VIBRATION PROBLEMS IN ENGINEERING

FIFTH EDITION

W. WEAVER, JR.
Professor Emeritus of Structural Engineering
Stanford University

The Late S. P. TIMOSHENKO
Professor Emeritus of Engineering Mechanics
Stanford University

The Late D. H. YOUNG
Professor Emeritus of Civil Engineering
Stanford University

A WILEY-INTERSCIENCE PUBLICATION

JOHN WILEY & SONS

New York / Chichester / Brisbane / Toronto / Singapore

***Library of Congress Cataloging in Publication Data*:**

Weaver, William, 1929–
Vibration problems in engineering/W. Weaver, Jr., S. P. Timoshenko, D. H. Young.—5th ed.
p. cm.
Timoshenko's name appears first on the earlier edition.
Includes bibliographies and index.
ISBN 0-471-63228-7
1. Vibration. I. Timoshenko, Stephen, 1878–1972. II. Young, D. H. (Donovan Harold), 1904–1979. III. Title.
TA355.W38 1989
620.3—dc20 89-31896
CIP

Printed in the United States of America

10 9 8 7 6 5 4 3 2

To

Stephen P. Timoshenko

and

Donovan H. Young,

two eminent scholars of Engineering Mechanics
who are no longer with us.

CONTENTS

PREFACE

The objectives of the fifth edition are not only to retain the useful portions of Timoshenko's classical work, but also to introduce more modern techniques for ease in computations. As before, we discuss one-degree, two-degree, multiple-degree, and infinite-degree of freedom systems, in that order. The chapter on nonlinear systems is updated and expanded to incorporate more numerical methods that have proven to be effective. Also, a new chapter on the finite-element method for discretized continua has been added at the end of the book. We emphasize matrix and numerical methods wherever feasible and describe computer programs that apply the theory in the text. Flowcharts for the programs appear in Appendix B. These programs are coded in FORTRAN for personal computers and recorded on diskettes for distribution to users. A diskette may be purchased from Paul R. Johnston, 838 Mesa Court, Palo Alto, CA 94306.

Material in this book is intended for engineering students in the last year of their undergraduate program or the first year of graduate studies. They should have a good understanding of calculus, including differential equations. Previous courses in statics, elementary dynamics, and mechanics of materials are also required. Although some exposure to structural analysis and theory of elasticity would be helpful, these subjects are not prerequisites for studying vibration theory. We assume that the student has some knowledge of matrix algebra and computer programming or is prepared to learn the essentials of these topics while using this text. The reader should find the explanations easy to understand, either as a student or practicing engineer.

The sequence of topics in the book has been arranged to draw the user forward from the simple to the complex. Chapter 1, dealing with linear systems having one degree of freedom, contains sections on free and forced vibrations for undamped and damped analytical models. Responses to initial conditions, arbitrary forcing functions, and support motions provide results that are used later for systems with multiple degrees of freedom. The last section (on step-by-step response calculations) has been completely re-written to describe only the best procedure for piecewise-linear forcing functions.

The discussion of nonlinear systems in Chapter 2 omits some outdated topics on approximate methods that were retained in the fourth edition. On the other hand, the section on numerical solutions is updated and expanded to give more information on recent developments.

In Chapter 3 the matrix format is introduced for both action equations of motion (involving stiffness influence coefficients) and displacement equations of motion (involving flexibility influence coefficients). Discussions in this chapter set the stage for multidegree systems in the next chapter. In addition, the topics of inertial and gravitational coupling are covered in a comprehensive manner, along with elasticity coupling and the influence of viscous damping.

Matrix equations of motion are generalized from two degrees of freedom to n degrees of freedom in Chapter 4, where the normal-mode method of dynamic analysis is developed and applied to a variety of problems. Normal-mode responses to initial conditions, applied actions, and support motions are covered, including rigid-body as well as vibrational modes. The iteration method for calculating frequencies and mode shapes is treated thoroughly, and modal truncation is emphasized. Various approaches are discussed for handling damping in multidegree systems, where modal damping is seen to be the simplest. The concluding section describes step-by-step calculations for the transient response of damped multidegree systems, subjected to piecewise-linear forcing functions.

Chapter 5, on continua with infinite degrees of freedom, required the least revision because of its classical nature. However, newly authored Chapter 6 describes the finite-element method for discretized continua. This approach is eminently suitable for solids and structures having arbitrary shapes and boundary conditions.

A bibliography on vibrations, appendices, and answers to problems close this work. Appendix A presents the two commonly used systems of units (SI and US) and gives material properties of significance in vibrational analysis. Appendix B documents computer programs that apply the matrix theory and numerical methods in the text. FORTRAN-oriented flowcharts display the logical steps in these programs clearly and completely.

I wish to thank Paul R. Johnston for coding the computer programs and dispensing them to interested users. Thanks are also due to Abdul R. Touqan for solving and organizing problem sets in the book. Patricia A. Krokel set a new example of excellence in typing parts of the revised

CHAPTER 1

SYSTEMS WITH ONE DEGREE OF FREEDOM

1.1 EXAMPLES OF ONE-DEGREE SYSTEMS

For vibrational analysis, many structures and machines may be idealized as systems having only one free displacement coordinate, or *one degree of freedom*. Figure 1.1 shows examples of analytical models for such idealized systems. In Fig. 1.1*a* the *mass m* may be either an attached concentration or some fraction of the distributed mass of the pole. The single translation $u(t)$ in the x direction is the only displacement coordinate required to describe the motion of the mass at any instant of time t. Similarly, the mass m attached to the center of the stretched wire in Fig. 1.1*b* may be considered to have only one degree of freedom. Another single-degree-of-freedom (SDOF) system appears in Fig. 1.1*c*, where $v(t)$ denotes the translation in the y direction of the mass attached to the beam.

A displacement coordinate may be rotational instead of translational, as indicated by the small angle $\phi(t)$ in Fig. 1.1*d*. Alternatively, an angular displacement may be represented symbolically by a curved arrow (see the figure). In this case the *mass moment of inertia I* of the disk (with respect to the axis of the shaft) acts in conjunction with the single rotational degree of freedom $\phi(t)$.

Topics of interest for linearly elastic SDOF systems consist of free and [illegible]rced harmonic motions without and with damping, response to arbitrary [illegible]arying loads or support motions, response spectra for dynamic loads, [illegible]y-step response calculations. Analytical models having nonlinear [illegible] and multiple degrees of freedom will be discussed in [illegible]apters.

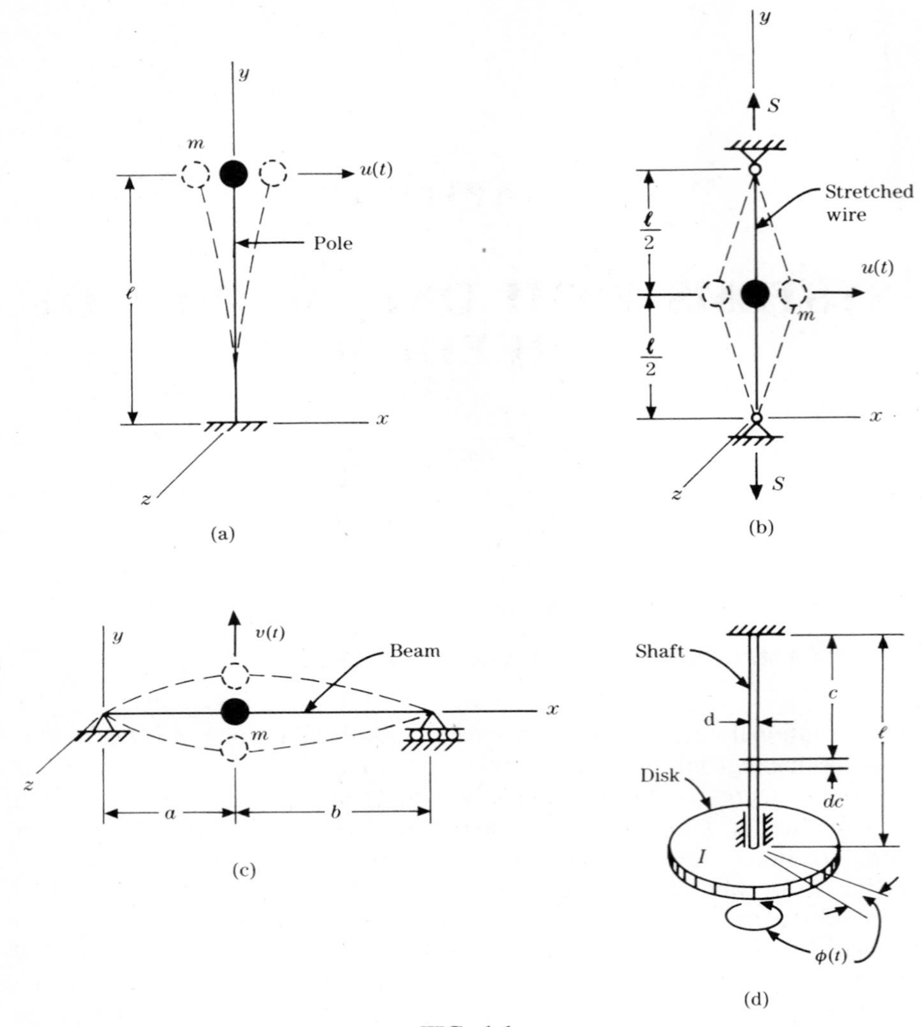

FIG. 1.1

1.2 UNDAMPED FREE TRANSLATIONAL VIBRATIONS

Figure 1.2*a* depicts a mass m of weight $W = mg$ that is suspended by a linearly elastic spring in the Earth's gravitational field. Here the symbol g represents the *gravitational constant* (acceleration due to gravity) at some point. If the arrangement is such that only vertical translation of the block is possible (and the mass of the spring is neglected), the system can be considered to have just one degree of freedom. The configuration of the system will be completely determined by the vertical translation u of the block from its static equilibrium position.

When the block is first attached to the spring, there will be a static

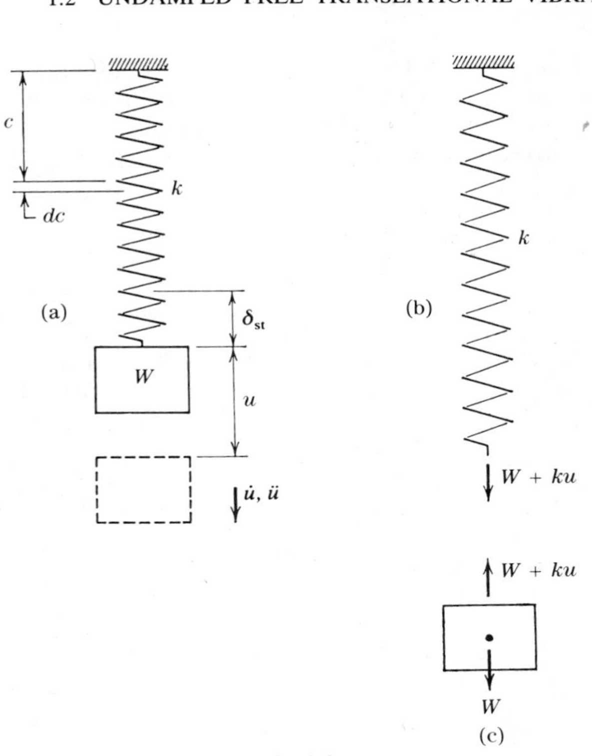

FIG. 1.2

deflection

$$\delta_{st} = \frac{W}{k} \tag{a}$$

in which the *spring constant k* denotes the force required to produce a unit change in length of the spring. For a *helical spring* with a closely wound series of n coils having a mean coil diameter D and a wire diameter d, the spring constant may be expressed [1] as†

$$k = \frac{Gd^4}{8nD^3} \tag{b}$$

where G represents the shearing modulus of elasticity of the wire.

Now let the block be displaced from its equilibrium position and released so that vibrations take place. Such vibrations, which are maintained by the elastic force in the spring alone, are called *free* or *natural vibrations*. If the

† Numbers in brackets indicate references at the end of the chapter.

displacement u is considered to be positive in the downward direction, the force in the spring corresponding to any position of the block will be $W + ku$, as shown in Fig. 1.2*b*. Knowing that the mass of the block is W/g and denoting its acceleration d^2u/dt^2 by $\ddot{u}$, we may apply *Newton's second law of motion* to obtain

$$\frac{W}{g}\ddot{u} = W - (W + ku) \tag{c}$$

The unbalanced forces acting on the block are shown in Fig. 1.2*c*. The weight W cancels on the right-hand side of Eq. (*c*), showing that this differential equation of motion for free vibrations of the system is independent of the gravitational field. In the discussion which follows, it is important to remember that the displacement u is measured from the static equilibrium position and that it is considered positive in the downward direction.

Introducing the notation

$$\omega^2 = \frac{k}{m} = \frac{kg}{W} = \frac{g}{\delta_{st}} \tag{d}$$

we can represent Eq. (*c*) in the following form:

$$\ddot{u} + \omega^2 u = 0 \tag{1.1}$$

This equation will be satisfied if we take $u = C_1 \cos \omega t$ or $u = C_2 \sin \omega t$, where C_1 and C_2 are arbitrary constants and ω is *angular frequency* (radians per second). By adding these solutions, we obtain the general solution of Eq. (1.1) as

$$u = C_1 \cos \omega t + C_2 \sin \omega t \tag{1.2}$$

It is seen that the vertical motion of the block has a vibratory character, because $\cos \omega t$ and $\sin \omega t$ are periodic functions which repeat themselves after an interval of time τ such that

$$\omega(\tau + 1) - \omega t = 2\pi \tag{e}$$

This interval of time is called the *period* of vibration. Its magnitude, from Eq. (*e*), is

$$\tau = \frac{2\pi}{\omega} \tag{f}$$

Or, by using notation (d),

$$\tau = 2\pi\sqrt{\frac{m}{k}} = 2\pi\sqrt{\frac{W}{kg}} = 2\pi\sqrt{\frac{\delta_{st}}{g}} \tag{1.3}$$

We see that the period of vibration depends only upon the magnitudes of the weight W and the spring constant k and is independent of the amount of displacement. If the deflection δ_{st} is determined theoretically or experimentally, the period τ can be calculated from Eq. (1.3).

The number of reciprocations per unit time (cycles per second) is called the *frequency* of vibration. Denoting frequency by f, we obtain

$$f = \frac{1}{\tau} = \frac{\omega}{2\pi} = \frac{1}{2\pi}\sqrt{\frac{k}{m}} = \frac{1}{2\pi}\sqrt{\frac{kg}{W}} = \frac{1}{2\pi}\sqrt{\frac{g}{\delta_{st}}} \tag{1.4}$$

In order to determine the constants of integration C_1 and C_2 in Eq. (1.2), we must consider the initial conditions. Assume that at the initial moment ($t = 0$) the block has a displacement u_0 from its position of equilibrium and that its initial velocity is $\dot{u}_0$. Substituting $t = 0$ into Eq. (1.2), we obtain

$$C_1 = u_0 \tag{g}$$

Taking the derivative of Eq. (1.2) with respect to time and substituting $t = 0$ therein, we have

$$C_2 = \frac{\dot{u}_0}{\omega} \tag{h}$$

The expression for the vibratory motion of the block is obtained by substituting into Eq. (1.2) the values of the constants from (g) and (h). Thus,

$$u = u_0 \cos \omega t + \frac{\dot{u}_0}{\omega} \sin \omega t \tag{1.5}$$

We see that the vibration consists of two parts: one is proportional to $\cos \omega t$ and depends on the initial displacement u_0 and the other is proportional to $\sin \omega t$ and depends on the initial velocity $\dot{u}_0$. Each of these parts can be represented graphically, as shown in Figs. 1.3a and 1.3b, by plotting displacements against time. The total displacement u of the vibrating block is obtained by adding the ordinates of the two curves to obtain the curve shown in Fig. 1.3c.

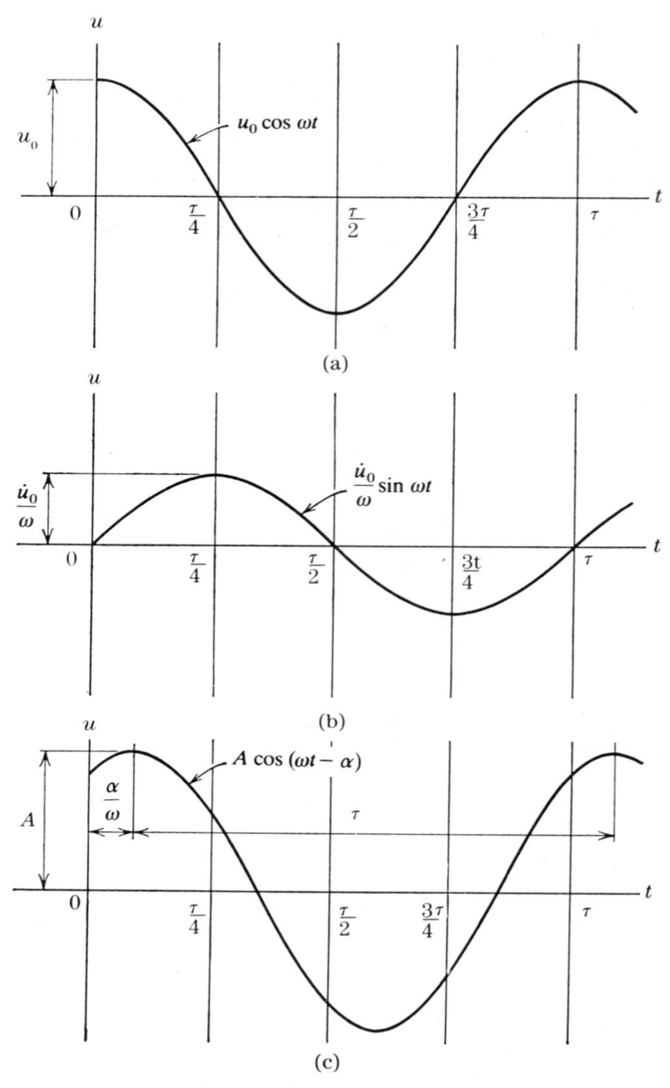

FIG. 1.3

Another method of representing vibrations is by means of *rotating vectors*. Imagine a vector $\overline{OP}$ (see Fig. 1.4) of magnitude x_0 rotating with constant angular velocity ω around a fixed point O. If at the initial moment $(t = 0)$ the vector $\overline{OP}$ coincides with the u axis, the angle that it makes with the same axis at any other time t is equal to ωt. The projection of this vector on the u axis is equal to $u_0 \cos \omega t$ and represents the first term of expression (1.5). Taking another vector $\overline{OQ}$ of magnitude $\dot{u}_0/\omega$ and perpendicular to

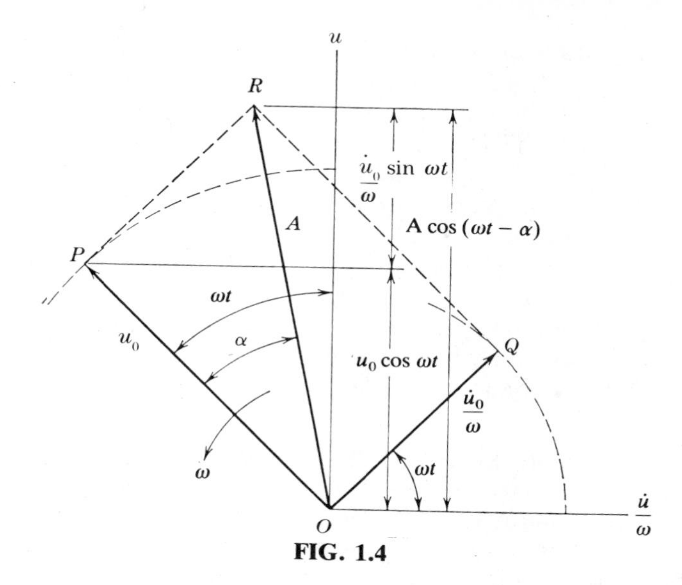

FIG. 1.4

the vector $\overline{OP}$, we observe that its projection on the u axis gives the second term of expression (1.5). The total displacement u of the oscillating weight is obtained by adding the projections on the u axis of the two perpendicular vectors $\overline{OP}$ and $\overline{OQ}$, rotating with the angular velocity ω.

The same result will be obtained if, instead of vectors $\overline{OP}$ and $\overline{OQ}$, we consider the vector $\overline{OR}$, equal to their vector sum, and take the projection of the resultant vector on the u axis. The magnitude A of this vector, from Fig. 1.4, is

$$A = \sqrt{u_0^2 + \left(\frac{\dot{u}_0}{\omega}\right)^2} \tag{i}$$

and the angle that it makes with the u axis is $\omega t - \alpha$, where

$$\alpha = \tan^{-1} \frac{\dot{u}_0}{\omega u_0} \tag{j}$$

From the above discussion, it is evident that Eq. (1.5) can be expressed in the equivlent form

$$u = A \cos(\omega t - \alpha) \tag{1.6}$$

where A and α, as represented by expressions (i) and (j), are new constants depending on the initial conditions of motion. It is seen that the addition of two simple harmonic motions, one proportional to $\cos \omega t$ and the other to $\sin \omega t$, is again a simple harmonic motion proportional to $\cos(\omega t - \alpha)$, as represented graphically in Fig. 1.3*c*. The maximum ordinate A of this curve, equal to the magnitude of vector $\overline{OR}$ in Fig. 1.4, represents the maximum displacement of the vibrating body from its position of equilibrium and is called the *amplitude of vibration*.

Due to the angle α between the two rotating vectors $\overline{OP}$ and $\overline{OR}$, the maximum ordinate of the curve in Fig. 1.3*c* is displaced with respect to the maximum ordinate of the curve in Fig. 1.3*a* by the amount α/ω. In such a case we may say that the total vibration, represented by the curve in Fig. 1.3*c*, lags the component of the motion given by the curve in Fig. 1.3*a*; and the angle α is called the *phase difference* or *phase angle* of these two vibrations. The coordinates u, $\dot{u}/\omega$ in Fig. 1.4 are said to define the *phase plane*, in which the motions are treated as rotating vectors.

Example 1. A simply supported steel beam of length $\ell = 120$ in. and flexural rigidity $EI = 20 \times 10^6$ lb-in.^{2} has a block of weight $W = 200$ lb dropped onto it at midspan from a height $h = \frac{1}{2}$ in., as shown in Fig. 1.5. Neglecting the distributed mass of the beam and assuming the block and beam do not separate after the initial contact, calculate the frequency and amplitude of the resulting free vibrations. Note that this example uses US units.†

Solution. Under the action of the load W resting at the center of the beam, the static deflection is

$$\delta_{st} = \frac{W\ell^3}{48EI} = \frac{(200)(120)^3}{(48)(20 \times 10^6)} = 0.36 \text{ in.}$$

Hence, from Eq. (1.4), the frequency of free vibration is

$$f = \frac{\omega}{2\pi} = \frac{1}{2\pi}\sqrt{\frac{g}{\delta_{st}}} = \frac{1}{2\pi}\sqrt{\frac{386}{0.36}} = \frac{32.8}{6.28} = 5.21 \text{ cps}$$

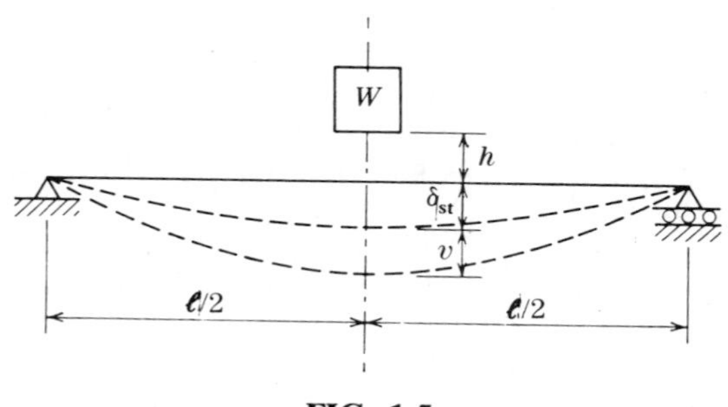

FIG. 1.5

† For a discussion of systems of units, see Appendix A.

In calculating the amplitude, we note that at the initial instant ($t = 0$) when the falling block first strikes the beam the initial displacement is

$$v_0 = -\delta_{st}$$

and the initial velocity is

$$\dot{v}_0 = \sqrt{2gh}$$

Hence, by Eq. (i) the amplitude is

$$A = \sqrt{(-\delta_{st})^2 + 2h\delta_{st}} = \sqrt{0.13 + 0.36} = \sqrt{0.49} = 0.70 \text{ in.}$$

Because this amplitude is measured from the static equilibrium position, it should be observed that the total deflection produced by the falling block is $A + \delta_{st} = 1.06$ in.

Example 2. In Fig. 1.6a, a block of weight W is suspended by two springs (having spring constants k_1 and k_2) connected *in series*. In Fig. 1.6b, the same block is supported by two springs (with constants k_1 and k_2), which are said to be connected *in parallel*. In each case find the *equivalent spring constant* k for the system.

Solution. For the case in Fig. 1.6a, each spring carries the same tension W, and their individual elongations are $\delta_1 = W/k_1$ and $\delta_2 = W/k_2$. Hence, the total static deflection of the weight is

$$\delta_{st} = \delta_1 + \delta_2 = \frac{W}{k_1} + \frac{W}{k_2}$$

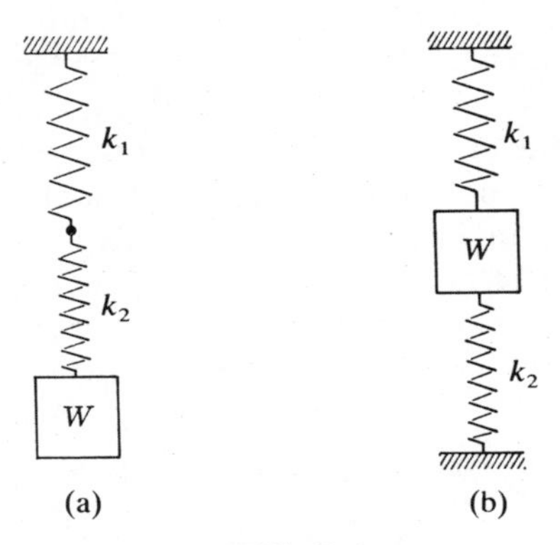

FIG. 1.6

Then from Eq. (*a*), the equivalent spring constant for the system is $k = W/\delta_{st}$, which becomes

$$k = \frac{k_1 k_2}{k_1 + k_2} \tag{k}$$

Using this value of k in Eq. (1.3), we can calculate the period of free vibration.

For the case in Fig. 1.6*b*, let S_1 be the tension in the upper spring and S_2 the compression in the lower spring due to the static action of the weight W. Then since each spring must have the same change in length, we have

$$\delta_{st} = \frac{S_1}{k_1} = \frac{S_2}{k_2} = \frac{W}{k} \tag{l}$$

Furthermore, a unit displacement of the block produces a restoring force of

$$k = k_1 + k_2 \tag{m}$$

which is the equivalent spring constant for the system. That is, for springs in parallel it is only necessary to add the individual spring constants to obtain the equivalent constant. Forces in the individual springs may be obtained from expressions (*l*) and (*m*) as

$$S_1 = \frac{k_1}{k_1 + k_2} W \qquad S_2 = \frac{k_2}{k_1 + k_2 W} \tag{n}$$

Example 3. A framed structure consists of a heavy rigid platform of mass $m = 20$ Mg, supported by four rigid vertical columns and braced laterally in each side panel by two diagonal steel wires, as shown in Fig. 1.7*a*. The columns (of height $h = 3$ m) are hinged at their ends; and each diagonal wire has a cross-sectional area of $A = 5 \times 10^4$ m^2 and is tensioned to a high stress. Neglecting all mass except that of the platform, find the period τ for free lateral vibrations of the structure, using SI units.†

Solution. Apply a force P in the x direction at the center of mass of the platform, as indicated in Fig. 1.7*b*. Due to this load, the change in tensile force in the diagonal AC will be $S = \sqrt{2}P/4$. The corresponding elongation of this diagonal is

$$\Delta = \frac{S\ell}{AE} = \frac{\sqrt{2}P\sqrt{2}h}{4AE} = \frac{Ph}{2AE}$$

† See Appendix A.

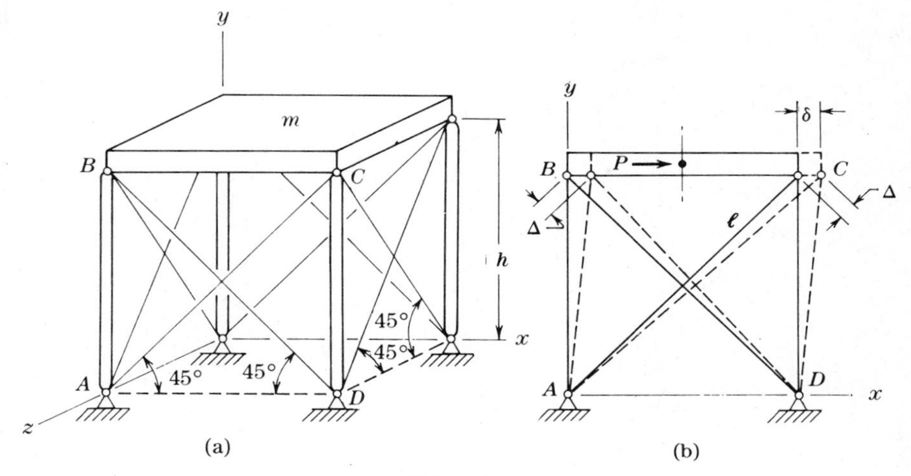

FIG. 1.7

and the diagonal BD shortens an equal amount. As a result of these length changes of the diagonals, we see that the platform has a lateral deflection $\delta = \sqrt{2}\Delta$. Thus, the spring constant for the structure becomes

$$k = \frac{P}{\delta} = \frac{\sqrt{2}AE}{h} = \frac{\sqrt{2}(5 \times 10^{-4})(207)}{3} = 48.8 \text{ MN/m}$$

in which $E = 207 \text{ GN/m}^2$ for steel. Substituting this value of k into Eq. (1.3), we obtain

$$\tau = 2\pi \sqrt{\frac{20}{48.8 \times 10^3}} = 0.127 \text{ sec}$$

It is left to the reader to show that in this example the horizontal force P need not be applied in the x direction. The same result is obtained if P is in the z direction or in any other direction in the horizontal plane.

Example 4. Assume that the block of weight W in Fig. 1.8 represents an elevator moving downward with constant velocity $\dot{u}_0$ and the spring consists of a steel cable. Determine the maximum stress in the cable if during motion the drum at the upper end suddenly locks. Let the weight $W = 10{,}000$ lb, $\ell = 60$ ft, the cross-sectional area of the cable $A = 2.5 \text{ in.}^2$, modulus of elasticity of the cable $E = 15 \times 10^6$ psi, and $\dot{u}_0 = 3$ ft/sec. The weight of the cable is to be neglected.

Solution. During the uniform motion of the elevator, the tensile force in the cable is equal to $W = 10{,}000$ lb; and the elongation of the cable at the instant of the accident is $\delta_{st} = W\ell/AE = 0.192$ in. Due to the initial velocity

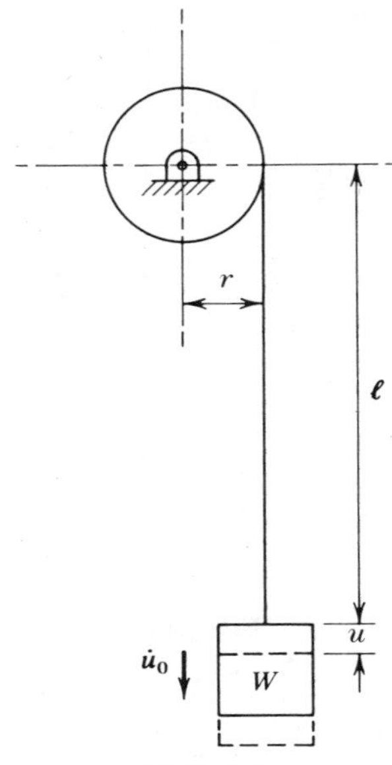

FIG. 1.8

$\dot{u}_0$, the elevator will not stop suddenly but will vibrate on the cable. Measuring time from the instant of the accident, we see that the displacement of the elevator from the position of equilibrium at that instant is zero, while its velocity is $\dot{u}_0$. From Eq. (1.5), we conclude that the amplitude of vibration will be equal to $\dot{u}_0/\omega$, where $\omega = \sqrt{g/\delta_{st}} = 44.8\ \text{sec}^{-1}$ and $\dot{u}_0 = 36$ in./sec. Hence, the maximum elongation of the cable is $\delta_{max} = \delta_{st} + \dot{u}_0/\omega = 0.192 + 36/44.8 = 0.192 + 0.803 = 0.995$ in., and the maximum stress is $\sigma_{max} = (10{,}000/2.5)(0.995/0.192) = 20{,}750$ psi. We see that due to the sudden stoppage of the drum, the stress in the cable increased by a factor of about 5.

1.3 ROTATIONAL VIBRATIONS

Now let us reconsider Fig. 1.1*d*, which shows an elastic shaft built in at its upper end and carrying at its lower end a solid circular disk. Such a system is called a *torsional pendulum*. If the disk is rotated through a small angle ϕ about the axis of the shaft and then released, the torque exerted upon it by the twisted shaft will set it in motion, and free *rotational vibrations* will ensue. During such vibrations, the torque exerted on the disk by the twisted shaft will be proportional to the angle ϕ and will always act in a direction opposite to the rotation of the disk. Thus, if I denotes the mass moment of inertia of the disk about the axis of the shaft, $\ddot{\phi}$ its angular acceleration, and

k_r the torque per unit angle of rotation (*rotational spring constant*), the differential equation of motion becomes

$$I\ddot{\phi} = -k_r\phi \tag{a}$$

Introducing the notation

$$\omega^2 = \frac{k_r}{I} \tag{b}$$

we may write Eq. (*a*) as

$$\ddot{\phi} + \omega^2\phi = 0 \tag{1.7}$$

This equation has the same form as Eq. (1.1) of the previous section; hence, its solution has the same form as Eq. (1.5), and we obtain

$$\phi = \phi_0 \cos \omega t + \frac{\dot{\phi}_0}{\omega} \sin \omega t \tag{1.8}$$

where ϕ_0 and $\dot{\phi}_0$ are the angular displacement and angular velocity, respectively, of the disk at the initial instant $t = 0$. Proceeding as in the previous section, we conclude from Eq. (1.8) that the period of rotational vibration is

$$\tau = \frac{2\pi}{\omega} = 2\pi\sqrt{\frac{I}{k_r}} \tag{1.9}$$

and its frequency is

$$f = \frac{1}{\tau} = \frac{1}{2\pi}\sqrt{\frac{k_r}{I}} \tag{1.10}$$

In the case of a shaft of circular cross section having length ℓ and diameter d, its *torsional spring constant* can be expressed by the formula [2]

$$k_r = \frac{GJ}{\ell} = \frac{\pi d^4 G}{32\ell} \tag{c}$$

where G is the modulus of elasticity in shear for the material. The symbol J in Eq. (*c*) represents the *torsion constant* of the cross section of the shaft, which is equal to the polar moment of inertia when the cross section is circular.

Furthermore, if the circular disk is homogeneous, with diameter D and weight W, its mass moment of inertia will be

$$I = \frac{WD^2}{8g} \tag{d}$$

With these values of the quantities k_r and I, the period and frequency of *torsional vibration* can be found from Eqs. (1.9) and (1.10).

In the more general case of a shaft of noncircular cross section or a body of irregular shape, the quantities k_r and I may be more difficult to calculate. However, they can always be determined by experiment if no formulas are available for their calculation. In order that the vibration be purely rotational, it is necessary that the axis of the shaft coincide with a principal axis of the body through its center of mass. Otherwise, it would be necessary to introduce restraints (in the form of bearings) to prevent other motions of the body. It should also be noted that rotational vibrations may occur in systems where no torsional deformations are involved (see Example 2 at the end of this section).

It has been assumed throughout the foregoing discussion that the shaft in Fig. 1.1d has a uniform cross section of diameter d. If the shaft consists of two parts having lengths ℓ_1 and ℓ_2 and diameters d_1 and d_2, respectively, the separate torsional spring constants k_{r1} and k_{r2} may be calculated from Eq. (c). Then, since the two portions of the shaft represent torsional springs connected in series, the equivalent spring constant can be obtained from Eq. (k) of the preceding section.

The case of a stepped shaft can also be handled in another way. If the shaft consisting of two parts is subjected to a twisting moment M, the total angle of twist in the shaft will be

$$\phi = \frac{M}{k_{r1}} + \frac{M}{k_{r2}} = \frac{32M\ell_1}{\pi d_1^4 G} + \frac{32M\ell_2}{\pi d_2^4 G} = \frac{32M}{\pi d_1^4 G}\left(\ell_1 + \ell_2 \frac{d_1^4}{d_2^4}\right)$$

It is seen that the angle of twist of a shaft with two diameters d_1 and d_2 is the same as that of a shaft of constant diameter d_1 and of a modified length L_1 given by the equation

$$L_1 = \ell_1 + \ell_2 \frac{d_1^4}{d_2^4} \tag{e}$$

The shaft of length L_1 and diameter d_1 has the same spring constant as the given shaft of two different diameters and is an *equivalent shaft* in this case.

Let us consider now the case of a shaft supported in frictionless bearings and carrying a rotating body at each end, as shown in Fig. 1.9. Such a case is of practical interest, because it may be considered to represent a propeller shaft with a propeller at one and a turbine rotor at the other.† If the two disks are twisted in opposite directions and then suddenly released, torsional vibrations will occur. From the principles of conservation of

† This was one of the earliest problems in which engineers found it necessary to make investigations of torsional vibrations. Rigid-body rotation of the whole system is ignored in this discussion.

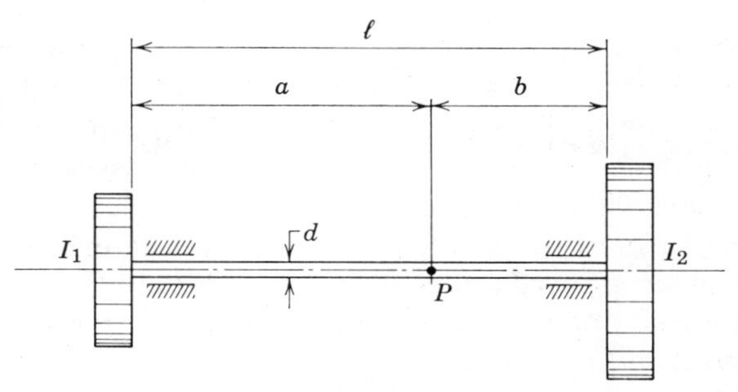

FIG. 1.9

angular momentum, it follows that the disks must always rotate in opposite directions during such vibrations. Thus, there is a certain intermediate cross section located at point P on the shaft (see Fig. 1.9) which remains stationary, and this cross section is called the *nodal section*. Its position is found from the fact that the two bodies must have the same period of vibration, because otherwise the condition that they must always rotate in opposite directions will not be fulfilled.

Applying Eq. (1.9) to each of the two subsystems on the two sides of the nodal section, we obtain

$$\sqrt{\frac{I_1}{k_{r1}}} = \sqrt{\frac{I_2}{k_{r2}}} \qquad \text{or} \qquad \frac{k_{r1}}{k_{r2}} = \frac{I_1}{I_2} \tag{f}$$

where k_{r1} and k_{r2} are the spring constants for the left- and the right-hand portions of the shaft, respectively. These quantities, as seen from Eq. (*c*), are inversely proportional to the lengths of the corresponding portions of the shaft, and from Eq. (*f*) it follows that

$$\frac{a}{b} = \frac{I_2}{I_1}$$

Then since $a + b = \ell$, we find

$$a = \frac{\ell I_2}{I_1 + I_2} \qquad b = \frac{\ell I_1}{I_1 + I_2} \tag{g}$$

Applying Eqs. (1.9) and (1.10) to the left-hand portion of the system, we obtain

$$\tau = 2\pi\sqrt{\frac{I_1}{k_{r1}}} = 2\pi\sqrt{\frac{32\ell I_1 I_2}{\pi d^4 G(I_1 + I_2)}} \tag{1.11}$$

$$f = \frac{1}{2\pi}\sqrt{\frac{\pi d^4 G(I_1 + I_2)}{32\ell I_1 I_2}} \tag{1.12}$$

From these formulas the period and the frequency of torsional vibrations can be calculated, provided that the dimensions of the shaft, the modulus G, and the mass moments of inertia of the bodies at the ends are known. The mass of the shaft is neglected in our present discussion; its effect on the period of vibration will be considered later in Sec. 1.5.

It can be seen from Eqs. (g) that if one of the rotating bodies has a very large mass moment of inertia in comparison with the other, the nodal cross section can be taken at the larger body, and the system with two bodies (Fig. 1.9) reduces to that with only one.

Example 1. Referring to Fig. 1.9, assume that two homogeneous disks at the ends of a steel shaft have weights $W_1 = 4.5$ kN and $W_2 = 9.0$ kN, and diameters $D_1 = 1.3$ m and $D_2 = 2.0$ m, while the shaft has length $\ell = 3.0$ m, diameter $d = 0.1$ m, and shear modulus $G = 79.6$ GPa. Compute the frequency of free torsional vibration of the system. In what proportion will this frequency be increased if along a length of 1.6 m the diameter of the shaft is increased from 0.1 m to 0.2 m?

Solution. Using the given numerical data in Eq. (d), we calculate the mass moments of inertia of the disks as

$$I_1 = \frac{(4.5)(1.3)^2}{(8)(9.80)} = 97.0\ \text{kg}\cdot\text{m}^2 \qquad I_2 = \frac{(9.0)(2.0)^2}{(8)(9.80)} = 459\ \text{kg}\cdot\text{m}^2$$

Substituting these values, together with the given data for the shaft into Eq. (1.12), we obtain

$$f = \frac{1}{2\pi}\sqrt{\frac{\pi(0.1)^4(79.6)(97+459)}{(32)(3)(97)(459)}} = 9.08\ \text{cps}$$

When the diameter of the shaft is increased from 0.1 m to 0.2 m over 1.6 m of its length, the equivalent length of 0.1 m diameter may be found from Eq. (e), as follows:

$$L_1 = 1.4 + 1.6\frac{(0.1)^4}{(0.2)^4} = 1.4 + 0.1 = 1.5\ \text{m}$$

Because this length is half of the original 3.0 m and the frequency is inversely proportional to the square root of the length (see Eq. (1.12)), we see that the frequency increases in the ratio $\sqrt{2}:1$.

Example 2. A flywheel consists of a heavy rim of weight W and mean radius R attached to a hub by four flexible prismatic spokes, as shown in Fig. 1.10a. If the hub is held fixed, find the period of free rotational vibration of the rim about its central axis through point O. Neglect the mass

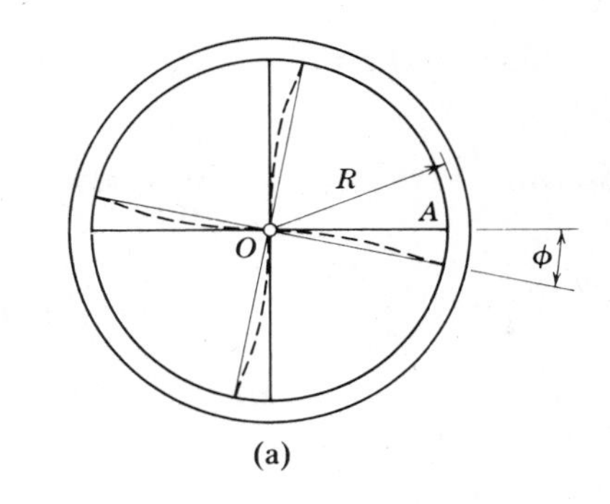

(a)

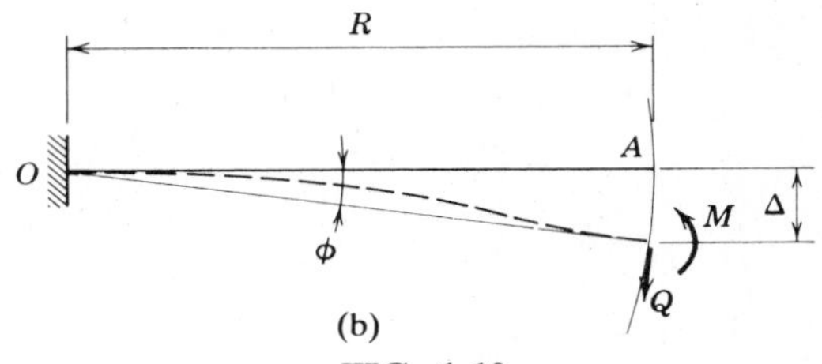

(b)

FIG. 1.10

of the spokes, and assume each spoke to have length R and bending rigidity B.

Solution. Let the rim have a small angle of rotation ϕ from its equilibrium position as shown. Each spoke behaves as a beam built in at the hub and constrained to move with the rim at the other end. At the outer end of a spoke, a shear force Q and a bending moment M act, as shown in Fig. 1.10*b*. Applying known formulas for stiffnesses of a beam, we have

$$Q = \frac{12B\Delta}{R^3} - \frac{6B\phi}{R^2} \tag{h}$$

$$M = \frac{6B\Delta}{R^2} - \frac{4B\phi}{R} \tag{i}$$

If the rim is assumed to be rigid, the tangent to the elastic line at the outer end of each spoke must be radial. Thus, the shear force Q and the bending moment M are related to the angle of rotation ϕ through the geometric condition that $\Delta \approx R\phi$. Substituting this relationship into Eqs. (*h*) and (*i*), we find

$$Q = \frac{6B\phi}{R^2} \quad \text{and} \quad M = \frac{2B\phi}{R} \tag{j}$$

Then the total moment acting on the rim will be

$$M_t = 4QR - 4M = \frac{16B\phi}{R} \qquad (k)$$

and the rotational spring constant in this case is seen to be

$$k_r = \frac{M_t}{\phi} = \frac{16B}{R} \qquad (l)$$

Substituting this value of k_r into Eq. (1.9) and noting that the mass moment of inertia of the flywheel rim is $I \approx WR^2/g$, we obtain

$$\tau = 2\pi\sqrt{\frac{WR^3}{16gB}} \qquad (m)$$

1.4 ENERGY METHOD

It is sometimes advantageous to use the principle of *conservation of energy* for vibrating systems in which there is no dissipation of energy. By this method the equation of motion for free vibrations of a one-degree-of-freedom system will be rederived, and the equality of maximum values of kinetic and potential energies during free vibration will be established.

Consider again the spring–mass system in Fig. 1.2*a* for the purpose of analyzing it on an energy basis. If we neglect the mass of the spring as before, the *kinetic energy* of the vibrating system is

$$\text{KE} = \frac{W}{g}\frac{\dot{u}^2}{2} \qquad (a)$$

The *potential energy* of the system in this case consists of two parts: (a) the potential energy of the weight W by virtue of its position below the equilibrium position as a datum; and (b) the *strain energy* stored in the spring because of the displacement u. The first of these two energies is simply

$$\text{PE} = -Wu \qquad (b)$$

To calculate the second quantity, we consider the diagram shown in Fig. 1.11, which is a plot of the spring force S as a function of the displacment u. In the static equilibrium position the force in the spring is W; and when the displacement is u, the force is $W + ku$. Thus, the strain energy stored in the

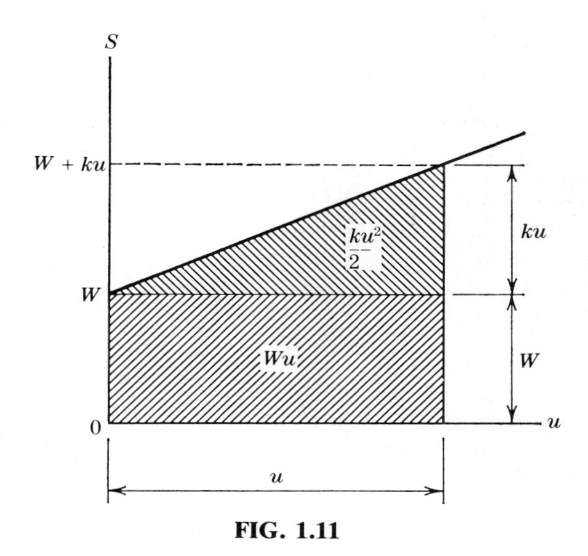

FIG. 1.11

spring during the displacement u is

$$\text{SE} = Wu + \frac{ku^2}{2} \tag{c}$$

Adding the energies (a), (b), and (c) and observing that their sum must be constant in accordance with the law of conservation of energy, we obtain

$$\frac{W}{g}\frac{\dot{u}^2}{2} + \frac{ku^2}{2} = \text{constant} \tag{d}$$

Because the expression in Eq. (d) is constant, its rate of change with respect to time must be equal to zero. Thus,

$$\frac{d}{dt}\left(\frac{W}{g}\frac{\dot{u}^2}{2} + \frac{ku^2}{2}\right) = 0 \tag{e}$$

Evaluating the derivatives in Eq. (e) and dividing by $\dot{u}$ results in the equation of motion obtained previously in Sec. 1.1.

$$\frac{W}{g}\ddot{u} + ku = 0 \tag{f}$$

If we are only interested in finding the natural frequency of a vibrating system, it is not necessary to consider the equation of motion at all. We may instead observe that the vibrating block in Fig. 1.2a is at rest in one of its extreme posititions, where its net potential energy is

$$(\mathrm{PE})_{\max} = \frac{ku_{\max}^2}{2} \tag{g}$$

while its kinetic energy is zero. On the other hand, when the block is passing through its static equilibrium position (at $u = 0$) with maximum velocity, its kinetic energy is

$$(\mathrm{KE})_{\max} = \frac{W}{g}\frac{\dot{u}_{\max}^2}{2} \tag{h}$$

while its potential energy is zero. Because the total energy remains constant, the maximum kinetic energy must be equal to the maximum potential energy. Thus,

$$(\mathrm{KE})_{\max} = (\mathrm{PE})_{\max} \tag{1.13}$$

This simple relationship is useful for the purpose of calculating the natural frequency or period of a vibrating system. For the system in Fig. 1.2*a*, equating expressions (*g*) and (*h*) gives:

$$\frac{W}{g}\frac{\dot{u}_{\max}^2}{2} = \frac{ku_{\max}^2}{2} \tag{i}$$

Assuming harmonic motion of the form given by Eq. (1.6),

$$u = A\cos(\omega t - \alpha) \qquad \dot{u} = -A\omega\sin(\omega t - \alpha)$$

we see that

$$\dot{u}_{\max} = \omega u_{\max} \tag{1.14}$$

Substituting this value for $\dot{u}_{\max}$ into Eq. (*i*), we obtain

$$\omega = \sqrt{\frac{k}{m}} = \sqrt{\frac{kg}{W}}$$

and the period of vibration is

$$\tau = 2\pi\sqrt{\frac{m}{k}} = 2\pi\sqrt{\frac{W}{kg}}$$

as obtained previously in Sec. 1.2. The use of Eq. (1.13) in calculating periods or frequencies is especially advantageous if, instead of a simple system as in Fig. 1.2*a*, we have a more complicated system involving several moving parts. Such cases are illustrated in the examples that follow.

Example 1. A displacement meter consists of a case enclosing a block of weight W, which is supported on a spring k_1, as shown in Fig. 1.12. Movement of the block relative to the case actuates a pointer BOA, which is pivoted at O and restrained by another spring k_2 as shown. Neglecting the masses of the two springs, calculate the period of free vibration of the system, assuming that it performs simple harmonic motion.

Solution. Let $\dot{u}_m$ be the maximum velocity of the block during vibration. Then the corresponding angular velocity of the arm BOA will be $\dot{u}_m/b$; and if I is the mass moment of inertia of this arm about point O, the total kinetic energy of the system in its equilibrium configuration will be

$$(\mathrm{KE})_{\max} = \frac{W}{g}\frac{\dot{u}_m^2}{2} + \frac{I}{b^2}\frac{\dot{u}_m^2}{2} \qquad (j)$$

When the system is in an extreme configuration, as defined by a vertical displacement u_m of the block, the spring k_2 will have an extension cu_m/b; and the total potential energy of the system will be

$$(\mathrm{PE})_{\max} = \frac{1}{2}k_1 u_m^2 + \frac{1}{2}k_2\left(\frac{c}{b}\right)^2 u_m^2 \qquad (k)$$

Equating expressions (j) and (k) in accordance with Eq. (1.13) and using for simple harmonic motion the relationship $u_m = \omega u_m$ from Eq. (1.14), we

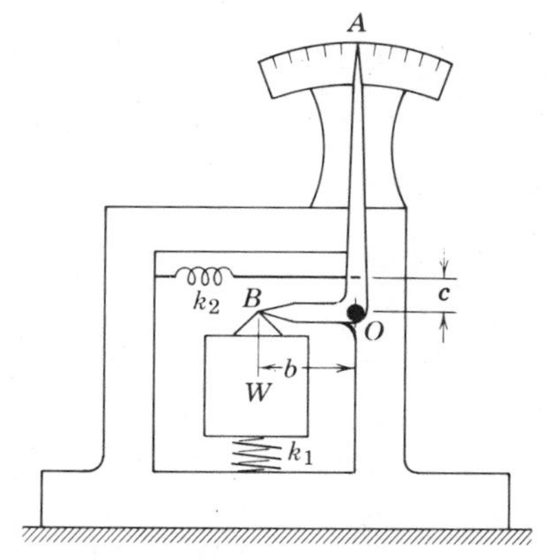

FIG. 1.12

find for the angular frequency

$$\omega = \sqrt{\frac{k_1 + (c/b)^2 k_2}{(W/g) + (I/b^2)}} \tag{l}$$

The period of vibration then is 2π divided by expression (l).

Example 2. An inverted pendulum consists of a ball of weight W at the end of a rigid bar OA of length ℓ, which is pinned at O and supported in a vertical position by a flexible spring, as shown in Fig. 1.13a. Neglecting the masses of the spring and the bar OA, determine the stabiity condition and angular frequency ω for small rotational oscillations of the pendulum in the plane of the figure.

Solution. Let ϕ_m (see Fig. 1.13b) be the amplitude of simple harmonic motion. For this extreme position the spring has an elongation of approximately $a\phi_m$, and the ball has dropped below its equilibrium position by the distance

$$\Delta = \ell(1 - \cos\phi_m) \approx \frac{1}{2}\ell\phi_m^2 \tag{m}$$

Hence, the potential energy of the system in the extreme position is approximately

$$(\text{PE})_{\max} = \frac{1}{2}ka^2\phi_m^2 - \frac{1}{2}W\ell\phi_m^2 \tag{n}$$

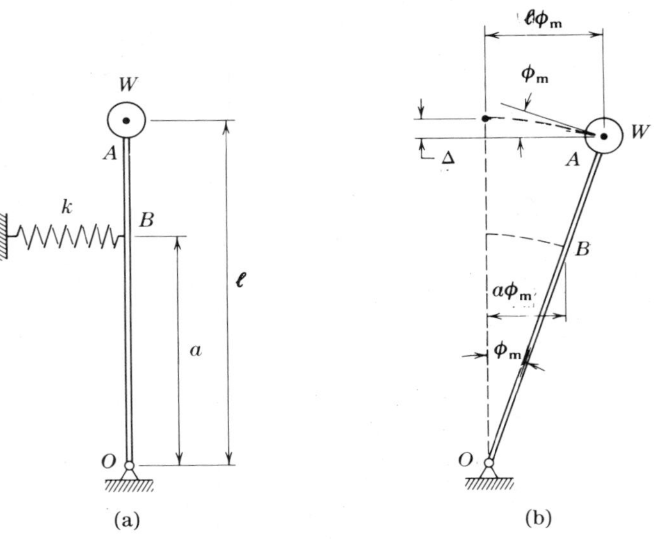

FIG. 1.13

In the vertical position (Fig. 1.13*a*), where the pendulum has angular velocity $\dot{\phi}_m$, the kinetic energy is simply $\frac{1}{2}I\dot{\phi}_m^2$, in which $I = W\ell^2/g$ represents the moment of inertia of the weight W about point O. Thus,

$$(\text{KE})_{\max} = \frac{W\ell^2}{2g}\dot{\phi}_m^2 \tag{o}$$

Equating expressions (n) and (o) using the relationship $\dot{\phi}_m = \omega\phi_m$ from Eq. (1.14), we find the angular frequency to be

$$\omega = \sqrt{\frac{g}{\ell}\left(\frac{ka^2}{W\ell} - 1\right)} \tag{p}$$

It is seen from Eq. (p) that we obtain a real value for ω only if

$$ka^2 > W\ell \tag{q}$$

If this condition is not satisfied, the vertical position of equilibrium of the pendulum is unstable.

Example 3. A solid circular cylinder of weight W and radius r rolls without slip on a cylindrical surface of radius a, as shown in Fig. 1.14. Assuming that the rolling cylinder performs simple harmonic motion, find its angular frequency ω for small amplitudes of displacement about the equilibrium position.

Solution. Consider the cylinder in an extreme position defined by the angle ϕ_m in Fig. 1.14. In this position the center of gravity of the cylinder has been raised through the height

$$(a - r)(1 - \cos\phi_m) \approx (a - r)\frac{\phi_m^2}{2} \tag{r}$$

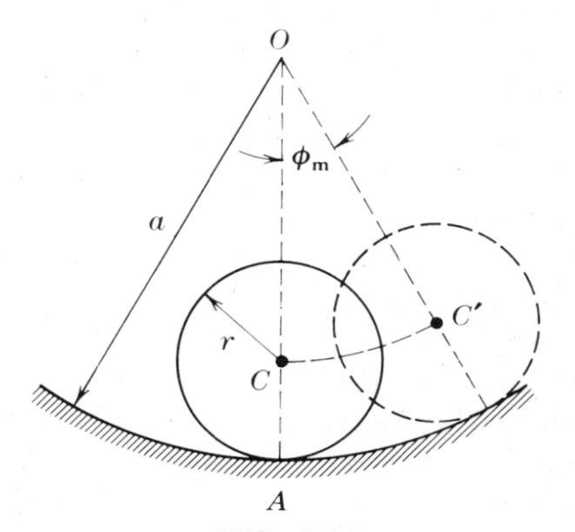

FIG. 1.14

above its equilibrium position, and the potential energy is

$$(\mathrm{PE})_{\max} = \frac{1}{2} W(a - r)\phi_{\mathrm{m}}^2 \tag{s}$$

In the middle position the point of contact A is the instantaneous center of rotation of the cylinder, and for the no-slip condition its instantaneous angular velocity about this point is

$$\dot{\theta}_{\mathrm{m}} = \frac{a - r}{r} \dot{\phi}_{\mathrm{m}} \tag{t}$$

Hence, the kinetic energy, which is $\frac{1}{2} I_A \dot{\theta}_{\mathrm{m}}^2$, becomes

$$(\mathrm{KE})_{\max} = \frac{1}{2} \frac{W}{g} \frac{3r^2}{2} \frac{(a - r)^2}{r^2} \dot{\phi}_{\mathrm{m}}^2 \tag{u}$$

Equating expressions (s) and (u) and using Eq. (1.14), we find for the angular frequency

$$\omega = \sqrt{\frac{2g}{3(a - r)}} \tag{v}$$

1.5 RAYLEIGH'S METHOD

In all of the cases considered previously, the problems were reduced to the case of vibration of a system with one degree of freedom by using simplifications. For instance, in the arrangement shown in Fig. 1.2, the mass of the spring was neglected in comparison with the mass of the block, while in the arrangement shown in Fig. 1.5 the mass of the beam was neglected. Again, in the case shown in Fig. 1.9, the mass moment of inertia of the shaft was neglected in comparison with those of the disks. Although these simplifications are accurate enough in many practical cases, there are technical problems in which a detailed consideration of the accuracy of such approximations becomes necessary. In order to determine the effect of such simplifications on the frequency of vibration, an approximate method developed by Lord Rayleigh [3] will now be discussed. In applying this method some assumption regarding the configuration of the system during vibration must be made. The frequency of vibration will then be found from a consideration of the conservation of energy in the system.

As a simple example of the application of Rayleigh's method, we reconsider the case shown in Fig. 1.2. If the mass of the spring is small in comparison with the mass of the block, the *mode of vibration* will not be substantially affected by the mass of the spring. With good accuracy it can be assumed that the displacement u_c of any point on the spring at a distance c from the fixed end (see Fig. 1.2*a*) is the same as in the case of a massless spring. Thus,

$$u_c = \frac{cu}{\ell} \tag{a}$$

where ℓ is the length of the spring in the equilibrium position.

If the displacements vary linearly as assumed above, the potential energy of the system will be the same as in the case of a massless spring, and only the kinetic energy of the system need not be reconsidered. Let w denote the weight of the spring per unit length. Then the mass of an infinitesimal element of the spring of length dc will be $w\,dc/g$, and its maximum kinetic energy becomes

$$\frac{w\,dc}{2g}\left(\frac{c\dot{u}_{\mathrm{m}}}{\ell}\right)^2$$

The complete kinetic energy of the spring will be

$$\frac{w}{2g}\int_0^{\ell}\left(\frac{c\dot{u}_{\mathrm{m}}}{\ell}\right)^2 dc = \frac{\dot{u}_{\mathrm{m}}^2}{2g}\left(\frac{w\ell}{3}\right) \tag{b}$$

This quantity must be added to the kinetic energy of the block; so that the energy equation (1.13) becomes

$$\frac{\dot{u}_{\mathrm{m}}^2}{2g}\left(W + \frac{w\ell}{3}\right) = \frac{ku_{\mathrm{m}}^2}{2} \tag{c}$$

Comparing this expression with Eq (*i*) of the preceding section, we may conclude that in order to estimate the effect of the mass of the spring on the period of natural vibration it is only necessary to add one-third of the weight of the spring to the weight W.

This conclusion, obtained on the assumption that the displacement of the spring varies linearly, can be used with sufficient accuracy even in cases where the weight of the spring is of the same order of magnitude as W. For instance, when $w\ell = 0.5W$, the error of the approximate solution is about 0.5%.† For $w\ell = W$, the error is about 0.8%; and for $w\ell = 2W$, the error is about 3%.

† A more detailed consideration of this problem is given in Sec. 5.5.

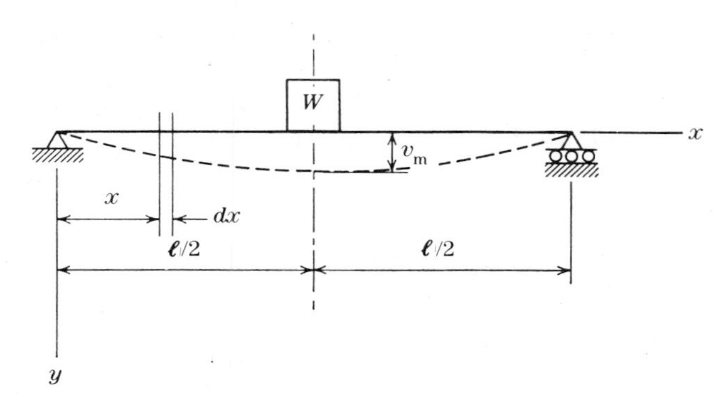

FIG. 1.15

As a second example, consider the case of vibration of a beam of uniform cross section loaded at the middle with a block of weight W (see Fig. 1.15). If the weight $w\ell$ of the beam is small in comparison with the load W, it can be assumed with sufficient accuracy that the deflection curve of the beam during vibration has the same shape as the static deflection curve for a concentrated load at the middle. Then, denoting by v_{m} the maximum displacement at the center of the beam during vibration, we express the displacement of any element located at a distance x from the support as

$$v = v_{\mathrm{m}}\left(\frac{3x\ell^2 - 4x^3}{\ell^3}\right) \tag{d}$$

The maximum kinetic energy of the beam itself will be

$$2\int_0^{\ell/2} \frac{w}{2g}\left(\dot{v}_{\mathrm{m}}\frac{3x\ell^2 - 4x^3}{\ell^3}\right)^2 dx = \frac{17}{35}w\ell\frac{\dot{v}_{\mathrm{m}}^2}{2g} \tag{e}$$

This kinetic energy of the vibrating beam must be added to the energy $W\dot{v}_{\mathrm{m}}^2/2g$ of the load concentrated at the middle in order to estimate the effect of the weight of the beam on the period of vibration. In this case the period of vibration will be the same as for a massless beam loaded at the middle by the weight

$$W' = W + \frac{17}{35}w\ell$$

Even in the extreme case when $W = 0$ and where the *equivalent weight* $(\frac{17}{35})w\ell$ is concentrated at the middle of the beam, the accuracy of the approximate method is sufficiently close for practical purposes. The deflection of the beam under the action of the equivalent load applied at the middle is

$$\delta_{st} = \frac{17}{35} w\ell\left(\frac{\ell^3}{48EI}\right)$$

Substituting this value into Eq. (1.3), we find that the period of the natural vibration is

$$\tau = 2\pi\sqrt{\frac{\delta_{st}}{g}} = 0.632\sqrt{\frac{w\ell^4}{EIg}} \tag{f}$$

The exact solution for this case is†

$$\tau = \frac{2}{\pi}\sqrt{\frac{w\ell^4}{EIg}} = 0.637\sqrt{\frac{w\ell^4}{EIg}} \tag{g}$$

It is seen that the error of the approximate solution for this limiting case is less than 1%.

As a third example, we shall consider a prismatic cantilever beam carrying a block of weight W at its free end, as shown in Fig. 1.16. Assume that during vibration the shape of the deflection curve of the beam is the same as the one produced by a load statically applied at the end. Denoting by v_m the maximum displacement of the load W, we may compute the kinetic energy of the distributed mass of the beam as follows:

$$\int_0^\ell \frac{w}{2g}\left(\dot{v}_m \frac{3x^2\ell - x^3}{2\ell^3}\right)^2 dx = \frac{33}{140} w\ell \frac{\dot{v}_m^2}{2g} \tag{h}$$

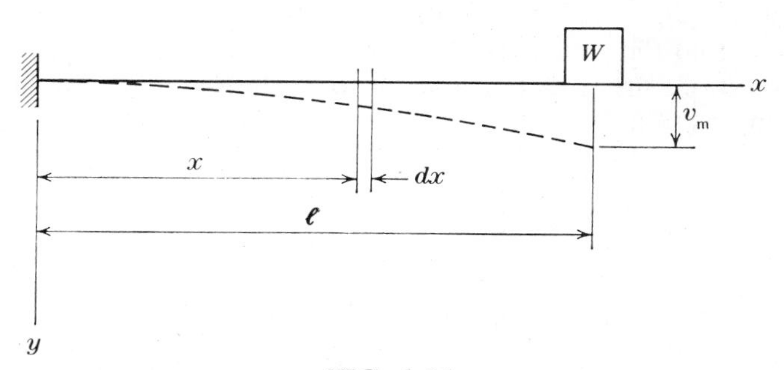

FIG. 1.16

† See Sec. 5.10.

The period of vibration for this case will be the same as for a massless cantilever beam loaded at the end by the weight

$$W' = W + \frac{33}{140} w\ell$$

The equivalent weight of $(\frac{33}{140})w\ell$ may be used even for cases where $w\ell$ is not small. Applying the result to the extreme case where $W = 0$, we obtain

$$\delta_{st} = \frac{33}{140} w\ell \left(\frac{\ell^3}{3EI}\right)$$

The resulting period of vibration will be

$$\tau = 2\pi \sqrt{\frac{\delta_{st}}{g}} = \frac{2\pi}{3.567} \sqrt{\frac{w\ell^4}{EIg}} \tag{i}$$

The exact solution for the same case is†

$$\tau = \frac{2\pi}{3.515} \sqrt{\frac{w\ell^4}{EIg}} \tag{j}$$

We see that the error of the approximate solution is about 1.5%.

In the two preceding cases of transverse vibrations of loaded beams, we assumed that the shapes of the beams during vibrations were the same as the deflection curves due to static application of the loads. In the limiting case of a massless beam this is the correct assumption; but when the beam has some distributed mass, it is only an approximation. In general, if we assume any reasonable shape for the elastic curve of the beam, we can expect to obtain a close appoximation to the true period of vibration. Of course, if the exact shape is chosen, we should obtain the exact period. To demonstrate this point, let us consider again the simply supported beam in Fig. 1.15 without the weight W. It is known for this case (see Sec. 5.10) that during vibration the exact shape of the elastic curve, relative to the equilibrium position, is

$$v = v_m \sin \frac{\pi x}{\ell} \tag{k}$$

As before, the symbol v_m in Eq. (k) is the maximum deflection at the center of the beam, and v is the deflection at any other point located at a distance x from the support.

† See Sec. 5.11.

The kinetic energy of the entire beam in the equilibrium position will be

$$(\mathrm{KE})_{\max} = \int_0^\ell \frac{w}{2g}\left(\dot{v}_{\mathrm{m}} \sin\frac{\pi x}{\ell}\right)^2 dx = \frac{w\ell}{2}\frac{\dot{v}_{\mathrm{m}}^2}{2g} \tag{l}$$

In order to compute the maximum potential energy of the beam relative to the equilibrium position, we will use the following expression [2] for flexural strain energy:

$$(\mathrm{PE})_{\max} = \frac{1}{2}\int_0^\ell EI\left(\frac{d^2v}{dx^2}\right)^2 dx \tag{m}$$

Upon substituting the second derivative of expression (k) into Eq. (m) and performing the indicated integration, we have

$$(\mathrm{PE})_{\max} = \frac{EI\pi^4}{4\ell^3} v_{\mathrm{m}}^2 \tag{n}$$

Finally, equating expressions (ℓ) and (n) in accordance with Eq. (1.13) and recalling that $\dot{v}_{\mathrm{m}} = \omega v_{\mathrm{m}}$, we find

$$\omega = \pi^2 \sqrt{\frac{EIg}{w\ell^4}} \tag{1.15}$$

The corresponding period is $\tau = 2\pi/\omega$, which yields the exact expression given previously as Eq. (g).

As mentioned above, any choice of function other than the exact shape will produce an approximate value of the true frequency or period. A very good choice of shape function for a vibrating beam consists of the deflection curve due to its own weight applied statically. In order to illustrate this idea, consider once more the case of an unloaded simply supported beam; and assume that its uniformly distributed weight w is applied statically, producing the following deflected shape:

$$v = v_{\mathrm{m}} \frac{16}{5\ell^4}(x^4 - 2\ell x^3 + \ell^3 x) \tag{o}$$

in which $v_{\mathrm{m}} = 5w\ell^4/384EI$ is the deflection at the center.

The kinetic energy of the beam in the equilibrium position is

$$(\mathrm{KE})_{\max} = \int_0^\ell \frac{w}{2g}\dot{v}^2\,dx = \frac{\omega^2}{2g}\int_0^\ell wv^2\,dx \tag{1.16}$$

Substituting expression (o) into Eq. (1.16) and integrating produces

$$(\mathrm{KE})_{\max} = 0.252 \frac{w\ell}{g} \omega^2 v_{\mathrm{m}}^2 \tag{p}$$

The maximum potential energy of the beam relative to the equilibrium position may be obtained by recognizing the fact that the external work of the statically applied distributed weight is equal to the flexural strain energy in the beam.† Thus,

$$(\mathrm{PE})_{\max} = \int_0^\ell \frac{1}{2} wv\, dx \tag{1.17}$$

Substituting Eq. (o) into Eq. (1.17) and integrating yields

$$(\mathrm{PE})_{\max} = 0.320 w\ell v_{\mathrm{m}} = 24.6 \frac{EI}{\ell^3} v_{\mathrm{m}}^2 \tag{q}$$

Equating expressions (p) and (q), we obtain

$$\omega = 9.87 \sqrt{\frac{EIg}{w\ell^4}} \tag{r}$$

Comparing this result to the exact expression given by Eq. (1.15), we find that it is accurate to three significant figures.

Equations (1.16) and (1.17) may be substituted into the energy equation (1.13) to give a general expression for ω^2 in this type of problem. Thus,

$$\omega^2 = \frac{g \int_0^\ell wv\, dx}{\int_0^\ell wv^2\, dx} \tag{1.18}$$

If w varies along the length of the beam, it must remain within the integrals of Eq. (1.18). However, for prismatic beams this term may be cancelled from the expression.

It must be noted that an elastic beam represents a system with an infinitely large number of degrees of freedom. It can, like a string, perform vibrations of various types. In using Rayleigh's method, the choosing of a definite shape for the deflection curve is equivalent to introducing some additional constraints that reduce the system to one having a single degree

† Equation (1.17) is equivalent to Eq. (m).

of freedom. Such additional constraints can only increase the stiffness of the system, that is, increase the frequency of vibration. Thus, in all cases considered above the approximate values of the frequencies as obtained by Rayleigh's method are somewhat higher than their exact values.

In the case of torsional vibrations (see Fig. 1.1d), the same approximate method can be used in order to calculate the effect of the inertia of the shaft on the frequency of the entire system. Let i denote the mass moment of inertia of the shaft per unit length. Then assuming that the mode of vibration is the same as in the case of a massless shaft, the angle of rotation of a cross section at a distance c from the fixed end of the shaft is $c\phi/\ell$; and the maximum kinetic energy of an infinitesimal element of the shaft will be

$$\frac{i\,dc}{2}\left(\frac{c\dot{\phi}_{\text{m}}}{\ell}\right)^2$$

The kinetic energy of the entire shaft is

$$\frac{i}{2}\int_0^{\ell}\left(\frac{c\dot{\phi}_{\text{m}}}{\ell}\right)^2 dc = \frac{\dot{\phi}_{\text{m}}^2}{2}\left(\frac{i\ell}{3}\right) \tag{s}$$

This kinetic energy must be added to the kinetic energy of the disk in order to estimate the effect of the mass of the shaft on the frequency of vibration for the system. Thus, the period of vibration will be the same as for a massless shaft having a disk at the end, for which the mass moment of inertia is

$$I' = I + \frac{i\ell}{3}$$

The torsional vibrations of shafts without attached bodies may be investigated in a manner analogous to that for transverse vibrations of beams. Following the same steps leading to Eq. (1.18) for beams, we establish a similar expression for shafts. Thus,

$$\omega^2 = \frac{\alpha\displaystyle\int_0^{\ell} i\phi\,dx}{\displaystyle\int_0^{\ell} i\phi^2\,dx} \tag{1.19}$$

in which ϕ represents the angle of twist at any point due to the application of a distributed torque that is taken to be numerically equal to αi per unit length of the shaft. The symbol α in Eq. (1.19) represents an angular acceleration that (for convenience) may be taken to be 1 rad/sec^2.

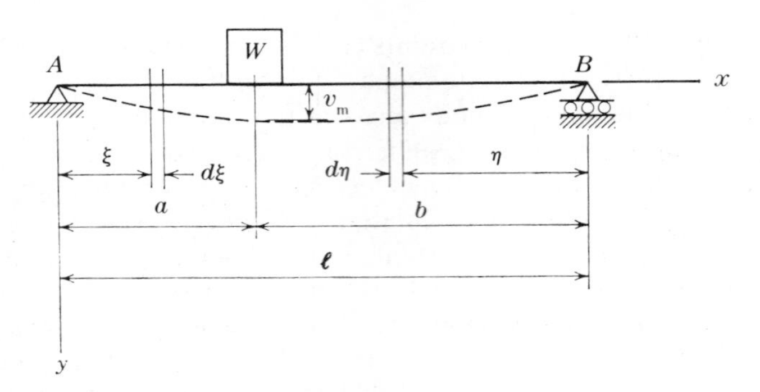

FIG. 1.17

Example 1. Determine the frequency of natural vibrations of the load W supported by a beam AB (Fig. 1.17) of constant cross section: (a) assuming that the weight of the beam can be neglected; (b) taking the weight of the beam into consideration and using Rayleigh's method.

Solution. If a and b are the distances of the load from the ends of the beam, the static deflection under the load is $\delta_{st} = Wa^2b^2/3\ell EI$. Taking for the spring constant the expression $k = 3\ell EI/a^2b^2$ and neglecting the mass of the beam, we find the angular frequency of vibration from the equation

$$\omega = \sqrt{\frac{g}{\delta_{st}}} = \sqrt{\frac{kg}{W}} = \sqrt{\frac{3\ell EIg}{Wa^2b^2}}$$

To take the mass of the beam into account, we consider the deflection curve of the beam under static action of the load W. The deflection at any point on the left portion of the beam at the distance ξ from the support A is

$$v_1 = \frac{Wb\xi}{6\ell EI}[a(\ell + b) - \xi^2]$$

For the deflection at any point to the right of the load W at the distance η from the support B we have

$$v_2 = \frac{Wa\eta}{6\ell EI}[b(\ell + a) - \eta^2]$$

Applying Rayleigh's method and assuming that during vibration the maximum velocity of any point of the left portion of the beam is given by the equation

$$\dot{v}_1 = \dot{v}_m \frac{v_1}{\delta_{st}} = \dot{v}_m \frac{\xi}{2a^2b}[a(\ell + b) - \xi^2]$$

in which $\dot{v}_{\mathrm{m}}$ is the maximum velocity of the load W, we find that the maximum kinetic energy of that portion is

$$\frac{w\dot{v}_{\mathrm{m}}^2}{2g}\int_0^a\left(\frac{v_1}{\delta_{\mathrm{st}}}\right)^2 d\xi=\frac{w\dot{v}_{\mathrm{m}}^2}{2g}\int_0^a\frac{\xi^2}{4a^4b^2}[a(\ell+b)-\xi^2]^2\,d\xi$$

$$=\dot{v}_{\mathrm{m}}^2\frac{wa}{2g}\left[\frac{\ell^2}{3b^2}+\frac{23a^2}{105b^2}-\frac{8a\ell}{15b^2}\right] \tag{t}$$

In the same manner, considering the right portion of the beam, we find that its maximum kinetic energy is

$$\dot{v}_{\mathrm{m}}^2\frac{wb}{2g}\left[\frac{(\ell+a)^2}{12a^2}+\frac{b^2}{28a^2}-\frac{b(\ell+a)}{10a^2}\right] \tag{u}$$

Thus, for this problem the equation of energy (1.13) becomes

$$\frac{(W+\alpha wa+\beta wb)}{2g}\dot{v}_{\mathrm{m}}^2=\frac{kv_{\mathrm{m}}^2}{2}$$

where α and β denote the quantities in the brackets of expressions (t) and (u). Using the relationship $\dot{v}_{\mathrm{m}}=\omega v_{\mathrm{m}}$, we obtain for the angular frequency of vibration the following formula:

$$\omega=\sqrt{\frac{3\ell EIg}{(W+\alpha aw+\beta bw)a^2b^2}} \tag{v}$$

Example 2. Referring to the prismatic cantilever beam in Fig. 1.16 and assuming that $W=0$, find the approximate period of free lateral vibration by the Rayleigh method. Make the assumption that the shape of the beam during vibration is that of the static deflection curve caused by the weight of the beam.

Solution. For a uniformly distributed load of intensity w on the beam, the static deflection at the distance x from the fixed end will be

$$v=\frac{v_{\mathrm{m}}}{3\ell^4}(x^4-4\ell x^3+6\ell^2x^2) \tag{w}$$

in which $v_{\mathrm{m}}=w\ell^4/8EI$ is the deflection at the free end. Substituting expression (w) into Eqs. (1.16) and (1.17) and integrating, we find

$$(\mathrm{KE})_{\max}=\frac{52w\ell}{405g}\omega^2v_{\mathrm{m}}^2 \tag{x}$$

and

$$(\mathrm{PE})_{\max} = \frac{w\ell}{5} v_{\mathrm{m}} = \frac{8EI}{5\ell^3} v_{\mathrm{m}}^2 \tag{y}$$

Equating expressions (x) and (y) results in

$$\omega = 3.530\sqrt{\frac{EIg}{w\ell^4}}$$

and the corresponding period of vibration is

$$\tau = \frac{2\pi}{3.530}\sqrt{\frac{w\ell^4}{EIg}} \tag{z}$$

The exact value for the fundamental period in this case is given by Eq. (j), and we see that the Rayleigh approximation is in error by about 0.5%.

As an extension of Rayleigh's method, let us consider the vibrational analysis of a weightless beam with several masses, as shown in Fig. 1.18. This type of analytical model may be associated with the idea of "lumping" the distributed mass of a beam at a succession of points along its length in order to approximate its dynamic characteristics. Of course, a series of weights may actually exist as a set of loads on the structure.

In either case, let W_1, W_2, and W_3 denote values of the weights on the beam, while v_1, v_2, and v_3 represent the corresponding static displacements in the y direction. The potential energy of deformation stored in the beam during bending is

$$(\mathrm{PE})_{\max} = \frac{1}{2} W_1 v_1 + \frac{1}{2} W_2 v_2 + \frac{1}{2} W_3 v_3 \tag{a'}$$

For the purpose of calculating the angular frequency of the *fundamental mode of vibration*,† we may write the kinetic energy of the system in its

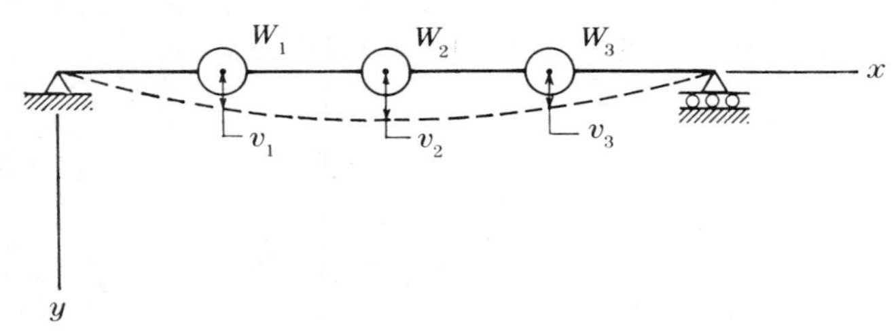

FIG. 1.18

† The fundamental mode has the lowest frequency.

equilibrium position as

$$(\text{KE})_{\max} = \frac{1}{2g} W_1 \dot{v}_1^2 + \frac{1}{2g} W_2 \dot{v}_2^2 + \frac{1}{2g} W_3 \dot{v}_3^2 \qquad (b')$$

From Eq. (1.14), we have

$$\dot{v}_1 = \omega v_1 \qquad \dot{v}_2 = \omega v_2 \qquad \dot{v}_3 = \omega \dot{v}_3 \qquad (c')$$

Thus, Eq. (b') may be rewritten as

$$(\text{KE})_{\max} = \frac{\omega^2}{2g} (W_1 v_1^2 + W_2 v_2^2 + W_3 v_3^2) \qquad (d')$$

When Eqs. (a') and (d') are equated, the following expression for ω^2 is obtained

$$\omega^2 = \frac{g(W_1 v_1 + W_2 v_2 + W_3 v_3)}{(W_1 v_1^2 + W_2 v_2^2 + W_3 v_3^2)} \qquad (e')$$

In general, for n masses on the beam expression (e') becomes:

$$\omega^2 = \frac{g \sum_{j=1}^{n} W_j v_j}{\sum_{j=1}^{n} W_j v_j^2} \qquad (1.20)$$

which is a discretized version of Eq. (1.18).

Equation (1.20) shows that for estimating the frequency or period of vibration of a beam with several masses only the weights $W_1, W_2, \ldots, W_n$ and the static deflections $v_1, v_2, \ldots, v_n$ are required. The latter quantities can be obtained easily by the methods of beam deflection theory. If the beam has a variable cross section or if the effect of the weight of the member itself is to be taken into account, it is necessary to divide the beam into several parts, the weights of which must be considered as concentrated loads.

Rotational vibrations of shafts with several rigid bodies attached may be handled in a manner analogous to that for beams. For this purpose the discretized version of Eq. (1.19) is

$$\omega^2 = \frac{\alpha \sum_{j=1}^{n} I_j \phi_j}{\sum_{j=1}^{n} I_j \phi_j^2} \qquad (1.21)$$

In this expression the symbol ϕ_j represents the rotation of the jth rigid body due to a set of static torques. Such a torque for the jth body is taken to be numerically equal to αI_j, where $\alpha = 1$ rad/sec^2.

An inherent features of Eqs. (1.18) through (1.21) is that potential and kinetic energies used in deriving them are referred to the static equilibrium positions of the systems. The static loads usually are not self-equilibrating, and appropriate restraints are required. On the other hand, vibrations of unrestrained systems may also be investigated by the Rayleigh method using imaginary restraints at known or estimated points of zero displacement. It should also be noted that all of the numerator terms in Eqs. (1.18) through (1.21) will be positive when the applied action is in the same sense as the corresponding displacement. This should always be the case in order to guarantee that the calculated frequency is an upper bound on the true frequency.

Example 3. Using the Rayleigh method, calculate the angular frequency of the fundamental mode of vibration of the beam with two masses shown in Fig. 1.19. The flexural rigidity of the beam is EI, and its distributed mass is to be neglected. Assume for simplicity that $W_1 = W_2 = W$.

Solution. In this case it will be assumed that during vibration the beam maintains a shape similar to the static deflection curve due to forces W_1 and W_2, acting in opposite directions, as shown in Fig. 1.19. The corresponding static deflections are found to be

$$v_1 = \frac{W_1 \ell^3}{48EI} + \frac{W_2 \ell^3}{32EI} = \frac{5W\ell^3}{96EI} \qquad v_2 = \frac{W_1 \ell^3}{32EI} + \frac{W_2 \ell^3}{8EI} = \frac{5W\ell^3}{32EI}$$

Substituting these values into Eq. (1.20), we obtain

$$\omega = \sqrt{\frac{192EIg}{25W\ell^3}}$$

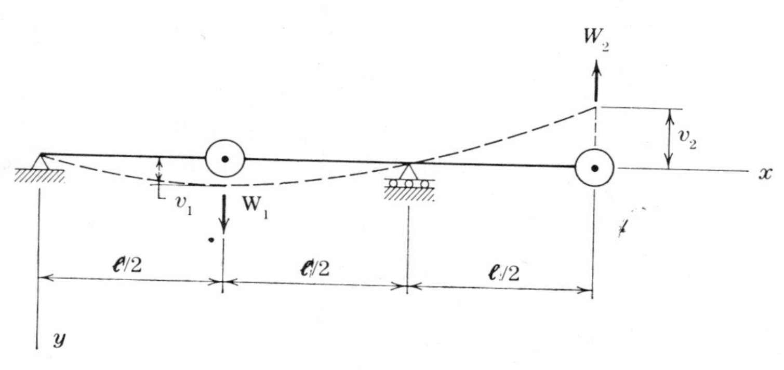

FIG. 1.19

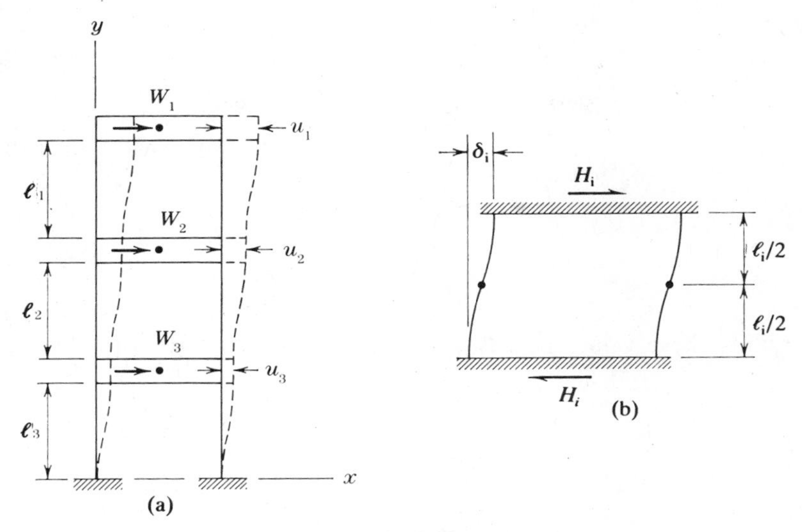

FIG. 1.20

Example 4. Figure 1.20*a* represents a simplified model of a three-story building, in which the floors are assumed to be rigid and the columns are assumed to be massless. Using the Rayleigh method, find the approximate fundamental period of free lateral vibration of the building. Assume for simplicity that $W_1 = W_2 = W_3 = W$, $\ell_1 = \ell_2 = \ell_3 = \ell$, and that each column has flexural rigidity EI.

Solution. We assume that during lateral vibrations the building maintains a shape similar to that produced by the floor weights W_1, W_2, and W_3 acting as horizontal forces at the floors. To calculate lateral deflections of the floors, we consider first the displacement δ_i of the ith floor from the top relative to the floor below it due to the action of the story shearing force H_i, as shown in Fig. 1.20*b*. Since each column has a shear force of magnitude $H_i/2$ and an inflection point at its midlength, we see that

$$\delta_i = 2\left[\frac{(H_i/2)(\ell/2)^3}{3EI}\right] = \frac{H_i\ell^3}{24EI} \qquad (f')$$

Now noting that $H_1 = W_1 = W$, $H_2 = W_1 + W_2 = 2W$, and $H_3 = W_1 + W_2 + W_3 = 3W$, we have from Eq. (f')

$$\delta_1 = \frac{W\ell^3}{24EI} \qquad \delta_2 = \frac{2W\ell^3}{24EI} \qquad \delta_3 = \frac{3W\ell^3}{24EI}$$

and the static deflections in Fig. 1.20*a* become

$$u_1 = \delta_3 + \delta_2 + \delta_1 = \frac{6W\ell^3}{24EI}$$

$$u_2 = \delta_3 + \delta_2 = \frac{5W\ell^3}{24EI}$$

$$u_3 = \delta_3 = \frac{3W\ell^3}{24EI}$$

Substituting these values into Eq. (1.20), together with $W_1 = W_2 = W_3 = W$, we obtain

$$\omega^2 = \frac{24EIg}{5W\ell^3}$$

and

$$\tau = \frac{2\pi}{\omega} = 2\pi\sqrt{\frac{5W\ell^3}{24EIg}}$$

Example 5. Suppose that a second weight equal to W is attached at the juncture of springs k_1 and k_2 in Fig. 1.6*a*, discussed earlier in Sec. 1.2. Determine approximately the angular frequency of the fundamental mode of this system by the Rayleigh method.

Solution. Due to both weights acting statically, the junction point displaces an amount $2W/k_1$, while the end point displaces an amount $2W/k_1 + W/k_2$. If we use a common denominator of k_1k_2 for these displacements, the first becomes $W(2k_2)/k_1k_2$ and the second is $W(2k_2 + k_1)/k_1k_2$. Substitution of these terms into Eq. (1.20) gives

$$\omega^2 = \frac{k_1k_2g[(2k_2) + (2k_2 + k_1)]}{W[(2k_2)^2 + (2k_2 + k_1)^2]}.$$

Thus,

$$\omega = \sqrt{\frac{k_1k_2g(k_1 + 4k_2)}{W(k_1^2 + 4k_1k_2 + 8k_2^2)}}$$

Example 6. Imagine that a second disk having a mass moment of inertia $2I$ is attached at midlength of the shaft in Fig. 1.1*d* (see Sec. 1.1). Using the Rayleigh method, estimate the angular frequency of the fundamental mode of rotational vibration.

Solution. To obtain rotational displacements, we apply torques that are numerically equal to $2I\alpha$ at midlength of the shaft and $I\alpha$ at the end. Such torques produce rotational displacements of $3I\alpha/2k_r$ at midlength and

$3I\alpha/2k_r + I\alpha/2k_r = 2I\alpha/k_r$ at the end (where k_r denotes the rotational stiffness of the entire shaft). Substitution of these terms into Eq. (1.21) produces

$$\omega^2 = \frac{\alpha[2I(3I\alpha/2k_r) + I(2I\alpha/k_r)]}{2I(3I\alpha/2k_r)^2 + I(2I\alpha/k_r)^2}$$

Hence,

$$\omega = \sqrt{\frac{10k_r}{17I}} = 0.767\sqrt{\frac{k_r}{I}}$$

1.6 FORCED VIBRATIONS: STEADY STATE

In Sec. 1.2 we studied the free vibrations of a spring-mass system and saw that its motion depends only upon the initial conditions and its physical characteristics k and $m = W/g$, which determine its natural frequency. If the system is subjected to other influences, such as time-varying forces or specified support motions, the dynamic response becomes more complicated. In many practical situations we encounter a *periodic forcing function* applied to the mass, and the response of the system to this condition is referred to as *forced vibrations*.

As an example of forced vibrations, consider the spring-suspended motor of weight W (Fig. 1.21), which is constrained to displace in the vertical

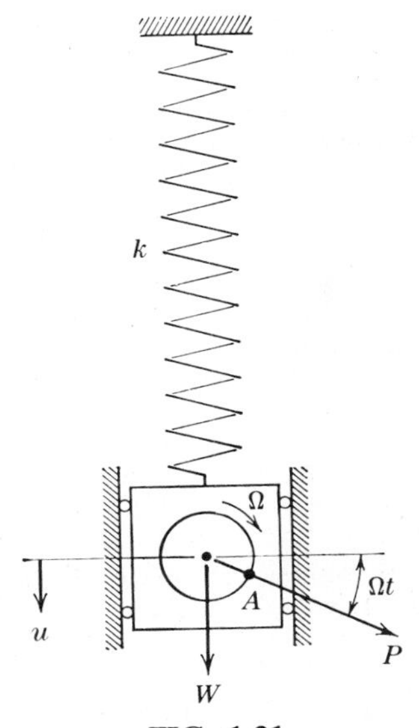

FIG. 1.21

direction only. This system has a natural angular frequency $\omega = \sqrt{kg/W}$, as discussed in Sec. 1.2. Suppose now that the motor runs at a constant angular speed Ω and that its rotor is slightly out of balance, as indicated in Fig. 1.21 by an eccentric mass at point A. This unbalance will create a rotating centrifugal force P, which will in turn produce forced vibrations of the system. In addition to the gravity force and the spring force, we now have the vertical component $P \sin \Omega t$ of the rotating force vector to consider. Thus, the equation of motion becomes

$$\frac{W}{g}\ddot{u} = W - (W + ku) + P \sin \Omega t \tag{a}$$

in which the term $P \sin \Omega t$ is called a *harmonic forcing function*. Introducing into Eq. (a) the notations

$$\omega^2 = \frac{k}{m} = \frac{kg}{W} \qquad \text{and} \qquad q = \frac{P}{m} = \frac{Pg}{W} \tag{b}$$

we obtain

$$\ddot{u} + \omega^2 u = q \sin \Omega t \tag{1.22}$$

A particular solution of this equation is obtained by assuming that u is proportional to $\sin \Omega t$, that is, by taking

$$u = C_3 \sin \Omega t \tag{c}$$

where C_3 is a constant, the magnitude of which must be chosen so as to satisfy Eq. (1.22). Substituting (c) into that equation, we find

$$C_3 = \frac{q}{\omega^2 - \Omega^2}$$

Thus, the required particular solution is

$$u = \frac{q \sin \Omega t}{\omega^2 - \Omega^2} \tag{d}$$

Adding this particular solution to the general solution (Eq. 1.2) of the homogeneous equation (1.1), we obtain

$$u = C_1 \cos \omega t + C_2 \sin \omega t + \frac{q \sin \Omega t}{\omega^2 - \Omega^2} \tag{1.23}$$

This expression contains two constants of integration and represents the total solution of Eq. (1.22).

The first two terms in Eq. (1.23) represent free vibrations, which were discussed before; and the third term, depending on the disturbing force, represents the forced vibration of the system. It is seen that this latter vibration has the same period $T = 2\pi/\Omega$ as that of the rotating force. Using notations (b) in Eq. (d) and ignoring the free vibrations,† we obtain so-called *steady-state forced vibrations*, defined by the equation

$$u = \left(\frac{P}{k}\sin\Omega t\right)\left(\frac{1}{1-\Omega^2/\omega^2}\right) \tag{1.24}$$

The factor $(P/k)\sin\Omega t$ is the deflection that the disturbing force $P\sin\Omega t$ would produce if it were acting statically, and the factor $1/(1-\Omega^2/\omega^2)$ accounts for the dynamical action of this force. The absolute value of the latter quantity is usually called the *magnification factor*:

$$\beta = \left|\frac{1}{1-\Omega^2/\omega^2}\right| \tag{e}$$

We see that β depends only on the *frequency ratio* Ω/ω, which is obtained by dividing the *impressed frequency* of the disturbing force by the *natural frequency* of free vibration of the system. In Fig. 1.22 values of the magnification factor β are plotted against the frequency ratio. It is seen that for small values of the ratio Ω/ω, that is, for the case where the frequency of the disturbing force is small in comparison with the frequency of free

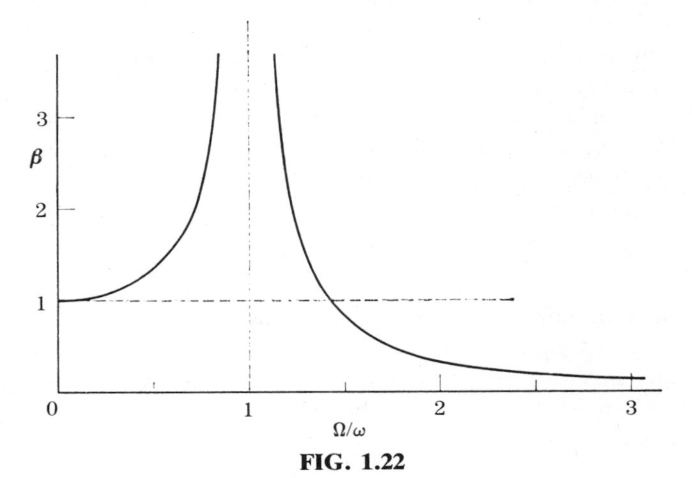

FIG. 1.22

† The effect of free vibrations in combination with forced vibrations will be discussed in the next section.

vibration, the magnification factor is approximately unity; and deflections are about the same as in the case of a statical action of the force $P \sin \Omega t$.

When the ratio Ω/ω approaches unity, the magnification factor and the *amplitude of forced vibration* rapidly increase and become infinite for $\Omega = \omega$, that is, for the case where the frequency of the disturbing force exactly coincides with the frequency of free vibration of the system. This is the *condition of resonance.* The infinite value obtained for the amplitude of forced vibrations indicates that if the periodic force acts on the vibrating system always at the proper time and in the proper direction the amplitude of vibration increases indefinitely, provided there is no dissipation of energy. In practical problems we always have dissipation of energy due to damping, and its effect on the amplitude of forced vibration will be discussed later (see Sec. 1.9).

When the frequency of the disturbing force increases beyond the frequency of free vibration, the magnification factor again becomes finite. Its absolute value diminishes with the increase of the ratio Ω/ω and approaches zero when this ratio becomes very large. Thus, when a periodic force of high frequency acts on the body, it produces vibrations of very small amplitude; and in many cases the body may be considered as remaining stationary.

Considering the sign of the expression $1/(1 - \Omega^2/\omega^2)$, one sees that for all cases where $\Omega < \omega$ this expression is positive; and the displacement of the vibrating mass is in the same direction as that of the disturbing force. On the other hand, in all cases where $\Omega > \omega$, the expression is negative; and the displacement of the mass is in the direction opposite to that of the force. For the first case the vibration is said to be *in phase* with the excitation, while in the latter case the response is said to be *out of phase.*

In the above discussion the disturbing force was taken to be proportional to $\sin \Omega t$, but the same conclusions will be reached if it is taken proportional to $\cos \Omega t$. Furthermore, it is also possible to produce forced vibrations by means of *periodic support motion* (or *ground motion*). For example, consider the spring-suspended weight in Fig. 1.23, and suppose that the upper end of the spring is given a simple harmonic motion

$$u_g = d \sin \Omega t \tag{f}$$

in the vertical direction. If we measure the displacement u of the suspended weight W from its equilibrium position (when $u_g = 0$), the elongation of the spring at any instant t will be $u - u_g + \delta_{st}$; and the corresponding force in the spring is $k(u - u_g) + W$. Thus, the equation of motion of the suspended weight becomes

$$\frac{W}{g} \ddot{u} = W - [W + k(u - u_g)] \tag{g}$$

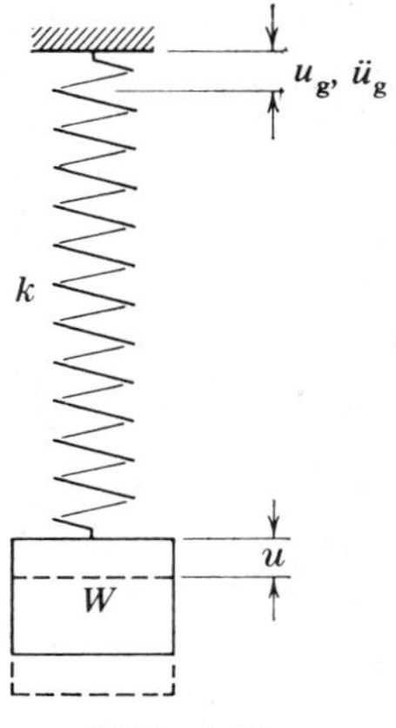

FIG. 1.23

Substituting for u_g its expression (f) and using the notations

$$\omega^2 = \frac{k}{m} = \frac{kg}{W} \qquad \text{and} \qquad q_g = \frac{kd}{m} = \frac{kgd}{W} \tag{h}$$

we obtain

$$\ddot{u} + \omega^2 u = q_g \sin \Omega t \tag{1.25}$$

which is mathematically the same as Eq. (1.22) obtained before. Thus, it may be concluded that giving the upper end of the spring a simple harmonic motion $d \sin \Omega t$ is equivalent to the direct application of a disturbing force $(kd)\sin \Omega t$. All previous statements about the solution of Eq. (1.22) hold also in this case, and we will again obtain steady-state forced vibrations defined by the equation

$$u = (d \sin \Omega t)\left(\frac{1}{1 - \Omega^2/\omega^2}\right) \tag{1.26}$$

The term $d \sin \Omega t$ in Eq. (1.26) may be considered to represent the motion of the mass when the support displacement occurs very slowly (or "statically"), and the factor $1/(1 - \Omega^2/\omega^2)$ accounts for the fact that the frequency of the support motion is nonzero. Thus, for any such problem we need only consider the displacement of the mass due to the static displacement of the support in order to calculate the steady-state response of the system.

In some cases it is more convenient to deal with ground accelerations than ground displacements because a measuring device called an *accelerometer* has been used to obtain information about the ground motion. For example, earthquake ground motions are typically measured and reported in terms of three orthonormal components (north-south, east-west, and vertical) of ground acceleration. Therefore, we shall re-examine the ground-motion problem by specifying a *periodic ground acceleration* instead of a periodic displacement.

Assume now that the upper end of the spring in Fig. 1.23 is subjected to the following condition of harmonic acceleration:

$$\ddot{u}_g = a \sin \Omega t \tag{i}$$

The rearranged form of Eq. (g) for this problem is

$$\frac{W}{g}\ddot{u} + k(u - u_g) = 0 \tag{j}$$

For the purpose of utilizing expression (i) in Eq. (j), the following change of coordinates is required:

$$u^* = u - u_g \qquad \ddot{u}^* = \ddot{u} - \ddot{u}_g \tag{k}$$

where the symbol u^* represents *relative displacement* of the mass with respect to ground. Substituting $u - u_g$ and $\ddot{u}$ from expressions (k) into Eq. (j) and rearranging, we obtain

$$\frac{W}{g}\ddot{u}^* + ku^* = -m\ddot{u}_g = -\frac{W}{g}\ddot{u}_g \tag{l}$$

Using the notations

$$\omega^2 = \frac{k}{m} = \frac{kg}{W} \qquad \text{and} \qquad q_g^* = -a \tag{m}$$

and substituting Eq. (i) into Eq. (l), we find

$$\ddot{u}^* + \omega^2 u^* = q_g^* \sin \Omega t \tag{1.27}$$

which is mathematically the same as Eqs. (1.22) and (1.25).

The solution of Eq. (1.27) follows the familiar pattern of the previous cases, and we may conclude that the response of the system in the *relative coordinate* (as defined by expressions k) is the same as that obtained by the application of a periodic force equal to $-ma \sin \Omega t = -(W/g)a \sin \Omega t$. In this case the steady-state forced vibration of the system relative to the moving support is given by

$$u^* = \left(-\frac{Wa}{kg}\sin \Omega t\right)\left(\frac{1}{1 - \Omega^2/\omega^2}\right) \tag{1.28}$$

Unless the initial ground displacement and velocity are available, the absolute motion of the mass cannot be computed. However, this fact is usually of no consequence, because the relative motion determines the force in the structure (which in this case is a simple spring).

Example 1. Determine the amplitude of forced torsional vibration of the shaft in Fig. 1.1*d* produced by a periodic torque $M \sin \Omega t$ if the free torsional vibration has the frequency $f = 10$ cps, the forcing frequency is $\Omega = 10\pi$ rad/sec, and the angle of twist produced by the torque M (acting statically) is equal to 0.01 rad.
Solution. The equation of motion in this case is (see Sec. 1.3)

$$\ddot{\phi} + \omega^2 \phi = \frac{M}{I} \sin \Omega t \qquad (n)$$

where ϕ is the angle of twist, and $\omega^2 = k_r/I$. The forced vibration is

$$\phi = \frac{M}{I(\omega^2 - \Omega^2)} \sin \Omega t = \frac{M}{k_r(1 - \Omega^2/\omega^2)} \sin \Omega t \qquad (o)$$

Noting that $M/k_r = 0.01$ and $\omega = 2\pi f = 20\,\pi$, we obtain the required amplitude

$$\phi_{\mathrm{m}} = \frac{0.01}{(1 - \frac{1}{4})} = 0.0133 \text{ rad}$$

Example 2. A wheel rolls along a wavy surface with constant horizontal speed v, as shown in Fig. 1.24. Determine the amplitude of the forced vertical vibrations of the load W attached to the axle of the wheel by a spring. Assume that the static deflection of the spring under the action of the load W is $\delta_{\mathrm{st}} = 3.86$ in., $v = 60$ ft/sec, and the wavy surface is defined by the equation $y = d \sin \pi x/\ell$, in which $d = 1$ in. and $\ell = 36$ in. Neglect the mass of the wheel.

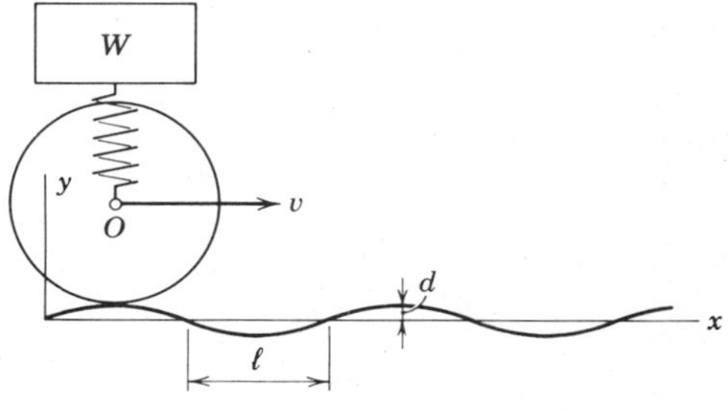

FIG. 1.24

Solution. Considering vertical vibrations of the load W on the spring, we find that the square of the angular frequency of these vibrations is $\omega^2 = g/\delta_{st} = 100\ \text{sec}^{-2}$. Due to the wavy surface, the center O of the rolling wheel oscillates vertically. Assuming that at the initial moment $(t = 0)$ the point of contact of the wheel is at $x = 0$ and putting $x = vt$, we specify these vertical oscillations by the equation $y = d \sin \pi vt/\ell$. The forced vibration of the load W is now obtained from Eq. (1.26) by substituting therein $d = 1$ in., $\Omega = \pi v/\ell = 20\pi\ \text{sec}^{-1}$, and $\omega^2 = 100\ \text{sec}^{-2}$. The amplitude of this forced vibration is $1/(4\pi^2 - 1) = 0.026$ in. At the given speed $v = 60$ ft/sec, the vertical oscillations of the wheel are transmitted to the load W only in a very small proportion. If we take the speed v of the wheel $\frac{1}{4}$ as great, we obtain $\Omega = 5\pi$; and the amplitude of forced vibration becomes $1/(\pi^2/4 - 1) = 0.68$ in. By further decrease in speed v we finally come to the condition of resonance (when $\pi v/\ell = \omega$), at which condition severe vibration of the load W will be produced.

Example 3. Refer to the braced platform in Example 3 of Sec. 1.2, and assume that a harmonic ground acceleration [as given by Eq. (i)] occurs in the x direction. Determine the steady-state amplitude of forced vibration of the platform if $a = 5\ \text{m/sec}^2$ and $\Omega = 40$ rad/sec.

Solution. Equation (1.28) gives the steady-state response of the system relative to ground, and the amplitude of this motion is

$$u_m^* = \frac{ma}{k}\left(\frac{1}{1 - \Omega^2/\omega^2}\right) \tag{p}$$

From the referenced example we have $m = 20$ Mg and $k = 48.8$ MN/m. Therefore,

$$\omega^2 = \frac{k}{m} = \frac{48.8}{20} = 2440\ \text{sec}^{-2}$$

The amplitude of the equivalent statically applied force is

$$ma = (20)(5) = 100\ \text{kN}$$

and the magnification factor in this case is

$$\beta = \frac{1}{1 - \Omega^2/\omega^2} = \frac{1}{1 - (1600/2440)} = 2.90$$

Finally, the amplitude of forced vibration in relative coordinates is obtained from Eq. (p) as

$$u_m^* = \frac{(100)(2.90)}{(48.8)} = 5.94\ \text{mm}$$

The change of stress in the bracing wires caused by this amount of displacement is

$$\frac{\sqrt{2}ku_{\mathrm{m}}^*}{4A} = \frac{\sqrt{2}(48.8)(5.94 \times 10^{-3})}{(4)(5 \times 10^{-4})} = 205 \text{ MPa}$$

Example 4. Suppose that a vertical periodic force $P \sin \Omega t$ is applied directly to the spring in Fig. 1.2*a* at a point located the distance c from the support. What will be the steady-state response of the weight W?
Solution. Consider the spring to be composed of two segments, as shown in Fig. 1.6*a*. The segment of length c has a spring constant designated by k_1, and the other segment has a spring constant represented by k_2. From Example 2, Sec. 1.2, we have the relationship

$$k = \frac{k_1 k_2}{k_1 + k_2} \tag{q}$$

At any instant the amount of force transmitted to the mass through the massless spring is

$$F(t) = \frac{k_2}{k_1 + k_2} P \sin \Omega t = \frac{k_1 k_2}{k_1 + k_2} \frac{P \sin \Omega t}{k_1} \tag{r}$$

The second factor in Eq. (r) represents the deflection of the mass when the forcing function is applied statically. Introducing the notation

$$\Delta_{\mathrm{st}} = \frac{P}{k_1} \tag{s}$$

we may write Eq. (r) as

$$F(t) = k\Delta_{\mathrm{st}} \sin \Omega t \tag{t}$$

The equivalent forcing function may be used in place of $P \sin \Omega t$ in Eq. (1.24) to obtain the steady-state response

$$x = (\Delta_{\mathrm{st}} \sin \Omega t)\left(\frac{1}{1 - \Omega^2/\omega^2}\right) \tag{u}$$

From this result we conclude that only the static displacement of the mass due to the forcing function need be considered, regardless of the point of application of the forcing function.

1.7 FORCED VIBRATIONS: TRANSIENT STATE

In the preceding section only the last term in Eq. (1.23), representing forced vibrations, was considered. In general, the application of a disturbing force also produces some free vibrations of the system, as represented by the first two terms of Eq. (1.23). Thus, the actual motion is a superposition of two harmonic motions having different amplitudes and different frequencies, resulting in a very complicated motion. However, because of damping, not considered in the derivation of Eq. (1.23), the free vibration disappears after a short time; and we are left only with the steady-state forced vibrations, which are constantly maintained by the action of the disturbing force. A particular case is illustrated graphically by the displacement curve in Fig. 1.25. On the dashed curve, representing forced vibrations with the angular frequency Ω, are superimposed free vibrations with a higher angular frequency ω and with decreasing amplitude due to damping. Thus, the complete motion is given by the solid curve, which gradually approaches the dashed curve as a steady-state condition. The early part of this motion, that is, the first few cycles in which free vibrations are present, is generally referred to as the *transient state*. It is sometimes of practical interest to study this motion in detail.

The amplitude of the free vibration can be found from Eq. (1.23) by taking into consideration the initial conditions. As in Sec. 1.2, we let $u = u_0$ and $\dot{u} = \dot{u}_0$ at time $t = 0$. Substituting these conditions into Eq. (1.23) and its first derivative with respect to time, we obtain

$$C_1 = u_0 \qquad \text{and} \qquad C_2 = \frac{\dot{u}_0}{\omega} - \frac{q\Omega/\omega}{\omega^2 - \Omega^2} \tag{a}$$

Substitution of these values into Eq. (1.23) produces

$$u = u_0 \cos \omega t + \frac{\dot{u}_0}{\omega} \sin \omega t + \frac{q}{\omega^2 - \Omega^2}\left(\sin \Omega t - \frac{\Omega}{\omega} \sin \omega t\right) \tag{1.29a}$$

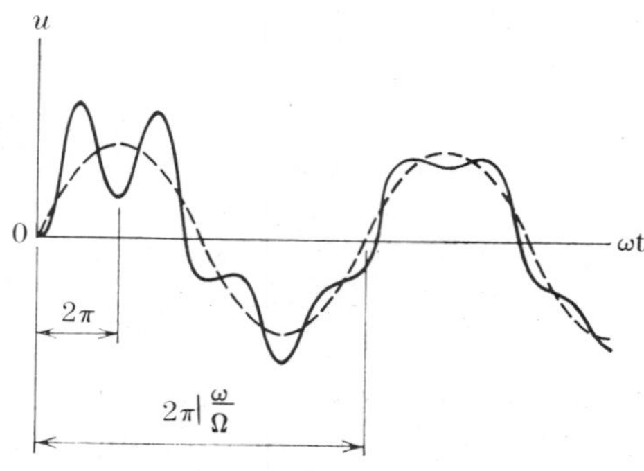

FIG. 1.25

If the initial conditions are taken to be $u_0 = \dot{u}_0 = 0$, this equation simplifies to:

$$u = \frac{q}{\omega^2 - \Omega^2}\left(\sin \Omega t - \frac{\Omega}{\omega} \sin \omega t\right) \tag{1.29b}$$

Equation (1.29b) represents the response to the forcing function $P \sin \Omega t$, and it is seen to consist of two parts. The first part is the steady-state response (discussed in the preceding section) proportional to $\sin \Omega t$, while the second part consists of a free vibration proportional to $\sin \omega t$. Their sum is not a harmonic motion even though it is composed of two harmonic functions, because the components have different frequencies.

If the forcing function is taken to be $P \cos \Omega t$ instead of $P \sin \Omega t$, the term $\cos \Omega t$ replaces $\sin \Omega t$ in Eq. (1.23). In this case the initial conditions result in the following constants of integration:

$$C_1 = u_0 - \frac{q}{\omega^2 - \Omega^2} \qquad \text{and} \qquad C_2 = \frac{\dot{u}_0}{\omega} \tag{b}$$

Substitution of these values into the solution gives

$$u = u_0 \cos \omega t + \frac{\dot{u}_0}{\omega} \sin \omega t + \frac{q}{\omega^2 - \Omega^2}(\cos \Omega t - \cos \omega t) \tag{1.30a}$$

If the initial conditions are taken to be $u_0 = \dot{u}_0 = 0$, this equation becomes

$$u = \frac{q}{\omega^2 - \Omega^2}(\cos \Omega t - \cos \omega t) \tag{1.30b}$$

In this case the free-vibration part of the response has the same amplitude as the steady-state part, regardless of the ratio Ω/ω.

Of special interest is the case when the frequency of the forcing function is equal to or very close to the frequency of free vibrations of the system, that is, Ω is close to ω. In exploring this case, let us introduce the notation

$$\omega - \Omega = 2\epsilon \tag{c}$$

where ϵ is a small quantity. Then rewrite Eq. (1.29b) (response due to the forcing function $P \sin \Omega t$) in the following equivalent form:†

$$u = \frac{q/\omega}{\omega^2 - \Omega^2}\left[\frac{\omega + \Omega}{2}(\sin \Omega t - \sin \omega t) + \frac{\omega - \Omega}{2}(\sin \Omega t + \sin \omega t)\right] \tag{d}$$

† This solution was contributed by C. C. Wang, Manager, Mechanical Engineering Dept., Interactive Technology, Inc., Santa Clara, California (private communication, 1970).

Substitution of trigonometric identities into Eq. (d) produces

$$u = \frac{q/\omega}{\omega^2 - \Omega^2}\left[(\omega + \Omega)\cos\frac{(\Omega + \omega)t}{2}\sin\frac{(\Omega - \omega)t}{2}\right.$$
$$\left. + (\omega - \Omega)\sin\frac{(\Omega + \omega)t}{2}\cos\frac{(\Omega - \omega)t}{2}\right] \qquad (e)$$

Using expression (c) in Eq. (e) and simplifying, we obtain

$$u = -\frac{q}{2\omega}\left[\frac{\sin \epsilon t}{\epsilon}\cos(\omega - \epsilon)t - \frac{\cos \epsilon t}{\omega - \epsilon}\sin(\omega - \epsilon)t\right] \qquad (f)$$

Evaluating this equation in the limit, we have†

$$\lim_{\epsilon \to 0} u = -\frac{q}{2\omega^2}(\omega t \cos \omega t - \sin \omega t) \qquad (1.31a)$$

In phase-angle form this expression becomes

$$u = -\frac{q}{\omega^2} A \cos(\omega t - \alpha) \qquad (g)$$

where

$$A = \frac{1}{2}\sqrt{(\omega t)^2 + 1} \qquad \text{and} \qquad \alpha = \tan^{-1}\frac{-1}{\omega t} \qquad (h)$$

Thus, in the limiting case when $\Omega = \omega$, the amplitude of vibration increases indefinitely with time, as shown in Fig. 1.26. The solid curve in that figure is a dimensionless plot of Eq. (1.31a), whereas the dashed curve is a similar plot of the first term only. It can be seen that after a short time the first term represents a good approximation to total response, as follows:

$$\lim_{\epsilon \to 0} u \approx -\frac{qt}{2\omega}\cos \omega t \qquad (1.31b)$$

The curves in Fig. 1.26 show that the system theoretically attains an infinite amplitude of forced vibration at resonance in the absence of damping, but this amplitude requires infinite time to build up. Thus, in the case of a machine designed to operate above resonance, no great difficult will be experienced in passing through the resonance condition, provided that this transition is made fairly rapidly. However, experiments show that if a

† Equation (1.31a) may also be obtained by application of L'Hôpital's rule to Eq. (1.29b).

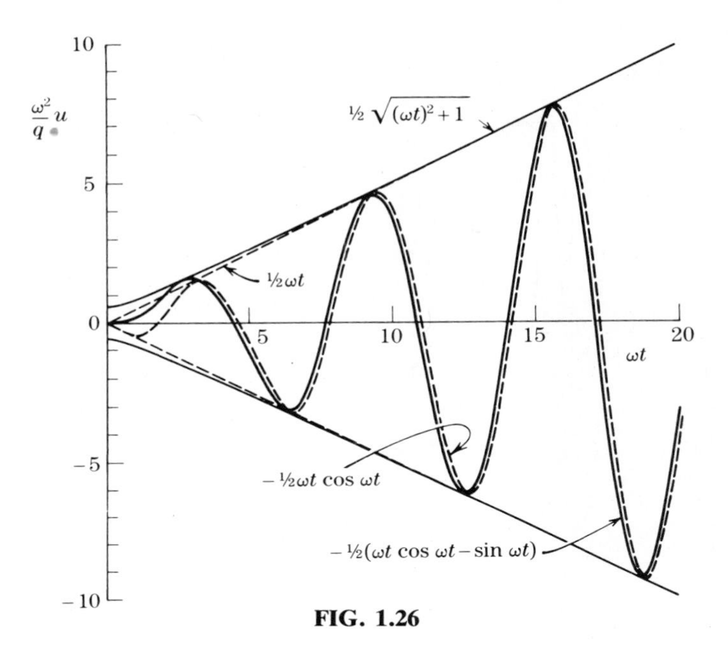

FIG. 1.26

vibrating system is allowed to reach a steady state just below resonance, it then becomes difficult to accelerate the machine through the resonance condition. Additional power supplied for this purpose increases the amplitude of vibration instead of the running speed of the machine.

When the frequency of the forcing function is close to (but not exactly equal to) that of the vibrating system, a phenomenon known as *beating* may be observed. Equation (f) represents this situation, and we may obtain a good approximation to the response by considering a simplified form of the first term, as follows:

$$u \approx -\frac{q \sin \epsilon t}{2\omega\epsilon} \cos \omega t \tag{1.32}$$

Because the quantity ϵ in Eq. (1.32) is small, the function $\sin \epsilon t$ varies slowly; and its period, equal to $2\pi/\epsilon$, is large. Therfore, Eq. (1.32) may be recognized as representing vibrations with period $2\pi/\omega$ and of variable amplitude equal to $(q/2\omega\epsilon) \sin \epsilon t$. This kind of vibration builds up and diminishes in a regular pattern of beats, as indicated in Fig. 1.27. The period of beating, equal to π/ϵ, increases as Ω approaches ω (or as $\epsilon \to 0$). At resonance, the period of beating becomes infinite, and the buildup is continuous, as shown in Fig. 1.26.

Example. The upper end of the spring in Fig. 1.23 has a uniform

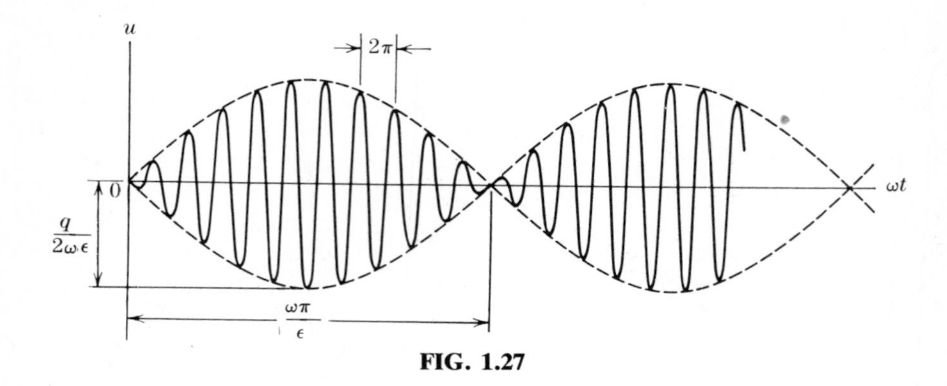

FIG. 1.27

downward velocity v_0 which, at a certain instant $t=0$, becomes a simple harmonic motion

$$u_g = \frac{v_0}{\Omega} \sin \Omega t \tag{i}$$

Determine the complete expression for the subsequent motion of the suspended weight W.
Solution. In this case the initial conditions of motion are

$$u_0 = 0 \qquad \dot{u}_0 = v_0 \tag{j}$$

Substituting these values into expressions (a), we have

$$C_1 = 0 \qquad \text{and} \qquad C_2 = \frac{v_0}{\omega} - \frac{q\Omega/\omega}{\omega^2 - \Omega^2} \tag{k}$$

Substituting these constants back into Eq. (1.23) and noting that in this case

$$q_g = \frac{kgd}{W} = \omega^2 d = \frac{\omega^2 v_0}{\Omega}$$

we obtain for the response

$$u = \frac{v_0/\Omega}{1 - \Omega^2/\omega^2}\left(\sin \Omega t - \frac{\Omega^3}{\omega^3} \sin \omega t\right) \tag{l}$$

1.8 FREE VIBRATIONS WITH VISCOUS DAMPING

In previous discussions of free and forced vibrations we did not consider the effects of dissipative forces, such as those due to friction or air resistance.

As a consequence, we found that the amplitude of free vibrations remains constant with time; but experience shows that the amplitude diminishes with time and that the vibrations are gradually damped out. In the case of forced vibrations, the theory indicates that the amplitude can grow without limit at resonance. However, we know that because of damping there is always some finite amplitude of steady-state response, even at resonance.

To bring an analytic discussion of vibrations into better agreement with actual conditions, *damping forces* must be taken into consideration. These damping forces may arise from several different sources, such as friction between dry sliding surfaces, friction between lubricated surfaces, air or fluid resistance, electrical damping, internal friction due to imperfect elasticity of materials, etc. Among all these sources of energy dissipation, the case in which the damping force is proportional to velocity, called *viscous damping*, is the simplest to deal with mathematically. For this reason resisting forces of a complicated nature are very often replaced, for purposes of analysis, by *equivalent viscous damping*. This equivalent damping is determined in such a manner as to produce the same dissipation of energy per cycle as that produced by the actual resisting forces. For example, damping due to internal friction can be treated by this approach.

We shall now consider the case of a spring–mass system that includes viscous damping, as indicated by the dashpot in Fig. 1.28. It is assumed that a viscous fluid in the dashpot resists motion in proportion to the velocity. In this case the differential equation of motion is

$$\frac{W}{g}\ddot{u} = W - (W + ku) - c\dot{u} \tag{a}$$

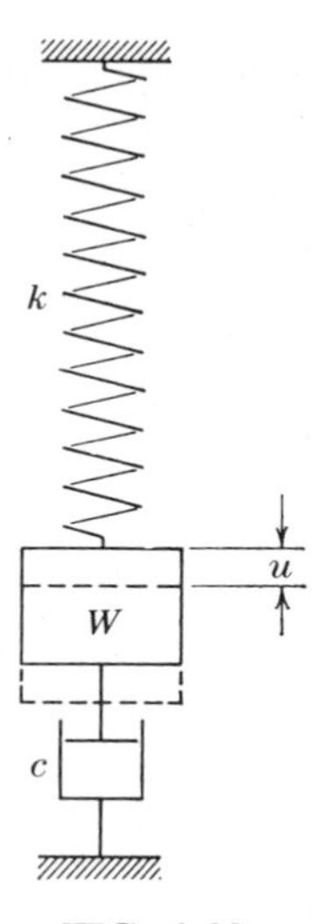

FIG. 1.28

The coefficient c in Eq. (a) represents the *coefficient of viscous damping,* or *damping constant*, which has dimensions of force per unit velocity. The minus sign on the damping force denotes the fact that it always acts in the direction opposite to the velocity. Dividing Eq. (a) by $m = W/g$ and using notations

$$\omega^2 = \frac{k}{m} = \frac{kg}{W} \qquad \text{and} \qquad 2n = \frac{c}{m} = \frac{cg}{W} \tag{b}$$

we obtain for *free vibrations with viscous damping* the following equation

$$\ddot{u} + 2n\dot{u} + \omega^2 u = 0 \tag{1.33}$$

In discussing this equation, we apply the usual method of solving linear differential equations with constant coefficients, and assume a solution in the form

$$u = Ce^{rt} \tag{c}$$

in which e is the base of natural logarithms, t is time, and r is a constant which must be determined from the condition that expression (c) satisfies Eq. (1.33). Substituting (c) into Eq. (1.33) we obtain

$$r^2 + 2nr + \omega^2 = 0$$

from which

$$r = -n \pm \sqrt{n^2 - \omega^2} \tag{d}$$

Let us consider first the case when the quantity n^2, depending on damping, is smaller than the quantity ω^2. In such a case the quantity

$$\omega_{\mathrm{d}}^2 = \omega^2 - n^2$$

is positive, and we obtain for r two complex roots:

$$r_1 = -n + i\omega_{\mathrm{d}} \qquad \text{and} \qquad r_2 = -n - i\omega_{\mathrm{d}}$$

Substituting these roots into expression (c), we obtain two solutions of Eq. (1.33). The sum or the difference of these two solutions multiplied by any constant will also be a solution. In this manner, we find

$$u_1 = \frac{C_1}{2}(e^{r_1 t} + e^{r_2 t}) = C_1 e^{-nt} \cos \omega_{\mathrm{d}} t$$

$$u_2 = \frac{C_2}{2i}(e^{r_1 t} - e^{r_2 t}) = C_2 e^{-nt} \sin \omega_{\mathrm{d}} t$$

Adding these solutions, we obtain the general solution of Eq. (1.33) in the following form:

$$u = e^{-nt}(C_1 \cos \omega_d t + C_2 \sin \omega_d t) \tag{1.34}$$

in which C_1 and C_2 are constants that must be determined from the initial conditions. The factor e^{-nt} in solution (1.34) decreases with time, and the vibrations originally generated will be gradually damped out.

The expression in the parentheses of Eq. (1.34) is of the same form that we had before for vibrations without damping [see Eq. (1.2)]. It represents a periodic function with the angular frequency ω_d, where

$$\omega_d = \sqrt{\omega^2 - n^2} \tag{e}$$

is called the *angular frequency of damped vibration*. The corresponding period is

$$\tau_d = \frac{2\pi}{\omega_d} = \frac{2\pi}{\omega} \frac{1}{\sqrt{1 - (n^2/\omega^2)}} \tag{f}$$

Comparing this expression with the period $\tau = 2\pi/\omega$, obtained before for vibrations without damping, we seen that the *period of damped vibration*, τ_d, is higher. However, if n is small in comparison with ω, this increase is a small quantity that can be neglected. Even if the *damping ratio* n/ω is as high as 0.2, the frequency ratio ω_d/ω is close to unity, as can be seen in Fig. 1.29, where the expression

$$\frac{\omega_d}{\omega} = \sqrt{1 - \frac{n^2}{\omega^2}}$$

(which is the equation of a circle) is plotted in the first quadrant.

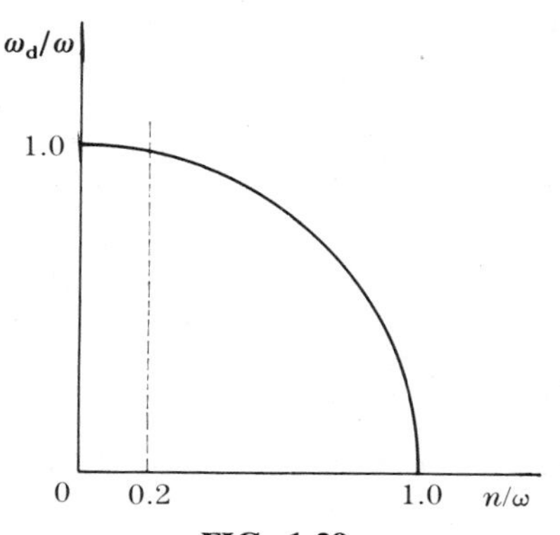

FIG. 1.29

To determine the constants C_1 and C_2 in Eq. (1.34), let us assume that at the initial instant ($t = 0$) the vibrating body is displaced from its position of equilibrium by the amount u_0 and has an initial velocity $\dot{u}_0$. Substituting these quantities into Eq. (1.34) and its first derivative with respect to time, we obtain

$$C_1 = u_0 \qquad \text{and} \qquad C_2 = \frac{\dot{u}_0 + nu_0}{\omega_d} \tag{g}$$

Substitution of expressions (g) into Eq. (1.34) yields

$$u = e^{-nt}\left(u_0 \cos \omega_d t + \frac{\dot{u}_0 + nu_0}{\omega_d} \sin \omega_d t\right) \tag{1.35}$$

The first term in this expression, proportional to $\cos \omega_d t$, depends only on the initial displacement u_0; and the second term, proportional to $\sin \omega_d t$, depends on both the initial displacement u_0 and the initial velocity $\dot{u}_0$.

Equation (1.35) can be written in the equivalent form

$$u = Ae^{-nt} \cos(\omega_d t - \alpha_d) \tag{1.36}$$

in which the maximum value is

$$A = \sqrt{C_1^2 + C_2^2} = \sqrt{u_0^2 + \frac{(\dot{u}_0 + nu_0)^2}{\omega_d^2}} \tag{h}$$

and

$$\alpha_d = \tan^{-1}\frac{C_2}{C_1} = \tan^{-1}\left(\frac{\dot{u}_0 + nu_0}{\omega_d u_0}\right) \tag{i}$$

We may regard Eq. (1.36) as representing a pseudoharmonic motion having an exponentially decreasing amplitude Ae^{-nt}, a phase angle α_d, and a period $\tau_d = 2\pi/\omega_d$. A graph of the motion is shown in Fig. 1.30. This displacement–time curve is tangent to the envelopes $\pm Ae^{-nt}$ at the points m_1, m_1', m_2, m_2', and so on, at instants of time separated by the interval $\tau_d/2$. Because the tangents at these points are not horizontal, the points of tangency do not coincide with the points of extreme displacement from the equilibrium position. If the damping ratio is small, the difference in these points may be neglected. In any case, however, the time interval between two consecutive extreme positions is constant and is equal to half the period $\tau_d/2$. To prove this statement, we differentiate Eq. (1.36) once with respect to time and obtain for the velocity of the vibrating body

$$\dot{u} = -Ae^{-nt}\omega_d \sin(\omega_d t - \alpha_d) - Ane^{-nt} \cos(\omega_d t - \alpha_d)$$

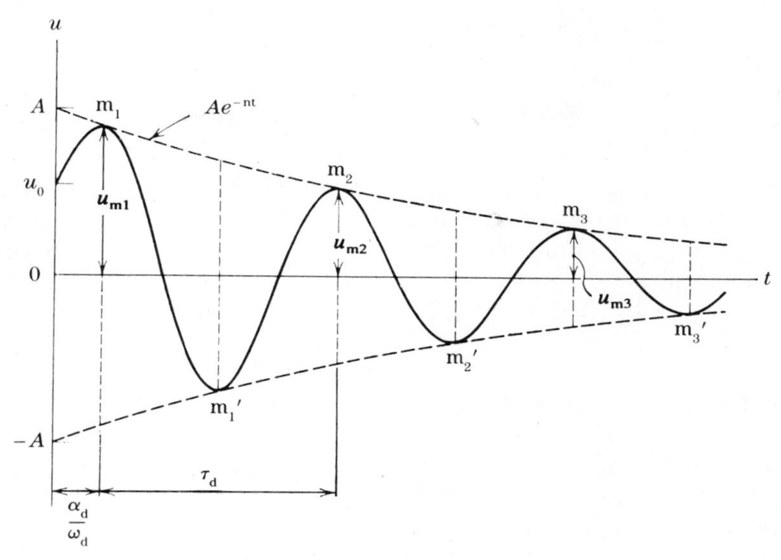

FIG. 1.30

Setting this velocity equal to zero, we find

$$\tan(\omega_d t - \alpha_d) = -\frac{n}{\omega_d}$$

Thus, points of extreme displacement (zero velocity) are separated by equal intervals of time $t = \pi/\omega_d = \tau_d/2$.

The rate of damping depends upon the damping ratio n/ω, and it is seen from Fig. 1.30 that the ratio of two successive amplitudes u_{mi} and $u_{m(i+1)}$ is

$$\frac{u_{mi}}{u_{m(i+1)}} = \frac{Ae^{-nt_i}}{Ae^{-n(t_i+\tau_d)}} = e^{n\tau_d} = e^{\delta} \tag{j}$$

The quantity $\delta = n\tau_d$ is called the *logarithmic decrement*, in accordance with the following expression:

$$\delta = \ln \frac{u_{mi}}{u_{m(i+1)}} = n\tau_d = \frac{2\pi n}{\omega_d} \approx \frac{2\pi n}{\omega} \tag{1.37}$$

Equation (1.37) can be used for an experimental determination of the damping coefficient n. It is only necessary to determine by experiment the ratio of two successive amplitudes of vibration. However, greater accuracy of the result is obtained if the ratio of two amplitudes j cycles apart is utilized. In this case Eq. (j) becomes

$$\frac{u_{mi}}{u_{m(i+j)}} = e^{jn\tau_d} \tag{k}$$

and the logarithmic decrement is

$$\delta = \frac{1}{j} \ln \frac{u_{mi}}{u_{m(i+j)}} \tag{l}$$

In the foregoing discussion of Eq. (1.33) we assumed that $n < \omega$. If $n > \omega$, both roots (d) become real and are negative. Substituting them into expression (c), we obtain two solutions of Eq. (1.33), and the general solution becomes

$$u = C_1 e^{r_1 t} + C_2 e^{r_2 t} \tag{1.38}$$

In this case the solution is not periodic and does not represent a vibratory motion. The viscous resistance is so large that when the body is displaced from its equilibrium position, it does not vibrate but only creeps gradually back to that position. In such a case, the system is said to be *overdamped,* and the motion is called *aperiodic.*

The constants in Eq. (1.38) may be evaluated by substituting $u = u_0$ and $\dot{u} = \dot{u}_0$ (at time $t = 0$) into the equation and its first derivative. These substitutions yield

$$C_1 + C_2 = u_0 \qquad r_1 C_1 + r_2 C_2 = \dot{u}_0$$

from which

$$C_1 = \frac{\dot{u}_0 - r_2 u_0}{r_1 - r_2} \qquad C_2 = \frac{r_1 u_0 - \dot{u}_0}{r_1 - r_2} \tag{m}$$

Thus, Eq. (1.38) becomes

$$u = \frac{\dot{u}_0 - r_2 u_0}{r_1 - r_2} e^{r_1 t} + \frac{r_1 u_0 - \dot{u}_0}{r_1 - r_2} e^{r_2 t} \tag{1.39}$$

The general appearance of a graph of Eq. (1.39) depends upon n, u_0, and $\dot{u}_0$.

Between the underdamped and overdamped cases lies the special case of $n = \omega$, which is the level of damping where the motion first loses its vibratory character. Using notations (b), we find for this case

$$c_{cr} = 2n \frac{W}{g} = 2\omega \frac{W}{g} = 2\sqrt{\frac{kW}{g}} = 2\sqrt{km} \tag{n}$$

in which the symbol c_{cr} represents *critical damping*. For the critically damped case of $n = \omega$, Eq. (d) shows that $r_1 = r_2 = -\omega$, and $\omega_d = 0$. Neither

Eq. (1.35) nor Eq. (1.39) constitutes the solution, which in this particular case of repeated roots takes the form

$$u = C_1 e^{-\omega t} + C_2 t e^{-\omega t} \tag{1.40}$$

Substituting the initial conditions into Eq. (1.40) and its first derivative, we find

$$C_1 = u_0 \qquad C_2 = \dot{u}_0 + n u_0 \tag{o}$$

Then the general solution becomes

$$u = e^{-\omega t}[u_0 + (\dot{u}_0 + n u_0)t] \tag{1.41}$$

Figure 1.31 shows a series of displacement-time curves represented by Eq. (1.41) for a fixed value of u_0 and several values of $\dot{u}_0$. For the upper two curves (indicated as 1 and 2 in the figure) the initial velocity $\dot{u}_0$ is positive, for curve 3 it is zero, and for curves 4 and 5 it is negative.

In the foregoing discussion, we always considered n as a positive quantity, that is, damping represented a resisting force. Thus, due to its action, energy is dissipated; the amplitude of vibration gradually diminishes; and the motion dies out. There are cases, however, in which energy is brought into the system during motion, and as a result of this the amplitude of vibration grows with time. In such situations, the notion of *negative damping* is sometimes used. From the solution (1.34), we see that if n is negative the factor e^{-nt} grows with time; and the vibrations gradually build up. The case of positive n (in which vibrations are dying out) represents a *stable motion*, while the case of negative n represents an *unstable motion.*

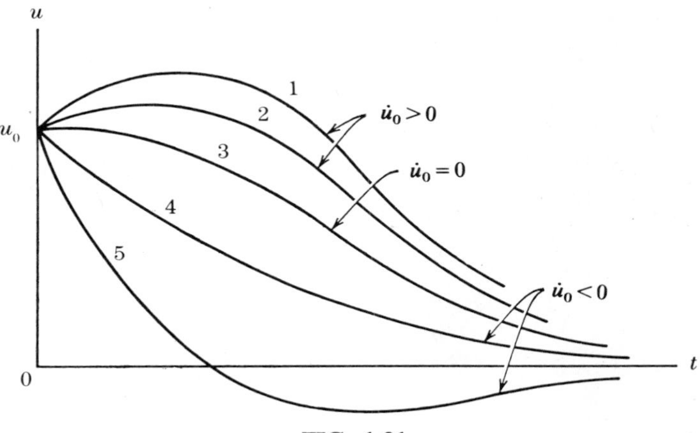

FIG. 1.31

Example 1. A body vibrating with viscous damping makes 10 complete oscillations per second and in 100 cycles its amplitude diminishes 10%. Calculate the logarithmic decrement, and determine the damping constant n and the damping ratio n/ω. In what proportion would the period of vibration be decreased if damping were removed?
Solution. From Eq. (l) the logarithmic decrement is

$$\delta = n\tau_d = \frac{1}{100}\ln\left(\frac{1.0}{0.9}\right) = 0.001054$$

The period $\tau_d = 0.1$ sec, and $\omega_d = 2\pi/\tau_d = 62.83\ \text{sec}^{-1}$. Thus, $n = 0.01054\ \text{sec}^{-1}$, and the damping ratio is approximately

$$\frac{n}{\omega} \approx \frac{n}{\omega_d} = \frac{0.01054}{62.83} = 0.000168$$

Equation (f) shows that the ratio of the period for the undamped case to the damped case is

$$\frac{\tau}{\tau_d} = \sqrt{1 - \frac{n^2}{\omega^2}} \approx \sqrt{1 - \frac{n^2}{\omega_d^2}} = \sqrt{1 - 2.82 \times 10^{-8}}$$

which is very close to unity.

Example 2. Determine the general nature of the displacement–time curve for the motion of a spring-suspended mass, released with initial displacement u_0 and without initial velocity, if the damping is greater than critical, i.e., $n > \omega$.
Solution. Substituting the initial conditions $u = u_0$ and $\dot{u}_0 = 0$ into Eq. (1.39), we find

$$u = \frac{u_0}{r_1 - r_2}(r_1 e^{r_2 t} - r_2 e^{r_1 t}) \tag{p}$$

Differentiating this equation with respect to time, we obtain the following expressions for velocity and acceleration:

$$\dot{u} = \frac{r_1 r_2 x_0}{r_1 - r_2}(e^{r_2 t} - e^{r_1 t}) \tag{q}$$

$$\ddot{u} = \frac{r_1 r_2 x_0}{r_1 - r_2}(r_2 e^{r_2 t} - r_1 e^{r_1 t}) \tag{r}$$

From expression (q), we see that the velocity is zero at $t = 0$ and $t = \infty$, and that it is negative for all intermediate values of t, since both r_1 and r_2 are negative. To find the time t_1 at which this negative velocity is a maximum, we set expression (r) equal to zero and find

$$t_1 = \frac{\ln(r_2/r_1)}{r_1 - r_2} \tag{s}$$

From expressions (p) through (s), we conclude that the displacement–curve has the general shape shown by curve 3 in Fig. 1.31. For any particular system with a given damping coefficient c, the exact details of the curve can be established by remembering that

$$r_1 = -n + \sqrt{n^2 - \omega^2}$$
$$r_2 = -n - \sqrt{n^2 - \omega^2}$$

where n and ω are defined by expressions (b).

1.9 FORCED VIBRATIONS WITH VISCOUS DAMPING

In the preceding section we discussed free vibrations of a spring-suspended mass with viscous damping. We shall now consider the case where, in addition to a spring force $-ku$ and a resisting force $-c\dot{u}$, there is also a harmonically varying force acting externally on the vibrating mass. As we have already seen in Sec. 1.6, such a disturbing force can result from an unbalanced motor running with constant angular velocity Ω. Thus, the rotating centrifugal force Q, acting as shown in Fig. 1.32, has a vertical component $Q \cos \Omega t$. Under such conditions, the equation of motion for the suspended mass m of the motor becomes

$$m\ddot{u} = -ku - c\dot{u} + Q \cos \Omega t \tag{a}$$

Dividing through by m and introducing the notations

$$\omega^2 = \frac{k}{m} \qquad 2n = \frac{c}{m} \qquad q = \frac{Q}{m} \tag{b}$$

we obtain

$$\ddot{u} + 2n\dot{u} + \omega^2 u = q \cos \Omega t \tag{1.42}$$

which is the differential equation of motion for *forced vibrations with viscous damping*. A particular solution of Eq. (1.42) can be taken in the form

$$u = M \cos \Omega t + N \sin \Omega t \tag{1.43}$$

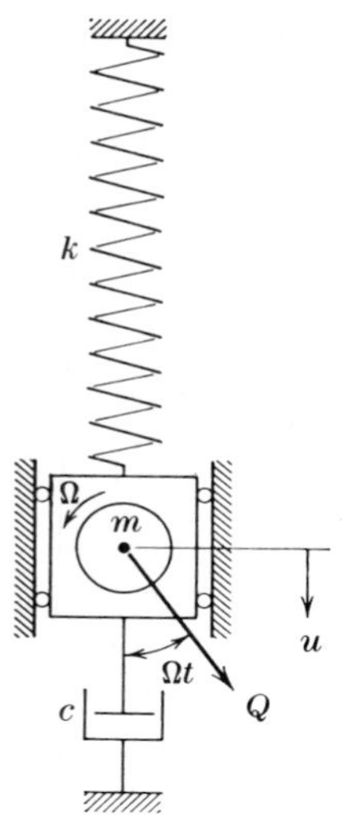

FIG. 1.32

where M and N are constants. To determine these constants, we substitute the trial solution (1.43) into Eq. (1.42) and obtain

$$(-\Omega^2 M + 2n\Omega N + \omega^2 M - q)\cos \Omega t + (-\Omega^2 N - 2n\Omega M + \omega^2 N)\sin \Omega t = 0$$

This equation can be satisfied for all values of t only if the expressions in the parentheses vanish. Thus, for calculating M and N we have two linear algebraic equations

$$-\Omega^2 M + 2n\Omega N + \omega^2 M = q$$
$$-\Omega^2 N - 2n\Omega M + \omega^2 N = 0$$

from which

$$M = \frac{q(\omega^2 - \Omega^2)}{(\omega^2 - \Omega^2)^2 + 4n^2\Omega^2}$$
$$N = \frac{q(2n\Omega)}{(\omega^2 - \Omega^2)^2 + 4n^2\Omega^2} \qquad (c)$$

Substituting these constants into Eq. (1.43), we obtain the particular solution of Eq. (1.42).

The total solution of Eq. (1.42) is obtained by adding the particular solution (1.43) to the general solution obtained as Eq. (1.34) in Sec. 1.8.

Thus, considering only subcritical damping, we have

$$u = e^{-nt}(C_1 \cos \omega_d t + C_2 \sin \omega_d t) + M \cos \Omega t + N \sin \Omega t \qquad (1.44)$$

The first two terms in Eq. (1.44) represent *damped free vibrations*, whereas the last two terms represent *damped forced vibrations*. The free vibrations have the period $\tau_d = 2\pi/\omega_d$, as discussed in the preceding section; and the forced vibrations have the period $T = 2\pi/\Omega$, which is identical with the period of the disturbing force that produces them. We see that due to the factor e^{-nt} the free vibrations gradually subside, leaving only the steady forced vibrations represented by the last two terms. These forced vibrations are maintained indefinitely by the action of the disturbing force and, therefore, are of great practical importance. We have already discussed such forced vibrations without damping in Sec. 1.6, but we shall now see how they are affected by damping.

The steady-state response (Eq. 1.43) may be written in the equivalent phase-angle form as

$$u = A \cos(\Omega t - \theta) \qquad (1.45)$$

where

$$A = \sqrt{M^2 + N^2} = \frac{q}{\sqrt{(\omega^2 - \Omega^2)^2 + 4n^2\Omega^2}} = \frac{q/\omega^2}{\sqrt{(1 - \Omega^2/\omega^2)^2 + 4n^2\Omega^2/\omega^4}} \qquad (d)$$

and

$$\theta = \tan^{-1}\left(\frac{N}{M}\right) = \tan^{-1}\left(\frac{2\Omega n}{\omega^2 - \Omega^2}\right) = \tan^{-1}\left(\frac{2n\Omega/\omega^2}{1 - \Omega^2/\omega^2}\right) \qquad (e)$$

Thus, we see that steady-state forced vibration with viscous damping is a simple harmonic motion having constant amplitude A [as given by expression (d)], phase angle θ [as given by expression (e)], and period $T = 2\pi/\Omega$.

Using the values of ω^2 and q from expressions (b) and introducing the symbol γ for the *damping ratio*

$$\gamma = \frac{n}{\omega} = \frac{c}{c_{cr}} \qquad (f)$$

we may substitute Eq. (d) into Eq. (1.45) to obtain

$$u = \frac{Q}{k} \beta \cos(\Omega t - \theta) \qquad (1.46)$$

in which the *magnification factor* is

$$\beta = \frac{1}{\sqrt{(1 - \Omega^2/\omega^2)^2 + (2\gamma\Omega/\omega)^2}} \qquad (1.47)$$

In addition, the expression for the phase angle [Eq. (*e*)] becomes

$$\theta = \tan^{-1}\left(\frac{2\gamma\Omega/\omega}{1-\Omega^2/\omega^2}\right) \tag{1.48}$$

From Eq. (1.46) we see that the amplitude of the steady-state forced vibration is obtained by multiplying the static-load displacement

$$u_{\mathrm{st}} = \frac{Q}{k} \tag{g}$$

by the magnification factor β. This factor depends not only upon the frequency ratio Ω/ω, but also upon the damping ratio γ.

Figure 1.33 shows the magnification factor β plotted against the ratio Ω/ω for various levels of damping. From these curves we see that when the impressed angular frequency Ω is small compared with the natural angular frequency ω, the value of the magnification factor β is not greatly different from unity. Thus, during vibration the displacements u of the suspended mass are approximately those which would be produced by a static action of the disturbing force $Q \cos \Omega t$.

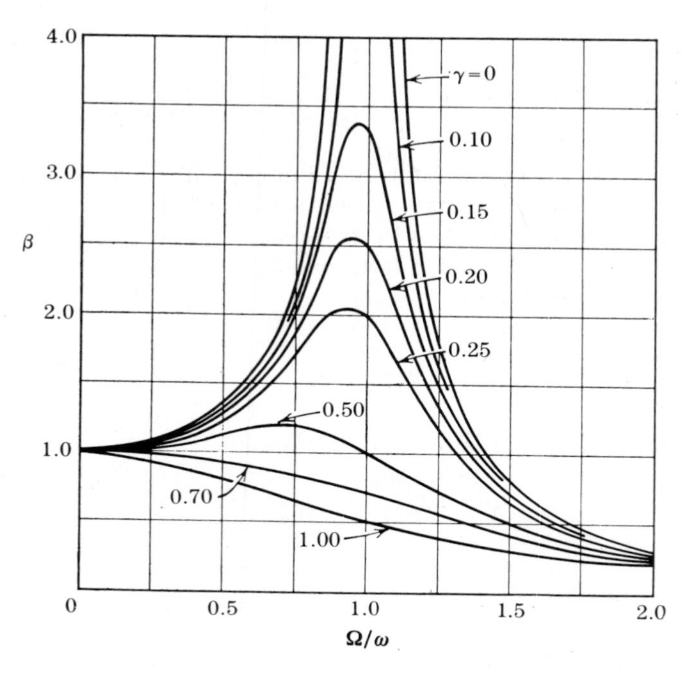

FIG. 1.33

When Ω is large compared with ω (that is, when the impressed frequency is much greater than the natural frequency), the value of the magnification factor tends towards zero, regardless of the amount of damping. This means that a high-frequency disturbing force produces practically no forced vibrations of a system has a low natural frequency. In both extreme cases ($\Omega \ll \omega$ and $\Omega \gg \omega$) we note that damping has only a secondary effect on the value of the magnification factor β. Thus, in these two cases of forced vibrations it is justifiable to neglect the effect of damping entirely and to use the equations previously developed in Sec. 1.6.

As the value of Ω approaches that of ω (that is, as Ω/ω approaches unity), the magnification factor grows rapidly, and its value at or near resonance is very sensitive to changes in the amount of damping. We should also note that the maximum value of β occurs for a value of Ω/ω that is slightly less than unity. Setting the derivative of β with respect to Ω/ω equal to zero, we find that the maximum occurs when

$$\frac{\Omega}{\omega} = \sqrt{1 - 2\gamma^2} \qquad (h)$$

For small damping ratios, the maximum value of β occurs very near to resonance, and it is permissible to take the value of β at resonance as the maximum. Then from Eq. (1.46) and expressions (b) and (f), the maximum amplitude A_m becomes approximately

$$A_m = \frac{Q}{k}\beta_{res} = \frac{Q}{k}\frac{1}{2\gamma} = \frac{Q}{\omega^2 m}\frac{1}{2n/\omega} = \frac{Q}{c\Omega} \qquad (i)$$

We see from this discussion that while damping has only a minor effect on the response of the system in the regions either well below or well above resonance, it is of great importance near the resonance condition and cannot be disregarded if we are to calculate meaningful results.

Let us now consider the phase relationship between the steady-state vibrations and the disturbing force that produces them. This is represented by the phase angle θ in Eq. (1.46), the value of which is given by expression (1.48). Since the disturbing force varies according to $\cos \Omega t$ and the forced vibrations according to $\cos(\Omega t - \theta)$, we say that the response lags the forcing function by the angle θ. That is, when the force Q in Fig. 1.32 is directed downward, the suspended mass on which it acts is not yet in its lowest position; but it arrives there θ/Ω sec later, by which time the force Q has advanced to a position where it makes the angle θ with the vertical. Expression (1.48) shows that the value of θ, like that of β, depends upon both the damping ratio and the frequency ratio. The curves in Fig. 1.34 show the variation in the phase angle θ with the ratio Ω/ω for several levels of damping. For the case of zero damping, the forced vibrations are exactly

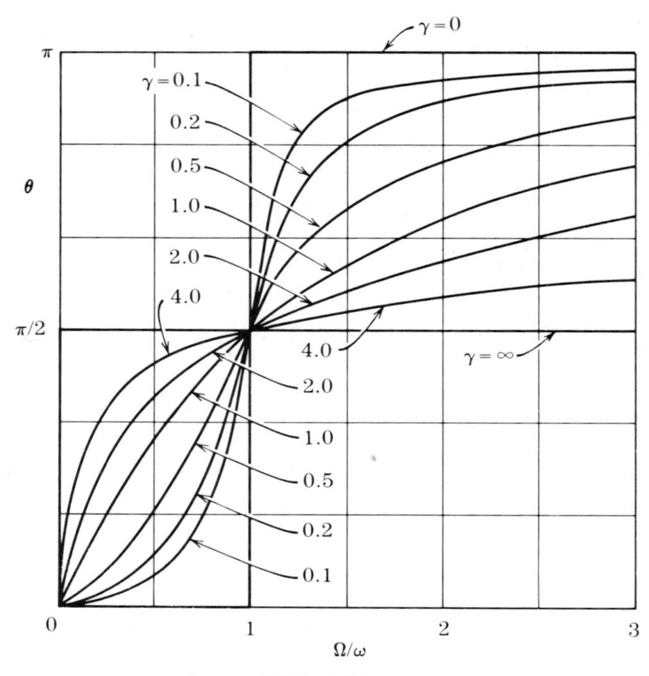

FIG. 1.34

in phase ($\theta = 0$) with the disturbing force for all values of $\Omega/\omega < 1$; and they are a full half cycle out of phase ($\theta = \pi$) for all values of $\Omega/\omega > 1$. Also for this condition, the phase angle is seen to be indeterminate at resonance ($\Omega = \omega$).

When damping is present, we note a continual change in θ as the ratio Ω/ω increases. Also, regardless of the amount of damping, we find $\theta = \pi/2$ at resonance. That is, at resonance the forced vibrations lag the disturbing force by one-quarter cycle. In Fig. 1.32, for example, the force q is directed downward when the vibrating mass passes through its middle position; and by the time the mass has moved to its lowest position, the force Q has rotated through the angle $\pi/2$ and acts horizontally to the right.

For values of Ω/ω either well below or well above resonance, we note that a small damping ratio has only a secondary effect upon the phase angle. That is, well below resonance the angle θ is practically zero, while well above resonance it is practically π. Thus, the effects of damping upon the phase angle can also be ignored except at or near resonance.

Example 1. Let the total weight of the unbalanced motor in Fig. 1.32 be $W = 1000$ lb and the mass of the eccentric weight be $m_1 = 1$ lb-sec^2/in., located at a radius of $r_1 = 1$ in. The speed of the motor is 600 rpm, the static-load deflection of the spring is $\delta_{st} = 0.01$ in., and the damping constant

c is 100 lb-sec/in. Determine the amplitude of steady-state forced vibrations at that speed and also at resonance, where $\Omega = \omega$.

Solution. $\Omega = 2\pi\left(\frac{600}{60}\right) = 20\pi \qquad \Omega^2 = 400\pi^2 \qquad \omega^2 = \frac{g}{\delta_{st}} = 38{,}600$

$$n = \frac{cg}{2W} = \frac{(100)(386)}{(2)(1000)} = 19.3 \qquad q = \frac{gm_1r_1}{W}\Omega^2 = (0.386)(400\pi^2)$$

From Eq. (d)

$$A = \frac{q}{\sqrt{(\omega^2 - \Omega^2)^2 + 4n^2\Omega^2}} = \frac{(0.386)(400\pi^2)}{\sqrt{(38{,}600 - 400\pi^2)^2 + (4)(19.3)^2(400\pi^2)}}$$
$$= 0.044 \text{ in.}$$

At resonance, $\Omega = \omega = \sqrt{38{,}600}$; and from Eq. ($i$) we have

$$A_{\max} = \frac{Q}{c\Omega} = \frac{m_1r_1\Omega}{c} = \frac{\sqrt{38{,}600}}{100} = 1.96 \text{ in.}$$

Example 2. For a damped system, consider the case of harmonic ground displacement (see Fig. 1.35), as given by

$$u_g = d \cos \Omega t \tag{j}$$

Derive expressions for the steady-state response of the system to this forcing function.

Solution. The equation of motion for this case becomes

$$m\ddot{u} = -c(\dot{u} - \dot{u}_g) - k(u - u_g) \tag{k}$$

in which

$$\dot{u}_g = -d\Omega \sin \Omega t \tag{l}$$

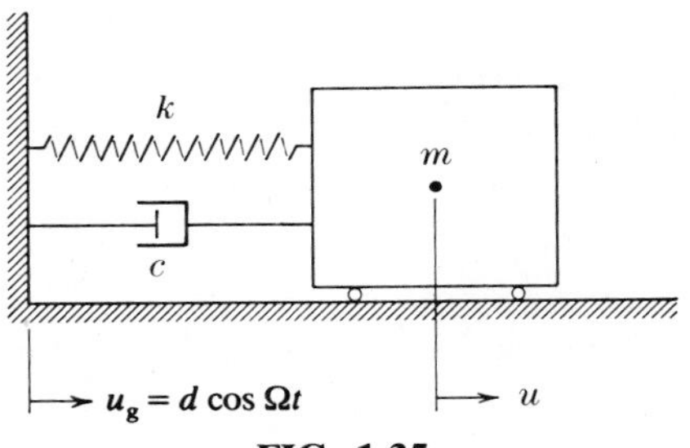

FIG. 1.35

Substituting expressions (*j*) and (*l*) into Eq. (*k*) and rearranging, we obtain

$$m\ddot{u} + c\dot{u} + ku = d(k \cos \Omega t - c\Omega \sin \Omega t) \tag{m}$$

Expressing the right-hand portion in phase-angle form gives:

$$m\ddot{u} + c\dot{u} + ku = Bd \cos(\Omega t - \phi) \tag{n}$$

in which

$$B = \sqrt{k^2 + c^2\Omega^2} \tag{o}$$

and

$$\phi = \tan^{-1}\left(\frac{-c\Omega}{k}\right) \tag{p}$$

In this case we see that the forcing function $Bd \cos(\Omega t - \phi)$ appears in place of $Q \cos \Omega t$, and the solution may be obtained from Eq. (1.46) by replacing Q with the product Bd and by including the phase angle ϕ in the result. Thus,

$$u = \frac{Bd}{k} \beta \cos(\Omega t - \phi - \theta) \tag{q}$$

Example 3. The transient response of a system with subcritical damping may be ascertained by introducing initial conditions into Eq. (1.44). Determine the free-vibrational response of such a system due to the forcing function $Q \cos \Omega t$.

Solution. Substituting $u = u_0$ and $\dot{u} = \dot{u}_0$ (at time $t = 0$) into Eq. (1.44) and its first derivative, we evaluate the constants of integration as

$$C_1 = u_0 - M \qquad \text{and} \qquad C_2 = \frac{\dot{u}_0 + n(u_0 - M) - N\Omega}{\omega_{\mathrm{d}}} \tag{r}$$

Substitution of these expressions into Eq. (1.44) gives

$$\begin{aligned} u &= e^{-nt}\left(u_0 \cos \omega_{\mathrm{d}} t + \frac{\dot{u}_0 + nu_0}{\omega_{\mathrm{d}}} \sin \omega_{\mathrm{d}} t\right) \\ &\quad + M\left[\cos \Omega t - e^{-nt}\left(\cos \omega_{\mathrm{d}} t + \frac{n}{\omega_{\mathrm{d}}} \sin \omega_{\mathrm{d}} t\right)\right] + N\left(\sin \Omega t - e^{-nt} \frac{\Omega}{\omega_{\mathrm{d}}} \sin \omega_{\mathrm{d}} t\right) \end{aligned} \tag{s}$$

If the initial conditions are taken to be $u_0 = \dot{u}_0 = 0$, the transient portion u_{tr} of the remaining response is seen to be

$$u_{tr} = -e^{-nt}\left(M \cos \omega_d t + \frac{Mn + N\Omega}{\omega_d} \sin \omega_d t\right) \tag{t}$$

Expressing this transient in phase-angle form, we have

$$u_{tr} = -e^{-nt} \frac{C}{\omega_d} \cos(\omega_d t - \psi) \tag{u}$$

where

$$C = \sqrt{(M\omega_d)^2 + (Mn + N\Omega)^2} \tag{v}$$

and

$$\psi = \tan^{-1}\left(\frac{Mn + N\Omega}{M\omega_d}\right) \tag{w}$$

1.10 EQUIVALENT VISCOUS DAMPING

As mentioned at the beginning of Sec. 1.8, various types of damping may be replaced by *equivalent viscous damping* [4], resulting in a linear differential equation for harmonic motion. The most important influence of damping occurs at or near resonance in the problem of forced vibrations. Therefore, let us consider the work done by the disturbing force per cycle during the steady-state response discussed in the previous section. In that case the work U_Q done by the force $Q \cos \omega t$ in one cycle is

$$U_Q = \int_0^T Q(\cos \Omega t)\dot{u}\, dt \tag{a}$$

The velocity $\dot{u}$ may be obtained by differentiating Eq. (1.45) with respect to time.

$$\dot{u} = -A\Omega \sin(\Omega t - \theta) \tag{b}$$

Substituting this expression into Eq. (*a*) and using a trigonometric identity, we have

$$U_Q = -QA\Omega \int_0^T (\cos \Omega t)(\sin \Omega t \cos \theta - \cos \Omega t \sin \theta)\, dt$$

Integration results in

$$U_Q = \pi QA \sin\theta \tag{c}$$

Similarly, the work U_c dissipated per cycle by the damping force $c\dot{u}$ is

$$U_c = \int_0^T c\dot{u}\dot{u}\, dt \tag{d}$$

Substitution of Eq. (b) into Eq. (d) gives

$$U_c = cA^2\Omega^2 \int_0^T \sin^2(\Omega t - \theta)\, dt$$

which upon integration becomes

$$U_c = \pi c A^2 \Omega \tag{e}$$

Thus, the input energy U_Q increases linearly with the amplitude A, while the dissipated energy U_c increases as the square of the amplitude. They will be equal where the two energy functions intersect (see Fig. 1.36), and we can find the steady-state amplitude by equating expressions (c) and (e). Hence,

$$A = \frac{Q \sin\theta}{c\Omega} \tag{f}$$

At resonance ($\Omega = \omega$) the phase angle θ (see Eq. 1.48) is $\pi/2$, and the value of A is maximum (for $c \ll c_{cr}$).

$$A_m = \frac{Q}{c\Omega} \tag{g}$$

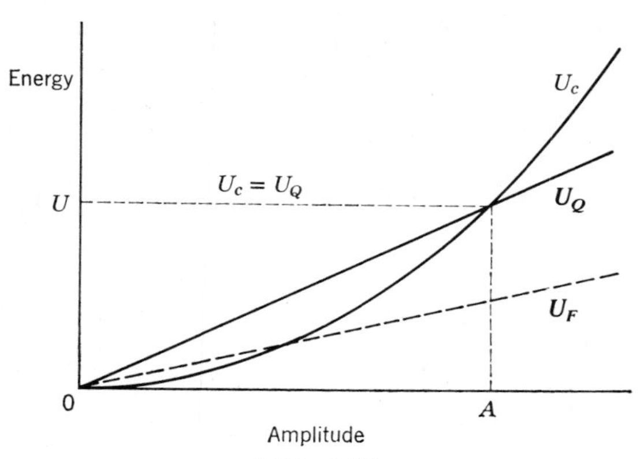

FIG. 1.36

This equation agrees with Eq. (i) obtained in Sec. 1.9 by another approach.

Equation (e) gives the energy dissipated by viscous damping in one cycle of forced vibration. This energy expression may be equated to that due to some other type of damping to obtain an *equivalent viscous damping constant*, c_{eq}. For example, consider *structural damping*, which is attributed to internal friction in structural materials (such as steel and aluminum alloys) that are not perfectly elastic. The energy dissipated per unit volume of material is represented by the hatched area with the *hysteresis loop* in Fig. 1.37. The loop is formed by stress–strain curves for increasing (or "loading") and decreasing (or "unloading") levels of stress and strain. Figure 1.37 shows a complete reversal of stress and strain, corresponding to one cycle of vibration. This damping mechanism dissipates energy approximately in proportion to the square of the strain amplitude [5], and the shape of the hysteresis loop is relatively independent of amplitude and strain rate.

Because amplitude of vibration is proportional to strain amplitude, the work U_s dissipated per cycle by structural damping may be written

$$U_s = sA^2 \tag{h}$$

in which s is a scalar representing the proportionality. Equating expressions (e) and (h) produces the equivalent viscous damping constant as

$$c_{eq} = \frac{s}{\pi\Omega} = \frac{\eta k}{\Omega} \tag{1.49}$$

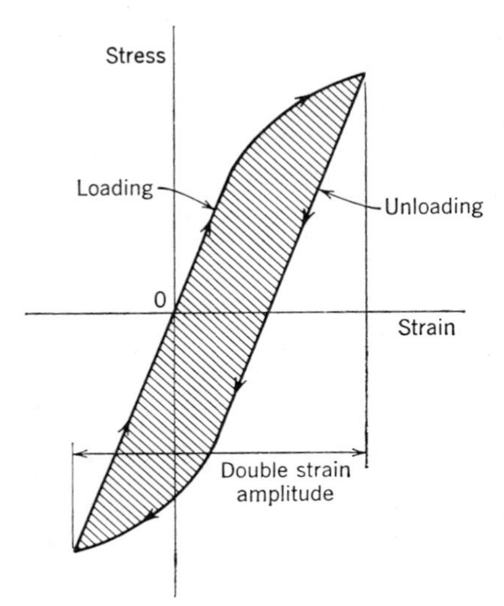

FIG. 1.37

The factor s/π has the units of k and is usually replaced by ηk (as in Eq. 1.49), where the dimensionless quantity

$$\eta = \frac{s}{\pi k} \tag{i}$$

is called the *structural damping factor*. This quantity may be related to the *equivalent viscous damping ratio,* γ_{eq}, by dividing Eq. (1.49) by the definition $c_{cr} = 2\omega m$ [see Eq. (n), Sec. 1.8] and using the notation $k = \omega^2 m$ [from Eq. (b), Sec. 1.9], as follows:

$$\gamma_{eq} = \frac{c_{eq}}{c_{cr}} = \frac{\omega}{2\Omega}\eta \tag{1.50}$$

When this definition for γ_{eq} is substituted into Eq. (1.47), the magnification factor for steady-state response becomes

$$\beta_s = \frac{1}{\sqrt{(1 - \Omega^2/\omega^2)^2 + \eta^2}} \tag{j}$$

Finally, at resonance we have $\gamma_{res} = \eta/2$, $\beta_{res} = 1/\eta$, and from Eqs. (1.49) and (g),

$$A_m = \frac{Q}{k\eta} \tag{k}$$

As a second example of determining equivalent viscous damping, consider Fig. 1.38, in which a body attached to a spring slides on a surface that offers frictional resistance to the motion. In the case of dry friction the laws of Coulomb [6] are usually applied, and it is assumed that the friction force F is proportional to the normal force N acting between the surfaces, so that

$$F = \mu N \tag{l}$$

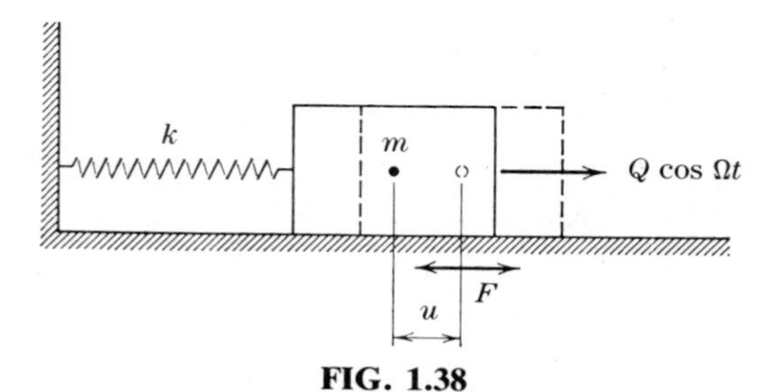

FIG. 1.38

where μ is the *coefficient of friction*. Experiments show that the coefficient of friction during motion is relatively constant (and smaller than the static value) for low velocities. Furthermore, cases involving rolling friction instead of sliding friction may also be handled with Eq. (*l*).

The friction force F in Fig. 1.38 always acts in a direction opposite to that of the velocity of the body, as in the case of a dashpot damper. However, the frictional resistance will be assumed to be constant, regardless of the velocity. This type of damping mechanism is referred to as *Coulomb damping*, and a rigorous solution [7] for the response of the system to a harmonic forcing function is more complex than that for the case of viscous damping. To determine the equivalent viscous damping constant to be used in place of the frictional resistance, we calculate the work U_F dissipated per cycle by the friction force F.

$$U_F = 4AF \tag{m}$$

Equating this expression to Eq. (*e*), we obtain

$$c_{eq} = \frac{4F}{\pi A \Omega} \tag{1.51}$$

In this case the magnitude of c_{eq} depends not only upon F and Ω, but also upon the amplitude A of the vibration. By dividing Eq. (1.51) by the definition $c_{cr} = 2\omega m$ and using the notation $k = \omega^2 m$, we write the equivalent viscous damping ratio as

$$\gamma_{eq} = \frac{c_{eq}}{c_{cr}} = \frac{2F\omega}{\pi A k \Omega} \tag{1.52}$$

When this formula for γ_{eq} is used, the amplitude of the steady-state forced vibration with equivalent viscous damping becomes

$$A = \frac{Q/k}{\sqrt{(1 - \Omega^2/\omega^2)^2 + (4F/\pi A k)^2}} \tag{n}$$

Solving for A, we obtain

$$A = \pm \frac{Q}{k} \frac{\sqrt{1 - (4F/\pi Q)^2}}{1 - \Omega^2/\omega^2} \tag{1.53}$$

The first factor on the right-hand side of Eq. (1.53) represents the static-load displacement, and the second is the magnification factor. We see that this factor has a real value only if

$$\frac{F}{Q} < \frac{\pi}{4} \tag{o}$$

In practical applications, where we are usually dealing with small frictional forces, this condition is satisfied. However, it is also observed that in all cases where condition (o) is satisfied, the magnification factor becomes infinite at resonance (where $\Omega = \omega$). This fact can be explained if we compare the dissipated energy U_F against the work U_Q done by the forcing function at resonance. Solving for F in condition (o) and substituting it into Eq. (m), we get

$$U_F < \pi QA \tag{p}$$

But Eq. (c) shows that the value of U_Q at resonance is πQA. Therefore, we conclude that

$$U_F < U_Q \tag{q}$$

Thus, the energy dissipated per cycle is less than the input energy. This fact is illustrated in Fig. 1.36, where the dashed line representing Eq. (m) is shown with a smaller slope than the solid line representing Eq. (c), provided that condition (o) is satisfied.

For a third example of the concept of equivalent viscous damping, we turn to the case of an oscillating body immersed in a fluid of low viscosity, such as air. If the mass of the body is small and its volume is large, the damping influence of the fluid resistance can be significant. Figure 1.39 depicts a light, hollow sphere subjected to forced vibrations in air, for which

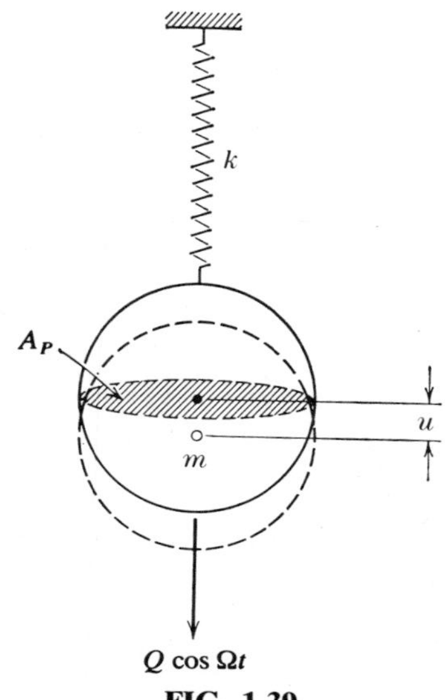

FIG. 1.39

case the resisting force due to the fluid may be approximated by [8]†

$$P = \frac{1}{2} \rho \dot{u}^2 C_D A_P \tag{r}$$

The symbol ρ in Eq. (r) represents the mass density of the fluid, C_D is the coefficient of drag, and A_P denotes the area of the body projected on a plane perpendicular to the direction of motion (see Fig. 1.39). In this case the resisting force is proportional to the square of the velocity and is always in the opposite direction. The work U_P dissipated per cycle by the force P is given by

$$U_P = 4 \int_0^{T/4} P\dot{u}\, dt \tag{s}$$

Substituting Eqs. (r) and (b) into Eq. (s) and letting $C_P = \frac{1}{2}\rho C_D A_P$, we have

$$U_P = 4C_P A^3 \Omega^3 \int_0^{T/4} \sin^3(\Omega t - \theta)\, dt$$

Integration results in

$$U_P = \frac{8}{3} C_P A^3 \Omega^3 \tag{t}$$

Setting this expression equal to Eq. (e) yields

$$c_{eq} = \frac{8C_P A\Omega}{3\pi} \tag{1.54}$$

Thus, the equivalent viscous damping in this case is directly proportional to C_P, A, and Ω. As before, we divide Eq. (1.54) by $c_{cr} = 2\omega m$ and use $k = \omega^2 m$ to obtain the equivalent viscous damping ratio as

$$\gamma_{eq} = \frac{c_{eq}}{c_{cr}} = \frac{4C_P A\Omega\omega}{3\pi k} \tag{1.55}$$

and the amplitude of steady-state forced vibration becomes

$$A = \frac{Q/k}{\sqrt{(1 - \Omega^2/\omega^2)^2 + (8C_P A\Omega^2/3\pi k)^2}} \tag{u}$$

† The drag coefficient C_D is not a constant, but varies with Reynold's number, which is a function of velocity. In the present discussion the use of an average value for C_D is implied.

Squaring Eq. (u) and rearranging terms produces the following fourth-degree polynomial:

$$\left(\frac{8C_P\Omega^2}{3\pi}\right)^2 A^4 + k^2\left(1 - \frac{\Omega^2}{\omega^2}\right)^2 A^2 - Q^2 = 0 \tag{1.56}$$

which must be solved by the quadratic formula to obtain A^2. Then the amplitude of vibration is calculated as $A = \sqrt{A^2}$.

In summary, an equivalent viscous damping constant can always be determined for any dissipative mechanism by equating the work of a hypothetical viscous damper to that of the actual mechanism. The work expression is based upon the velocity of the system [Eq. (b)] for steady-state response to a harmonic forcing function, and the equivalent viscous damping constant is determined as

$$c_{\text{eq}} = \frac{1}{\pi A^2 \Omega} \int_0^T R\dot{u}\, dt = \frac{U_R}{\pi A^2 \Omega} \tag{1.57}$$

where the symbol R denotes the resisting force. A simplified dynamic analysis of the system may then be performed, using the value of c_{eq} so obtained. Furthermore, more than one type of damping may be handled simultaneously. For example, with a combination of Coulomb damping and viscous damping, we find from Eq. (1.51)

$$c_{\text{eq}} = \frac{4F}{\pi A \Omega} + c \tag{v}$$

Proceeding with this value of c_{eq} as before, we obtain the following equation for determining the amplitude A of forced vibrations:

$$\left[\left(1 - \frac{\Omega^2}{\omega^2}\right)^2 + \left(2\gamma\frac{\Omega}{\omega}\right)^2\right]A^2 + \frac{16F\gamma\Omega}{\pi k \omega} A + \left(\frac{4F}{\pi k}\right)^2 - \frac{Q^2}{k^2} = 0 \tag{w}$$

which must be solved by the quadratic formula.

1.11 GENERAL PERIODIC FORCING FUNCTIONS

In all previous discussions of forced vibrations we assumed simple harmonic forcing functions proportional to $\sin \Omega t$ or $\cos \Omega t$. In general, it is possible to encounter periodic forcing functions that are more complicated. The response of one-degree-of-freedom systems to such functions will be discussed in this section.

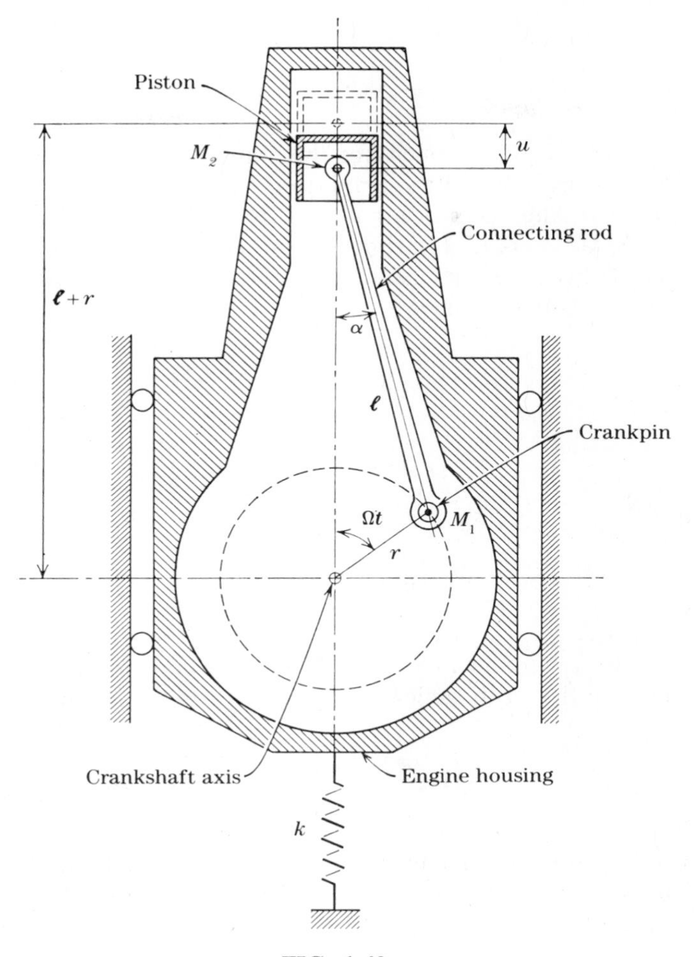

FIG. 1.40

Consider, for example, the one-cylinder engine shown in Fig. 1.40. When such an engine has reciprocating parts that are not balanced, it will generate a periodic force that causes vibrations of the whole system. In studying such forced vibrations, we will need to know the exact nature of the disturbing force; and of particular importance is its period in relation to the natural period of the system.

In the analysis of the disturbing force, the mass of the connecting rod can be replaced with sufficient accuracy by two masses, one at the crankpin and the second at the piston. All other unbalanced masses in motion can readily be reduced to the same two points, so that finally only two masses M_1 and M_2 need be taken into consideration (see Fig. 1.40). If we take downward forces to be positive, the vertical component of the inertia force of the mass

M_1 is

$$F_1 = -M_1\Omega^2 r \cos \Omega t \tag{a}$$

in which Ω is the angular velocity of the crankshaft, r is the radius of the crank, and Ωt is the angle of the crank to the vertical axis.

The motion of the reciprocating mass M_2 is more complicated. Let u denote the displacement of M_2 from the top-dead-center position, and let α be the angle between the connecting rod and the vertical axis. From the geometry of the figure, we have

$$u = \ell(1 - \cos \alpha) + r(1 - \cos \Omega t) \tag{b}$$

and

$$r \sin \Omega t = \ell \sin \alpha \tag{c}$$

From Eq. (c),

$$\sin \alpha = \frac{r}{\ell} \sin \Omega t$$

The length ℓ of the connecting rod is usually several times larger than the crank radius r so that with sufficient accuracy it can be assumed that

$$\cos \alpha = \sqrt{1 - \frac{r^2}{\ell^2} \sin^2 \Omega t} \approx 1 - \frac{r^2}{2\ell^2} \sin^2 \Omega t$$

which includes only the first two terms of the binomial expansion of the actual expression. Substituting this approximation into Eq. (b), we obtain

$$u = r(1 - \cos \Omega t) + \frac{r^2}{2\ell} \sin^2 \Omega t \tag{d}$$

By differentiating this equation with respect to time, we find the velocity of the reciprocating mass M_2 to be

$$\dot{u} = r\Omega \sin \Omega t + \frac{r^2\Omega}{2\ell} \sin 2\Omega t$$

and a second differentiation yields its acceleration; so the inertia force for mass M_2 is

$$F_2 = -M_2\Omega^2 r\left(\cos \Omega t + \frac{r}{\ell} \cos 2\Omega t\right) \tag{e}$$

Combining this result with Eq. (*a*), we obtain the complete expression for the disturbing force as

$$F(t) = -(M_1 + M_2)\Omega^2 r \cos \Omega t - \frac{r}{\ell} M_2 \Omega^2 r \cos 2\Omega t \tag{f}$$

It will be noted that this expression consists of two terms, one having an angular frequency equal to the speed of the machine and another having twice that frequency. From this fact it can be concluded that in the case under consideration we have two critical speeds of the engine: the first when the number of revolutions of the machine per second is equal to the natural frequency $f = 1/\tau$ of the system, and the second when the number of revolutions of the machine is half this frequency. By a suitable choice of the spring stiffness k it is always possible to make the natural frequency sufficiently remote from such critical speeds and to eliminate in this manner the possibility of large-amplitude vibrations.

It must be remembered that expression (*e*) for the inertia force of the reciprocating mass was obtained by taking only the first two terms of the binomial expansion of $\cos \alpha$. A more accurate solution will also contain harmonics, resulting in lower critical speeds than those found above. However, these are usually of no practical importance, because the corresponding forces are too small to produce substantial vibrations of the system.

In general, a periodic forcing function of any kind can be represented in the form of a *trigonometic* (or *Fourier*) *series*, as follows:

$$\begin{aligned} F(t) &= a_0 + a_1 \cos \Omega t + a_2 \cos 2\Omega t + \cdots + b_1 \sin \Omega t + b_2 \sin 2\Omega t + \cdots \\ &= a_0 + \sum_{i=1}^{\infty} (a_i \cos i\Omega t + b_i \sin i\Omega t) \end{aligned} \tag{1.58}$$

The period of the disturbing force is $T = 2\pi/\Omega$, and the quantities a_0, a_i, and b_i are constants to be determined.

In order to calculate any of the coefficients of Eq. (1.58), provided $F(t)$ is known, the following procedure may be used. Assume that the coefficient a_i is desired; then both sides of the equation can be multiplied by $\cos i\Omega t\, dt$ and integrated from $t = 0$ to $t = T$. It can be shown that

$$\int_0^T a_0 \cos i\Omega t\, dt = 0 \qquad \int_0^T a_j \cos j\Omega t \cos i\Omega t\, dt = 0$$

$$\int_0^T b_j \sin j\Omega t \cos i\Omega t\, dt = 0 \qquad \int_0^T a_i \cos^2 i\Omega t\, dt = \frac{a_i}{2} T = a_i \frac{\pi}{\Omega}$$

where i and j denote integer numbers 1, 2, 3, ... By using these formulas, we obtain from Eq. (1.58)

$$a_i = \frac{2}{T}\int_0^T F(t)\cos i\Omega t\,dt \tag{1.59a}$$

In the same manner, by multiplying Eq. (1.58) by $\sin i\Omega t\,dt$ and integrating, we obtain

$$b_i = \frac{2}{T}\int_0^T F(t)\sin i\Omega t\,dt \tag{1.59b}$$

Finally, multiplying Eq. (1.58) by dt and integrating from $t = 0$ to $t = T$, we have

$$a_0 = \frac{1}{T}\int_0^T F(t)\,dt \tag{1.59c}$$

It is seen that by using formulas (1.59), the coefficients of Eq. (1.58) can be calculated if $F(t)$ is known analytically. If $F(t)$ is given graphically, with no analytical expression available, some approximate numerical method for calculating the integrals (1.59) must be used.

Assuming that the forcing function is represented in the form of a trigonometric series, we write the equation for damped forced vibrations as

$$m\ddot{u} + c\dot{u} + ku = a_0 + a_1\cos\Omega t + a_2\cos 2\Omega t + \cdots$$
$$+ b_1\sin\Omega t + b_2\sin 2\Omega t + \cdots \tag{1.60}$$

The general solution of this equation will consist of two parts, one for free vibrations and one for forced vibrations. The free vibrations will gradually diminish due to damping. The forced vibrations will be obtained in the case of a linear equation by superimposing the steady-state forced vibrations produced by every term of the series (1.58). These latter vibrations can be found in the same manner as that explained in Sec. 1.9, and it can be concluded that large forced vibrations may occur when the period of one of the terms of series (1.58) coincides with the period of the natural vibrations of the system, that is, if the period T of the disturbing force is equal to, or a multiple of, the period τ_d.

Example 1. For the system shown in Fig. 1.40, the following numerical data are given.

Weight of piston, $W_p = 6.00$ lb
Weight of connecting rod, $W_c = 3.00$ lb
$M_1 g = \frac{2}{3}W_c = 2.00$ lb
$M_2 g = W_p + \frac{1}{3}W_c = 7.00$ lb

Total weight of engine, $W = 500$ lb
Speed of engine $= 600$ rpm
Crank radius, $r = 8$ in.
Length of connecting rod, $\ell = 24$ in.
Spring stiffness, $k = 11{,}500$ lb per in.

Neglecting damping, find the maximum displacement of the engine from its equilibrium position during steady-state forced vibrations of the system. Assume that the crankshaft is perfectly balanced.

Solution. We being by calculating the natural angular frequency of vibration of the system.

$$\omega = \sqrt{\frac{kg}{W}} = \sqrt{\frac{(11{,}500)(386)}{(500)}} = \sqrt{8878} = 94.3 \text{ sec}^{-1}$$

Also,

$$\Omega = \frac{(600)(2\pi)}{(60)} = 20\pi = 62.83 \text{ sec}^{-1}$$

and we find

$$\frac{\Omega}{\omega} = \frac{62.83}{94.3} = \frac{2}{3} \qquad \frac{2\Omega}{\omega} = \frac{4}{3}$$

From these ratios, we see that the disturbing force proportional to $\cos \Omega t$ will be working below resonance, while that proportional to $\cos 2\Omega t$ will be working above resonance. Neglecting higher frequency components of the disturbing force, we have only to superimpose the effects of the inertia forces represented by Eqs. (a) and (e) and combined in (f) above. Writing these terms in the form

$$\begin{aligned} P_1 \cos \Omega t &= -(M_1 + M_2)\Omega^2 r \cos \Omega t \\ P_2 \cos 2\Omega t &= -\frac{r}{\ell} M_2 \Omega^2 r \cos 2\Omega t \end{aligned} \tag{g}$$

we have

$$P_1 = -(M_1 + M_2)\Omega^2 r = -\left(\frac{2+7}{386}\right)(400\pi^2)(8) = -736 \text{ lb}$$

$$P_2 = -\frac{r}{\ell} M_2 \Omega^2 r = -\left(\frac{8}{24}\right)\left(\frac{7}{386}\right)(400\pi^2)(8) = -191 \text{ lb}$$

Returning now to Eq. (1.24) in Sec. 1.6, we find that the undamped forced vibrations produced separately by the two disturbing forces [Eqs. (*g*)] are

$$u_1 = \frac{P_1}{k}\left(\frac{1}{1-\Omega^2/\omega^2}\right)\cos\Omega t = \frac{-736}{11{,}500}\left(\frac{1}{1-4/9}\right)\cos\Omega t = -0.115\cos\Omega t$$

$$u_2 = \frac{P_2}{k}\left(\frac{1}{1-4\Omega^2/\omega^2}\right)\cos 2\Omega t = \frac{-191}{11{,}500}\left(\frac{1}{1-16/9}\right)\cos 2\Omega t = 0.0214\cos 2\Omega t$$

For a maximum displacement, we take $\Omega t = \pi$; then

$$(u_1 + u_2)_m = 0.115 + 0.0214 = 0.136 \text{ in.}$$

Example 2. A one-degree-of-freedom system is subjected to a disturbing force $F(t)$ that varies with time according to the diagram shown in Fig. 1.41. Neglecting damping, find the steady-state forced vibrations that will be produced if the mass m and the spring constant k are such that $\Omega/\omega = 0.9$.

Solution. We begin by making a harmonic analysis of the given force, which we assume can be represented by the trigonometric series (1.58). For this purpose, we have Eqs. (1.59) defining the coefficients a_0, a_i, and b_i of the series.

Considering first Eq. (1.59c), we see that $\int_0^{2\pi/\Omega} F(t)\,dt$ is simply the area under the given saw-tooth diagram in Fig. 1.41 between the ordinates $t = 0$ and $t = T = 2\pi/\Omega$. Clearly, this area is zero, and hence $a_0 = 0$.

Considering next Eq. (1.59a), we see that each ordinate of the diagram in Fig. 1.41 must be multiplied by $\cos i\Omega t$ and then integrated from $t = 0$ to $t = 2\pi/\Omega$. Now from the antisymmetry of $F(t)$ with respect to $t = \pi/\Omega$, together with the symmetry of $\cos i\Omega t$ with respect to the same time, we conclude again that the integral in Eq. (1.59a) vanishes; and $a_i = 0$.

Finally, considering Eq. (1.59b), we see that each ordinate of $F(t)$ in Fig. 1.41 must be multiplied by $\sin i\Omega t$ and integrated from $t = 0$ to $t = 2\pi/\Omega$. In

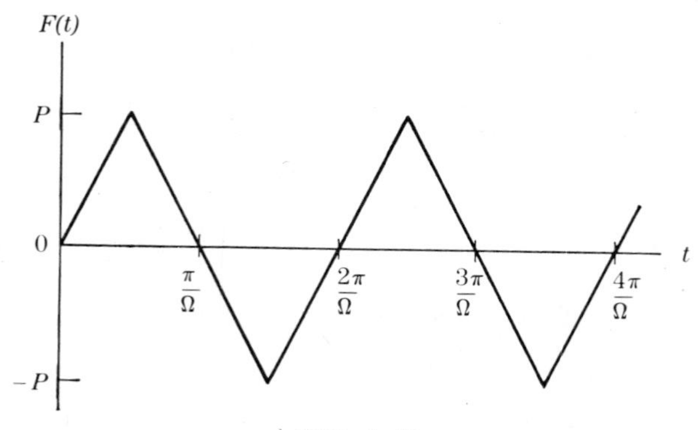

FIG. 1.41

this case $F(t)$ from $t=0$ to $t=\pi/\Omega$ is symmetric about $t=\pi/2\Omega$; while from $t=\pi/\Omega$ to $t=2\pi/\Omega$, it is symmetric about $t=3\pi/2\Omega$. However, when i is an even integer, corresponding parts of $\sin i\Omega t$ are antisymmetric about $t=\pi/2\Omega$ and $t=3\pi/2\Omega$, respectively. Thus, for $i=2, 4, 6, \ldots$, we conclude that $b_i=0$.

When i is an odd integer, both $F(t)$ and $\sin i\Omega t$ are antisymmetric about the ordinate $t=\pi/\Omega$, and Eq. (1.59b) gives

$$b_i = \frac{\Omega}{\pi}\int_0^{2\pi/\Omega} F(t)\sin i\Omega t\, dt = \frac{4\Omega}{\pi}\int_0^{\pi/2\Omega} F(t)\sin i\Omega t\, dt \tag{h}$$

Referring to Fig. 1.41, we see that in the interval from $t=0$ to $t=\pi/2\Omega$

$$F(t) = \frac{2P\Omega t}{\pi}$$

Substituting this expression into Eq. (h), we obtain

$$b_i = \frac{8P\Omega^2}{\pi^2}\int_0^{\pi/2\Omega} t\sin i\Omega t\, dt = \frac{8P}{i^2\pi^2}\int_0^{i\pi/2} u\sin u\, du$$

Integrating and substituting limits results in

$$b_i = \frac{8P}{i^2\pi^2}\sin\frac{i\pi}{2} = \frac{8P}{i^2\pi^2}(-1)^{(i-1)/2} \tag{i}$$

where $i=1, 3, 5, 7, \ldots$.

Using $a_0=0$, $a_i=0$, and expression (i) for b_i, we find that the trigonometric series (1.58) becomes

$$F(t) = \frac{8P}{\pi^2}\left(\sin\Omega t - \frac{1}{3^2}\sin 3\Omega t + \frac{1}{5^2}\sin 5\Omega t - \cdots\right) \tag{j}$$

Therefore, to represent the saw-tooth diagram in Fig. 1.41 by a trigonometric series, we need only to superimpose sine curves with an odd number of full waves in the interval $t=0$ to $t=2\pi/\Omega$. Furthermore, we see that the series (j) converges rapidly; so only the first term is of practical importance. That is, the saw-tooth disturbing force produces approximately the same effect as a sinusoidal disturbing force of slightly smaller amplitude given by

$$F(t) = \frac{8P}{\pi^2}\sin\Omega t \tag{k}$$

To judge the insignificance of the second term, we note that for $\Omega/\omega = 0.9$ the magnification factor is

$$\beta_3 = \frac{1}{1-(3\Omega/\omega)^2} = -0.159$$

Thus, the amplitude of forced vibration produced by the second term is only $0.159/3^2 = 0.0177$ times that produce statically by the force $8P/\pi^2$, while the magnification factor for the first term is

$$\beta_1 = \frac{1}{1-\Omega^2/\omega^2} = 5.26$$

We conclude that using the approximate expression (k) results in an error of less than 0.4%, and the approximate solution for the response is

$$u = \frac{8P\beta_1}{\pi^2 k} \sin \Omega t \qquad (l)$$

1.12 ARBITRARY FORCING FUNCTIONS

In the preceding section we considered the general case of a periodic forcing funtion that could be represented in the form of a Fourier series. However, for the case of an *arbitrary disturbing force* with no periodic character, a somewhat different approach to the problem must be taken.

Let us now consider the differential equation of motion for a damped one-degree system (see Fig. 1.42*a*), subjected to an arbitrary forcing function $Q = F(t')$.

$$m\ddot{u} = -c\dot{u} - ku + Q \qquad (a)$$

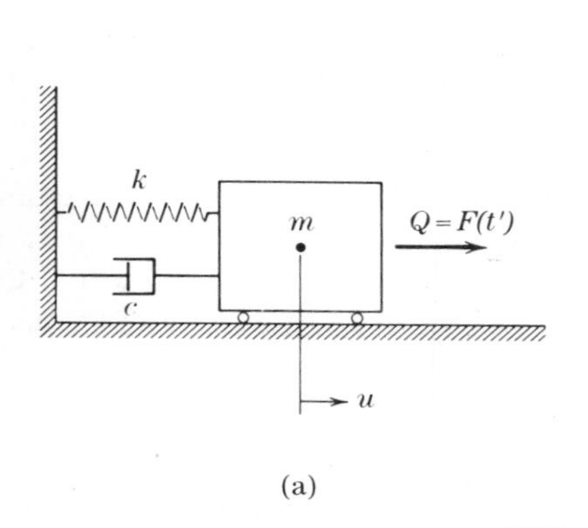

(a)

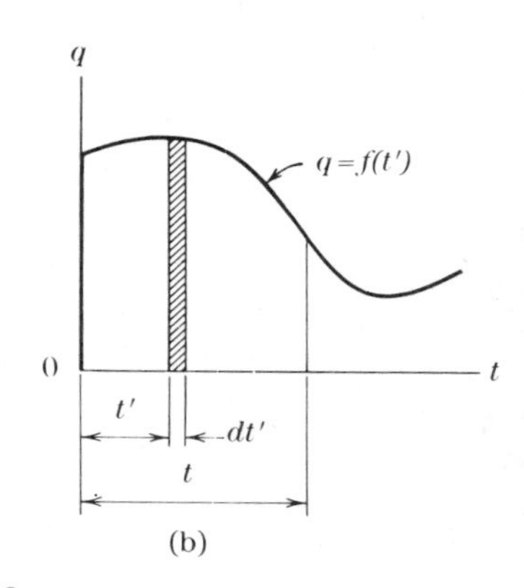

(b)

FIG. 1.42

Dividing Eq. (a) by the mass m and rearranging terms, we have

$$\ddot{u} + 2n\dot{u} + \omega^2 u = q \tag{1.61}$$

where

$$q = \frac{Q}{m} = \frac{F(t')}{m} = f(t') \tag{b}$$

which is the disturbing force per unit of mass. In dealing with Eq. (1.61), we assume that the force q is expressed as a function of the dummy time variable t', as indicated in Fig. 1.42b. Then at any instant t' we may calculate an *incremental impulse* $q\,dt'$, represented by the hatched strip in the diagram. This impulse imparts to each unit of mass an instantaneous increase in velocity (or *incremental velocity*) equal to

$$d\dot{u} = q\,dt' \tag{c}$$

regardless of what other forces (such as the spring force) may be acting upon it, and regardless of its displacement and velocity at the instant t'. Treating this increment of velocity as if it were an initial velocity (at the instant t') and using Eq. (1.35), Sec. 1.8, we conclude that the *incremental displacement* of the system at any later time t will be

$$du = e^{-n(t-t')}\frac{q\,dt'}{\omega_{\mathrm{d}}}\sin\omega_{\mathrm{d}}(t-t') \tag{d}$$

Since each incremental impulse $q\,dt'$ between $t'=0$ and $t'=t$ has such an effect, we obtain, as a result of the continuous action of the disturbing force q, the total displacement

$$u = \frac{e^{-nt}}{\omega_{\mathrm{d}}}\int_0^t e^{nt'} q \sin\omega_{\mathrm{d}}(t-t')\,dt' \tag{1.62}$$

This mathematical form is referred to as *Duhamel's integral.*

Expression (1.62) represents the complete displacement produced by the disturbing force q acting during the time interval 0 to t. It includes both steady-state and transient terms and is especially useful in studying the response of a vibratory system to any kind of disturbing force. If the function $q = f(t')$ cannot be expressed analytically, the integral in Eq. (1.62) can always be evaluated approximately by a suitable numerical method of integration. To take account of the effect of initial displacement u_0 and initial velocity $\dot{u}_0$ at $t=0$, it is only necessary to add to the results for Eq. (1.62) the solution for such initial conditions given by Eq. (1.35), Sec. 1.8.

Thus, the total solution is

$$u = e^{-nt}\left[u_0 \cos \omega_d t + \frac{\dot{u}_0 + nu_0}{\omega_d} \sin \omega_d t + \frac{1}{\omega_d}\int_0^t e^{nt'} q \sin \omega_d (t - t')\, dt'\right] \tag{1.63}$$

If damping is neglected, we have $n = 0$ and $\omega_d = \omega$, in which case Eq. (1.62) reduces to

$$u = \frac{1}{\omega}\int_0^t q \sin \omega(t - t')\, dt' \tag{1.64}$$

In this instance, if we also consider the effect of initial displacement u_0 and initial velocity $\dot{u}_0$ at $t = 0$, Eq. (1.63) without damping becomes

$$u = u_0 \cos \omega t + \frac{\dot{u}_0}{\omega} \sin \omega t + \frac{1}{\omega}\int_0^t q \sin \omega(t - t')\, dt' \tag{1.65}$$

As an example of the application of Eq. (1.64), let us assume that a constant force Q_1 (see Fig. 1.43*a*) is suddenly applied to the mass in Fig. 1.42*a*. This condition of dynamic loading is called a *step function*. In this case we have $q_1 = Q_1/m$ = constant, and Eq. (1.64) becomes

$$u = \frac{q_1}{\omega}\int_0^t \sin \omega(t - t')\, dt' \tag{e}$$

This integral is readily evaluated, as follows:

$$u = \frac{q_1}{\omega^2}(1 - \cos \omega t) = \frac{Q_1}{k}(1 - \cos \omega t) \tag{1.66}$$

From this result we see that a suddenly applied constant force produces free

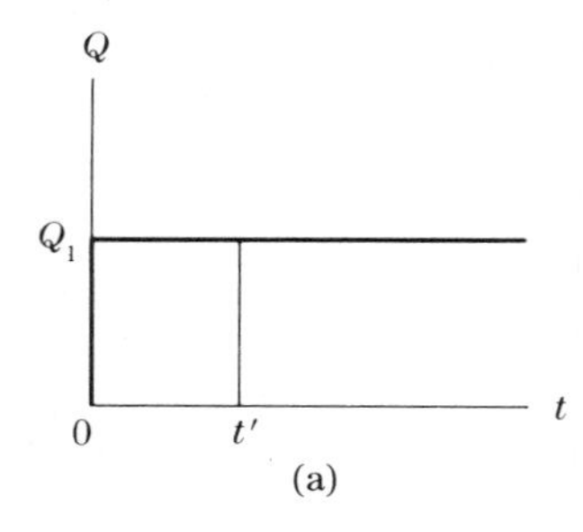

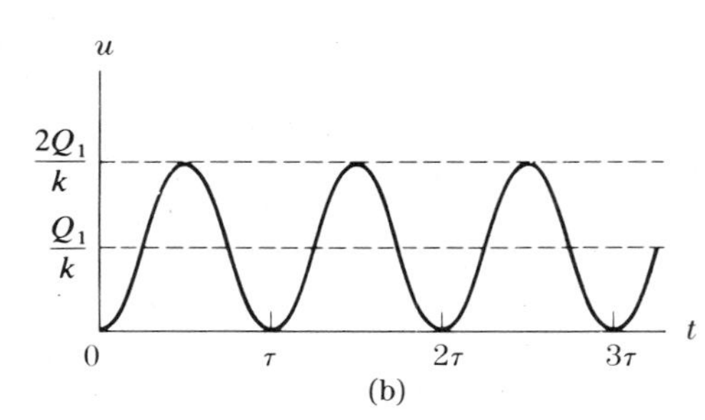

FIG. 1.43

vibrations of amplitude Q_1/k superimposed on a static displacement of the same magnitude Q_1/k (see Fig. 1.43*b*). Thus, the maximum deflection due to a suddenly applied force is twice as large as that produced by the same force acting statically.

In the case discussed above, the constant Q_1 acts for an indefinite time. If it acts only for a period of time t_1, we have the *rectangular impulse* shown in Fig. 1.44*a*. During the time when the force is nonzero, the response of the system is the same as that given by Eq. (1.66). The response subsequent to time t_1 may be obtained by evaluating the Duhamel integral for the two ranges 0 to t_1 and t_1 to t. Only the integration over the first range will produce nonzero results, because the forcing function is zero in the second range. Altogether, the solution for this case is summarized as follows:

$$\text{for} \quad 0 \leqq t \leqq t_1 \qquad u = \frac{Q_1}{k}(1 - \cos \omega t) \qquad \text{(1.66 repeated)}$$

$$\text{for} \quad t \leqq t \qquad u = \frac{Q_1}{k}[\cos \omega(t - t_1) - \cos \omega t] \qquad (1.67)$$

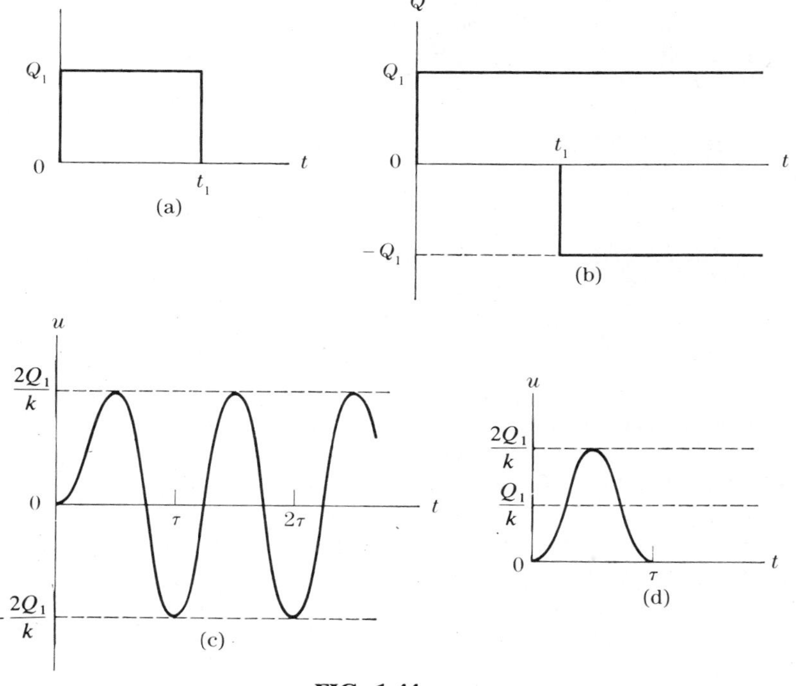

FIG. 1.44

The same results may be obtained by considering the retangular impulse in Fig. 1.44*a* to consist of the sum of two step functions, as indicated in Fig. 1.44*b*. The first step function (of magnitude Q_1) begins at time $t = 0$, while the second step function (of magnitude $-Q_1$) begins at time $t = t_1$.

A third method for determining the result given as Eq. (1.67) involves evaluating the displacement and velocity of the system at time t_1, using Eq. (1.66). Thus,

$$u_{t_1} = \frac{Q_1}{k}(1 - \cos \omega t_1) \tag{f}$$

$$\dot{u}_{t_1} = \frac{Q_1 \omega}{k} \sin \omega t_1 \tag{g}$$

If these two quantities are treated as initial displacement and initial velocity at the time t_1, the ensuing free-vibration response may be calculated from

$$u = u_{t_1} \cos \omega(t - t_1) + \frac{\dot{u}_{t_1}}{\omega} \sin \omega(t - t_1) \tag{h}$$

Substitution of expressions (f) and (g) into Eq. (h), followed by simple trigonometric manipulation, yields the same result as that in Eq. (1.67).

The amplitude of the free-vibration response subsequent to the rectangular impulse may be determined from

$$A = \sqrt{u_{t_1}^2 + \left(\frac{\dot{u}_{t_1}}{\omega}\right)^2} \tag{i}$$

Using expressions (f) and (g) in Eq. (i) and simplifying, we find

$$A = \frac{Q_1}{k}\sqrt{2(1 - \cos \omega t_1)} = \frac{2Q_1}{k} \sin(\omega t_1/2) = \frac{2Q_1}{k} \sin(\pi t_1/\tau) \tag{j}$$

From the last form of Eq. (j) we see that the amplitude of free vibration depends upon the ratio t_1/τ, where τ is the natural period of the system. By taking the duration of the rectangular impulse as $t_1 = \tau/2$, we obtain the amplitude $A = 2Q_1/k$, and the response is as shown in Fig. 1.44*c*. In this case the force Q_1 acts through the displacement from 0 to A and does positive work on the system. After removal of the force in the extreme position, the system (without damping) retains this energy; and we have free vibrations resulting from the initial displacement $2Q_1/k$ at time t_1.

Considering another special case, let us specify the duration of the impulse to be $t_1 = \tau$. From Eq. (j) we obtain the amplitude $A = 0$, and the response is as shown in Fig. 1.44*d*. In this instance the constant force Q_1 does positive work from 0 to A and an equal amount of negative work from A back to 0. Therefore, the net work is zero, and the system remains at rest when the force is removed.

Now suppose that a series of step functions with alternating signs act as shown in Fig. 1.45*a* to produce the sequence of rectangular pulses appearing in Fig. 1.45*b*. The time difference between successive step functions is specified to be $\tau/2$, so that the impulsive action is always in phase with the

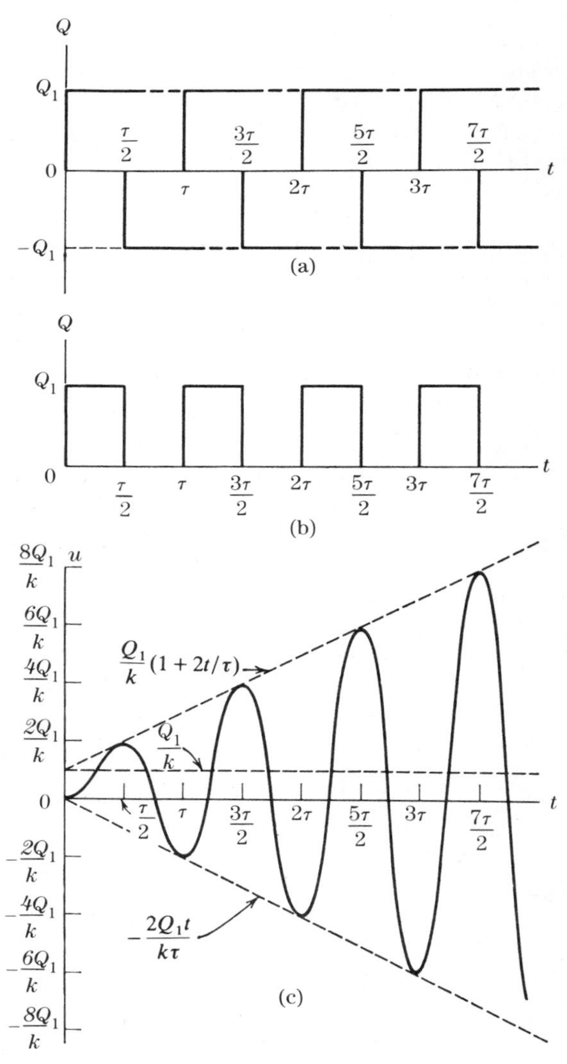

FIG. 1.45

velocity and will do positive work in each cycle of vibration. From the principle of superposition, we may conclude that the amplitude of free vibration after n rectangular pulses is

$$A_n = \frac{2nQ_1}{k} \tag{k}$$

That is, in every cycle of vibration the amplitude increases by $2Q_1/k$, and the response builds up indefinitely. Figure 1.45c shows a curve representing this buildup over the first few cycles of vibration. We may conclude from this demonstration that any periodic forcing function in resonance with the system will cause large amplitudes of forced vibrations if it does positive net work in each cycle. Furthermore, the use of the Duhamel integral for calculating the reponse to general periodic forcing functions represents an alternative method to that discussed in Sec. 1.11, where the dynamic loads were decomposed into Fourier series.

Example 1. Determine the undamped response of a one-degree-of-freedom system to the linearly increasing force, known as a *ramp funtion*, given in Fig. 1.46a. The rate of increase of the force Q per unit of time is δQ.

Solution. The forcing function in this example, expressed in terms of δQ and t', is simply

$$Q = \delta Q\, t' \tag{l}$$

and the force per unit mass is

$$q = \frac{\delta Q}{m} t' \tag{m}$$

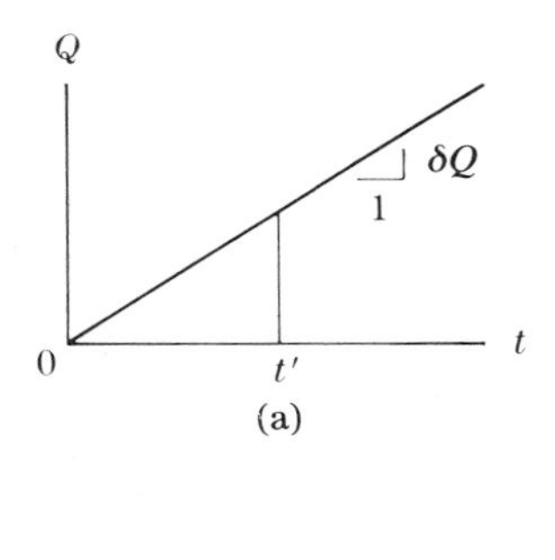

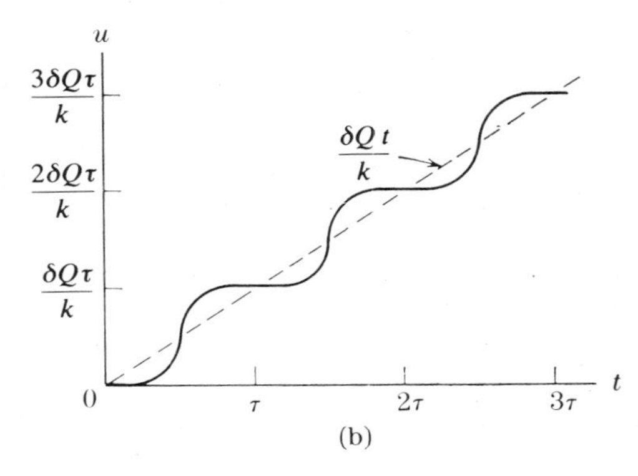

FIG. 1.46

Applying Eq. (1.64) to this case, we have

$$u = \frac{\delta Q}{m\omega}\int_0^t t' \sin \omega(t - t')\, dt'$$

which, when integrated by parts, yields

$$u = \frac{\delta Q}{k}\left(t - \frac{1}{\omega}\sin \omega t\right) \tag{n}$$

From this result we see that the response to a ramp function consists of the sum of a linearly increasing static displacement $\delta Qt/k$ and a free vibration of amplitude $\delta Q/k\omega$, as shown in Fig. 1.46*b*. The velocity at any time t is given by the first derivative of expression (n) with respect to time. That is,

$$\dot{u} = \frac{\delta Q}{k}(1 - \cos \omega t) = \frac{\delta Q}{k}\left(1 - \cos\frac{2\pi t}{\tau}\right) \tag{o}$$

Thus, the velocity is zero at times $t = 0,\ \tau,\ 2\tau,\ 3\tau$, etc.; so the slope of the displacement curve in Fig. 1.46*b* is zero at those times. Furthermore, the velocity is always positive and has a maximum value of $2\delta Q/k$ at times $t = \tau/2,\ 3\tau/2,\ 5\tau/2$, etc.

The right-hand side of expression (o) has the same mathematical form as the right-hand side of Eq. (1.66) for the step function. This follows from the fact that the ramp function is proportional to the time instead of constant with time. We may also conclude that a parabolic variation of the forcing function produces a velocity function of the same form as the right-hand side of Eq. (n) and an acceleration function of the same form as the right-hand side of Eq. (o).

Example 2. Let us rederive the expression for undamped forced vibrations of a one-degree system subjected to a harmonic function. This case was previously discussed and developed in detail in Secs. 1.6 and 1.7. Assume that the forcing function is given as

$$Q = P \sin \Omega t' \tag{p}$$

The force per unit mass is

$$q = q_{\mathrm{m}} \sin \Omega t' \tag{q}$$

in which the maximum value of q is $q_{\mathrm{m}} = P/m$.

Substitution of expression (q) into Eq. (1.64) gives

$$u = \frac{q_{\mathrm{m}}}{\omega} \int_0^t \sin \Omega t' \sin \omega(t - t')\, dt' \qquad (r)$$

Using a trigonometric identity for the product within the integral of Eq. (r), we obtain

$$u = \frac{q_{\mathrm{m}}}{2\omega} \int_0^t \{\cos[\Omega t' - \omega t + \omega t'] - \cos[\Omega t' + \omega t - \omega t']\}\, dt'$$

$$= \frac{q_{\mathrm{m}}}{2\omega} \int_0^t \{\cos[(\Omega + \omega)t' - \omega t] - \cos[(\Omega - \omega)t' + \omega t]\}\, dt'$$

which can be integrated directly. Performing the integrations and simplifying, we find

$$u = \frac{P}{k}\left(\sin \Omega t - \frac{\Omega}{\omega} \sin \omega t\right)\left(\frac{1}{1 - \Omega^2/\omega^2}\right) \qquad (s)$$

Upon comparison of this result with Eq. (1.29b) in Sec. 1.7, we see that they are the same. The first factor in Eq. (s) is the static displacement of the system due to a constant load P; the terms in the second factor represent the steady-state and transient responses of the system; and the third factor is β, the magnification factor for zero damping. Note that the transient part of the response is included in the results from the Duhamel integral, without specific reference to initial conditions.

Example 3. For the system shown in Fig. 1.42a, determine the damped response to the step function given in Fig. 1.43a.
Solution. In this case the force per unit mass, $q_1 = Q_1/m$, is substituted into Eq. (1.62) in order to evaluate the damped response.

$$u = \frac{q_1 e^{-nt}}{\omega_{\mathrm{d}}} \int_0^t e^{nt'} \sin \omega_{\mathrm{d}}(t - t')dt' \qquad (t)$$

Equation (t) may be integrated by parts and put into the form

$$u = \frac{Q_1}{k}\left[1 - e^{-nt}\left(\cos \omega_{\mathrm{d}} t + \frac{n}{\omega_{\mathrm{d}}} \sin \omega_{\mathrm{d}} t\right)\right] \qquad (u)$$

This result represents the sum of a static displacement Q_1/k and a damped free vibration (see Sec. 1.8) of amplitude

$$Ae^{-nt} = \frac{Q_1}{k} e^{-nt} \sqrt{1 + \left(\frac{n}{\omega_{\mathrm{d}}}\right)^2} \qquad (v)$$

and having the phase angle

$$\alpha_d = \tan^{-1} \frac{n}{\omega_d} \tag{w}$$

When damping is put to zero, Eq. (u) becomes the same as Eq. (1.66); the amplitude A becomes equal to Q_1/k; and the phase angle α_d becomes zero.

1.13 ARBITRARY SUPPORT MOTIONS

In certain practical problems the response of a vibratory system is due to support motion instead of a directly applied disturbing force. Forced vibrations caused by harmonic support displacements and accelerations were discussed in Secs. 1.6 and 1.9 for undamped systems and for systems with viscous damping. In this section we will deal with cases where the specified support motions are arbitrary functions of time.

Consider the damped one-degree system in Fig. 1.47*a*, and let the displacement of ground, u_g, be given as an analytical function of time. The equation of motion for this case becomes

$$m\ddot{u} = -c(\dot{u} - \dot{u}_g) - k(u - u_g) \tag{a}$$

Rearranging Eq. (a), we have

$$m\ddot{u} + c\dot{u} + ku = ku_g + c\dot{u}_g \tag{b}$$

If the expression for u_g can be differentiated with respect to time, we have two analytical forcing functions on the right-hand side of Eq. (b). The first of these is equivalent to a disturbing force of magnitude ku_g applied directly

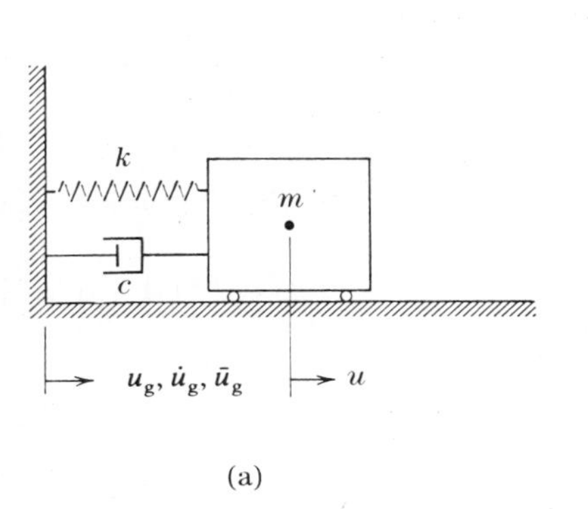

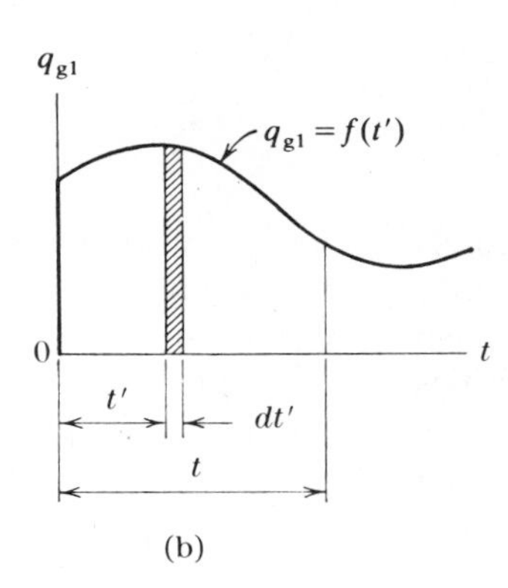

FIG. 1.47

to the mass, while the second is a similar force of magnitude $c\dot{u}_g$. Dividing both sides of Eq. (*b*) by the mass m, we obtain

$$\ddot{u} + 2n\dot{u} + \omega^2 u = q_g = q_{g1} + q_{g2} \tag{1.68}$$

in which

$$q_{g1} = \omega^2 u_g = \omega^2 F(t') = f(t') \tag{c}$$

represents the equivalent force per unit mass due to the support displacement u_g, and the quantity q_{g2} has the definition

$$q_{g2} = 2n\dot{u}_g = \frac{2n}{\omega^2}\dot{q}_{g1} \tag{d}$$

As with applied forcing functions, we assume that the displacement u_g and the corresponding force q_{g1} are expressed as functions of the dummy time variable t' (see Fig. 1.47*b*).

From this point the analysis proceeds in a manner similar to that for an applied forcing function. However, in this case the incremental impulse has two parts, and the expression for incremental velocity at time t' becomes

$$d\dot{u} = (q_{g1} + q_{g2})\, dt' \tag{e}$$

the first term of which is represented by the hatched strip in Fig. 1.47*b*. At any later time t, the incremental displacement is

$$du = e^{-n(t-t')} \frac{1}{\omega_d}(q_{g1} + q_{g2}) \sin \omega_d(t - t')\, dt' \tag{f}$$

Due to the continuous effect of the support motion, the total displacement of the mass becomes

$$u = u_1 + u_2 = \frac{e^{-nt}}{\omega_d} \int_0^t e^{nt'}(q_{g1} + q_{g2}) \sin \omega_d(t - t')\, dt' \tag{1.69}$$

which is a somewhat more complicated expression for the Duhamel integral than that given previously as Eq. (1.62) in Sec. 1.12.

If damping is neglected, we have $n = 0$ and $\omega_d = \omega$; and Eq. (1.69) simplifies to

$$u = \frac{1}{\omega} \int_0^t q_{g1} \sin \omega(t - t')\, dt' = \omega \int_0^t u_g \sin \omega(t - t')\, dt' \tag{1.70}$$

the first of which is mathematically the same as Eq. (1.64) in Sec. 1.12.

Next, let us consider the case of specified ground acceleration, $\ddot{u}_g$. As was done in Sec. 1.6 for forced vibrations, we will use the following change of coordinates

$$u^* = u - u_g \qquad \dot{u} = \dot{u} - \dot{u}_g \qquad \ddot{u}^* = \ddot{u} - \ddot{u}_g \tag{g}$$

where the symbol u^* denotes relative displacement of the mass with respect to ground. Substituting $u - u_g$, $\dot{u} - \dot{u}_g$, and $\ddot{u}$ from Eqs. (g) into Eq. (a) and rearranging, we obtain

$$m\ddot{u}^* + c\dot{u}^* + ku^* = -m\ddot{u}_g \tag{h}$$

The term on the right-hand side of this equation is equivalent to a disturbing force of magnitude $-m\ddot{u}_g$ applied directly to the mass. When Eq. (h) is divided by m, it may be written as

$$\ddot{u}^* + 2n\dot{u}^* + \omega^2 u^* = q_g^* \tag{1.71}$$

where

$$q_g^* = -\ddot{u}_g = -f(t') \tag{i}$$

represents the forcing function in the relative coordinate due to ground acceleration.

Equation (1.71) is mathematically the same as Eq. (1.61) in Sec. 1.12, and we may conclude that the response of the system in the relative coordinate follows the familiar pattern of the previous cases. In this instance, the Duhamel integral for damped response relative to ground is

$$u^* = \frac{e^{-nt}}{\omega_d} \int_0^t e^{nt'} q_g^* \sin \omega_d(t - t')\, dt' \tag{1.72}$$

If damping is neglected, Eq. (1.72) reduces to the simpler form

$$u^* = \frac{1}{\omega} \int_0^t q_g^* \sin \omega(t - t')\, dt' \tag{1.73}$$

In addition, the absolute response of the system may be calculated if initial conditions of ground displacement and velocity are given.

Example 1. Assume that the support in Fig. 1.47a suddenly moves to the right, in accordance with the displacement step function given in Fig. 1.48. Determine the undamped response of the system to this instantaneous support displacement.

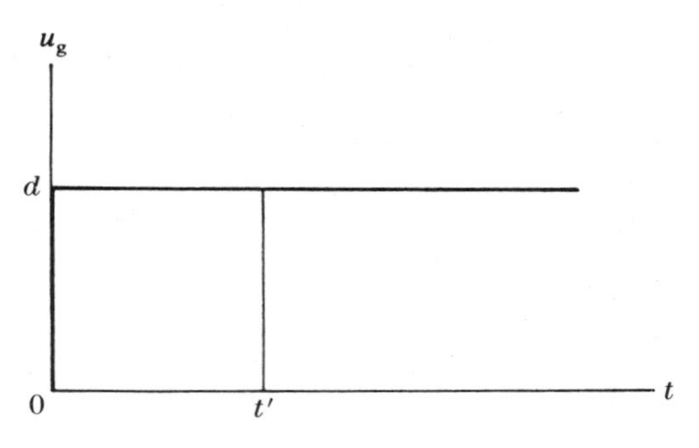

FIG. 1.48

Solution. In this case we have from Eq. (*c*) $q_{g1} = \omega^2 d = \text{constant}$, and Eq. (1.70) gives for the undamped response

$$u = \omega d \int_0^t \sin \omega(t - t')\, dt' = d(1 - \cos \omega t) \tag{j}$$

which is the same as Eq. (1.66), except that Q_1/k is replaced by the constant d. Thus, we see that the response consists of free vibrations of amplitude d superimposed on a static displacement of the same magnitude.

Example 2. As an illustration of the use of Eq. (1.69), let us consider the ramp function for support displacement given in Fig. 1.49. The slope of the straight line in the figure is δd per unit of time. Develop an expression for the damped response of the system in Fig. 1.47*a* caused by this support motion.

Solution. The support displacement in terms of t' and δd is

$$u_g = t'\, \delta d \tag{k}$$

and the forcing function q_{g1} in Eq. (*c*) becomes

$$q_{g1} = \omega^2 t'\, \delta d \tag{l}$$

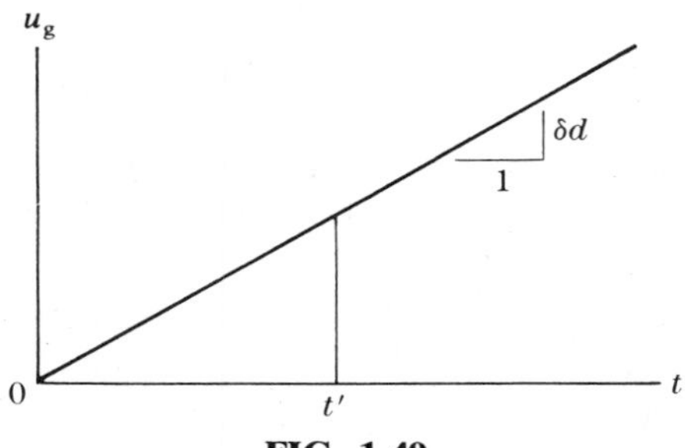

FIG. 1.49

The second part of the equivalent force per unit mass, given by Eq. (*d*), is

$$q_{g2} = 2n\,\delta d \tag{m}$$

Substituting the expressions for q_{g1} and q_{g2} from Eqs. (*l*) and (*m*) into Eq. (1.69), we have

$$u = u_1 + u_2 = \frac{\delta d\, e^{-nt}}{\omega_d} \int_0^t e^{nt'}(\omega^2 t' + 2n) \sin \omega_d (t - t')\, dt' \tag{n}$$

Performing the indicated integrations, we obtain for the first part

$$u_1 = \frac{\delta d}{\omega^2} \left\{ \omega^2 t - 2n + e^{-nt} \left[2n \cos \omega_d t + \frac{1}{\omega_d} (n^2 - \omega_d^2) \sin \omega_d t \right] \right\} \tag{o}$$

and for the second part

$$u_2 = 2n\,\delta d \left[1 - e^{-nt} \left(\cos \omega_d t + \frac{n}{\omega_d} \sin \omega_d t \right) \right] \tag{p}$$

The result for the first part consists of a response of the type shown in Fig. 1.46*b*, but with the oscillation gradually diminishing. When damping is put equal to zero, Eq. (*o*) becomes the same as Eq. (*n*) in Sec. 1.12, except that δd replaces $\delta Q/k$. In addition, the result in Eq. (*p*) for the second part is the same as Eq. (*u*) in Sec. 1.12, except that $2n\,\delta d$ replaces Q_1/k.

Example 3. Suppose that the ramp function in Fig. 1.49 specifies ground acceleration, $\ddot{u}_g$, instead of ground displacement, and that the slope of the line is δa per unit of time. Assuming initial conditions of ground displacement u_{g0} and ground velocity $\dot{u}_{g0}$ are known at time $t = 0$, determine the absolute undamped response of a one-degree system at time t due to this ground motion.

Solution. In this case the forcing function q_g^*, given by Eq. (*i*), becomes

$$q_g^* = -t'\,\delta a \tag{q}$$

Substituting this expression into Eq. (1.73), we have

$$u^* = -\frac{\delta a}{\omega} \int_0^t t' \sin \omega (t - t')\, dt' \tag{r}$$

Integration of Eq. (*r*) yields the undamped response in the relative coordinate as

$$u^* = -\frac{\delta a}{\omega^2} \left(t - \frac{1}{\omega} \sin \omega t \right) \tag{s}$$

and

$$\dot{u}^* = -\frac{\delta a}{\omega^2}(1 - \cos \omega t) \tag{t}$$

The total solution consists of the sum of the support motion and the relative motion. Thus, from Eqs. (*g*) and the initial conditions we find that the absolute velocity is

$$\begin{aligned} \dot{u} &= \dot{u}_g + \dot{u}^* = \dot{u}_{g0} + \int_0^t \ddot{u}_g \, dt' + \dot{u}^* \\ &= \dot{u}_{g0} + \delta a\left[\frac{t^2}{2} - \frac{1}{\omega^2}(1 - \cos \omega t)\right] \end{aligned} \tag{u}$$

and the absolute displacement is

$$\begin{aligned} u &= u_g + u^* = u_{g0} + \dot{u}_{g0}t + \int_0^t \dot{u}_g \, dt' + u^* \\ &= u_{g0} + \dot{u}_{g0}t + \delta a\left[\frac{t^3}{6} - \frac{1}{\omega^2}\left(t - \frac{1}{\omega}\sin \omega t\right)\right] \end{aligned} \tag{v}$$

Example 4. An undamped elevator of weight W (see Fig. 1.50) is suspended by a flexible cable of cross-sectional area A and modulus of elasticity E. It is traveling downward at a constant velocity $\dot{u}_0$ when brakes are applied to the hoist, causing an angular deceleration equal to a/r, where r is the radius of the drum. Under this condition, the cable will cease to unwind at the time $\dot{u}_0/a$ (measuring time from $t = 0$ when the brakes are initially applied). Find the displacement u of the elevator during the time interval $0 \leqq t \leqq \dot{u}_0/a$, assuming that the length of free cable is ℓ at time $t = 0$ and neglecting the change in length during braking.

Solution. This example represents a case of constant support acceleration of magnitude $-a$, and Eq. (1.73) for undamped relative response gives

$$u^* = \frac{a}{\omega}\int_0^t \sin \omega(t - t') \, dt' = \frac{a}{\omega^2}(1 - \cos \omega t) \tag{w}$$

As in the previous example, we may add this relative response to the motion at the top of the cable to obtain the total response.

$$u = u_g + u^* = \dot{u}_0 t - a\left[\frac{t^2}{2} - \frac{1}{\omega^2}(1 - \cos \omega t)\right] \tag{x}$$

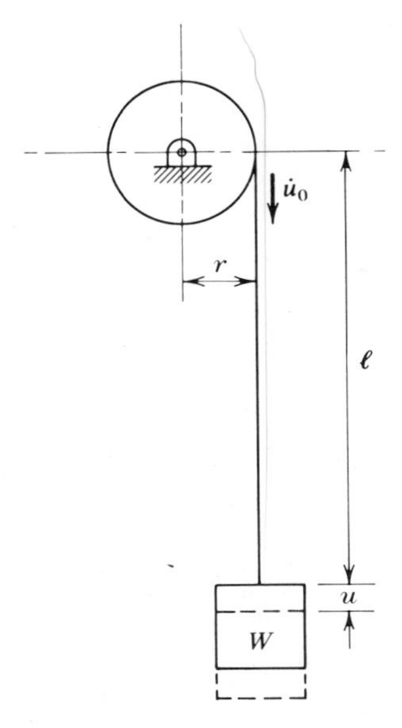

FIG. 1.50

Of course, in an actual elevator the design is such that the brakes are applied more smoothly, and any tendency to oscillate is suppressed by suitable dampers.

1.14 RESPONSE SPECTRA

The forcing functions discussed in Secs 1.12 and 1.13 cause vibrational responses in elastic systems, and the maximum values of these responses may be less than, equal to, or greater than the corresponding static responses. In general, the maximum response depends upon the characteristics of the system and the nature of the loading. For an undamped one-degree system, the natural period (or frequency) is the characteristic that determines its response to a given forcing function. In addition, the shape and duration of the forcing function itself play important roles in the response. Plots of maximum response values against selected parameters of the system or of the forcing function are called *response spectra*. Such diagrams are of interest in design because they provide the possibility of predicting the ratio of the maximum dynamic stress in a structure to the corresponding static stress. The time at which the maximum response of a system occurs is also of interest, and plots of this variable accompany the response spectra discussed in this section.

Let us reconsider the rectangular impulse shown in Fig. 1.51*a*, which was discussed extensively in Sec. 1.12. In that section we learned that a

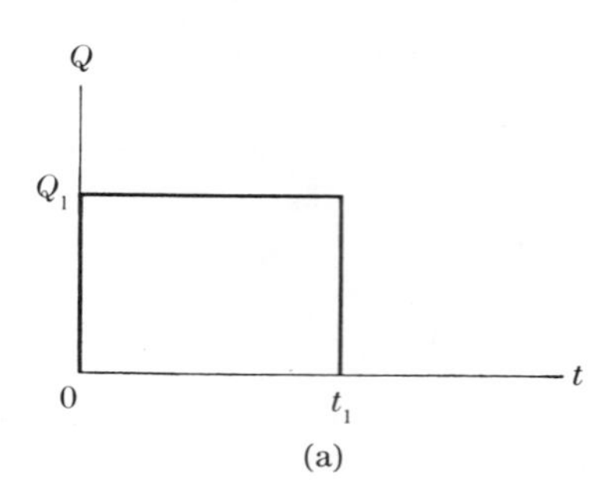

(a)

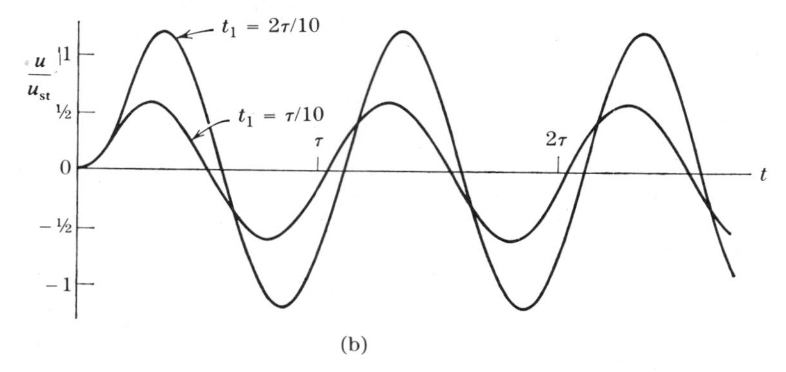

(b)

FIG. 1.51

rectangular impulse of duration $t_1 = \tau/2$ is just sufficient to cause a maximum response of $u_m = 2Q_1/k$. This maximum value is the same as that caused by a suddenly applied force Q_1 of indefinite duration (step function). Thus, a duration of the rectangular impulse in excess of $\tau/2$ will always cause a maximum response of the system equal to twice the static-load response. Using the notation

$$u_{st} = \frac{Q_1}{k} \tag{a}$$

we may state for this case:

$$\frac{u_m}{u_{st}} = 2 \qquad (t_1 \geqq \tau/2) \tag{b}$$

If the duration of the rectangular impulse is less than $\tau/2$, the maximum response will be less than $2u_{st}$. Examples of this type appear in Fig. 1.51*b*, where response curves for $t_1 = \tau/10$ and $t_1 = 2\tau/10$ are plotted. In all such cases the maximum displacement occurs after the impulsive action has terminated, because the velocity at time t_1 is positive [see Eq. (g), Sec. 1.12]. Hence, to find the maximum value of the response and the time at which it occurs, we must examine Eq. (1.67), for which $t_1 \leqq t$. That equation may be written in dimensionless form as

$$\frac{u}{u_{\mathrm{st}}} = \cos \omega(t - t_1) - \cos \omega t \tag{c}$$

Differentiating this expression with respect to time, we obtain

$$\frac{\dot{u}}{u_{\mathrm{st}}} = \omega[\sin \omega t - \sin \omega(t - t_1)] \tag{d}$$

By setting the term in brackets equal to zero, we find an expression for the time t_{m} at which the maximum displacement occurs.

$$\sin \omega t_{\mathrm{m}} = \sin \omega(t_{\mathrm{m}} - t_1) \tag{e}$$

Thus,

$$\omega t_{\mathrm{m}} = \frac{\pi}{2} + \frac{\omega t_1}{2} \tag{f}$$

Equation (f) shows that t_{m} is linearly related to t_1. Furthermore, since the range of interest for ωt_1 is $0 \leqq \omega t_1 \leqq \pi$, the corresponding range of ωt_{m} is $\pi/2 \leqq \omega t_{\mathrm{m}} \leqq \pi$. Substitution of the expression for ωt_{m} from Eq. (f) in place of ωt in Eq. (c) yields

$$\frac{u_{\mathrm{m}}}{u_{\mathrm{st}}} = 2 \sin \frac{\omega t_1}{2} = \sqrt{2(1 - \cos \omega t_1)} \tag{g}$$

Equation (g) is the same as Eq. (j) in Sec. 1.12, which was obtained by a different approach. Thus, the response spectrum for the rectangular impulse may be summarized as follows:

$$\text{for} \quad 0 \leqq \frac{t_1}{\tau} \leqq \frac{1}{2} \qquad \frac{u_{\mathrm{m}}}{u_{\mathrm{st}}} = 2 \sin \frac{\pi t_1}{\tau} \tag{1.74a}$$

$$\frac{t_{\mathrm{m}}}{\tau} = \frac{1}{4}\left(1 + \frac{2t_1}{\tau}\right) \tag{1.74b}$$

$$\text{for} \quad \frac{1}{2} \leqq \frac{t_1}{\tau} \qquad \frac{u_{\mathrm{m}}}{u_{\mathrm{st}}} = 2 \tag{1.74c}$$

$$\frac{t_{\mathrm{m}}}{\tau} = \frac{1}{2} \tag{1.74d}$$

Diagrams of these dimensionless expressions are given in Figs. 1.52*a* and 1.52*b*, where u_m/u_{st} and t_m/τ are plotted against t_1/τ. From Eq. (1.74a) we see that if the duration of the impulse is less than $\tau/6$, the dynamic response is less than that due to the load applied statically. On the other hand, if the duration of the impulse is between $\tau/6$ and $\tau/2$, the value of u_m/u_{st} is between 1 and 2. Of course, if $t_1 \geqq \tau/2$, the value of u_m/u_{st} is always equal to 2.

At this point it is of interest to recognize that plots of magnification factors for forced vibrations represent response spectra, as defined in this section. Figure 1.33 in Sec. 1.9 contains a family of curves for $\beta = u_m/u_{st}$ plotted against the frequency ratio Ω/ω. It should be recalled that these curves constitute only the steady-state part of the response and that a different curve is obtained for each level of damping. If the transient parts of the forced vibrations were to be included, the response spectra in Fig. 1.33 would be somewhat higher; but this effect is of little significance.

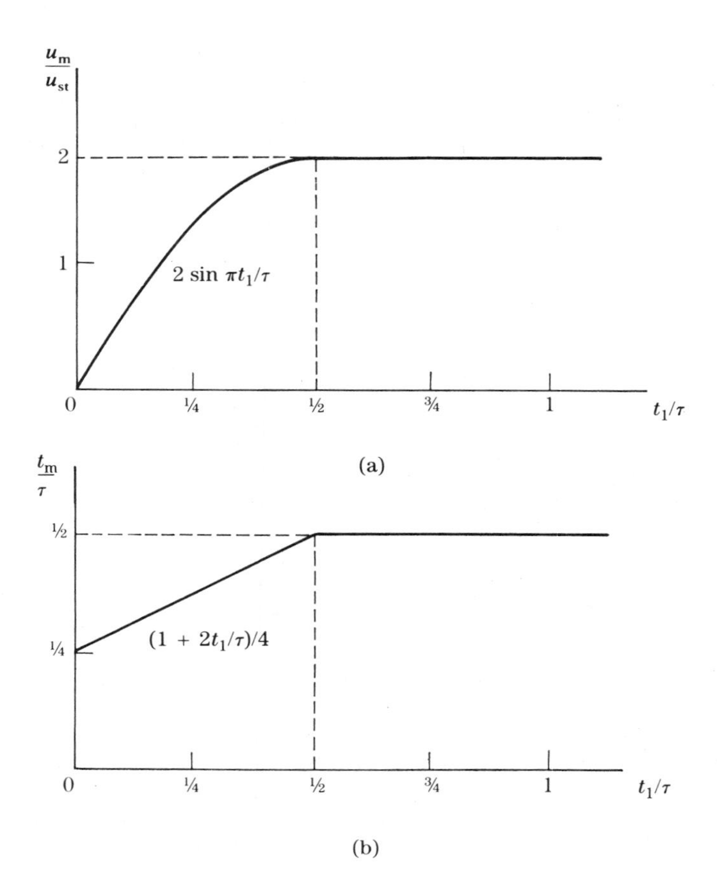

FIG. 1.52

Furthermore, while damping is of great importance in the problem of forced vibrations, it is often omitted as a consideration in response spectra due to impulsive excitations. Low values of damping have little effect upon such response maxima, which usually occur before much energy is dissipated. However, a family of damped response spectra can always be constructed for any forcing function, with a different curve for each level of damping. This may be accomplished for simple cases by deriving the appropriate analytical functions, but for complicated cases a numerical approach must be applied.

Example 1. Figure 1.53*a* shows a forcing function that increases linearly from zero to Q_1 in the time t_1 and is constant thereafter. The response of an undamped one-degree system to this excitation is (see Prob. 1.13-2)

$$u = \frac{Q_1}{k}\left(\frac{t}{t_1} - \frac{\sin \omega t}{\omega t_1}\right) \qquad (0 \leqq t \leqq t_1) \tag{h}$$

$$u = \frac{Q_1}{k}\left[1 + \frac{\sin \omega(t - t_1) - \sin \omega t}{\omega t_1}\right] \qquad (t_1 \leqq t) \tag{i}$$

Determine the response spectrum and the corresponding time function for this case.

Solution. By inspection of Eqs. (*h*) and (*i*), we see that the maximum response will always ocur after time t_1. Thus, only Eq (*i*) is of interest here, and it may be expressed in the dimensionless form

$$\frac{u}{u_{\text{st}}} = \frac{1}{\omega t_1}[\omega t_1 + \sin \omega(t - t_1) - \sin \omega t] \tag{j}$$

Differentiating Eq. (*j*) with respect to time, we obtain

$$\frac{\dot{u}}{u_{\text{st}}} = \frac{1}{t_1}[\cos \omega(t - t_1) - \cos \omega t] \tag{k}$$

Setting the term in brackets equal to zero yields an expression for the time t_{m}.

$$\cos \omega t_{\text{m}} = \cos \omega(t_{\text{m}} - t_1) \tag{l}$$

Hence

$$\omega t_{\text{m}} = \pi + \frac{\omega t_1}{2} \tag{m}$$

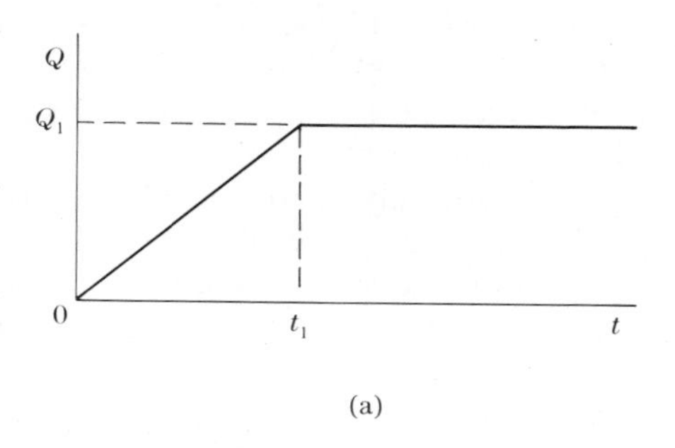
Q
Q_1
0
t_1
t

(a)

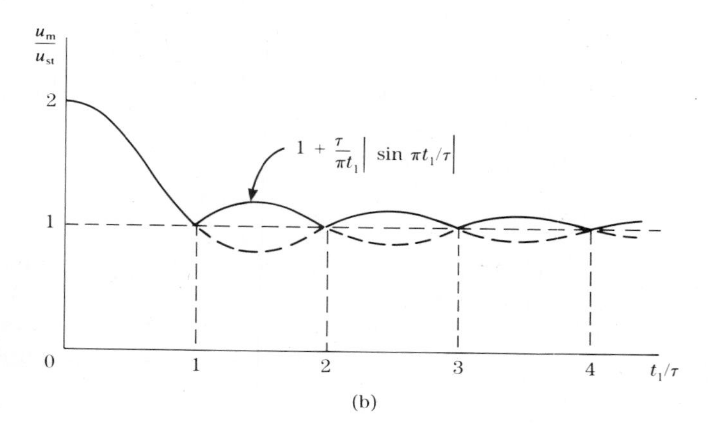
$\frac{u_m}{u_{st}}$
2
1
0
$1 + \frac{\tau}{\pi t_1}\left|\sin \pi t_1/\tau\right|$
1
2
3
4
t_1/τ

(b)

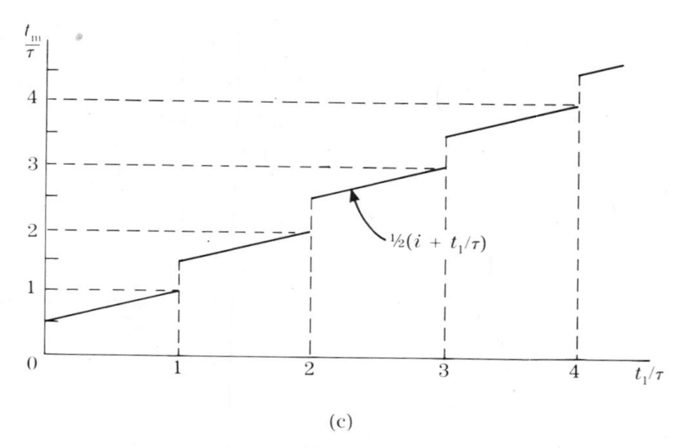
$\frac{t_m}{\tau}$
4
3
2
1
0
$\frac{1}{2}(i + t_1/\tau)$
1
2
3
4
t_1/τ

(c)

FIG. 1.53

As in the previous case, t_m is linearly related to t_1. In addition, we conclude from Eq. (m) that the range of ωt_m is $\pi \leqq \omega t_m$. Substitution of expression (m) into Eq. (j) produces

$$\frac{u_m}{u_{st}} = 1 + \frac{2}{\omega t_1} \sin \frac{\omega t_1}{2} = 1 \pm \frac{1}{\omega t_1} \sqrt{2(1 - \cos \omega t_1)} \qquad (n)$$

These expressions represent both maxima and minima, which depend upon the value of ωt_1. Summarizing for maxima in the range $t_1 \leqq t$, we have

$$\frac{u_m}{u_{st}} = 1 + \frac{\tau}{\pi t_1} |\sin \pi t_1/\tau| \qquad (o)$$

Because of half-cycle discontinuities of $|\sin \pi t_1/\tau|$ in Eq. (o), we must also write the time formula (m) as

$$\frac{t_m}{\tau} = \frac{(i + t_1/\tau)}{2} \qquad (p)$$

where $i = 1, 2, 3, \ldots$ denotes each half cycle.

Figures 1.53*b* and 1.53*c* contain plots of Eqs. (o) and (p), respectively. We see from this response spectrum (Fig. 1.53*b*) that the highest value of $u_m/u_{st} = 2$ occurs when $t_1 = 0$, which corresponds to the case of the step function. For $t_1 \leqq \tau/4$ the value of u_m/u_{st} is approximately 2; and the result is not much different from that for the step funtion. Since a zero rise time is physically impossible, it is well to recognize that a small but finite rise time gives practically the same results. For $t_1 \geqq \tau$ the value of u_m does not exceed u_{st} by very much, and with a large rise time the loading is essentially static.

Example 2. Consider the case of a rectangular impulse (Fig. 1.51*a*) applied to a damped one-degree system (Fig. 1.42*a*). We may synthesize this forcing function as the sum of a step function (of magnitude Q_1) beginning at time $t = 0$ and a second step function (of magnitude $-Q_1$) beginning at time $t = t_1$. Thus, the dimensionless response (for $t_1 \leqq t$) of the damped system becomes (see Example 3, Sec. 1.12)

$$\frac{u}{u_{st}} = e^{-n(t-t_1)}\left[\cos \omega_d(t - t_1) + \frac{n}{\omega_d} \sin \omega_d(t - t_1)\right]$$
$$- e^{-nt}\left(\cos \omega_d t + \frac{n}{\omega_d} \sin \omega_d t\right) \qquad (q)$$

Derive expressions for the response spectrum and the time for maximum response.

Solution. Equation (q) may be simplified by using trigonometric identities and rearranging terms to obtain

$$\frac{u}{u_{\mathrm{st}}} = e^{-nt}(A \cos \omega_{\mathrm{d}} t + B \sin \omega_{\mathrm{d}} t) \tag{r}$$

in which

$$A = e^{nt_1}\left(\cos \omega_{\mathrm{d}} t_1 - \frac{n}{\omega_{\mathrm{d}}} \sin \omega_{\mathrm{d}} t_1\right) - 1 \tag{s}$$

and

$$B = e^{nt_1}\left(\sin \omega_{\mathrm{d}} t_1 + \frac{n}{\omega_{\mathrm{d}}} \cos \omega_{\mathrm{d}} t_1\right) - \frac{n}{\omega_{\mathrm{d}}} \tag{t}$$

Differentiating Eq. (r) with respect to time and setting the result equal to zero produces

$$t_{\mathrm{m}} = \frac{1}{\omega_{\mathrm{d}}} \tan^{-1}\left(\frac{\omega_{\mathrm{d}} B - nA}{\omega_{\mathrm{d}} A + nB}\right) \tag{u}$$

Also

$$\sin \omega_{\mathrm{d}} t_{\mathrm{m}} = \frac{\omega_{\mathrm{d}} B - nA}{C} \qquad \cos \omega_{\mathrm{d}} t_{\mathrm{m}} = \frac{\omega_{\mathrm{d}} A + nB}{C} \tag{v}$$

where

$$C = \sqrt{(\omega_{\mathrm{d}}^2 + n^2)(A^2 + B^2)}$$

Substituting expressions (v) into Eq. (r) and combining terms, we obtain

$$\frac{u_{\mathrm{m}}}{u_{\mathrm{st}}} = e^{-nt_{\mathrm{m}}}\sqrt{1 + e^{2nt_1} - 2e^{nt_1} \cos \omega_{\mathrm{d}} t_1} \tag{w}$$

When the damping constant is set equal to zero, Eq. (w) becomes the same as Eq. (g) for the undamped case.

The examples treated in this section lead to explicit expressions for t_{m}/τ and $u_{\mathrm{m}}/u_{\mathrm{st}}$, but it should be mentioned that these are exceptional cases. In general, it is difficult to identify the time range within which the maximum response occurs. In addition, the equation containing t_{m}/τ may be transcendental, in which case it cannot be solved explicitly. Under these circumstances, values of $u_{\mathrm{m}}/u_{\mathrm{st}}$ and t_{m}/τ must be obtained by exhaustive calculations, using a succession of values for the time ratio t_1/τ. For each value of t_1/τ, the expression for u/u_{st} in terms of t/τ may be calculated on a computer; and a value of $u_{\mathrm{m}}/u_{\mathrm{st}}$ as well as t_{m}/τ may be obtained therefrom.

1.15 STEP-BY-STEP RESPONSE CALCULATIONS

In many practical problems the forcing functions are not analytical expressions but are represented by a series of points on a diagram or a list of numbers in a table. In such cases it may be feasible to replace the data with certain formulas by curve fitting methods and then to use those formulas in the Duhamel integral. However, a more general method for evaluating the response consists of using some simple interpolation function in a repetitive series of calculations. The latter approach will be discussed in this section for *piecewise-linear interpolation functions.*

Assume that the damped one-degree system in Fig. 1.54 is subjected to a time-varying force $Q(t)$ that is approximated by a series of straight lines, as shown in Fig. 1.55. For a particular interpolating line in the time interval $t_j \leqq t \leqq t_{j+1}$, the response of the system may be written as the sum of three parts, as follows:

$$u = u_1 + u_2 + u_3 \tag{1.75}$$

Using the definition $t' = t - t_j$, we have for the first part

$$u_1 = e^{-nt'}\left(u_j \cos \omega_d t' + \frac{\dot{u}_j + nu_j}{\omega_d} \sin \omega_d t'\right) \tag{1.76a}$$

This equation contains the free-vibrational motion of the system due to the displacement u_j and the velocity $\dot{u}_j$ at time $t = t_j$ (the beginning of the interval). The formula for this portion of the response is drawn from Eq. (1.35) in Sec. 1.8.

The other two parts of the response in Eq. (1.75) are associated with the straight-line forcing function in Fig. 1.55. The one caused by the rectangular impulse of magnitude Q_j is

$$u_2 = \frac{Q_j}{k}\left[1 - e^{-nt'}\left(\cos \omega_d t' + \frac{n}{\omega_d} \sin \omega_d t'\right)\right] \tag{1.76b}$$

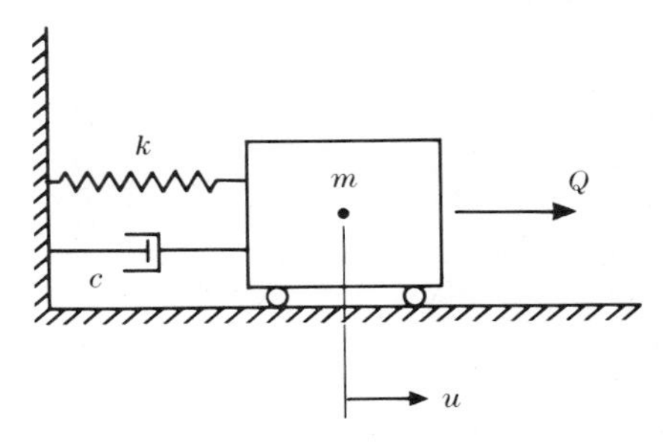

FIG. 1.54

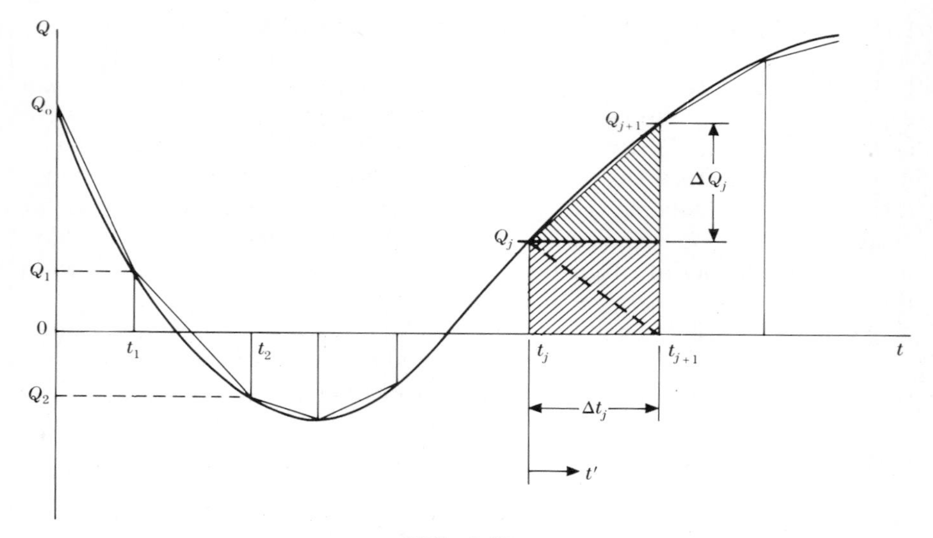

FIG. 1.55

which comes from Eq. (u) in Example 3 of Sec. 1.12. On the other hand, the effect of the triangular impulse of magnitude $\Delta Q_j = Q_{j+1} - Q_j$ becomes

$$u_3 = \frac{\Delta Q_j}{\Delta t_j k \omega^2}\left[\omega^2 t' - 2n + e^{-nt'}\left(2n \cos \omega_d t' - \frac{\omega_d^2 - n^2}{\omega_d} \sin \omega_d t'\right)\right] \tag{1.76c}$$

This formula is available as the solution of Prob. 1.12-7.

By differentiating Eqs. (1.75) and (1.76) with respect to time, we can also find velocity expressions in three parts. Thus,

$$\dot{u} = \dot{u}_1 + \dot{u}_2 + \dot{u}_3 \tag{1.77}$$

and

$$\dot{u}_1 = e^{-nt'}\left[-\left(u_j \omega_d + n \frac{\dot{u}_j + n u_j}{\omega_d}\right) \sin \omega_d t' + \dot{u}_j \cos \omega_d t'\right] \tag{1.78a}$$

and

$$\dot{u}_2 = \frac{Q_j \omega^2}{k \omega_d} e^{-nt'} \sin \omega_d t' \tag{1.78b}$$

Also,

$$\dot{u}_3 = \frac{\Delta Q_j}{\Delta t_j k}\left[1 - e^{-nt'}\left(\cos \omega_d t' + \frac{n}{\omega_d} \sin \omega_d t'\right)\right] \tag{1.78c}$$

At the end of the time interval Δt_j, the displacement expressions in Eqs.

(1.76) become

$$(u_1)_{j+1} = e^{-n\Delta t_j}\left(u_j \cos \omega_d \Delta t_j + \frac{\dot{u}_j + nu_j}{\omega_d} \sin \omega_d \Delta t_j\right) \tag{1.79a}$$

$$(u_2)_{j+1} = \frac{Q_j}{k}\left[1 - e^{-n\Delta t_j}\left(\cos \omega_d \Delta t_j + \frac{n}{\omega_d} \sin \omega_d \Delta t_j\right)\right] \tag{1.79b}$$

$$(u_3)_{j+1} = \frac{\Delta Q_j}{\Delta t_j k\omega^2}\left[\omega^2 \Delta t_j - 2n + e^{-n\Delta t_j}\left(2n \cos \omega_d \Delta t_j - \frac{\omega_d^2 - n^2}{\omega_d} \sin \omega_d \Delta t_j\right)\right] \tag{1.79c}$$

In addition, the velocity expressions in Eqs. (1.78) are rewritten as

$$(\dot{u}_1)_{j+1} = e^{-n\Delta t_j}\left[-\left(u_j\omega_d + n\frac{\dot{u}_j + nu_j}{\omega_d}\right) \sin \omega_d \Delta t_j + \dot{u}_j \cos \omega_d \Delta t_j\right] \tag{1.80a}$$

$$(\dot{u}_2)_{j+1} = \frac{Q_j\omega^2}{k\omega_d} e^{-n\Delta t_j} \sin \omega_d \Delta t_j \tag{1.80b}$$

$$(\dot{u}_3)_{j+1} = \frac{\Delta Q_j}{\Delta t_j k}\left[1 - e^{-n\Delta t_j}\left(\cos \omega_d \Delta t_j + \frac{n}{\omega_d} \sin \omega_d \Delta t_j\right)\right] \tag{1.80c}$$

Equations (1.79) and (1.80) constitute *recurrence formulas* that may be used to calculate the damped response at the end of step j and to provide initial conditions at the beginning of step $j + 1$.

If damping is neglected, Eqs. (1.79) for displacements simplify, as follows:

$$(u_1)_{j+1} = u_j \cos \omega\Delta t_j + \frac{\dot{u}_j}{\omega} \sin \omega\Delta t_j \tag{1.81a}$$

$$(u_2)_{j+1} = \frac{Q_j}{k}(1 - \cos \omega\Delta t_j) \tag{1.81b}$$

$$(u_3)_{j+1} = \frac{\Delta Q_j}{\Delta t_j k\omega}(\omega\Delta t_j - \sin \omega\Delta t_j) \tag{1.81c}$$

and Eqs. (1.80) for velocities become

$$(\dot{u}_1)_{j+1} = -u_j\omega \sin \omega\Delta t_j + \dot{u}_j \cos \omega\Delta t_j \tag{1.82a}$$

$$(\dot{u}_2)_{j+1} = \frac{Q_j}{k} \omega \sin \omega\Delta t_j \tag{1.82b}$$

$$(\dot{u}_3)_{j+1} = \frac{\Delta Q_j}{\Delta t_j k}(1 - \cos \omega\Delta t_j) \tag{1.82c}$$

Equations (1.81) and (1.82) are simple enough for hand calculations to obtain approximate results.

Of course, we need not take the shaded impulse in Fig. 1.55 as the sum of a rectangle and a triangle. Alternatively, it could be divided into two triangles, as indicated by the dashed diagonal line in the figure. Then it would be possible to express the second and third parts of the response in terms of Q_j and Q_{j+1}. Furthermore, if the time step Δt_j is constant, the coefficients of u_j, $\dot{u}_j$, Q_j, and ΔQ_j (or Q_{j+1}) all become constants for both the displacement and the velocity expressions. Hence, these coefficients need be computed only once and then used repetitively throughout the numerical solution [9].

Because step-by-step response calculations are tedious, we provide a flowchart for a computer program in Appendix B to handle the numerical operations. The program named LINFORCE computes the response of a damped one-degree system to a piecewise-linear forcing function, using Eqs. (1.79) and (1.80).

Example 1. Figure 1.56*a* shows a forcing function $Q = Q_1 \sin \Omega t$ that is applied to an undamped one-degree system. The function is discretized by piecewise-linear interpolation into 20 equal time steps of duration $\Delta t = T/20$. Using the method of this section, calculate approximately the response of the system. Assume that the initial conditions are $u_0 = \dot{u}_0 = 0$, that the values of Q_1 and k are both unity, and that the frequency ratio is $\Omega/\omega = 0.9$.

From Eq. (1.29b) in Sec. 1.7, the exact solution to this problem (with zero damping) is

$$u = \frac{Q_1}{k}\left(\sin \Omega t - \frac{\Omega}{\omega} \sin \omega t\right)\beta \tag{a}$$

in which the magnification factor β has the value

$$\beta = \left| \frac{1}{1-(\Omega/\omega)^2} \right| = \frac{1}{1-(0.9)^2} = 5.263 \tag{b}$$

An approximate solution is found by applying Eqs. (1.81) and (1.82) recursively in 20 specified time steps. Results of such calculations (by hand or computer) are summarized in Table 1.1. Also given in the table are exact displacements obtained from Eq. (a). As expected, the approximate displacements are slightly less than their exact counterparts, because linear interpolation of the sine curve is imperfect. Decreasing the step size would, of course, lead to exact values (except for roundoff errors).

The approximate undamped response in Table 1.1 is plotted as the solid curve in Fig. 1.56*b*. Also shown in this figure is a dashed curve that results from recursively applying Eqs. (1.79) and (1.80) with the damping ratio $\gamma = n/\omega = 0.05$.

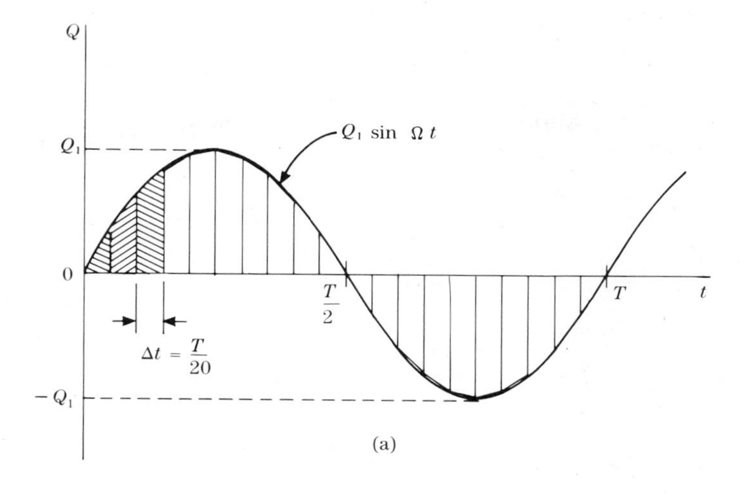

(a)

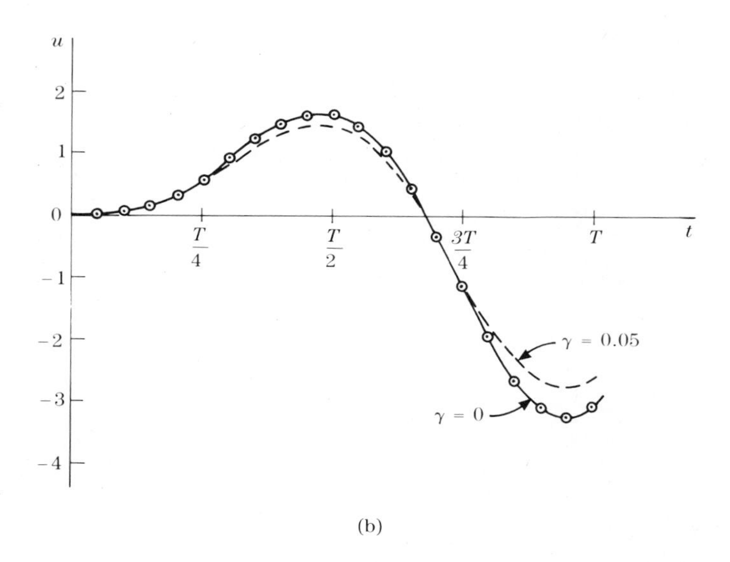

(b)

FIG. 1.56

Example 2. Figure 1.57*a* gives a series of plotted points simulating a *blast load* that impinges upon an undamped one-degree structure, such as the building frame in Fig. 1.20. Note that the blast force rises quickly to the maximum value Q_1 and then diminishes more slowly (and even becomes negative for a while). In this case we have 16 equal time steps, each of which has the value $\Delta t = \tau/30$. Apply the method of this section to find the approximate response of the structure. Let the values of both Q_1 and k be unity, and assume that the initial conditions are $u_0 = \dot{u}_0 = 0$.

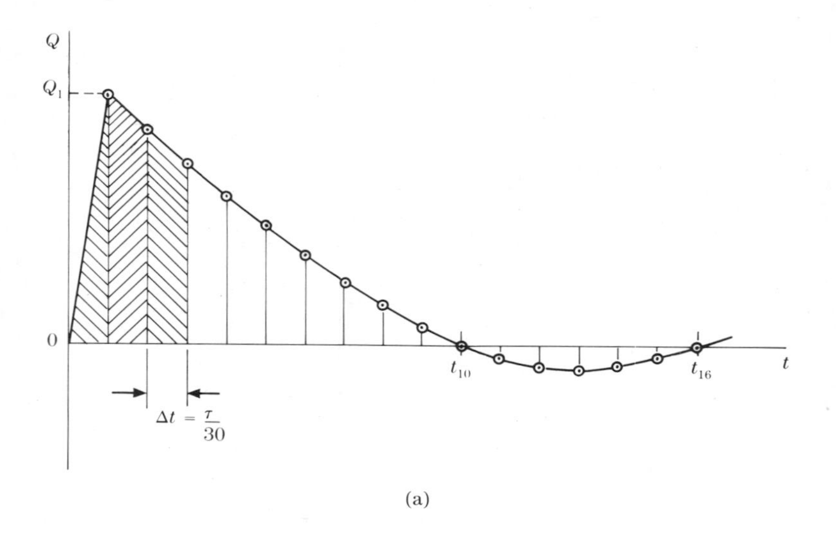

(a)

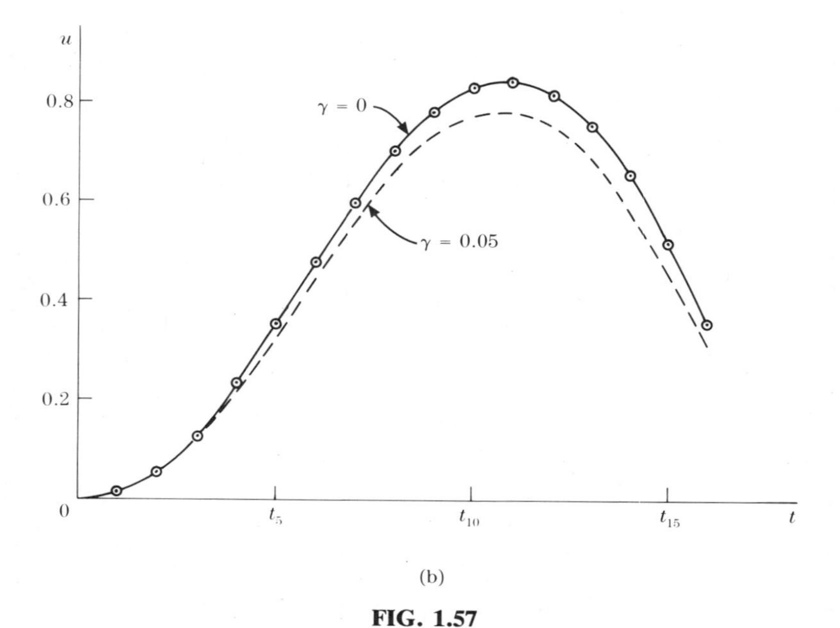

(b)

FIG. 1.57

After using Eqs. (1.81) and (1.82) recursively to obtain the time history of the response, we then list the calculated displacements in Table 1.2. For this problem there are no exact results with which to compare these numbers. Nevertheless, we have great confidence in their validity. Figure 1.57b shows a plot of the time history of undamped response (the solid

TABLE 1.1 Solution for Example 1

j	Q_j	u_j	Exact	j	Q_j	u_j	Exact
1	0.309	0.006	0.006	11	−0.309	1.407	1.418
2	0.588	0.048	0.049	12	−0.588	1.000	1.001
3	0.809	0.154	0.156	13	−0.809	0.404	0.407
4	0.951	0.338	0.341	14	−0.951	−0.338	−0.341
5	1.000	0.593	0.598	15	−1.000	−1.151	−1.161
6	0.951	0.896	0.903	16	−0.951	−1.945	−1.961
7	0.809	1.203	1.213	17	−0.809	−2.616	−2.634
8	0.588	1.461	1.473	18	−0.588	−3.068	−3.094
9	0.309	1.613	1.626	19	−0.309	−3.220	−3.246
10	0	1.607	1.620	20	0	−3.020	−3.045

TABLE 1.2 Solution for Example 2

j	Q_j	u_j	j	Q_j	u_j
1	1.000	0.007	9	0.070	0.780
2	0.850	0.050	10	0	0.830
3	0.720	0.127	11	−0.050	0.844
4	0.590	0.230	12	−0.080	0.819
5	0.475	0.350	13	−0.100	0.755
6	0.360	0.474	14	−0.080	0.654
7	0.250	0.594	15	−0.050	0.521
8	0.155	0.699	16	0	0.363

curve), which appears to be quite reasonable. Also appearing in the figure is a dashed curve that gives the effect of using Eqs. (1.79) and (1.80) with $\gamma = 0.05$.

REFERENCES

[1] Bowes, W. H., Russell, L. T., and Suter, G. T., *Mechanics of Engineering Materials,* Wiley, New York, 1984.

[2] Gere, J. M., and Timoshenko, S. P., *Mechanics of Materials,* 3rd ed., PWS-Kent, Boston, MA, 1990.

[3] Rayleigh, J. W. S., *Theory of Sound,* 2nd ed., Vol. 1, Macmillan, London, 1894, Sec. 88 (reprinted by Dover, New York, 1945).

[4] Jacobsen, L. S., "Steady Forced Vibrations as Influenced by Damping," *Trans. ASME,* **52,** 1930, pp. APM 169–181.

[5] Kimball, A. L., "Vibration Damping, Including the Case of Solid Damping," *Trans. ASME,* **51,** 1929, pp. APM 227–236.

[6] Coulomb, C. A., *Théorie des machines simples,* Paris, 1821.

[7] Den Hartog, J. P., "Forced Vibrations with Combined Coulomb and Viscous Damping," *Trans. ASME,* **53,** 1931, pp. APM 107–115.

[8] Vennard, J. K., and Street, R. L., *Elementary Fluid Mechanics,* 5th ed., Wiley, New York, 1975.

[9] Craig, R. R., *Structural Dynamics,* Wiley, New York, 1981.

PROBLEMS

1.2-1. The helical spring in Fig. 1.2 has a mean coil diameter $D = 1$ in., a wire diameter $d = 0.1$ in., and contains 20 coils. The modulus of elasticity of the wire in shear is $G = 12 \times 10^6$ psi, and the suspended weight is $W = 30$ lb. Calculate the period of free vibration.

1.2-2. A simply supported beam having flexural rigidity $EI = 350\ \text{kN} \cdot \text{m}^2$ has a clear span $\ell_1 = 2$ m between supports and an overhang $\ell_2 = 1$ m at one end, as shown in the figure. Neglecting the distributed mass of the beam, find the natural frequency f of free vibration of a block of weight $W = 2700$ kN at the end of the overhang.

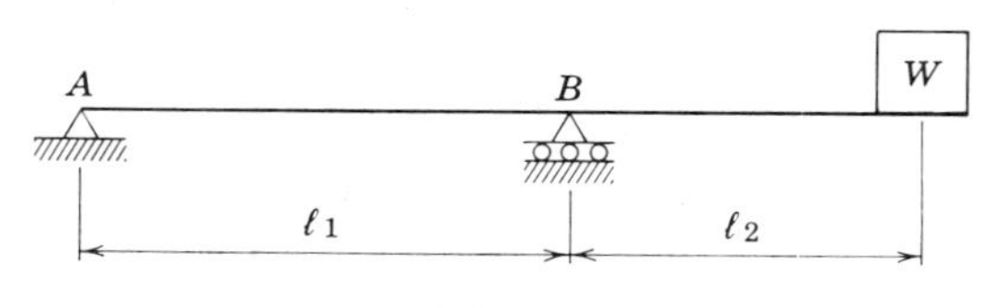

PROB. 1.2-2

1.2-3. A beam AB having flexural rigidity $EI = 30 \times 10^6\ \text{lb-in.}^2$ is supported at points A and B by springs, each having a spring constant $k = 300$ lb/in., as shown in the figure. Neglecting the distributed mass of the beam, compute the period of free vibration of the block of weight $W = 1000$ lb, located at point C. Let $a = 7$ ft and $b = 3$ ft.

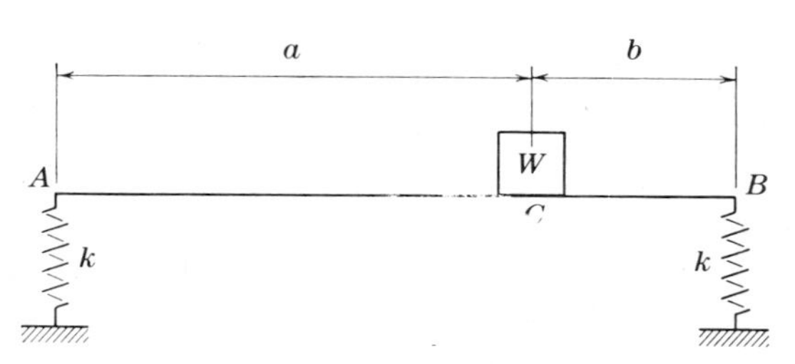

PROB. 1.2-3

1.2-4. A water tank of weight $W = 670$ kN is supported by four vertical pipe columns built in at the ends, as shown in the figure. Each column has flexural rigidity $EI = 6$ Mn · m². Calculate the period of free vibration of the tank in the horizontal direction. Neglect the distributed mass of the columns.

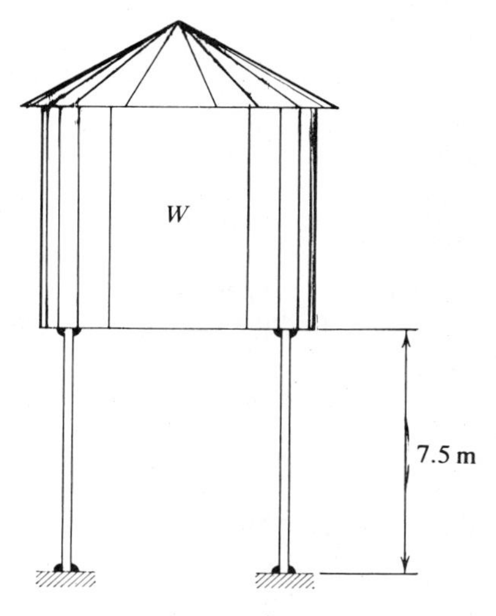

PROB. 1.2-4

1.2-5. To reduce the maximum dynamical stress occurring under the conditions of Example 4 in Sec. 1.2, assume that a short spring having a spring constant $k = 2000$ lb/in. is inserted between the lower end of the cable and the elevator. Calculate the maximum stress that will result in this case when the upper end of the cable is suddenly stopped. Take the same numerical data as given in Example 4.

1.2-6. A portal frame consists of a 24 in. steel I-beam 20 ft long rigidly connected by welding to two relatively flexible columns, as shown in the figure. Each column is a channel section having cross-sectional area $A = 4.02$ in.², least radius of gyration $r = 0.62$ in., and $E = 30 \times 10^6$ psi. Calculate the natural period of lateral vibration in the plane of the frame:

(a) Assuming complete fixity at A and B

(b) Assuming hinges at A and B.

Neglect bending of the I-beam and mass of the columns.

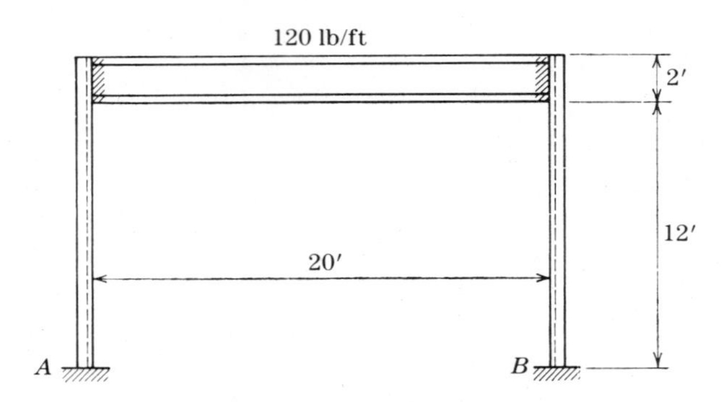

PROB. 1.2-6

1.2-7. A two-span continuous steel beam ($I = 15 \times 10^{-6}\,\text{m}^4$) carries a motor of weight $W = 55$ kN at the middle of the span BC, as shown in the figure. Compute the natural frequency f of free vertical vibration of the motor, neglecting the distributed mass of the beam.

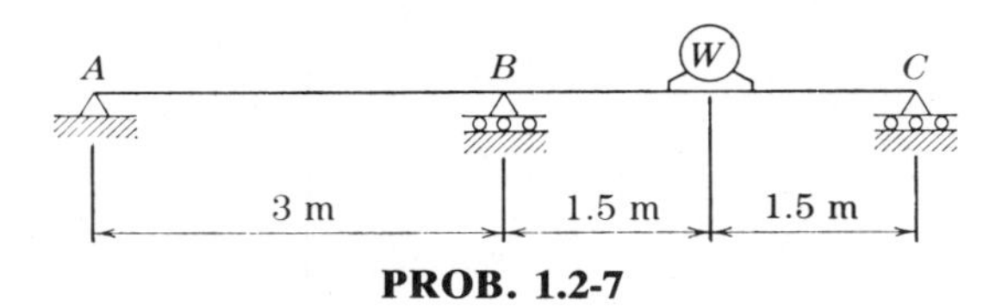

PROB. 1.2-7

1.2-8. A small ball of mass m is attached to the midpoint of a tightly stretched wire of length 2ℓ, as shown in the figure. The wire cannot resist bending and is subjected to a high initial tension S. Set up the differential equation of motion for small lateral vibrations of the ball; and show that if the tension in the wire can be assumed to remain constant, the motion will be simple harmonic. In this case what is the period of vibration?

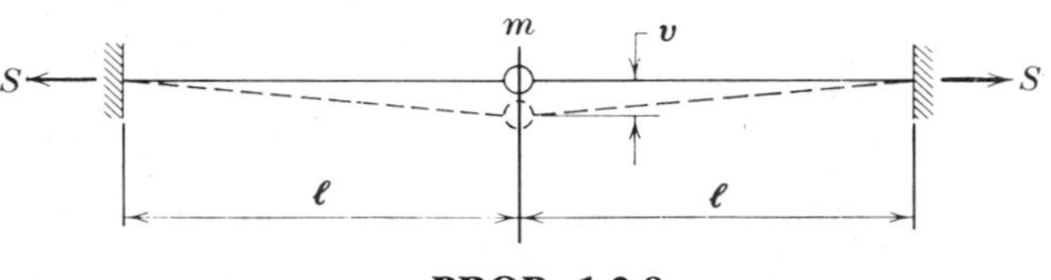

PROB. 1.2-8

1.2-9. A weight W is suspended by three springs k_1, k_2, k_3, connected in series in a manner similar to that for two springs in Fig. 1.6a. Derive the equivalent spring constant for the system.

1.3-1. Determine the frequency of torsional vibration of a horizontal bar AB of weight $W = 18$ N and length $a = 0.5$ m, which is suspended at its midpoint on a vertical steel wire of length $\ell = 0.5$ m and diameter $d = 3$ mm. Assume that the bar is slender but rigid, neglect the mass of the wire, and take its shear modulus to be $G = 83$ GPa.

1.3-2. The figure shows a device that is quite useful for the experimental determination of mass moments of inertia of irregularly shaped bodies. It consists of two parallel plates connected in such a manner that the whole assemblage acts as a rigid body attached to a vertical shaft and inside of which any body of limited size can be placed. When empty (Fig. *a*), this torsional pendulum has an observed period τ_0. When carrying a body of known moment of inertia I_1 that oscillates with it (Fig. *b*), the pendulum has a period τ_1; and when carrying a body of unknown moment of inertia I_2 (Fig. *c*), the pendulum oscillates with a period of τ_2. Find the moment of inertia I_2 of the last body with respect to the axis of rotation, that is, the axis of the shaft.

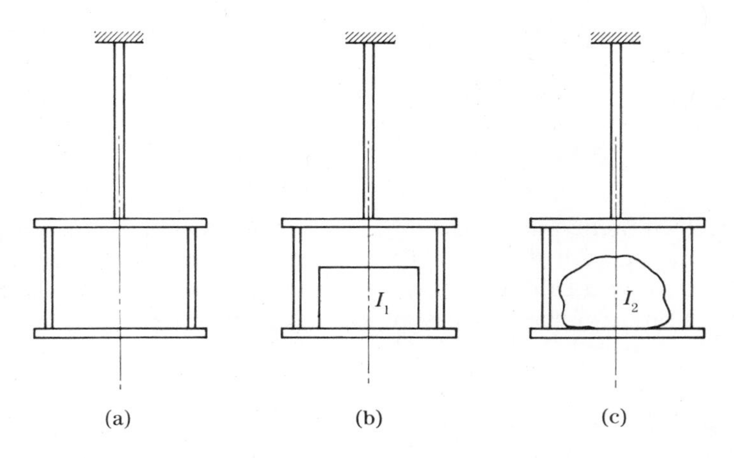

PROB. 1.3-2

1.3-3. A slender prismatic bar AB of weight W and length ℓ is supported in a horizontal position by a hinge at A and by a spring of constant k at B, as shown in the figure. For small values of angular displacement ϕ of the bar in the vertical plane, find the period of rotational oscillation. Neglect the mass of the spring, and consider the bar to be rigid.

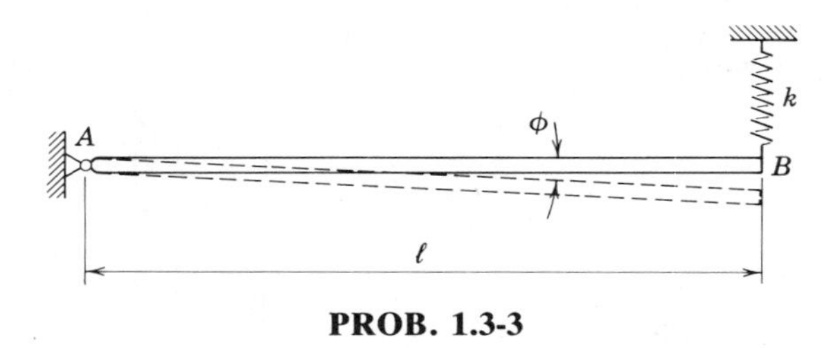

PROB. 1.3-3

1.3-4. A slender but rigid prismatic bar AB of weight W and length ℓ is hinged at A and supported in a horizontal position by a vertical spring attached to it at C (see figure). For small amplitudes of rotational oscillation of the bar in a vertical plane, calculate the period τ if the spring has constant k and negligible mass.

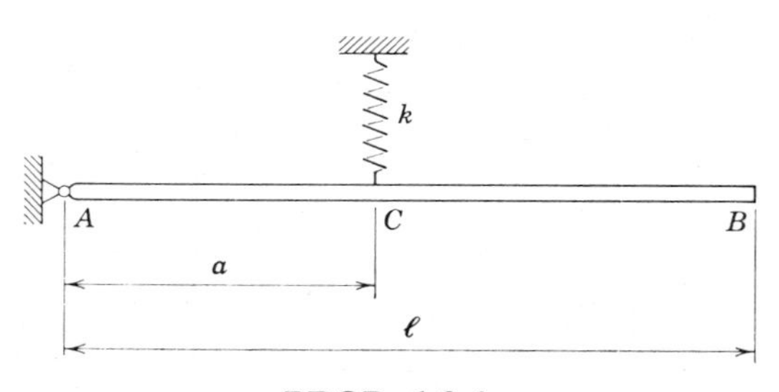

PROB. 1.3-4

1.3-5. Determine the frequency of torsional vibration of the disk shown in the figure if the ends of the shaft are built in at A and B. The two portions of the shaft have the same diameter d but different lengths ℓ_1 and ℓ_2. The disk has moment of inertia I.

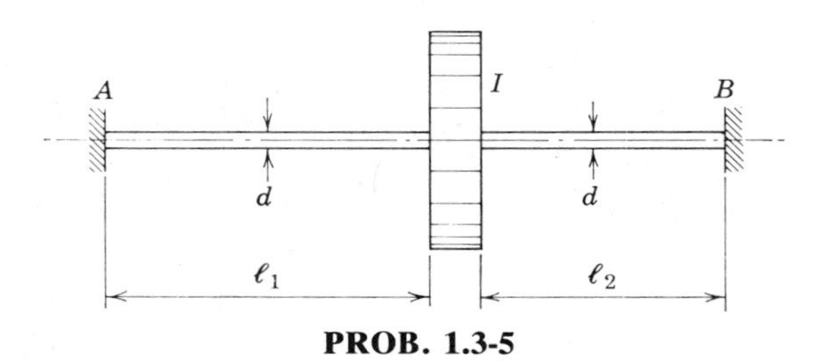

PROB. 1.3-5

1.3-6. Determine the equivalent length L_1 of a straight shaft having the same torsional rigidity C_1 as the journals of the crankshaft shown in the figure. The crank webs CE and DF have flexural rigidity B.

Assume that the bearings at A and B have sufficient clearance to allow free lateral deflection of C and D during twist of the crankshaft. The crankpin EF has torsional rigidity C_2 and throw-radius r.

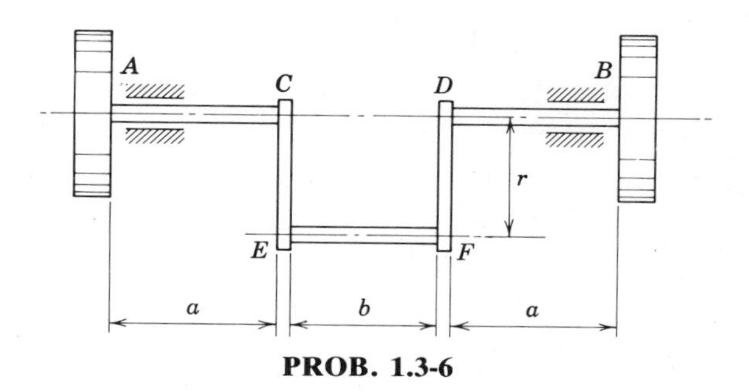

PROB. 1.3-6

1.3-7. A circular steel rim of weight W and mean radius r is attached to a fixed hub of radius r_0 by n radial spokes, each of which carries a high initial tension S_0 (see figure). Determine the period of rotational oscillation of the rim, assuming that the tension in each spoke remains constant for small amplitudes of vibration. The spokes are pinned at the ends and cannot sustain bending.

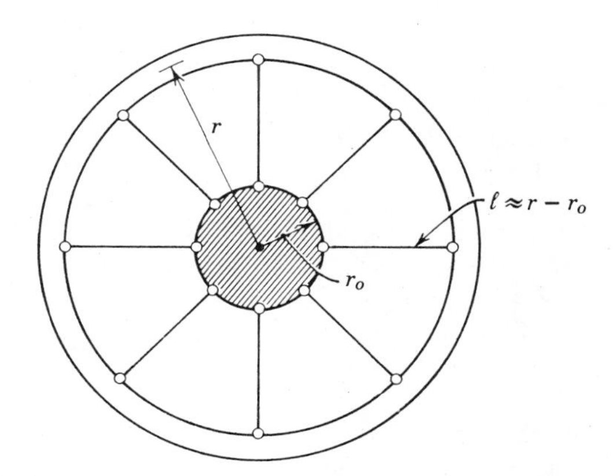

PROB. 1.3-7

1.3-8. The braced platform of Example 3 in Sec. 1.2 has the mass $m = 20$ Mg distributed uniformly over the platform. Calculate the period of rotational vibration of the system about a vertical axis through its center of mass.

1.4-1. The figure represents a heavy pendulum having an axis of rotation that makes an angle β with the vertical. Determine the frequency of small oscillations considering only the weight W of the ball, which is assumed to be concentrated at its mass center C. Neglect friction in the bearings.

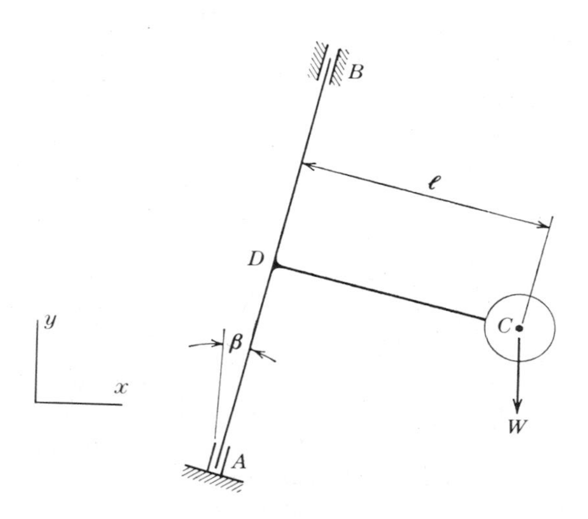

PROB. 1.4-1

1.4-2. Calculate the natural frequency of free vibrations of the system in Fig. 1.12 for the following numerical data: $W = 5$ lb, $k_1 = 2$ lb/in., $k_2 = 10$ lb/in., $b = 4$ in., and $c = 2$ in. Treat the arm BOA as a uniform slender rod of total weight $W' = 0.4$ lb, and assume OA to be 12 in. long.

1.4-3. When the system in Fig. 1.13*a* carries a weight $W_1 = 9$ N at the top end of the vertical bar, the observed frequency is 90 cpm. With a weight $W_2 = 18$ N, the observed frequency is 45 cpm. What weight W_3 at the top will just bring the system to a condition of unstable equilibrium? Neglect the weight of the bar.

1.4-4. Determine the angular frequency ω for the system in Fig. 1.13*a* if the vertical bar has a total weight $w\ell$ uniformly distributed along its length.

1.4-5. For recording vertical vibrations, the instrument shown in the figure is used, in which a rigid frame AOB carrying the weight W can rotate about an axis through O perpendicular to the plane of the figure. Determine the angular frequency of small vertical vibrations of the weight, neglecting the masses of the frame and the springs.

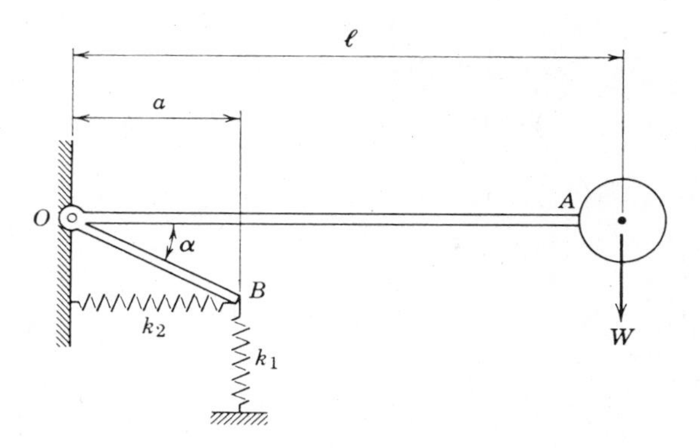

PROB. 1.4-5

1.4-6. A prismatic bar AB of weight W, suspended on two identical vertical wires (see figure), performs small rotational oscillations in the horizontal plane about the central axis. Determine the angular frequency of these oscillations.

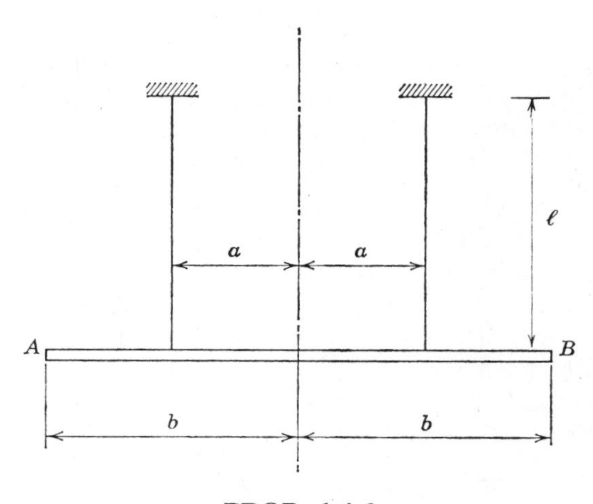

PROB. 1.4-6

1.4-7. A semicircular segment of a cylinder oscillates by rolling without sliding on a horizontal plane (see figure). Determine the angular frequency of small vibrations if r is the radius of the cylinder, c is the distance to the center of gravity, and $i^2 = Ig/W$ is the square of the radius of gyration about the centroidal axis.

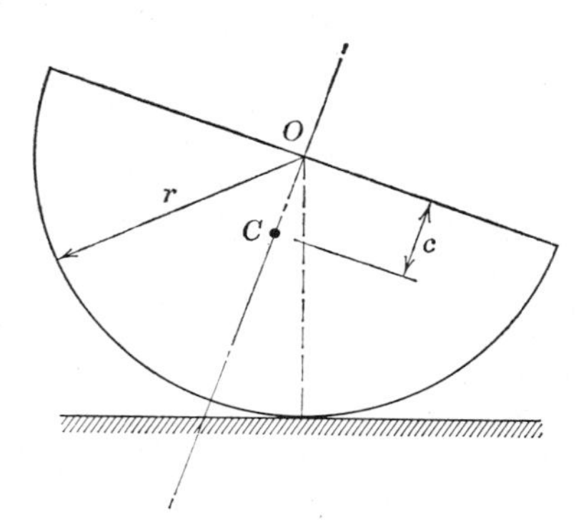

PROB. 1.4-7

1.5-1. Referring to the cantilever beam in Fig. 1.16, calculate the period of lateral vibration for the limiting case where $W = 0$, that is, for a uniform beam without load at the end, using Rayleigh's method. Assume that the shape of the beam during vibration is given by the expression $v = v_m[1 - \cos(\pi x/2\ell)]$, where v_m is the deflection at the free end. Note that this assumed shape function gives an approximate period that is in error by about 4%.

1.5-2. If the beam in Fig. 1.15 has both ends built-in instead of simply supported, what fraction of its total weight should be added to the weight W at midspan in calculating the natural period of lateral vibration? Assume a configuration of the beam during vibration corresponding to a static deflection curve under the influence of the load W.

1.5-3. Using the Rayleigh method, calculate the period of free lateral vibration of a uniform beam of weight $w\ell$ and flexural rigidity EI if it has built-in ends as in the preceding problem. Assume that during vibration the beam maintains a configuration identical in shape with a full wave of a cosine curve. That is, with the left end of the beam as an origin, the dynamic deflection curve will be represented by the equation $v = (v_m/2)[1 - \cos(2\pi x/\ell)]$, where v_m is the deflection at the center of the beam.

1.5-4. For the frame given earlier in Prob. 1.2-6, assume that each vertical column has a weight of 20 lb/ft and is pinned at the bottom end. Then calculate the natural period of lateral vibration of the frame corrected for the mass of the columns. Use the same data and shape function as in Prob. 1.2-6b.

1.5-5. What portion of the uniformly distributed weight of the beam ABC in the figure should be added to the weight W at the free end in calculating the natural frequency of lateral vibration? Assume a static deflection curve due to the load at C.

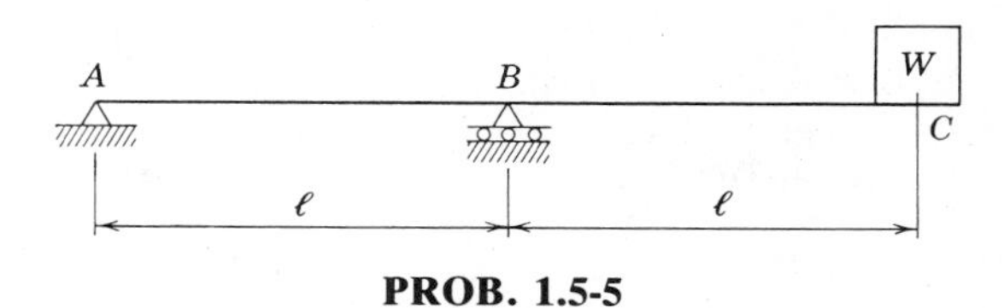

PROB. 1.5-5

1.5-6. Suppose that the spring in Fig. 1.2*a* (discussed in Sec. 1.2) hangs under its own weight w per unit length and that the load W is removed. Using the Rayleigh method, estimate the angular frequency of the fundamental mode of vibration of the spring. For this purpose, use the deflection pattern of the spring due to its own weight applied statically.

1.5-7. Consider the weight W suspended by two springs in series, as shown in Fig. 1.6*a* and discussed earlier in Sec. 1.2. Let the symbols ℓ_1 and w_1 represent the length and the weight per unit length of the spring having stiffness k_1, and let ℓ_2 and w_2 be the corresponding terms for the spring of stiffness k_2. Assume a configuration of the vibrating system to be that due to the weight W applied statically, and calculate the term necessary to correct for the weights of the springs.

1.5-8. Determine the expression to be added to I in the figure to correct for the mass moment of inertia of the stepped shaft. Let the symbols i_1 and i_2 denote the mass moments of inertia per unit lengths of parts 1 and 2 of the shaft, and let k_{r1} and k_{r2} represent their rotational stiffnesses. Assume a variation of twist corresponding to that due to a static torque applied to the disk.

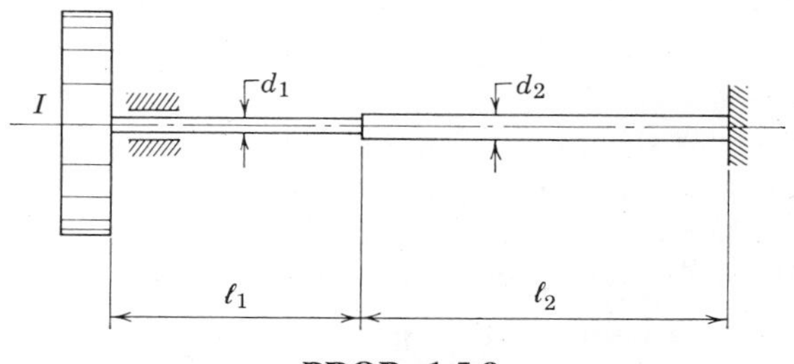

PROB. 1.5-8

1.5-9. Referring to Prob. 1.3-7 of Sec. 1.3, calculate the frequency of rotational oscillation of the rim of the wheel in the figure. Correct for the mass of the n radial spokes, assuming that each spoke has a weight W_s.

1.5-10. Referring to Example 2 of Sec. 1.3, calculate the frequency of rotational vibration of the wheel in Fig. 1.10*a* corrected for the mass of the radial spokes. Assume each spoke to have a mass wR/g uniformly distributed along its length.

1.5-11. Assume that the weights W_1, W_2, and W_3 in Fig. 1.18 represent the distributed weight of a prismatic beam lumped at midspan and at the quarter points. Let each of the weights be of magnitude $w\ell/4$, and compute an approximate value of the fundamental period by Rayleigh's method.

1.5-12. The figure shows a lumped-mass analytical model for a cantilever beam of uniform section. By Rayleigh's method estimate the fundamental period.

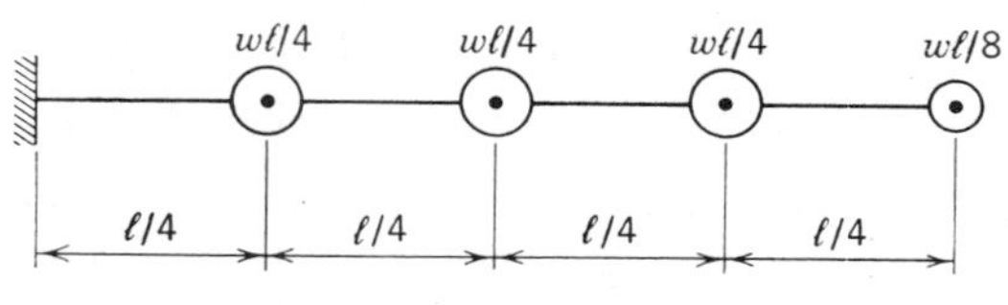

PROB. 1.5-12

1.5-13. A lumped-inertia analytical model for a uniform shaft with fixed ends appears in the figure. Estimate the angular frequency of its first mode of rotational vibration.

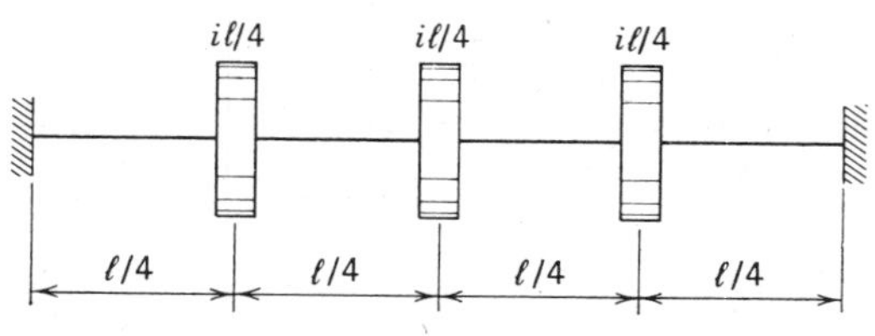

PROB. 1.5-13

1.5-14. Consider the system in Prob, 1.5-12 to represent a lumped-mass analytical model for axial vibrations of a prismatic bar with cross-sectional area A. Using Rayleigh's method, estimate the angular frequency of its first mode of axial vibration.

1.6-1. If the upper end of the spring in Fig. 1.23 has a vertical harmonic motion with amplitude $d = 1$ in. and angular frequency $\Omega = 180\,\text{sec}^{-1}$, find the amplitude of forced vibration of the suspended load W. Assume that this weight has a static deflection $\delta_{st} = 3$ in.

1.6-2. In Fig. 1.21 the suspended weight W has a static deflection $\delta_{st} = 2.5$ cm. What amplitude of forced vibration will be produced by a periodic force $P \cos \Omega t$ if $P = 9$ N, $\Omega = 10\pi\,\text{sec}^{-1}$, and $W = 45$ N?

1.6-3. A standard 8 in. steel I-beam with clear span $\ell = 12$ ft and $I = 57.6$ in.^{4} is simply supported at its ends, as shown in the figure. At midspan it carries an electric motor of weight $W = 1000$ lb that runs at 1800 rpm. Due to unbalance, the rotor sets up a rotating centrifugal force $P = 500$ lb. What amplitude of steady-state forced vibrations will be produced? Neglect the mass of the beam.

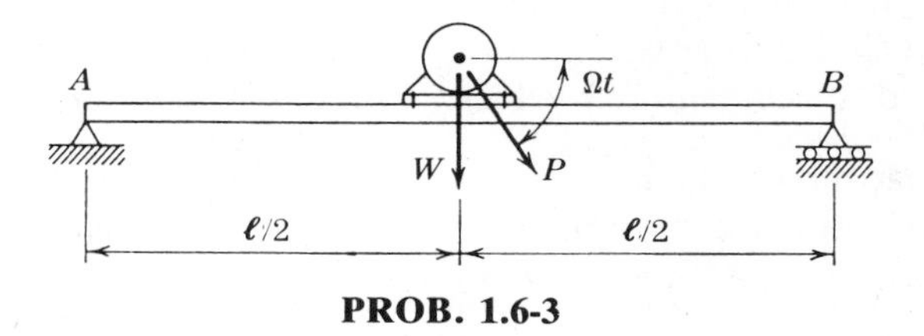

PROB. 1.6-3

1.6-4. A steel beam having cross-sectional moment of inertia $I = 1.7 \times 10^{-6}$ m^4 is supported as shown in the figure and carries at its free end a weight $W = 2.7$ kN. Compute the amplitude of steady-state forced vibrations of the weight W if the support at A performs small vertical oscillations defined by the expression $v_A = d \sin \Omega t$, where $d = 3$ mm and $\Omega = 30$ sec^{-1}. The support B does not move, and the mass of the beam is to be neglected.

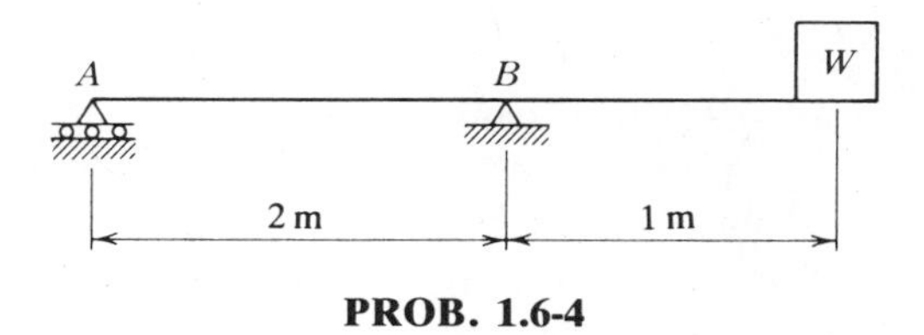

PROB. 1.6-4

1.6-5. A weight $W = 12$ kips is supported at the middle of a simple beam (see figure) consisting of two 6 in. steel channels ($I = 2 \times 17.4 = 34.8$ in.^{4}) placed back-to-back. Neglecting the mass of the beam, compute the amplitude of steady-state forced vibrations of W if a periodic moment $M = M_1 \cos \Omega t$ acts on one end of the beam as shown. Assume that $\Omega = 0.90\omega$ and that $M_1 = 10{,}000$ lb-in.

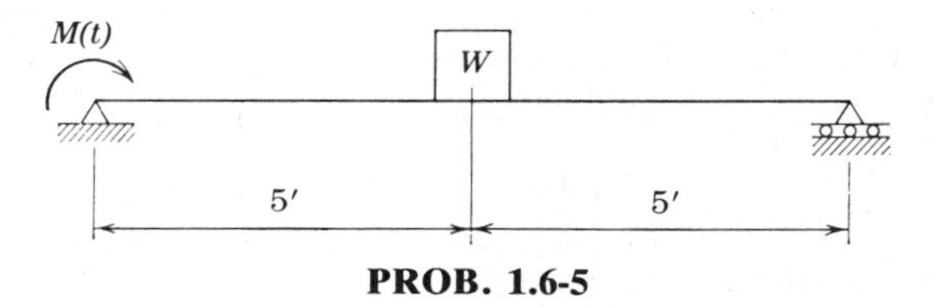

PROB. 1.6-5

1.6-6. Assume that the motor in Prob. 1.6-3 is stopped but that the supports of the beam at A and B accelerate vertically in accordance with Eq. (i) in Sec. 1.6. Determine the steady-state amplitude of forced vibration of the motor relative to ground if $a = 40$ in./sec^2 and $\Omega = 60$ rad/sec.

1.7-1. For the system shown in Fig. 1.21, the weight $W = 10$ lb and the spring constant $k = 10$ lb/in. Assume a force $P \sin \Omega t$, where $P = 2$ lb and $\Omega = 10\pi$ sec^{-1}, and initial conditions $u_0 = \dot{u}_0 = 0$ when $t = 0$. Find the displacement and velocity of the weight W at time $t = 1$ sec.

1.7-2. What displacement and velocity will the weight W of the preceding problem have at time $t = 1$ sec if the force is $P \cos \Omega t$ instead of $P \sin \Omega t$? Assume all other data to be the same as in Prob. 1.7-1.

1.7-3. Derive an expression similar to Eq. (1.31a) but due to the forcing function $P \cos \Omega t$ instead of $P \sin \Omega t$.

1.7-4. Make a dimensionless plot of the response curve from Prob. 1.7-3 similar to that in Fig. 1.26.

1.8-1. A body of weight $W = 10$ lb is supported by a spring with constant $k = 10$ lb/in. and has connected to it a dashpot damper (Fig. 1.28) that is so adjusted as to produce a resistance of 0.01 lb at a velocity of 1 in./sec. In what ratio will the amplitude of vibration be reduced after 10 cycles?

1.8-2. A weight $W = 9$ N hangs by a spring for which $k = 175$ N/m and is subject to damping such that $n = \sqrt{5}\omega/2$. If released from rest with an initial displacement $u_0 = 5$ cm, what maximum negative velocity will it attain in its return to the equilibrium position?

1.8-3. A spring-suspended mass that is critically damped has a spring constant $k = 1$ lb/in. and a weight $W = 3.86$ lb. If the weight is given an initial displacement $u_0 = 1$ in. and an initial velocity $\dot{u} = -12$ in./sec, at what time t after release will it reach the equilibrium position $u = 0$? In such a case, how far does the weight overshoot the equilibrium position, that is, what is the numerically largest negative displacement that it attains?

1.8-4. A spring-suspended weight $W = 2$ lb vibrates with an observed period $\pi_d = \frac{1}{2}$ sec, and damping is such that after 10 complete cycles the amplitude has decreased from $u_1 = 2$ in. to $u_{11} = 1$ in. Calculate the coefficient of viscous damping c.

1.8-5. A spring-mass system has a natural frequency of vibration f when there is no damping. Calculate the frequency f_d when the coefficient of viscous damping $c = c_{cr}/2$.

1.9-1. A simply supported beam carries an electric motor of weight $W = 2000$ lb at midspan, as given in Prob. 1.6-3. The stiffness of the beam is such that the static deflection at midspan is $\delta_{st} = 0.10$ in., and viscous damping is such that after 10 cycles of free vibration the amplitude is reduced to half of its initial value. The motor runs at 600 rpm and, due to unbalance in the rotor, sets up a centrifugal force $Q = 500$ lb at this speed. Neglecting the distributed mass of the beam, find the amplitude of steady-state forced vibrations.

1.9-2. A rotating machine having weight $W = 72$ kN is supported at the middle of two parallel simply supported steel beams, each having a clear span $\ell = 4$ m and cross-sectional moment of inertia $I = 25 \times 10^{-6}\ \text{m}^4$. The machine runs at 300 rpm, and its rotor is out of balance to the extent of 180 N at a radius of 25 cm. What will be the amplitude of steady-state forced vibrations if the equivalent viscous damping for the system is 10% of critical?

1.9-3. Referring to Fig. 1.33, write the equation of the locus of the peaks of the plotted curves for all values of the damping ratio.

1.9-4. For a damped system subjected to the forcing function $Q \sin \Omega t$, derive the expression for the steady-state response in phase-angle form.

1.9-5. For a damped system subjected to the support motion $u_g = d \sin \Omega t$, derive the expression for the steady-state response in phase-angle form.

1.9-6. Assume a harmonic ground acceleration of the type $\ddot{u}_g = a \cos \Omega t$, and derive the damped steady-state response of the system in phase-angle form.

1.9-7. Derive the steady-state response (in phase-angle form) of a damped system subjected to the ground acceleration $\ddot{u}_g = a \sin \Omega t$.

1.9-8. For a system with subcritical damping, determine the transient response due to the forcing function $Q \sin \Omega t$. Give the solution in a form similar to Eq. (t) in Example 3 of Sec. 1.9.

1.11-1. Using data from Example 1 in Sec. 1.11, construct the displacement-time curve $u = f(t)$ for steady-state forced vibrations of the system in Fig. 1.40.

1.11-2. Expand the force $F(t)$ represented graphically in the figure into a trigonometric series.

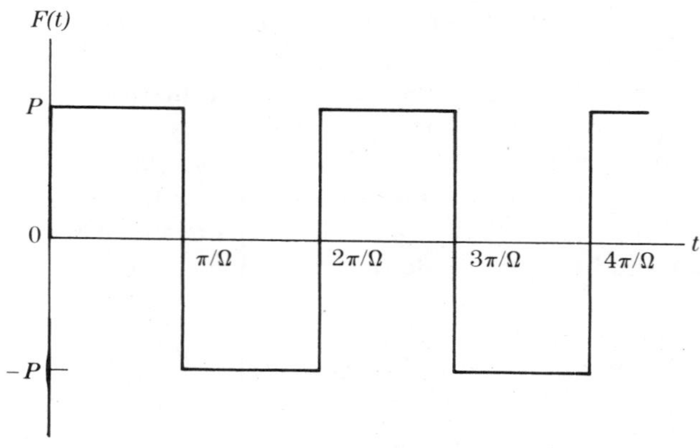

PROB. 1.11-2

1.11-3. Expand the force $F(t)$ represented graphically in the figure into a trigonometric series.

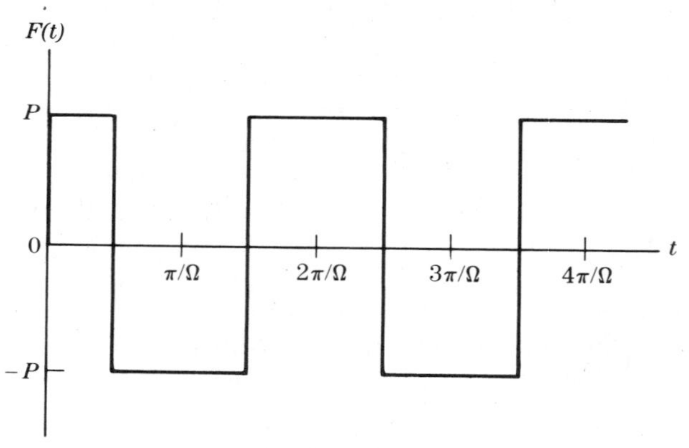

PROB. 1.11-3

1.11-4. Expand the force $F(t)$ represented graphically in the figure into a trigonometric series.

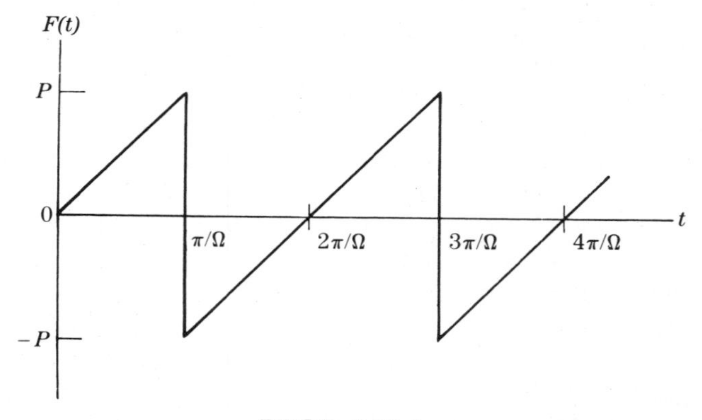

PROB. 1.11-4

1.11-5. Expand the force $F(t)$ represented graphically in the figure into a trigonometric series.

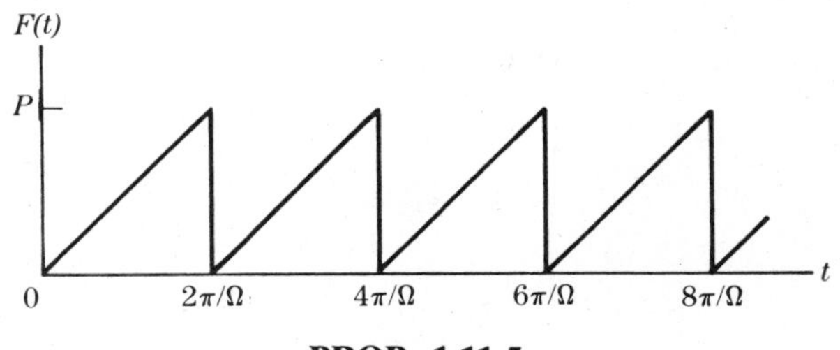

PROB. 1.11-5

1.11-6. Derive the general expression for steady-state forced vibrations of a damped one-degree system subjected to the forcing function expressed in Eq. (1.58).

1.12-1. Determine the undamped response of a one-degree system to the forcing function in the figure.

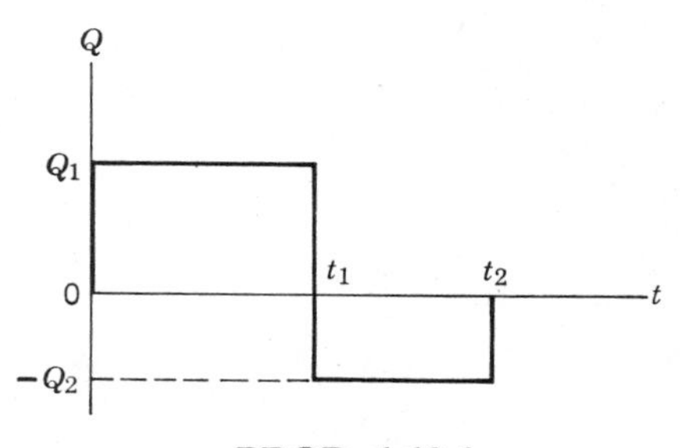

PROB. 1.12-1

1.12-2. Determine the undamped response of a one-degree system to the forcing function in the figure.

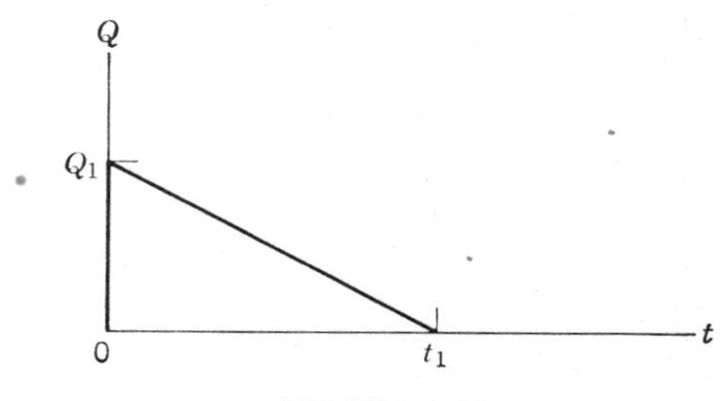

PROB. 1.12-2

1.12-3. Determine the undamped response of a one-degree system to the forcing function in the figure.

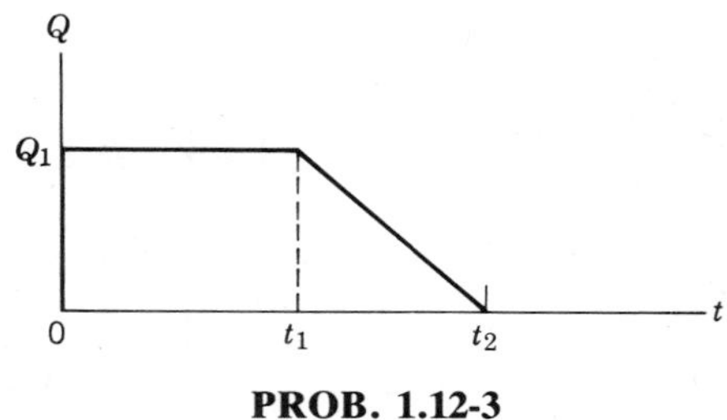

PROB. 1.12-3

1.12-4. Determine the undamped response of a one-degree system to the forcing function in the figure.

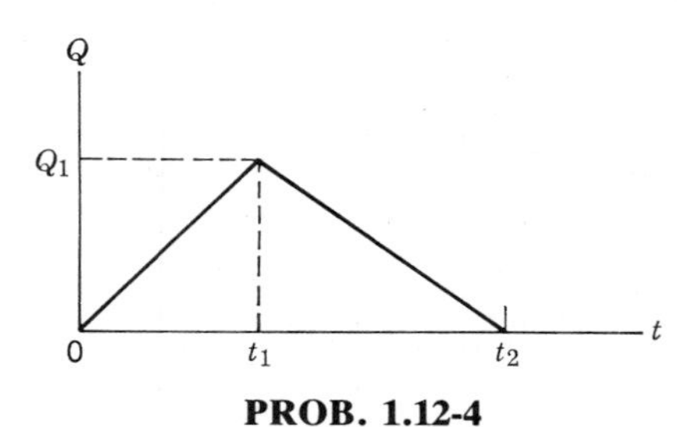

PROB. 1.12-4

1.12-5. Determine the undamped response of a one-degree system to the parabolic forcing function $Q = Q_1(1 - t^2/t_1^2)$ in the figure.

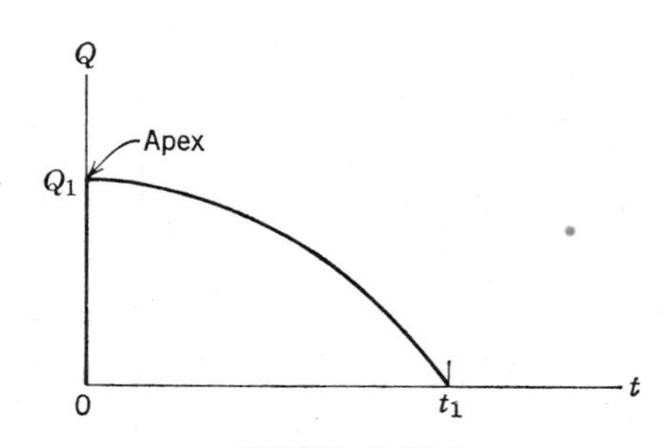

PROB. 1.12-5

1.12-6. Determine the undamped response of a one-degree system to the parabolic forcing function $Q = Q_1(t - t_1)^2/t_1^2$ in the figure.

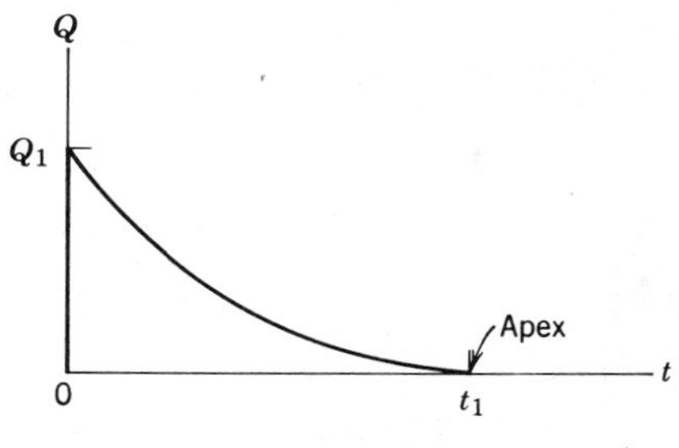

PROB. 1.12-6

1.12-7. For the one-degree system in Fig. 1.42*a*, determine the damped response to the ramp function in Fig. 1.46*a*.

1.13-1. Determine the undamped response of a one-degree system to the support displacement in the figure.

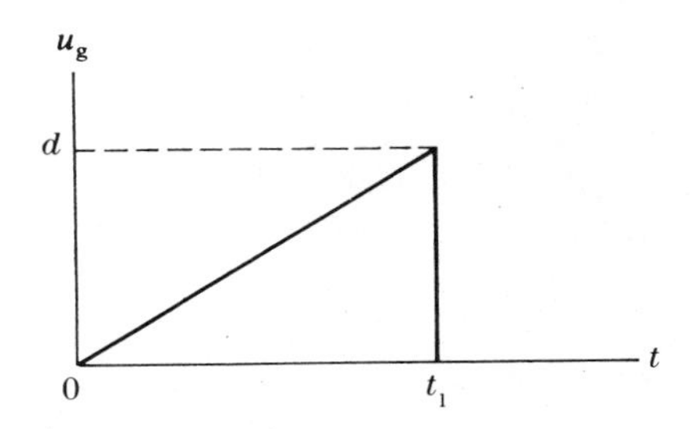

PROB. 1.13-1

1.13-2. Determine the undamped response of a one-degree system to the support acceleration in the figure.

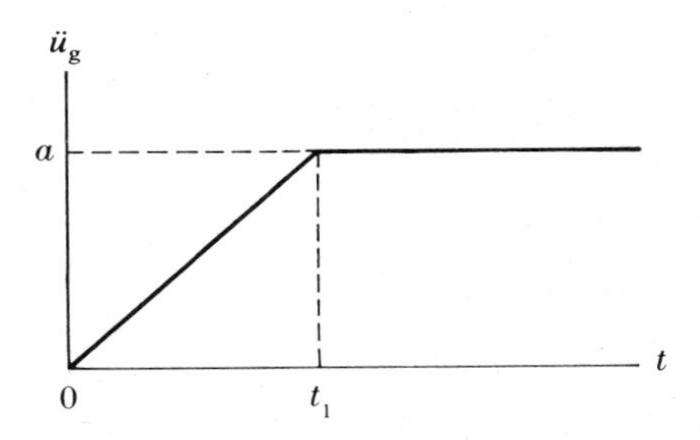

PROB. 1.13-2

1.13-3. Determine the undamped response of a one-degree system to the support displacement in the figure.

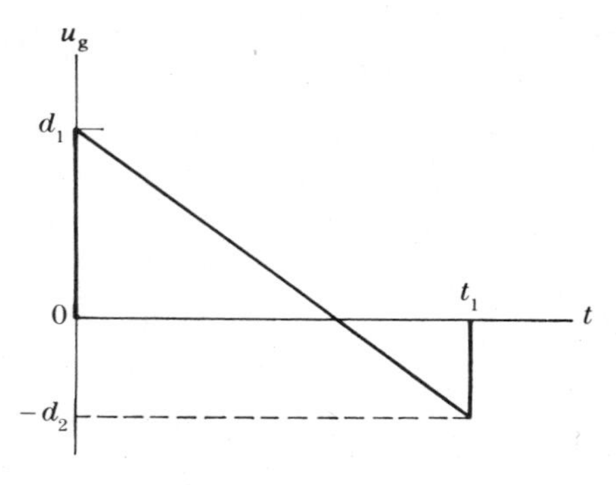

PROB. 1.13-3

1.13-4. Determine the undamped response of a one-degree system to the support acceleration in the figure.

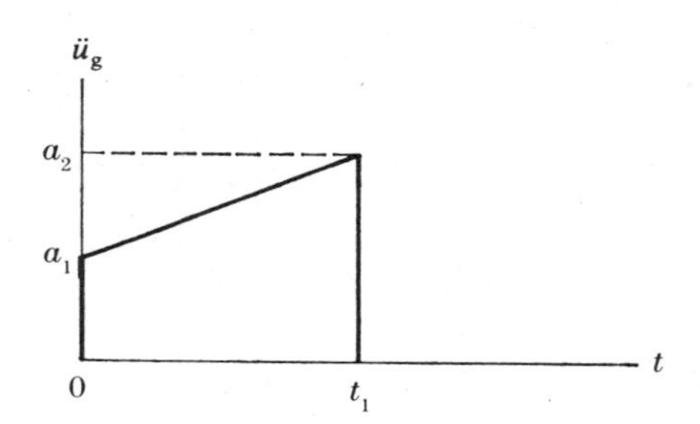

PROB. 1.13-4

1.13-5. Determine the undamped response of a one-degree system to the parabolic function $u_g = d[1 - (t - t_1)^2/t_1^2]$ for support displacement given in the figure.

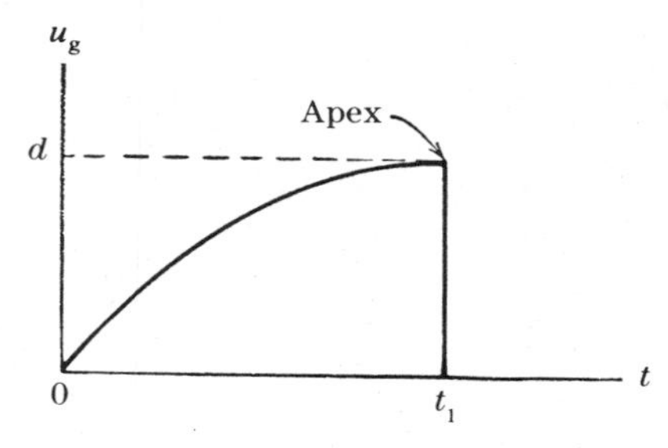

PROB. 1.13-5

1.13-6. Determine the undamped response of a one-degree system to the parabolic function $\ddot{u}_g = at^2/t_1^2$ for support acceleration given in the figure.

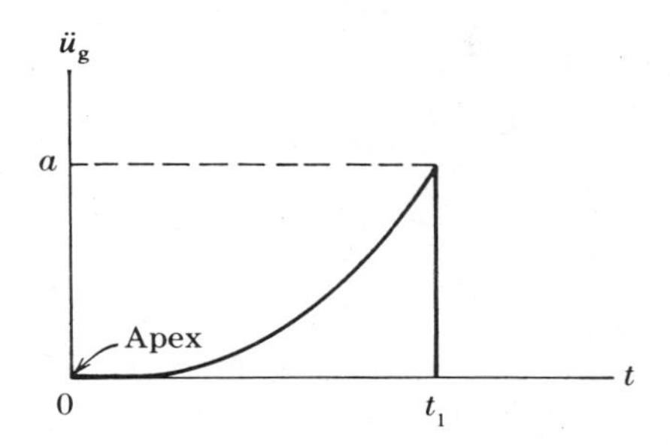

PROB. 1.13-6

1.13-7. Refer to the undamped system in Fig. 1.50, and use the results of Example 4 in Sec. 1.13 to find the amplitude of free vibration that the elevator will have after the hoist has stopped.

1.14-1. Plot the response spectrum u_m/u_{st} and the time for maximum response t_m/τ against t_1/τ for the forcing function in the figure. (See Prob. 1.13-1 for response formulas.)

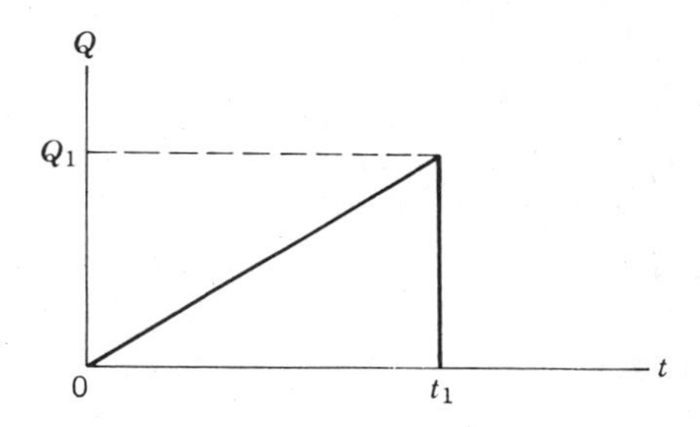

PROB. 1.14-1

1.14-2. Repeat Prob. 1.14-1 for the forcing function in the figure. (See Prob. 1.12-4 for response formulas.)

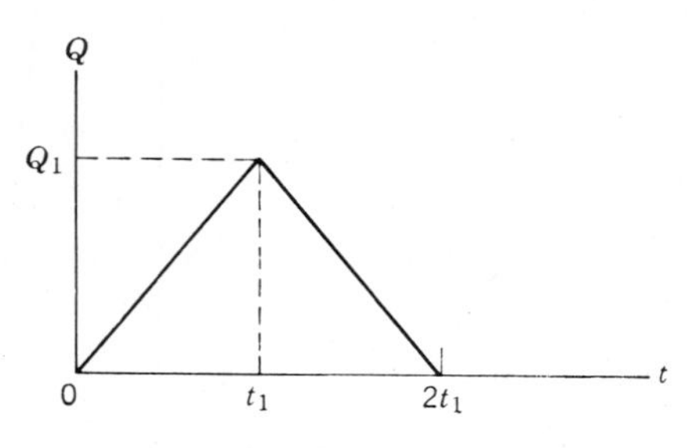

PROB. 1.14-2

1.14-3. Repeat Prob. 1.14-1 for the forcing function in the figure. (See Prob. 1.13-4 for response formulas.)

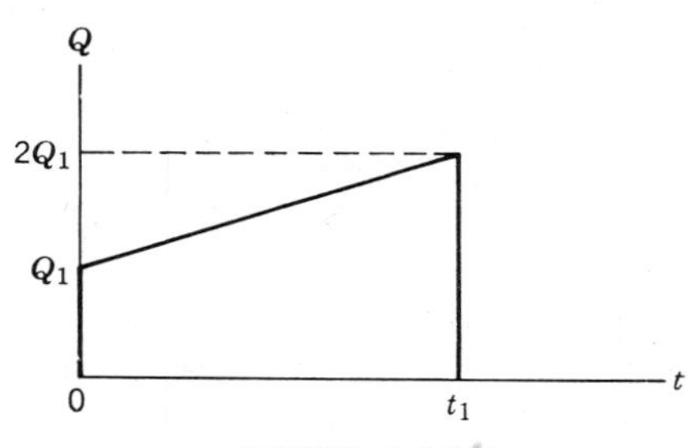

PROB. 1.14-3

1.14-4. Repeat Prob. 1.14-1 for the forcing function in the figure. (See Prob. 1.12-3 for response formulas.)

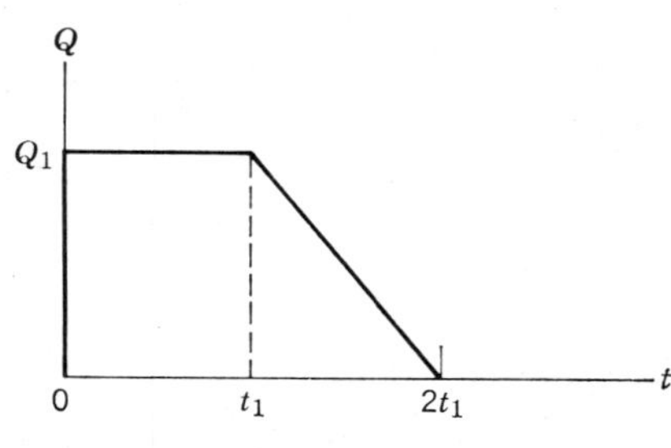

PROB. 1.14-4

1.14-5. Repeat Prob. 1.14-1 for the parabolic forcing function $Q = Q_1 t^2/t_1^2$, in the figure. (See Prob. 1.13-6 for response formulas.)

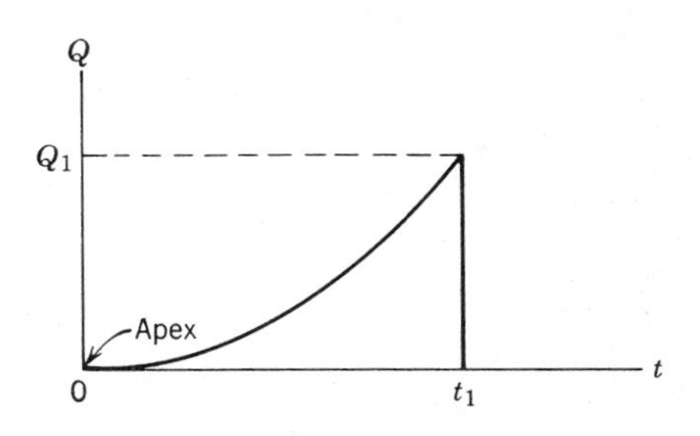

PROB. 1.14-5

1.14-6. Repeat Prob. 1.14-1 for the parabolic forcing function $Q = Q_1(1 - t^2/t_1^2)$ in the figure. (See prob. 1.12-5 for response formulas.)

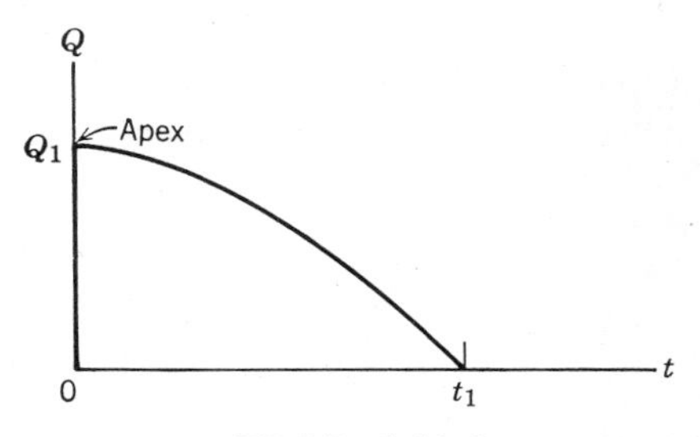

PROB. 1.14-6

1.15-1. For a step function of magnitude $Q = Q_1$, calculate the undamped response of a one-degree system, using the recursive procedure in Sec. 1.15 with 10 equal time steps of duration $\Delta t = \tau/10$.

1.15-2. Assume that a ramp function $Q = Q_1 t/t_1$ is applied to an undamped one-degree system. With the step-by-step procedure, find the response of the system for 10 equal time steps of duration $\Delta t = \tau/10 = t_1$.

1.15-3. Confirm the approximate results for Example 1 of Sec. 1.15 in Table 1.1.

1.15-4. Confirm the results for Example 2 of Sec. 1.15 in Table 1.2.

1.15-5. Using 20 equal time steps with $\Delta t = \tau/20$, determine the undamped response of a one-degree system to the forcing function in the figure.

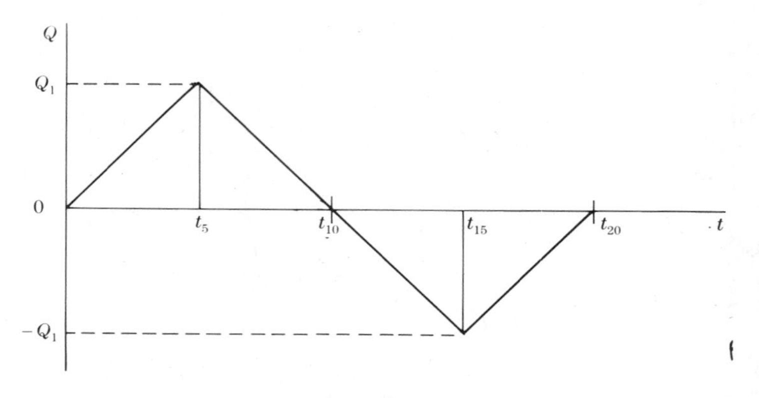

PROB. 1.15-5

1.15-6. Calculate the undamped response of a one-degree system to the forcing function in the figure, using 20 equal time steps with $\Delta t = \tau/20$.

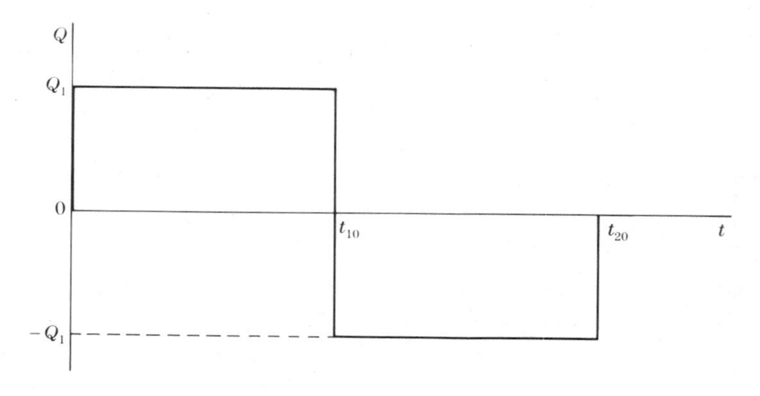

PROB. 1.15-6

1.15-7. Divide the triangular impulse in the figure into 10 equal time steps $\Delta t = \tau/30$, and find the undamped response when it is applied to a one-degree system.

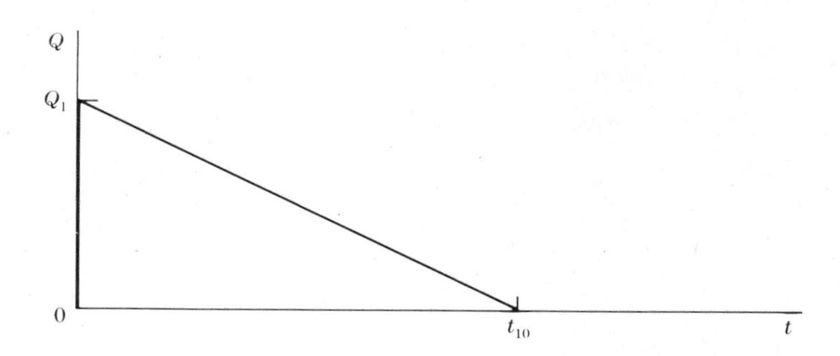

PROB. 1.15-7

1.15-8. Determine the undamped response of a one-degree system to the parabolic forcing function $Q = Q_1 t^2/t_{10}^2$ in the figure, using 10 equal time steps of duration $\Delta t = \tau/30$.

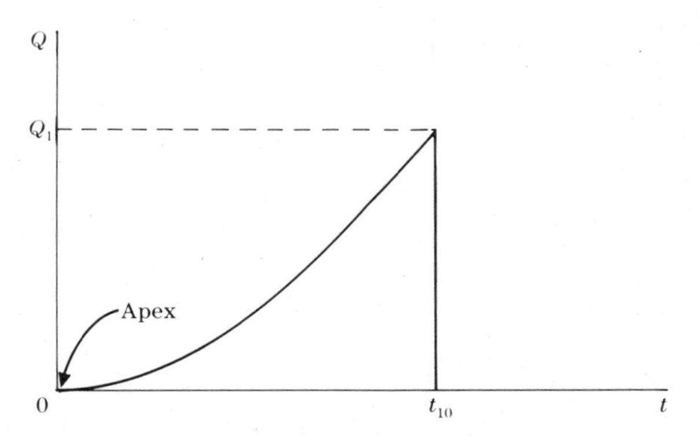

PROB. 1.15-8

1.15-9. The parabolic forcing function in the figure has the formula $Q = Q_1(1 - t^2/t_{10}^2)$. With 10 equal time steps of duration $\Delta t = \tau/25$, obtain the undamped response of a one-degree system.

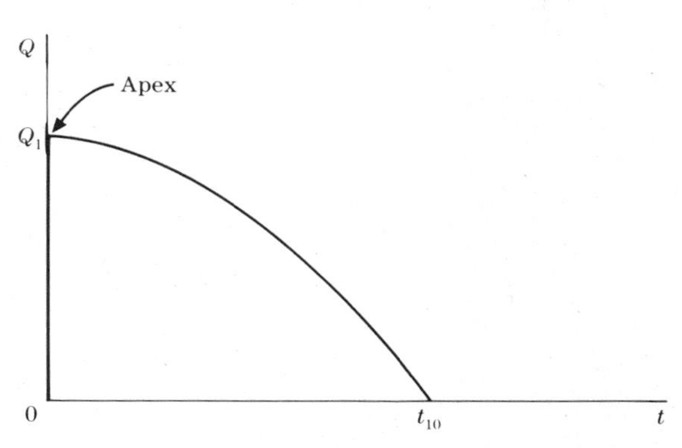

PROB. 1.15-9

1.15.10. Find the undamped response of a one-degree system to the parabolic forcing function $Q = Q_1[1 - (t - t_{10})^2/t_{10}^2]$ in the figure, using 10 equal time steps of duration $\Delta t = \tau/25$.

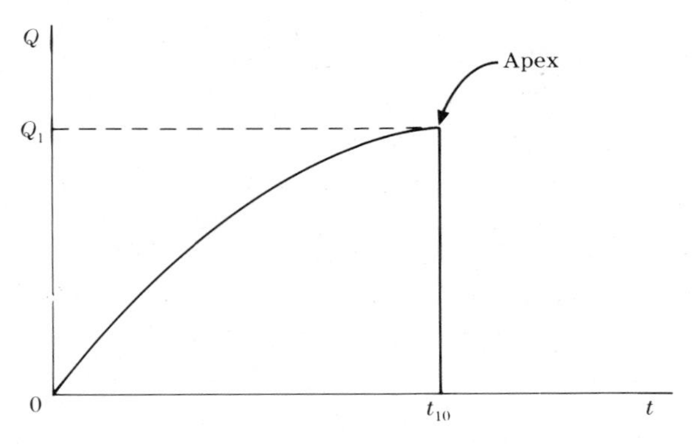

PROB. 1.15-10

CHAPTER 2

SYSTEMS WITH NONLINEAR CHARACTERISTICS

2.1 EXAMPLES OF NONLINEAR SYSTEMS

When discussing the vibrational properties of one-degree-of-freedom systems in Chapter 1, we always assumed that the force in a spring was proportional to its deformation. We also recognized the fact that viscous damping, where the damping force is proportional to velocity, is much easier to handle than other causes of energy dissipation. To avoid mathematical difficulty, the concept of equivalent viscous damping was introduced in Sec. 1.10. In addition, the mass was always taken to be unvarying with time. Consequently, the equation of motion for such a system is a linear, second-order ordinary differential equation with constant coefficients, as follows:

$$m\ddot{u} + c\dot{u} + ku = F(t) \tag{2.1}$$

This equation represents many practical problems very well and plays a central role in linear vibration theory. However, there are also many physical systems for which linear differential equations with constant coefficients are insufficient to describe the motion, and the analysis of such systems requires a discussion of nonlinear differential equations.

If we omit the possibility of variable mass, the general form of the equation of motion for a one-degree system may be expressed as

$$m\ddot{u} + F(u, \dot{u}, t) = 0 \tag{2.2}$$

We will refer to systems with nonlinear characteristics as *nonlinear systems*, and we will refer to their motions as *nonlinear vibrations*, or *nonlinear response*. It should be recognized at the outset that the principle of superposition, used many times in Chapter 1, does not apply to nonlinear systems. For example, if the magnitude of a forcing function is doubled, the response of a nonlinear system is not necessarily doubled. In general, nonlinear vibrations are not harmonic, and their frequencies vary with amplitude.

One important type of nonlinearity arises when the restoring force of a spring is not proportional to its deformation. Figure 2.1*a* shows the static load–displacement curve for a nonlinearly elastic "*hardening spring*," where the slope increases as the load increases. The dashed line in the figure is tangent to the curve at the origin, and its slope k represents the initial stiffness of the spring. Similarly, Fig. 2.1*b* depicts the load–displacement curve for a nonlinearly elastic "*softening spring*," where the slope decreases as the load increases. In each of these figures the curve is symmetric with respect to the origin, and the spring is said to have a *symmetric restoring force*. If the load–displacement curve is not symmetric with respect to the origin, the spring is said to have an *unsymmetric restoring force*.

An example of a system having the characteristics of a hardening spring with a symmetric restoring force appears in Fig. 2.2*a*. A small mass m is attached to the middle of a stretched wire AB of length 2ℓ, which is subjected to an initial tensile force denoted by the symbol S. When the mass is displaced laterally a distance x from its equilibrium position, a restoring force is developed by the wire, as indicated in Fig. 2.2*b*. Thus, the system may perform free vibrations, for which the equation of motion is

$$m\ddot{u} + 2\left(S + \frac{AE\Delta}{\ell}\right)\sin\theta = 0 \qquad (a)$$

The symbols A, E, and Δ in Eq. (a) represent the cross-sectional area of

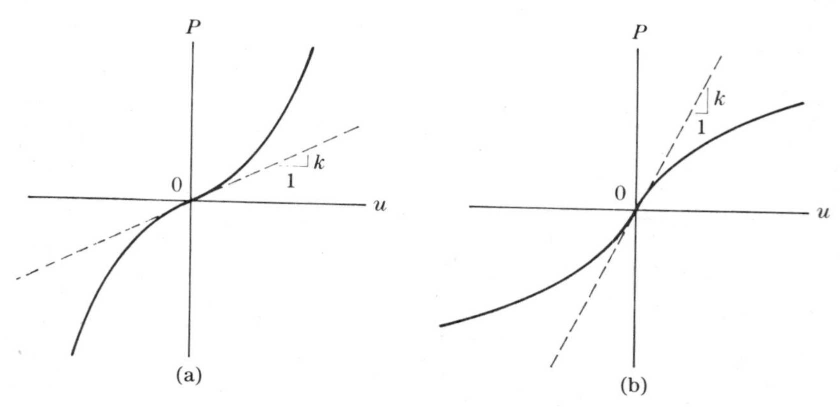

FIG. 2.1

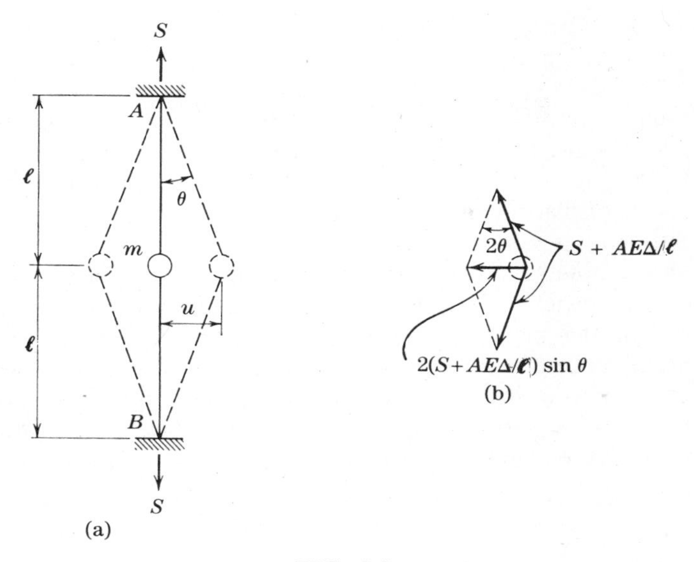

FIG. 2.2

the wire, its modulus of elasticity, and the change in the length ℓ due to the displacement u; and the angle θ denotes the inclination of the wire from the vertical. Inspection of the geometry in Fig. 2.2*a* shows that

$$\Delta = \sqrt{\ell^2 + u^2} - \ell \qquad \sin\theta = \frac{u}{\sqrt{\ell^2 + u^2}} \tag{b}$$

When these expressions are substituted into Eq. (*a*), we obtain

$$m\ddot{u} + 2\left[S + \frac{AE(\sqrt{\ell^2 + u^2} - \ell)}{\ell}\right]\frac{u}{\sqrt{\ell^2 + u^2}} = 0 \tag{2.3a}$$

This exact nonlinear equation of motion may be replaced with a simpler (but less accurate) equation by using the approximate relationships

$$\Delta \approx \frac{u^2}{2\ell} \qquad \sin\theta \approx \frac{u}{\ell} \tag{c}$$

Substitution of these expressions into Eq. (*a*) yields

$$m\ddot{u} + \frac{2S}{\ell}u + \frac{AE}{\ell^3}u^3 = 0 \tag{2.3b}$$

This differential equation for approximating the motion contains a term with

u raised to the third power and is, therefore, still nonlinear. If the initial tensile force S is high and the displacement u is small, the cubic term in Eq. (2.3b) may be neglected. Under these conditions the motion of the mass is approximately harmonic, as represented by the remaining terms. Otherwise, the cubic term must be considered, and the restoring force is of the type represented by Fig. 2.1*a*. Because the slope of the load–displacement curve increases with displacement, the frequency of free vibration will increase with amplitude.

It should be noted that the nonlinearity of the system in Fig. 2.2*a* arises from the geometric considerations for large displacements, not from nonlinearity of the material in the wire. Another example of *geometric nonlinearity* is represented by the *simple pendulum* of weight W and length L in Fig. 2.3. In a position displaced by the angle ϕ from the vertical, the pendulum has a restoring moment about the pivot point C equal to $WL \sin \phi$. Thus, the equation for rotational motion about the pivot becomes

$$I\ddot{\phi} + WL \sin \phi = 0 \tag{d}$$

Substituting into this expression $I = WL^2/g$ for the mass moment of inertia, we have

$$\ddot{\phi} + \frac{g}{L} \sin \phi = 0 \tag{2.4a}$$

For small amplitudes $\sin \phi$ is approximately equal to the angle ϕ, and the

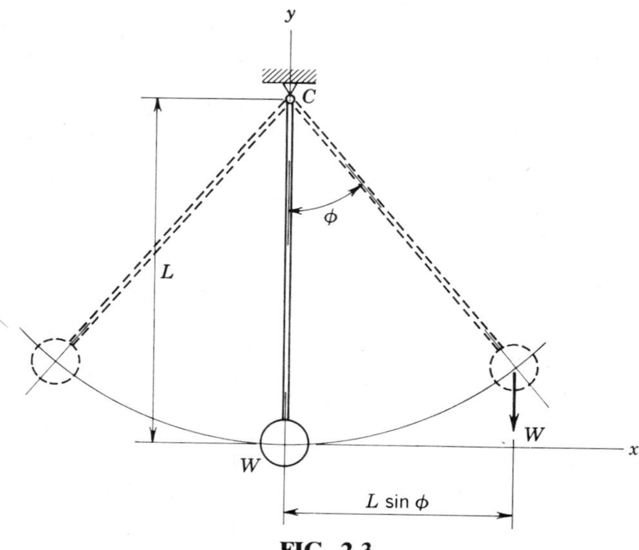

FIG. 2.3

motion can be considered as simple harmonic. If the amplitude is not small, the restoring moment is proportional to sin ϕ, which can be approximated by a power series. Substitution of the first two terms of the series into Eq. (2.4a) gives

$$\ddot{\phi} + \frac{g}{L}\left(\phi - \frac{\phi^3}{6}\right) = 0 \tag{2.4b}$$

In this case we see that the slope of the curve for restoring moment versus ϕ decreases as the rotation increases; so the frequency of oscillation decreases with amplitude.

Comparison of Eqs. (2.3b) and (2.4b) reveals the fact that the nonlinear terms in these two expressions would tend to compensate each other in a combined system of the type shown in Fig. 2.4. That is, by restraining the pendulum with a horizontal stretched wire AB attached to the bar of the pendulum at point D (and perpendicular to the plane of oscillation), a better approximation to *isochronic oscillations* may be obtained.

Figures 2.5*b* and 2.6*b* show diagrams of piecewise-linear restoring forces that might be considered as approximations to the continuous curves in Figs. 2.1*a* and 2.1*b*. Actually, they correspond to the linearly elastic *discontinuous systems* in Figs. 2.5*a* and 2.6*a*. Although the springs in these figures are linearly elastic, the motions of the masses are not continuous functions. Such systems will be treated in detail later in this chapter (see Sec. 2.5). However, at this point it is well to recognize the possibility of approximating a nonlinear restoring force as a piecewise-linear function by simulating it with a succession of straight-line segments.

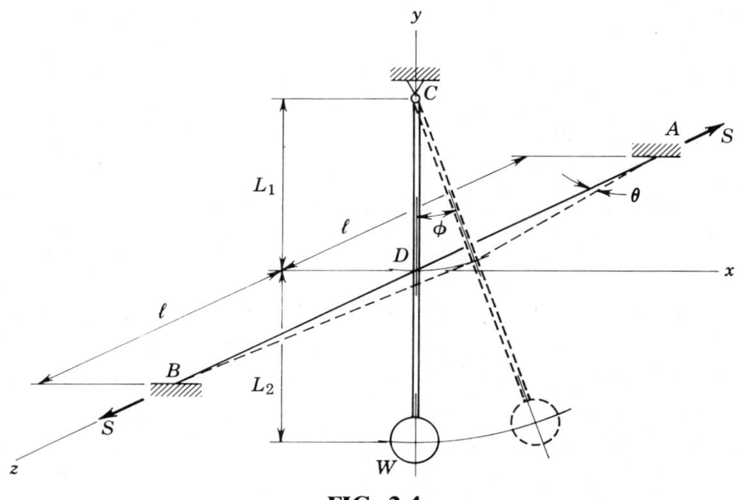

FIG. 2.4

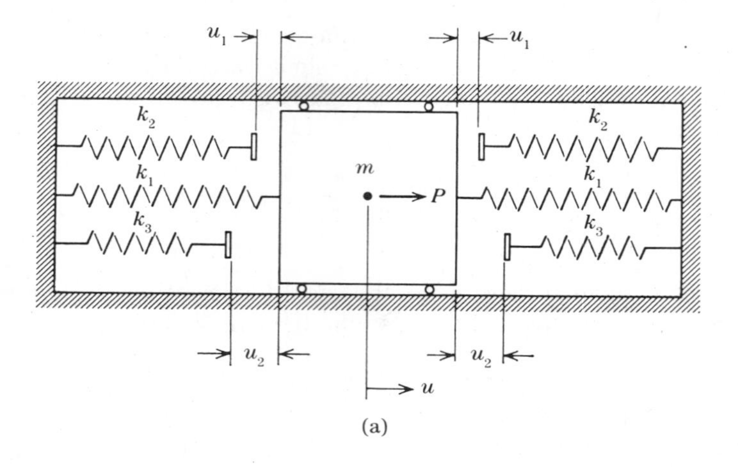

(a)

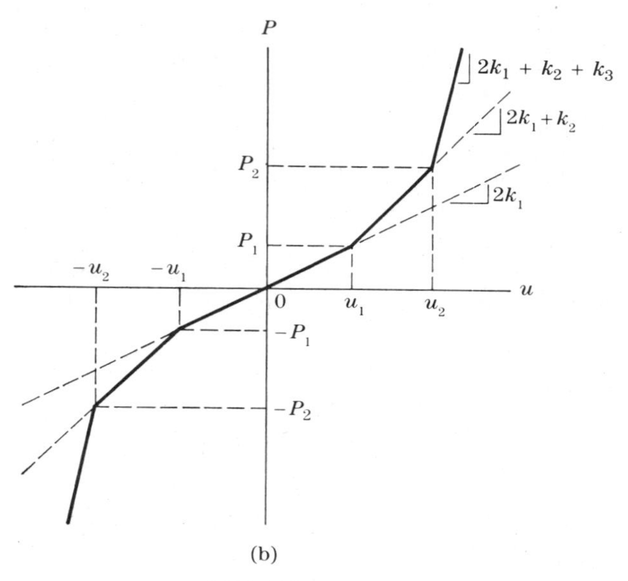

(b)

FIG. 2.5

If dynamic loads deform a structure or the parts of a machine beyond the elastic range of the material, the resulting motion is called *inelastic response.* Although excursions beyond the elastic range are not usually permitted under normal operating conditions, the extent of permanent damage in a structure or a machine subjected to extreme conditions is often of interest to the design engineer. For example, a building subjected to severe blast or earthquake loading will probably be deformed inelastically.

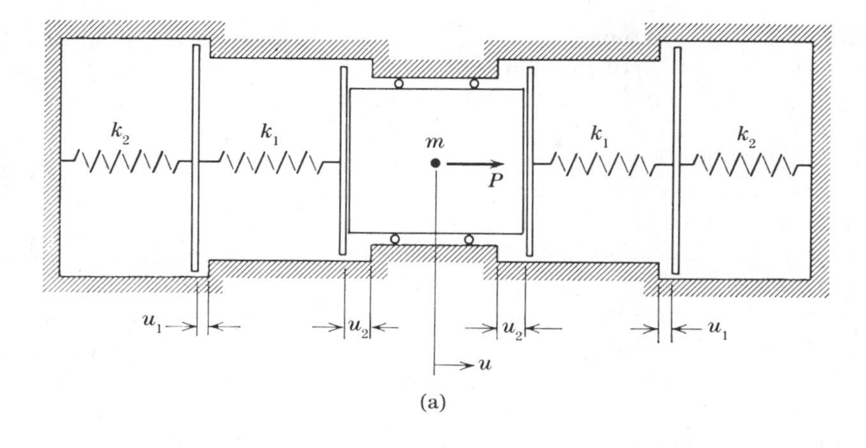

(a)

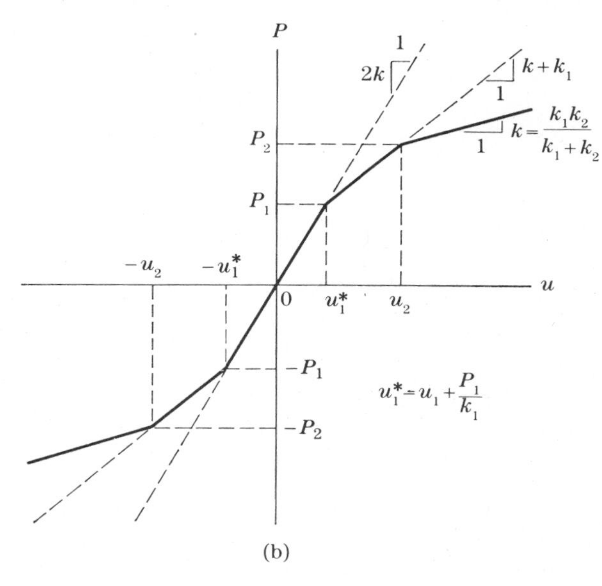

(b)

FIG. 2.6

Figure 2.7*a* depicts an idealized two-dimensional steel frame in a rectangular building, subjected to a lateral load P applied at the roof level. If the flexural rigidities of the columns are less than that of the beam and the load is increased indefinitely, so-called *plastic hinges* will form at the ends of the columns. A plot of the load P against the displacement u is linear up to the value of P_{y1} (see the line labeled ① in Fig. 2.7*b*), where yielding of the material begins. Subsequently, it curves (see the curve labeled ②) in a manner that is similar to the diagram in Fig. 2.1*b* for a softening spring. Upon unloading, the material rebounds elastically, as indicated by part ③ of the plot in Fig. 2.7*b*. That portion of the diagram is represented by another straight line parallel to the elastic-loading line (part ① of the plot). If a reverse loading is then applied, the parts in Fig. 2.7*b* labeled ④ and ⑤ result; and a subsequent unloading produces line ⑥. Experiments show that if the maximum positive and negative forces P_{m1} and $-P_{m2}$ (the ordinates of points B and E on the diagram) are numerically equal, the *hysteresis loop* formed by cyclic loading is symmetric with respect to the origin [1–3].

The curved portions of Fig. 2.7*b* are often replaced by straight lines approximating the true behavior. Figure 2.7*c* illustrates such a simplified

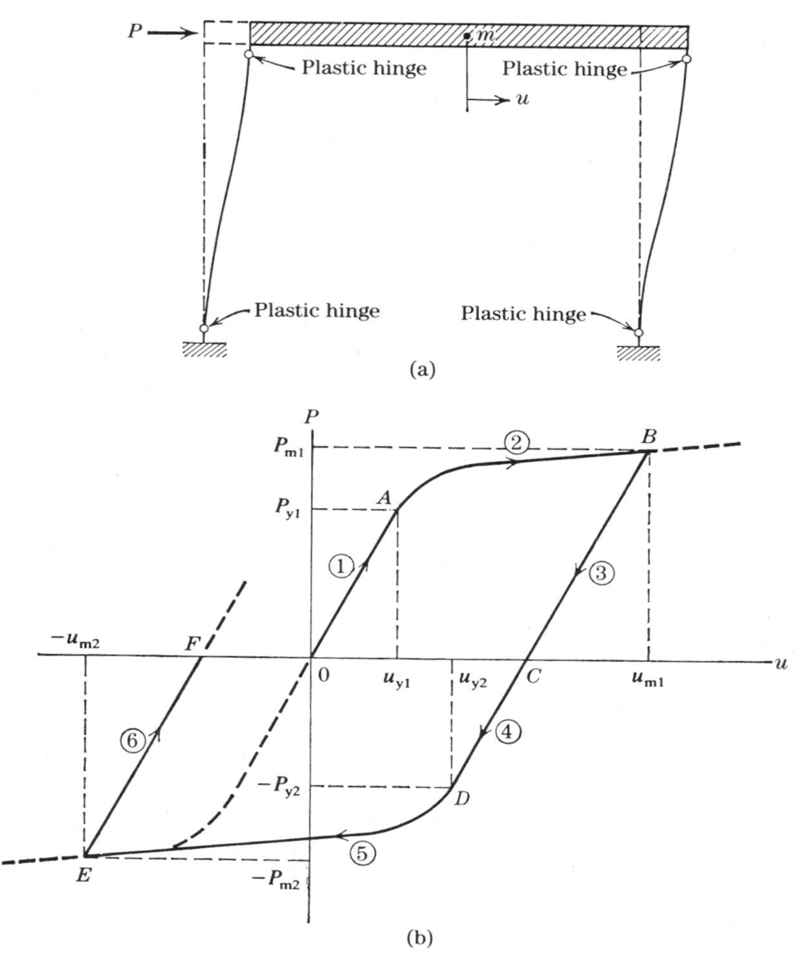

FIG. 2.7

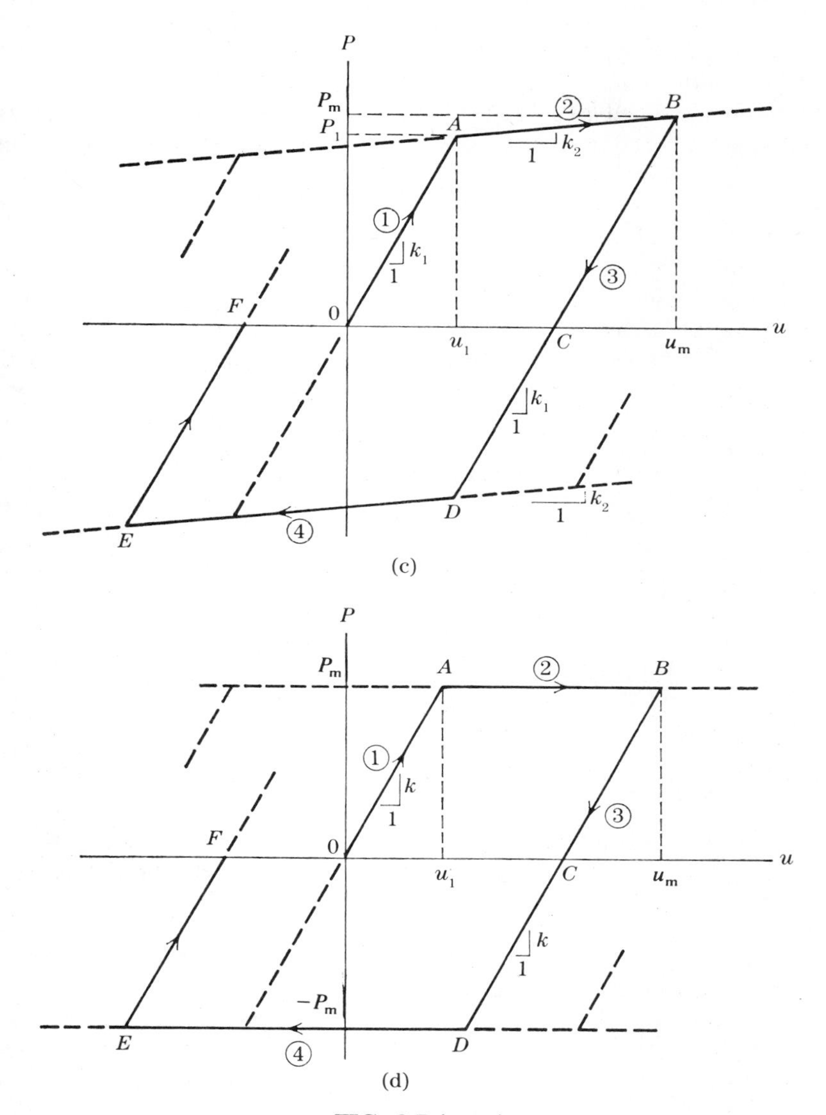

FIG. 2.7 (cont.)

load–displacement diagram, called a *bilinear inelastic restoring force.* It consists of two parallel lines (labeled ② and ④ in the figure) for inelastic behavior and a family of parallel lines (of which those labeled ① and ③ are representative) for elastic behavior. If the slopes of lines ② and ④ are zero, as in Fig. 2.7*d*, the diagram represents an *elastoplastic restoring force.* That is, the plot of P against u consists of straight-line segments, where the behavior is assumed to be either perfectly elastic or perfectly plastic. For example, let us reconsider the frame in Fig. 2.7*a* and suppose that the load P increases to the value P_m (see point A in Fig. 2.7*d*). If the plastic hinges in Fig. 2.7*a* are assumed to form instantaneously, the displacement increases without a corresponding increase of the load, as shown by the horizontal line from A to B in Fig. 2.7*d*. A decrease in the load causes a decrease of displacement in accordance with line ③ in Fig. 2.7*d*, and so on.

The hysteresis loop inherent in elastoplastic analyses of beams and frames represents a discretized, macroscopic form of structural damping, which was discussed earlier in Sec. 1.10. In this case, all of the dissipated energy is tacitly assumed to be lost at the plastic hinges, while energy is conserved in the rest of the structure. Let us refer to this dissipative mechanism as *elastoplastic damping*, which represents a particular case of *hysteretic damping*. It is a piecewise-linear function of the displacement and has certain characteristics in common with *Coulomb* (or *frictional*) *damping*,

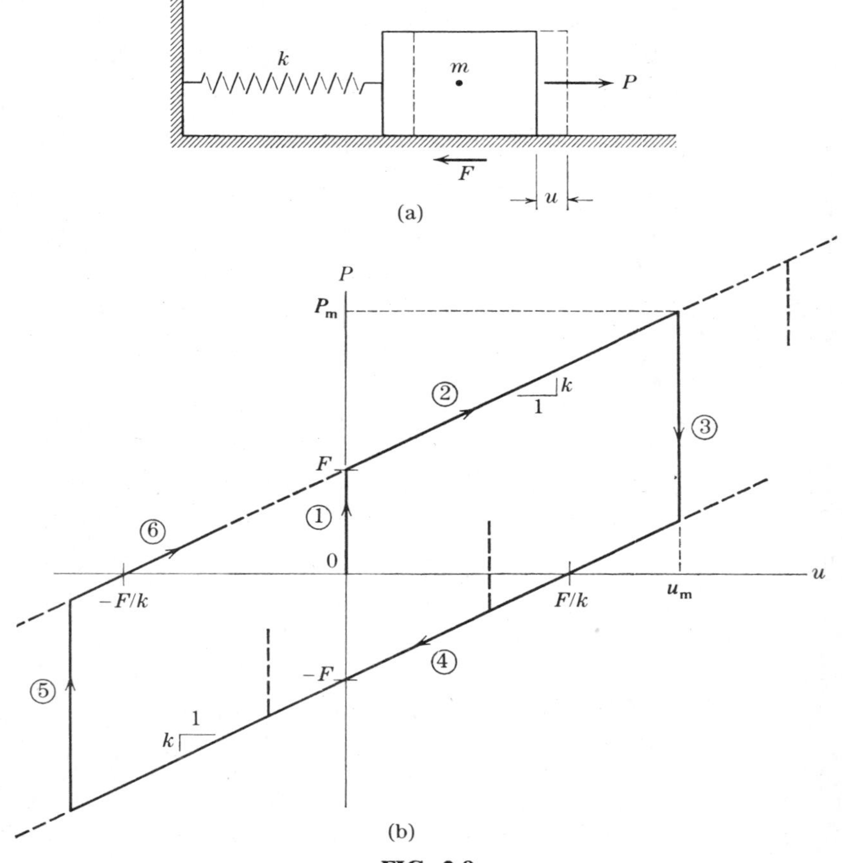

FIG. 2.8

which is another type of hysteretic damping that was discussed in Sec. 1.10. Figure 2.8*a* shows a one-degree system with such a frictional resisting force F, and the hysteresis loop for this case appears in Fig. 2.8*b*. If the system has no residual frictional force when P is first applied, the plot of P against u follows lines ① and ② in Fig. 2.8*b*. A subsequent decrease and reversal of loading are indicated by lines ③ and ④, and a reverse unloading followed by reloading conform to lines ⑤ and ⑥ in the figure. In this case the bilinear diagram consists of the inclined lines ② and ④ (having slopes that are both equal to the spring constant k) and a family of vertical lines, of which ③ and ⑤ are representative.

It is interesting to note that if the slopes of the inclined lines (types ① and ③) in Fig. 2.7*d* are taken to be infinite and if the slopes of the inclined lines (types ② and ④) in Fig. 2.8*b* are taken to be zero, the two problems represented by these figures become mathematically the same. The former problem specializes to the case of the *rigid-plastic restoring force,* for which the deformation of the spring in its elastic range is assumed to be small in comparison with that in the plastic range of the material. On the other hand, the latter problem becomes that of a mass without a spring, the motion of which is inhibited only by friction.

As was mentioned in Sec. 1.10, all known types of dissipative mechanisms, except viscous damping, result in nonlinear vibrations. For example, fluid (or "velocity-squared") damping causes a term proportional to $\dot{u}\,|\dot{u}|$ to appear in the equation of motion for a rapidly moving body immersed in a fluid. This type of nonlinearity and many others may be analyzed by the methods of this chapter. However, it will always be assumed that the mass, damping, and stiffness characteristics of the vibrating system do not vary with time and that the forcing function is independent of the displacement, the velocity, and the acceleration. These possibilities are associated with problems in stability of oscillation, viscoelasticity, and "self-excited" vibrations [4, 5] that are considered to be outside the scope of this book.

2.2 DIRECT INTEGRATION FOR VELOCITY AND PERIOD

Let us consider the free vibrations of an undamped system with a nonlinearly elastic, symmetric restoring force. The equation of motion for this case is

$$m\ddot{u} + F(u) = 0 \tag{a}$$

or

$$\ddot{u} + \omega^2 f(u) = 0 \tag{2.5}$$

in which the term $\omega^2 f(u) = F(u)/m$ represents the restoring force per unit mass as a function of the displacement u. We may express the acceleration in Eq. (2.5) as the derivative of the velocity in the following manner:

$$\ddot{u} = \frac{d\dot{u}}{dt} = \frac{d\dot{u}}{du}\frac{du}{dt} = \frac{d\dot{u}}{du}\dot{u} = \frac{1}{2}\frac{d(\dot{u})^2}{du} \tag{b}$$

Substituting the last expression into Eq. (2.5), we have

$$\frac{1}{2}\frac{d(\dot{u})^2}{du} + \omega^2 f(u) = 0 \tag{c}$$

Assuming that the restoring force $\omega^2 f(u)$ per unit mass is given by the curve in Fig. 2.9 and that the velocity corresponding to u_m in an extreme position is zero, we may integrate Eq. (*c*) to obtain

$$\frac{1}{2}\dot{u}^2 = -\omega^2 \int_{u_m}^{u} f(u')\,du' = \omega^2 \int_{u}^{u_m} f(u')\,du' \tag{d}$$

Thus, for any position of the vibrating system its kinetic energy per unit mass is equal to the potential energy represented by the hatched area under the curve in Fig. 2.9. Of course, the maximum kinetic energy is attained in the equilibrium position, and from Eq. (1.13) of Sec. 1.4 we have

$$(\mathrm{KE})_{\max} = \frac{1}{2}\dot{u}_m^2 = \omega^2 \int_0^{u_m} f(u')\,du' = (\mathrm{PE})_{\max} \tag{2.6}$$

Equation (*d*) yields an expression for the velocity $\dot{u}$ of the vibrating mass

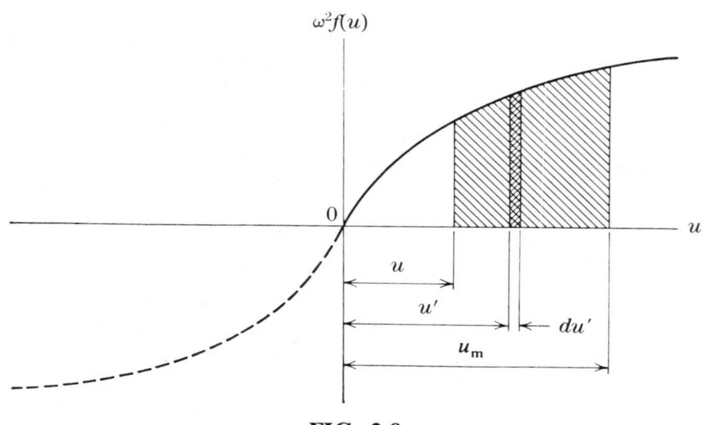

FIG. 2.9

in any position as

$$\dot{u} = \frac{du}{dt} = \pm\omega\sqrt{2\int_u^{u_m} f(u')\,du'} \tag{e}$$

from which the time for any portion of a cycle may be obtained by a second integration. Hence, the time for a full cycle becomes

$$\tau = \frac{4}{\omega}\int_0^{u_m} \frac{du}{\sqrt{2\int_u^{u_m} f(u')\,du'}} \tag{2.7}$$

Thus, if an analytical expression for the restoring force is given, the natural period of the system may be found by evaluating the integrals in Eq. (2.7). Furthermore, the relationship between the velocity $\dot{u}_m$ in the equilibrium position and the displacement u_m in an extreme position may be obtained from Eq. (2.6). This expression is useful for finding the maximum velocity of a nonlinear system that is displaced initially and then allowed to vibrate freely. Conversely, it may be used to calculate the maximum displacement due to an initial velocity. Such an initial velocity might be imparted to the mass by an impulse that is of short duration in comparison with the period of the system.

We shall now consider some particular cases, beginning with a restoring force proportional to any odd power of u. That is,

$$f(u) = u^{2n-1} \tag{f}$$

where n is a positive integer, and the load–displacement curves are symmetric with respect to the origin. Using expression (f) in Eq. (2.6) and integrating, we find

$$\dot{u}_m = \pm\frac{\omega u_m^n}{\sqrt{n}} \tag{g}$$

which becomes $\dot{u}_m = \pm\omega u_m$ for $n = 1$, $\dot{u}_m = \pm 0.707\omega u_m^2$ for $n = 2$, and so on. Next, we substitute expression (f) into Eq. (2.7) and integrate to obtain

$$\tau = \frac{4\sqrt{n}}{\omega}\int_0^{u_m} \frac{du}{\sqrt{u_m^{2n} - u^{2n}}} \tag{2.8a}$$

For the case of a linear restoring force ($n = 1$), the second integration results in

$$\tau = \frac{4}{\omega}\int_0^{u_m}\frac{du}{\sqrt{u_m^2 - u^2}} = \frac{4}{\omega}\int_0^1 \frac{d\xi}{\sqrt{1-\xi^2}} = \frac{4}{\omega}\left[\cos^{-1}\xi\right]_1^0 = \frac{2\pi}{\omega} \tag{h}$$

where $\xi = u/u_m$. When $n = 2$, the restoring force is proportional to u^3, and Eq. (2.8a) yields

$$\tau = \frac{4\sqrt{2}}{\omega}\int_0^{u_m}\frac{du}{\sqrt{u_m^4 - u^4}} = \frac{4\sqrt{2}}{\omega u_m}\int_0^1\frac{d\xi}{\sqrt{1-\xi^4}} \tag{i}$$

The numerical value of the latter integral in Eq. (*i*) is known† and is equal to $1.8541/\sqrt{2}$. Therefore, the expression for the natural period becomes

$$\tau = \frac{7.4164}{\omega u_m} \tag{2.8b}$$

In this case the period of vibration is inversely proportional to the amplitude. A plot of Eq. (2.8b) appears in Fig. 2.10; it pertains to the system in Fig. 2.2*a* when the initial tension *S* in the wire is equal to zero.

If the initial tension in the wire of Fig. 2.2*a* is nonzero, we have the more general case of vibration in which the restoring force per unit mass takes the form

$$\omega^2 f(u) = \omega^2(u + \alpha u^3) \tag{j}$$

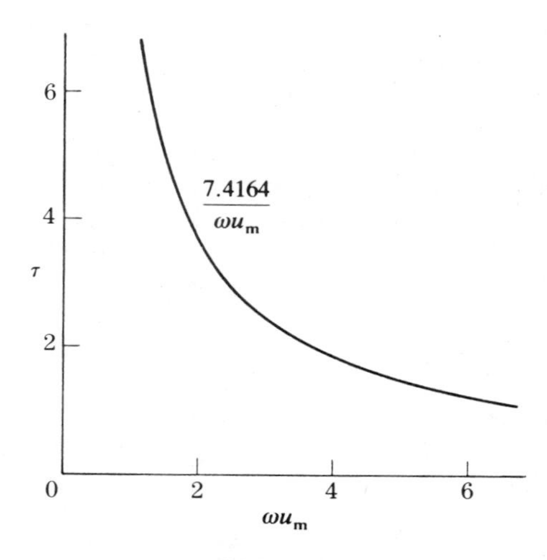

FIG. 2.10

† See standard tables of functions for numerical results.

where $\omega^2 = 2S/m\ell$ and $\alpha = AE/2S\ell^2$. In this case Eq. (2.6) yields

$$\dot{u}_{\mathrm{m}} = \pm\omega u_{\mathrm{m}}\sqrt{1 + \alpha u_{\mathrm{m}}^2/2} \tag{k}$$

which reduces to $\dot{u}_{\mathrm{m}} = \pm\omega u_{\mathrm{m}}$ when $\alpha = 0$. For the purpose of calculating the period of free vibrations, we substitute expression (j) into Eq. (2.7) and obtain

$$\tau = \frac{4}{\omega}\int_0^{u_{\mathrm{m}}} \frac{du}{\sqrt{(u_{\mathrm{m}}^2 - u^2) + \alpha(u_{\mathrm{m}}^4 - u^4)/2}}$$

or

$$\tau = \frac{4}{\omega}\int_0^{u_{\mathrm{m}}} \frac{du}{\sqrt{(u_{\mathrm{m}}^2 - u^2)[1 + \alpha(u_{\mathrm{m}}^2 + u^2)/2]}}$$

To convert the elliptic integral on the right-hand side of the latter expression to standard form, we use the notations

$$\xi = \frac{u}{u_{\mathrm{m}}} \quad \text{and} \quad v = \alpha u_{\mathrm{m}}^2 \tag{l}$$

Then

$$\tau = \frac{4}{\omega}\int_0^1 \frac{d\xi}{\sqrt{(1 - \xi^2)[1 + v(1 + \xi^2)/2]}}$$

Or

$$\tau = \frac{4}{\omega}\sqrt{\frac{2}{v}}\int_0^1 \frac{d\xi}{\sqrt{(1 - \xi^2)[(2 + v)/v + \xi^2]}} \tag{m}$$

In tables of elliptic integrals we find the form

$$\int_0^1 \frac{d\xi}{\sqrt{(a^2 - \xi^2)(b^2 + \xi^2)}} = \frac{1}{c}F\left(\frac{a}{c}, \phi\right) \tag{n}$$

where $F(a/c, \phi)$ represents the elliptic integral of the first kind. The symbols c and ϕ in Eq. (n) have the following meanings:

$$c^2 = a^2 + b^2 \quad \text{and} \quad \sin^2\phi = \frac{c^2}{a^2(b^2 + 1)}$$

Comparing the integrals in Eqs. (m) and (n), we conclude that

$$a^2 = 1 \quad \text{and} \quad b^2 = \frac{2 + v}{v}$$

Therefore

$$c = \sqrt{\frac{2(1 + v)}{v}} \quad \text{and} \quad \phi = \sin^{-1}(1) = \frac{\pi}{2}$$

As the result, Eq. (m) becomes

$$\tau = \frac{4}{\omega} \frac{1}{\sqrt{1+v}} F\left(\sqrt{\frac{v}{2(1+v)}}, \frac{\pi}{2}\right) \tag{2.9}$$

If the deviation of the spring characteristic from linearity is very small, we may set α (and v) equal to zero. Then Eq. (2.9) yields expression (h), corresponding to the case of a linear restoring force. On the other hand, if α (and v) is very large, the first term in Eq. (j) may be neglected. Consequently, the quantity $1 + v$ in Eq. (2.9) becomes approximately equal to v, and we obtain for τ the expression

$$\tau = \frac{7.4164}{\omega u_{\mathrm{m}} \sqrt{\alpha}} \tag{o}$$

Equation (o) is the same as Eq. (2.8b), except for the appearance of $\sqrt{\alpha}$ [due to the fact that the cubic term in Eq. (j) is $\omega^2 \alpha u^3$ instead of $\omega^2 u^3$]. For any intermediate case between these two extremes, it is necessary to calculate the numerical value of $\sqrt{v/2(1+v)}$ and to obtain the corresponding value of the elliptic integral from the tables.

In the above derivation we dealt with the case of a "hardening spring" [see Eq. (j)], where the slope of the restoring force increases with the displacement. Let us now consider the case of a "softening spring" and take the following expression for the restoring force:

$$\omega^2 f(u) = \omega^2(u - \alpha u^3) \tag{p}$$

Proceeding in the same manner as before, we obtain the counterparts of Eqs. (k) and (m) as

$$\dot{u}_{\mathrm{m}} = \pm \omega u_{\mathrm{m}} \sqrt{1 - \alpha u_{\mathrm{m}}^2/2} \tag{q}$$

and

$$\tau = \frac{4}{\omega} \sqrt{\frac{2}{v}} \int_0^1 \frac{d\xi}{\sqrt{(1-\xi^2)[(2-v)/v - \xi^2]}} \tag{r}$$

In the tables we find the form

$$\int_0^1 \frac{d\xi}{\sqrt{(a^3 - \xi^2)(b^2 - \xi^2)}} = \frac{1}{b} F\left(\frac{a}{b}, \phi\right) \tag{s}$$

where $\sin\phi = 1/a$. Comparing the integrals in Eqs. (r) and (s), we conclude that

$$a^2 = 1 \qquad b^2 = \frac{2-v}{v} \qquad \phi = \frac{\pi}{2}$$

Therefore, Eq. (r) becomes

$$\tau = \frac{4}{\omega}\sqrt{\frac{2}{2-v}}\,F\left(\sqrt{\frac{v}{2-v}}, \frac{\pi}{2}\right) \tag{2.10}$$

Again, for any value of α (and v), we can readily calculate values of τ from Eq. (2.10) using tables of elliptic integrals.

Another example of a symmetric restoring force for which a rigorous solution can be obtained is represented by the pendulum in Fig. 2.3. In this case the equation of motion is [see Eq. (2.4a), Sec. 2.1]

$$\ddot{\phi} + \omega^2 \sin \phi = 0$$

where $\omega^2 = g/L$. The counterparts of Eqs. (2.6) and (2.7) for rotational oscillations are

$$(\mathrm{KE})_{\max} = \frac{1}{2}\dot{\phi}_{\mathrm{m}}^2 = \omega^2 \int_0^{\phi_{\mathrm{m}}} f(\psi)\, d\psi = (\mathrm{PE})_{\max} \tag{2.11a}$$

and

$$\tau = \frac{4}{\omega}\int_0^{\phi_{\mathrm{m}}} \frac{d\phi}{\sqrt{2\int_{\phi}^{\phi_{\mathrm{m}}} f(\psi)\, d\psi}} \tag{2.11b}$$

For the pendulum these equations become

$$\dot{\phi}_{\mathrm{m}} = \pm\omega\sqrt{2(1 - \cos \phi_{\mathrm{m}})} \tag{t}$$

and

$$\tau = \frac{4}{\omega}\int_0^{\phi_{\mathrm{m}}} \frac{d\phi}{\sqrt{2(\cos \phi - \cos \phi_{\mathrm{m}})}} = \frac{2}{\omega}\int_0^{\phi_{\mathrm{m}}} \frac{d\phi}{\sqrt{\sin^2(\phi_{\mathrm{m}}/2) - \sin^2(\phi/2)}} \tag{u}$$

Introducing the notation $s = \sin(\phi_{\mathrm{m}}/2)$ and a new variable θ, such that

$$\sin(\phi/2) = s \sin \theta = \sin(\phi_{\mathrm{m}}/2) \sin \theta \tag{v}$$

we find

$$d\phi = \frac{2s \cos \theta \, d\theta}{\sqrt{1 - s^2 \sin^2 \theta}} \tag{w}$$

Substituting expressions (v) and (w) into Eq. (u) and observing from Eq. (v) that θ varies from 0 to $\pi/2$ while ϕ varies from 0 to ϕ_{m}, we obtain

$$\tau = \frac{4}{\omega}\int_0^{\pi/2} \frac{d\theta}{\sqrt{1 - s^2 \sin^2 \theta}} = \frac{4}{\omega} F\left(s, \frac{\pi}{2}\right) \tag{2.12}$$

which is in the standard form of the elliptic integral of the first kind. Thus, the numerical value of the integral in Eq. (2.12) corresponding to any value of s may be taken from the tables.

When the maximum rotational displacement ϕ_m of the pendulum is small, the value of s is also small; and we can neglect the quantity $s^2 \sin^2 \theta$ in Eq. (2.12). The integral then becomes equal to $\pi/2$, and we obtain the natural period $\tau = 2\pi/\omega$ for the pendulum with small rotations.

Example 1. Suppose that a package containing a mass m, constrained by springs within the package, is dropped from a height h onto a concrete floor. The restoring force exerted by the springs upon the mass is known from experiments to be approximately

$$F(u) = \alpha(u)^5 \tag{x}$$

where u is the displacement of the mass relative to the package. Determine the maximum displacement of the mass relative to the package, assuming inelastic impact of the package with the floor.

Solution. At the instant of impact the dropped package has kinetic energy per unit mass equal to gh. Dividing Eq. (x) by m, we have $\omega^2 f(u) = \alpha u^5/m$; and Eq. (2.6) yields

$$(\text{KE})_{\text{max}} = gh = \alpha u_{\text{m}}^6/6m = (\text{PE})_{\text{max}}$$

Thus,

$$u_{\text{m}} = \left(\frac{6mgh}{\alpha}\right)^{1/6} \tag{y}$$

Example 2. Derive an expression for the initial velocity required for the mass in Prob. 2.1–5 (see Sec. 2.1) to "snap through" from a position with $\theta < \pi/2$ to a position with $\theta > \pi/2$.

Solution. Using energy considerations, we see that the initial velocity must be at least that which gives an initial kinetic energy equal to the potential energy in the spring when it is in the vertical position ($\theta = \pi/2$). The change in length of the spring in the vertical position is

$$\Delta = \ell(1 - \sin\theta)$$

and the potential energy stored in the spring is

$$(\text{PE})_{\text{vert}} = k\Delta^2/2 = k\ell^2(1 - \sin\theta)^2/2$$

Equating this expression to the initial kinetic energy of the mass, we find

$$\dot{u}_0 \geqq \ell(1 - \sin\theta)\sqrt{k/m} \tag{z}$$

as the criterion for "snap-through."

2.3 APPROXIMATE METHODS FOR FREE VIBRATIONS

In general, the exact solution of a nonlinear differential equation is unknown, and only an approximate solution can be obtained. In any case, we may simulate a nonlinear vibration with appropriately selected functions of time that satisfy certain specified criteria. Two well-known methods that involve such approximate simulations are described in this section.

Method of Successive Approximations

If the deviation of the spring characteristic from linearity is comparatively small, the equation of motion for free vibrations of an undamped one-degree system can be represented in the following form:

$$\ddot{u} + \omega^2 u + \alpha f(u) = 0 \tag{2.13}$$

in which α is a small factor, and $f(u)$ is a polynomial of u with the lowest power of u not smaller than 2. For systems in which the load–displacement curve is symmetric with respect to the origin, we have

$$f(u) = \sum_{i=1}^{n} \pm u\,|u^i| \tag{a}$$

A commonly occurring case of this kind is obtained by substituting only the second positive term of expression (a) into Eq. (2.13). Then the equation of motion becomes

$$\ddot{u} + \omega^2 u + \alpha u^3 = 0 \tag{2.14}$$

One method for simulating the motions of such *quasi-linear systems* consists of determining periodical solutions by successive approximations [6].

Let us assume that at time $t = 0$ the initial conditions of the system are $u_0 = u_{\mathrm{m}}$ and $\dot{u}_0 = 0$. The resulting harmonic motion of a linear system replacing the actual system is represented by

$$u = u_{\mathrm{m}} \cos \omega_1 t \tag{b}$$

in which the symbol ω_1 denotes the angular frequency of the substitute system. Equation (b) may be considered as a first approximation to the solution of Eq. (2.14) for the given initial conditons. Since the factor α is small, we may assume that the angular frequency ω_1 is not substantially different from the frequency ω for a linear system. Then we can write

$$\omega^2 = \omega_1^2 + (\omega^2 - \omega_1^2) \tag{c}$$

where $(\omega^2 - \omega_1^2)$ is a small quantity. Substituting expression (c) into Eq. (2.14), we obtain

$$\ddot{u} + \omega_1^2 u + (\omega^2 - \omega_1^2)u + \alpha u^3 = 0 \tag{d}$$

The first approximation (b) for u may now be substituted into the last two terms of Eq. (d), which are small, to give

$$\ddot{u} + \omega_1^2 u = -u_{\mathrm{m}}(\omega^2 - \omega_1^2)\cos\omega_1 t - \alpha u_{\mathrm{m}}^3 \cos^3 \omega_1 t$$

Or, by using the identity

$$\cos^3 \omega_1 t = (3\cos\omega_1 t + \cos 3\omega_1 t)/4$$

we find

$$\ddot{u} + \omega_1^2 u = -\left(\omega^2 - \omega_1^2 + \frac{3\alpha u_{\mathrm{m}}^2}{4}\right)u_{\mathrm{m}}\cos\omega_1 t - \frac{\alpha u_{\mathrm{m}}^3}{4}\cos 3\omega_1 t \tag{e}$$

This equation has the same mathematical form as that for an undamped one-degree system subjected to harmonic forcing functions. However, the first term on the right-hand side of the equation represents a forcing function having the same frequency ω_1 as that of the substitute system. It will cause a response that builds up indefinitely with time and violates the condition of free vibrations with constant amplitude. To eliminate this spurious resonance, we must set the coefficient of $\cos\omega_1 t$ equal to zero. This step represents an essential feature in the method at hand because it yields an approximate value for ω_1. Thus,

$$\omega_1^2 = \omega^2 + \frac{3\alpha u_{\mathrm{m}}^2}{4} \tag{2.15}$$

Whereas the term ω^2 may be considered to be a first approximation of ω_1^2, Eq. (2.15) represents a second approximation, consisting of the first augmented by the term $3\alpha u_{\mathrm{m}}^2/4$.

From the remaining parts of Eq. (e) we find that the total solution consists of

$$u = C_1\cos\omega_1 t + C_2\sin\omega_1 t + \frac{\alpha u_{\mathrm{m}}^3}{32\omega_1^2}\cos 3\omega_1 t \tag{f}$$

To satisfy the given initial conditions ($u_0 = u_{\mathrm{m}}$, $\dot{u}_0 = 0$), the integration constants must be

$$C_1 = u_{\mathrm{m}} - \frac{\alpha u_{\mathrm{m}}^3}{32\omega_1^2} \qquad \text{and} \qquad C_2 = 0$$

Therefore, the second approximation of the motion becomes

$$u = u_{\mathrm{m}} \cos \omega_1 t + \frac{\alpha u_{\mathrm{m}}^3}{32\omega_1^2}(\cos 3\omega_1 t - \cos \omega_1 t) \tag{2.16}$$

Thus, the correcting term in the second approximation contains a higher harmonic proportional to $\cos 3\omega_1 t$, which is shown graphically in Fig. 2.11*a*. Of course, the magnitude of the deviation from a simple cosine curve depends upon the magnitude of the factor α. It should also be remembered that the angular frequency ω_1 increases with amplitude, as indicated by the plot of Eq. (2.15) in Fig. 2.11*b*.

If a third approximation of the motion is desired, the second approximation (Eq. 2.16) may be substituted into Eq. (*d*) and the process repeated. However, the trigonometric manipulations become rather involved, and a more orderly technique is desired. Toward this end, we observe that Eqs. (2.15) and (2.16) may be written in the following forms:

$$\begin{aligned} \omega^2 &= \omega_1^2 + \alpha c_1 \\ u &= \phi_0 + \alpha\phi_1 \end{aligned} \tag{g}$$

That is, expressions for the second approximations of the frequency and the displacement contain the quantity α to the first power. Then for further approximations we will take additional terms in the series

$$u = \phi_0 + \alpha\phi_1 + \alpha^2\phi_2 + \alpha^3\phi_3 + \cdots \tag{2.17a}$$

and

$$\omega^2 = \omega_1^2 + \alpha c_1 + \alpha^2 c_2 + \alpha^3 c_3 + \cdots \tag{2.17b}$$

which both contain powers of the small quantity α. In these series the symbols ϕ_0, ϕ_1, ϕ_2, etc., represent unknown functions of time; and c_1, c_2, c_3, etc., are constants that must be chosen to eliminate conditions of resonance, as was explained above in the calculation of the second approximation. By increasing the number of terms in the series, we can calculate as many successive approximations as desired. In the following discussion, we limit our calculations by omitting all the terms containing α in a power higher than the third. Substituting Eqs. (2.17a) and (2.17b) into Eq. (2.14), we obtain

$$\begin{aligned} &\ddot{\phi}_0 + \alpha\ddot{\phi}_1 + \alpha^2\ddot{\phi}_2 + \alpha^3\ddot{\phi}_3 \\ &\qquad + (\omega_1^2 + \alpha c_1 + \alpha^2 c_2 + \alpha^3 c_3)(\phi_0 + \alpha\phi_1 + \alpha^2\phi_2 + \alpha^3\phi_3) \\ &\qquad\qquad + \alpha(\phi_0 + \alpha\phi_1 + \alpha^2\phi_2 + \alpha^3\phi_3)^3 = 0 \end{aligned} \tag{h}$$

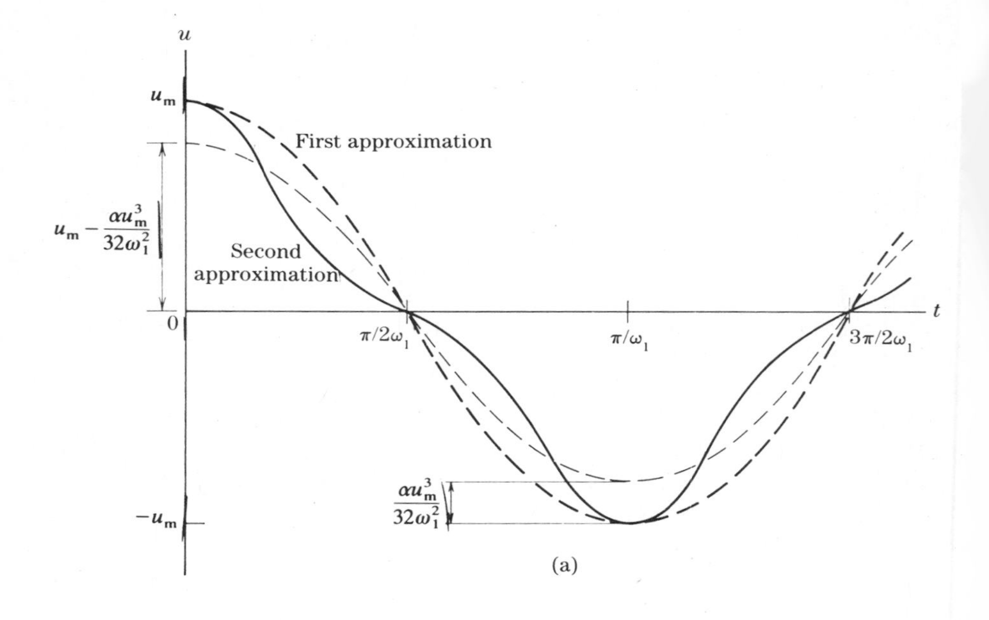

(a)

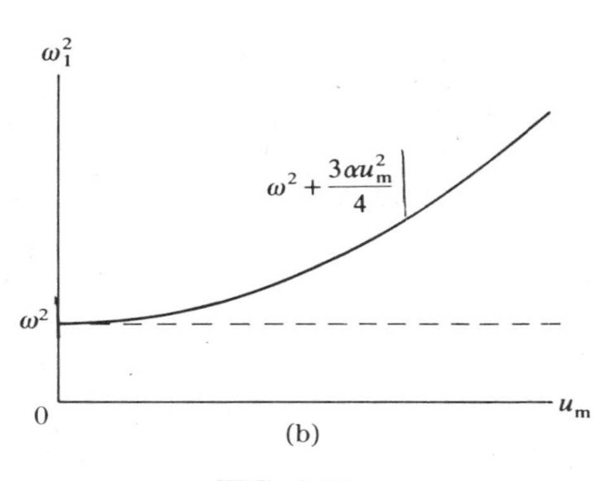

(b)

FIG. 2.11

After performing the indicated algebraic operations and neglecting all of the terms containing α to a power higher than the third, we can represent Eq. (h) in the following form:

$$\begin{aligned}
&\ddot{\phi}_0 + \omega_1^2\phi_0 + \alpha(\ddot{\phi}_1 + \omega_1^2\phi_1 + c_1\phi_0 + \phi_0^3) \\
&\quad + \alpha^2(\ddot{\phi}_2 + \omega_1^2\phi_2 + c_2\phi_0 + c_1\phi_1 + 3\phi_0^2\phi_1) \\
&\quad + \alpha^3(\ddot{\phi}_3 + \omega_1^2\phi_3 + c_3\phi_0 + c_2\phi_1 + c_1\phi_2 + 3\phi_0^2\phi_2 + 3\phi_0\phi_1^2) = 0 \qquad (i)
\end{aligned}$$

This equation must hold for any value of the small quantity α, which means that the factor for each of the three powers of α must be zero. Thus, Eq. (i)

may be separated into the following system of equations:

$$
\begin{aligned}
&\ddot{\phi}_0 + \omega_1^2\phi_0 = 0 \\
&\ddot{\phi}_1 + \omega_1^2\phi_1 = -c_1\phi_0 - \phi_0^3 \\
&\ddot{\phi}_2 + \omega_1^2\phi_2 = -c_2\phi_0 - c_1\phi_1 - 3\phi_0^2\phi_1 \\
&\ddot{\phi}_3 + \omega_1^2\phi_3 = -c_3\phi_0 - c_2\phi_1 - c_1\phi_2 - 3\phi_0^2\phi_2 - 3\phi_0\phi_1^2
\end{aligned}
\tag{j}
$$

Taking the same initial conditions as before (that is, for $t=0$, $u_0 = u_{\mathrm{m}}$ and $\dot{u}_0 = 0$) and substituting them into eq. (2.17a) and its derivative, we obtain

$$
\begin{aligned}
&\phi_0(0) + \alpha\phi_1(0) + \alpha^2\phi_2(0) + \alpha^3\phi_3(0) = u_{\mathrm{m}} \\
&\dot{\phi}_0(0) + \alpha\dot{\phi}_1(0) + \alpha^2\dot{\phi}_2(0) + \alpha^3\dot{\phi}_3(0) = 0
\end{aligned}
$$

Again, because these equations must hold for any magnitude of α, we have

$$
\begin{aligned}
&\phi_0(0) = u_{\mathrm{m}} \qquad && \dot{\phi}_0(0) = 0 \\
&\phi_1(0) = 0 && \dot{\phi}_1(0) = 0 \\
&\phi_2(0) = 0 && \dot{\phi}_2(0) = 0 \\
&\phi_3(0) = 0 && \dot{\phi}_3(0) = 0
\end{aligned}
\tag{k}
$$

Considering the first of Eqs. (j) and the corresponding initial conditions represented by the first row of the system (k), we find as before

$$
\phi_0 = u_{\mathrm{m}} \cos \omega_1 t \tag{l}
$$

Substituting this first approximation into the right-hand side of the second of Eqs. (j), we obtain

$$
\begin{aligned}
\ddot{\phi}_1 + \omega_1^2\phi_1 &= -c_1 u_{\mathrm{m}} \cos \omega_1 t - u_{\mathrm{m}}^3 \cos^3 \omega_1 t \\
&= -\left(c_1 u_{\mathrm{m}} + \frac{3u_{\mathrm{m}}^3}{4}\right) \cos \omega_1 t - \frac{u_{\mathrm{m}}^3}{4} \cos 3\omega_1 t
\end{aligned}
$$

To eliminate the condition of resonance, we will choose the constant c_1 so as to make the first term on the right-hand side of the equation equal to zero. Thus, we find

$$
c_1 = -\frac{3u_{\mathrm{m}}^2}{4} \tag{m}
$$

The general solution for ϕ_1 then becomes

$$\phi_1 = C_1 \cos \omega_1 t + C_2 \sin \omega_1 t + \frac{u_{\mathrm{m}}^3}{32\omega_1^2} \cos 3\omega_1 t$$

To satisfy the initial conditions given by the second row of the system (k), we find

$$C_1 = -\frac{u_{\mathrm{m}}^3}{32\omega_1^2} \qquad \text{and} \qquad C_2 = 0$$

Thus,

$$\phi_1 = \frac{u_{\mathrm{m}}^3}{32\omega_1^2}(\cos 3\omega_1 t - \cos \omega_1 t) \tag{n}$$

If we limit our calculations to the second approximation and substitute expressions (l), (m), and (n) into Eqs. (2.17a) and (2.17b), we obtain

$$u = u_{\mathrm{m}} \cos \omega_1 t + \frac{\alpha u_{\mathrm{m}}^3}{32\omega_1^2}(\cos 3\omega_1 t - \cos \omega_1 t) \tag{o}$$

where

$$\omega_1^2 = \omega^2 + \frac{3\alpha u_{\mathrm{m}}^2}{4} \tag{p}$$

These results coincide exactly with expressions that were previously obtained (see Eqs. 2.15 and 2.16).

To obtain the third approximation, we substitute the expressions (l), (m), and (n) into the right-hand side of the third of Eqs. (j) and obtain

$$\begin{aligned}\ddot{\phi}_2 + \omega_1^2 \phi_2 = & -c_2 u_{\mathrm{m}} \cos \omega_1 t \\ & + 3u_{\mathrm{m}}^2\left(\frac{1}{4} - \cos^2 \omega_1 t\right)\left[\frac{u_{\mathrm{m}}^3}{32\omega_1^2}(\cos 3\omega_1 t - \cos \omega_1 t)\right]\end{aligned}$$

By usng formulas for trigonometric functions of multiple angles, we can write this equation in the following form:

$$\ddot{\phi}_2 + \omega_1^2 \phi_2 = -u_{\mathrm{m}}\left(c_2 - \frac{3u_{\mathrm{m}}^4}{128\omega_1^2}\right) \cos \omega_1 t - \frac{3u_{\mathrm{m}}^5}{128\omega_1^2} \cos 5\omega_1 t$$

Again, to eliminate the condition of resonance, we put

$$c_2 = \frac{3u_{\mathrm{m}}^4}{128\omega_1^2} \tag{q}$$

Then the general solution for ϕ_2 becomes

$$\phi_2 = C_1 \cos \omega_1 t + C_2 \sin \omega_1 t + \frac{u_m^5}{1024\omega_1^4} \cos 5\omega_1 t$$

By using the third row of the system (k), we find that the constants of integration are

$$C_1 = -\frac{u_m^5}{1024\omega_1^4} \qquad \text{and} \qquad C_2 = 0$$

Thus, we obtain

$$\phi_2 = \frac{u_m^5}{1024\omega_1^4}(\cos 5\omega_1 t - \cos \omega_1 t) \tag{r}$$

and the third approximation to the response becomes

$$u = u_m \cos \omega_1 t + \frac{\alpha u_m^3}{32\omega_1^2}(\cos 3\omega_1 t - \cos \omega_1 t) + \frac{\alpha^2 u_m^5}{1024\omega_1^4}(\cos 5\omega_1 t - \cos \omega_1 t) \tag{2.18}$$

where ω_1 is now determined by the expression

$$\omega_1^2 = \omega^2 + \frac{3\alpha u_m^2}{4} - \frac{3\alpha^2 u_m^4}{128\omega^2} \tag{2.19}$$

To obtain the fourth approximation, we substitute the expressions for ϕ_0, ϕ_1, ϕ_2, c_1, and c_2 into the last of Eqs. (j); and proceeding as before, we find

$$u = u_m \cos \omega_1 t + \frac{\alpha u_m^3}{32\omega_1^2}(\cos 3\omega_1 t - \cos \omega_1 t) + \frac{\alpha^2 u_m^5}{1024\omega_1^4}(\cos 5\omega_1 t - \cos \omega_1 t)$$
$$+ \frac{\alpha^3 u_m^7}{32{,}768\omega_1^6}(\cos 7\omega_1 t - 6 \cos 3\omega_1 t + 5 \cos \omega_1 t) \tag{2.20}$$

in which ω_1 is determined by the approximation

$$\omega_1^2 = \omega^2 + \frac{3\alpha u_m^2}{4} - \frac{3\alpha^2 u_m^4}{128\omega^2} + \frac{9\alpha^3 u_m^6}{512\omega^4} \tag{2.21}$$

In summary, the method of successive approximations involves the simulation of nonlinear free vibrations by a series of functions obtained by assuming the form of a first approximation, such as eq. (b), and then by solving recursively a set of equations of the type (j), subject to the initial conditions exemplified by Eqs. (k). An approximation selected by this method inherently satisfies the equation of motion only at the instants when the vibrating system is in an extreme or middle position. Although there is theoretically no limit on the number of successive approximations that can be obtained, the second approximation should ordinarily suffice for practical purposes.

Ritz Averaging Method

Another technique for approximating nonlinear vibrations by a series solution involves making the average value of the virtual work per cycle vanish. This approach, known as the Ritz averaging method [7], can produce a better solution than does the method of successive approximations with the same number of terms. Furthermore, the averaging method is not restricted to quasi-linear systems, and it may be applied to forced vibrations (see the next section) as well as free vibrations.

Let us consider the free vibrations of an undamped one-degree system, for which the equation of motion may be written as

$$\ddot{u} + f(u) = 0 \tag{2.22}$$

in which the terms on the left represent the inertia force and the restoring force per unit mass. By D'Alembert's principle, Eq. (2.22) may be considered to be an equation of dynamic equilibrium in which these two forces balance each other. If the system is subjected to a virtual displacement δu, the work done by these forces must be equal to zero. Thus,

$$[\ddot{u} + f(u)]\,\delta u = 0 \tag{s}$$

In applying the Ritz method, we assume as an approximate solution for the free vibrations a series

$$u = a_1\phi_1(t) + a_2\phi_2(t) + a_3\phi_3(t) + \cdots = \sum_{i=1}^{n} a_i\phi_i(t) \tag{2.23}$$

in which $\phi_1(t)$, $\phi_2(t)$, and so on, are selected functions of time and a_1, a_2, etc., are weighting factors determined in such a manner that the virtual work per cycle vanishes. Virtual displacements are taken in the form

$$\delta u_i = \delta a_i\phi_i(t) \tag{t}$$

and by integrating the work over one cycle we obtain

$$\sum_{i=1}^{n} \int_0^{\tau} [\ddot{u} + f(u)]\,\delta a_i\phi_i(t)\,dt = 0 \tag{u}$$

It follows that

$$\begin{aligned} &\int_0^{\tau} [\ddot{u} + f(u)]\phi_1(t)\,dt = 0 \\ &\int_0^{\tau} [\ddot{u} + f(u)]\phi_2(t)\,dt = 0 \\ &\qquad \cdots \\ &\int_0^{\tau} [\ddot{u} + f(u)]\phi_n(t)\,dt = 0 \end{aligned} \tag{2.24}$$

Equations (2.24) represent n algebraic equations that can be solved simultaneously to determine the values of $a_1, a_2, \ldots, a_n$.

As an example, let us reconsider the quasi-linear system for which Eq. (2.14) is the equation of motion. If we take as a first approximation of the free vibration

$$u = a_1\phi_1(t) = a_1 \cos \omega_1 t \tag{v}$$

the first of Eqs. (2.24) gives

$$\int_0^{\tau} [-\omega_1^2 a_1 \cos \omega_1 t + \omega^2 a_1 \cos \omega_1 t + \alpha a_1^3 \cos^3 \omega_1 t] \cos \omega_1 t\,dt = 0$$

Observing that

$$\int_0^{\tau} \cos^2 \omega_1 t\,dt = \frac{1}{\omega_1}\int_0^{2\pi} \cos^2 \omega_1 t\,d(\omega_1 t) = \frac{\pi}{\omega_1}$$

and that

$$\int_0^{\tau} \cos^4 \omega_1 t\,dt = \frac{1}{\omega_1}\int_0^{2\pi} \cos^4 \omega_1 t\,d(\omega_1 t) = \frac{3\pi}{4\omega_1}$$

we find

$$\omega_1^2 = \omega^2 + \frac{3\alpha a_1^2}{4} \tag{2.25}$$

This expression is of the same form as Eq. (2.15) in the method of successive approximations, and it determines the value of a_1 in terms of ω, ω_1, and α. Solving for a_1 in Eq. (2.25) and substituting the result into Eq. (v) yields

$$u = 2\sqrt{\frac{\omega_1^2 - \omega^2}{3\alpha}} \cos \omega_1 t \tag{2.26}$$

As a more refined approximation satisfying the condition of symmetry in this example, we can take the two-term series

$$u = a_1\phi_1(t) + a_2\phi_2(t) = a_1 \cos \omega_1 t + a_2 \cos 3\omega_1 t \qquad (w)$$

Substitution of expression (w) into the first two of Eqs. (2.24), followed by integration of individual terms, yields two simultaneous cubic equations that must be solved numerically for a_1 and a_2. Although this part of the analysis is cumbersome, the difficulty is only an algebraic one.

2.4 FORCED NONLINEAR VIBRATIONS

In the preceding sections of this chapter we considered only free vibrations of nonlinear systems. We shall now discuss the steady-state response of a nonlinear system subjected to a periodic forcing function, using the Ritz averaging method to obtain an approximate solution.

Suppose that a system has damping in proportion to some function $f_1(\dot{u})$ of the velocity and a restoring force proportional to some other function $f_2(u)$ of the displacement. If a periodic forcing function $mf_3(t)$ acts upon the mass, its equation of motion may be written as

$$\ddot{u} + 2nf_1(\dot{u}) + \omega^2 f_2(u) = f_3(t) \qquad (2.27)$$

in which the terms represent the inertial force, the damping force, the restoring force, and the applied force per unit mass. By the Ritz averaging method we assume for the steady-state vibrations an approximate solution in the form of a series, as given by Eq. (2.23) in Sec. 2.3. In order that the virtual work per cycle of forced vibrations shall vanish, we have the conditions

$$\int_0^\tau [\ddot{u} + 2nf_1(\dot{u}) + \omega^2 f_2(u) - f_3(t)]\phi_i(t)\, dt = 0 \qquad (i = 1, 2, \ldots, n) \qquad (2.28)$$

Let us consider first the particular case in which there is no damping and for which the equation of motion is

$$\ddot{u} + \omega^2(u \pm \mu u^3) = q \cos \Omega t \qquad (2.29)$$

This expression is known as *Duffing's equation* because it was studied extensively by Duffing [8] in his book on vibrations. Forced vibrations of this type are symmetric with respect to the equilibrium position, and without damping the response will be either in phase or 180° out of phase with the disturbing force. As a first approximation, we can assume

$$u = a_1\phi_1(t) = a_1 \cos \Omega t \qquad (a)$$

Then the first of Eqs. (2.28) becomes

$$\int_0^{\tau} [-\Omega^2 a_1 \cos \Omega t + \omega^2(a_1 \cos \Omega t \pm \mu a_1^3 \cos^3 \Omega t) - q \cos \Omega t] \cos \Omega t \, dt = 0$$

Integration and rearrangement of terms results in

$$\omega^2 a_1 \pm \frac{3\omega^2 \mu a_1^3}{4} = \Omega^2 a_1 + q \tag{2.30}$$

For any given values of the parameters ω^2, μ, and q, Eq. (2.30) represents two approximate relationships between the amplitude a_1 and the impressed frequency Ω for the steady-state forced vibrations. A more convenient form for plotting response spectra (similar to Fig. 1.22 in Sec. 1.6) in the case of a hardening spring is

$$\frac{3\mu a_1^3}{4} = \left(\frac{\Omega^2}{\omega^2} - 1\right) a_1 + \frac{q}{\omega^2} \tag{2.31}$$

and for the case of a softening spring is

$$\frac{3\mu a_1^3}{4} = \left(1 - \frac{\Omega^2}{\omega^2}\right) a_1 - \frac{q}{\omega^2} \tag{2.32}$$

A method for constructing plots of a_1 against Ω/ω from Eqs. (2.31) and (2.32) is given in the following discussion.

Equation (2.31) may be envisioned as representing the intersection of the cubic function of a_1 on the left-hand side and the linear function of a_1 on the right-hand side. Figure 2.12*a* shows the cubic function as well as a family of linear functions for several values of the frequency ratio Ω/ω. The inclined line labeled ① in Fig. 2.12*a*, corresponding to the vertical line $\Omega/\omega = 0$ in Fig. 2.12*b*, intersects the cubic curve at point A. The value of a_1 for that intersection is plotted in Fig. 2.12*b* as the point A'. In Fig. 2.12*a* the line labeled ②, for which $0 < \Omega/\omega < 1$, intersects the curve at point B; and the corresponding point in Fig. 2.12*b* is indicated as B'. The horizontal line labeled ③ in Fig. 2.12*a* would represent resonance for a linear system, but in this case it merely produces another pair of points C and C' on the diagrams. As the slopes of the lines in Fig. 2.12*a* increase, a condition is reached where the line ④ not only intersects the upper branch of the curve at D, but is also tangent to the lower branch at E. The corresponding points D' and E' on the response spectrum in Fig. 2.12*b* both occur at the *critical frequency* ($\Omega_{cr} \geqq \omega$), where the spectrum has an infinite slope (at point E'). Steeper lines in Fig. 2.12*a*, such as the one labeled ⑤, intersect the cubic

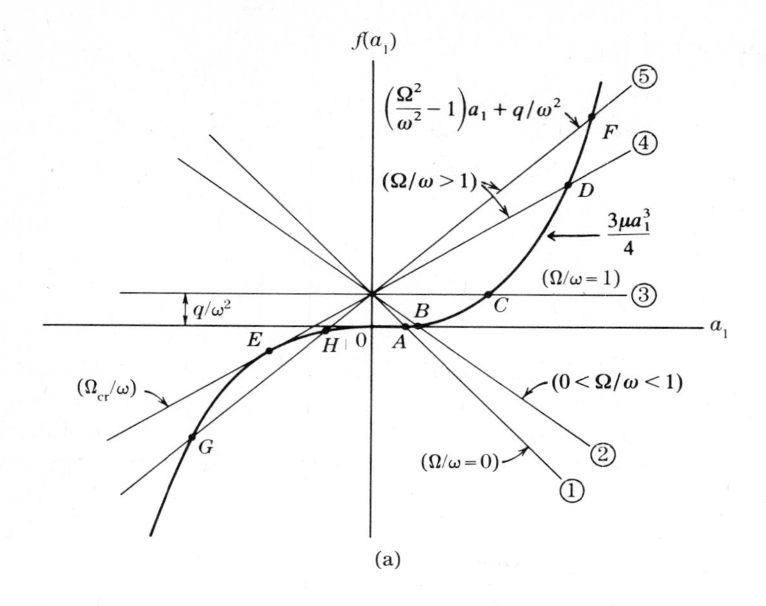

(a)

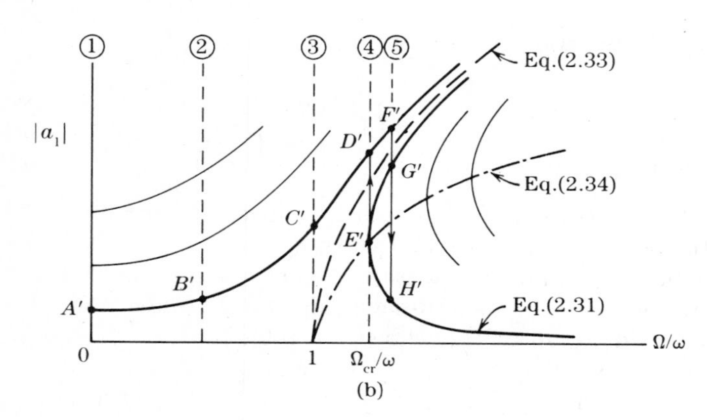

(b)

FIG. 2.12

curve at three points, such as F, G, and H; and their counterparts (F', G', and H') are indicated on the response spectrum. In this manner the solid curve in Fig. 2.12*b* is constructed as a graphical representation of Eq. (2.31).

The response spectrum for the hardening spring has a hyperbolic asymptote, which is shown by the dashed curve in Fig. 2.12*b*. It corresponds to the case of nonlinear free vibrations, as obtained by setting q equal to zero in Eq. (2.31). Thus, the amplitude–frequency relationship for free vibrations becomes

$$\frac{3\mu a_1^2}{4} = \frac{\Omega^2}{\omega^2} - 1 \tag{2.33}$$

In which Ω now represents the angular frequency of free nonlinear vibrations. Moreover, a family of response spectra (indicated by the light curves) similar to the solid curve in Fig. 2.12*b* can be generated for various values of the load q [9]. A locus of critical points of type E' (where the slope is infinite) for such spectra is shown by the dot-dash curve in Fig. 2.12*b*. The equation of this curve may be obtained as

$$\frac{9\mu a_1^2}{4} = \frac{\Omega^2}{\omega^2} - 1 \tag{2.34}$$

by differentiation of Eq. (2.31).

Returning our attention to Eq. (2.32) for the case of a softening spring, we may plot a typical response spectrum using a graphical technique similar to that described above. The cubic curve for this case is shown in Fig. 2.13*a* together with a family of lines representing the right-hand side of Eq. (2.31) for several values of Ω/ω. Lines ① and ② both have three intersections with the cubic curve; line ③ has one intersection and is also tangent to the curve; and each of lines ④ and ⑤ has a single point of intersection with the curve. Points A–J in Fig. 2.13*a* have their counterparts indicated as points A'–J' in Fig. 2.13*b*. In this case the response spectrum has a vertical tangent at point H', and the critical frequency occurs when the impressed frequency is less than that for resonance in a linear system ($\Omega_{cr} \leqq \omega$).

By setting $q = 0$ in Eq. (2.32), we obtain the equation for the dashed curve in Fig. 2.13*b* as

$$\frac{3\mu a_1^2}{4} = 1 - \frac{\Omega^2}{\omega^2} \tag{2.35}$$

which is an ellipse representing the case of free vibrations, The locus of critical points of type H' for a family of response spectra also appears in the figure as the dot-dash curve. Its equation is found to be

$$\frac{9\mu a_1^2}{4} = 1 - \frac{\Omega^2}{\omega^2} \tag{2.36}$$

by differentiation of Eq. (2.32).

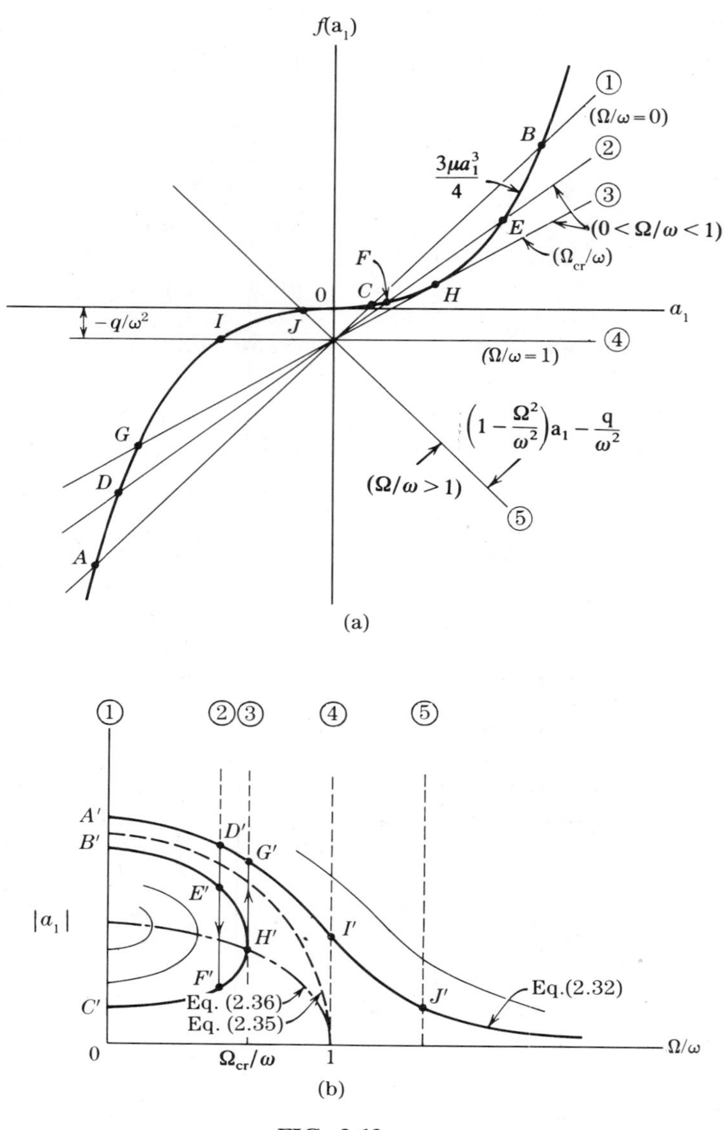

FIG. 2.13

The types of response spectra shown in Figs. 2.12*b* and 2.13*b* represent mathematical models of the *drop-jump phenomenon* observed in experiments upon nonlinear mechanical systems subjected to harmonic forcing functions [8]. In the case of the hardening spring, a gradual increase of the forcing frequency (starting with $\Omega = 0$) will result in steady-state amplitudes determined by the left-hand branch of the curve in Fig. 2.12*b* until some point such as F' is reached. Due to extraneous disturbances, the amplitude will drop suddenly from point F' to point H' on the curve, at which time the phase angle abruptly changes from 0° to 180°. A subsequent increase of the impressed frequency then causes the response to follow the ever-diminishing portion of the right-hand branch of the response spectrum. On the other hand, if the forcing frequency is slowly decreasing from a high value (greater than Ω_{cr}), the steady-state amplutide will gradually increase until the critical point E' is reached. Then the amplitude will jump from E' to D', and the phase angle abruptly changes from 180° to 0°. A subsequent decrease in the impressed frequency then causes the response to diminish along the path $D'C'B'A'$. We see that when the forcing frequency is decreasing, the amplitude must jump from point E' to point D' because there is only one solution for the case of $\Omega < \Omega_{cr}$.

It has been shown [10] that the dashed curve and the dot-dash curve in Fig. 2.12*b* delineate a region of instability, and that points on the curve $E'G'$ represent conditions that cannot be observed in physical systems. Thus, the point E' divides the right-hand branch of the spectrum into the unstable upper part $E'G'$ and the stable lower part $E'H'$.

In the case of a softening spring, slowly increasing the forcing frequency from $\Omega = 0$ to $\Omega > \Omega_{cr}$ will cause the steady-state amplitude to follow the path $C'F'H'G'I'J'$ in Fig. 2.13*b*. On the other hand, if we gradually decrease the impressed frequency from $\Omega > \Omega_{cr}$ to zero, the response follows the path $J'I'G'D'F'C'$. In the first instance there is a jump from point H' to point G', where the phase angle shifts from 0° to 180°. In the latter instance we have an abrupt drop from point D' to point F', where the phase angle reverses from 180° to 0°. In this case the unstable region is enclosed by the vertical line $\Omega/\omega = 0$ and the curves representing Eqs. (2.35) and (2.36). Thus, the point H' divides the left-hand branch of the spectrum into the unstable upper part $B'E'H'$ and the stable lower part $C'F'H'$.

Let us now consider the problem of Duffing's equation with viscous damping, where the damping force is proportional to velocity. In this case the equation of motion may be written as

$$\ddot{u} + 2n\dot{u} + \omega^2(u \pm \mu u^3) = q \cos \Omega t \tag{2.37}$$

The steady-state forced vibration will involve a phase angle ψ, and we assume as a first approximation

$$u = c_1 \cos(\Omega t - \psi) = a_1 \cos \Omega t + b_1 \sin \Omega t \tag{b}$$

where $c_1^2 = a_1^2 + b_1^2$, and $\tan \psi = b_1/a_1$. For determining the two constants a_1 and b_1 by the Ritz averaging method, we use two equations of the form

(2.28), which in his case become

$$\int_0^\tau [\ddot{u} + 2n\dot{u} + \omega^2(u \pm \mu u^3) - q \cos \Omega t] \cos \Omega t \, dt = 0$$

$$\int_0^\tau [\ddot{u} + 2n\dot{u} + \omega^2(u \pm \mu u^3) - q \cos \Omega t] \sin \Omega t \, dt = 0$$

Substituting Eq. (b) for u into these expressions and integrating, we obtain the following equations:

$$-a_1\Omega^2 + 2n\Omega b_1 + \omega^2 a_1 \pm \frac{3\omega^2 \mu a_1 c_1^2}{4} - q = 0 \qquad (c)$$

$$-b_1\Omega^2 - 2n\Omega a_1 + \omega^2 b_1 \pm \frac{3\omega^2 \mu b_1 c_1^2}{4} = 0 \qquad (d)$$

Then by using $a_1 = c_1 \cos \psi$ and $b_1 = c_1 \sin \psi$, we have

$$2n\Omega c_1 \sin \psi + \left(-\Omega^2 + \omega^2 \pm \frac{3\omega^2 \mu c_1^2}{4}\right) c_1 \cos \psi - q = 0 \qquad (e)$$

$$-2n\Omega c_1 \cos \psi + \left(-\Omega^2 + \omega^2 \pm \frac{3\omega^2 \mu c_1^2}{4}\right) c_1 \sin \psi = 0 \qquad (f)$$

Multiplying Eq. (e) by $\cos \psi$ and Eq. (f) by $\sin \psi$ and adding, we obtain

$$-\Omega^2 + \omega^2 \pm \frac{3\omega^2 \mu c_1^2}{4} = \frac{q}{c_1} \cos \psi \qquad (g)$$

On the other hand, multiplying Eq. (e) by $\sin \psi$ and Eq. (f) by $\cos \psi$ and subtracting, we find

$$2n\Omega = \frac{q}{c_1} \sin \psi \qquad (h)$$

By squaring and adding Eqs. (g) and (h) we have

$$\left(-\Omega^2 + \omega^2 \pm \frac{3\omega^2 \mu c_1^2}{4}\right)^2 + 4n^2\Omega^2 = \left(\frac{q}{c_1}\right)^2 \qquad (2.38)$$

and dividing Eq. (h) by Eq. (g) yields

$$\psi = \tan^{-1}\left(\frac{2n\Omega}{-\Omega^2 + \omega^2 \pm 3\omega^2 \mu c_1^2/4}\right) \qquad (2.39)$$

These two equations relate the amplitude c_1 and the phase angle ψ to the forcing frequency Ω for any given values of ω^2, μ. n, and q. If the damping factor n is set equal to zero, the phase angle becomes 0 or π, $b_1 = 0$, $c_1 = a_1$, and Eq. (2.38) becomes the same as Eq. (2.30), derived previously for the undamped case.

For the purpose of plotting response spectra, Eq. (2.38) can be rearranged and separated into two expressions, as follows:

$$\frac{3\mu c_1^3}{4} = \left(\frac{\Omega^2}{\omega^2} - 1\right)c_1 + \frac{q}{\omega^2}\sqrt{1 - \frac{(2n\Omega c_1)^2}{q^2}} \tag{2.40}$$

and

$$\frac{3\mu c_1^3}{4} = \left(1 - \frac{\Omega^2}{\omega^2}\right)c_1 - \frac{q}{\omega^2}\sqrt{1 - \frac{(2n\Omega c_1)^2}{q^2}} \tag{2.41}$$

where the first applies to the case of the hardening spring and the second to the softening spring. For zero damping these equations become the same as Eqs. (2.31) and (2.32). The right-hand sides of Eqs. (2.40) and (2.41) are no longer straight lines; so the graphical construction of response spectra is now more complicated than for the undamped cases. However, the general shapes of the resulting curves are similar, as indicated in Figs. 2.14 and 2.15 for the hardening and softening springs, respectively.

The dashed curve in Fig. 2.14 has the same definition as before (see Eq. 2.33), and the locus of points where response spectra intersect this free-vibration curve may be found by solving Eqs. (2.33) and (2.40) simultaneously. In this manner we find

$$(c_1)_{\text{res}} = \frac{q/2n\omega}{\Omega_{\text{res}}/\omega} = \frac{q}{2n\Omega_{\text{res}}} \tag{2.42}$$

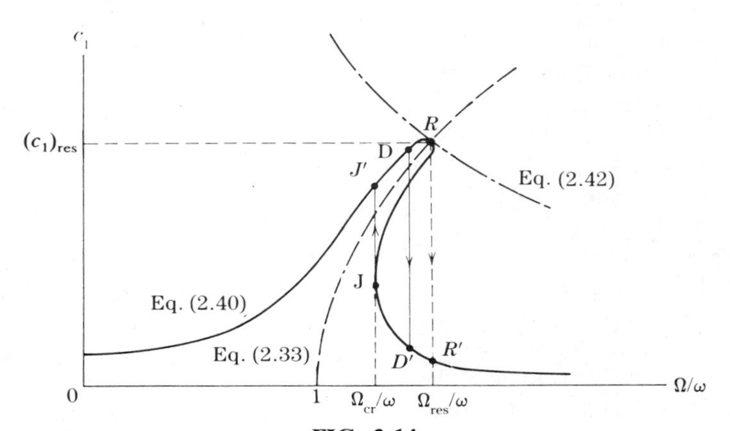

FIG. 2.14

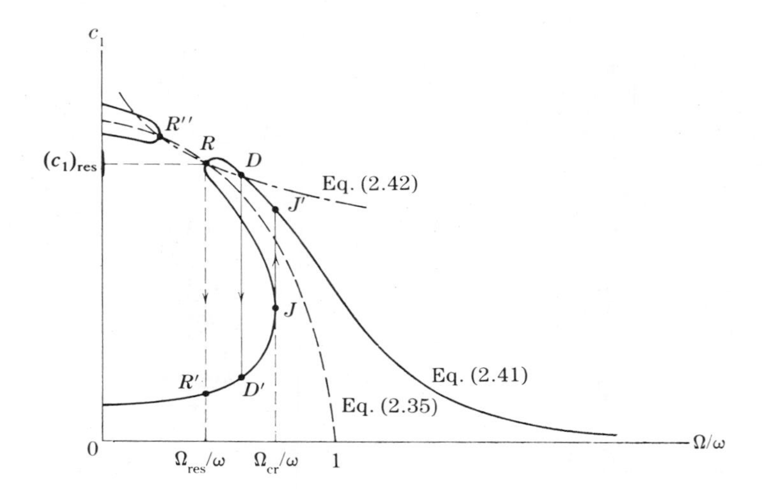

FIG. 2.15

in which Ω_{res} denotes the *resonance frequency* and $(c_1)_{res}$ is the *resonance amplitude.* Equation (2.42) represents a family of hyperbolas in the plane of c_1 versus Ω/ω (see the dot-dash line in Fig. 2.14). Thus, for any particular value of the quantity $q/2n\omega$, the hyperbola given by Eq. (2.42) may be constructed. Its intersection with the free-vibration curve determines approximately the maximum possible amplitude of steady-state forced vibrations that can be attained. If this maximum value is the only information desired, there is no need to construct the entire response spectrum; and the difficulty of plotting Eq. (2.40) is avoided.

Similarly, the dashed curve in Fig. 2.15 also has the same definition as before (see Eq. 2.35), and the locus of points where the response spectra intersect this free-vibration curve may be found by solving Eqs. (2.35) and (2.41) simultaneously. The resulting expression is the same as that in Eq. (2.42) given above, and the dot-dash curve in Fig. 2.15 exemplifies such a hyperbola. In this case there may be two intersections of the loci, indicating the existence of an upper branch of the response spectrum that has no physical meaning.

In lightly damped systems the theoretical condition of resonance, represented by the points labeled R in Figs. 2.14 and 2.15, may not actually be attained because of the drop-jump phenomenon. In each of these figures the possibility of a drop is indicated as occurring from point D to point D', whereas a jump always occurs from point J to point J'. Extraneous disturbances may cause drops to occur prematurely, thereby precluding the possibility of true resonances. Without such disturbing influences, the drop in each figure will be approximately from point R to point R' on the spectrum.

For damped systems the phase angle ψ varies continuously from 0 to π as the forcing frequency Ω varies from 0 to ∞. At resonance the phase angle is theoretically $\pi/2$, but in actuality it changes abruptly when a drop or a jump occurs. Such a change is from a value slightly below (or above) $\pi/2$ to a value slightly above (or below) $\pi/2$.

The Ritz averaging method has been applied successfully to a variety of problems involving free and forced vibrations of nonlinear systems. Excellent accuracy has been attained [9] using single-term approximations for systems with symmetric nth-order and piecewise-linear restoring forces. Duffing's equation for forced vibrations represents only one example of the former type. For systems with unsymmetric restoring forces, at least two-term approximations are required; and the algebraic difficulties increase rapidly.

2.5 PIECEWISE-LINEAR SYSTEMS

As was mentioned in Sec. 2.1, certain vibrational systems exhibit piecewise-linear characteristics. They are often simple to analyze and in some cases are amenable to exact solutions. Included in this category are systems with discontinuous linearly elastic springs, multilinear inelastic material behavior (including elastoplasticity), and Coulomb (or frictional) damping. In this section we will discuss their free-vibrational responses to initial conditions of displacement and velocity, forced vibrations caused by periodic functions, and transient responses due to arbitrary forcing functions.

Figure 2.16*a* shows a vibrational system with a mass located in a gap between two linearly elastic springs. If we measure the motions of the mass from the middle position, the static load–displacement diagram takes the form given in Fig. 2.16*b*. In this case of a *contact problem,* the period of free vibrations depends upon the magnitude of the gap as well as other parameters. Assume that at time $t = 0$ the initial displacement of the mass is zero but that its initial velocity is $\dot{u}_0$. The time required to cross the clearance u_1 is given by

$$t_1 = \frac{u_1}{\dot{u}_0} \tag{a}$$

After crossing the clearance, the mass contacts the right-hand spring; and subsequent motion is harmonic until the mass rebounds from the spring at a later time t_2. The time during which the velocity changes from $\dot{u}_0$ to zero is equal to one-quarter of the natural period of the mass m attached to a spring having a stiffness equal to k. Thus, the time t_m at which the maximum displacement occurs is

$$t_m = t_1 + \frac{\pi}{2\omega} = \frac{u_1}{\dot{u}_0} + \frac{\pi}{2}\sqrt{\frac{m}{k}} \tag{b}$$

Hence, the complete period of vibration of the actual system is

$$\tau = 4t_m = \frac{4u_1}{\dot{u}_0} + \frac{2\pi}{\omega} = \frac{4u_1}{\dot{u}_0} + 2\pi\sqrt{\frac{m}{k}} \tag{2.43}$$

Furthermore, the maximum displacement of the mass due to the initial velocity is equal to the sum of the clearance u_1 and the amplitude of the harmonic motion mentioned above. That is,

$$u_m = u_1 + \frac{\dot{u}_0}{\omega} = u_1 + \dot{u}_0\sqrt{\frac{m}{k}} \tag{2.44}$$

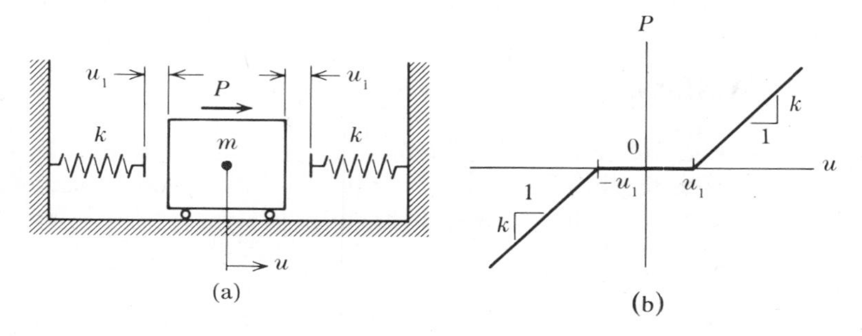

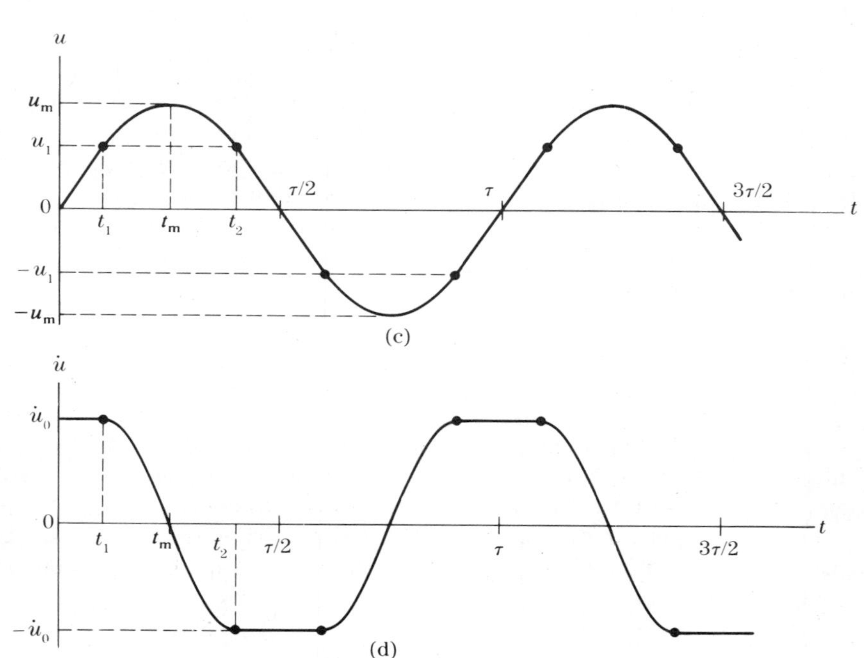

FIG. 2.16

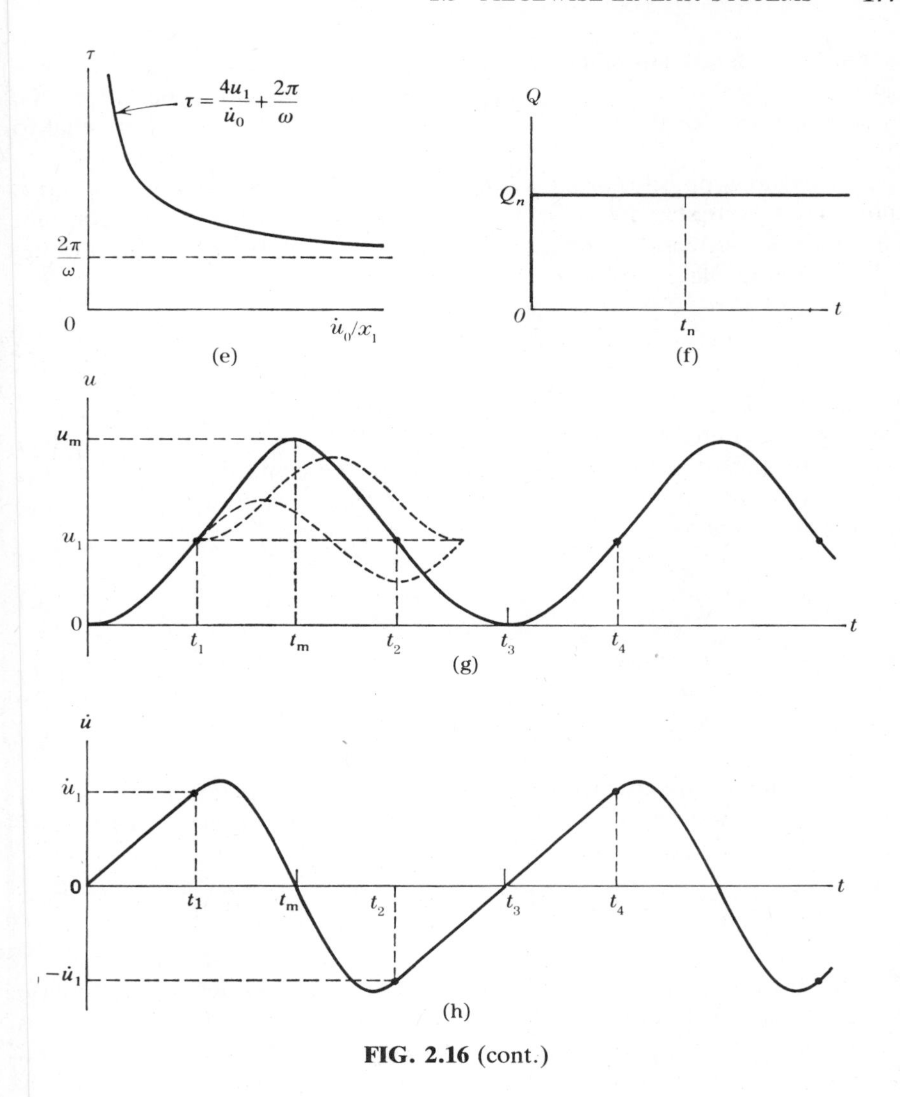

FIG. 2.16 (cont.)

Time histories of the displacement and the velocity for the undamped free vibrations of this system appear in Figs. 2.16*c* and 2.16*d*. Note that the velocity in the latter figure is constant whenever the mass is not in contact with one of the springs.

For particular values of the clearance u_1, the mass m, and the spring constant k, the period τ in Eq. (2.43) depends only upon the initial velocity $\dot{u}_0$. As the value of $\dot{u}_0$ approaches zero, the period tends to infinity; and as the velocity becomes large, the period approaches $2\pi/\omega$. Figure 2.16*e* contains a plot of this variation, and it is evident that such a system will tend to resonate with any periodic forcing function having a period greater than or equal to $2\pi/\omega$. However, the amplitude of such forced vibrations will always have an upper limit, except in the case where the period of the forcing function (or one of its harmonic components) has a period equal to $2\pi/\omega$.

Now let us suppose that the system in Fig. 2.16*a* is initially at rest and is subjected to the step function shown in Fig. 2.16*f*. Due to the constant force Q_n, the acceleration of the mass within the gap is $q_n = Q_n/m$; and the velocity and displacement are given by

$$\dot{u} = q_n t \qquad \text{and} \qquad u = q_n t^2/2 \tag{c}$$

The latter expression is represented by the parabola from $t = 0$ to $t = t_1$ in the displacement-time diagram plotted in Fig. 2.16*g*, and the former expression appears as the straight line within the same time range for the velocity plot in Fig. 2.16*h*. In this case the time t_1 required for the mass to cross the clearance u_1 is

$$t_1 = \sqrt{2u_1/q_n} \tag{d}$$

and the velocity at time t_1 is

$$\dot{u}_1 = q_n t_1 = \sqrt{2q_n u_1} \tag{e}$$

When the mass contacts the right-hand spring, it has the initial velocity $\dot{u}_1$, and the response of the system in the time range $t_1 \leqq t \leqq t_2$ can be written as

$$u = u_1 + \frac{\dot{u}_1}{\omega} \sin \omega(t - t_1) + \frac{Q_n}{k}[1 - \cos \omega(t - t_1)] \tag{f}$$

in which the second term on the right-hand side is due to the initial velocity, $\dot{u}_1$, while the last term is caused by the forcing function. These two terms are indicated as dashed curves in the time range $t_1 \leqq t \leqq t_2$ in Fig. 2.16*g*, and the sum of all three terms is shown as the bell-shaped solid curve. By differentiation of expression (*f*), we find that the maximum displacement u_m occurs at time t_m as follows:

$$t_m = t_1 + \frac{1}{\omega} \tan^{-1}\left(-\frac{\dot{u}_1/\omega}{Q_n/k} \right) \tag{2.45}$$

and that its magnitude is

$$u_m = u_1 + \frac{Q_n}{k} + \sqrt{\left(\frac{Q_n}{k}\right)^2 + \left(\frac{\dot{u}_1}{\omega}\right)^2} \tag{2.46}$$

Because the bell-shaped curve is symmetric with respect to the time t_m, it descends to u_1 again at time t_2, given by

$$t_2 = 2t_m - t_1 \tag{g}$$

Then the mass loses contact with the right-hand spring and follows a parabolic trajectory within the time range $t_2 \leqq t \leqq t_4$, as shown in Fig. 2.16*g*. The parabola is symmetric with respect to the time t_3, and the times of interest here are

$$t_3 = t_2 + t_1 \qquad \text{and} \qquad t_4 = t_2 + 2t_1 \tag{h}$$

At time t_4 the mass again contacts the right-hand spring, and the previous displacement pattern is repeated. The corresponding time history of velocity is given in Fig. 2.16*h*, and we see that the plot is linear whenever the mass is not in contact with the spring.

If the constant force Q_n were suddenly removed from the system at time t_n (see Fig. 2.16*f*), we would have a rectangular impulse instead of the simple step function. At the time t_n the system has some displacement u_n and some velocity $\dot{u}_n$ that may be ascertained from the diagrams in Figs. 2.16*g* and 2.16*h*. With these quantities as initial conditions, the subsequent free-vibration response may be deduced in a manner similar to that which resulted in the plots of Figs. 2.16*c* and 2.16*d*.

As a second example of a *piecewise-linear elastic system,* we shall consider the symmetric arrangement in Fig. 2.17*a*. This system is similar to the one in Fig. 2.16*a*, but it has additional springs to provide a restoring force for any nonzero displacement from the middle position. The static load–displacement diagram for this case is shown in Fig. 2.17*b*, in which the line through the origin has a slope k_1 and the steeper lines have slopes equal to k_2. If the displacement of the system never exceeds $\pm u_1$, the motion will be simple harmonic; but if the displacement is numerically larger than u_1, the motion becomes more complicated.

To study the free-vibrational characteristics of this system, let us assume that at time $t = 0$ the initial displacement of the mass is zero but that its initial velocity is $\dot{u}_0 > \omega_1 u_1$, where $\omega_1 = \sqrt{k_1/m}$. A velocity of this magnitude will cause excursions of the mass beyond the transition point at u_1, resulting in the time history of displacement plotted in Fig. 2.17*c*. In the time range $0 \leqq t \leqq t_1$ the displacement of this system is

$$u = \frac{\dot{u}_0}{\omega_1} \sin \omega_1 t \tag{i}$$

and the velocity is given by

$$\dot{u} = \dot{u}_0 \cos \omega_1 t \tag{j}$$

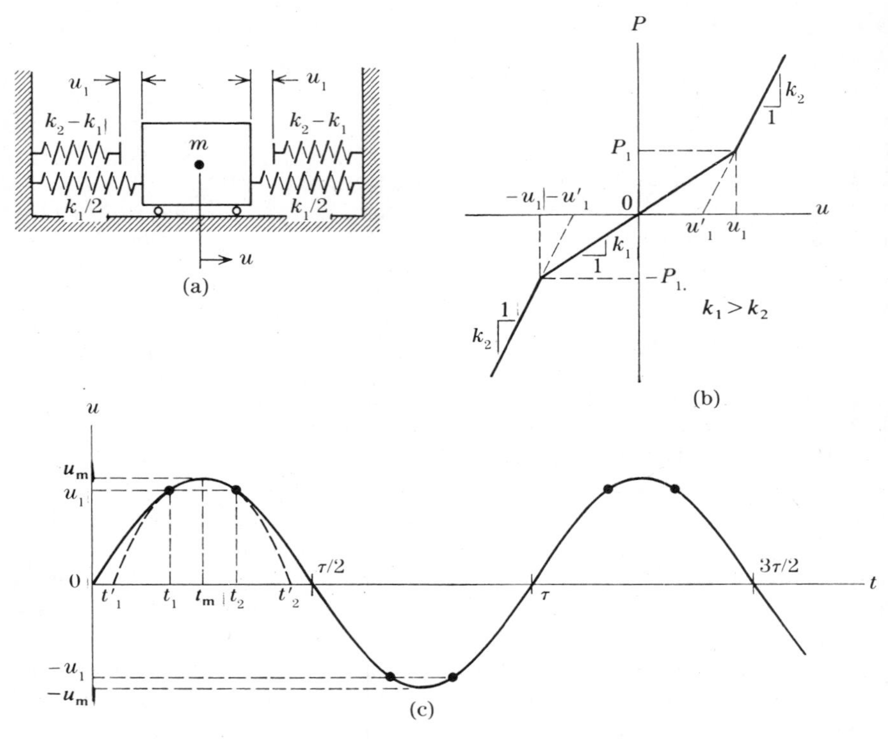

FIG. 2.17

At the time t_1 the mass has traveled to the limit u_1 of the first elastic range, and we have from Eq. (i)

$$t_1 = \frac{1}{\omega_1} \sin^{-1}\left(\frac{\omega_1 u_1}{\dot{u}_0}\right) \tag{k}$$

and the corresponding velocity given by Eq. (j) is

$$\dot{u}_1 = \dot{u}_0 \sqrt{1 - \left(\frac{\omega_1 u_1}{\dot{u}_0}\right)^2} \tag{l}$$

Then the mass contacts the upper right-hand spring in Fig. 2.17a, and the equation of motion becomes

$$m\ddot{u} + k_1 u + (k_2 - k_1)(u - u_1) = 0$$

or

$$m\ddot{u} + k_2 u = (k_2 - k_1)u_1 \tag{m}$$

The right-hand term in Eq. (m) is a constant that may be considered as a pseudostep function applied to the system with spring constant k_2. By this approach, the total response may be calculated as the sum of the effects of initial conditions at time t_1 and the influence of the pseudostep function. Thus, we have

$$
\begin{aligned}
u &= u_1 \cos \omega_2(t - t_1) + \frac{\dot{u}_1}{\omega_2} \sin \omega_2(t - t_1) + \frac{k_2 - k_1}{k_2} u_1[1 - \cos \omega_2(t - t_1)] \\
&= \left(1 - \frac{k_1}{k_2}\right) u_1 + \frac{k_1}{k_2} u_1 \cos \omega_2(t - t_1) + \frac{\dot{u}_1}{\omega_2} \sin \omega_2(t - t_1) \qquad (n)
\end{aligned}
$$

and the expression for velocity is

$$
\dot{u} = -\frac{k_1}{k_2} \omega_2 u_1 \sin \omega_2(t - t_1) + \dot{u}_1 \cos \omega_2(t - t_1) \qquad (o)
$$

The symbol ω_2 in Eqs. (n) and (o) represents the angular frequency $\omega_2 = \sqrt{k_2/m}$ for the harmonic motion in the second elastic range. From Eq. (o) we see that the time for maximum response is

$$
t_{\mathrm{m}} = t_1 + \frac{1}{\omega_2} \tan^{-1}\left(\frac{k_2 \dot{u}_1}{k_1 \omega_2 u_1}\right) \qquad (p)
$$

The first term in Eq. (n), given by

$$
u_1' = \left(1 - \frac{k_1}{k_2}\right) u_1 \qquad (q)
$$

represents the point where the steeper line in Fig. 2.17b crosses the positive u axis. Furthermore, the coefficient of the cosine term in Eq. (n) denotes the initial displacement

$$
\frac{k_1}{k_2} u_1 = u_1 - u_1'
$$

relative to the crossing point. The dashed curve in Fig. 2.17c may be imagined as extensions of the displaced harmonic motion in the second elastic range. A half-cycle of this motion begins at time t_1' and ends at time t_2'. These times are

$$
t_1' = t_{\mathrm{m}} - \frac{\pi}{2\omega_2} \qquad \text{and} \qquad t_2' = t_{\mathrm{m}} + \frac{\pi}{2\omega_2}
$$

and the time t_2 for the second point of tangency of the dashed and solid curve is

$$t_2 = t_1 + \frac{2}{\omega_2}\tan^{-1}\left(\frac{k_2\dot{u}_1}{k_1\omega_2 u_1}\right) \tag{r}$$

As in the previous case, the complete period of vibration may be calculated as

$$\tau = 4t_{\mathrm{m}} = 4t_1 + \frac{4}{\omega_2}\tan^{-1}\left(\frac{k_2\dot{u}_1}{k_1\omega_2 u_1}\right) \tag{2.47}$$

Furthermore, Eq. (n) shows that the maximum displacement u_{m} of the mass due to the given initial velocity is equal to the sum of u_1' and the amplitude of the displaced harmonic motion in the second elastic range. That is,

$$u_{\mathrm{m}} = \left(1 - \frac{k_1}{k_2}\right)u_1 + \sqrt{\left(\frac{k_1}{k_2}u_1\right)^2 + \left(\frac{\dot{u}_1}{\omega_2}\right)^2} \tag{2.48}$$

Of course, expression (l) must be substituted for $\dot{u}_1$ in Eqs. (2.47) and (2.48) in order to express them in terms of the initial velocity $\dot{u}_0$.

If the system in Fig. 2.17a is subjected to the harmonic forcing function $Q \sin \Omega t$, its equation of motion must be written for three ranges of u, as follows:

$$\text{for } -u_1 \leqq u \leqq u_1 \qquad m\ddot{u} + k_1 u = Q \sin \Omega t \tag{s}$$

$$\text{for } u_1 \leqq u \qquad m\ddot{u} + k_2 u = Q \sin \Omega t + (k_2 - k_1)u_1 \tag{t}$$

$$\text{for } u \leqq -u_1 \qquad m\ddot{u} + k_2 u = Q \sin \Omega t - (k_2 - k_1)u_1 \tag{u}$$

While these equations may be used to calculate the transient response of the system, they are not suitable for evaluating the steady-state response. Klotter [11] studied this case of forced vibrations using the Ritz averaging method with a one-term approximation. A family of response spectra for the stiffness ratio of $k_1/k_2 = \frac{1}{2}$ are presented in Fig. 2.18. To make the diagrams dimensionless, values of u/u_1 are plotted against Ω^2/ω^2 for several values of the load parameter $\zeta = Q/k_1u_1$.

As an illustration of an *inelastic piecewise-linear system*, let us turn our attention to the system in Fig. 2.19a, where a mass m is attached to the end of a vertical flexural member. It is assumed that the application of a horizontal load P causes only small displacements u and that the system behaves elastoplastically, as discussed previously in Sec. 2.1. That is, the static load–displacement diagram (Fig. 2.19b) has a slope equal to some nonzero constant k until a plastic hinge forms instantaneously near the base

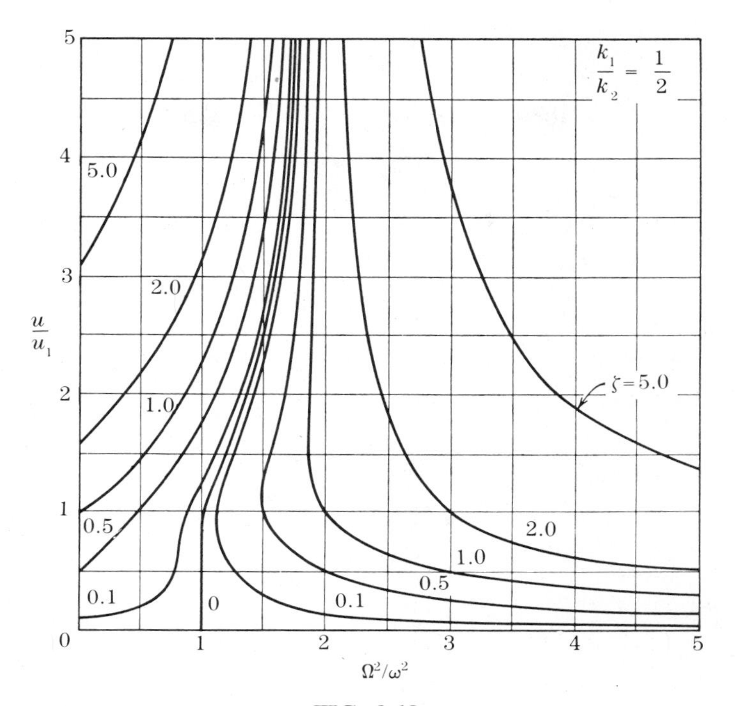

FIG. 2.18

of the member, at which time the ordinate of the diagram is equal to the maximum value P_m and the slope of the diagram becomes zero. Let us suppose that at time $t = 0$ the initial displacement of the mass is zero but that its initial velocity is $\dot{u}_0 > \omega u_1$, where $\omega = \sqrt{k/m}$ and u_1 is the displacement at time t_1 when the plastic hinge first forms. In the time range $0 \leqq t \leqq t_1$, the equations of displacement and velocity are

$$u = \frac{\dot{u}_0}{\omega} \sin \omega t \qquad \text{and} \qquad \dot{u} = \dot{u}_0 \cos \omega t \tag{v}$$

These functions are shown by the curves labeled ① in Figs. 2.19*c* and 2.19*d*, which contain time histories of displacement and velocity. Substituting $u_1 = P_m/k$ into Eqs. (v), we obtain

$$t_1 = \frac{1}{\omega} \sin^{-1}\left(\frac{P_m \omega}{k\dot{u}_0}\right) \qquad \text{and} \qquad \dot{u}_1 = \dot{u}_0 \sqrt{1 - \left(\frac{P_m \omega}{k\dot{u}_0}\right)^2} \tag{w}$$

After the plastic hinge has formed, the displacement and velocity of the

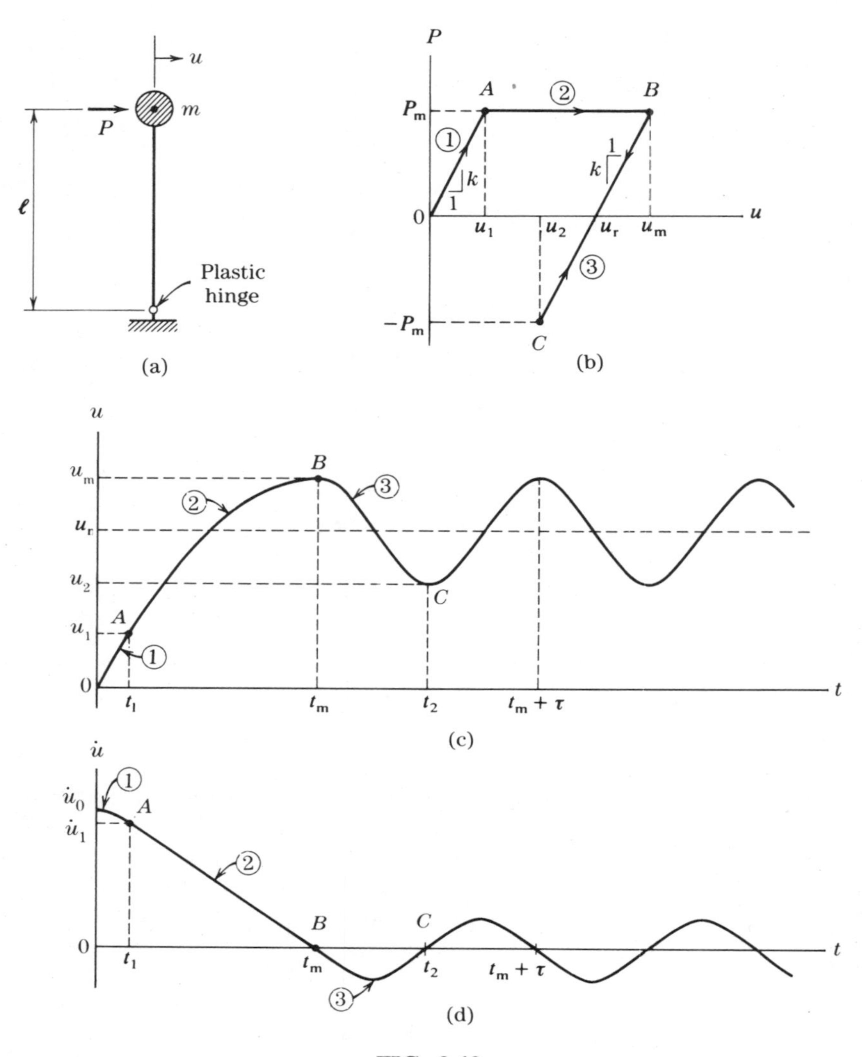

FIG. 2.19

mass in the time range $t_1 \leqq t \leqq t_m$ are given by the expressions

$$u = u_1 + \dot{u}_1(t - t_1) - \frac{P_m(t - t_1)^2}{2m}$$

and $$\tag{x}$$

$$\dot{u} = \dot{u}_1 - \frac{P_m(t - t_1)}{m}$$

These formulas (a parabola and a straight line) are plotted as the parts labeled ② in Figs. 2.19*c* and 2.19*d*. From Eqs. (x) we see that the maximum displacement u_m occurs at time

$$t_m = t_1 + \frac{m\dot{u}_1}{P_m} = t_1 + \frac{1}{\omega}\sqrt{\left(\frac{k\dot{u}_0}{P_m\omega}\right)^2 - 1} \tag{2.49}$$

and that its magnitude is

$$u_m = u_1 + \dot{u}_1(t_m - t_1) - \frac{P_m(t_m - t_1)^2}{2m} = \frac{P_m}{2k}\left[\left(\frac{k\dot{u}_0}{P_m\omega}\right)^2 + 1\right] \tag{2.50}$$

Subsequent to time t_m, the plastic hinge is no longer present; and the mass rebounds on an elastic branch of the load–displacement diagram, as indicated by the line labeled ③ in Fig. 2.19*b*. The free-vibration response in the time range $t_m \leqq t$ is the simple harmonic motion represented by the expression

$$u = u_m - \frac{P_m}{k}[1 - \cos \omega t] = u_r + \frac{P_m}{k}\cos \omega t \tag{y}$$

in which the *residual displacement* u_r, due to permanent deformation of the material, is

$$u_r = u_m - \frac{P_m}{k} = \frac{P_m}{2k}\left[\left(\frac{k\dot{u}_0}{P_m\omega}\right)^2 - 1\right] \tag{2.51}$$

Finally, the velocity in that time range is

$$\dot{u} = -\frac{P_m\omega}{k}\sin \omega t \tag{z}$$

Equations (y) and (z) are plotted as the curves labeled ③ in Figs. 2.19*c* and 2.19*d*. From the second form of expression (y), we see that the new equilibrium position for the *residual oscillation* is the residual displacement u_r (see Fig. 2.19*c*).

If the system in Fig. 2.19*a* is subjected to an impulsive loading, its response may be determined in a manner similar to that for an initial velocity described above. In particular, a rectangular impulse of magnitude Q_n and duration t_n may be considered; and by making a large number of solutions with various values for the parameters, we can construct the response spectra [12] shown in Fig. 2.20. These curves represent the maximum response ratio u_m/u_1 plotted against t_m/τ for various levels of the ratio P_m/Q_n, including the case of elastic response discussed in Sec. 1.14 (see Fig. 1.52*a*). When the lateral displacement u becomes significant in the

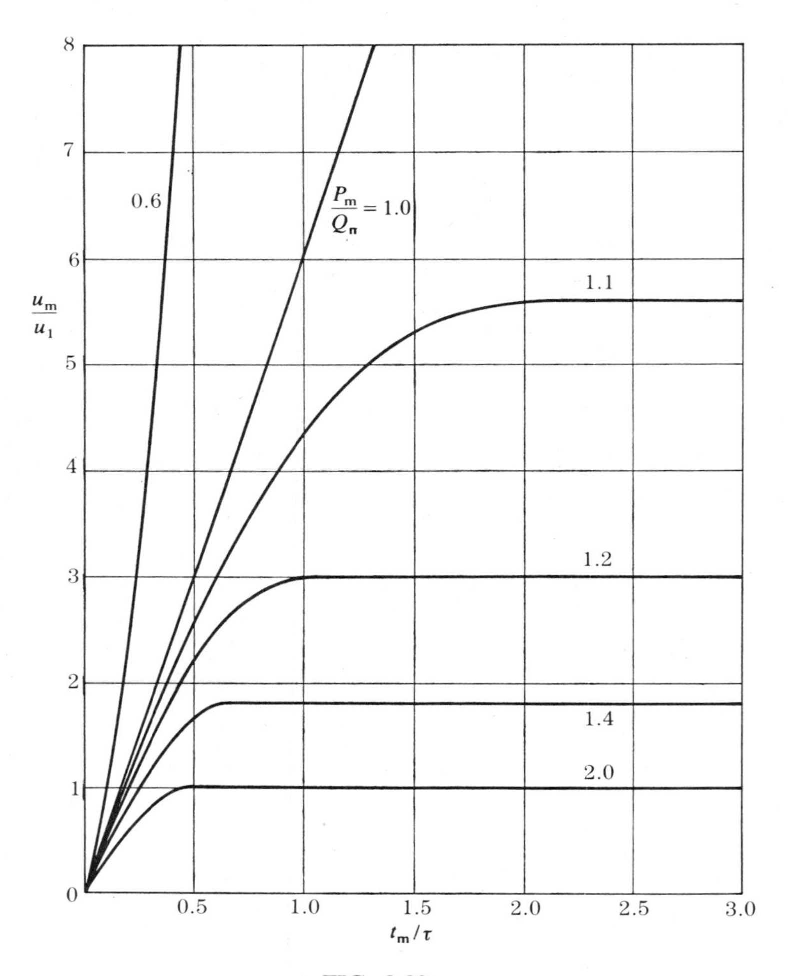

FIG. 2.20

geometry of Fig. 2.19*a*, the effect of the gravity-induced moment mgu should be taken into account as a quantity augmenting the moment $P\ell$ (due to the horizontal force).

To conclude our discussion of piecewise-linear problems, we shall examine a system with *Coulomb* (or *frictional*) *damping*, which was also described briefly in Sec. 2.1. As a preliminary matter, we recognize the fact that the block in Fig. 2.21*a* has more than one position of static equilibrium. In fact, it has an infinite number of such positions in the displacement range $-\Delta \leqq u \leqq \Delta$, where the symbol $\Delta = F/k$ denotes the position for which the frictional force F and the spring force $k\Delta$ are equal. Furthermore, the frictional force F always acts in the direction opposite to that of the velocity when the system is in motion. Therefore, we must write two differential

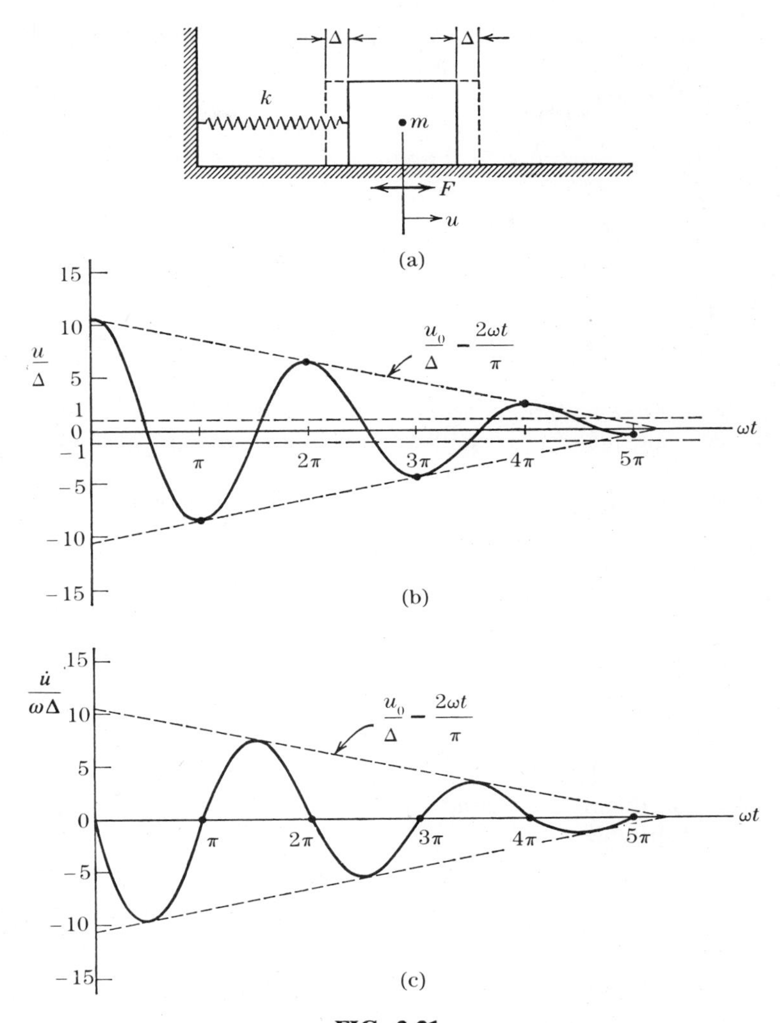

FIG. 2.21

equations for the free vibrations of this system. When the block in Fig. 2.21*a* is moving to the right, we have

$$m\ddot{u} + ku = -F \qquad (\dot{u} > 0) \tag{a'}$$

and when it is moving to the left, we have

$$m\ddot{u} + ku = F \qquad (\dot{u} < 0) \tag{b'}$$

Equations (a') and (b') may be written more succinctly as

$$m\ddot{u} + ku = -F \operatorname{sgn}(\dot{u}) \tag{c'}$$

in which the function $\mathrm{sgn}(\dot{u})$ yields a sign corresponding to that of the velocity $\dot{u}$.

Suppose that the block in Fig. 2.21*a* is displaced to the right by an amount $u_0 \gg \Delta$ and released with zero initial velocity. While the block is moving to the left, equation (*b'*) applies, and its solution is

$$u = u_0 \cos \omega t + \frac{F}{k}(1 - \cos \omega t) = \Delta + (u_0 - \Delta) \cos \omega t \tag{2.52a}$$

During this time the velocity is given by

$$\dot{u} = -\omega(u_0 - \Delta) \sin \omega t \tag{2.52b}$$

Thus, the motion within the time range $(0 \leqq t \leqq \pi/\omega)$ is harmonic, and $\omega = \sqrt{k/m}$ represents the angular frequency of this motion. At time $t = \pi/\omega$, the maximum negative displacement is $-(u_0 - 2\Delta)$, and the sense of the velocity changes from minus to plus. Then in the next time range $(\pi/\omega \leqq t \leqq 2\pi/\omega)$ the block moves to the right, and we have from Eq. (*a'*) the solution

$$u = -(u_0 - 2\Delta) \cos \omega t - \frac{F}{k}(1 - \cos \omega t) = -\Delta - (u_0 - 3\Delta) \cos \omega t \tag{2.53a}$$

while the velocity is

$$\dot{u} = \omega(u_0 - 3\Delta) \sin \omega t \tag{2.53b}$$

Thus, the motion in the second time range is also harmonic and has the same angular frequency ω.

Inspection of Eqs. (2.52a) and (2.53a) shows that the former represents an oscillation of amplitude $u_0 - \Delta$ about the right-hand equilibrium position (at $u = \Delta$), while the latter is an oscillation of amplitude $u_0 - 3\Delta$ about the left-hand equilibrium position (at $u = -\Delta$). Thus, in the time π/ω the numerical value of the maximum displacement decreases by the amount 2Δ, and in the time $2\pi/\omega$ it decreases by the amount 4Δ. By continuing the analysis, we find that the amplitude of the motion decreases by the amount 2Δ in every half-cycle until the amplitude becomes less than Δ. Then the block stops in one of its extreme positions within the displacement range $-\Delta \leqq u \leqq \Delta$.

Figures 2.21*b* and 2.21*c* contain dimensionless time histories of the displacement and the velocity of the block in Fig. 2.21*a* due to releasing it with the initial conditions $u_0 = 10.5\Delta$ and $\dot{u}_0 = 0$. The amplitudes of these plots diminish arithmetically in accordance with the expressions

$$\frac{u_{\mathrm{m}}}{\Delta} = \frac{\dot{u}_{\mathrm{m}}}{\omega\Delta} = \pm\left(\frac{u_0}{\Delta} - \frac{2\omega t}{\pi}\right) \tag{2.54}$$

These functions are indicated as the inclined dashed lines enveloping the curves in Figs. 2.21*b* and 2.21*c*. For the assumed initial conditions the mass comes to rest in the position $u = -0.5\Delta$ after oscillating for $2\frac{1}{2}$ cycles. Because the damping force changes abruptly at times $t = \pi/\omega$, $2\pi/\omega$, etc., the slopes of the curves in Fig. 2.21*c* are discontinuous at these times.

Although the frictional force changes direction in every half-cycle, the

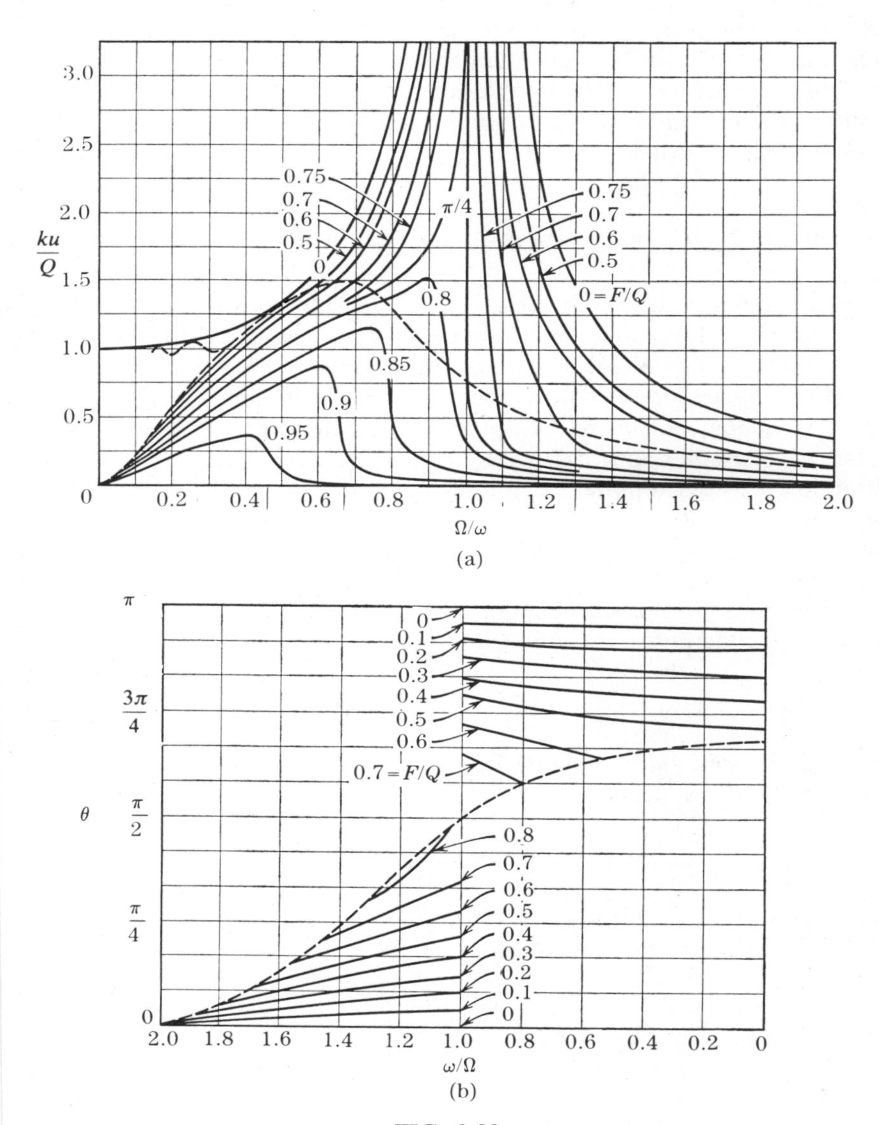

FIG. 2.22

system in Fig. 2.21*a* is not difficult to analyze for transient response due to impulsive excitations. However, its steady-state response to a periodic forcing function such as $Q \cos \Omega t$ is rather complicated. An approximate solution based on the concept of equivalent viscous damping was given in Sec. 1.10, but a rigorous solution is also available [13]. Diagrams of the magnification factor ku/Q and the phase angle θ for the forced vibrations of a system with frictional damping are presented in Figs. 2.22*a* and 2.22*b*. Each curve in these figures represents a different level of damping, as determined by the ratio F/Q. The dashed curve in Fig. 2.22*a* indicates the limit above which a nonstop oscillatory motion occurs. Below that limit the motion has intermittent stops during which the frictional force has some passive value P in the range $-F \leqq P \leqq F$.

2.6 NUMERICAL SOLUTIONS FOR NONLINEAR SYSTEMS

Using numerical integration methods, we can always solve approximately the equations of motion for nonlinear systems. Many of the well-known numerical methods involve extrapolation or interpolation formulas for responses tht are applied in a series of small but finite time steps. One effective technique consists of applying predictor and corrector formulas iteratively. Another approach uses direct linear extrapolation for equations of motion, expressed in incremental form. In this seciton we discuss both of these methods and also examine their stability and accuracy. Although the applications in this section have only one degree of freedom, we can extend any numerical method to systems with multiple degrees of freedom.

The general form of the equation of motion for a one-degree system with nonlinear characteristics may be expressed as

$$\ddot{u} = f(t, u, \dot{u}) \tag{2.55}$$

To begin the solution, we can evaluate the initial acceleration (at time $t = 0$) from Eq. (2.55), which yields

$$\ddot{u}_0 = f(0, u_0, \dot{u}_0) \tag{2.56}$$

The desired solution of Eq. (2.55) at any subsequent time t will be written in the symbolic form

$$u = F(t) \tag{2.57}$$

Figure 2.23 shows a graph of the numerical solution for the response, represented as a smooth curve in the u–t plane, even though it may actually have slight discontinuities. The symbols $u_0, u_1, u_2, \ldots, u_{j-1}, u_j, u_{j+1}, \ldots$ denote values of u at the time stations $t_0, t_1, t_2, \ldots, t_{j-1}, t_j, t_{j+1}, \ldots$ and

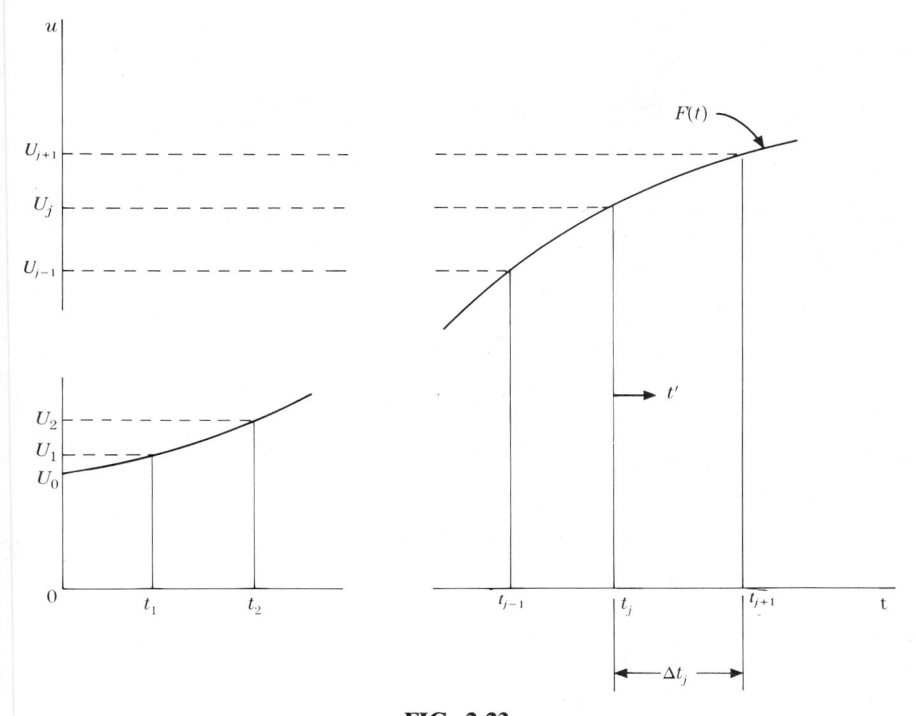

FIG. 2.23

so on. The time interval Δt_j between times t_j and t_{j+1} is usually taken to be of uniform duration Δt, although such a specialization is not necessary. The objective of the numerical integration process is to calculate the response at time t_{j+1} from previous responses by using approximation formulas.

Iterative Predictor–Corrector Methods

The technique we shall now explain is called the *predictor-corrector method* [14]. In each time step an *explicit* formula (a *predictor*) is first used to estimate the response at the end of the step. This is followed by one or more applications of an *implicit* formula (a *corrector*) to improve the results. Such an iterative procedure is often required for nonlinear problems, where physical properties can change in each cycle of iteration.

By an approach that we shall refer to as the *average-acceleration method,* the velocity $\dot{u}_{j+1}$ for a one-degree system at time t_{j+1} is approximated as

$$\dot{u}_{j+1} = \dot{u}_j + \tfrac{1}{2}(\ddot{u}_j + \ddot{u}_{j+1})\,\Delta t_j \tag{2.58}$$

in which $\dot{u}_j$ is the velocity at the preceding time station t_j (see Fig. 2.23). This formula, known to numerical analysts as the *trapezoidal rule,* implies that the acceleration in the step is taken to be the average of $\ddot{u}_j$ and $\ddot{u}_{j+1}$. Similarly, the displacement u_{j+1} at the end of the step is approximated by the trapezoidal rule with the expression

$$u_{j+1} = u_j + \tfrac{1}{2}(\dot{u}_j + \dot{u}_{j+1})\,\Delta t_j \tag{2.59}$$

where the velocity in the step is taken to be the average of $\dot{u}_j$ and $\dot{u}_{j+1}$. Substitution of Eq. (2.58) into Eq. (2.59) yields

$$u_{j+1} = u_j + \dot{u}_j\,\Delta t_j + \tfrac{1}{4}(\ddot{u}_j + \ddot{u}_{j+1})(\Delta t_j)^2 \tag{2.60}$$

When applying this method, we do not use Eq. (2.60) directly; but Eqs. (2.58) and (2.59) are used in succession. Because the value of $\ddot{u}_{j+1}$ is not known in advance, the approximation is said to be implicit; so the solution must be iterative within each step. The following recurrence equations represent the ith iteration of the jth step:

$$(\dot{u}_{j+1})_i = Q_j + \tfrac{1}{2}(\ddot{u}_{j+1})_{i-1}\,\Delta t_j \qquad (i > 1) \tag{2.61}$$

$$(u_{j+1})_i = R_j + \tfrac{1}{2}(\dot{u}_{j+1})_i\,\Delta t_j \qquad (i \geqq 1) \tag{2.62}$$

$$(\ddot{u}_{j+1})_i = f[t_{j+1}, (u_{j+1})_i, (\dot{u}_{j+1})_i] \qquad (i \geqq 1) \tag{2.63}$$

where

$$Q_j = \dot{u}_j + \tfrac{1}{2}\ddot{u}_j\,\Delta t_j \tag{2.64}$$

and

$$R_j = u_j + \tfrac{1}{2}\dot{u}_j\,\Delta t_j \tag{2.65}$$

This iterative procedure is not self-starting because it requires a supplementary formula for determining the first estimate of $\dot{u}_{j+1}$ in each time step. After evaluating the initial acceleration from Eq. (2.56), we may start the iteration for the first step by approximating $\dot{u}_1$ with *Euler's extrapolation formula,* as follows:

$$(\dot{u}_1)_1 = \dot{u}_0 + \ddot{u}_0\,\Delta t_0 \qquad (j = 0; i = 1) \tag{2.66}$$

Then the first approximations for u_1 and $\ddot{u}_1$ are obtained from Eqs. (2.62) and (2.63). All subsequent iterations for the first time step involve the repetitive use of Eqs. (2.61), (2.62), and (2.63).

To start the iteration in the jth time step, we may again apply Euler's formula to determine a first estimate of $\dot{u}_{j+1}$ as

$$(\dot{u}_{j+1})_1 = \dot{u}_j + \ddot{u}_j\,\Delta t_j \qquad (i = 1) \tag{2.67}$$

Both Eqs. (2.66) and (2.67) imply constant values of the accelerations within the steps. To improve the accuracy of the results for the first iteration of the jth step, we can use the slightly more elaborate formula that is valid only for uniform time steps

$$(\dot{u}_{j+1})_1 = \dot{u}_{j-1} + 2\ddot{u}_j\,\Delta t \qquad (i = 1) \tag{2.68}$$

This expression spans two equal time steps from t_{j-1} to t_{j+1} (see Fig. 2.23) and utilizes the midpoint acceleration at time t_j.

Equations (2.67) and (2.68) are called explicit predictors because they provide estimates of $\dot{u}_{j+1}$ in terms of previous values of $\dot{u}$ and $\ddot{u}$. On the other hand, Eq. (2.58) is referred to as an implicit corrector that yields an improved value of $\dot{u}_{j+1}$ after an estimation of $\ddot{u}_{j+1}$ has been obtained. The method described here involves one application of a predictor, followed by repetitive applications of the corrector.

An iterative type of solution requires some criterion for stopping or changing the step size, such as a limit on the number of iterations. A convenient method for measuring the *rate of convergence* is to control the number of significant figures in u_{j+1}, as follows:

$$|(u_{j+1})_i - (u_{j+1})_{i-1}| < \epsilon_u\,|(u_{j+1})_i| \tag{2.69}$$

where ϵ_u is some small number selected by the analyst. For example, an *accuracy* of approximately four digits may be specified by taking $\epsilon_u = 0.0001$. That level of accuracy is used in the numerical examples of this section.

Another implicit approach for approximating responses is known as the *linear-acceleration method.* As indicated by its name, this technique has the assumption that the acceleration varies linearly within each time step. Thus, an expression for $\ddot{u}$ during the step Δt_j may be written

$$\ddot{u}(t') = \ddot{u}_j + (\ddot{u}_{j+1} - \ddot{u}_j)\frac{t'}{\Delta t_j} \tag{2.70}$$

where t' is measured from the beginning of the step (see Fig. 2.23). If the acceleration varies linearly, the corresponding velocity and displacement will vary quadratically and cubically with time. Therefore,

$$\dot{u}(t') = \dot{u}_j + \ddot{u}_j t' + (\ddot{u}_{j+1} - \ddot{u}_j)\frac{(t')^2}{2\,\Delta t_j} \tag{2.71}$$

and

$$u(t') = u_j + \dot{u}_j t' + \ddot{u}_j\frac{(t')^2}{2} + (\ddot{u}_{j+1} - \ddot{u}_j)\frac{(t')^3}{6\,\Delta t_j} \tag{2.72}$$

At the end of the step the velocity and displacement become

$$\dot u_{j+1} = \dot u_j + \tfrac{1}{2}(\ddot u_j + \ddot u_{j+1})\,\Delta t_j \tag{2.73}$$

and

$$u_{j+1} = u_j + \dot u_j\,\Delta t_j + \tfrac{1}{6}(2\ddot u_j + \ddot u_{j+1})(\Delta t_j)^2 \tag{2.74}$$

Equation (2.73) is the same as Eq. (2.58) of the average-acceleration method, but Eq. (2.74) is slightly different from its counterpart in Eq. (2.60).

We will apply the linear-acceleration method in a manner analogous to that for the average-acceleration approach. Because Eq. (2.73) is the same as Eq. (2.58), the recurrence expression for the ith iteration of $\dot u_{j+1}$ is the same as that in Eq. (2.61). To obtain a direct relationship between u_{j+1} and $\dot u_{j+1}$, we solve for $\ddot u_{j+1}$ in Eq. (2.73) and substitute the result into Eq. (2.74), which yields

$$u_{j+1} = u_j + \tfrac{1}{3}(2\dot u_j + \dot u_{j+1})\,\Delta t_j + \tfrac{1}{6}u_j(\Delta t_j)^2 \tag{2.75}$$

Thus, we form the recurrence equation for the ith iteration of u_{j+1} as

$$(u_{j+1})_i = R_j^* + \tfrac{1}{3}(\dot u_{j+1})_i\,\Delta t_j \qquad (i \geqq 1) \tag{2.76}$$

where

$$R_j^* = u_j + \tfrac{2}{3}\dot u_j\,\Delta t_j + \tfrac{1}{6}\ddot u_j(\Delta t_j)^2 \tag{2.77}$$

The formulas given earlier [see Eqs. (2.66) and (2.67) or (2.68)] may be used again to start the iteration in each step.

It is well known that the linear-acceleration method is somewhat more accurate than the average-acceleration method. However, it has been shown [15] that the former technique is only *conditionally stable.* Therefore, the solution diverges if the time step is too large. On the other hand, the average-acceleration method is *unconditionally stable,* although less accurate. The topics of stability and accuracy are discussed further at the end of this section.

Example 1. To demonstrate the preditor-corrector methods, let us first consider a linear one-degree system subjected to a step force P_1, starting from rest. In this case the equation of motion is

$$m\ddot u + c\dot u + ku = P_1 \tag{a}$$

Rearranging Eq. (a) into the form of Eq. (2.55) produces

$$\ddot u = \frac{1}{m}(P_1 - ku - c\dot u) \tag{b}$$

For zero damping ($c = 0$), we can express the mass m in terms of the stiffness k and the period τ as

$$m = \frac{k}{\omega^2} = k\left(\frac{\tau}{2\pi}\right)^2 \tag{c}$$

At time $t_0 = 0$, we have $u_0 = \dot{u}_0 = 0$, so the initial acceleration is

$$\ddot{u}_0 = \frac{P_0}{m} = \frac{P_1}{m} \tag{d}$$

We shall calculate the approximate response of the system using 20 uniform time steps of duration $\Delta t = \tau/20$.

To apply the average-acceleration method, we start the first iteration in the first time step using Eq. (2.66) to estimate the velocity at time $t_1 = \Delta t = \tau/20$, as follows:

$$(\dot{u}_1)_1 = \dot{u}_0 + \ddot{u}_0\,\Delta t = 0 + \left(\frac{P_1}{m}\right)\left(\frac{\tau}{20}\right) = 0.05\,\frac{P_1\tau}{m} \tag{e}$$

Then the displacement at time t_1 is found from Eq. (2.62) to be

$$(u_1)_1 = R_0 + \frac{1}{2}(\dot{u}_1)_1\,\Delta t = 0 + \frac{0.05}{2k}\left(\frac{P_1\tau^2}{20}\right)\left(\frac{2\pi}{\tau}\right)^2 = 0.04935\,\frac{P_1}{k} \tag{f}$$

Next, we obtain the acceleration at time t_1 from Eq. (2.63) [or Eq. (b)] as

$$(\ddot{u}_1)_1 = \frac{1}{m}(P_1 - ku_1)_1 = \frac{P_1}{m}(1 - 0.04935) = 0.9507\,\frac{P_1}{m} \tag{g}$$

For the second iteration in the first time step, Eqs. (2.61), (2.62), and (2.63) yield

$$(\dot{u}_1)_2 = (1 + 0.9507)\,\frac{P_1\tau}{40m} = 0.04877\,\frac{P_1\tau}{m}$$

$$(u_1)_2 = 0 + \frac{0.04877}{2k}\left(\frac{P_1\tau^2}{20}\right)\left(\frac{2\pi}{\tau}\right)^2 = 0.04813\,\frac{P_1}{k}$$

$$(\ddot{u}_1)_2 = \frac{P_1}{m}(1 - 0.04813) = 0.9519\,\frac{P_1}{m}$$

Third iteration:

$$(\dot{u}_1)_3 = (1 + 0.9519)\frac{P_1\tau}{40m} = 0.04880\frac{P_1\tau}{m}$$

$$(u_1)_3 = 0 + \frac{0.04880}{2k}\left(\frac{P_1\tau^2}{20}\right)\left(\frac{2\pi}{\tau}\right)^2 = 0.04816\frac{P_1}{k}$$

$$(\ddot{u}_1)_3 = \frac{P_1}{m}(1 - 0.04816) = 0.9518\frac{P_1}{m}$$

Fourth iteration:

$$(\dot{u}_1)_4 = (1 + 0.9518)\frac{P_1\tau}{40m} = 0.04880\frac{P_1\tau}{m}$$

$$(u_1)_4 = 0 + \frac{0.04880}{2k}\left(\frac{P_1\tau^2}{20}\right)\left(\frac{2\pi}{\tau}\right)^2 = 0.04816\frac{P_1}{k}$$

$$(\ddot{u}_1)_4 = \frac{P_1}{m}(1 - 0.04816) = 0.9518\frac{P_1}{m}$$

TABLE 2.1 Response for Example 1 Using Iteration Methods[a]

	Average-acceleration Method		*Linear-acceleration Method*	
j	n_i	Approx. u	n_i	Approx. u
1	4	0.04816	4	0.04855
2	4	0.1880	4	0.1895
3	4	0.4061	4	0.4091
4	4	0.6813	4	0.6861
5	4	0.9873	3	0.9936
6	4	1.294	3	1.302
7	4	1.573	3	1.581
8	3	1.797	3	1.803
9	3	1.944	3	1.948
10	3	2.000	3	2.000
11	3	1.959	3	1.955
12	3	1.827	3	1.818
13	3	1.614	3	1.601
14	4	1.343	3	1.326
15	4	1.038	3	1.019
16	4	0.7300	4	0.7105
17	4	0.4478	4	0.4299
18	4	0.2187	4	0.2047
19	4	0.06497	4	0.05665
20	5	0.00126	5	0.00025.

[a] Tabulated values to be multiplied by P_1/k.

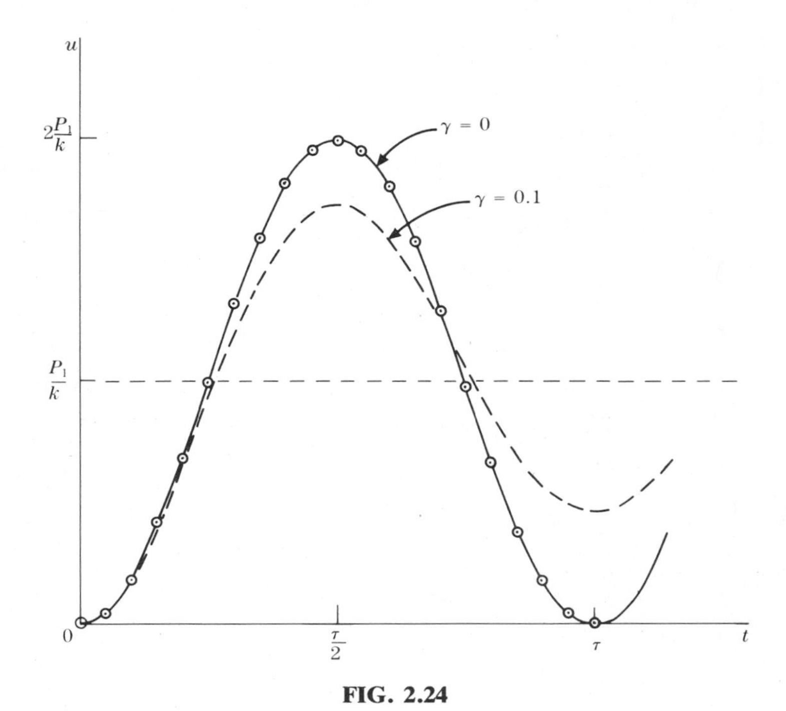

FIG. 2.24

In the fourth iteration, we see that the response has converged to within four significant digits. Results for 20 time steps are given in Table 2.1, along with the number of iterations n_i required in each step. Also, Fig. 2.24 shows plots of the approximate results for both zero damping and a damping ratio of $\gamma = 0.1$. For the scale used in the figure, the plotted curves are indistinguishable from exact responses.

Next, we apply the linear-acceleration method, using Eqs. (2.76) and (2.77) in place of Eqs. (2.62) and (2.65). In this case, the approximate responses are calculated somewhat more accurately and with fewer iterations. Results for this second analysis also appear in Table 2.1. Having tested these methods with a linear problem, we shall now apply them to nonlinear examples.

Example 2. The simple pendulum in Fig. 2.3 (see Sec. 2.1) has the nonlinear equation of motion

$$\ddot{\phi} + \omega^2 \sin \phi = 0 \tag{2.4a}$$

where $\omega^2 = g/L$. If we let L be numerically equal to g, then $\omega^2 = 1$; additionally, we will take the initial conditons to be $\phi_0 = \pi/2$ and $\dot{\phi}_0 = 0$. In Sec. 2.2 the exact expression for the period of the pendulum was derived as

TABLE 2.2 Results for Example 2

j	t_j (sec)	n_i	Approx. ϕ (rad) Average Method	Linear Method
0	0	—	1.5708	1.5708
1	0.1	2	1.5536	1.5536
2	0.2	2	1.5021	1.5021
3	0.3	2	1.4163	1.4163
4	0.4	3	1.2967	1.2966
5	0.5	3	1.1442	1.1440
6	0.6	3	0.9608	0.9603
7	0.7	3	0.7496	0.7487
8	0.8	3	0.5154	0.5140
9	0.9	3	0.2646	0.2627
10	1.0	4	0.0051	0.0025
11	1.1	2	−0.2546	−0.2577
12	1.2	3	−0.5059	−0.5093
13	1.3	3	−0.7409	−0.7444
14	1.4	3	−0.9530	−0.9564
15	1.5	3	−1.1376	−1.1407
16	1.6	3	−1.2913	−1.2939
17	1.7	3	−1.4123	−1.4142
18	1.8	2	−1.4994	−1.5007
19	1.9	2	−1.5522	−1.5529
20	2.0	2	−1.5708	−1.5708

Eq. (2.12). Using the initial condition $\phi_0 = \phi_m = \pi/2$, we obtain from a table of elliptic integrals $K = F(k, \pi/2) = 1.8541$. Hence, the quarter-period is $\tau/4 = K/\omega = 1.8541$ sec. Alternatively, if we take $\omega = K = 1.8541$, then $\tau/4 = 1$ sec; and this relationship will be used because it is simpler.

Thus, the equation to be solved numerically is

$$\ddot{\phi} = -\omega^2 \sin \phi = -3.4377 \sin \phi \tag{h}$$

and the initial conditions give

$$\ddot{\phi}_0 = -\omega^2 \sin \pi/2 = -3.4377 \tag{i}$$

Table 2.2 contains the results for 20 time steps (with $\Delta t = 0.1$ sec), using the average-acceleration and the linear-acceleration methods. The value of the angle ϕ should be equal to zero at time t_{10}, and the linear-acceleration method produces the smaller of the two approximate values of ϕ_{10}. However, both methods yield the correct final value of $\phi_{20} = -1.5708$ rad. A plot of the approximate values of ϕ against time is shown in Fig. 2.25.

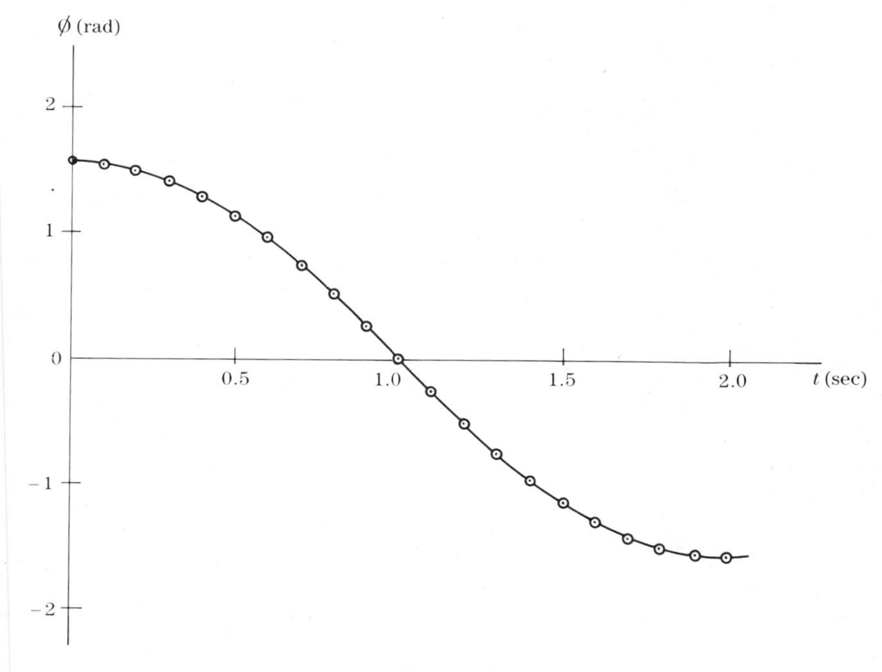

FIG. 2.25

Example 3. As a second nonlinear example, consider the following equation of motion for a system with a hardening spring:

$$m\ddot{u} + c\dot{u} + k(u + \alpha u^3) = Q(t) \tag{j}$$

Or

$$\ddot{u} + 2n\dot{u} + \omega^2(u + \alpha u^3) = q(t) \tag{k}$$

From Prob. 2.2-1 (see Sec. 2.2) we will take

$$m = 100 \text{ lb-sec}^2/\text{in.} \qquad k = 400 \text{ lb/in.}$$

$$\omega^2 = 4 \text{ sec}^{-2} \qquad \alpha = 2 \text{ in.}^{-2} \qquad c = Q(t) = 0$$

For these values of the parameters, the equation to be solved numerically from Prob. 2.2-1 becomes

$$\ddot{u} = -\omega^2(u + \alpha u^3) = -4(u + 2u^3) \tag{l}$$

Substitution of the initial conditions $u_0 = 0$ and $\dot{u}_0 = 10$ in./sec into Eq. (l) yields

$$\ddot{u}_0 = -4(0 + 0) = 0 \tag{m}$$

Results for 20 time steps (with $\Delta t = 0.025$ sec), using both methods, are given in Table 2.3; and a plot of the approximate values of u against time appears in Fig. 2.26. The maximum displacement of approximately 2.12 in. occurs in the vicinity of time $t_{12} = 0.30$ sec, as it should.

While the iterative examples in this section can be solved using a desk calculator, the numerical operations become rather tedious; and the use of a digital computer is preferred. Special-purpose computer programs named **AVAC1**, **AVAC2A**, and **AVAC3A** were used for the calculations by the average-acceleration method in Examples 1, 2, and 3, respectively. These programs are treated as modifications of Program **LINFORCE** in Appendix B. They can easily be converted to programs named **LINAC1**, **LINAC2A**, and **LINAC3A** for the linear-acceleration method by changing only a few lines in each program. Furthermore, most of the numerical integration problems at the end of the chapter can be solved with variations of these same programs.

Direct Linear Extrapolation Methods

If linear equations of motion are converted to incremental form, they can be solved numerically as approximations for nonlinear problems, in which the

TABLE 2.3 Results for Example 3

j	t_j (sec)	n_i	Approx. u (in.)	
			Average Method	Linear Method
0	0	—	0	0
1	0.025	3	0.2498	0.2499
2	0.050	3	0.4988	0.4990
3	0.075	3	0.7457	0.7461
4	0.100	3	0.9884	0.9890
5	0.125	3	1.2234	1.2243
6	0.150	3	1.4457	1.4472
7	0.175	3	1.6490	1.6511
8	0.200	2	1.8256	1.8282
9	0.225	2	1.9673	1.9702
10	0.250	3	2.0665	2.0694
11	0.275	3	2.1171	2.1196
12	0.300	3	2.1158	2.1175
13	0.325	3	2.0628	2.0632
14	0.350	3	1.9614	1.9603
15	0.375	3	1.8178	1.8151
16	0.400	2	1.6397	1.6355
17	0.425	3	1.4353	1.4298
18	0.450	3	1.2122	1.2058
19	0.475	3	0.9768	0.9696
20	0.500	3	0.7339	0.7263

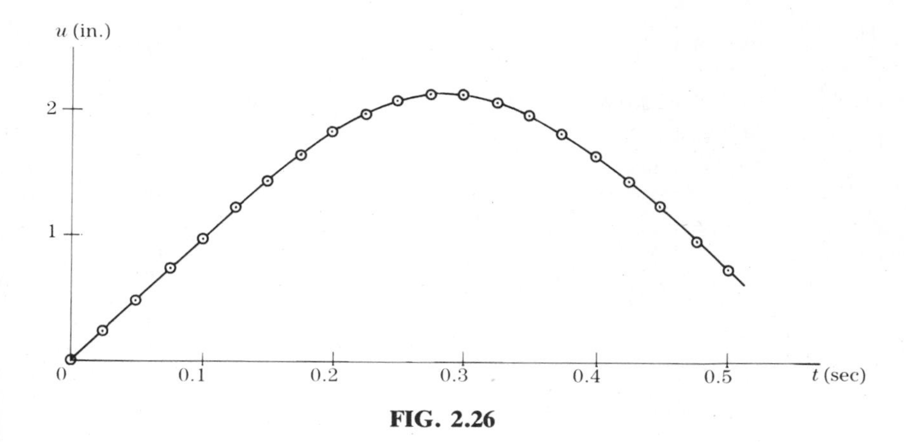

FIG. 2.26

mass, damping, and stiffness coefficients are variables. It is feasible to set up such equations for incremental accelerations, velocities, or displacements. Here we will use the *incremental displacement* Δu_j as the unknown and solve a *pseudostatic problem* for each time step.

At time t_j (see Fig. 2.23), the damped equation of motion for a one-degree system may be written as

$$m_j \ddot{u}_j + c_j \dot{u}_j + k_j u_j = P_j \tag{2.78}$$

similarly, at time $t_{j+1} = t_j + \Delta t_j$, the equation of motion becomes:

$$m_j(\ddot{u}_j + \Delta\ddot{u}_j) + c_j(\dot{u}_j + \Delta\dot{u}_j) + k_j(u_j + \Delta u_j) = P_j + \Delta P_j \tag{2.79}$$

This expression implies that the variable coefficients m_j, c_j, and k_j are assumed to remain constant within the step, which is true as $\Delta t_j \rightarrow 0$. Subtraction of Eq. (2.78) from Eq. (2.79) produces the *incremental equation of motion*

$$m_j \,\Delta\ddot{u}_j + c_j \,\Delta\dot{u}_j + k_j \,\Delta u_j = \Delta P_j \tag{2.80}$$

This equation will be used in conjunction with *Newmark's generalized acceleration method,* which is explained in the following discussion.

In his 1959 paper, Newmark [15] generalized certain numerical integration methods that had been in use up to that time. He presented equations for approximating the velocity and displacement of a one-degree system at time t_{j+1}, as follows:

$$\dot{u}_{j+1} = \dot{u}_j + [(1-\gamma)\ddot{u}_j + \gamma\ddot{u}_{j+1}]\,\Delta t_j \tag{2.81}$$

$$u_{j+1} = u_j + \dot{u}_j\,\Delta t_j + [(\tfrac{1}{2} - \beta)\ddot{u}_j + \beta\ddot{u}_{j+1}](\Delta t_j)^2 \tag{2.82}$$

The parameter γ in Eq. (2.81) produces *numerical* (or *algorithmic*) *damping* within the time step Δt_j. If γ is taken to be less than $\frac{1}{2}$, an artificial negative damping results. On the other hand, if γ is greater than $\frac{1}{2}$, such damping is positive. To avoid numerical damping altogether, the value of γ must be equal to $\frac{1}{2}$; and Eq. (2.81) becomes the trapezoidal rule.

The parameter β in Eq. (2.82) controls the variation of acceleration within the time step. For this reason the technique is referred to as Newmark's generalized acceleration method (or Newmark's β method). For example, if we take $\beta = 0$, Eq. (2.82) becomes

$$u_{j+1} = u_j + \dot{u}_j \Delta t_j + \tfrac{1}{2}\ddot{u}_j(\Delta t_j)^2 \tag{2.83}$$

Use of this formula is known as the *constant-acceleration method,* because the acceleration $\ddot{u}_j$ at the beginning of the time step Δt_j is taken to be constant within the step. Equation (2.83) also corresponds to a truncated Taylor series that results from using Euler's formula [see Eq. (2.67)] for velocity and the trapezoidal rule for displacement.

If we let $\beta = \frac{1}{4}$, Eq. (2.82) yields

$$u_{j+1} = u_j + \dot{u}_j \Delta t_j + \tfrac{1}{4}(\ddot{u}_j + \ddot{u}_{j+1})(\Delta t_j)^2 \tag{2.84}$$

This expression is the same as that in Eq. (2.60) for the *average-acceleration method.* When we take $\beta = \frac{1}{6}$, Eq. (2.82) produces

$$u_{j+1} = u_j + \dot{u}_j \Delta t_j + \tfrac{1}{6}(2\ddot{u}_j + \ddot{u}_{j+1})(\Delta t_j)^2 \tag{2.85}$$

In this case the formula is identical to Eq. (2.74) for the *linear-acceleration method.*

Now let us cast the Newmark-β method into the form described previously for direct linear extrapolation. For this purpose, Eq. (2.81) is restated in the incremental form

$$\begin{aligned} \Delta\ddot{u}_j &= [(1-\gamma)\ddot{u}_j + \gamma\ddot{u}_{j+1}]\,\Delta t_j \\ &= (\ddot{u}_j + \gamma\,\Delta\ddot{u}_j)\,\Delta t_j \end{aligned} \tag{2.86}$$

Similarly, Eq. (2.82) is rewritten as

$$\begin{aligned} \Delta u_j &= \dot{u}_j \Delta t_j + [(\tfrac{1}{2}-\beta)\ddot{u}_j + \beta\ddot{u}_{j+1}](\Delta t_j)^2 \\ &= \dot{u}_j \Delta t_j + (\tfrac{1}{2}\ddot{u}_j + \beta\,\Delta\ddot{u}_j)(\Delta t_j)^2 \end{aligned} \tag{2.87}$$

Solving for $\Delta\ddot{u}_j$ in Eq. (2.87) gives us

$$\Delta\ddot{u}_j = \frac{1}{\beta(\Delta t_j)^2}\Delta u_j - \frac{1}{\beta\,\Delta t_j}\dot{u}_j - \frac{1}{2\beta}\ddot{u}_j \tag{2.88}$$

Substituting Eq. (2.88) into Eq. (2.86), we find

$$\Delta \dot{u}_j = \frac{\gamma}{\beta\, \Delta t_j} \Delta u_j - \frac{\gamma}{\beta} \dot{u}_j - \left(\frac{\gamma}{2\beta} - 1\right) \Delta t_j \ddot{u}_j \tag{2.89}$$

For convenience in later operations, we define the terms

$$\hat{Q}_j = \frac{1}{\beta\, \Delta t_j} \dot{u}_j + \frac{1}{2\beta} \ddot{u}_j \tag{2.90}$$

$$\hat{R}_j = \frac{\gamma}{\beta} \dot{u}_j + \left(\frac{\gamma}{2\beta} - 1\right) \Delta t_j \ddot{u}_j \tag{2.91}$$

Now rewrite Eqs. (2.88) and (2.89) as

$$\Delta \ddot{u}_j = \frac{1}{\beta (\Delta t_j)^2} \Delta u_j - \hat{Q}_j \tag{2.92}$$

$$\Delta \dot{u}_j = \frac{\gamma}{\beta\, \Delta t_j} \Delta u_j - \hat{R}_j \tag{2.93}$$

then substitute Eqs. (2.92) and (2.93) into the incremental equation of motion [see Eq. (2.80)] and collect terms to obtain

$$\hat{k}_j\, \Delta u_j = \Delta \hat{P}_j \tag{2.94}$$

in which

$$\hat{k}_j = k_j + \frac{1}{\beta (\Delta t_j)^2} m_j + \frac{\gamma}{\beta\, \Delta t_j} c_j \tag{2.95}$$

and

$$\Delta \hat{P}_j = \Delta P_j + m_j \hat{Q}_j + c_j \hat{R}_j \tag{2.96}$$

We solve the pseudostatic problem in Eq. (2.94) for the incremental displacement Δu_j and substitute it into Eqs. (2.92) and (2.93) to find the incremental acceleration and velocity $\Delta \ddot{u}_j$ and $\Delta \dot{u}_j$. Finally, the total values of displacement, velocity, and acceleration are determined from

$$u_{j+1} = u_j + \Delta u_j \tag{2.97}$$

$$\dot{u}_{j+1} = \dot{u}_j + \Delta \dot{u}_j \tag{2.98}$$

$$\ddot{u}_{j+1} = \ddot{u}_j + \Delta \ddot{u}_j \tag{2.99}$$

which completes the calculations for the jth time step.

If convergence to the true solution is poor by this approach, we can simply decrease the size of the time step until satisfactory accuracy is achieved. Alternatively, we could make a second, third, etc., set of calculations in each step, using average values of m_j, c_j, and k_j within the step.

In order to improve control of numerical damping, Hilber et al. [16] introduced a parameter α into the equation of motion at time t_{j+1} as follows:

$$m_j\ddot{u}_{j+1} + c_j\dot{u}_{j+1} + (1+\alpha)k_ju_{j+1} - \alpha k_ju_j = P_{j+1} \tag{2.100}$$

Subtracting a similar equation of motion at time t_j from Eq. (2.100) produces the incremental equation

$$m_j\,\Delta\ddot{u}_j + c_j\,\Delta\dot{u}_j + (1+\alpha)k_j\,\Delta u_j - \alpha k_j\,\Delta u_{j-1} = \Delta P_j \tag{2.101}$$

Now substitute Eqs. (2.92) and (2.93) from Newmark's method into Eq. (2.101) and collect terms to find

$$\hat{k}_{\alpha j}\,\Delta u_j = \Delta\hat{P}_{\alpha j} \tag{2.102}$$

in which

$$\hat{k}_{\alpha j} = \hat{k}_j + \alpha k_j \tag{2.103}$$

and

$$\Delta\hat{P}_{\alpha j} = \Delta\hat{P}_j + \alpha k_j\,\Delta u_{j-1} \tag{2.104}$$

Expressions for $\hat{k}_j$ and $\Delta\hat{P}_j$ were derived previously as Eqs. (2.95) and (2.96). For the first time step, we take $\Delta u_{j-1} = 0$ in Eq. (2.104).

Numerical Stability and Accuracy

In order to study the stability and accuracy of various one-step numerical integration procedures, we may cast them into *operator form* [17], as follows:

$$\mathbf{U}_{j+1} = \mathbf{A}\mathbf{U}_j + \mathbf{L}P_{j+1} \tag{2.105}$$

The symbol $\mathbf{U}_j$ in Eq. (2.105) represents a column vector [18] containing the three response quantities u_j, $\dot{u}_j$, and $\ddot{u}_j$ at the time station t_j. That is,

$$\mathbf{U}_j = \{u_j, \dot{u}_j, \ddot{u}_j\} \tag{2.106}$$

And the vector $\mathbf{U}_{j+1}$ is similarly defined at time t_{j+1} to be

$$\mathbf{U}_{j+1} = \{u_{j+1}, \dot{u}_{j+1}, \ddot{u}_{j+1}\} \tag{2.107}$$

The coefficient matrix $\mathbf{A}$ in Eq. (2.105) is a 3×3 array called the *amplification matrix* that we shall examine to answer questions about stability and accuracy. Finally, the symbol $\mathbf{L}$ denotes a column vector called the *load operator,* which is multiplied by the load P_{j+1} at time t_{j+1}. If there is no loading, Eq. (2.105) simplifies to

$$\mathbf{U}_{j+1} = \mathbf{A}\mathbf{U}_j \tag{2.108}$$

for free-vibrational response.

To investigate the stability of a numerical algorithm, we apply *spectral decomposition* [18] to the amplification matrix $\mathbf{A}$, as follows:

$$\mathbf{A} = \mathbf{\Phi}\,\boldsymbol{\lambda}\,\mathbf{\Phi}^{-1} \tag{2.109}$$

In this equation $\boldsymbol{\lambda}$ is the *spectral matrix* of $\mathbf{A}$, containing eigenvalues λ_1, λ_2, and λ_3 in diagonal positions; and $\mathbf{\Phi}$ is the 3×3 *modal matrix* of $\mathbf{A}$, with eigenvectors $\mathbf{\Phi}_1$, $\mathbf{\Phi}_2$, and $\mathbf{\Phi}_3$ listed columnwise. If we start at time $t_0 = 0$ and take n_j time steps using Eq. (2.108), we have

$$\mathbf{U}_{n_j} = \mathbf{A}^{n_j}\mathbf{U}_0 \tag{2.110}$$

where the vector $\mathbf{U}_0$ contains initial conditions, and vector $\mathbf{U}_{n_j}$ gives the response values at time t_{n_j}. Raising the decomposed form of matrix $\mathbf{A}$ in Eq. (2.110) to the power n_j yields

$$\mathbf{A}^{n_j} = \mathbf{\Phi}\,\boldsymbol{\lambda}^{n_j}\mathbf{\Phi}^{-1} \tag{2.111}$$

Now let us define the *spectral radius* of matrix $\mathbf{A}$ as

$$(r)_{\mathbf{A}} = \max |\lambda_i| \qquad (i = 1, 2, 3) \tag{2.112}$$

Then Eq. (2.111) shows that we must have

$$(r)_{\mathbf{A}} \leqq 1 \tag{2.113}$$

in order to keep the numerical solution from growing without bound. This condition is known as the *stability criterion* for a given method. By applying this criterion to the constant-acceleration method [see Eq. (2.83)], we find the critical time step for this conditionally stable approach to be

$$(\Delta t)_{\text{cr}} = \frac{\tau}{\pi} = 0.318\tau \tag{2.114}$$

For the linear-acceleration method, the critical time step is

$$(\Delta t)_{cr} = \sqrt{3}\,\frac{\tau}{\pi} = 0.551\tau \tag{2.115}$$

On the other hand, the spectral radius for the average-acceleration method is always unity. Therefore, it has no critical time step and is said to be unconditionally stable.

The matter of accuracy of a numerical integration procedure is closely related to that of stability. Figure 2.27 shows the undamped response of a one-degree system to an initial displacement u_0. The curve labeled 1 is the exact result, and those labeled 2, 3, and 4 represent various approximations. Curve 2 demonstrates an amplitude increase (AI) that implies an unstable algorithm. Curve 3 shows no amplitude change, and curve 4 depicts an amplitude decrease (AD). Because of the stability criterion in Eq. (2.113), only curves of types 3 and 4 are admissible approximations. Curve 3 may be considered to be the result for the average-acceleration method, which has a spectral radius of unity. Other admissible algorithms are represented by curve 4, which implies a spectral radius less than unity. Thus, one important type of error to be considered is *amplitude suppression,* as exhibited by curve 4.

All of the approximate responses in Fig. 2.27 also show *period elongation* (PE), which is a second type of error introduced by any numerical algorithm. Both the amplitude suppression and the period elongation may be made negligible by using sufficiently small time steps. Newmark [15] recommended a time step of duration equal to $\frac{1}{5}$ or $\frac{1}{6}$ of τ. However, a more commonly-used time step is $\Delta t = \tau/10$.

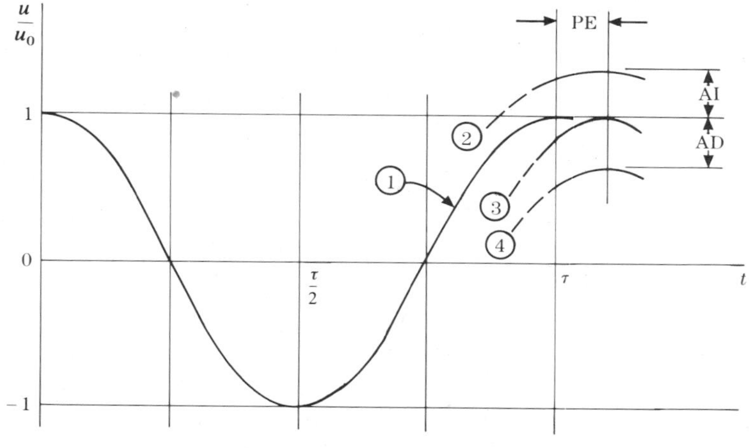

FIG. 2.27

At first glance, the average-acceleration method appears to be the best choice among the implicit approaches, because it has no amplitude suppression and the least period elongation [17]. However, a small amount of amplitude suppression is considered to be desirable for reducing or eliminating unwanted responses of higher modes in multidegree systems. But if the spectral radius of an amplification matrix is too small, the response of the system will be unduly suppressed. Thus, we conclude that the best choice of algorithm is the Newmark-β method with the Hilber-α modification, as described in this section, Probably, the optimum selection of parameters for this approach is to let $\alpha = -0.1$, $\beta = 0.3025$, and $\gamma = 0.6$ [19].

REFERENCES

[1] Ramberg, W., and Osgood, W. R., "Description of Stress-Strain Curves by Three Parameters," NACA Tech. Note No. 902, July, 1943.

[2] Goel, S. C., and Berg, G. V., "Inelastic Earthquake Response of Tall Steel Frames," *Jour. Struc. Div.*, ASCE, **94** (ST8), August, 1968, pp. 1907–1934.

[3] Popov, E. P., and Pinkney, B. R., "Cyclic Yield Reversal in Steel Building Joints," *Jour. Struct. Div.*, ASCE, **95** (ST3), March, 1969, pp. 327–353.

[4] Den Hartog, J. P., *Mechanical Vibrations,* 4th edn., McGraw-Hill, New York, 1956.

[5] Belvins, R. D., *Flow-Induced Vibration,* Van Nostrand-Reinhold, New York, 1977.

[6] Kryloff, A. N., and Bogoliuboff, N., "Introduction to Nonlinear Mechanics," *Ann. Math. Studies* **11,** Princeton University Press, Princeton, NJ, 1943.

[7] Ritz, W., "Über eine neue Methode zur Lösung gewisser Variations–probleme der mathematischen Physik," *Crelles Jour. f. d. reine u. ang. Math.*, No. 135, 1909, pp. 1–61.

[8] Duffing, G., *Erzwungene Schwingungen bei verländerlicher Eigenfrequenz,* Braunsweig, F. Vieweg u. Sohn, 1918.

[9] Klotter, K., "Nonlinear Vibration Problems Treated by the Averaging Method of W. Ritz," *Proc. 1st U.S. Natl. Congr. Appl. Mech.*, 1951, pp. 125–131.

[10] Klotter, K., and Pinney, E., "A Comprehensive Stability Criterion for the Forced Vibrations of Nonlinear Systems," *Jour. Appl. Mech.*, **20,** 1953, pp. 9–12.

[11] Klotter, K., "Nonlinear Vibration Problems Treated by the Averaging Method of W. Ritz," Technical Report No. 17, Parts I and II, Division of Engineering Mechanics, Stanford University, Stanford, CA, 1951.

[12] U.S. Army Corps of Engineers, "Design of Structures to Resist the Effects of Atomic Weapons," Manual EM 1110–345–415, 1957.

[13] Den Hartog, J. P., "Forced Vibrations with Combined Coulomb and Viscous Damping," *Trans ASME,* **53,** 1931, pp. APM 107–115.

[14] Morino, L., Leech, J. W., and Witmer, E. A., "Optimal Predictor-Corrector Method for Systems of Second-Order Differential Equations," *AIAA J.*, **12** (10), 1974, pp. 1343–1347.

[15] Newmark, N. M., "A Method of Computation for Structural Dynamics," *ASCE J. Eng. Mech. Div.*, **85** (EM3), 1959, pp. 67–94.

[16] Hilber, H. M., Hughes, T. J. R., and Taylor, R. L., "Improved Numerical Dissipation for Time Integration Algorithms in Structural Mechanics," *Earthquake Eng. Struct. Dyn.*, **5** (3), 1977, pp. 283–292.

[17] Bathe, K. J., *Finite Element Procedures in Engineering Analysis*, Prentice-Hall, Englewood Cliffs, NJ, 1982.

[18] Gere, J. M., and Weaver, W., *Jr.*, *Matrix Algebra for Engineers*, 2nd edn., Wadsworth, Belmont, CA, 1983.

[19] Weaver, W., Jr., and Johnston, P. R., *Structural Dynamics by Finite Elements*, Prentice-Hall, Englewood Cliffs, NJ, 1987.

PROBLEMS

2.1-1. The figure shows a small mass m that is restrained by four linearly elastic springs, each of which has a stiffness constant k and an unstressed length ℓ. Determine

(a) The nonlinear equation of motion for large displacements of the mass in the x direction.

(b) An approximate nonlinear equation of motion for large displacements.

(c) An approximate linear equation of motion for small displacements.

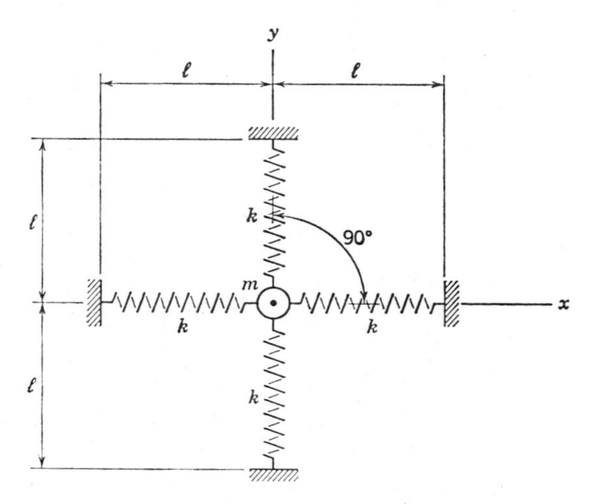

PROB. 2.1-1

2.1-2. The system in the figure is the same as that in Prob. 2.1-1, but the angle between each spring and the x axis is 45°. Determine

(a) The nonlinear equation of motion for large displacements of the mass in the x direction.

(b) An approximate linear equation of motion for small displacements.

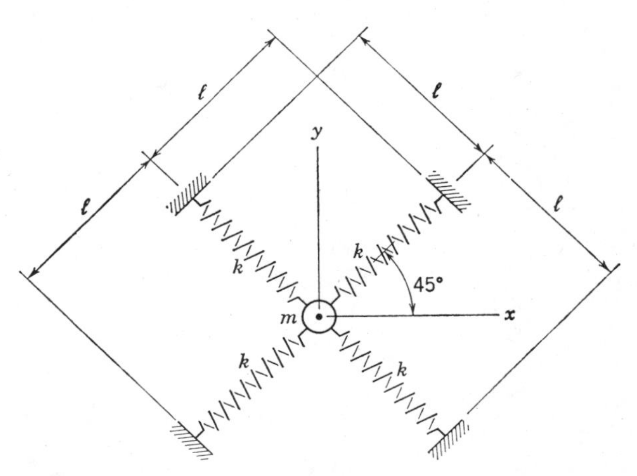

PROB. 2.1-2

2.1-3. For the restrained pendulum in the figure determine

(a) The nonlinear equation of motion for large rotations.

(b) An approximate linear equation of motion for small rotations.

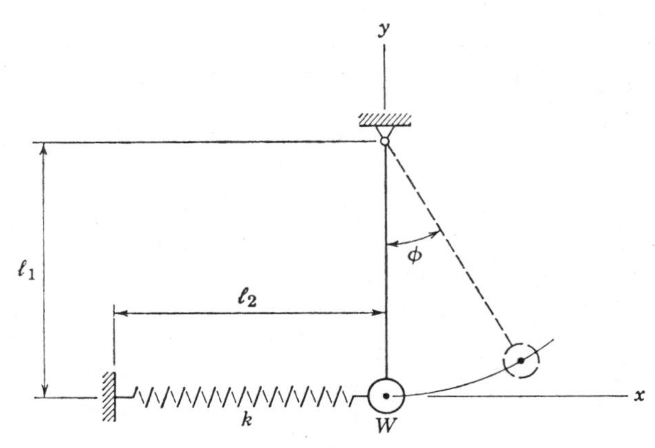

PROB. 2.1-3

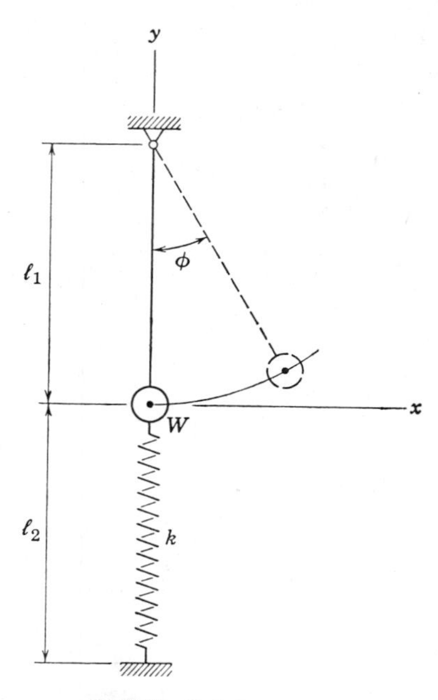

PROB. 2.1-4

2.1-4. For the restrained pendulum in the figure determine

(a) The nonlinear equation of motion for large rotations.

(b) An approximate nonlinear equation of motion for large rotations.

(c) An approximate linear equation of motion for small rotations.

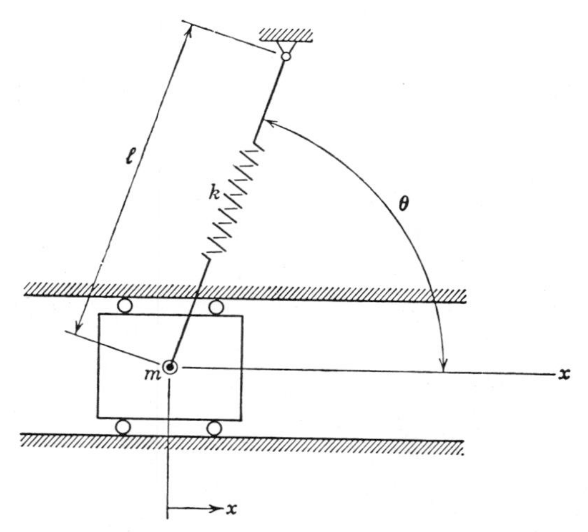

PROB. 2.1-5

2.1-5. The figure shows a mass that is restrained by an inclined spring having an undeformed length equal to ℓ. In one of its stable equilibrium positions the spring is inclined to the horizontal by the angle θ, as indicated in the figure. Write the nonlinear equation of motion for this system associated with large displacements in the x direction.

2.1-6. An inverted pendulum has a linearly elastic rotational spring at the pivot, as indicated in the figure. Determine the nonlinear equation of motion for large rotations of this system.

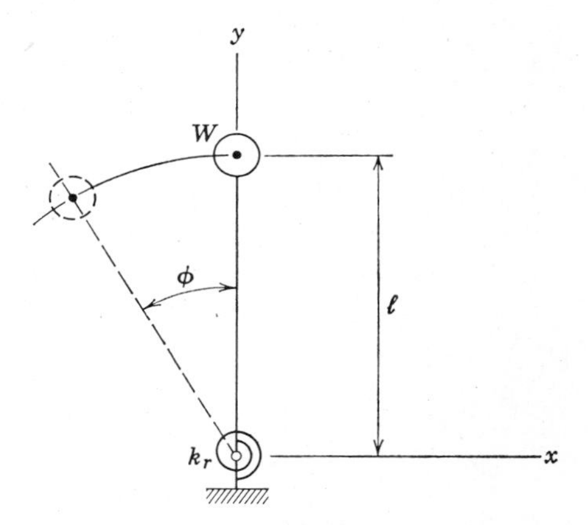

PROB. 2.1-6

2.2-1. The figure shows a shock-absorbing bumper installed at a truck-loading dock. The bumper has a hardening spring that provides a restoring force given by $F(u) = k(u + \alpha u^3)$, where $k = 400\ \text{lb/in.}$

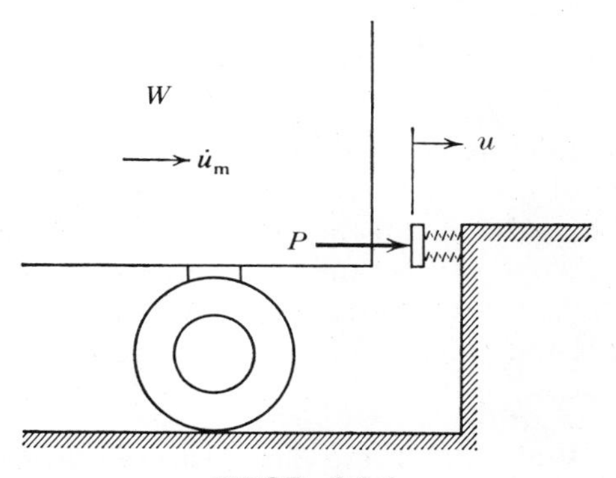

PROB. 2.2-1

and $\alpha = 2\ \text{in.}^{-2}$. If the weight of a truck is $W = 38{,}600$ lb and its velocity when backing into the dock is 10 in./sec, determine the maximum displacement of the bumper as well as its maximum restoring force and the time (measured from $t = 0$ at impact) when it occurs. Assume that the mass of the bumper is small with respect to that of the vehicle and that the two remain in contact after impact.

2.2-2. Suppose that the springs in the bumper in Prob. 2.2-1 are replaced by another set which provide a restoring force as given by the tangent function in the figure. Let $k = 400$ lb/in. as before, but the limiting displacement at which the resistance becomes infinite is $u_1 = 10$ in. Determine the maximum displacement u_m and the maximum restoring force P_m.

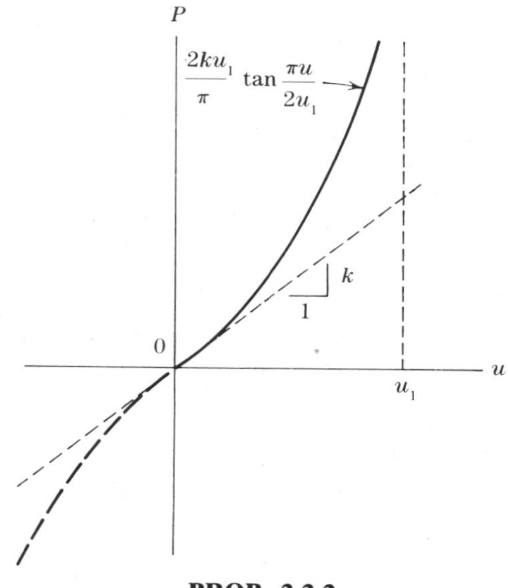

PROB. 2.2-2

2.2-3. Let the hardening springs in the bumper in Prob. 2.2-1 be replaced by a set of softening springs which provide a restoring force as given by the hyperbolic tangent function in the figure. In this case assume that the slope of the curve at the origin is $k = 1000$ lb/in. and that the ultimate resisting capacity of the bumper is $P_1 = 100{,}000$ lb. Calculate the values of u_m and P_m caused by the truck striking the bumper.

2.2-4. Repeat Prob. 2.2-3 for the restoring force represented by the exponential function in the figure.

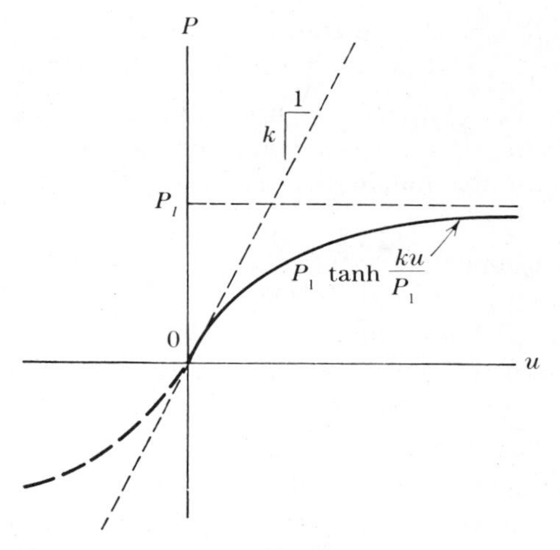

PROB. 2.2-3

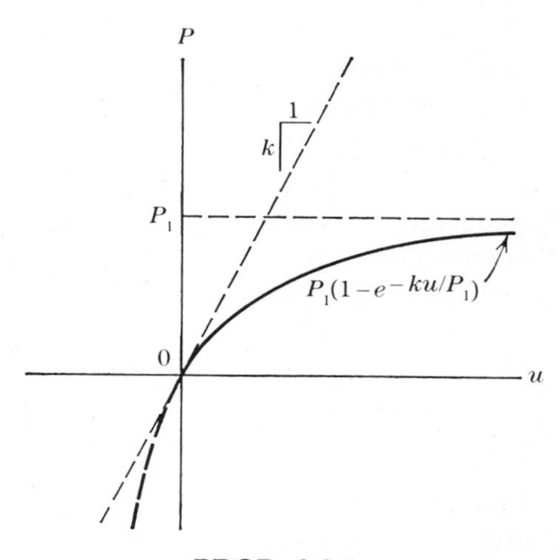

PROB. 2.2-4

2.5-1. For the system in Fig. 2.16*a*, assume that a rectangular impulse of magnitude Q_n terminates at time t_n (see Fig. 2.16*f*). Let the time t_n be equal to the time required for the velocity $\dot{u}$ in Fig. 2.16*h* to reach its first maximum value. Determine this time t_n and the amplitude of the resulting free vibrations.

2.5-2. Let the system in Fig. 2.17*a* be subjected to a step function of magnitude $Q_n = k_1 u_1$, beginning at time $t = 0$. Determine the maximum response u_m and the time t_m at which it occurs.

2.5-3. Suppose that the impulse in Prob. 2.5-2 is discontinued precisely at the time t_n when the transition point ($u = u_1$) is reached. Determine the time t_n and the amplitude of the remaining free vibrations.

2.5-4. Assume that the system in Fig. 2.19*a* is subjected to a step function of magnitude $Q_n = ku_1/1.5$, beginning at time $t = 0$. Derive expressions for the maximum response u_m and the time t_m at which it occurs.

2.5-5. Assume that the system in Fig. 2.19*a* is subjected to a rectangular impulse of magnitude $Q_n = ku_1$, beginning at time $t = 0$. If the impulse is discontinued at the time t_n when the plastic hinge forms, determine expressions for the maximum response u_m and the residual displacement u_r.

2.5-6. The block in Fig. 2.21*a* is displaced from the unstressed position by the amount $u_0 = 10$ in. and then released without initial velocity. Its weight is $W = 2$ lb, and the spring constant is $k = 1$ lb/in. If the coefficient of friction is $\frac{1}{4}$, how long will the body vibrate?

2.5-7. Suppose that the spring constant for the system in Fig. 2.21*a* is given as $k = 4W$, where W is the weight of the block. If the amplitude of free vibration diminishes from 25 in. to 22.5 in. during 10 cycles, what is the coefficient of friction?

2.5-8. The mass shown in Fig. (*a*) is restrained by springs having unequal stiffnesses, and the static load-displacement diagram in Fig. (*b*) is unsymmetric with respect to the origin. Assuming that at time $t = 0$ the initial conditions are $u_0 = 0$ and $\dot{u}_0 \neq 0$, construct time histories of displacement and velocity for one complete cycle of free vibration.

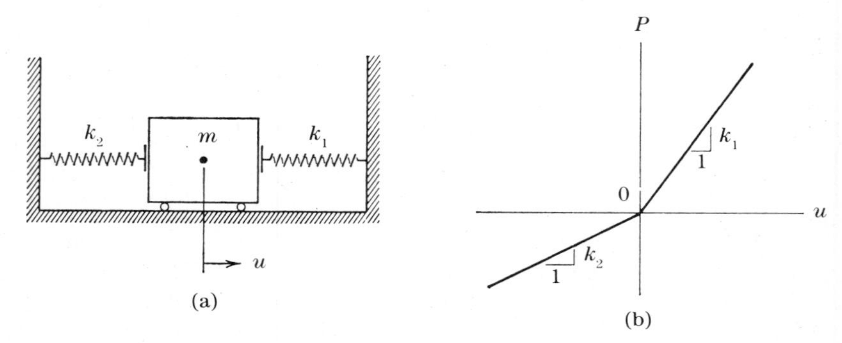

PROB. 2.5-8

2.5-9. Each of the springs in Fig. (*a*) is precompressed by the amount P_1, as indicated by the static load-displacement diagram in Fig. (*b*). For the initial conditions $u_0 = 0$ and $\dot{u}_0 \neq 0$, construct time histories of

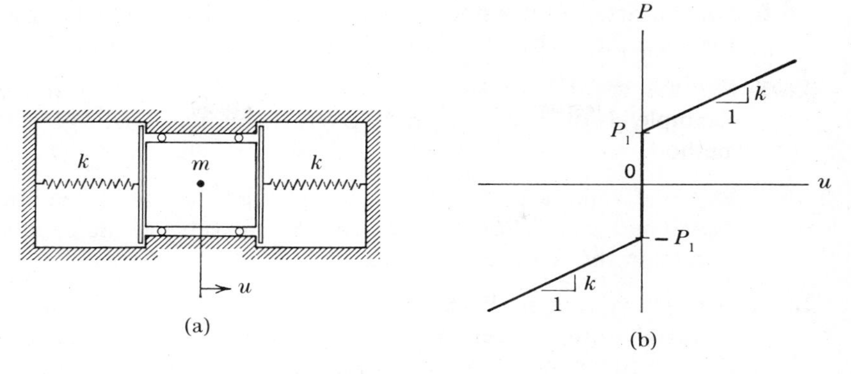

PROB. 2.5-9

displacement and velocity representing one complete cycle of free vibration.

2.5-10. Assume that a piecewise-linear elastic system has the static load-displacement diagram shown in the figure. Construct time histories of displacement and velocity representing one complete cycle of free vibration due to the initial conditions $u_0 = 0$ and $\dot{u}_0 > \omega_1 u_1$.

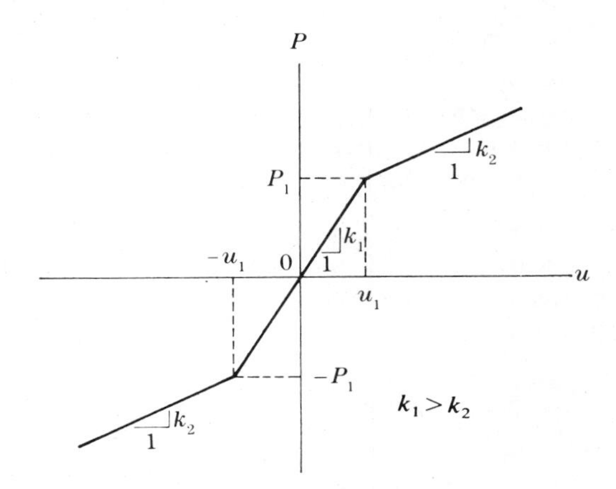

PROB. 2.5-10

2.5-11. Refer to the bilinear hysteretic diagram in Fig. 2.7*c* (discussed in Sec. 2.1) and assume that $k_1 = 5k_2$. If the system to which this diagram applies is subjected to a step function of magnitude $Q_n = k_1 u_1$, construct the time history of the displacement, including one cycle of residual oscillation.

2.5-12. Repeat Prob. 2.5-11, assuming that $Q_n = 2k_1 u_1$.

2.6-1. For the first 10 time steps, confirm the iterative results for Example 1 of Sec. 2.6 in Table 2.1 by the average-acceleration method.

2.6-2. For the second 10 time steps, confirm the iterative results for Example 1 of Sec. 2.6 in Table 2.1 by the linear-acceleration method.

2.6-3. Repeat Example 2 in Sec. 2.6, but change the initial conditions to $\phi_0 = 0$ and $\dot{\phi}_0 = 2.618$ rad/sec. All other features of the problem are to remain the same as in the example.

2.6-4. For the system in Prob. 2.1-1 (see Sec. 2.1), determine by the iterative average-acceleration method the free-vibrational response to the initial conditions $u_0 = \ell/4$ and $\dot{u}_0 = v_0 = \dot{v}_0 = 0$. Use the exact equation of motion and the following values for the parameters of the system: $k = 10$ lb/in.; $m = 8$ lb-sec^2/in.; $\ell = 4$ in. Take 10 equal time steps of duration $\Delta t = 0.1$ sec, and plot the results.

2.6-5. For the system in Prob. 2.1-6 (see Sec. 2.1), determine by the iterative linear-acceleration method the free-vibrational response to the initial conditions $\phi_0 = 0$ and $\dot{\phi}_0 = 10.64$ rad/sec. Use the exact equation of motion and the following values for the parameters of the system: $W = 5$ lb; $\ell = 10$ in.; $k_r = 100$ lb-in./rad. Take 20 equal time steps of duration $\Delta t = 0.025$ sec, and plot the results.

2.6-6. Repeat Example 3 in Sec. 2.6, but change the spring characteristic from that of Prob. 2.2-1 to that of Prob. 2.2-2. Determine by the average-acceleration method the maximum displacement and the time at which it occurs.

2.6-7. Repeat Example 3 in Sec. 2.6, but change the spring characteristic from that of Prob. 2.2-1 to that of Prob. 2.2-4. Determine by the linear-acceleration method the maximum displacement and the time at which it occurs.

2.6-8. Solve Prob. 2.5-2 (see Sec. 2.5) by the iterative average-acceleration method,† using the following values for the parameters of the system: $m = 1$ lb-sec^2/in.; $k_1 = \pi^2$ lb/in.; $k_2 = 4k_1$; $u_1 = 1$ in.

2.6-9. Solve Prob. 2.5-4 (see Sec. 2.5) by the iterative linear-acceleration method,† using the following values for the parameters of the system: $m = 1$ lb-sec^2/in.; $k = \pi^2$ lb/in.; $u_1 = 1$ in.

2.6-10. Solve Prob. 2.5-11 (see Sec. 2.5) by the iterative linear-acceleration method,† using the following values for the parameters of the system: $m = 1$ lb-sec^2/in.; $k_1 = \pi^2$ lb/in.; $u_1 = 1$ in.

† Problems 2.6-8, 2.6-9, and 2.6-10 are piecewise-linear, and the value of k changes abruptly from one constant to another at various stages in the analyses.

CHAPTER 3

SYSTEMS WITH TWO DEGREES OF FREEDOM

3.1 EXAMPLES OF TWO-DEGREE SYSTEMS

In Chapters 1 and 2 we dealt only with systems having one degree of freedom. In this chapter and the next we shall discuss systems having several degrees of freedom, the simplest of which are *systems with two degrees of freedom.* The configuration of such a vibrating system is completely defined by two coordinates, or displacements, and two differential equations are required to describe its motions.

Figure 3.1*a* shows two masses m_1 and m_2 that are connected to ground and to each other by springs having constants k_1 and k_2, respectively. It is assumed that the masses can move only in the x direction and that there is no friction or other type of damping in the system. As coordinates defining the motions of this sytem, we take the displacements u_1 and u_2 of the masses from their positions of static equilibrium (corresponding to no deformations in the springs). Forcing functions $Q_1 = F_1(t)$ and $Q_2 = F_2(t)$, applied to masses m_1 and m_2, also appear in Figure 3.1*a*. In a displaced position during motion the forces exerted on the masses by the springs will be as shown in Fig. 3.1*b*. Using Newton's second law, we obtain the equations of motion for m_1 and m_2 as

$$m_1\ddot{u}_1 = -k_1u_1 + k_2(u_2 - u_1) + Q_1 \qquad (a)$$

$$m_2\ddot{u}_2 = -k_2(u_2 - u_1) + Q_2 \qquad (b)$$

If $u_1 > u_2$, these expressions are unaltered, because in that case the compressive spring force $k_2(u_1 - u_2)$ acting on each mass has a negative sign in Eq. (*a*) and a positive sign in Eq. (*b*). Rearranging terms in these

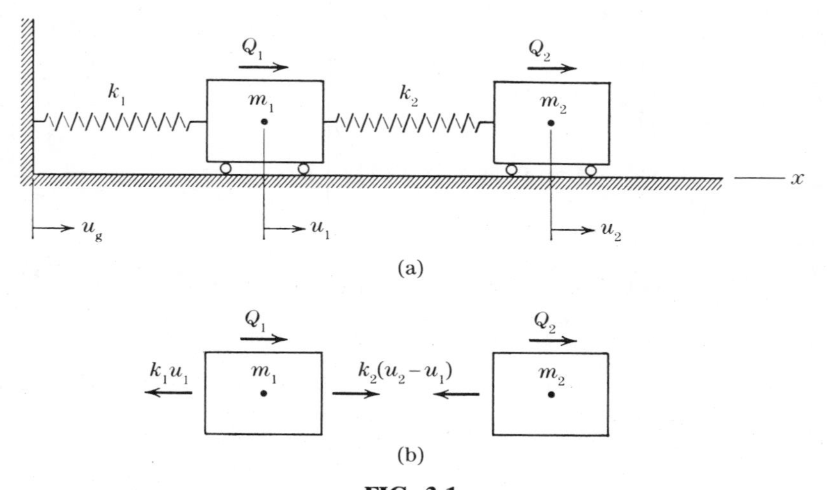

FIG. 3.1

equations yields

$$m_1\ddot{u}_1 + (k_1 + k_2)u_1 - k_2 u_2 = Q_1 \tag{3.1a}$$

$$m_2\ddot{u}_2 - k_2 u_1 + k_2 u_2 = Q_2 \tag{3.1b}$$

Thus, we have two simultaneous linear second-order differential equations with constant coefficients.

To investigate free vibrations for this system, we set Q_1 and Q_2 equal to zero and obtain the homogeneous equations

$$m_1\ddot{u}_1 + (k_1 + k_2)u_1 - k_2 u_2 = 0 \tag{c}$$

$$m_2\ddot{u}_2 - k_2 u_1 + k_2 u_2 = 0 \tag{d}$$

In a manner analogous to that for a system with one degree of freedom, we assume solutions of the form

$$u_1 = A\sin(\omega t + \phi) \tag{e}$$

$$u_2 = B\sin(\omega t + \phi) \tag{f}$$

These expressions mply that in a *natural mode of vibration* both masses follow the same harmonic function, having the angular frequency ω and the phase angle ϕ. The symbols A and B represent maximum values, or *amplitudes,* of the vibratory motions. Substitution of Eqs. (e) and (f) into Eqs. (c) and (d) produces the following algebraic equations that must be satisfied:

$$(k_1 + k_2 - \omega^2 m_1)A - k_2 B = 0 \tag{g}$$

$$-k_2 A + (k_2 - \omega^2 m_2)B = 0 \tag{h}$$

One possible solution of these equations is that $A = B = 0$, which corresponds to the equilibrium position and yields no information about vibrations. The equations can have nonzero solutions only if the determinant of the coefficients of A and B is equal to zero. Thus,

$$\begin{vmatrix} (k_1 + k_2 - \omega^2 m_1) & -k_2 \\ -k_2 & (k_2 - \omega^2 m_2) \end{vmatrix} = 0 \tag{i}$$

Expansion of this determinant results in

$$(k_1 + k_2 - \omega^2 m_1)(k_2 - \omega^2 m_2) - k_2^2 = 0 \tag{j}$$

or,

$$m_1 m_2 \omega^4 - [m_1 k_2 + m_2(k_1 + k_2)]\omega^2 + k_1 k_2 = 0 \tag{k}$$

This expression, which is quadratic in ω^2, is called the *frequency equation* (or *characteristic equation*) for the system. It has two roots (called *characteristic values*) that may be determined by the quadratic formula

$$\omega_{1,2}^2 = \frac{-b \mp \sqrt{b^2 - 4ac}}{2a} \tag{l}$$

where

$$a = m_1 m_2 \qquad b = -[m_1 k_2 + m_2(k_1 + k_2)] \qquad c = k_1 k_2 \tag{m}$$

Because the expression under the radical sign in Eq. (*l*) is always positive, both of the roots ω_1^2 and ω_2^2 are real. It is also apparent that the square-root term is less than $-b$ and that both roots are positive. Furthermore, Eq. (*l*) is written in such a manner that $\omega_1 < \omega_2$. Thus, the characteristic equation yields two angular frequencies of vibration that depend only upon the physical constants of the system.

Substituting the characteristic values ω_1^2 and ω_2^2 into the homogeneous algebraic equations (*g*) and (*h*), we find that we cannot obtain actual values for A and B. However, those equations can be used to provide values for the ratios $r_1 = A_1/B_1$ and $r_2 = A_2/B_2$, corresponding to ω_1^2 and ω_2^2, respectively. Thus,

$$r_1 = \frac{A_1}{B_1} = \frac{k_2}{k_1 + k_2 - \omega_1^2 m_1} = \frac{k_2 - \omega_1^2 m_2}{k_2} \tag{n}$$

$$r_2 = \frac{A_2}{B_2} = \frac{k_2}{k_1 + k_2 - \omega_2^2 m_1} = \frac{k_2 - \omega_2^2 m_2}{k_2} \tag{o}$$

These *amplitude ratios* represent the shapes of the two natural modes of vibration (also called *principal modes*) of the system. They have dual definitions by virtue of Eq. (l), and their values depend only upon the physical constants m_1, m_2, k_1, and k_2.

Using the smaller angular frequency ω_1 and the corresponding amplitude ratio r_1 in Eqs. (e) and (f), we have

$$u_1' = r_1 B_1 \sin(\omega_1 t + \phi_1) \tag{p}$$

$$u_2' = B_1 \sin(\omega_1 t + \phi_1) \tag{q}$$

These expressions completely describe the *first mode* of vibration, which is also called the *fundamental mode*. It consists of a simple harmonic motion of both masses with the angular frequency ω_1 and the phase angle ϕ_1. At any time during such motion, the displacement ratio u_1'/u_2' is the same as the amplitude ratio r_1. In each cycle of the vibration both masses pass twice through their equilibrium positions, and they reach their extreme positions simultaneously. Our analysis reveals no limitation on the phase angle; but by definition in Eqs. (e) and (f), it must be the same for both masses.

Substitution of the larger angular frequency ω_2 and the corresponding amplitude ratio r_2 into Eqs. (e) and (f) yields

$$u_1'' = r_2 B_2 \sin(\omega_2 t + \phi_2) \tag{r}$$

$$u_2'' = B_2 \sin(\omega_2 t + \phi_2) \tag{s}$$

which describe the *second mode* of vibration. This simple harmonic motion of the two masses occurs at the angular frequency ω_2 and with the common phase angle ϕ_2. In this case the displacement ratio is always $u_1''/u_2'' = r_2$.

The general solution of Eqs. (c) and (d) consists of the sum of the principal-mode solutions in Eqs. (p), (q), (r), and (s). Thus, we have

$$u_1 = u_1' + u_1'' = r_1 B_1 \sin(\omega_1 t + \phi_1) + r_2 B_2 \sin(\omega_2 t + \phi_2) \tag{t}$$

$$u_2 = u_2' + u_2'' = B_1 \sin(\omega_1 t + \phi_1) + B_2 \sin(\omega_2 t + \phi_2) \tag{u}$$

These expressions contain four arbitrary constants (B_1, B_2, ϕ_1, and ϕ_2) that may be obtained by satisfying the four initial conditions of displacement and velocity of the two masses at time $t = 0$. Equations (t) and (u) represent rather complicated motions that are not periodic unless the natural frequencies ω_1 and ω_2 happen to be commensurate. The system performs a pure harmonic motion only if carefully started in one of its principal modes.

When dealing with any system having two degrees of freedom, we will always be able to determine its frequencies and mode shapes in a manner analogous to that shown above for the system in Fig. 3.1a. Because the

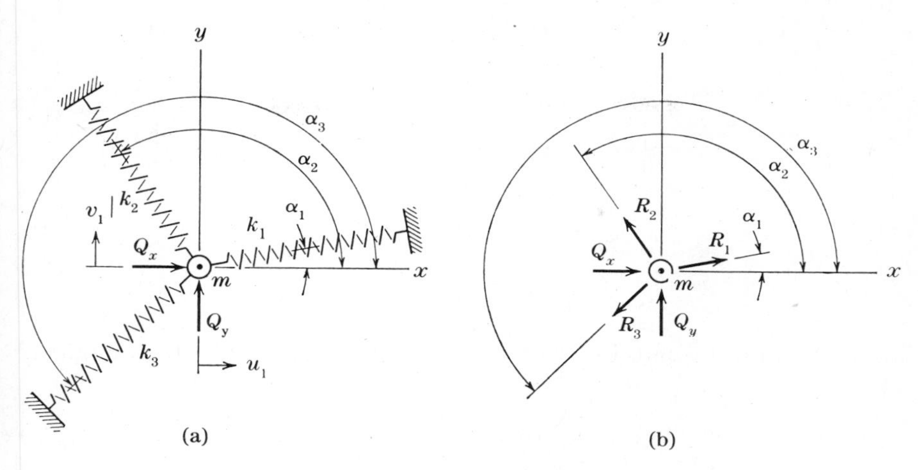

FIG. 3.2

equations of motion for all multi-degree systems have a similar mathematical form, further solutions of these equations will be deferred for the present. They will be handled systematically by matrix methods later in this chapter and in Chapter 4.

As a second example of a system with two degrees of freedom, consider the spring-suspended mass in Fig. 3.2*a*. The three springs shown in the figure all lie in the same plane, and they have constants k_1, k_2, and k_3. It is assumed that the mass is restrained to displace only in the plane of the springs (the x–y plane); where its motion may be defined by the x and y components (u_1 and v_1) of its translation relative to the equilibrium position. Forcing functions Q_x and Q_y, acting in the x and y directions, are also shown in Fig. 3.2*a*. If we consider only small displacements, the restoring forces R_1, R_2, and R_3 (see Fig. 3.2*b*) exerted on the mass by the springs may be considered to have the same inclinations as the springs in the equilibrium position. With this assumption the equations of motion for the mass become

$$m\ddot{u}_1 = \sum_{i=1}^{3} R_i \cos \alpha_i + Q_x \tag{v}$$

and

$$m\ddot{v}_1 = \sum_{i=1}^{3} R_i \sin \alpha_i + Q_y \tag{w}$$

in which

$$R_i = -k_i(u_1 \cos \alpha_i + v_1 \sin \alpha_i) \tag{x}$$

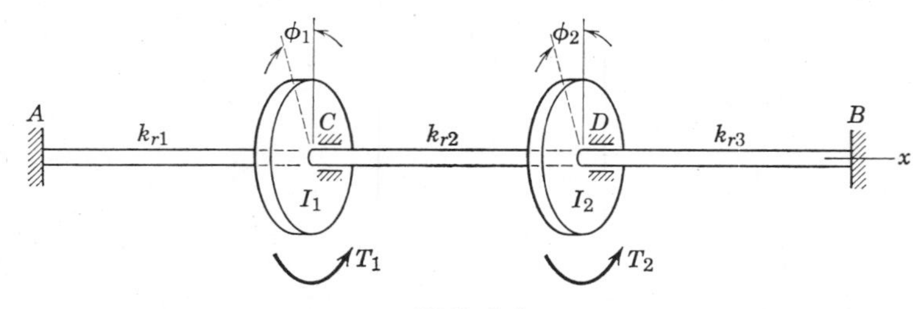

FIG. 3.3

Substituting Eq. (x) into Eqs. (v) and (w) and rearranging terms, we obtain

$$m\ddot{u}_1 + \sum_{i=1}^{3} k_i(u_1 \cos^2 \alpha_i + v_1 \sin \alpha_i \cos \alpha_i) = Q_x \tag{3.2a}$$

$$m\ddot{v}_1 + \sum_{i=1}^{3} k_i(u_1 \sin \alpha_i \cos \alpha_i + v_1 \sin^2 \alpha_i) = Q_y \tag{3.2b}$$

Figure 3.3 shows a third example, consisting of two disks mounted on a shaft that is fixed at points A and B and restrained against lateral translation at points C and D. The thee segments of the shaft have rotational spring constants k_{r1}, k_{r2}, and k_{r3}, as indicated in the figure. Also shown are the rotational degrees of freedom ϕ_1 and ϕ_2 of the disks, their mass moments of inertia I_1 and I_2, and the applied torques T_1 and T_2. In this case the equations of rotational motion are

$$I_1\ddot{\phi}_1 = -k_{r1}\phi_1 + k_{r2}(\phi_2 - \phi_1) + T_1 \tag{y}$$

$$I_2\ddot{\phi}_2 = -k_{r2}(\phi_2 - \phi_1) - k_{r3}\phi_2 + T_2 \tag{z}$$

which upon rearrangement become

$$I_1\ddot{\phi}_1 + (k_{r1} + k_{r2})\phi_1 - k_{r2}\phi_2 = T_1 \tag{3.3a}$$

$$I_2\ddot{\phi}_2 - k_{r2}\phi_1 + (k_{r2} + k_{r3})\phi_2 = T_2 \tag{3.3b}$$

As a final example we consider the spring-connected pair of simple pendulums in Fig. 3.4. They have the same length ℓ and mass m, and the hinges at A and B allow them to swing freely only in the plane of the figure. The angles θ_1 and θ_2 define the configuration of the system during motion; and the forcing functions P_1 and P_2 act in the horizontal direction. Assuming small displacements of the system, we write the equations of motion as

$$m\ell^2\ddot{\theta}_1 = -mg\ell\theta_1 + kh^2(\theta_2 - \theta_1) + P_1\ell \tag{a'}$$

$$m\ell^2\ddot{\theta}_2 = -mg\ell\theta_2 - kh^2(\theta_2 - \theta_1) + P_2\ell \tag{b'}$$

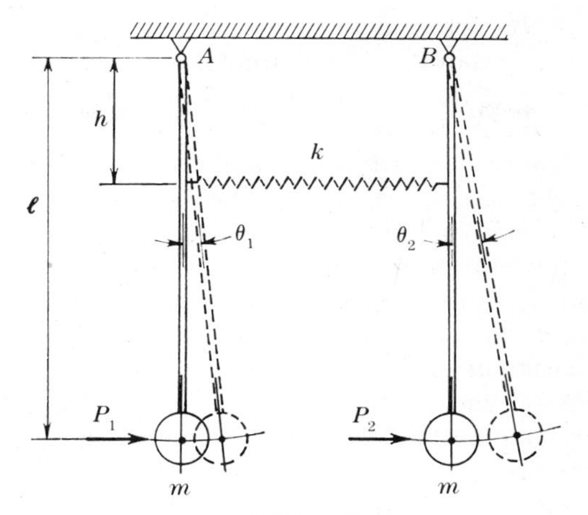

FIG. 3.4

Rearranging terms in these expressions, we have

$$m\ell^2\ddot{\theta}_1 + (kh^2 + mg\ell)\theta_1 - kh^2\theta_2 = P_1\ell \tag{3.4a}$$

$$m\ell^2\ddot{\theta}_2 - kh^2\theta_1 + (kh^2 + mg\ell)\theta_2 = P_2\ell \tag{3.4b}$$

It is apparent that the equations of motion for each of the examples are of similar form. This similarity will be utilized in the next section.

3.2 ACTION EQUATIONS: STIFFNESS COEFFICIENTS

Equations (3.1a) and (3.1b) may be rewritten in *matrix form* [1] as follows:

$$\begin{bmatrix} m_1 & 0 \\ 0 & m_2 \end{bmatrix}\begin{bmatrix} \ddot{u}_1 \\ \ddot{u}_2 \end{bmatrix} + \begin{bmatrix} k_1 + k_2 & -k_2 \\ -k_2 & k_2 \end{bmatrix}\begin{bmatrix} u_1 \\ u_2 \end{bmatrix} = \begin{bmatrix} Q_1 \\ Q_2 \end{bmatrix} \tag{3.5}$$

The same relationships can also be stated more concisely as

$$\mathbf{M}\ddot{\mathbf{D}} + \mathbf{S}\mathbf{D} = \mathbf{Q} \tag{3.6}$$

where it is understood that $\mathbf{D}$, $\ddot{\mathbf{D}}$, and $\mathbf{Q}$ are the column matrices

$$\mathbf{D} = \begin{bmatrix} u_1 \\ u_2 \end{bmatrix} \qquad \ddot{\mathbf{D}} = \begin{bmatrix} \ddot{u}_1 \\ \ddot{u}_2 \end{bmatrix} \qquad \mathbf{Q} = \begin{bmatrix} Q_1 \\ Q_2 \end{bmatrix} \tag{a}$$

and the coefficient matrices are

$$\mathbf{S} = \begin{bmatrix} S_{11} & S_{12} \\ S_{21} & S_{22} \end{bmatrix} = \begin{bmatrix} k_1 + k_2 & -k_2 \\ -k_2 & k_2 \end{bmatrix} \qquad \mathbf{M} = \begin{bmatrix} m_1 & 0 \\ 0 & m_2 \end{bmatrix} \tag{b}$$

Equation (3.6) will be referred to as a set of *action equations of motion* expressed in matrix notation. This terminology is adopted because it is representative of a large class of equations of motion, wherein the terms are either forces or moments (more generally, *actions*). The *stiffness matrix* **S** consists of *stiffness influence coefficients,* and the *mass matrix* **M** contains m_1 and m_2 in diagonal positions. While the mass matrix is diagonal for simple problems, there are many vibrational systems for which it is not.

We shall now concentrate upon the characteristics of the stiffness matrix **S** and the development of its contents in an organized fashion. Any term S_{ij} in this array is an action corresponding to a displacement of type i, resulting from a unit displacement of type j. We develop all such terms by inducing unit displacements for each of the *displacement coordinates* (one at a time) and calculating the required holding actions. Figures 3.5*a* and 3.5*b* illustrate this process for the first example of the preceding section. In Fig. 3.5*a* a unit displacement $u_1 = 1$ is induced, while $u_2 = 0$. The static holding forces required for this condition are labeled S_{11} and S_{21} (slashes on action vectors serve as reminders that they are holding actions). The symbol S_{11} represents the holding action of type 1 required for a unit displacement of type 1, and S_{21} denotes the action of type 2 required for a unit displacement of type 1. Their values are $S_{11} = k_1 + k_2$ and $S_{21} = -k_2$, and they constitute the first column of the stiffness matrix. Terms for the second column of **S** are obtained from Fig. 3.5*b,* which shows a unit displacement of $u_2 = 1$ (while $u_1 = 0$). In this case the holding forces are $S_{12} = -k_2$ and $S_{22} = k_2$, which are the actions of types 1 and 2 required for a unit displacement of type 2. For linearly elastic systems with small displacements, the stiffness matrices are always symmetric [2]; and in this case we see that $S_{12} = S_{21} = -k_2$.

When the equations of motion for the second example in the preceding

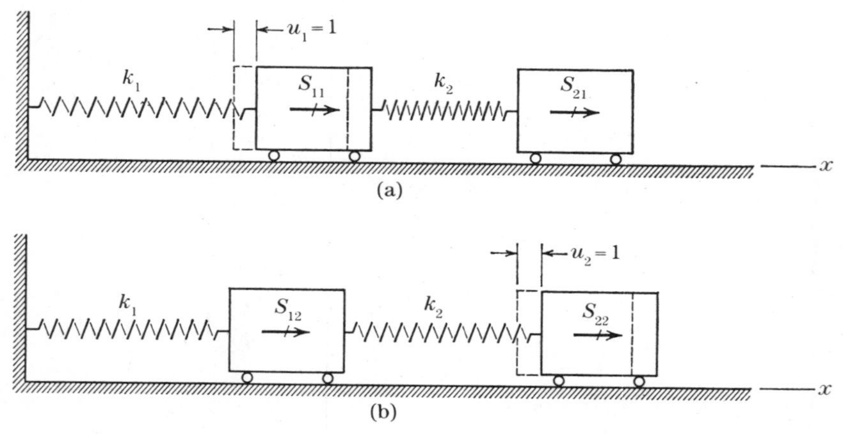

FIG. 3.5

section [see Fig. 3.2 and Eqs. (3.2a,b)] are cast into matrix form, we obtain

$$\begin{bmatrix} m & 0 \\ 0 & m \end{bmatrix}\begin{bmatrix} \ddot{u}_1 \\ \ddot{v}_1 \end{bmatrix} + \sum_{i=1}^{3} k_i \begin{bmatrix} \cos^2 \alpha_i & \sin \alpha_i \cos \alpha_i \\ \sin \alpha_i \cos \alpha_i & \sin^2 \alpha_i \end{bmatrix}\begin{bmatrix} u_1 \\ v_1 \end{bmatrix} = \begin{bmatrix} Q_x \\ Q_y \end{bmatrix} \tag{3.7}$$

The first column of the stiffness matrix in Eq. (3.7) may be obtained directly from the condition $u_1 = 1$ (while $v_1 = 0$) and the second column from the condition $v_1 = 1$ (while $u_1 = 0$).

Similarly, the equations of motion for the third example [see Fig. 3.3 and Eqs. (3.3a, b)] in the matrix format become

$$\begin{bmatrix} I_1 & 0 \\ 0 & I_2 \end{bmatrix}\begin{bmatrix} \ddot{\phi}_1 \\ \ddot{\phi}_2 \end{bmatrix} + \begin{bmatrix} k_{r1} + k_{r2} & -k_{r2} \\ -k_{r2} & k_{r2} + k_{r3} \end{bmatrix}\begin{bmatrix} \phi_1 \\ \phi_2 \end{bmatrix} = \begin{bmatrix} T_1 \\ T_2 \end{bmatrix} \tag{3.8}$$

In this case the displacements are rotations, and the corresponding actions are torques, or moments. The coefficient matrix containing I_1 and I_2 will still be referred to as the "mass matrix," which is, of course, a misnomer. As before, the terms in the stiffness matrix may be derived from the conditions $\phi_1 = 1$ (for the first column) and $\phi_2 = 1$ (for the second column).

Finally, the matrix form of the equations of motion for the last example [see Fig. 3.4 and Eqs. (3.4a, b)] may be written as

$$\begin{bmatrix} m\ell^2 & 0 \\ 0 & m\ell^2 \end{bmatrix}\begin{bmatrix} \ddot{\theta}_1 \\ \ddot{\theta}_2 \end{bmatrix} + \begin{bmatrix} kh^2 + mg\ell & -kh^2 \\ -kh^2 & kh^2 + mg\ell \end{bmatrix}\begin{bmatrix} \theta_1 \\ \theta_2 \end{bmatrix} = \begin{bmatrix} P_1\ell \\ P_2\ell \end{bmatrix} \tag{3.9}$$

However, in this case we have a mixture of stiffness and gravitational restoring actions. If the influence coefficients for these two types of restoring actions are segregated into separate arrays, we have

$$\mathbf{S}^* = \mathbf{S} + \mathbf{G} \tag{3.10}$$

where

$$\mathbf{S} = \begin{bmatrix} kh^2 & -kh^2 \\ -kh^2 & kh^2 \end{bmatrix} \qquad \mathbf{G} = \begin{bmatrix} mg\ell & 0 \\ 0 & mg\ell \end{bmatrix} \tag{c}$$

The former array contains the usual stiffness influence coefficients, whereas the latter contains *gravitational influence coefficients,* which are defined as actions required for unit displacements in the presence of gravity. In the absence of gravity, the terms in the *gravitational matrix* **G** are zero.

3.3 DISPLACEMENT EQUATIONS: FLEXIBILITY COEFFICIENTS

For statically determinate systems it is sometimes more convenient to work with *displacement equations of motion* instead of action equations. In this approach we write expressions for the *displacement coordinates* (translations or rotations) of a system in terms of its flexibilities. For this purpose we introduce the notation

$$\delta = \frac{1}{k} \tag{a}$$

which will be referred to as the flexibility constant of a spring that has a *stiffness constant* k (previously called the spring constant). With this notation we determine the flexibility constants of the two springs in Fig. 3.1*a* to be $\delta_1 = 1/k_1$ and $\delta_2 = 1/k_2$.

Suppose that the forces Q_1 and Q_2, acting upon the masses in Fig. 3.1*a*, were to be applied statically (so that no inertia forces arise). Under this condition the displacements of the masses, in terms of the flexibility constants δ_1 and δ_2, would be

$$(u_1)_{st} = \delta_1(Q_1 + Q_2) \tag{b}$$

$$(u_2)_{st} = \delta_1(Q_1 + Q_2) + \delta_2 Q_2 \tag{c}$$

These expressions may be written in matrix form as follows:

$$\begin{bmatrix} u_1 \\ u_2 \end{bmatrix}_{st} = \begin{bmatrix} \delta_1 & \delta_1 \\ \delta_1 & \delta_1 + \delta_2 \end{bmatrix} \begin{bmatrix} Q_1 \\ Q_2 \end{bmatrix} \tag{d}$$

Such displacement-action relationships can also be stated more concisely as

$$\mathbf{D}_{st} = \mathbf{FQ} \tag{e}$$

where the symbol $\mathbf{F}$ denotes the *flexibility matrix*

$$\mathbf{F} = \begin{bmatrix} F_{11} & F_{12} \\ F_{21} & F_{22} \end{bmatrix} = \begin{bmatrix} \delta_1 & \delta_1 \\ \delta_1 & \delta_1 + \delta_2 \end{bmatrix} \tag{f}$$

This array contains *flexibility influence coefficients,* which are defined as displacements due to unit values of the actions corresponding to the displacements.

Terms in the flexibility matrix may be developed in a manner that is complementary to that used for obtaining stiffnesses. Any element F_{ij} in the flexibility matrix is a displacement of type i due to a unit action of type j. We determine all such elements by applying unit actions corresponding to each of the displacement coordinates (one at a time) and calculating the resulting displacements. Figure 3.6*a* and 3.6*b* illustrate this process for the system in Fig. 3.1*a*. In Fig. 3.6*a*, a unit force $Q_1 = 1$ is applied statically to the mass m_1, while no force is applied to m_2. The resultng static

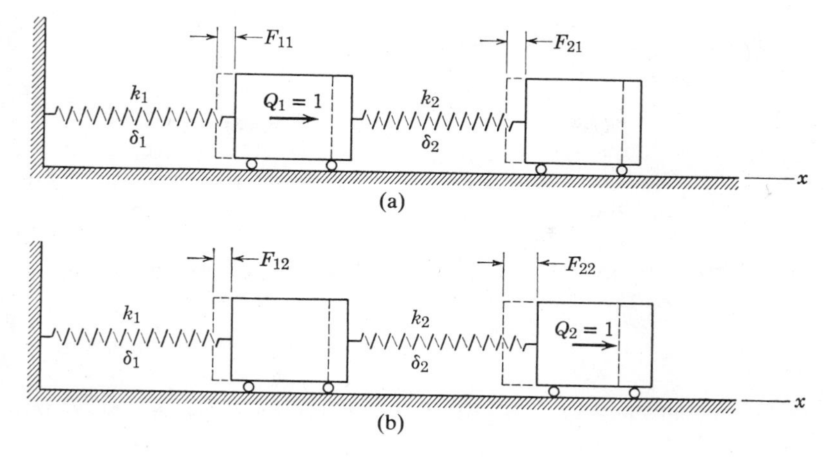

FIG. 3.6

displacements are labeled F_{11} and F_{21} in the figure. The symbol F_{11} represents the displacement of type 1 due to a unit action of type 1, and F_{21} is the displacement of type 2 due to a unit action of type 1. Using notation (a), we determine $F_{11} = F_{21} = \delta_1 = 1/k_1$, and these terms constitute the first column of the flexibility matrix. Terms for the second column of **F** are obtained from Fig. 3.6b, which shows a unit force $Q_2 = 1$ applied statically to the mass m_2, while none is applied to m_1. In this case the flexibilities are $F_{12} = \delta_1 = 1/k_1$ and $F_{22} = \delta_1 + \delta_2 = (k_1 + k_2)/k_1k_2$, which are the displacements of types 1 and 2 due to a unit action of type 2. As with the stiffness matrix, the flexibility matrix for a linearly elastic system is always symmetric† (an inherent characteristic for the inverse of a symmetric matrix), and in this case we have $F_{12} = F_{21} = \delta_1$.

Now let the forces Q_1 and Q_2 be applied dynamically, in which case inertia forces $-m_1\ddot{u}_1$ and $-m_2\ddot{u}_2$ must also be considered. We rewrite Eq. (d) as

$$\begin{bmatrix} u_1 \\ u_2 \end{bmatrix} = \begin{bmatrix} \delta_1 & \delta_1 \\ \delta_1 & \delta_1 + \delta_2 \end{bmatrix} \begin{bmatrix} Q_1 - m_1\ddot{u}_1 \\ Q_2 - m_2\ddot{u}_2 \end{bmatrix} \tag{g}$$

If the masses and accelerations are placed into separate arrays, Eq. (g) takes the expanded form

$$\begin{bmatrix} u_1 \\ u_2 \end{bmatrix} = \begin{bmatrix} \delta_1 & \delta_1 \\ \delta_1 & \delta_1 + \delta_2 \end{bmatrix} \left(\begin{bmatrix} Q_1 \\ Q_2 \end{bmatrix} - \begin{bmatrix} m_1 & 0 \\ 0 & m_2 \end{bmatrix} \begin{bmatrix} \ddot{u}_1 \\ \ddot{u}_2 \end{bmatrix} \right) \tag{3.11}$$

† Symmetry of the flexibility matrix is due to *Maxwell's reciprocal theorem*, presented by J. C. Maxwell in 1864.

which can also be represented succinctly as

$$\mathbf{D} = \mathbf{F}(\mathbf{Q} - \mathbf{M}\ddot{\mathbf{D}}) \tag{3.12}$$

This expression states that the dynamic displacements are equal to the product of the flexibility matrix and the actions in the problem. Both the applied actions and the inertial actions are included within the parentheses on the right-hand side.

In order to compare this method with that in the preceding section, we solve Eq. (3.6) for $\mathbf{D}$, as follows:

$$\mathbf{D} = \mathbf{S}^{-1}(\mathbf{Q} - \mathbf{M}\ddot{\mathbf{D}}) \tag{h}$$

In Eq. (h) it is implied that the stiffness matrix $\mathbf{S}$ is nonsingular and that its inverse $\mathbf{S}^{-1}$ exists. Comparing Eqs. (3.12) and (h), we see that

$$\mathbf{F} = \mathbf{S}^{-1} \tag{3.13}$$

which is a relationship that applies whenever $\mathbf{F}$ and $\mathbf{S}$ correspond to the same coordinates for the same system. For example, if we invert matrix $\mathbf{F}$ in Eq. (f) and use the relationship (a), we find

$$\mathbf{F}^{-1} = \frac{1}{\delta_1\delta_2}\begin{bmatrix} \delta_1 + \delta_2 & -\delta_1 \\ -\delta_1 & \delta_1 \end{bmatrix} = \begin{bmatrix} k_1 + k_2 & -k_2 \\ -k_2 & k_2 \end{bmatrix} \tag{i}$$

which is the stiffness matrix for the system in Fig. 3.1a [see Eq. (b), Sec. 3.2]. Of course, if the stiffness matrix of a system is singular, the corresponding flexibility matrix does not exist.

Because the system in Fig. 3.1a is statically determinate, its flexibility matrix is easily derived, which is usually not true for statically indeterminate systems. The majority of vibrational systems are more easily analyzed using action equations with stiffness coefficients, but many cases also occur where the complementary approach is more convenient. The following examples illustrate the use of flexibility influence coefficients.

Example 1. Figure 3.7a shows a cantilever beam with masses m_1 and m_2 located at midlength and at the free end, respectively. We shall assume that the beam is prismatic and has a flexural rigidity of EI. Considering only small displacements associated with flexural deformations, we take the translations v_1 and v_2 in the y direction as the displacement coordinates. The displacement equations of motion are to be set up for this problem, using flexibility influence coefficients.

Solution. To derive the required flexibility coefficients, we first apply a unit force $Q_1 = 1$, as shown in Fig. 3.7b, and evaluate the terms

$$F_{11} = \frac{\ell^3}{24EI} \qquad F_{21} = \frac{5\ell^3}{48EI} \tag{j}$$

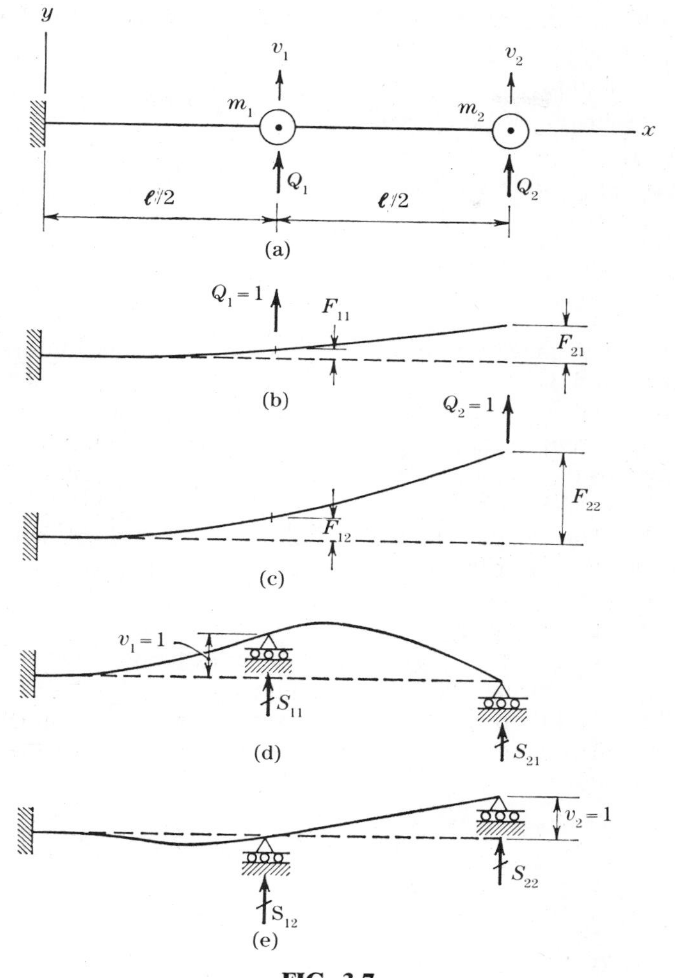

FIG. 3.7

Then we apply a unit force $Q_2 = 1$, as given in Fig. 3.7*c*, to obtain

$$F_{12} = \frac{5\ell^3}{48EI} \qquad F_{22} = \frac{\ell^3}{3EI} \tag{k}$$

Thus, the flexibility matrix is

$$\mathbf{F} = \frac{\ell^3}{48EI}\begin{bmatrix} 2 & 5 \\ 5 & 16 \end{bmatrix} \tag{l}$$

and the displacement equations of motion in the matrix format become

$$\begin{bmatrix} v_1 \\ v_2 \end{bmatrix} = \frac{\ell^3}{48EI}\begin{bmatrix} 2 & 5 \\ 5 & 16 \end{bmatrix}\left(\begin{bmatrix} Q_1 \\ Q_2 \end{bmatrix} - \begin{bmatrix} m_1 & 0 \\ 0 & m_2 \end{bmatrix}\begin{bmatrix} \ddot{v}_1 \\ \ddot{v}_2 \end{bmatrix}\right) \tag{m}$$

Inversion of the flexibility matrix produces

$$\mathbf{S} = \mathbf{F}^{-1} = \frac{48EI}{7\ell^3}\begin{bmatrix} 16 & -5 \\ -5 & 2 \end{bmatrix} \tag{n}$$

This inverse array may also be obtained directly by the procedure indicated in Figs. 3.7*d* and 3.7*e*. However, the direct development of stiffnesses for his type of problem is much more difficult than the determination of flexibilities. Therefore, if stiffnesses are desired, they are more conveniently obtained by inversion of the flexibility matrix, as in this example.

Example 2. The simple frame in Fig. 3.8*a* consists of two prismatic members with flexural rigidities EI. A mass m is connected to the frame at its free end, and the small displacements u_1 and v_1 at that point (due to flexural deformations) are of the same order of magnitude. Write action equations of motion for this system, using u_1 and v_1 as displacement coordinates and omitting gravitational influences.
Solution. As in the preceding example, the flexibilities are much easier to derive than the stiffnesses. Figures 3.8*b* and 3.8*c* show the effects of unit loads $Q_x = 1$ and $Q_y = 1$, applied one at a time. The resulting flexibility matrix is

$$\mathbf{F} = \frac{\ell^3}{6EI}\begin{bmatrix} 8 & 3 \\ 3 & 2 \end{bmatrix} \tag{o}$$

By inversion we obtain

$$\mathbf{S} = \mathbf{F}^{-1} = \frac{6EI}{7\ell^3}\begin{bmatrix} 2 & -3 \\ -3 & 8 \end{bmatrix} \tag{p}$$

Then the action equations of motion may be written as

$$\begin{bmatrix} m & 0 \\ 0 & m \end{bmatrix}\begin{bmatrix} \ddot{u}_1 \\ \ddot{v}_1 \end{bmatrix} + \frac{6EI}{7\ell^3}\begin{bmatrix} 2 & -3 \\ -3 & 8 \end{bmatrix}\begin{bmatrix} u_1 \\ v_1 \end{bmatrix} = \begin{bmatrix} Q_x \\ Q_y \end{bmatrix} \tag{q}$$

Example 3. As a third example of computing flexibilities, we consider the two rigid pendulums in Fig. 3.9*a* that are connected by a torsion bar having rotational stiffness k_r. Displacement equations of motion are to be set up for small rotations (θ_1 and θ_2) of the pendulums about the x axis.

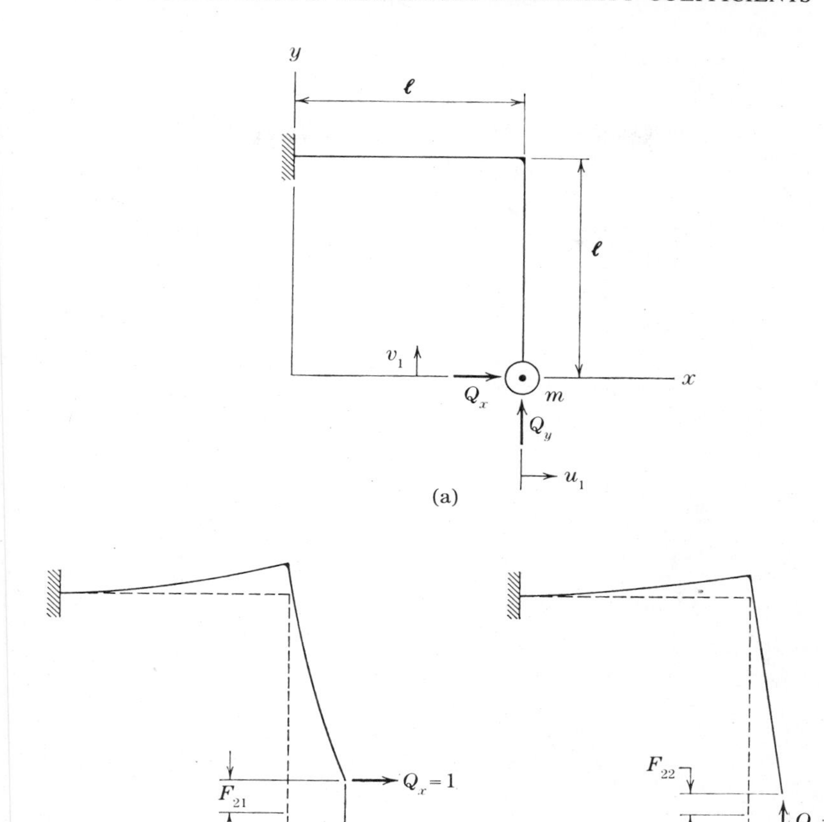

FIG. 3.8

Solution. Because stiffness (and gravitational) coefficients are easy to derive for this system, they will be written directly as [see Eq. (3.10)]

$$\mathbf{S}^* = \mathbf{S} + \mathbf{G} = \begin{bmatrix} k_r & -k_r \\ -k_r & k_r \end{bmatrix} + \begin{bmatrix} mg\ell & 0 \\ 0 & mg\ell \end{bmatrix} \tag{r}$$

We see that the stiffness matrix **S** is singular, and the flexibility matrix $\mathbf{F} = \mathbf{S}^{-1}$ does not exist. However, the inverse of $\mathbf{S}^*$ does exist, and we evaluate it to obtain

$$\mathbf{F}^* = (\mathbf{S}^*)^{-1} = \frac{1}{mg\ell(2k_r + mg\ell)} \begin{bmatrix} k_r + mg\ell & k_r \\ k_r & k_r + mg\ell \end{bmatrix} \tag{s}$$

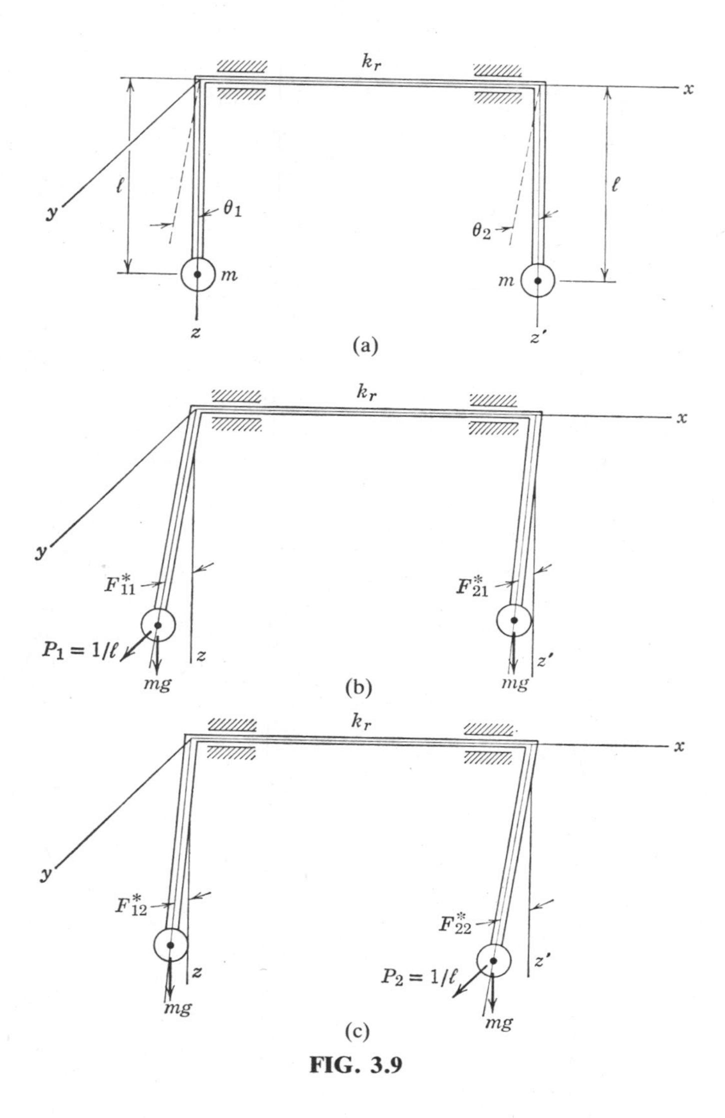

FIG. 3.9

The terms in the array $\mathbf{F}^*$ cannot be segregated into flexibility and gravitational influence coefficients; so they will be referred to as *pseudoflexibilities.* They may be derived directly by applying unit moments (or the equivalent effects in the forms of forces $P_1 = 1/\ell$ and $P_2 = 1/\ell$), as indicated in Figs. 3.9*b* and 3.9*c*. For the former figure we write the moment equilibrium condition

$$mg\ell F_{11}^* + mg\ell F_{21}^* = P_1\ell = 1 \tag{t}$$

and the rotational compatibility condition

$$F_{11}^* - F_{21}^* = \frac{mg\ell F_{21}^*}{k_r} \tag{u}$$

When Eqs. (t) and (u) are solved simultaneously, we find

$$F_{11}^* = \frac{k_r + mg\ell}{mg\ell(2k_r + mg\ell)} \qquad F_{21}^* = \frac{k_r}{mg\ell(2k_r + mg\ell)} \tag{v}$$

which are the same as the terms in the first column of the array in Eq. (s). Similarly, the terms in the second column of $\mathbf{F}^*$ may be derived directly from the situation in Fig. 3.9c. Then the displacement equations for this example, expressed in terms of $\mathbf{F}^*$, become

$$\begin{bmatrix} \theta_1 \\ \theta_2 \end{bmatrix} = \mathbf{F}^* \left(\begin{bmatrix} P_1\ell \\ P_2\ell \end{bmatrix} - \begin{bmatrix} m\ell^2 & 0 \\ 0 & m\ell^2 \end{bmatrix} \begin{bmatrix} \ddot{\theta}_1 \\ \ddot{\theta}_2 \end{bmatrix} \right) \tag{w}$$

3.4 INERTIAL AND GRAVITATIONAL COUPLING

For most of the two-degree systems considered so far in this chapter, the mass and gravitational matrices have been diagonal. Coupling terms in the equations of motion often appear only as off-diagonal elements in the stiffness or flexibility matrices. This type of coupling will be referred to as *elasticity coupling,* because the terms are associated with either the stiffness or flexibility properties of elastic elements. It is also possible to generate off-diagonal terms in the mass and gravitational matrices, depending upon how the equations of motion are written. Terms of the first type often occur in equations of motion for systems containing rigid bodies and will be called *inertial coupling,* while those of the second type will be referred to as *gravitational coupling.*

To demonstrate how inertial coupling can arise, we shall write action equations of motion for the system in Fig. 3.10a for several choices of displacement coordinates. The rigid bar of mass m is supported at points A and D by springs having stiffness constants k_1 and k_2. It is restrained against translation in the x direction and can move only in the x–y plane. Point C represents the center of mass of the bar, and I_C denotes its mass moment of inertia about a z axis (not shown) through C. the point labeled B is that for which

$$k_1\ell_4 = k_2\ell_5 \tag{a}$$

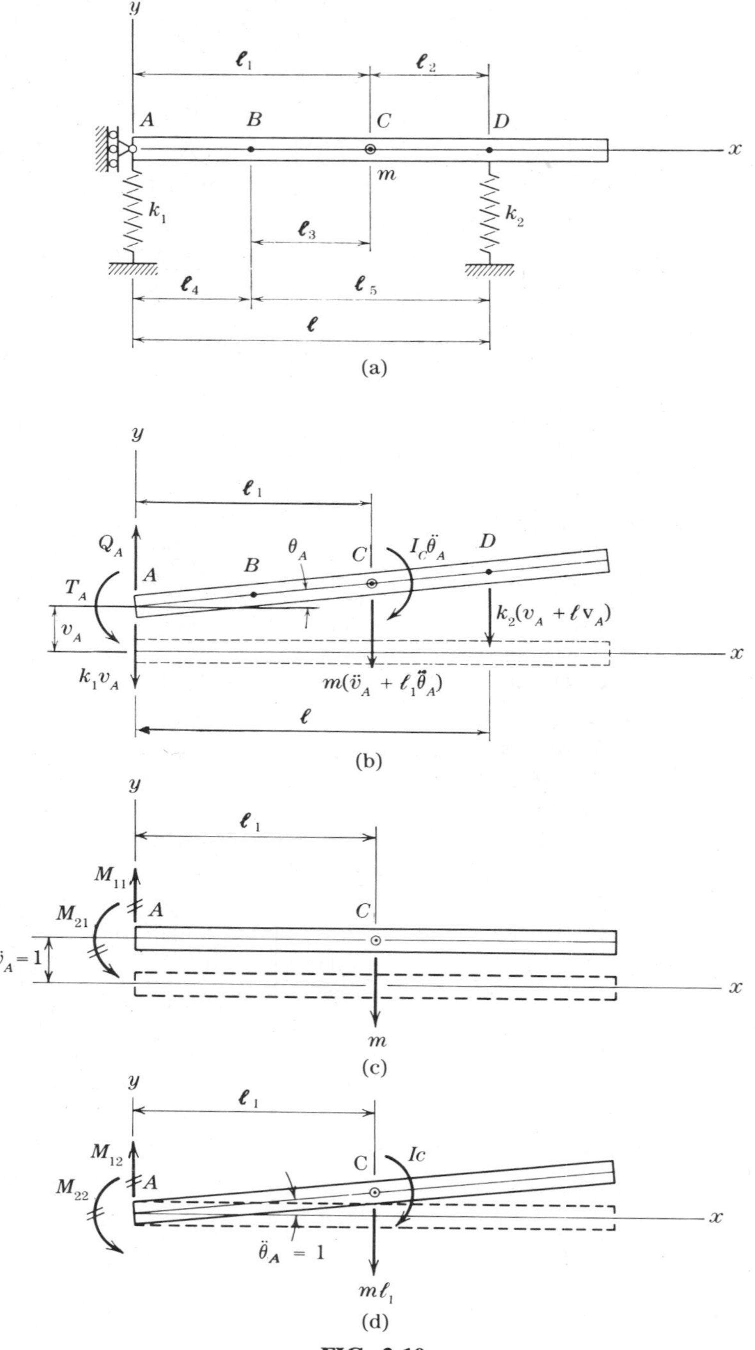

y
ℓ_1
ℓ_2
A
B
C
D
x
m
k_1
k_2
ℓ_3
ℓ_4
ℓ_5
ℓ
(a)
y
ℓ_1
Q_A
θ_A
C
$I_C\ddot{\theta}_A$
D
A
B
T_A
v_A
$k_2(v_A + \ell v_A)$
x
$k_1 v_A$
$m(\ddot{v}_A + \ell_1\ddot{\theta}_A)$
ℓ
(b)
y
ℓ_1
M_{11}
A
C
M_{21}
$\ddot{v}_A = 1$
x
m
(c)
y
ℓ_1
M_{12}
C
Ic
M_{22}
A
x
$\ddot{\theta}_A = 1$
$m\ell_1$
(d)

FIG. 3.10

A force at B in the y direction produces translation without rotation, and a moment produces rotation without translation.

Figure 3.10*b* shows one choice of displacement coordinates for this system, which are taken to be v_A (y translation of point A) and θ_A (rotation of the bar about point A). Also shown on the figure are applied actions (Q_A and T_A) at A, spring forces at A and D, and inertial actions at C. When the last type of actions are shown on a free-body diagram, the body may be considered to be in a state of *dynamic equilibrium*. Then we apply *d'Alembert's principle* to obtain the force equilibrium equation in the y direction as

$$m(\ddot{v}_A + \ell_1\ddot{\theta}_A) + k_1 v_A + k_2(v_A + \ell\theta_A) = Q_A \tag{b}$$

For the second equilibrium equation, we calculate moments with respect to point A and write

$$m(\ddot{v}_A + \ell_1\ddot{\theta}_A)\ell_1 + I_C\ddot{\theta}_A + k_2(v_A + \ell\theta_A)\ell = T_A \tag{c}$$

Equations (b) and (c) in matrix form become

$$\begin{bmatrix} m & m\ell_1 \\ m\ell_1 & I_C + m\ell_1^2 \end{bmatrix}\begin{bmatrix} \ddot{v}_A \\ \ddot{\theta}_A \end{bmatrix} + \begin{bmatrix} k_1 + k_2 & k_2\ell \\ k_2\ell & k_2\ell^2 \end{bmatrix}\begin{bmatrix} v_A \\ \theta_A \end{bmatrix} = \begin{bmatrix} Q_A \\ T_A \end{bmatrix} \tag{3.14a}$$

which exhibit both inertial coupling and elasticity coupling.

As a second choice of displacement coordinates for this system, we take v_B and θ_B (the y translation of point B and the rotation of the bar about point B) and the corresponding applied actions Q_B and T_B. Proceeding as before, we obtain the action equations of motion in matrix form as

$$\begin{bmatrix} m & m\ell_3 \\ m\ell_3 & I_C + m\ell_3^2 \end{bmatrix}\begin{bmatrix} \ddot{v}_B \\ \ddot{\theta}_B \end{bmatrix} + \begin{bmatrix} k_1 + k_2 & 0 \\ 0 & k_1\ell_4^2 + k_2\ell_5^2 \end{bmatrix}\begin{bmatrix} v_B \\ \theta_B \end{bmatrix} = \begin{bmatrix} Q_B \\ T_B \end{bmatrix} \tag{3.14b}$$

in which there is inertial coupling but no elasticity coupling.

For the third choice of coordinates, we use the center of mass (point C) as the reference point for rigid-body motions of the bar. In this case the displacement coordinates are v_C and θ_C (y translation and rotation of the bar at point C), and the corresponding applied actions are Q_C and T_C. The equations of motion for this reference point become

$$\begin{bmatrix} m & 0 \\ 0 & I_C \end{bmatrix}\begin{bmatrix} \ddot{v}_C \\ \ddot{\theta}_C \end{bmatrix} + \begin{bmatrix} k_1 + k_2 & k_2\ell_2 - k_1\ell_1 \\ k_2\ell_2 - k_1\ell_1 & k_1\ell_1^2 + k_2\ell_2^2 \end{bmatrix}\begin{bmatrix} v_C \\ \theta_C \end{bmatrix} = \begin{bmatrix} Q_C \\ T_C \end{bmatrix} \tag{3.14c}$$

which have elasticity coupling but no inertial coupling. Thus, it is apparent that the coupling in a set of equations of motion depends upon the choice of displacement coordinates.

The terms in any mass matrix, such as that in Eq. (3.14*a*), may be considered to be *inertial influence coefficients,* defined as actions required for unit accelerations.

$$\mathbf{M} = \begin{bmatrix} M_{11} & M_{12} \\ M_{21} & M_{22} \end{bmatrix} = \begin{bmatrix} m & m\ell_1 \\ m\ell_1 & I_C + m\ell_1^2 \end{bmatrix} \tag{d}$$

A typical term M_{ij} in a mass matrix is an action of type i required for a unit (instantaneous) acceleration of type j. This definition is analogous to that for a stiffness-influence coefficient, and the formal procedure for determining columns of **M** is similar to that described previously for deriving columns of **S**. Figures 3.10*c* and 3.10*d* illustrate the process when point A is used as the reference for motions of the rigid bar. The actions M_{11} and M_{21}, required for a unit acceleration $\ddot{v}_A = 1$ (while $\ddot{\theta}_A = 0$), appear in Fig. 3.10*c*; and the actions M_{12} and M_{22} required for a unit acceleration $\ddot{\theta}_A = 1$ (while $\ddot{v}_A = 0$) are indicated in Fig. 3.10*d*. (As visual aids, the accelerations are indicated as if they were displacements and double slashes on the arrows at point A serve as reminders that the actions are those required for unit accelerations.) From conditions of dynamic equilibrium, we see that the inertial influence coefficients are $M_{11} = m$, $M_{21} = M_{12} = m\ell_1$, and $M_{22} = I_C + m\ell_1^2$, as given in Eq. (*d*).

It is also possible to derive *inverse inertial influence coefficients,* defined as accelerations due to unit actions, that are analogous to flexibility influence coefficients. The inverse matrix $\mathbf{M}^{-1}$ exists if the array **M** is nonsingular, and solving for the accelerations $\ddot{\mathbf{D}}$ in Eq. (3.6) results in

$$\ddot{\mathbf{D}} = \mathbf{M}^{-1}(\mathbf{Q} - \mathbf{SD}) \tag{3.15}$$

which are acceleration equations of motion. Such equations are comparable to the displacement equations of motion (see Eq. 3.12) discussed in the preceding article. However, this concept is of secondary interest and will not be considered further in this article.

To demonstrate the occurrence of gravitational coupling, let us reconsider the spring-connected pair of pendulums in Fig. 3.4. Their equations of motion (Eqs. 3.4a and 3.4b), as previously written, have elasticity coupling only However, if Eq. (3.4b) is added to Eq. (3.4a) and the result is paired with Eq. (3.4b), we obtain the equally valid set

$$\begin{bmatrix} m\ell^2 & m\ell^2 \\ 0 & m\ell^2 \end{bmatrix}\begin{bmatrix} \ddot{\theta}_1 \\ \ddot{\theta}_2 \end{bmatrix} + \left(\begin{bmatrix} 0 & 0 \\ -kh^2 & kh^2 \end{bmatrix} + \begin{bmatrix} mg\ell & mg\ell \\ 0 & mg\ell \end{bmatrix}\right)\begin{bmatrix} \theta_1 \\ \theta_2 \end{bmatrix} = \begin{bmatrix} T_1 + T_2 \\ T_2 \end{bmatrix} \tag{e}$$

where $T_1 = P_1\ell$ and $T_2 = P_2\ell$. The first equation in the new set expresses dynamic moment equilibrium about point A for the whole system in Fig. 3.4, while the second equation represents moment equilibrium about point B for the right-hand pendulum only. By this linear combination of the original equations, we have introduced off-diagonal terms in both the mass matrix and the gravitational matrix. At the same time, the symmetry of the stiffness matrix has been destroyed. Equation (*e*) may also be visualized as the result of premultiplying Eq. (3.9) in Sec. 3.2 by the transposed operator $\mathbf{A}^{\mathrm{T}}$, where

$$\mathbf{A} = \begin{bmatrix} 1 & 0 \\ 1 & 1 \end{bmatrix} \qquad \mathbf{A}^{\mathrm{T}} = \begin{bmatrix} 1 & 1 \\ 0 & 1 \end{bmatrix} \tag{f}$$

and the superscript T denotes transposition. Thus, in succinct matrix notation we restate Eq. (*e*) as

$$\mathbf{A}^{\mathrm{T}}\mathbf{M}\ddot{\mathbf{D}} + \mathbf{A}^{\mathrm{T}}(\mathbf{S} + \mathbf{G})\mathbf{D} = \mathbf{A}^{\mathrm{T}}\mathbf{T} \tag{g}$$

Symmetry in all of the coefficient matrices will be restored if we insert before $\ddot{\mathbf{D}}$ and $\mathbf{D}$ in Eq. (*g*) the identity matrix

$$\mathbf{I} = \mathbf{A}\mathbf{A}^{-1} \tag{h}$$

where $\mathbf{A}^{-1}$ is the inverse of the operator $\mathbf{A}$.

$$\mathbf{A}^{-1} = \begin{bmatrix} 1 & 0 \\ -1 & 1 \end{bmatrix} \tag{i}$$

Then Eq. (*g*) becomes

$$\mathbf{A}^{\mathrm{T}}\mathbf{M}\mathbf{A}\mathbf{A}^{-1}\ddot{\mathbf{D}} + \mathbf{A}^{\mathrm{T}}(\mathbf{S} + \mathbf{G})\mathbf{A}\mathbf{A}^{-1}\ddot{\mathbf{D}} = \mathbf{A}^{\mathrm{T}}\mathbf{T} \tag{j}$$

or

$$\mathbf{M}_{\mathrm{A}}\ddot{\mathbf{D}}_{\mathrm{A}} + (\mathbf{S}_{\mathrm{A}} + \mathbf{G}_{\mathrm{A}})\mathbf{D}_{\mathrm{A}} = \mathbf{T}_{\mathrm{A}} \tag{3.16}$$

where

$$\begin{aligned} \mathbf{D}_{\mathrm{A}} &= \mathbf{A}^{-1}\mathbf{D} = \begin{bmatrix} \theta_1 \\ -\theta_1 + \theta_2 \end{bmatrix} & \mathbf{T}_{\mathrm{A}} &= \mathbf{A}^{\mathrm{T}}\mathbf{T} = \begin{bmatrix} T_1 + T_2 \\ T_2 \end{bmatrix} \\ \ddot{\mathbf{D}}_{\mathrm{A}} &= \mathbf{A}^{-1}\ddot{\mathbf{D}} = \begin{bmatrix} \ddot{\theta}_1 \\ -\ddot{\theta}_1 + \ddot{\theta}_2 \end{bmatrix} & \mathbf{M}_{\mathrm{A}} &= \mathbf{A}^{\mathrm{T}}\mathbf{M}\mathbf{A} = m\ell^2\begin{bmatrix} 2 & 1 \\ 1 & 1 \end{bmatrix} \\ \mathbf{S}_{\mathrm{A}} &= \mathbf{A}^{\mathrm{T}}\mathbf{S}\mathbf{A} = kh^2\begin{bmatrix} 0 & 0 \\ 0 & 1 \end{bmatrix} & \mathbf{G}_{\mathrm{A}} &= \mathbf{A}^{\mathrm{T}}\mathbf{G}\mathbf{A} = mg\ell\begin{bmatrix} 2 & 1 \\ 1 & 1 \end{bmatrix} \end{aligned} \tag{k}$$

Equation (3.16) represents an alternative set of action equations of motion, wherein the *generalized actions* are $\mathbf{T}_{\mathrm{A}}$; and the corresponding *generalized displacements* are $\mathbf{D}_A$. Changing the coordinates (from $\mathbf{D}$ to $\mathbf{D}_A$) in this manner is referred to as a *coordinate transformation.* Symmetry in the transformed coefficient matrices is maintained by virtue of the fact that *congruence transformations,* such as $\mathbf{M}_A = \mathbf{A}^T\mathbf{M}\mathbf{A}$, and so on, produce symmetric results. In the new coordinates we see that the equations have both inertial and gravitational coupling but that there is no elasticity coupling.

Example 1. Figure 3.11 shows a rigid body of mass m connected to the end of a cantilever beam. Let I_C represent the mass moment of inertia of the body about a z axis through its center of mass (point C). This point is located on the x axis at the distance b from the end of the beam. We shall assume that the beam is prismatic and has a flexural rigidity of EI. Considering only small displacements in the x–y plane due to flexural deformations, we will treat the system as one having two degrees of freedom and write displacement equations of motion.

Solution. If we choose the translation v_B and the rotation θ_B (for point B on the rigid body) as displacement coordinates, the flexibility coefficients are easily obtained. In addition, the terms in the mass matrix are similar to those in Eq. (d), except that ℓ_1 is replaced by the distance b. Thus, we can write the displacement equations of motion for reference point B as

$$\begin{bmatrix} v_B \\ \theta_B \end{bmatrix} = \frac{\ell}{6EI}\begin{bmatrix} 2\ell^2 & 3\ell \\ 3\ell & 6 \end{bmatrix}\left(\begin{bmatrix} Q_B \\ T_B \end{bmatrix} - \begin{bmatrix} m & mb \\ mb & I_C + mb^2 \end{bmatrix}\begin{bmatrix} \ddot{v}_B \\ \ddot{\theta}_B \end{bmatrix}\right) \tag{l}$$

which have both elasticity and inertial coupling.

On the other hand, if point C is used as the reference point for rigid-body motions, we obtain the following displacement equations:

$$\begin{bmatrix} v_C \\ \theta_C \end{bmatrix} = \frac{\ell}{6EI}\begin{bmatrix} 2(\ell^2 + 3\ell b + 3b^2) & 3(\ell + 2b) \\ 3(\ell + 2b) & 6 \end{bmatrix}\left(\begin{bmatrix} Q_C \\ T_C \end{bmatrix} - \begin{bmatrix} m & 0 \\ 0 & I_C \end{bmatrix}\begin{bmatrix} \ddot{v}_C \\ \ddot{\theta}_C \end{bmatrix}\right) \tag{m}$$

which have a more complicated elasticity coupling but no inertial coupling.

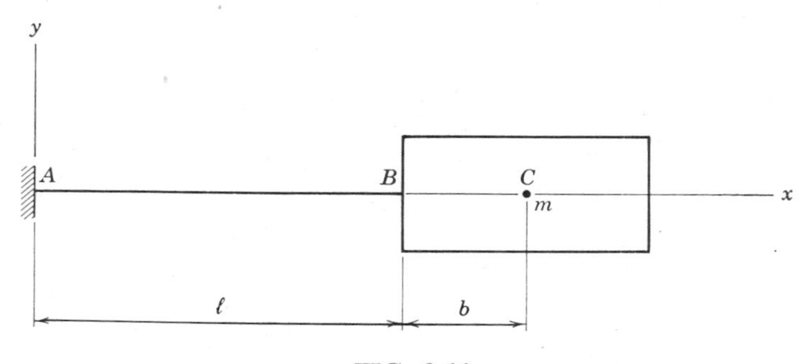

FIG. 3.11

Example 2. Consider the *double compound pendulum* shown in Fig. 3.12*a*, consisting of two rigid bodies hinged together at point B and hinged to the support at point A. In the presence of gravity, this system may oscillate in the x–y plane, and we take the small rotations θ_1 and θ_2 as displacement coordinates. The bodies have masses m_1 and m_2: their centers of masses are located at points C_1 and C_2; and their mass moments of

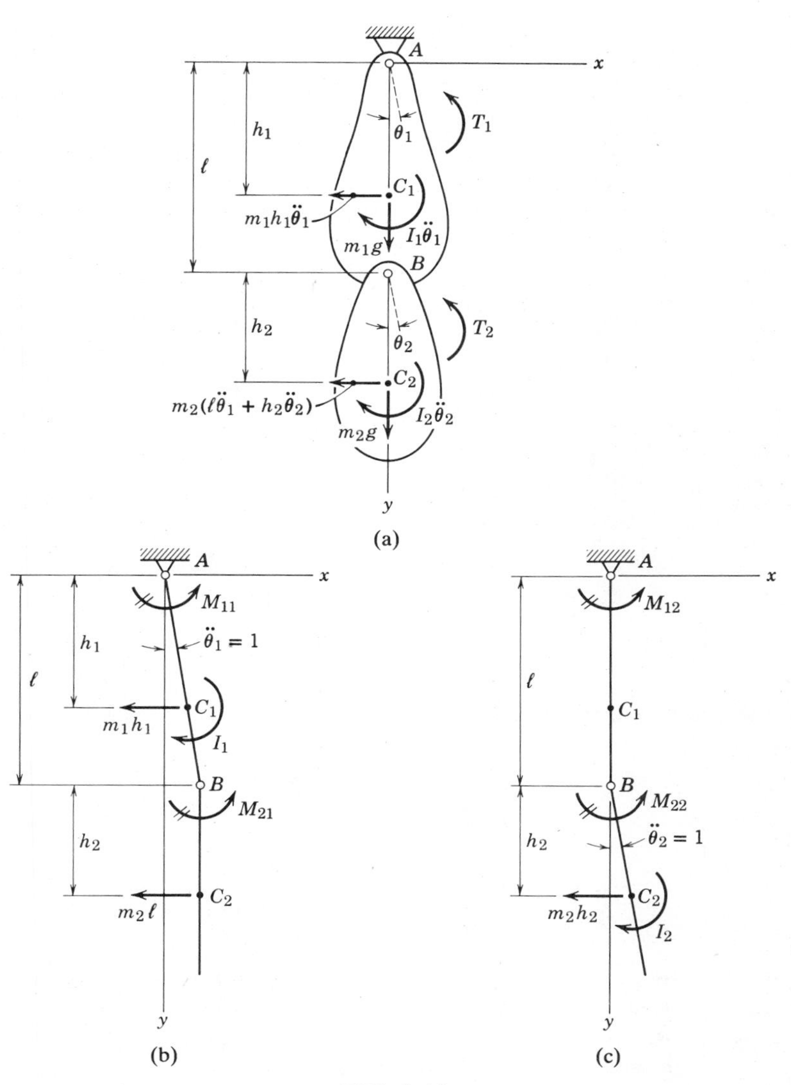

FIG. 3.12

inertia about z axes through those points are denoted by I_1 and I_2. Action equations of motion are to be written for this system.

Solution. Using d'Alembert's principle, we write a moment equation of dynamic equilibrium about point A for the whole system (see Fig. 3.12a) to obtain

$$I_1\ddot{\theta}_1 + I_2\ddot{\theta}_2 + m_1 h_1^2 \ddot{\theta}_1 + m_2(\ell\ddot{\theta}_1 + h_2\ddot{\theta}_2)(\ell + h_2) + m_1 g h_1 \theta_1 + m_2 g(\ell\theta_1 + h_2\theta_2) = T_1 + T_2 \qquad (n)$$

Similarly, the moment-equilibrium condition about point B for the second body yields

$$I_2\ddot{\theta}_2 + m_2(\ell\ddot{\theta}_1 + h_2\ddot{\theta}_2)h_2 + m_2 g h_2 \theta_2 = T_2 \qquad (o)$$

Putting Eqs. (n) and (o) into matrix form, we have

$$\begin{bmatrix} I_1 + m_1 h_1^2 + m_2\ell(\ell + h_2) & I_2 + m_2 h_2(\ell + h_2) \\ m_2\ell h_2 & I_2 + m_2 h_2^2 \end{bmatrix} \begin{bmatrix} \ddot{\theta}_1 \\ \ddot{\theta}_2 \end{bmatrix} + \begin{bmatrix} (m_1 h_1 + m_2\ell)g & m_2 h_2 g \\ 0 & m_2 h_2 g \end{bmatrix} \begin{bmatrix} \theta_1 \\ \theta_2 \end{bmatrix} = \begin{bmatrix} T_1 + T_2 \\ T_2 \end{bmatrix} \qquad (p)$$

Equation (p) resembles Eq. (e) because the generalized actions do not correspond to the displacement coordinates, and the coefficient matrices are unsymmetric. However, if we subtract Eq. (o) from Eq. (n) and pair the result with Eq. (o), we obtain the set

$$\mathbf{M}\ddot{\mathbf{D}} + \mathbf{G}\mathbf{D} = \mathbf{T} = \begin{bmatrix} T_1 \\ T_2 \end{bmatrix} \qquad (q)$$

where $\mathbf{M}$ and $\mathbf{G}$ are the following symmetric matrices:

$$\mathbf{M} = \begin{bmatrix} I_1 + m_1 h_1^2 + m_2\ell^2 & m_2\ell h_2 \\ m_2\ell h_2 & I_2 + m_2 h_2^2 \end{bmatrix} \qquad (r)$$

and

$$\mathbf{G} = \begin{bmatrix} (m_1 h_1 + m_2\ell)g & 0 \\ 0 & m_2 h_2 g \end{bmatrix} \qquad (s)$$

The first equation now represents dynamic moment equilibrium about point A for the first body only.

The symmetric arrays $\mathbf{M}$ and $\mathbf{G}$ may be generated directly as inertial and gravitational influence coefficients, respectively. Figures 3.12b and 3.12c show schematically the conditions $\ddot{\theta}_1 = 1$ (while $\ddot{\theta}_2 = 0$) and $\ddot{\theta}_2 = 1$ (while

$\ddot{\theta}_1 = 0$) that are required for developing the terms in **M**. From the first of these figures we see that

$$M_{21} = m_2 \ell h_2 \tag{t}$$

and

$$M_{11} = I_1 + m_1 h_1^2 + m_2 \ell(\ell + h_2) - M_{21}$$
$$= I_1 + m_1 h_1^2 + m_2 \ell^2 \tag{u}$$

which constitute the first column of **M** [see Eq. (*r*)]. From Fig. 3.12*c* we obtain the terms for the second column as

$$M_{22} = I_2 + m_2 h_2^2 \tag{v}$$

and

$$M_{12} = I_2 + m_2 h_2(\ell + h_2) - M_{22} = m_2 \ell h_2 \tag{w}$$

The elements in **G** may be generated in similar fashion, using unit displacements instead of accelerations.

3.5 UNDAMPED FREE VIBRATIONS

In Sec. 3.1 we derived expressions for the natural frequencies and mode shapes for undamped free vibrations of the two-mass system in Fig. 3.1*a*. That task will now be repeated in a more formal manner that yields expressions applicable to all vibratory systems with two degrees of freedom. Both action equations and displacement equations will be considered, and the determination of arbitrary constants from initial conditions of displacement and velocity will be covered. In addition, several special topics related to free vibrations will be discussed and illustrated with examples.

If there are no loads applied to the two-mass system in Fig. 3.1*a*, the action equations of motion [Eq. (3.6)] become

$$\mathbf{M\ddot{D}} + \mathbf{SD} = \mathbf{0} \tag{3.17}$$

where the symbol **0** denotes a null load matrix. For the discussion in this section, we shall consider only diagonal mass matrices (as is the case for the system in Fig. 3.1*a*) and write Eq. (3.17) in expanded form as

$$\begin{bmatrix} M_{11} & 0 \\ 0 & M_{22} \end{bmatrix} \begin{bmatrix} \ddot{u}_1 \\ \ddot{u}_2 \end{bmatrix} + \begin{bmatrix} S_{11} & S_{12} \\ S_{21} & S_{22} \end{bmatrix} \begin{bmatrix} u_1 \\ u_2 \end{bmatrix} = \begin{bmatrix} 0 \\ 0 \end{bmatrix} \tag{3.18}$$

For this homogeneous set of equations, we assume harmonic solutions of the form introduced previously in Sec. 3.1. Thus,

$$u_1 = u_{m1} \sin(\omega t + \phi) \tag{a}$$

$$u_2 = u_{m2} \sin(\omega t + \phi) \tag{b}$$

The symbols u_{m1} and u_{m2} in Eqs. (*a*) and (*b*) represent maximum values, or amplitudes, of the vibratory motions.

Substituting Eqs. (*a*) and (*b*) and their second derivatives into Eq. (3.18) results in the following algebraic equations that must be satisfied:

$$-\omega^2 M_{11} u_{m1} + S_{11} u_{m1} + S_{12} u_{m2} = 0 \tag{c}$$

$$-\omega^2 M_{22} u_{m2} + S_{21} u_{m1} + S_{22} u_{m2} = 0 \tag{d}$$

Or,

$$\begin{bmatrix} S_{11} - \omega^2 M_{11} & S_{12} \\ S_{21} & S_{22} - \omega^2 M_{22} \end{bmatrix} \begin{bmatrix} u_{m1} \\ u_{m2} \end{bmatrix} = \begin{bmatrix} 0 \\ 0 \end{bmatrix} \tag{e}$$

In order to have nonzero solutions for the displacements, the determinant of the coefficient matrix in Eq. (*e*) must be equal to zero. Thus,

$$\begin{vmatrix} S_{11} - \omega^2 M_{11} & S_{12} \\ S_{21} & S_{22} - \omega^2 M_{22} \end{vmatrix} = 0 \tag{f}$$

Expanding this determinant produces

$$(S_{11} - \omega^2 M_{11})(S_{22} - \omega^2 M_{22}) - S_{12}^2 = 0 \tag{g}$$

Or,

$$M_{11} M_{22} (\omega^2)^2 - (M_{11} S_{22} + M_{22} S_{11})\omega^2 + S_{11} S_{22} - S_{12}^2 = 0 \tag{h}$$

This characteristic equation is quadratic in ω^2, and its roots represent the characteristic values for the system. Solving Eq. (*h*) by the quadratic formula, we obtain

$$\omega_{1,2}^2 = \frac{-b \mp \sqrt{b^2 - 4ac}}{2a} \tag{3.19}$$

where

$$a = M_{11} M_{22} \qquad b = -(M_{11} S_{22} + M_{22} S_{11}) \qquad c = S_{11} S_{22} - S_{12}^2 = |\mathbf{S}| \tag{i}$$

Expansion of the term $b^2 - 4ac$ in Eq. (3.19) using expressions (*i*) shows that it is always positive; therfore, both of the roots ω_1^2 and ω_2^2 are real. Furthermore, if the determinant of **S** (equal to the constant c) is not negative, the square-root term will be smaller than or equal to b; so the roots ω_1^2 and ω_2^2 are both positive (or zero).

By substituting the characteristic values ω_1^2 and ω_2^2 into the homogeneous equations (c) and (d), we may evolve the solutions as the amplitude ratios, r_1 and r_2.

$$r_1 = \frac{u_{m1,1}}{u_{m2,1}} = \frac{-S_{12}}{S_{11} - \omega_1^2 M_{11}} = \frac{S_{22} - \omega_1^2 M_{22}}{-S_{21}} \tag{3.20a}$$

$$r_2 = \frac{u_{m1,2}}{u_{m2,2}} = \frac{-S_{12}}{S_{11} - \omega_2^2 M_{11}} = \frac{S_{22} - \omega_2^2 M_{22}}{-S_{21}} \tag{3.20b}$$

Both of the definitions for each amplitude ratio are valid by virtue of Eq. (g). As is the case for homogeneous algebraic equations in general, such solutions may be obtained only within arbitrary constants. That is, the absolute values of the amplitudes cannot be determined, but only their ratios, or mode shapes. The second subscript (1 or 2) on the amplitudes in Eqs. (3.20a) and (3.20b) denote the natural modes (or principal modes) of vibration, corresponding to the roots ω_1^2 and ω_2^2. As in Sec. 3.1, the solution [Eq. (3.19)] of the characteristic equation is written in such a manner that $\omega_1 < \omega_2$. The lower value is the angular frequency of the first mode (or fundamental mode), and the larger value is that for the second mode.

To demonstrate the calculation of frequencies and mode shapes, let the masses for the system in Fig. 3.1a be $m_1 = m_2 = m$ and the stiffness constants of the springs be $k_1 = k_2 = k$. Then $M_{11} = M_{22} = m$; $S_{11} = 2k$; $S_{12} = S_{21} = -k$; $S_{22} = k$; and Eq. (3.19) yields

$$\omega_1^2 = \frac{3 - \sqrt{5}}{2} \frac{k}{m} = 0.382 \frac{k}{m} \tag{j}$$

$$\omega_2^2 = \frac{3 + \sqrt{5}}{2} \frac{k}{m} = 2.618 \frac{k}{m} \tag{k}$$

Substitution of each of these roots into either of the definitions for amplitude ratios in Eqs. (3.20a) and (3.20b) produces the following mode shapes:

$$\text{first mode:} \qquad r_1 = \frac{u_{m1,1}}{u_{m2,1}} = \frac{2}{1 + \sqrt{5}} = \frac{-1 + \sqrt{5}}{2} = 0.618 \tag{l}$$

$$\text{second mode:} \qquad r_2 = \frac{u_{m1,2}}{u_{m2,2}} = \frac{2}{1 - \sqrt{5}} = \frac{-1 - \sqrt{5}}{2} = -1.618 \tag{m}$$

Of course, the amplitudes of these two mode shapes may be scaled as desired, but their ratios remain constant. Figure 3.13a shows the two-mass system vibrating in its fundamental mode, where the amplitudes are normalized with respect to that of the second mass. Similarly, the shape of

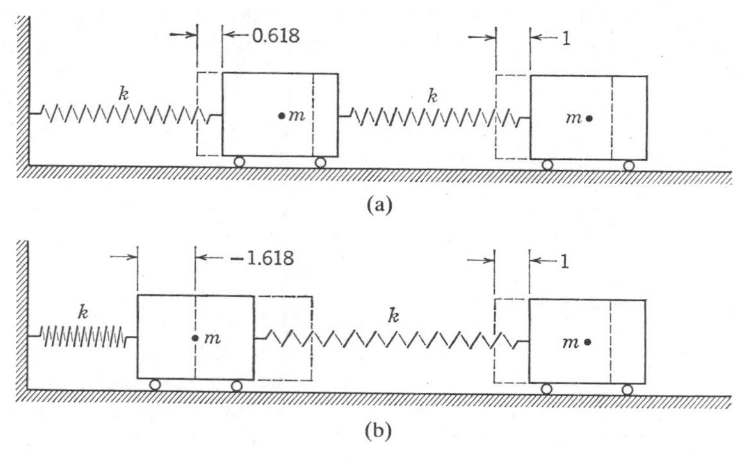

FIG. 3.13

the second mode of vibration is depicted in Fig. 3.13*b*, where the amplitudes are again normalized with respect to that of the second mass.

If we work with displacement equations of motion instead of action equations, Eq. (3.17) is replaced by

$$\mathbf{FM\ddot{D}} + \mathbf{D} = \mathbf{0} \tag{3.21}$$

In expanded form, this set of equations may be written as

$$\begin{bmatrix} F_{11}M_{11} & F_{12}M_{22} \\ F_{21}M_{11} & F_{22}M_{22} \end{bmatrix} \begin{bmatrix} \ddot{u}_1 \\ \ddot{u}_2 \end{bmatrix} + \begin{bmatrix} u_1 \\ u_2 \end{bmatrix} = \begin{bmatrix} 0 \\ 0 \end{bmatrix} \tag{3.22}$$

Substituting Eqs. (*a*) and (*b*) into Eq. (3.22), we obtain the algebraic equations

$$-\omega^2 F_{11}M_{11}u_{m1} - \omega^2 F_{12}M_{22}u_{m2} + u_{m1} = 0 \tag{n}$$

$$-\omega^2 F_{21}M_{11}u_{m1} - \omega^2 F_{22}M_{22}u_{m2} + u_{m2} = 0 \tag{o}$$

Or,

$$\begin{bmatrix} F_{11}M_{11} - \lambda & F_{12}M_{22} \\ F_{21}M_{11} & F_{22}M_{22} - \lambda \end{bmatrix} \begin{bmatrix} u_{m1} \\ u_{m2} \end{bmatrix} = \begin{bmatrix} 0 \\ 0 \end{bmatrix} \tag{p}$$

in which $\lambda = 1/\omega^2$. For nontrivial solutions, the determinant of the coefficient matrix in Eq. (*p*) must be set equal to zero, producing

$$(F_{11}M_{11} - \lambda)(F_{22}M_{22} - \lambda) - F_{12}^2 M_{11}M_{22} = 0 \tag{q}$$

or,

$$\lambda^2 - (F_{11}M_{11} + F_{22}M_{22})\lambda + (F_{11}F_{22} - F_{12}^2)M_{11}M_{22} = 0 \tag{r}$$

Expression (r) is the characteristic equation corresponding to the homogeneous displacement equations of motion, and its roots are the reciprocals of the squares of the angular frequencies. We calculate the roots either from

$$\lambda_{1,2} = \frac{-d \pm \sqrt{d^2 - 4e}}{2} \tag{3.23a}$$

or from

$$\omega_{1,2}^2 = \frac{1}{\lambda_{1,2}} = \frac{2}{-d \pm \sqrt{d^2 - 4e}} \tag{3.23b}$$

where

$$\begin{aligned} d &= -(F_{11}M_{11} + F_{22}M_{22}) \\ e &= (F_{11}F_{22} - F_{12}^2)M_{11}M_{22} = |\mathbf{F}|\, M_{11}M_{22} \end{aligned} \tag{s}$$

In Eqs. (3.23a) and (3.23b) the value λ_1 (the larger value of λ) corresponds to ω_1^2 (the smaller value of ω^2), and the value λ_2 (the smaller value of λ) corresponds to ω_2^2 (the larger value of ω^2). Both of the roots (and their reciprocals will be real and positive if the determinant of $\mathbf{F}$ is positive.

Upon substituting the characteristic values λ_1 and λ_2 into the homogeneous equations [see Eq. (p)], we obtain dual definitions for the amplitude ratios r_1 and r_2 as

$$r_1 = \frac{u_{\mathrm{m}1,1}}{u_{\mathrm{m}2,1}} = \frac{-F_{12}M_{22}}{F_{11}M_{11} - \lambda_1} = \frac{F_{22}M_{22} - \lambda_1}{-F_{21}M_{11}} \tag{3.24a}$$

$$r_2 = \frac{u_{\mathrm{m}1,2}}{u_{\mathrm{m}2,2}} = \frac{-F_{12}M_{22}}{F_{11}M_{11} - \lambda_2} = \frac{F_{22}M_{22} - \lambda_2}{-F_{21}M_{11}} \tag{3.24b}$$

which are all valid by virtue of Eq. (q).

If we again assume equal masses and equal spring constants for the system in Fig. 3.1a, the flexibility influence coefficients are $F_{11} = \delta$, $F_{12} = F_{21} = \delta$, and $F_{22} = 2\delta$ (where $\delta = 1/k$). Then Eq. (3.23a) gives the values

$$\lambda_1 = \frac{(3 + \sqrt{5})m\delta}{2} = \frac{m\delta}{0.382} \tag{t}$$

$$\lambda_2 = \frac{(3 - \sqrt{5})m\delta}{2} = \frac{m\delta}{2.618} \tag{u}$$

which are the reciprocals of those in Eqs. (j) and (k). Substitution of Eqs. (t) and (u) into Eqs. (3.24a) and (3.24b) yields the following amplitude ratios:

first mode: $$r_1 = \frac{u_{m1,1}}{u_{m2,1}} = \frac{2}{1+\sqrt{5}} = \frac{-1+\sqrt{5}}{2} = 0.618$$

second mode: $$r_2 = \frac{u_{m1,2}}{u_{m2,2}} = \frac{2}{1-\sqrt{5}} = \frac{-1-\sqrt{5}}{2} = -1.618$$

which are the same as Eqs. (l) and (m).

Having determined the vibrational characteristics of a given system, we can write the complete solution for free vibrations by superposition of the natural modes, as follows:

$$u_1 = r_1 u_{m2,1} \sin(\omega_1 t + \phi_1) + r_2 u_{m2,2} \sin(\omega_2 t + \phi_2) \tag{v}$$

$$u_2 = u_{m2,1} \sin(\omega_1 t + \phi_1) + u_{m2,2} \sin(\omega_2 t + \phi_2) \tag{w}$$

In the first of these expressions, $r_1 u_{m2,1}$ and $r_2 u_{m2,2}$ are used in place of $u_{m1,1}$ and $u_{m1,2}$, respectively. Equations (v) and (w) may also be written in the equivalent forms

$$u_1 = r_1(C_1 \cos \omega_1 t + C_2 \sin \omega_1 t) + r_2(C_3 \cos \omega_2 t + C_4 \sin \omega_2 t) \tag{3.25a}$$

$$u_2 = C_1 \cos \omega_1 t + C_2 \sin \omega_1 t + C_3 \cos \omega_2 t + C_4 \sin \omega_2 t \tag{3.25b}$$

Velocity expressions are obtained by differentiating Eqs. (3.25a) and (3.25b) with respect to time.

$$\dot{u}_1 = -\omega_1 r_1(C_1 \sin \omega_1 t - C_2 \cos \omega_1 t) - \omega_2 r_2(C_3 \sin \omega_2 t - C_4 \cos \omega_2 t) \tag{3.25c}$$

$$\dot{u}_2 = -\omega_1(C_1 \sin \omega_1 t - C_2 \cos \omega_1 t) - \omega_2(C_3 \sin \omega_2 t - C_4 \cos \omega_2 t) \tag{3.25d}$$

The four arbitrary constants C_1–C_4 in Eqs. (3.25a) through (3.25d) may be determined from the four initial conditions of displacement and velocity. For a two-degree system at time $t = 0$, these conditions will be denoted by the symbols u_{01}, $\dot{u}_{01}$, u_{02}, and $\dot{u}_{02}$. Substituting the initial conditions into Eqs. (3.25a) through (3.25d), we evaluate the constants as

$$\begin{aligned} C_1 &= \frac{u_{01} - r_2 u_{02}}{r_1 - r_2} & \quad C_2 &= \frac{\dot{u}_{01} - r_2 \dot{u}_{02}}{\omega_1(r_1 - r_2)} \\ C_3 &= \frac{r_1 u_{02} - u_{01}}{r_1 - r_2} & \quad C_4 &= \frac{r_1 \dot{u}_{02} - \dot{u}_{01}}{\omega_2(r_1 - r_2)} \end{aligned} \tag{3.26}$$

To demonstrate the use of these expressions, we will calculate the response of the system in Fig. 3.1a to the initial conditions $u_{01} = u_{02} = 1$ and $\dot{u}_{01} = \dot{u}_{02} = 0$. The values of ω_1, ω_2, r_1, and r_2 are already available [see Eqs.

(j), (k), (l), and (m)] for the system with equal masses and equal spring constants. Substituting the known values and conditions into Eqs. (3.26), we find the constants to be $C_1 = 1.171$, $C_2 = 0$, $C_3 = -0.171$, and $C_4 = 0$. Then from Eqs. (3.25a) and (3.25b) the response of the system becomes

$$u_1 = 0.724 \cos \omega_1 t + 0.276 \cos \omega_2 t \tag{x}$$

$$u_2 = 1.171 \cos \omega_1 t - 0.171 \cos \omega_2 t \tag{y}$$

In this case there are only cosine terms in the response, because the initial velocities are zero. If the initial displacements were zero and the initial velocities were nonzero, only sine terms would appear in the response. Furthermore, both of the natural modes of the system contribute terms to the response, except in cases where the initial conditions coincide with a possible motion of one of those modes. For example, if the initial displacements exactly match the pattern of the first mode ($u_{01}/u_{02} = r_1$), while $\dot{u}_{01} = \dot{u}_{02} = 0$, the response becomes

$$u_1 = u_{01} \cos \omega_1 t \qquad u_2 = u_{02} \cos \omega_1 t$$

which consists of pure first-mode response.

In summary, the assumption of natural modes of vibration in the form of Eqs. (a) and (b) enables us to convert homogeneous differential equations for free vibrations [such as Eqs. (3.17) or (3.21)] to a set of algebraic equations. Setting the determinant of their coefficient matrix equal to zero, we obtain the characteristic equation and evaluate the frequencies and mode shapes. In this manner the form of the solution is ascertained, but the amounts of the modes contributing to the total response must be determined from the initial conditions.

For the purpose of discussing several special situations related to free vibrations, we will consider again the spring-connected pair of simple pendulums in Fig. 3.4. For this system the terms in the diagonal mass matrix [see Eq. (3.9), Sec. 3.2] are $M_{11} = M_{22} = m\ell^2$. In addition, the array $\mathbf{S}^*$, containing gravitational terms, replaces the matrix $\mathbf{S}$; so we have $S_{11}^* = S_{22}^* = kh^2 + mg\ell$, and $S_{12}^* = S_{21}^* = -kh^2$. Using these terms, we obtain from Eq. (3.19) the natural angular frequencies as

$$\omega_1 = \sqrt{\frac{g}{\ell}} \qquad \omega_2 = \sqrt{\frac{g}{\ell} + \frac{2kh^2}{m\ell^2}} \tag{z}$$

Then the amplitude ratios may be found from Eqs. (3.20a) and (3.20b) to be simply

$$r_1 = 1 \qquad r_2 = -1 \tag{a'}$$

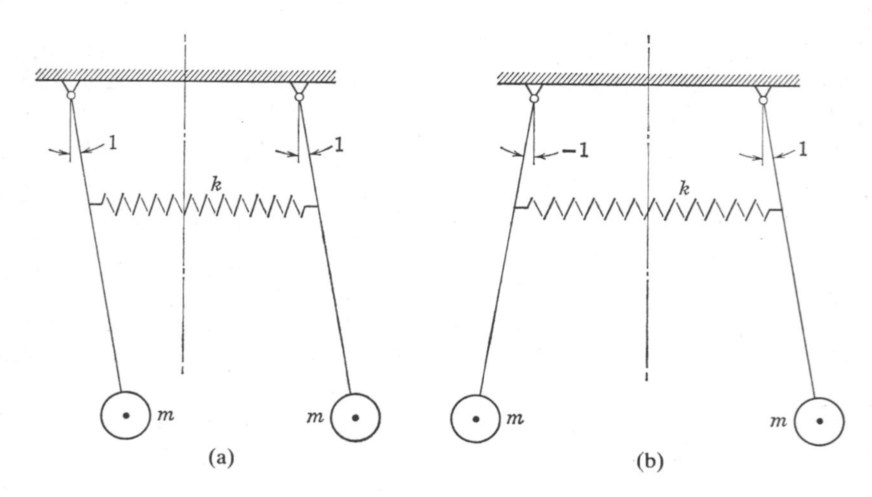

FIG. 3.14

The two natural modes of vibration are shown in Figs. 3.14*a* and 3.14*b*, where the amplitudes are normalized with respect to that of the right-hand pendulum. In the first mode the pendulums swing in the same direction and at the same amplitude (as if they were one), without deforming the spring. In the second mode they swing in opposite directions at equal amplitudes, and the spring is periodically stretched and compressed.

This system is symmetric with respect to a vertical plane midway between the pendulums. We see from Fig. 3.14*b* that the second natural mode is symmetric with respect to that plane, and it is called a *symmetric mode of vibration*. Half the system may be used to represent this mode by restraining the midpoint of the spring against motion (in which case the effective stiffness constant is $2k$ for half the spring). On the other hand, the first natural mode in Fig. 3.14*a* is antisymmetric with respect to the plane of symmetry; and it is referred to as an *antisymmetric mode of vibration*. In this case half the system can be used if the midpoint of the spring is allowed to displace freely across the plane of symmetry (so that the effective stiffness constant of half the spring is zero). In general, a vibratory system having a single plane of symmetry will possess only symmetric and antisymmetric modes with respect to that plane, and two reduced systems may be analyzed instead. One of the reduced systems must be restrained at the plane of symmetry to allow only symmetric patterns of displacements, while the other is allowed to displace only in antisymmetric configurations.

If the spring-connected pendulums are not subjected to a gravitational field, the matrix $\mathbf{S}^*$ reverts to $\mathbf{S}$, which is singular. In this situation the roots in Eqs. (z) become

$$\omega_1 = 0 \qquad \omega_2 = \frac{h}{\ell}\sqrt{\frac{2k}{m}} \tag{b'}$$

Now the first mode of the system consists of a rigid-body motion that encounters no resistance. The natural frequency of such a *rigid-body mode* is zero, and its period is infinite. Characteristic equations having only positive roots are referred to as positive-definite, while those with one or more zero roots are called positive-semidefinite. For this reason vibrational systems with one or more rigid-body modes are sometimes referred to as *semidefinite systems.*

As another consideration for the pair of pendulums, suppose that gravity is present but that the stiffness of the connecting spring is zero. In this case the second angular frequency in Eqs. (z) becomes the same as the first, and we have *repeated roots.* The pendulums are able to oscillate independently at the same frequency, and there is no inherent relationship between their amplitudes.

On the other hand, if the spring connecting the pendulums has a small (but nonzero) stiffness constant, the two parts of the system are said to be *lightly coupled.* In such a case the frequency of the second mode will be only slightly higher than that of the first mode [see Eqs. (z)]. Suppose that we start vibrations of the system by the initial conditions $\theta_{01} = \theta_0$, $\theta_{02} = 0$, and $\dot{\theta}_{01} = \dot{\theta}_{02} = 0$. Using Eqs. (3.26), we evaluate $C_1 = -C_3 = \theta_0/2$ and $C_2 = C_4 = 0$; and from Eqs. (3.25a) and (3.25b) the response is found to be

$$\theta_1 = \frac{\theta_0}{2}(\cos\omega_1 t + \cos\omega_2 t) = \theta_0 \cos\frac{(\omega_1 - \omega_2)t}{2}\cos\frac{(\omega_1 + \omega_2)t}{2} \qquad (c')$$

$$\theta_2 = \frac{\theta_0}{2}(\cos\omega_1 t - \cos\omega_2 t) = -\theta_0 \sin\frac{(\omega_1 - \omega_2)t}{2}\sin\frac{(\omega_1 + \omega_2)t}{2} \qquad (d')$$

When the frequencies ω_1 and ω_2 are close to one another, each of the displacements θ_1 and θ_2 consists of the product of a trigonometric function of low frequency $(\omega_1 - \omega_2)/2$ and one of high frequency $(\omega_1 + \omega_2)/2$. Consequently, the phenomenon called *beating*† will occur. At the beginning of the response, the left-hand pendulum will oscillate with amplitude θ_0, while the right-hand pendulum is stationary. Gradually, the amplitude of the former decreases, whereas that of the latter increases. At time $t = \pi/(\omega_1 - \omega_2)$ the left-hand pendulum is stationary, while the right-hand pendulum oscillates with amplitude θ_0. Then the vibration of the former begins to increase while that of the latter decreases, until at time $t = 2\pi/(\omega_1 - \omega_2)$ the initial conditions are again attained. The phenomenon repeats itself indefinitely if there is no damping in the system. As the spring stiffness is made smaller and smaller, the period of the beats increases. Of course, when $k = 0$ there is no interaction at all between the pendulums, and the modal relationships are undefined.

Example 1. Referring to the system in Fig. 3.3, assume that each portion of the shaft has the same rotational stiffness constant k_r and that $I_2 = 2I_1$. If

† For a previous discussion of beating, see Sec. 1.7.

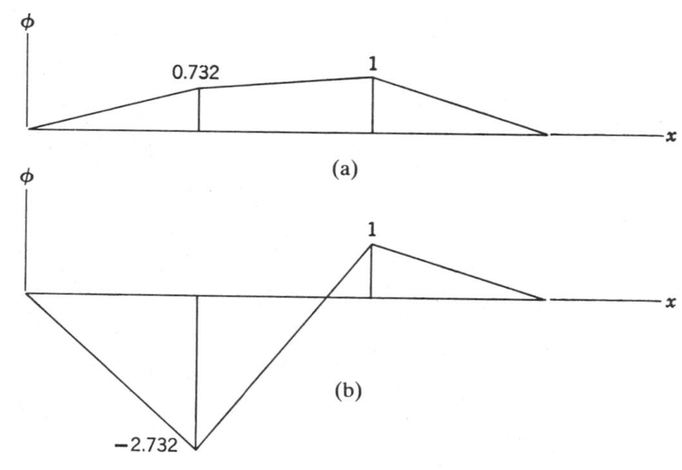

FIG. 3.15

the whole system is rotating at a constant angular velocity $\dot{\phi}_0$ when the shaft is suddenly stopped at points A and B, determine the resulting free-vibrational response.

Solution. From Eq. (3.8) in Sec. 3.2 we see that $M_{11} = I_1$, $M_{22} = 2I_1$, $S_{11} = S_{22} = 2k_r$, and $S_{12} = S_{21} = -k_r$. Using these values in Eq. (3.19), we determine the characteristic values to be

$$\omega_1^2 = \frac{3-\sqrt{3}}{2}\frac{k_r}{I_1} = 0.634\frac{k_r}{I_1}$$
$$\omega_2^2 = \frac{3+\sqrt{3}}{2}\frac{k_r}{I_1} = 2.366\frac{k_r}{I_1} \qquad (e')$$

and from Eqs. (3.20a) and (3.20b) the amplitude ratios are

$$r_1 = \frac{2}{1+\sqrt{3}} = 0.732 \qquad r_2 = \frac{2}{1-\sqrt{3}} = -2.732 \qquad (f')$$

The shapes of the first and second modes are indicated in Figs. 3.15*a* and 3.15*b*, respectively, where the amplitudes are normalized with respect to that of the right-hand disk.

Using Eqs. (3.26), we evaluate the constants $C_1 = C_3 = 0$, $C_2 = 1.352\dot{\phi}_0\sqrt{I_1/k_r}$, and $C_4 = -0.0502\dot{\phi}_0\sqrt{I_1/k_r}$. Then from Eqs. (3.25a) and (3.25b) the response is found to be

$$\phi_1 = (0.992 \sin \omega_1 t + 0.137 \sin \omega_2 t)\dot{\phi}_0\sqrt{\frac{I_1}{k_r}} \qquad (g')$$

$$\phi_2 = (1.352 \sin \omega_1 t - 0.0502 \sin \omega_2 t)\dot{\phi}_0\sqrt{\frac{I_1}{k_r}} \qquad (h')$$

Example 2. The system in Fig. 3.10*a* of the preceding section may be considered to be a simplified representation of an automobile supported on its front and rear springs. To avoid inertial coupling, we shall use the center of mass at point C as a reference for bouncing and pitching motions. Assume the following values for the properties of the automobile:

$$mg = 3220 \text{ lb} \qquad k_1 = 2000 \text{ lb/ft} \qquad k_2 = 2500 \text{ lb/ft}$$

$$I_C = 1500 \text{ lb-sec}^2\text{-ft} \qquad \ell_1 = 4 \text{ ft} \qquad \ell_2 = 6 \text{ ft}$$

Determine the frequencies and mode shapes of the system, and calculate the free-vibrational response to an initial vertical translation Δ without rotation ($v_{0C} = \Delta$; $\theta_{0C} = 0$; $\dot{v}_{0C} = \dot{\theta}_{0C} = 0$).

Solution. Referring to Eq. (3.14c) of the preceding section, we see that $M_{11} = m = 100 \text{ lb-sec}^2/\text{ft}$; $M_{22} = I_C = 1500 \text{ lb-sec}^2\text{-ft}$; $S_{11} = k_1 + k_2 = 4500 \text{ lb/ft}$, $S_{12} = S_{21} = k_2\ell_2 - k_1\ell_1 = 7000 \text{ lb}$, and $S_{22} = k_1\ell_1^2 + k_2\ell_2^2 = 122{,}000 \text{ lb-ft}$. With these values Eqs. (3.19), (3.20a), and (3.20b) yield

$$\omega_1 = 6.13 \text{ rad/sec} \qquad \omega_2 = 9.42 \text{ rad/sec}$$

$$(\tau_1 \approx 1.02 \text{ cps} \qquad \tau_2 \approx 0.67 \text{ cps})$$

$$r_1 = -9.40 \text{ ft/rad} = -1.97 \text{ in./degree}$$

$$r_2 = 1.59 \text{ ft/rad} = 0.333 \text{ in./degree}$$

Figures 3.16*a* and 3.16*b* show the first and second mode shapes, normalized with respect to rotations. They are equivalent to rigid-body rotations about node points O' and O'', which are located at distances 9.40 ft to the right and 1.59 ft to the left of point C. In its first mode the automobile is mostly bouncing, whereas in its second mode it is mostly pitching.

Due to the given initial conditions, the constants from Eqs. (3.26) are $C_1 = -C_3 = -\Delta/10.99$ and $C_2 = C_4 = 0$. Finally, the response of the automobile is found from Eqs. (3.25a) and (3.25b) to be

$$v_C = (0.856 \cos \omega_1 t + 0.145 \cos \omega_2 t)\Delta \qquad (i')$$

$$\theta_C = -0.0911(\cos \omega_1 t - \cos \omega_2 t)\Delta \qquad (j')$$

which is a complicated nonperiodic combination of pitching and bouncing.

3.6 UNDAMPED FORCED VIBRATIONS

We shall now consider harmonic excitations of systems with two degrees of freedom. For example, suppose that the two-mass system in Fig. 3.1*a* is subjected to the sinusoidal forcing functions

$$Q_1 = P_1 \sin \Omega t \qquad Q_2 = P_2 \sin \Omega t \qquad (a)$$

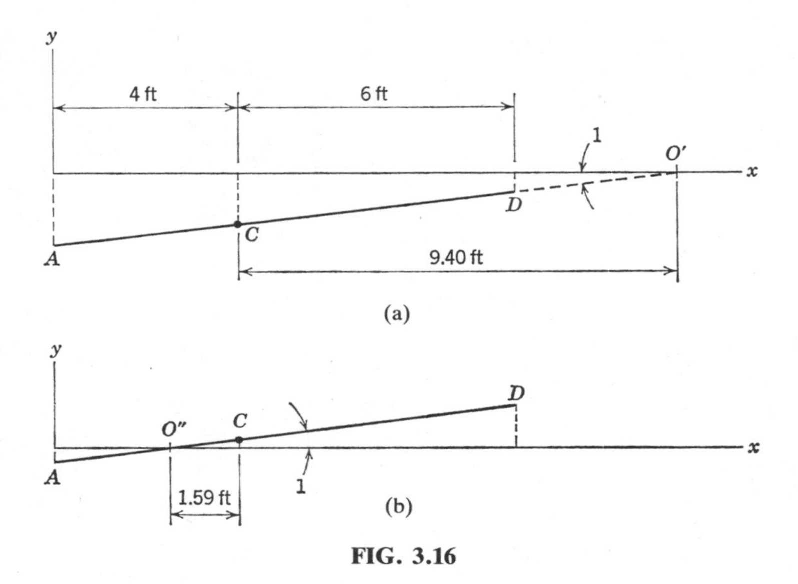

FIG. 3.16

which have the same angular frequency Ω but different magnitudes P_1 and P_2. In this case the action equations of motion [see Eq. (3.6)] become

$$\mathbf{M}\ddot{\mathbf{D}} + \mathbf{S}\mathbf{D} = \mathbf{P} \sin \Omega t \tag{3.27}$$

where

$$\mathbf{P} = \begin{bmatrix} P_1 \\ P_2 \end{bmatrix}$$

In this section we shall again consider only cases with diagonal mass matrices and write Eq. (3.27) in the expanded form

$$\begin{bmatrix} M_{11} & 0 \\ 0 & M_{22} \end{bmatrix} \begin{bmatrix} \ddot{u}_1 \\ \ddot{u}_2 \end{bmatrix} + \begin{bmatrix} S_{11} & S_{12} \\ S_{21} & S_{22} \end{bmatrix} \begin{bmatrix} u_1 \\ u_2 \end{bmatrix} = \begin{bmatrix} P_1 \\ P_2 \end{bmatrix} \sin \Omega t \tag{3.28}$$

A particular solution of these equations may be taken as

$$u_1 = A_1 \sin \Omega t \qquad u_2 = A_2 \sin \Omega t$$

Or

$$\mathbf{D} = \mathbf{A} \sin \Omega t \tag{b}$$

where the amplitudes of *steady-state responses* are

$$\mathbf{A} = \begin{bmatrix} A_1 \\ A_2 \end{bmatrix}$$

Substituting Eq. (*b*) into Eq. (3.28) produces the following algebraic equations:

$$\begin{bmatrix} S_{11} - \Omega^2 M_{11} & S_{12} \\ S_{21} & S_{22} - \Omega^2 M_{22} \end{bmatrix} \begin{bmatrix} A_1 \\ A_2 \end{bmatrix} = \begin{bmatrix} P_1 \\ P_2 \end{bmatrix} \tag{c}$$

Solving for **A**, we obtain

$$\mathbf{A} = \mathbf{BP} \tag{d}$$

in which the matrix **B** is the inverse of the coefficient matrix in Eq. (*c*). Thus,

$$\mathbf{B} = \begin{bmatrix} B_{11} & B_{12} \\ B_{21} & B_{22} \end{bmatrix} = \frac{1}{C} \begin{bmatrix} S_{22} - \Omega^2 M_{22} & -S_{12} \\ -S_{21} & S_{11} - \Omega^2 M_{11} \end{bmatrix} \tag{e}$$

and

$$C = (S_{11} - \Omega^2 M_{11})(S_{22} - \Omega^2 M_{22}) - S_{12}^2 \tag{f}$$

The terms in **B** are influence coefficients (also called *transfer functions*) that may be defined as amplitudes of steady-state response due to unit harmonic forcing functions. Substituting Eq. (*d*) into Eq. (*b*) gives the final form of the solution as

$$\mathbf{D} = \mathbf{BP} \sin \Omega t \tag{3.29}$$

which represents simple harmonic motions of the two masses at the angular frequency Ω.

For slowly varying disturbing forces (i.e., when $\Omega \to 0$), the matrix **B** becomes the inverse of the stiffness matrix, which is the flexibility matrix. Comparing the expression for C in Eq. (*f*) with the characteristic equation [Eq. (*g*) in Sec. 3.5], we conclude that when $\Omega = \omega_1$ or $\Omega = \omega_2$ the amplitudes become infinitely large. Thus, for a two-degree system there are two conditions of resonance, one corresponding to each of the two natural frequencies of free vibration.

The ratio of the amplitudes A_1 and A_2 from Eq. (*d*) is

$$\frac{A_1}{A_2} = -\frac{(S_{22} - \Omega^2 M_{22})P_1 - S_{12}P_2}{S_{21}P_1 - (S_{11} - \Omega^2 M_{11})P_2} \tag{g}$$

When $P_2 = 0$ and $\Omega = \omega_1$ or $\Omega = \omega_2$, this ratio approaches the second forms of r_1 or r_2, given by Eqs. (3.20a) and (3.20b) in Sec. 3.5. On the other hand, when $P_1 = 0$, the ratio approaches the first forms of r_1 or r_2 at the resonance conditions. More generally, if we divide the numerator and denominator of Eq. (*g*) by $-S_{12}$, we see that

$$\frac{A_1}{A_2} = \frac{r_i P_1 + P_2}{P_1 + P_2/r_i} = r_i \qquad (i = 1, 2) \tag{3.30}$$

This result means that for each condition of resonance the forced vibrations are in the corresponding principal mode.

To construct a response spectrum for the steady-state amplitudes of a two-degree system, we must assume specific values for the parameters of a problem. Thus, for the two-mass system in Fig. 3.1*a*, we take $m_1 = 2m$, $m_2 = m$, and $k_1 = k_2 = k$. For convenience in plotting, we introduce the notation

$$\omega_0^2 = \frac{k_1}{m_1} = \frac{k}{2m} \tag{h}$$

and calculate the characteristic values of the system (see Eq. 3.19) in terms of ω_0^2 as

$$\omega_1^2 = 0.586\omega_0^2 \qquad \omega_2^2 = 3.414\omega_0^2 \tag{i}$$

When matrix **B** is evaluated in terms of ω_0^2, it becomes

$$\mathbf{B} = \frac{k}{C}\begin{bmatrix} 1 - \Omega^2/2\omega_0^2 & 1 \\ 1 & 2(1 - \Omega^2/2\omega_0^2) \end{bmatrix} \tag{j}$$

where

$$C = k^2[2(1 - \Omega^2/2\omega_0^2)^2 - 1] \tag{k}$$

In this case the units for all of the terms in the **B** matrix are the same; so it can be made dimensionless by the simple operation of multiplying it by k. Hence, we let

$$\boldsymbol{\beta} = k\mathbf{B} \tag{l}$$

and rewrite Eq. (3.29) as

$$\mathbf{D} = \boldsymbol{\beta}(\mathbf{P}/k) \sin \Omega t \tag{m}$$

Equation (m) is analogous to Eq. (1.24) in Sec. 1.6, and the $\boldsymbol{\beta}$ matrix may be considered to be an *array of magnification factors* (without taking absolute values).

Figure 3.17 shows a dimensionless plot of the magnification factors

$$\beta_{11} = \frac{1 - \Omega^2/2\omega_0^2}{2(1 - \Omega^2/2\omega_0^2)^2 - 1} \tag{n}$$

$$\beta_{21} = \frac{1}{2(1 - \Omega^2/2\omega_0^2)^2 - 1} \tag{o}$$

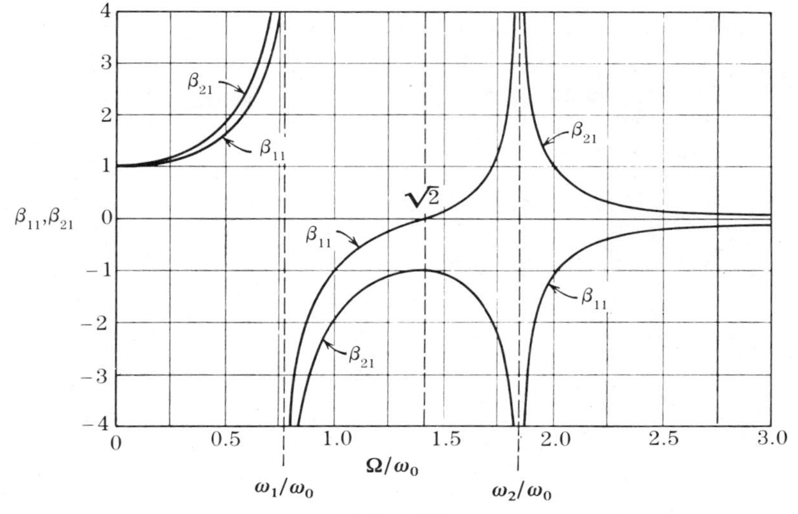

FIG. 3.17

which are associated with the function $(P_1/k)\sin\Omega t$. Both of these factors are unity when $\Omega = 0$; and as Ω increases they are both positive, indicating that the masses are vibrating in phase with the disturbing force $P_1 \sin \Omega t$. As Ω approaches the value of the first natural frequency ω_1, both factors approach infinity. When Ω becomes slightly greater than ω_1, both factors become negative, indicating that the masses are out of phase with the force but are still in phase with each other. With further increase in Ω, both factors diminish until (when $\Omega = \sqrt{2}\,\omega_0$) the factor β_{11} becomes zero, while β_{21} has the value -1. When Ω exceeds the value $\sqrt{2}\,\omega_0$, β_{11} is positive, while β_{21} remains negative. This means that the masses are out of phase with each other but that the first mass is in phase with the force again. The factors become infinite a second time when $\Omega = \omega_2$; and as Ω increases well beyond ω_2, the motions of the masses tend toward zero.

The fact that β_{11} becomes zero when $\Omega = \sqrt{2}\,\omega_0$ is of particular interest. At this frequency the first mass is stationary while the second mass moves in opposition to the forcing function at the amplitude $-P_1/k$. We see from the definition of β_{11} in Eq. (*e*) that it becomes zero when

$$\Omega = \sqrt{S_{22}/M_{22}} \tag{p}$$

which is equal to $\sqrt{k_2/m_2} = \sqrt{2}\,\omega_0$ for the two-mass system. To show how this condition can be used advantageously, consider the motor of mass m_1 supported on a beam of stiffness k_1 in Fig. 3.18*a*. the rotating force vector P_1 due to unbalance of the rotor can cause heavy vibrations of the system when the angular frequency is at the critical value $\Omega_{cr} = \sqrt{k_1/m_1}$. To

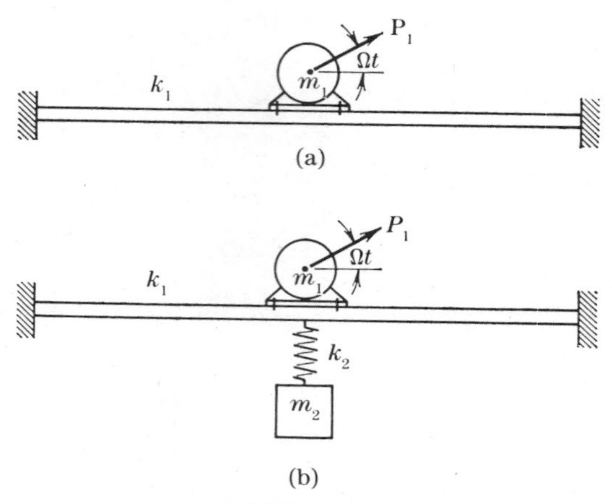

FIG. 3.18

suppress these forced vibrations, we attach an auxiliary mass m_2 with a spring of stiffness k_2, as shown in Fig. 3.18*b*. If m_2 and k_2 are selected so that $\sqrt{k_2/m_2} = \Omega_{cr}$, we have a two-degree system for which the vibrations of the motor vanish, while the auxiliary mass vibrates at the amplitude $-P_1/k_2$. Such an auxiliary system is called a *dynamic damper*, because it can prevent vibrations of constant-speed machinery in the absence of actual damping. To design the "damper" we first select the spring k_2 so that the amplitude $-P_1/k_2$ is of reasonable magnitude and then select the mass so that $\sqrt{k_2/m_2} = \Omega_{cr}$. To be effective at speeds other than Ω_{cr}, the dynamic damper requires actual damping [3].

As discussed previously in Sec. 1.6, forced vibrations may also result from *periodic ground motions*. For example, suppose that the ground in Fig. 3.1*a* translates in the x direction in accordance with the simple harmonic function $u_g = d \sin \Omega t$, where d is the displacement amplitude. In this case the action equations of motion are

$$m_1\ddot{u}_1 = -k_1(u_1 - u_g) + k_2(u_2 - u_1) \tag{q}$$

$$m_2\ddot{u}_2 = -k_2(u_2 - u_1) \tag{r}$$

These equations may be written in matrix form as

$$\mathbf{M}\ddot{\mathbf{D}} + \mathbf{S}\mathbf{D} = \mathbf{P}_g \sin \Omega t \tag{3.31}$$

The array $\mathbf{P}_g$ in Eq. (3.31) consists of the maximum forces transmitted to the masses through the springs, due to the ground displacements. In this case only one of the terms in $\mathbf{P}_g$ is nonzero, and we have

$$\mathbf{P}_g = \begin{bmatrix} k_1 d \\ 0 \end{bmatrix} \tag{s}$$

On the other hand, suppose that horizontal ground accelerations are specified as $\ddot{u}_g = a \sin \Omega t$, where a is the acceleration amplitude. In this case we change coordinates to the relative displacements

$$u_1^* = u_1 - u_g \qquad u_2^* = u_2 - u_g \tag{t}$$

The corresponding accelerations are

$$\ddot{u}_1^* = \ddot{u}_1 - \ddot{u}_g \qquad \ddot{u}_2^* = \ddot{u}_2 - \ddot{u}_g \tag{u}$$

Then the action equations in the relative coordinates may be written as

$$\mathbf{M\ddot{D}}^* + \mathbf{SD}^* = \mathbf{P}_g^* \sin \Omega t \tag{3.32}$$

For the two-mass system in Fig. 3.1*a*, the array $\mathbf{P}_g^*$ in Eq. (3.32) becomes

$$\mathbf{P}_g^* = -\begin{bmatrix} m_1 \\ m_2 \end{bmatrix} a \tag{v}$$

Thus, we can always put problems of forced vibrations due to support motions into the same mathematical form as that for applied actions corresponding to the displacement coordinates. In addition, it is always possible to calculate *equivalent loads* [2], corresponding to the displacement coordinates, due to applied actions that do not correspond to those coordinates.

If we analyze forced vibrations with displacement equations instead of action equations, Eq. (3.27) is replaced by

$$\mathbf{FM\ddot{D}} + \mathbf{D} = \mathbf{FP} \sin \Omega t \tag{3.33}$$

Or, in expanded form, we have

$$\begin{bmatrix} F_{11}M_{11} & F_{12}M_{22} \\ F_{21}M_{11} & F_{22}M_{22} \end{bmatrix} \begin{bmatrix} \ddot{u}_1 \\ \ddot{u}_2 \end{bmatrix} + \begin{bmatrix} u_1 \\ u_2 \end{bmatrix} = \begin{bmatrix} F_{11} & F_{12} \\ F_{21} & F_{22} \end{bmatrix} \begin{bmatrix} P_1 \\ P_2 \end{bmatrix} \sin \Omega t \tag{3.34}$$

Substitution of Eq. (*b*) into Eq. (3.34) yields

$$\begin{bmatrix} 1 - \Omega^2 F_{11}M_{11} & -\Omega^2 F_{12}M_{22} \\ -\Omega^2 F_{21}M_{11} & 1 - \Omega^2 F_{22}M_{22} \end{bmatrix} \begin{bmatrix} A_1 \\ A_2 \end{bmatrix} = \begin{bmatrix} F_{11} & F_{12} \\ F_{21} & F_{22} \end{bmatrix} \begin{bmatrix} P_1 \\ P_2 \end{bmatrix} \tag{w}$$

In this case the solution for $\mathbf{A}$ may be written as

$$\mathbf{A} = \mathbf{EFP} \tag{x}$$

where the matrix $\mathbf{E}$ is the inverse of the coefficient matrix on the left-hand side of Eq. (w). Thus,

$$\mathbf{E} = \begin{bmatrix} E_{11} & E_{12} \\ E_{21} & E_{22} \end{bmatrix} = \frac{1}{H} \begin{bmatrix} 1 - \Omega^2 F_{22} M_{22} & \Omega^2 F_{12} M_{22} \\ \Omega^2 F_{21} M_{11} & 1 - \Omega^2 F_{11} M_{11} \end{bmatrix} \tag{y}$$

where

$$H = (1 - \Omega^2 F_{11} M_{11})(1 - \Omega^2 F_{22} M_{22}) - \Omega^4 F_{12}^2 M_{11} M_{22} \tag{z}$$

The terms in $\mathbf{E}$ are influence coefficients that may be defined as amplitudes of steady-state response due to unit harmonic displacements of the masses. Substituting Eq. (x) into Eq. (b) produces the solution as

$$\mathbf{D} = \mathbf{EFP} \sin \Omega t \tag{3.35}$$

Comparing Eq. (3.35) with Eq. (3.29), we conclude that

$$\mathbf{EF} = \mathbf{B} \tag{3.36a}$$

Therefore,

$$\mathbf{E} = \mathbf{BS} \tag{3.36b}$$

Although both $\mathbf{B}$ and $\mathbf{S}$ are symmetric, their product $\mathbf{E}$ is usually unsymmetric.

We may also introduce the notation

$$\mathbf{\Delta}_{st} = \mathbf{FP} \tag{3.37}$$

Using this identifier, we rewrite Eq. (3.33) as

$$\mathbf{FM\ddot{D}} + \mathbf{D} = \mathbf{\Delta}_{st} \sin \Omega t \tag{3.38}$$

Then the solution in Eq. (3.35) becomes

$$\mathbf{D} = \mathbf{E\Delta}_{st} \sin \Omega t \tag{3.39}$$

which is analogous to Eq. (u) in Example 4 of Sec. 1.6. the matrix $\mathbf{\Delta}_{st}$ consists of terms that are defined as displacements of the masses due to static application of the maximum values of the forcing functions. As formed in Eq. (3.37), the terms in $\mathbf{\Delta}_{st}$ arise from applied actions corresponding to the displacement coordinates; but similar terms may also arise from actions of other types, or from support motions.

Harmonic translations of ground for the system in Fig. 3.1a are particularly easy to analyze using Eq. (3.38). If the ground sisplacement is given by $u_g = d \sin \Omega t$ as before, the array $\mathbf{\Delta}_{st}$ becomes simply

$$\mathbf{\Delta}_{st} = \begin{bmatrix} 1 \\ 1 \end{bmatrix} d \tag{a'}$$

which represents rigid-body motion of the system. On the other hand, the ground accelerations given by $\ddot{u}_g = a \sin \Omega t$ are more difficult to handle. Equation (3.38) in relative coordiantes [see Eqs. (t) and (u)] becomes

$$\mathbf{FM\ddot{D}}^* + \mathbf{D}^* = \mathbf{\Delta}^*_{st} \sin \Omega t \tag{3.40}$$

where

$$\mathbf{\Delta}^*_{st} = \mathbf{FP}^*_g \tag{b'}$$

Example 1. Assume that the two-mass system in Fig. 3.1a is subjected to the sinusoidal ground displacement $u_g = d \sin \Omega t$. As was done for the plot of the response spectrum in Fig. 3.17, we take $m_1 = 2m$, $m_2 = m$, and $k_1 = k_2 = k$. The steady-state response of the system is to be calculated by both the action-equation method and the displacement-equation technique.
Solution. Having already determined the matrices **B** [see Eqs. (j) and (k)] and $\mathbf{P}_g$ [see Eq. (s)] for this system, we can immediately substitute them into Eq. (3.29) to obtain the results as

$$u_1 = \frac{(1 - \Omega^2 m/k) d \sin \Omega t}{2(1 - \Omega^2 m/k)^2 - 1} \tag{c'}$$

$$u_2 = \frac{d \sin \Omega t}{2(1 - \Omega^2 m/k)^2 - 1} \tag{d'}$$

Next, we develop the matrix **E** [see Eqs. (y) and (z)] using the flexibilities $F_{11} = F_{12} = F_{21} = \delta$ and $F_{22} = 2\delta$.

$$\mathbf{E} = \frac{1}{H} \begin{bmatrix} 1 - 2\Omega^2 m\delta & \Omega^2 m\delta \\ 2\Omega^2 m\delta & 1 - 2\Omega^2 m\delta \end{bmatrix} \tag{e'}$$

where

$$H = (1 - 2\Omega^2 m\delta)^2 - 2\Omega^4 m^2 \delta^2 = 2(1 - \Omega^2 m\delta)^2 - 1 \tag{f'}$$

Substitution of **E** and $\mathbf{\Delta}_{st}$ [see Eq. (a')] into Eq. (3.39) yields

$$u_1 = \frac{(1 - \Omega^2 m\delta) d \sin \Omega t}{2(1 - \Omega^2 m\delta)^2 - 1} \tag{g'}$$

$$u_2 = \frac{d \sin \Omega t}{2(1 - \Omega^2 m\delta)^2 - 1} \tag{h'}$$

Since $\delta = 1/k$, we see that Eqs. (g') and (h') are the same as Eqs. (c') and (d').

Example 2. Suppose that the frame in Fig. 3.8a (see Example 2 of Sec. 3.3) is subjected to a torque $T = T_m \cos \Omega t$ applied in the z sense at its upper right-hand corner. Determine steady-state response of the mass (attached to the free end of the frame) due to this excitation.

Solution. We shall analyze this system using the displacement-equation approach and recall that the flexibilities are $F_{11} = 4\ell^3/3EI$, $F_{12} = F_{21} = \ell^3/2EI$, and $F_{22} = \ell^3/3EI$. In addition, the mass terms are $M_{11} = M_{22} = m$. Using these values in Eqs. (y) and (z), we develop the **E** matrix as

$$\mathbf{E} = \frac{1}{H}\begin{bmatrix} 1 - \Omega^2 m\ell^3/3EI & \Omega^2 m\ell^3/2EI \\ \Omega^2 m\ell^3/2EI & 1 - 4\Omega^2 m\ell^3/3EI \end{bmatrix} \qquad (i')$$

where

$$H = 1 - \frac{5\Omega^2 m\ell^3}{3EI} + \frac{7\Omega^4 m^2 \ell^6}{36(EI)^2} \qquad (j')$$

The maximum torque T_m applied statically to the upper right-hand corner of the frame causes the mass to translate an amount $T_m\ell^2/EI$ in the x direction and an amount $T_m\ell^2/2EI$ in the y direction. Therefore, the matrix $\mathbf{\Delta}_{st}$ for this problem becomes

$$\mathbf{\Delta}_{st} = \begin{bmatrix} 2 \\ 1 \end{bmatrix} \frac{T_m\ell^2}{2EI} \qquad (k')$$

Substitution of these arrays into Eq. (3.39), with sin Ωt replaced by cos Ωt, produces the results

$$u_1 = \left(12 - \frac{\Omega^2 m\ell^3}{EI}\right)\frac{T_m\ell^2 \cos \Omega t}{12EIH} \qquad (l')$$

$$v_1 = \left(3 - \frac{\Omega^2 m\ell^3}{EI}\right)\frac{T_m\ell^2 \cos \omega t}{6EIH} \qquad (m')$$

3.7 FREE VIBRATIONS WITH VISCOUS DAMPING

Figure 3.19*a* shows a two-mass system with dashpot dampers, having viscous damping constants c_1 and c_2. If there are no loads applied to the system, the action equations of motion are (see Fig. 3.19*b*)

$$m_1\ddot{u}_1 = -c_1\dot{u}_1 + c_2(\dot{u}_2 - \dot{u}_1) - k_1u_1 + k_2(u_2 - u_1) \qquad (a)$$

$$m_2\ddot{u}_2 = -c_2(\dot{u}_2 - \dot{u}_1) - k_2(u_2 - u_1) \qquad (b)$$

In concise matrix notation these equations become

$$\mathbf{M\ddot{D}} + \mathbf{C\dot{D}} + \mathbf{SD} = \mathbf{0} \qquad (3.41)$$

where

$$\mathbf{C} = \begin{bmatrix} C_{11} & C_{12} \\ C_{21} & C_{22} \end{bmatrix} = \begin{bmatrix} c_1 + c_2 & -c_2 \\ -c_2 & c_2 \end{bmatrix} \qquad \dot{\mathbf{D}} = \begin{bmatrix} \dot{u}_1 \\ \dot{u}_2 \end{bmatrix} \qquad (c)$$

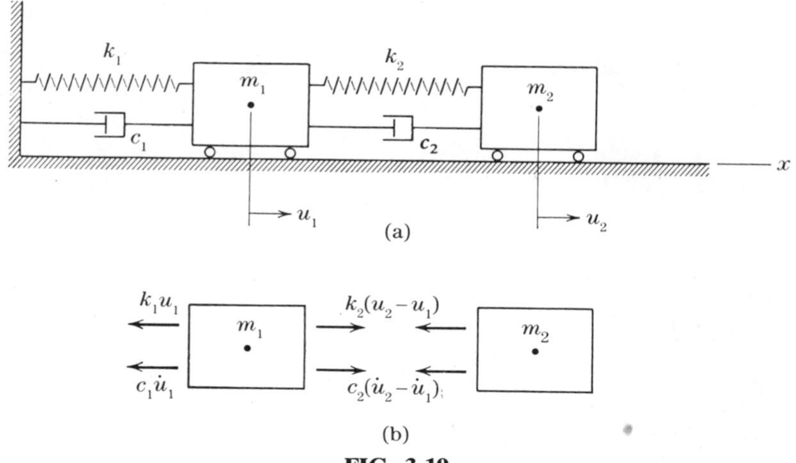

FIG. 3.19

and the other matrices in Eq. (3.41) are as defined previously. The *damping matrix* **C** contains *damping influence coefficients* that may be defined as actions required for unit velocities. That is, any term C_{ij} in an array of viscous damping coefficients is an action of type i equilibrating damping actions associated with a unit velocity of type j. This definition is analogous to those for stiffness and inertial influence coefficients, and the procedure for deriving columns of **C** is similar to those described previously for determining columns of **S** and **M**. If such a procedure is followed, the damping matrix will always be symmetric.

Because velocity terms are present in Eq. (3.41), the solution of the homogeneous differential equations is more complicated than in Sec. 3.5 for the undamped case. Here we shall seek solutions of the genral form

$$u_1 = A_1 e^{st} \tag{d}$$

$$u_2 = A_2 e^{st} \tag{e}$$

Substitution of Eqs. (*d*) and (*e*) and their derivatives into Eq. (3.41) results in the following algebraic equations to be satisfied:

$$\begin{bmatrix} M_{11}s^2 + C_{11}s + S_{11} & C_{12}s + S_{12} \\ C_{21}s + S_{21} & M_{22}s^2 + C_{22}s + S_{22} \end{bmatrix} \begin{bmatrix} A_1 \\ A_2 \end{bmatrix} = \begin{bmatrix} 0 \\ 0 \end{bmatrix} \tag{f}$$

For nontrivial solutions the determinant of the coefficient matrix in Eq. (*f*) must be equal to zero. Hence, we obtain the characteristic equation as

$$(M_{11}s^2 + C_{11}s + S_{11})(M_{22}s^2 + C_{22}s + S_{22}) - (C_{12}s + S_{12})^2 = 0 \tag{g}$$

Or,

$$M_{11}M_{22}s^4 + (M_{11}C_{22} + M_{22}C_{11})s^3 + (M_{11}S_{22} + C_{11}C_{22} + M_{22}S_{11} - C_{12}^2)s^2 \\ +(C_{11}S_{22} + C_{22}S_{11} - 2C_{12}S_{12})s + S_{11}S_{22} - S_{12}^2 = 0 \quad (h)$$

Equation (h) simplifies somewhat if the actual values of $M_{11} = m_1$, $M_{22} = m_2$, etc., for the system in Fig. 3.19*a* are used. When such values are substituted into Eq. (h), it becomes

$$m_1m_2s^4 + [m_1c_2 + m_2(c_1 + c_2)]s^3 + [m_1k_2 + c_1c_2 + m_2(k_1 + k_2)]s^2 \\ +(c_1k_2 + c_2k_1)s + k_1k_2 = 0 \quad (i)$$

This expression must be solved by some numerical method for extracting roots of a polynomial, but the general form of the solution is known and will be discussed in detail.

Because all of the coefficients in Eq. (i) are positive, nonzero roots of this fourth degree polynomial can be neither real and positive nor complex with positive real parts [4]. The remaining possibilities are that they may be either real and negative or complex with negative real parts. If damping is small, the system can vibrate freely; and all nonzero roots will be complex. They occur in conjugate pairs that may be expressed as

$$s_{11} = -n_1 + i\omega_{d1} \qquad s_{12} = -n_1 - i\omega_{d1} \tag{3.42a}$$

and

$$s_{21} = -n_2 + i\omega_{d2} \qquad s_{22} = -n_2 - i\omega_{d2} \tag{3.42b}$$

The symbols n_1 and n_2 represent positive numbers attributable to damping, while ω_{d1} and ω_{d2} denote damped angular frequencies. Substituting each of these roots into Eq. (f), we obtain the corresponding amplitude ratios

$$r_{jk} = \frac{-C_{12}s_{jk} - S_{12}}{M_{11}s_{jk}^2 + C_{11}s_{jk} + S_{11}} = \frac{M_{22}s_{jk}^2 + C_{22}s_{jk} + S_{22}}{-C_{21}s_{jk} - S_{21}} \tag{3.43}$$

where $j = 1$ or 2 and $k = 1$ or 2. The resulting ratios r_{11}, r_{12} and r_{21}, r_{22} are complex conjugate pairs. Then the complete solution may be written as

$$u_1 = r_{11}A_{11}e^{s_{11}t} + r_{12}A_{12}e^{s_{12}t} + r_{21}A_{21}e^{s_{21}t} + r_{22}A_{22}e^{s_{22}t} \tag{3.44a}$$

$$u_2 = A_{11}e^{s_{11}t} + A_{12}e^{s_{12}t} + A_{21}e^{s_{21}t} + A_{22}e^{s_{22}t} \tag{3.44b}$$

in which the coefficients A_{11}, A_{12} and A_{21}, A_{22} are complex conjugate pairs that must be determined from initial conditions.

Proceeding as for the one-degree system in Sec. 1.8, we can convert Eqs. (3.44a) and (3.44b) to equivalent trigonometric expressions. We rewrite the first two terms of u_2 in Eq. (3.44b) as

$$A_{11}e^{s_{11}t} + A_{12}e^{s_{12}t} = e^{-n_1t}(C_1 \cos \omega_{d1}t + C_2 \sin \omega_{d1}t)$$

where

$$\cos \omega_{d1}t = \frac{e^{i\omega_{d1}t} + e^{-i\omega_{d1}t}}{2} \qquad \sin \omega_{d1}t = \frac{e^{i\omega_{d1}t} - e^{-i\omega_{d1}t}}{2i}$$

and

$$C_1 = A_{11} + A_{12} \qquad C_2 = i(A_{11} - A_{12}) \tag{j}$$

which are real constants. The corresponding terms in Eq. (3.44a) will be converted to trigonometric form by introducing the notations

$$r_{11} = a + ib \qquad r_{12} = a - ib \tag{k}$$

representing the first pair of complex conjugate amplitude ratios. Then we rewrite the first two terms of u_1 in the equivalent form

$$r_{11}A_{11}e^{s_{11}t} + r_{12}A_{12}e^{s_{12}t} = e^{-n_1t}[(C_1a - C_2b)\cos \omega_{d1}t + (C_1b + C_2a)\sin \omega_{d1}t]$$

Similarly, the last two terms of u_1 and u_2 can be converted to trigonometric form, using the real constants

$$C_3 = A_{21} + A_{22} \qquad C_4 = i(A_{21} - A_{22}) \tag{l}$$

and introducing the notations

$$r_{21} = c + id \qquad r_{22} = c - id \tag{m}$$

In this form the total solution becomes

$$u_1 = e^{-n_1t}(r_1C_1 \cos \omega_{d1}t + r_1'C_2 \sin \omega_{d1}t) + e^{-n_2t}(r_2C_3 \cos \omega_{d2}t + r_2'C_4 \sin \omega_{d2}t) \tag{3.45a}$$

$$u_2 = e^{-n_1t}(C_1 \cos \omega_{d1}t + C_2 \sin \omega_{d1}t) + e^{-n_2t}(C_3 \cos \omega_{d2}t + C_4 \sin \omega_{d2}t) \tag{3.45b}$$

where the real amplitude ratios are

$$r_1 = \frac{C_1a - C_2b}{C_1} \qquad r_1' = \frac{C_1b + C_2a}{C_2}$$
$$r_2 = \frac{C_3c - C_4d}{C_3} \qquad r_2' = \frac{C_3d + C_4c}{C_4} \tag{n}$$

Equations (3.45a) and (3.45b) are similar in many ways to their counterparts for undamped vibrations [see Eqs. (3.25a) and (3.25b) in Sec. 3.5]. However, they differ from the simpler expressions in several important aspects. The amplitudes of vibration diminish with time in accordance with the factors $e^{-n_1 t}$ and $e^{-n_2 t}$, and they eventually die out completely. In addition, the damped angular frequencies ω_{d1} and ω_{d2} are not the same as those for the undamped case. Furthermore, there are four amplitude ratios in the damped case, whereas only two ratios characterize the undamped modes of vibration. As a consequence, the first part of the expression for u_1 in Eq. (3.45a) is not in phase with the first part of the expression for u_2 in Eq. (3.45b). Similarly, the second part of the expression for u_1 is not in phase with the second part of the expression for u_2. We see the phase differences clearly by writing those parts in phase-angle form, as follows:

$$u_1 = B_1' e^{-n_1 t} \cos(\omega_{d1} t - \alpha_{d1}') + B_2' e^{-n_2 t} \cos(\omega_{d2} t - \alpha_{d2}') \tag{3.46a}$$

$$u_2 = B_1 e^{-n_1 t} \cos(\omega_{d1} t - \alpha_{d1}) + B_2 e^{-n_2 t} \cos(\omega_{d2} t - \alpha_{d2}) \tag{3.46b}$$

where

$$\begin{aligned} B_1 &= \sqrt{C_1^2 + C_2^2} \qquad & B_2 &= \sqrt{C_3^2 + C_4^2} \\ B_1' &= B_1\sqrt{a^2 + b^2} \qquad & B_2' &= B_2\sqrt{c^2 + d^2} \end{aligned} \tag{o}$$

and

$$\begin{aligned} \alpha_{d1} &= \tan^{-1}\left(\frac{C_2}{C_1}\right) \qquad & \alpha_{d2} &= \tan^{-1}\left(\frac{C_4}{C_3}\right) \\ \alpha_{d1}' &= \tan^{-1}\left(\frac{r_1' C_2}{r_1 C_1}\right) \qquad & \alpha_{d2}' &= \tan^{-1}\left(\frac{r_2' C_4}{r_2 C_3}\right) \end{aligned} \tag{p}$$

Thus, principal modes of vibration, as defined in Sec. 3.5, do not exist for the damped two-degree system under consideration. The natural modes that do exist have phase relationships that complicate the analysis. The topics of principal modes and damping will be discussed further in Chapter 4 for systems with multiple degrees of freedom.

If the coefficients of viscous damping are very small, the characteristic equation (h) tends toward that for the undamped case; and we make the following approximations:

$$\begin{aligned} \omega_{d1} &\approx \omega_1 \qquad & \omega_{d2} &\approx \omega_2 \\ r_1' &\approx r_1 \qquad & r_2' &\approx r_2 \end{aligned} \tag{q}$$

With these assumptions, Eqs. (3.45a) and (3.45b) are simplified to

$$\begin{aligned} u_1 \approx\; & r_1 e^{-n_1 t}(C_1 \cos \omega_1 t + C_2 \sin \omega_1 t) \\ & + r_2 e^{-n_2 t}(C_3 \cos \omega_2 t + C_4 \sin \omega_2 t) \end{aligned} \tag{3.47a}$$

$$\begin{aligned} u_2 \approx\; & e^{-n_1 t}(C_1 \cos \omega_1 t + C_2 \sin \omega_1 t) \\ & + e^{-n_2 t}(C_3 \cos \omega_1 t + C_4 \sin \omega_2 t) \end{aligned} \tag{3.47b}$$

To evaluate the constants C_1 through C_4, we substitute the initial conditions u_{01}, u_{02}, $\dot{u}_{01}$, and $\dot{u}_{02}$ (at time $t = 0$) into Eqs. (3.47a), (3.47b), and their derivatives. The resulting formulas for these constants are

$$C_1 \approx \frac{u_{01} - r_2 u_{02}}{r_1 - r_2} \qquad C_2 \approx \frac{\dot{u}_{01} + n_1 u_{01} - r_2(\dot{u}_{02} + n_1 u_{02})}{\omega_1(r_1 - r_2)}$$
$$C_3 \approx \frac{r_1 u_{02} - u_{01}}{r_1 - r_2} \qquad C_4 \approx \frac{r_1(\dot{u}_{02} + n_2 u_{02}) - (\dot{u}_{01} + n_2 u_{01})}{\omega_2(r_1 - r_2)} \tag{r}$$

On the other hand, if damping is very high, all roots of the characteristic equation will be real and negative. In this case the solution is not vibratory and can be represented as

$$u_1 = r_1 D_1 e^{-v_1 t} + r_2 D_2 e^{-v_2 t} + r_3 D_3 e^{-v_3 t} + r_4 D_4 e^{-v_4 t} \tag{3.48a}$$

$$u_2 = D_1 e^{-v_1 t} + D_2 e^{-v_2 t} + D_3 e^{-v_3 t} + D_4 e^{-v_4 t} \tag{3.48b}$$

where $v_1 - v_4$ are positive numbers. In addition, the constants D_1 through D_4 and r_1 through r_4 are real.

It is also possible to have two roots that are real and negative, while the other two are complex conjugates with negative real parts. In such a case the solution may be written as

$$u_1 = e^{-nt}(r_1 C_1 \cos \omega_d t + r_1' C_2 \sin \omega_d t) + r_3 C_3 e^{-v_3 t} + r_4 C_4 e^{-v_4 t} \tag{3.49a}$$

$$u_2 = e^{-nt}(C_1 \cos \omega_d t + C_2 \sin \omega_d t) + C_3 e^{-v_3 t} + C_4 e^{-v_4 t} \tag{3.49b}$$

Damped free vibrations can also be analyzed using displacement equations of motion instead of action equations. By that approach the differential equations become

$$\mathbf{F}(\mathbf{M}\ddot{\mathbf{D}} + \mathbf{C}\dot{\mathbf{D}}) + \mathbf{D} = \mathbf{0} \tag{3.50}$$

The details of this method are similar to those for the action equations of motion but will not be discussed here.

3.8 FORCED VIBRATIONS WITH VISCOUS DAMPING

Let the system in Fig. 3.19*a* be subjected to a general harmonic excitation, given in complex form as

$$\mathbf{Q} = \mathbf{P}e^{i\Omega t} = \mathbf{P}(\cos \Omega t + i \sin \Omega t) \tag{a}$$

where the vector **P** has the same meaning as in Eq. (3.27) of Sec. 3.6. then Eq. (3.41) in Sec. 3.7 becomes

$$\mathbf{M}\ddot{\mathbf{D}} + \mathbf{C}\dot{\mathbf{D}} + \mathbf{S}\mathbf{D} = \mathbf{P}e^{i\Omega t} \tag{3.51}$$

Considering only steady-state forced vibrations, we assume solutions in complex form to be

$$\mathbf{D} = \mathbf{A}e^{i\Omega t} \tag{b}$$

Substitution of Eq. (b) and its derivatives into Eq. (3.51) yields the following algebraic matrix equation:

$$(\mathbf{S} - \Omega^2\mathbf{M} + i\Omega\mathbf{C})\mathbf{A} = \mathbf{P} \tag{c}$$

Solving for **A** in Eq. (c), we obtain

$$\mathbf{A} = \mathbf{B}^*\mathbf{P} \tag{d}$$

Substitution of Eq. (d) into Eq. (b) shows the solution to be

$$\mathbf{D} = \mathbf{B}^*\mathbf{P}e^{i\Omega t} \tag{3.52}$$

which represents harmonic motions of the two masses at the angular frequency Ω.

From Eqs. (c) and (d), we observe that the definition of matrix $\mathbf{B}^*$ is

$$\mathbf{B}^* = (\mathbf{S} - \Omega^2\mathbf{M} + i\Omega\mathbf{C})^{-1} \tag{e}$$

This array is similar to matrix **B** in Sec. 3.6, but $\mathbf{B}^*$ includes imaginary parts due to damping. When **M** is diagonal, the expanded form of $\mathbf{B}^*$ becomes

$$B^* = \begin{bmatrix} B_{11}^* & B_{12}^* \\ B_{21}^* & B_{22}^* \end{bmatrix} = \frac{1}{C^*}\begin{bmatrix} S_{22} - \Omega^2 M_{22} + i\Omega C_{22} & -S_{12} - i\Omega C_{12} \\ -S_{21} - i\Omega C_{21} & S_{11} - \Omega^2 M_{11} + i\Omega C_{11} \end{bmatrix} \tag{f}$$

where

$$C^* = (S_{11} - \Omega^2 M_{11} + i\Omega C_{11})(S_{22} - \Omega^2 M_{22} + i\Omega C_{22}) - (S_{12} + i\Omega C_{12})^2 \tag{g}$$

The terms in $\mathbf{B}^*$ are influence coefficients called *complex transfer functions*. The complex numbers in such an array represent amplitudes and phases of steady-state damped response due to unit harmonic forcing functions.

Using standard relationships for complex numbers, we may express the solution given by Eq. (3.52) in terms of real amplitudes and phase angles, as follows:

$$u_1 = \frac{\sqrt{a^2+b^2}}{\sqrt{g^2+h^2}} P_1 \cos(\Omega t - \theta_1) + \frac{\sqrt{c^2+d^2}}{\sqrt{g^2+h^2}} P_2 \cos(\Omega t - \theta_2) \qquad (3.53a)$$

$$u_2 = \frac{\sqrt{c^2+d^2}}{\sqrt{g^2+h^2}} P_1 \cos(\Omega t - \theta_2) + \frac{\sqrt{e^2+f^2}}{\sqrt{g^2+h^2}} P_2 \cos(\Omega t - \theta_3) \qquad (3.53b)$$

in which

$$\begin{aligned} &a = S_{22} - \Omega^2 M_{22} \qquad b = \Omega C_{22} \qquad c = S_{12} \\ &d = \Omega C_{12} \qquad e = S_{11} - \Omega^2 M_{11} \qquad f = \Omega C_{11} \\ &g = (S_{11} - \Omega^2 M_{11})(S_{22} - \Omega^2 M_{22}) - S_{12}^2 - \Omega^2(C_{11}C_{22} - C_{12}^2) \\ &h = \Omega[C_{11}(S_{22} - \Omega^2 M_{22}) + C_{22}(S_{11} - \Omega^2 M_{11}) - 2C_{12}S_{12}] \end{aligned} \qquad (h)$$

and

$$\begin{aligned} &\theta_1 = \tan^{-1}\left(\frac{ah - bg}{ag + bh}\right) \qquad \theta_2 = \tan^{-1}\left(\frac{ch - dg}{cg + dh}\right) \\ &\theta_3 = \tan^{-1}\left(\frac{eh - fg}{eg + fh}\right) \end{aligned} \qquad (i)$$

At this point we can see that including viscous damping becomes quite complicated, even for a system with only two degrees of freedom. In Chapter 4 the matter of damping will be reconsidered for systems with multiple degrees of freedom. For such problems we shall use simplifying assumptions about phase relationships that allow practical calculations yielding good accuracy.

REFERENCES

[1] Gere, J. M., and Weaver, W., Jr., *Matrix Algebra for Engineers,* 2nd edn., Wadsworth, Belmont, CA, 1983.

[2] Weaver, W., Jr., and Gere, J. M., *Matrix Analysis of Framed Structures,* 2nd edn., Van Nostrand-Reinhold, New York, 1980.

[3] Ormondroyd, J., and Den Hartog, J. P., "Theory of Dynamic Vibration Absorber," *Trans. ASME,* **50,** 1928, pp. APM 241–254.

[4] Rayleigh, J. W. S., *Theory of Sound,* 2nd edn., Vol. 1, Dover, New York, 1945.

PROBLEMS

3.2-1. For the two-mass system shown in the figure determine the stiffness matrix **S**, and write the action equations of motion in matrix form.

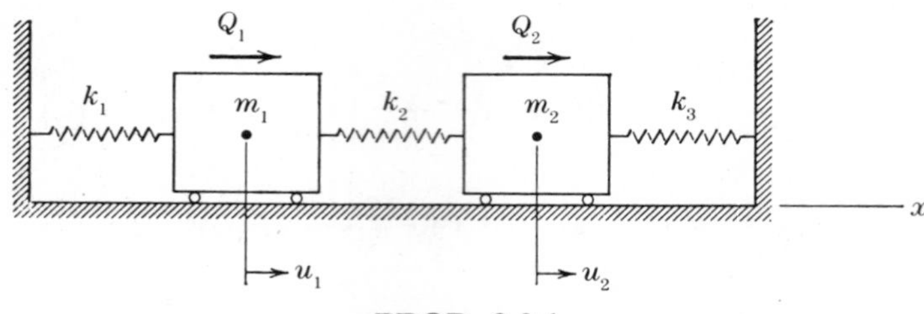

PROB. 3.2-1

3.2-2. Let the spring constants in Fig. 3.2 be $k_1 = k_2 = k_3 = k$, and assume the following values for the angles: $\alpha_1 = 0°$, $\alpha_2 = 60°$, and $\alpha_3 = 210°$. Determine the stiffness matrix of the suspended mass in terms of k.

3.2-3. The figure shows a double pendulum with restraining springs attached to the masses m_1 and m_2. Use the small horizontal translations u_1 and u_2 of the masses as displacement coordinates. Develop the stiffness matrix **S** and the gravitational matrix **G** for this sytem, and write the action equations of motion in matrix form.

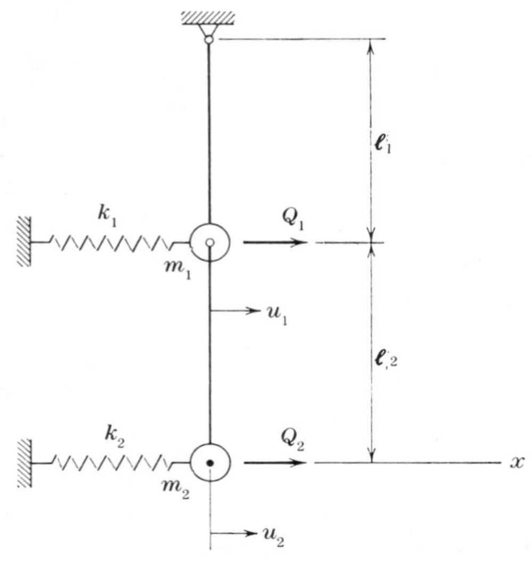

PROB. 3.2-3

3.2-4. The double pendulum in the figure has rotational springs across both of the hinges. Using the small horizontal translations u_1 and u_2 of the masses as displacement coordinates, determine the matrices **S** and **G**, and write the action equations of motion in matrix form.

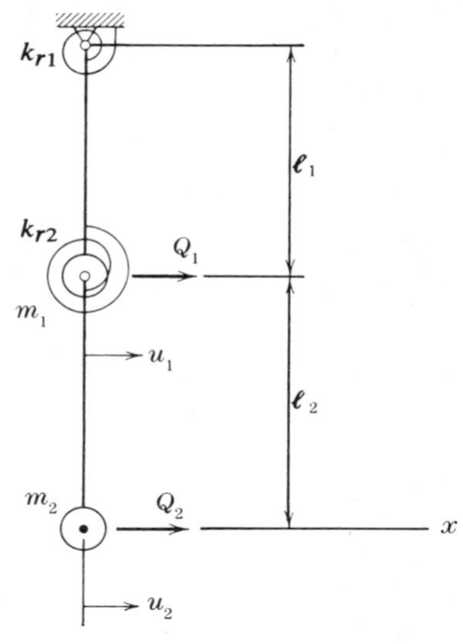

PROB. 3.2-4

3.2-5. For the two-story building frame in the figure determine the stiffness matrix **S**, and write the action equations of motion in matrix form. For this purpose assume that the girders are rigid, and use the small horizontal translations u_1 and u_2 as displacement coordinates. Columns in the frame are prismatic, having flexural rigidities of EI_1 in the lower story and EI_2 in the upper story.

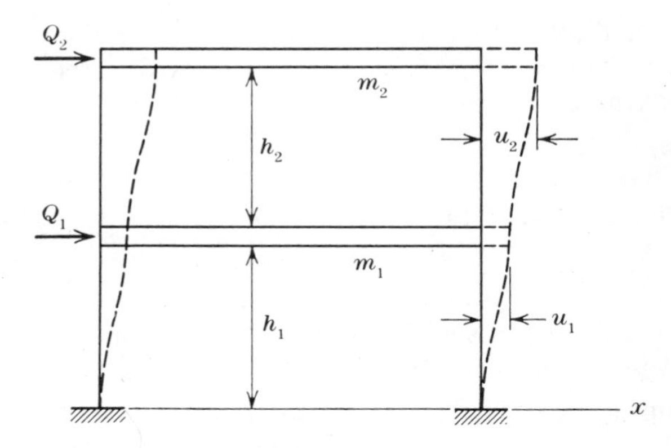

PROB. 3.2-5

3.2-6. A rigid prismatic bar is supported vertically on a roller and restrained against lateral motion by horizontal springs at top and bottom (see figure). Let the symbols ℓ, A, and ρ represent the length, cross-

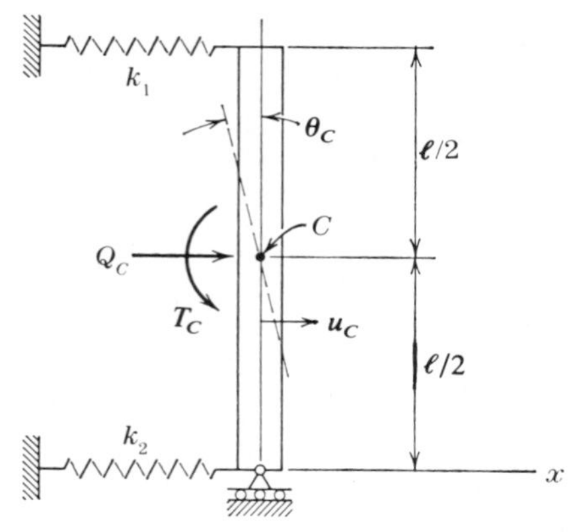

PROB. 3.2-6

sectional area, and mass density of the bar. Develop the stiffness, gravitational, and mass matrices for this system, using the small displacements u_C and θ_C of its center of mass (point C) as displacement coordinates. Write the action equations of motion in matrix form, including the horizontal force Q_C and torque T_C applied at point C.

3.3-1. For the two-mass system of Prob. 3.2-1, determine flexibility coefficients by applying unit forces to masses m_1 and m_2, one at a time. Write the displacement equations of motion in matrix form, and check the relationship $\mathbf{S} = \mathbf{F}^{-1}$.

3.3-2. Apply the displacement-equation method to the system in Fig. 3.3 of Sec. 3.1. Derive the flexibility coefficients directly, and check the relationship $\mathbf{S} = \mathbf{F}^{-1}$.

3.3-3. Reconsider the spring-connected pair of pendulums in Fig. 3.4 of Sec. 3.1, and determine the pseudoflexibility matrix $\mathbf{F}^*$ by inversion of $\mathbf{S}^*$. In addition, derive the elements of $\mathbf{F}^*$ directly by applying unit actions corresponding to the displacement coordinates θ_1 and θ_2.

3.3-4. For the two-story frame in Prob. 3.2-5, determine flexibilities by applying unit loads. Write the displacement equations of motion in matrix form, and check the relationship $\mathbf{S} = \mathbf{F}^{-1}$.

3.3-5. Develop the flexibility matrix $\mathbf{F}_C$ for the system of Prob. 3.2-6 in the absence of gravity. Invert $\mathbf{F}_C$ to obtain $\mathbf{S}_C$; then add $\mathbf{G}_C$ to produce $\mathbf{S}_C^*$; and invert $\mathbf{S}_C^*$ to determine $\mathbf{F}_C^*$.

3.3-6. The simply supported beam shown in the figure has masses m_1 and m_2 located at its third points. Assume that the beam is prismatic and has a flexural rigidity of EI. Using the translations v_1 and v_2 as displacement coordinates, derive the flexibility coefficients and write the displacement equations of motion in matrix form.

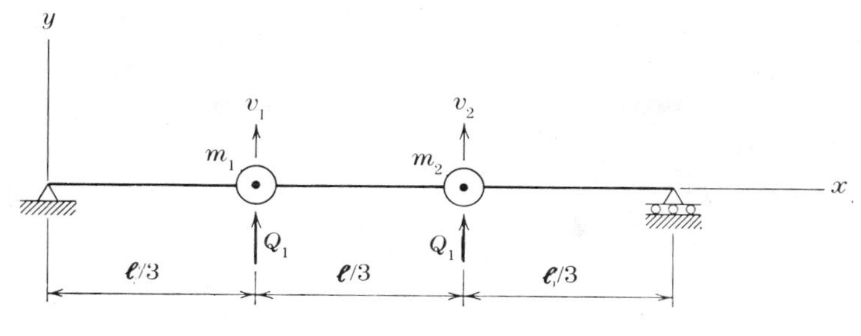

PROB. 3.3-6

3.3-7. The figure shows an overhanging prismatic beam (flexural rigidity of EI) with two masses m_1 and m_2. Determine the flexibility matrix $\mathbf{F}$; invert it to obtain the stiffness matrix $\mathbf{S} = \mathbf{F}^{-1}$; and write the action equations of motion in matrix form.

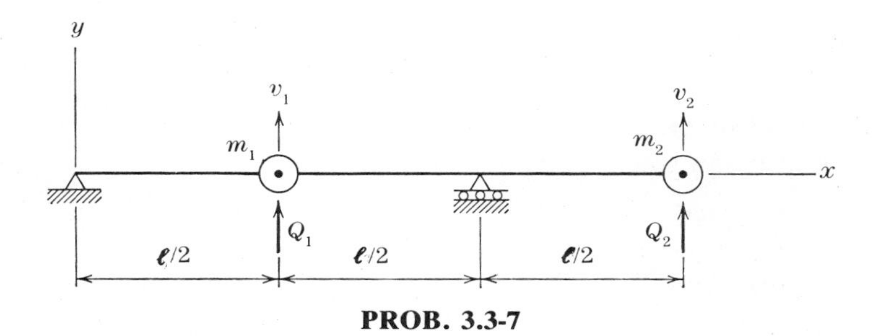

PROB. 3.3-7

3.3-8. Assume that each member of the horizontal frame in the figure is prismatic, with flexural rigidity EI and torsional rigidity GJ. Derive the flexibility matrix for the vertical translations v_1 and v_2, invert it, and set up the action equations of motion in matrix form.

3.5-1. For the system in Prob. 3.2-1 (see Sec. 3.2), assume that $m_1 = m_2 = m$ and that $k_1 = k_2 = k_3 = k$. Determine the characteristic values ω_1^2 and ω_2^2 and the modal ratios r_1 and r_2. Let the initial conditions be $u_{01} = u_{02} = \Delta$ and $\dot{u}_{01} = \dot{u}_{02} = 0$, and calculate the free-vibrational response.

3.5-2. Using the parameters specified in Prob. 3.2-2 (see Sec. 3.2) for the system in Fig. 3.2, calculate the characteristic values ω_1^2 and ω_2^2; and determine the modal ratios r_1 and r_2. In addition, evaluate the response of the system to the initial conditions $u_{01} = \Delta$, $v_{01} = 0$, $\dot{u}_{01} = \dot{v}_{01} = 0$.

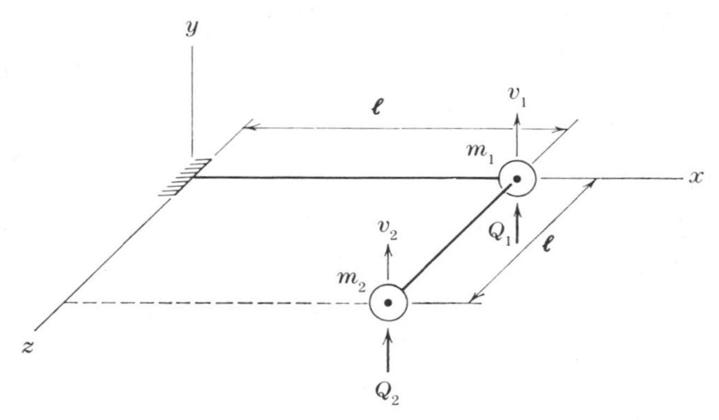

PROB. 3.3-8

3.5-3. Let $m_1 = m_2 = m$, $\ell_1 = \ell_2 = \ell$, and $k_1 = k_2 = 0$ for the system in Prob. 3.2-3 (see Sec. 3.2), and determine ω_1^2, ω_2^2, r_1, and r_2. Then calculate the response to the initial conditions $u_{01} = u_{02} = 0$ and $\dot{u}_{01} = \dot{u}_{02} = \dot{u}_0$.

3.5-4. For the two-story frame in Prob. 3.2-5 (see Sec. 3.2), let $m_1 = 2m$; $m_2 = m$; $h_1 = h_2 = h$; and $EI_1 = EI_2 = EI$. Evaluate ω_1^2, ω_2^2, r_1, and r_2; and determine the free-vibrational response due to the sudden release of a static load $(Q_1)_{st}$ at the lower level.

3.5-5. Assume that $m_1 = m_2 = m$ for the system in Fig. 3.7*a* (see Example 1, Sec. 3.3), and calculate λ_1, λ_2, r_1, and r_2 from Eqs. (3.23a), (3.24a), and (3.24b). Then find the response due to suddenly releasing a static load $(Q_2)_{st}$ at the free end of the beam.

3.5-6. Determine λ_1, λ_2, r_1, and r_2 for the system in Fig. 3.8*a* (see Example 2 of Sec. 3.3). Due to impact, the mass acquires an initial velocity $\dot{u}_0$ in the x direction ($\dot{u}_{01} = \dot{u}_0$; $\dot{v}_{01} = 0$; $u_{01} = v_{01} = 0$). Calculate the free-vibrational response to these initial conditions.

3.5-7. Let $m_1 = m_2 = m$ for the system in Prob. 3.3-6 (see Sec. 3.3), and evaluate λ_1, λ_2, r_1, and r_2. Suppose that the beam is dropped upon its supports from a height h and is retained by the supports thereafter. Determine the free-vibrational response to this initial condition.

3.5-8. Assuming that the system in Prob. 3.3-8 (see Sec. 3.3) has equal masses $m_1 = m_2 = m$ and that $R = EI/GJ = \frac{1}{3}$, find the values of λ_1, λ_2, r_1, and r_2. In addition, calculate the response to the sudden release of a static load $(Q_1)_{st}$ applied at the first mass.

3.5-9. Assuming that the mass matrix **M** for a two-degree system is filled instead of diagonal, derive expressions for ω_1^2, ω_2^2, r_1, and r_2 using action equations with stiffness coefficients.

3.5-10. Repeat Prob. 3.5-9 using displacement equations with flexibility coefficients.

3.5-11. Solve Example 2 of Sec. 3.5 using point A as the reference point for rigid-body motions. Action equations for this reference point were derived as Eq. (3.14a) in Sec. 3.4. (Expressions from Prob. 3.5-9 are required to solve this problem.)

3.5-12. For the system in Fig. 3.11 (see Example 1 of Sec. 3.4), let $b = \ell/3$ and $I_C = 2mb^2$; and calculate λ_1, λ_2, r_1, and r_2. Use point B as the reference for rigid-body motions, for which displacement equations are given as Eq. (l) in Sec. 3.4. (Expressions from Prob. 3.5-10 are required to solve this problem.)

CHAPTER 4

SYSTEMS WITH MULTIPLE DEGREES OF FREEDOM

4.1 INTRODUCTION

The concepts introduced in the preceding chapter for systems with two degrees of freedom will be extended in this chapter to *systems with multiple degrees of freedom.* In this category we include all systems having more than one degree of freedom but fewer han an infinite number of degrees of freedom. The configuration of such a vibrating system is determined by a finite number of displacement coordinates. If there are n degrees of freedom with which mass terms are associated, then n differential equations are required to describe the motions of the system.

In Chapter 3 free vibrations of two-degree systems and their response to harmonic excitations were handled without much difficulty, except for damped cases. Additional complications arise in multidegree systems because the number of terms increases rapidly with the number of degrees of freedom. Of course, matrix formulations prove to be very effective for the purpose of manipulating large numbers of terms. More important than this consideration, however, is the fact that systems subjected to arbitrary forcing functions become extremely difficult to analyze in the original coordinates, especially in the presence of damping. These difficulties will be avoided by using a more suitable set of coordinates.

If the principal modes of vibration for a multidegree system are used as generalized coordinates, the equations of undamped motion become uncoupled. In these coordinates each equation may be solved as if it pertained to a system with only one degree of freedom. This approach, known as the *normal-mode method of dynamic analysis*, is developed in this chapter and applied to problems of general interest. Undamped systems are treated first, and special considerations required for damped systems are discussed in the latter parts of the chapter.

4.2 FREQUENCIES AND MODE SHAPES FOR UNDAMPED SYSTEMS

For a system with n degrees of freedom, the action equations for undamped free vibrations [see Eq. (3.17) in Sec. 3.5] take the general form

$$\begin{bmatrix} M_{11} & M_{12} & M_{13} & \cdots & M_{1n} \\ M_{21} & M_{22} & M_{23} & \cdots & M_{2n} \\ M_{31} & M_{32} & M_{33} & \cdots & M_{3n} \\ \cdots & \cdots & \cdots & \cdots & \cdots \\ M_{n1} & M_{n2} & M_{n3} & \cdots & M_{nn} \end{bmatrix} \begin{bmatrix} \ddot{u}_1 \\ \ddot{u}_2 \\ \ddot{u}_3 \\ \cdots \\ \ddot{u}_n \end{bmatrix} + \begin{bmatrix} S_{11} & S_{12} & S_{13} & \cdots & S_{1n} \\ S_{21} & S_{22} & S_{23} & \cdots & S_{2n} \\ S_{31} & S_{32} & S_{33} & \cdots & S_{3n} \\ \cdots & \cdots & \cdots & \cdots & \cdots \\ S_{n1} & S_{n2} & S_{n3} & \cdots & S_{nn} \end{bmatrix} \begin{bmatrix} u_1 \\ u_2 \\ u_3 \\ \cdots \\ u_n \end{bmatrix} = \begin{bmatrix} 0 \\ 0 \\ 0 \\ \cdots \\ 0 \end{bmatrix} \tag{4.1}$$

or

$$\mathbf{M}\ddot{\mathbf{D}} + \mathbf{S}\mathbf{D} = \mathbf{0} \qquad \text{(3.17 repeated)}$$

We assume that in a natural mode of vibration all masses follow the harmonic function

$$\mathbf{D}_i = \mathbf{\Phi}_{\mathrm{M}i} \sin(\omega_i t + \phi_i) \tag{a}$$

in which ω_i and ϕ_i are the angular frequency and phase angle of the ith mode. The symbol $\mathbf{D}_i$ in Eq. (a) denotes the column matrix (or vector) of displacements for the ith mode, and $\mathbf{\Phi}_{\mathrm{M}i}$ represents the corresponding vector of maximum values, or amplitudes. That is,

$$\mathbf{D}_i = \begin{bmatrix} u_1 \\ u_2 \\ u_3 \\ \cdots \\ u_n \end{bmatrix}_i \qquad \mathbf{\Phi}_{\mathrm{M}i} = \begin{bmatrix} u_{\mathrm{m}1} \\ u_{\mathrm{m}2} \\ u_{\mathrm{m}3} \\ \cdots \\ u_{\mathrm{m}n} \end{bmatrix}_i$$

Substitution of Eq. (a) into Eq. (3.17) produces a set of algebraic equations that may be stated as

$$\mathbf{H}_i \mathbf{\Phi}_{\mathrm{M}i} = \mathbf{0} \tag{4.2}$$

where $\mathbf{H}_i$ is the *characteristic matrix*

$$\mathbf{H}_i = \mathbf{S} - \omega_i^2 \mathbf{M} \tag{4.3}$$

For nontrivial solutions of Eq. (4.2), the determinant of the characteristic matrix is set equal to zero, giving the following general form of the *characteristic equation*:

$$|\mathbf{H}_i| = \begin{vmatrix} S_{11} - \omega_i^2 M_{11} & S_{12} - \omega_i^2 M_{12} & S_{13} - \omega_i^2 M_{13} & \cdots & S_{1n} - \omega_i^2 M_{1n} \\ S_{21} - \omega_i^2 M_{21} & S_{22} - \omega_i^2 M_{22} & S_{23} - \omega_i^2 M_{23} & \cdots & S_{2n} - \omega_i^2 M_{2n} \\ S_{31} - \omega_i^2 M_{31} & S_{32} - \omega_i^2 M_{32} & S_{33} - \omega_i^2 M_{33} & \cdots & S_{3n} - \omega_i^2 M_{3n} \\ \cdots & \cdots & \cdots & \cdots & \cdots \\ S_{n1} - \omega_i^2 M_{n1} & S_{n2} - \omega_i^2 M_{n2} & S_{n3} - \omega_i^2 M_{n3} & \cdots & S_{nn} - \omega_i^2 M_{nn} \end{vmatrix} = 0 \tag{4.4}$$

Expansion of this determinant yields a polynomial in which the term of highest order is $(\omega_i^2)^n$. If the polynomial cannot be factored, its n roots $\omega_1^2, \omega_2^2, \ldots, \omega_i^2, \ldots, \omega_n^2$ may be found by a numerical procedure. Such roots, which were referred to previously as characteristic values, are also called *eigenvalues* (the prefix *eigen* means *own* in German). If $\mathbf{M}$ is positive-definite† and $\mathbf{S}$ is either positive-definite or positive-semidefinite, the eigenvalues will all be real and either positive or zero. However, they are not necessarily distinct (that is, different from another). The subject of repeated roots is discussed later in Sec. 4.7.

Vectors of modal amplitudes, any one of which is represented by $\mathbf{\Phi}_{\mathrm{M}i}$, are called *characteristic vectors*, or *eigenvectors*. If the eigenvalues of a system have been calculated as the roots of the characteristic equation (4.4), the eigenvectors may be evaluated (to within arbitrary constants) from the homogeneous algebraic equations (4.2). Because there are n characteristic values, there will also be n corresponding modal vectors. For a distinct eigenvalue (not a repeated root), $n - 1$ of the amplitudes in the eigenvector may be expressed in terms of the remaining amplitude by solving $n - 1$ simultaneous equations. However, we shall see that such extensive calculations are not necessary if the formal definition for the inverse of $\mathbf{H}_i$ is considered, as follows:

$$\mathbf{H}_i^{-1} = \frac{\mathbf{H}_i^{\mathrm{a}}}{|\mathbf{H}_i|} = \frac{(\mathbf{H}_i^{\mathrm{c}})^{\mathrm{T}}}{|\mathbf{H}_i|} \tag{b}$$

The symbol $\mathbf{H}_i^{\mathrm{a}}$ in Eq. (*b*) denotes the *adjoint matrix* of $\mathbf{H}_i$, which is defined as the transpose of the *cofactor matrix* $\mathbf{H}_i^{\mathrm{c}}$. Of course, the inverse of $\mathbf{H}_i$ does not actually exist because the determinant $|\mathbf{H}_i|$ is zero [see Eq. (4.4)]. For the purpose of this discussion, however, we may rewrite Eq. (*b*) as

$$\mathbf{H}_i \mathbf{H}_i^{\mathrm{a}} = |\mathbf{H}_i| \, \mathbf{I} = \mathbf{0} \tag{c}$$

† A matrix of real numbers is positive-definite if all of its leading principal minors are positive. If some of these minors are zero, it is said to be positive-semidefinite [1].

Comparing Eq. (c) with Eq. (4.2), we conclude that the eigenvector $\mathbf{\Phi}_{\mathrm{M}i}$ is proportional to any nonzero column of the adjoint matrix $\mathbf{H}_i^{\mathrm{a}}$. Since the eigenvector can be scaled arbitrarily, it may be taken to be either equal to such a column or normalized as desired.†

When displacement equations of motion are used instead of action equations, Eq. (4.1) is replaced by [See Eq. (3.21) in Sec. 3.5]:

$$\begin{bmatrix} F_{11} & F_{12} & F_{13} & \cdots & F_{1n} \\ F_{21} & F_{22} & F_{23} & \cdots & F_{2n} \\ F_{31} & F_{32} & F_{33} & \cdots & F_{3n} \\ \cdots & \cdots & \cdots & \cdots & \cdots \\ F_{n1} & F_{n2} & F_{n3} & \cdots & F_{nn} \end{bmatrix} \begin{bmatrix} M_{11} & M_{12} & M_{13} & \cdots & M_{1n} \\ M_{21} & M_{22} & M_{23} & \cdots & M_{2n} \\ M_{31} & M_{32} & M_{33} & \cdots & M_{3n} \\ \cdots & \cdots & \cdots & \cdots & \cdots \\ M_{n1} & M_{n2} & M_{n3} & \cdots & M_{nn} \end{bmatrix} \begin{bmatrix} \ddot{u}_1 \\ \ddot{u}_2 \\ \ddot{u}_3 \\ \cdots \\ \ddot{u}_n \end{bmatrix} + \begin{bmatrix} u_1 \\ u_2 \\ u_3 \\ \cdots \\ u_n \end{bmatrix} = \begin{bmatrix} 0 \\ 0 \\ 0 \\ \cdots \\ 0 \end{bmatrix} \tag{4.5}$$

or

$$\mathbf{F}\mathbf{M}\ddot{\mathbf{D}} + \mathbf{D} = \mathbf{0} \qquad \text{(3.21 repeated)}$$

Substituting Eq. (a) into Eq. (3.21), we obtain the algebraic equations

$$\mathbf{L}_i \mathbf{\Phi}_{\mathrm{M}i} = \mathbf{0} \tag{4.6}$$

In this expression $\mathbf{L}_i$ represents the characteristic matrix, defined as

$$\mathbf{L}_i = \mathbf{F}\mathbf{M} - \lambda_i \mathbf{I} \tag{4.7}$$

where $\lambda_i = 1/\omega_i^2$. For nontrivial solutions of Eq. (4.6), the determinant of $\mathbf{L}_i$ is set equal to zero, producing the general form of the characteristic equation for this case as

$$\left| \begin{bmatrix} F_{11} & F_{12} & F_{13} & \cdots & F_{1n} \\ F_{21} & F_{22} & F_{23} & \cdots & F_{2n} \\ F_{31} & F_{32} & F_{33} & \cdots & F_{3n} \\ \cdots & \cdots & \cdots & \cdots & \cdots \\ F_{n1} & F_{n2} & F_{n3} & \cdots & F_{nn} \end{bmatrix} \begin{bmatrix} M_{11} & M_{12} & M_{13} & \cdots & M_{1n} \\ M_{21} & M_{22} & M_{23} & \cdots & M_{2n} \\ M_{31} & M_{32} & M_{33} & \cdots & M_{3n} \\ \cdots & \cdots & \cdots & \cdots & \cdots \\ M_{n1} & M_{n2} & M_{n3} & \cdots & M_{nn} \end{bmatrix} - \lambda_i \begin{bmatrix} 1 & 0 & 0 & \cdots & 0 \\ 0 & 1 & 0 & \cdots & 0 \\ 0 & 0 & 1 & \cdots & 0 \\ \cdots & \cdots & \cdots & \cdots & \cdots \\ 0 & 0 & 0 & \cdots & 1 \end{bmatrix} \right| = 0 \tag{4.8}$$

Expansion of the determinant in Eq. (4.8) yields a polynomial of order n, from which the roots $\lambda_1, \lambda_2, \ldots, \lambda_i, \ldots, \lambda_n$ may be obtained. These roots constitute the eigenvalues for this case, and the eigenvectors can be determined by substituting these values into Eq. (4.6). Alternatively, a typical eigenvector may be taken to be any column of the adjoint matrix $\mathbf{L}_i^{\mathrm{a}}$, scaled as desired.

† If submatrices of $\mathbf{H}_i$ are uncoupled, analyze the parts separately.

From previous discussions in Chapter 3, it is evident that the stiffness matrix **S** in Eq. (4.3) can be replaced or augmented by a gravitational matrix **G** [see Eq. (3.10) in Sec. 3.2]. Similarly, the flexibility matrix **F** in Eq. (4.7) can be replaced by an array of pseudoflexibilities associated with gravitational influences (see Example 3, Sec. 3.3). In any case, the calculations are simplified considerably if the mass matrix **M** is diagonal instead of filled. Determination of natural frequencies and mode shapes will now be illustrated by specific examples of multidegree systems.

Example 1. Figure 4.1*a* shows three masses that are connected to each other and to ground by three springs. The displacement coordinates u_1, u_2, and u_3 define the motions of this three-degree-of-freedom system. For

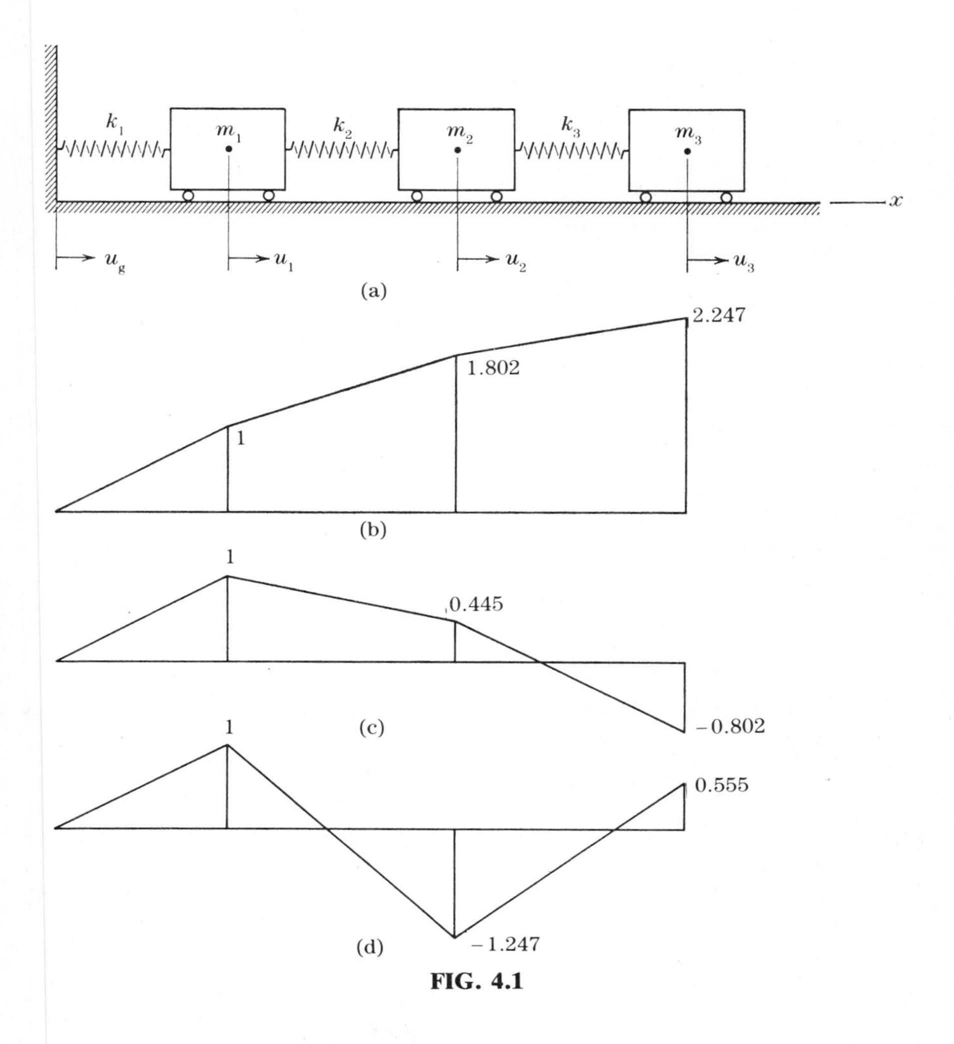

FIG. 4.1

simplicity, let $m_1 = m_2 = m_3 = m$ and $k_1 = k_2 = k_3 = k$. Determine the characteristic values and principal mode shapes using the action-equation approach.

Solution. The mass matrix for this system is the diagonal array

$$\mathbf{M} = \begin{bmatrix} m_1 & 0 & 0 \\ 0 & m_2 & 0 \\ 0 & 0 & m_3 \end{bmatrix} = m \begin{bmatrix} 1 & 0 & 0 \\ 0 & 1 & 0 \\ 0 & 0 & 1 \end{bmatrix} \tag{d}$$

and the stiffness matrix is

$$\mathbf{S} = \begin{bmatrix} k_1 + k_2 & -k_2 & 0 \\ -k_2 & k_2 + k_3 & -k_3 \\ 0 & -k_3 & k_3 \end{bmatrix} = k \begin{bmatrix} 2 & -1 & 0 \\ -1 & 2 & -1 \\ 0 & -1 & 1 \end{bmatrix} \tag{e}$$

We use the terms in these arrays to form the characteristic matrix in Eq. (4.3) as

$$\mathbf{H}_i = \begin{bmatrix} 2k - \omega_i^2 m & -k & 0 \\ -k & 2k - \omega_i^2 m & -k \\ 0 & -k & k - \omega_i^2 m \end{bmatrix} \tag{f}$$

Setting the determinant of $\mathbf{H}_i$ equal to zero [see Eq. (4.4)] and collecting terms, we obtain the following characteristic equation:

$$(\omega_i^2)^3 - 5\left(\frac{k}{m}\right)(\omega_i^2)^2 + 6\left(\frac{k}{m}\right)^2(\omega_i^2) - \left(\frac{k}{m}\right)^3 = 0 \tag{g}$$

This cubic equation cannot be factored, but its roots may be determined by trial and error to be

$$\omega_1^2 = 0.198\frac{k}{m} \qquad \omega_2^2 = 1.555\frac{k}{m} \qquad \omega_3^2 = 3.247\frac{k}{m} \tag{h}$$

To determine the mode shape corresponding to the lowest eigenvalue, we substitute the value of ω_1^2 into Eq. (4.2) and solve for $u_{\mathrm{m}2,1}$ and $u_{\mathrm{m}3,1}$ in terms of $u_{\mathrm{m}1,1}$. The results of this solution are

$$u_{\mathrm{m}2,1} = 1.802u_{\mathrm{m}1,1} \qquad u_{\mathrm{m}3,1} = 2.247u_{\mathrm{m}1,1} \tag{i}$$

Similarly, substitution of the values of ω_2^2 and ω_3^3 into Eq. (4.2) leads to the solutions

$$u_{\mathrm{m}2,2} = 0.445u_{\mathrm{m}1,2} \qquad u_{\mathrm{m}3,2} = -0.802u_{\mathrm{m}1,2} \tag{j}$$

and

$$u_{m2,3} = -1.247 u_{m1,3} \qquad u_{m3,3} = 0.555 u_{m1,3} \tag{k}$$

Alternatively, the adjoint $\mathbf{H}_i^a$ is developed from Eq. (f), as follows:

$$\mathbf{H}_i^a = \begin{bmatrix} (2k-\omega_i^2 m)(k-\omega_i^2 m)-k^2 & k(k-\omega_i^2 m) & k^2 \\ k(k-\omega_i^2 m) & (2k-\omega_i^2 m)(k-\omega_i^2 m) & k(2k-\omega_i^2 m) \\ k^2 & k(2k-\omega_i^2 m) & (2k-\omega_i^2 m)^2-k^2 \end{bmatrix} \tag{l}$$

Substitution of the value of ω_1^2 into Eq. (l) yields

$$\mathbf{H}_i^a = k^2 \begin{bmatrix} 0.445 & 0.802 & 1.000 \\ 0.802 & 1.445 & 1.802 \\ 1.000 & 1.802 & 2.247 \end{bmatrix}$$

The third column of this matrix (divided by k^2) provides the first eigenvector normalized with respect to the amplitude of the first mass. Thus, we have†

$$\mathbf{\Phi}_{M1} = \{1.000, 1.802, 2.247\} \tag{m}$$

which checks with Eqs. (i). Of course, we need generate only one column of the adjoint matrix to determine this result because all of the columns are proportional to $\mathbf{\Phi}_{M1}$.

In a similar manner, the eigenvectors $\mathbf{\Phi}_{M2}$ and $\mathbf{\Phi}_{M3}$ are determined by substituting the values of ω_2^2 and ω_3^2 into the third column of $\mathbf{H}_i^a$ in Eq. (l). The resulting vectors are

$$\mathbf{\Phi}_{M2} = \{1.000, 0.445, -0.802\} \tag{n}$$

and

$$\mathbf{\Phi}_{M3} = \{1.000, -1.247, 0.555\} \tag{o}$$

which check with Eqs. (j) and (k). The three mode shapes in Eqs. (m), (n), and (o) are represented by the ordinates in Figs. 4.1b, 4.1c, and 4.1d, respectively.

Suppose now that the stiffness constant k_1 for the first spring in Fig. 4.1a is zero. In this case the system will be free to translate as a rigid body as well as to vibrate. The stiffness coefficient S_{11} changes from $2k$ to k, and the corresponding term in the characteristic matrix becomes $H_{i11} = k - \omega_i^2 m$.

† To save space, column vectors may be written in a row enclosed by braces and with commas separating the terms.

Consequently, the characteristic equation simplifies to the factored expression

$$\omega_i^2\left(\omega_i^2 - \frac{k}{m}\right)\left(\omega_i^2 - \frac{3k}{m}\right) = 0 \tag{p}$$

from which we obtain

$$\omega_1^2 = 0 \qquad \omega_2^2 = \frac{k}{m} \qquad \omega_3^2 = \frac{3k}{m} \tag{q}$$

where the zero root corresponds to the rigid-body mode.

The third column of the adjoint matrix $\mathbf{H}_i^{\mathrm{a}}$ in Eq. (l) changes to

$$\mathbf{H}_{i,3}^{\mathrm{a}} = \begin{bmatrix} k^2 \\ k(k - \omega_i^2 m) \\ (k - \omega_i^2 m)(2k - \omega_i^2 m) - k^2 \end{bmatrix} \tag{r}$$

Successive substitution of the eigenvalues for this semidefinite system into Eq. (r) produces the following eigenvectors:

$$\mathbf{\Phi}_{\mathrm{M1}} = \begin{bmatrix} 1 \\ 1 \\ 1 \end{bmatrix} \qquad \mathbf{\Phi}_{\mathrm{M2}} = \begin{bmatrix} 1 \\ 0 \\ -1 \end{bmatrix} \qquad \mathbf{\Phi}_{\mathrm{M3}} = \begin{bmatrix} 1 \\ -2 \\ 1 \end{bmatrix} \tag{s}$$

The principal mode shapes for this sytem are easily visualized by inspection of these vectors.

Example 2. Three masses are attached to a tighly stretched wire, as shown in Fig. 4.2a. The tensile force T in the wire is assumed to be large, so that for small lateral displacements of the particles it does not change appreciably. Characteristic values and vectors of this system are to be found by the displacement-equation approach, taking $m_1 = m_2 = m_3 = m$ and $\ell_1 = \ell_2 = \ell_3 = \ell_4 = \ell$.

Solution. The mass matrix is the same as in Example 1, and the flexibility matrix is derived as

$$\mathbf{F} = \frac{\ell}{4T}\begin{bmatrix} 3 & 2 & 1 \\ 2 & 4 & 2 \\ 1 & 2 & 3 \end{bmatrix} \tag{t}$$

Using these terms, we form the characteristic matrix in Eq. (4.7) as

$$\mathbf{L}_i = \begin{bmatrix} 3\alpha - \lambda_i & 2\alpha & \alpha \\ 2\alpha & 4\alpha - \lambda_i & 2\alpha \\ \alpha & 2\alpha & 3\alpha - \lambda_i \end{bmatrix} \tag{u}$$

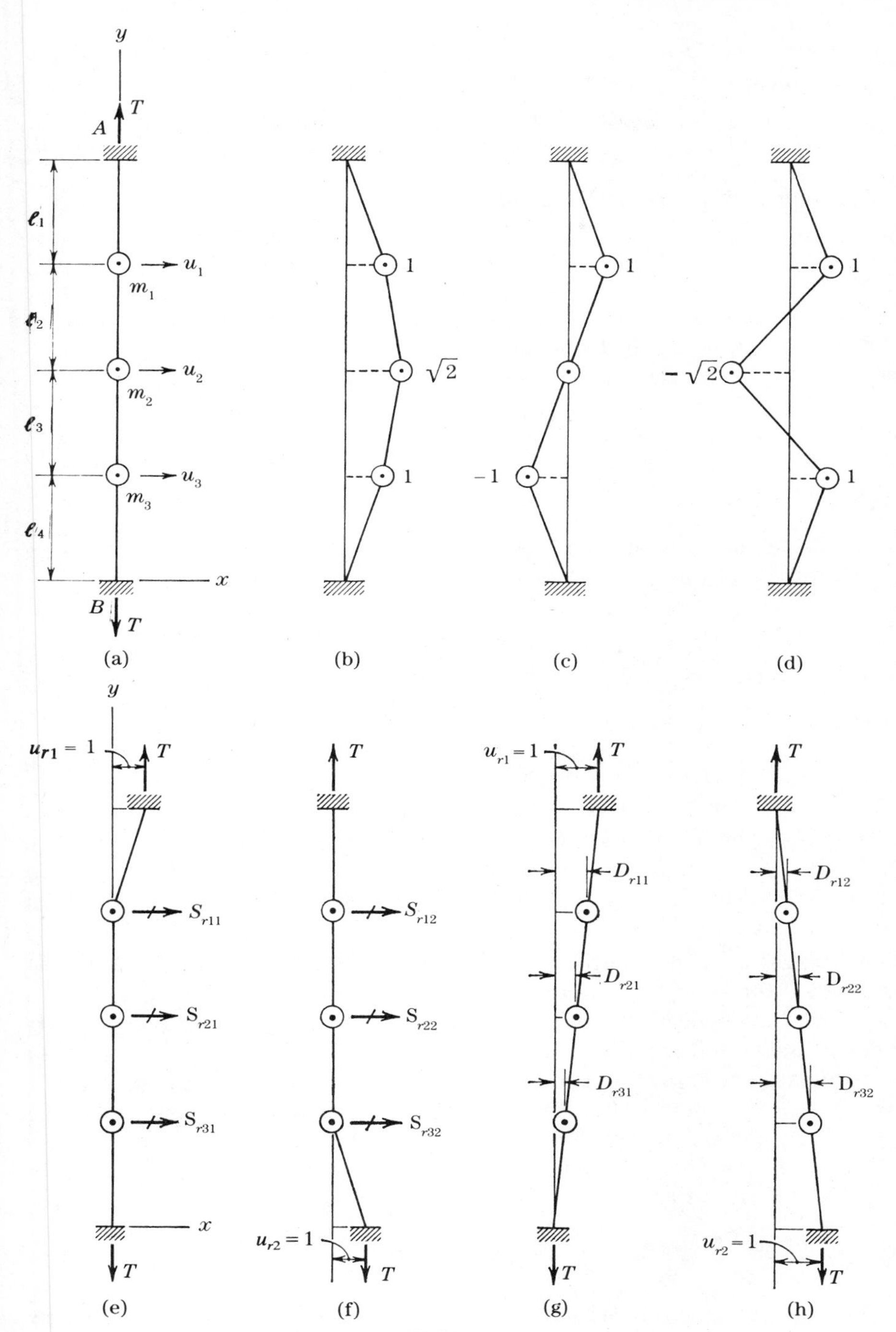

y
T
A
ℓ_1
ℓ_2
ℓ_3
ℓ_4
u_1
u_2
u_3
m_1
m_2
m_3
x
B
T
1
$\sqrt{2}$
1
1
-1
1
$-\sqrt{2}$
1
(a)
(b)
(c)
(d)
y
$u_{r1} = 1$
T
S_{r11}
S_{r21}
S_{r31}
x
T
T
S_{r12}
S_{r22}
S_{r32}
$u_{r2} = 1$
T
$u_{r1} = 1$
T
D_{r11}
D_{r21}
D_{r31}
T
T
D_{r12}
D_{r22}
D_{r32}
$u_{r2} = 1$
T
(e)
(f)
(g)
(h)

FIG. 4.2

where $\alpha = \ell m/4T$. Then the determinant of $\mathbf{L}_i$ is set equal to zero to obtain the characteristic equation

$$(\lambda_i - 2\alpha)(\lambda_i^2 - 8\alpha\lambda_i + 8\alpha^2) = 0 \tag{v}$$

The roots of Eq. (v), in descending order, are

$$\lambda_1 = 2(2+\sqrt{2})\alpha \qquad \lambda_2 = 2\alpha \qquad \lambda_3 = 2(2-\sqrt{2})\alpha \tag{w}$$

To find the mode shapes for this system, we develop only the first column of the adjoint matrix of $\mathbf{L}_i$, as follows:

$$\mathbf{L}_{i,1}^{\mathrm{a}} = \begin{bmatrix} (4\alpha - \lambda_i)(3\alpha - \lambda_i) - 4\alpha^2 \\ -2\alpha(3\alpha - \lambda_i) + 2\alpha^2 \\ 4\alpha^2 - \alpha(4\alpha - \lambda_i) \end{bmatrix} \tag{x}$$

Successive substitution of the eigenvalues into Eq. (x) and (normalization with respect to $u_{\mathrm{m}1}$ in each mode) yields the eigenvectors

$$\mathbf{\Phi}_{\mathrm{M}1} = \begin{bmatrix} 1 \\ \sqrt{2} \\ 1 \end{bmatrix} \qquad \mathbf{\Phi}_{\mathrm{m}2} = \begin{bmatrix} 1 \\ 0 \\ -1 \end{bmatrix} \qquad \mathbf{\Phi}_{\mathrm{M}3} = \begin{bmatrix} 1 \\ -\sqrt{2} \\ 1 \end{bmatrix} \tag{y}$$

These mode shapes appear in Figs. 4.2b, 4.2c, and 4.2d, respectively.

The two methods described in this section for obtaining characteristic values and characteristic vectors can both be expressed as

$$\mathbf{A}\mathbf{\Phi}_{\mathrm{M}i} = \lambda_i \mathbf{\Phi}_{\mathrm{M}i} \tag{z}$$

where $\mathbf{A}$ is a square matrix of real numbers. Equation (z) is known as the *standard form of the eigenvalue problem.* For linearly elastic vibratory systems the coefficient matrix $\mathbf{A}$ can always be developed as a symmetric, positive-definite (or positive-semidefinite) array by an appropriate choice of coordinates. Such a form is often desirable when the eigenvalue problem is to be solved numerically, and techniques for attaining symmetry in the coefficient matrix are given in the following discussion.

Equation (4.6) from the displacement-equation method may be written in terms of $\mathbf{F}$ and $\mathbf{M}$ as

$$\mathbf{F}\mathbf{M}\mathbf{\Phi}_{\mathrm{M}i} = \lambda_i \mathbf{\Phi}_{\mathrm{M}i} \tag{4.9}$$

This equation is in the standard form of Eq. (z), but the coefficient matrix $\mathbf{FM}$ is unsymmetric. The product of $\mathbf{F}$ and $\mathbf{M}$ lacks symmetry even when $\mathbf{M}$

is a diagonal matrix, except for the special case when all terms on the diagonal are equal. To attain symmetry in the coefficient matrix, a change of coordinates is required. If $\mathbf{M}$ is positive-definite, it may be factored by the *Cholesky square-root method* [2] into

$$\mathbf{M} = \mathbf{U}^{\mathrm{T}}\mathbf{U} \tag{4.10}$$

where $\mathbf{U}$ is an upper-triangular matrix, and $\mathbf{U}^{\mathrm{T}}$ is its transpose. By substituting Eq. (4.10) into Eq. (4.9) and premultiplying both sides of the latter by $\mathbf{U}$, we obtain

$$\mathbf{UFU}^{\mathrm{T}}\mathbf{U}\mathbf{\Phi}_{\mathrm{M}i} = \lambda_i \mathbf{U}\mathbf{\Phi}_{\mathrm{M}i}$$

This equation can be rewritten as

$$\mathbf{F}_{\mathrm{U}}\mathbf{\Phi}_{\mathrm{U}i} = \lambda_i \mathbf{\Phi}_{\mathrm{U}i} \tag{4.11}$$

where

$$\mathbf{\Phi}_{\mathrm{U}i} = \mathbf{U}\mathbf{\Phi}_{\mathrm{M}i} \tag{4.12a}$$

or

$$\mathbf{\Phi}_{\mathrm{M}i} = \mathbf{U}^{-1}\mathbf{\Phi}_{\mathrm{U}i} \tag{4.12b}$$

and

$$\mathbf{F}_{\mathrm{U}} = \mathbf{UFU}^{\mathrm{T}} \tag{4.13}$$

The symbol $\mathbf{\Phi}_{\mathrm{U}i}$, defined by Eq. (4.12a), represents modal amplitudes transformed to a new set of coordinates, where the generalized mass matrix is the identity matrix. In these coordinates the generalized flexibility matrix $\mathbf{F}_{\mathrm{U}}$, given by Eq. (4.13), is symmetric because it results from a congruence transformation of the symmetric matrix $\mathbf{F}$.

Thus, Eq. (4.11) is in the standard form of the eigenvalue problem with a symmetric, positive-definite coefficient matrix. It is evident that this transformed equation has the same eigenvalues λ_i as the original equation (4.9). However, the eigenvectors $\mathbf{\Phi}_{\mathrm{U}i}$ are not the same as $\mathbf{\Phi}_{\mathrm{M}i}$. After solving for the eigenvectors $\mathbf{\Phi}_{\mathrm{U}i}$ in the generalized coordinates, we may transform them to the original coordinates using Eq. (4.12b).

The transformations given by Eqs. (4.12a) and (4.12b) are simplified if the mass matrix is diagonal. In this case the factorization of $\mathbf{M}$ in Eq. (4.10) results in

$$\mathbf{U} = \mathbf{U}^{\mathrm{T}} = \mathbf{M}^{1/2} \qquad \mathbf{U}^{-1} = (\mathbf{U}^{-1})^{\mathrm{T}} = \mathbf{M}^{-1/2} \tag{4.14}$$

In these expressions the symbol $\mathbf{M}^{1/2}$ signifies a diagonal matrix containing terms that are equal to the square roots of those in $\mathbf{M}$, and $\mathbf{M}^{-1/2}$ denotes a diagonal matrix containing the reciprocals of those in $\mathbf{M}^{1/2}$. Then the relationships in Eqs. (4.12a), (4.12b), and (4.13) become

$$\mathbf{\Phi}_{\mathrm{U}i} = \mathbf{M}^{1/2}\mathbf{\Phi}_{\mathrm{M}i} \tag{4.15a}$$

or

$$\mathbf{\Phi}_{\mathrm{M}i} = \mathbf{M}^{-1/2}\mathbf{\Phi}_{\mathrm{U}i} \tag{4.15b}$$

and

$$\mathbf{F}_{\mathrm{U}} = \mathbf{M}^{1/2}\mathbf{F}\mathbf{M}^{1/2} \tag{4.16}$$

Equation (4.2) for the action-equation approach may also be transformed to the generalized coordinates. We first write it in terms of **S** and **M** as

$$\mathbf{S}\mathbf{\Phi}_{\mathrm{M}i} = \omega_i^2\mathbf{M}\mathbf{\Phi}_{\mathrm{M}i} \tag{4.17}$$

which differs from the standard form in Eq. (z) because of the presence of **M** on the right-hand side. Equation (4.17) represents the *eigenvalue problem in nonstandard form*, having two symmetric coefficient matrices. It could be converted to the standard form by premultiplying both sides with $\mathbf{M}^{-1}$, but the resulting coefficient matrix $\mathbf{M}^{-1}\mathbf{S}$ would be unsymmetric. To avoid this loss of symmetry, we substitute Eq. (4.12b) into Eq. (4.17) and premultiply both sides of the latter by $(\mathbf{U}^{-1})^{\mathrm{T}}$ to obtain

$$(\mathbf{U}^{-1})^{\mathrm{T}}\mathbf{S}\mathbf{U}^{-1}\mathbf{\Phi}_{\mathrm{U}i} = \omega_i^2(\mathbf{U}^{-1})^{\mathrm{T}}\mathbf{M}\mathbf{U}^{-1}\mathbf{\Phi}_{\mathrm{U}i} \tag{4.18}$$

Use of the factored form of **M** from Eq. (4.10) in the right-hand side of Eq. (4.18) yields the result

$$\mathbf{S}_{\mathrm{U}}\mathbf{\Phi}_{\mathrm{U}i} = \omega_i^2\mathbf{M}_{\mathrm{U}}\mathbf{\Phi}_{\mathrm{U}i} = \omega_i^2\mathbf{\Phi}_{\mathrm{U}i} \tag{4.19}$$

where

$$\mathbf{S}_{\mathrm{U}} = (\mathbf{U}^{-1})^{\mathrm{T}}\mathbf{S}\mathbf{U}^{-1} = \mathbf{F}_{\mathrm{U}}^{-1} \tag{4.20}$$

Here we see clearly the transformation of **M** to the identity matrix, as follows:

$$\mathbf{M}_{\mathrm{U}} = (\mathbf{U}^{-1})^{\mathrm{T}}\mathbf{M}\mathbf{U}^{-1} = (\mathbf{U}^{\mathrm{T}})^{-1}\mathbf{U}^{\mathrm{T}}\mathbf{U}\mathbf{U}^{-1} = \mathbf{I} \tag{4.21}$$

Equation (4.19) is now in the standard form of the eigenvalue problem with a symmetric coefficient matrix. As indicated by Eq. (4.20), the generalized stiffness matrix $\mathbf{S}_{\mathrm{U}}$ is the inverse of the generalized flexibility matrix, given by Eq. (4.13). Of course, the converse of this statement is true only when **S** (and hence $\mathbf{S}_{\mathrm{U}}$) is positive-definite. When the mass matrix is diagonal, the transformation of $\mathbf{S}$ in Eq. (4.20) simplifies to

$$\mathbf{S}_{\mathrm{U}} = \mathbf{M}^{-1/2}\mathbf{S}\mathbf{M}^{-1/2} \tag{4.22}$$

Example 3. Suppose that the second mass in Fig. 4.2*a* is given the value $m_2 = 4m$, while $m_1 = m_3 = m$, as in Example 2. In this case the product $\mathbf{FM}$ in Eq. (4.9) becomes

$$\mathbf{FM} = \frac{\ell}{4T}\begin{bmatrix} 3 & 2 & 1 \\ 2 & 4 & 2 \\ 1 & 2 & 3 \end{bmatrix}\begin{bmatrix} 1 & 0 & 0 \\ 0 & 4 & 0 \\ 0 & 0 & 1 \end{bmatrix} m = \alpha\begin{bmatrix} 3 & 8 & 1 \\ 2 & 16 & 2 \\ 1 & 8 & 3 \end{bmatrix} \tag{a'}$$

This array is unsymmetric becasue the second column of $\mathbf{F}$ has been scaled differently from the others. Instead of using this form, we can preserve the inherent symmetry in $\mathbf{F}$ by first forming the diagonal matrix $\mathbf{M}^{1/2}$ (and its inverse), as follows

$$\mathbf{M}^{1/2} = \sqrt{m}\begin{bmatrix} 1 & 0 & 0 \\ 0 & 2 & 0 \\ 0 & 0 & 1 \end{bmatrix} \qquad \mathbf{M}^{-1/2} = \frac{1}{\sqrt{m}}\begin{bmatrix} 1 & 0 & 0 \\ 0 & \frac{1}{2} & 0 \\ 0 & 0 & 1 \end{bmatrix} \tag{b'}$$

Then the array $\mathbf{M}^{1/2}$ is applied as a transformation operator, as indicated in Eq. (4.16), to obtain

$$\mathbf{F}_{\mathrm{U}} = \mathbf{M}^{1/2}\mathbf{F}\mathbf{M}^{1/2} = \alpha\begin{bmatrix} 3 & 4 & 1 \\ 4 & 16 & 4 \\ 1 & 4 & 3 \end{bmatrix} \tag{c'}$$

which has the desired symmetry. The transformed matrix $\mathbf{F}_{\mathrm{U}}$ would be used to solve the eigenvalue problem in the standard form given by Eq. (4.11). Finally, Eq. (4.15b) provides the means of determining the eigenvectors in the original coordinates, using the operator $\mathbf{M}^{-1/2}$.

4.3 PRINCIPAL AND NORMAL COORDINATES

To examine certain inherent relationships among principal modes of vibration, we shall consider modes i and j of the eigenvalue problem for action equations [see Eq. (4.17) in the preceding section], as follows:

$$\mathbf{S\Phi}_{\mathrm{M}i} = \omega_i^2\mathbf{M\Phi}_{\mathrm{M}i} \tag{a}$$

$$\mathbf{S\Phi}_{\mathrm{M}j} = \omega_j^2\mathbf{M\Phi}_{\mathrm{M}j} \tag{b}$$

Premultiplication of the first of these expressions by $\mathbf{\Phi}_{\mathrm{M}j}^{\mathrm{T}}$ and postmultiplication of the transpose of the second by $\mathbf{\Phi}_{\mathrm{M}i}$ produces

$$\mathbf{\Phi}_{\mathrm{M}j}^{\mathrm{T}}\mathbf{S}\mathbf{\Phi}_{\mathrm{M}i} = \omega_i^2 \mathbf{\Phi}_{\mathrm{M}j}^{\mathrm{T}}\mathbf{M}\mathbf{\Phi}_{\mathrm{M}i} \tag{c}$$

$$\mathbf{\Phi}_{\mathrm{M}j}^{\mathrm{T}}\mathbf{S}\mathbf{\Phi}_{\mathrm{M}i} = \omega_j^2 \mathbf{\Phi}_{\mathrm{M}j}^{\mathrm{T}}\mathbf{M}\mathbf{\Phi}_{\mathrm{M}i} \tag{d}$$

The left-hand side of Eqs. (*c*) and (*d*) are equal, so that subtraction of the second equation from the first yields the relationship

$$(\omega_i^2 - \omega_j^2)\mathbf{\Phi}_{\mathrm{M}j}^{\mathrm{T}}\mathbf{M}\mathbf{\Phi}_{\mathrm{M}i} = 0 \tag{e}$$

On the other hand, if we divide both sides of Eq. (*c*) by ω_i^2 and both sides of Eq. (*d*) by ω_j^2, the right-hand sides become equal. Then subtraction gives

$$\left(\frac{1}{\omega_i^2} - \frac{1}{\omega_j^2}\right)\mathbf{\Phi}_{\mathrm{M}j}^{\mathrm{T}}\mathbf{S}\mathbf{\Phi}_{\mathrm{M}i} = 0 \tag{f}$$

To satisfy Eqs. (*e*) and (*f*) when $i \neq j$ and the eigenvalues are distinct ($\omega_i^2 \neq \omega_j^2$), the following relationships must hold:

$$\mathbf{\Phi}_{\mathrm{M}j}^{\mathrm{T}}\mathbf{M}\mathbf{\Phi}_{\mathrm{M}i} = \mathbf{\Phi}_{\mathrm{M}i}^{\mathrm{T}}\mathbf{M}\mathbf{\Phi}_{\mathrm{M}j} = 0 \tag{4.23}$$

and

$$\mathbf{\Phi}_{\mathrm{M}j}^{\mathrm{T}}\mathbf{S}\mathbf{\Phi}_{\mathrm{M}i} = \mathbf{\Phi}_{\mathrm{M}i}^{\mathrm{T}}\mathbf{S}\mathbf{\Phi}_{\mathrm{M}j} = 0 \tag{4.24}$$

These expressions represent *orthogonality relationships* among the principal modes of vibration. From Eq. (4.23) we see that the eigenvectors are orthogonal with respect to **M**, and Eq. (4.24) shows that they are also orthogonal with respect to **S**.

For the case when $i = j$, Eqs. (*e*) and (*f*) yield

$$\mathbf{\Phi}_{\mathrm{M}i}^{\mathrm{T}}\mathbf{M}\mathbf{\Phi}_{\mathrm{M}i} = M_{\mathrm{P}i} \tag{4.25}$$

and

$$\mathbf{\Phi}_{\mathrm{M}i}^{\mathrm{T}}\mathbf{S}\mathbf{\Phi}_{\mathrm{M}i} = S_{\mathrm{P}i} \tag{4.26}$$

where $M_{\mathrm{P}i}$ and $S_{\mathrm{P}i}$ are constants that depend upon how the eigenvector $\mathbf{\Phi}_{\mathrm{M}i}$ is normalized. For operational efficiency, we place all of the eigenvectors column-wise into an $n \times n$ *modal matrix* of the form

$$\mathbf{\Phi}_{\mathrm{M}} = [\mathbf{\Phi}_{\mathrm{M1}} \quad \mathbf{\Phi}_{\mathrm{M2}} \quad \mathbf{\Phi}_{\mathrm{M3}} \quad \cdots \quad \mathbf{\Phi}_{\mathrm{M}n}] \tag{4.27}$$

Then we can state Eqs. (4.23) and (4.25) collectively as

$$\mathbf{\Phi}_{\mathrm{M}}^{\mathrm{T}}\mathbf{M}\mathbf{\Phi}_{\mathrm{M}} = \mathbf{M}_{\mathrm{P}} \tag{4.28}$$

where $\mathbf{M}_{\mathrm{P}}$ is a diagonal array that will be referred to as a *principal mass matrix*. Similarly, Eqs. (4.24) and (4.26) are combined into

$$\mathbf{\Phi}_{\mathrm{M}}^{\mathrm{T}}\mathbf{S}\mathbf{\Phi}_{\mathrm{M}} = \mathbf{S}_{\mathrm{P}} \tag{4.29}$$

in which $\mathbf{S}_{\mathrm{P}}$ is another diagonal array that will be called a *principal stiffness matrix*. Equations (4.28) and (4.29) represent *diagonalization* of matrices $\mathbf{M}$ and $\mathbf{S}$. If either of them is already diagonal, the operation in either Eq. (4.28) or Eq. (4.29) merely scales the values on the diagonal.

In order to take advantage of the diagonalization process, let us reconsider the action equations of motion for free vibrations of an undamped multidegree system, as follows:

$$\mathbf{M}\ddot{\mathbf{D}} + \mathbf{S}\mathbf{D} = \mathbf{0} \tag{4.30}$$

Premultiplication of Eq. (4.30) by $\mathbf{\Phi}_{\mathrm{M}}^{\mathrm{T}}$ and insertion of $\mathbf{I} = \mathbf{\Phi}_{\mathrm{M}}\mathbf{\Phi}_{\mathrm{M}}^{-1}$ before $\ddot{\mathbf{D}}$ and $\mathbf{D}$ produces

$$\mathbf{\Phi}_{\mathrm{M}}^{\mathrm{T}}\mathbf{M}\mathbf{\Phi}_{\mathrm{M}}\mathbf{\Phi}_{\mathrm{M}}^{-1}\ddot{\mathbf{D}} + \mathbf{\Phi}_{\mathrm{M}}^{\mathrm{T}}\mathbf{S}\mathbf{\Phi}_{\mathrm{M}}\mathbf{\Phi}_{\mathrm{M}}^{-1}\mathbf{D} = \mathbf{0}$$

which can be restated as

$$\mathbf{M}_{\mathrm{P}}\ddot{\mathbf{D}}_{\mathrm{P}} + \mathbf{S}_{\mathrm{P}}\mathbf{D}_{\mathrm{P}} = \mathbf{0} \tag{4.31}$$

The displacement and acceleration vectors in this equation are defined to be

$$\mathbf{D}_{\mathrm{P}} = \mathbf{\Phi}_{\mathrm{M}}^{-1}\mathbf{D} \tag{4.32}$$

and

$$\ddot{\mathbf{D}}_{\mathrm{P}} = \mathbf{\Phi}_{\mathrm{M}}^{-1}\ddot{\mathbf{D}} \tag{4.33}$$

By virtue of Eqs. (4.28) and (4.29), the generalized mass and stiffness matrices in Eq. (4.31) are both diagonal. The generalized displacements $\mathbf{D}_{\mathrm{P}}$ given by Eq. (4.32) are called *principal coordinates*, for which the equations of motion (4.31) have neither inertial nor elasticity coupling. From Eq. (4.32) we find that the original coordinates are related to the principal coordinates by the operation

$$\mathbf{D} = \mathbf{\Phi}_{\mathrm{M}}\mathbf{D}_{\mathrm{P}} \tag{4.34}$$

Also, from Eq. (4.33) we have

$$\ddot{\mathbf{D}} = \mathbf{\Phi}_{\mathrm{M}}\ddot{\mathbf{D}}_{\mathrm{P}} \tag{4.35}$$

Recalling the definition of the modal matrix in Eq. (4.27), we see that the generalized displacements $\mathbf{D}_\mathrm{P}$ operate in Eq. (4.34) as multipliers of the modal columns in $\mathbf{\Phi}_\mathrm{M}$ to produce values of the actual displacements $\mathbf{D}$. Thus, the shape functions for the principal coordinates of a multidegree system are its natural modes of vibration.

The eigenvalue problem in Eq. (*a*) may be restated in comprehensive fashion by expanding $\mathbf{\Phi}_{\mathrm{M}i}$ to $\mathbf{\Phi}_\mathrm{M}$ (see Eq. 4.27), which results in

$$\mathbf{S\Phi}_\mathrm{M} = \mathbf{M\Phi}_\mathrm{M}\boldsymbol{\omega}^2 \tag{g}$$

The symbol $\boldsymbol{\omega}^2$ in Eq. (*g*) represents a diagonal matrix with values of ω_i^2 on the diagonal, as follows:

$$\boldsymbol{\omega}^2 = \begin{bmatrix} \omega_1^2 & 0 & 0 & \cdots & 0 \\ 0 & \omega_2^2 & 0 & \cdots & 0 \\ 0 & 0 & \omega_3^2 & \cdots & 0 \\ \cdots & \cdots & \cdots & \cdots & \cdots \\ 0 & 0 & 0 & \cdots & \omega_n^2 \end{bmatrix} \tag{4.36}$$

This array, sometimes called the *spectral matrix*, will be referred to as the *eigenvalue matrix*, or *matrix of characteristic values*. It postmultiplies the matrix $\mathbf{\Phi}_\mathrm{M}$ in Eq. (*g*), so that a typical modal column $\mathbf{\Phi}_{\mathrm{M}i}$ is scaled by the corresponding eigenvalue, ω_i^2. Premultiplying Eq. (*g*) by $\mathbf{\Phi}_\mathrm{M}^\mathrm{T}$ and using the relationships (4.28) and (4.29), we obtain

$$\mathbf{S}_\mathrm{P} = \mathbf{M}_\mathrm{P}\boldsymbol{\omega}^2 \tag{4.37}$$

Hence,

$$S_{\mathrm{P}i} = M_{\mathrm{P}i}\omega_i^2 \tag{h}$$

Thus, in principal coordinates the ith principal stiffness is equal to the ith principal mass multiplied by the ith eigenvalue.

Because the modal vectors may be scaled arbitrarily, the principal coordinates are not unique. In fact, there is an infinite number of sets of such generalized displacements, but the most common choice is that for which the mass matrix is transformed to the identity matrix. We state this condition by specifying that $M_{\mathrm{P}i}$ in Eq. (4.25) must be equal to unity, as follows:

$$\mathbf{\Phi}_{\mathrm{N}i}^\mathrm{T}\mathbf{M\Phi}_{\mathrm{N}i} = M_{\mathrm{P}i} = 1 \tag{i}$$

in which

$$\mathbf{\Phi}_{\mathrm{N}i} = \mathbf{\Phi}_{\mathrm{M}i}/C_i \tag{j}$$

Under this condition, the scaled eigenvector $\mathbf{X}_{\mathrm{N}i}$ is said to be *normalized with respect to the mass matrix*. The scalar C_i in Eq. (j) is computed as

$$C_i = \pm\sqrt{\mathbf{\Phi}_{\mathrm{M}i}^{\mathrm{T}}\mathbf{M}\mathbf{\Phi}_{\mathrm{M}i}} = \pm\sqrt{\sum_{j=1}^{n} \Phi_{\mathrm{M}ji}\left(\sum_{k=1}^{n} M_{jk}\Phi_{\mathrm{M}ki}\right)} \tag{4.38}$$

If the mass matrix is diagonal, this expression simplifies to

$$C_i = \pm\sqrt{\sum_{j=1}^{n} (M_j\Phi_{\mathrm{M}ji}^2)} \tag{4.39}$$

When all of the vectors in the modal matrix are normalized in this manner, we change the subscript M to N and use the symbol $\mathbf{\Phi}_{\mathrm{N}}$ in place of $\mathbf{\Phi}_{\mathrm{M}}$. Then the principal mass matrix, given by Eq. (4.28), becomes

$$\mathbf{\Phi}_{\mathrm{N}}^{\mathrm{T}}\mathbf{M}\mathbf{\Phi}_{\mathrm{N}} = \mathbf{M}_{\mathrm{N}} = \mathbf{I} \tag{4.40}$$

Furthermore, the principal stiffness matrix, from Eqs. (4.29) and (4.37), is seen to be

$$\mathbf{\Phi}_{\mathrm{N}}^{\mathrm{T}}\mathbf{S}\mathbf{\Phi}_{\mathrm{N}} = \mathbf{S}_{\mathrm{N}} = \boldsymbol{\omega}^2 \tag{4.41}$$

Or, for the ith mode

$$\mathbf{\Phi}_{\mathrm{N}i}^{\mathrm{T}}\mathbf{S}\mathbf{\Phi}_{\mathrm{N}i} = S_{\mathrm{N}i} = \omega_i^2 \tag{k}$$

Thus, when the eigenvectors are normalized with respect to $\mathbf{M}$, the stiffnesses in principal coordinates are equal to the eigenvalues. This particular set of principal coordinates is known as *normal coordinates.*

To illustrate the use of normal coordinates for action equations, let us reconsider the three masses on a stretched wire in Fig. 4.2*a*. From Example 2 in the preceding section, we have the eigenvectors for this system, which become the columns of the modal matrix

$$\mathbf{\Phi}_{\mathrm{M}} = \begin{bmatrix} 1 & 1 & 1 \\ \sqrt{2} & 0 & -\sqrt{2} \\ 1 & -1 & 1 \end{bmatrix} \tag{l}$$

To normalize this array with respect to $\mathbf{M}$, which is equal to $m\mathbf{I}$, we calculate from Eq. (4.39) the scalars

$$C_1 = \sqrt{m(1)^2 + m(\sqrt{2})^2 + m(1)^2} = 2\sqrt{m}$$
$$C_2 = \sqrt{m(1)^2 + m(0)^2 + m(-1)^2} = \sqrt{2m}$$
$$C_3 = \sqrt{m(1)^2 + m(-\sqrt{2})^2 + m(1)^2} = 2\sqrt{m}$$

Dividing the columns of $\mathbf{\Phi}_{\mathrm{M}}$ by these values produces

$$\mathbf{\Phi}_{\mathrm{N}} = \frac{1}{2\sqrt{m}} \begin{bmatrix} 1 & \sqrt{2} & 1 \\ \sqrt{2} & 0 & -\sqrt{2} \\ 1 & -\sqrt{2} & 1 \end{bmatrix} \tag{m}$$

The stiffness matrix for this system is

$$\mathbf{S} = \frac{T}{\ell} \begin{bmatrix} 2 & -1 & 0 \\ -1 & 2 & -1 \\ 0 & -1 & 2 \end{bmatrix} \tag{n}$$

Substitution of Eqs. (m) and (n) into Eq. (4.41) yields

$$\mathbf{S}_{\mathrm{N}} = \frac{T}{m\ell} \begin{bmatrix} 2-\sqrt{2} & 0 & 0 \\ 0 & 2 & 0 \\ 0 & 0 & 2+\sqrt{2} \end{bmatrix} \tag{o}$$

which contains the values of ω_1^2, ω_2^2, and ω_3^2 on the diagonal (see the answers for Prob. 4.2-1). Of course, the eigenvalues in Eq. (o) were already available; so the advantages of transforming stiffnesses to normal coordinates are not obvious. These advantages will become apparent for the response calculations in succeeding sections.

As an alternative to working with the equations of motion in the form of Eq. (4.30), we could premultiply by $\mathbf{M}^{-1}$ and convert them to *acceleration equations*, as follows:

$$\ddot{\mathbf{D}} + \mathbf{M}^{-1}\mathbf{S}\mathbf{D} = \mathbf{0} \tag{4.42}$$

This equation can be transformed to principal coordinates by substituting Eq. (4.34) for $\mathbf{D}$ and Eq. (4.35) for $\ddot{\mathbf{D}}$. Then premultiplication by $\mathbf{\Phi}_{\mathrm{M}}^{-1}$ gives

$$\ddot{\mathbf{D}}_{\mathrm{P}} + \mathbf{\Phi}_{\mathrm{M}}^{-1}\mathbf{M}^{-1}\mathbf{S}\mathbf{\Phi}_{\mathrm{M}}\mathbf{D}_{\mathrm{P}} = \mathbf{0} \tag{p}$$

Upon inserting the identity matrix $\mathbf{I} = (\mathbf{\Phi}_{\mathrm{M}}^{-1})^{\mathrm{T}}\mathbf{\Phi}_{\mathrm{M}}^{\mathrm{T}}$ before $\mathbf{S}$, the coefficient matrix in Eq. (p) becomes

$$\mathbf{\Phi}_{\mathrm{M}}^{-1}\mathbf{M}^{-1}(\mathbf{\Phi}_{\mathrm{M}}^{-1})^{\mathrm{T}}\mathbf{\Phi}_{\mathrm{M}}^{\mathrm{T}}\mathbf{S}\mathbf{\Phi}_{\mathrm{M}} = \mathbf{M}_{\mathrm{P}}^{-1}\mathbf{S}_{\mathrm{P}} = \boldsymbol{\omega}^2 \tag{q}$$

Hence, the matrix equation for accelerations in principal coordinates may be written as

$$\ddot{\mathbf{D}}_{\mathrm{P}} + \boldsymbol{\omega}^2\mathbf{D}_{\mathrm{P}} = \mathbf{0} \tag{4.43}$$

which can also be obtained by premultiplying Eq. (4.31) with $\mathbf{M}_\mathrm{P}^{-1}$. Since the same result is obtained by either approach, we can avoid using Eq. (4.42), which requires the development of $\mathbf{M}^{-1}$. Of course, there is no difficulty in calculating the inverse of $\mathbf{M}$ when it is diagonal, but the determination of $\mathbf{M}^{-1}$ is more complicated when the array is filled.

On the other hand, $\mathbf{M}_\mathrm{P}$ is always diagonal; so obtaining its inverse is a trivial matter. This fact proves to be advantageous for the purpose of generating the inverse of the modal matrix, which is sometimes needed. A formula for this inverse is determined by postmultiplying Eq. (4.28) by $\mathbf{\Phi}_\mathrm{M}^{-1}$ and premultiplying it by $\mathbf{M}_\mathrm{P}^{-1}$, which yields

$$\mathbf{\Phi}_\mathrm{M}^{-1} = \mathbf{M}_\mathrm{P}^{-1}\mathbf{\Phi}_\mathrm{M}^\mathrm{T}\mathbf{M} \tag{4.44a}$$

If the eigenvectors are normalized with respect to $\mathbf{M}$, the principal masses are all unity, and Eq. (4.44a) becomes

$$\mathbf{\Phi}_\mathrm{N}^{-1} = \mathbf{\Phi}_\mathrm{N}^\mathrm{T}\mathbf{M} \tag{4.44b}$$

Formulas similar to Eqs. (4.44a) and (4.44b) can be derived from Eq. (4.29), involving stiffnesses and principal stiffnesses. However, unless the eigenvectors are normalized with respect to $\mathbf{S}$, Eqs. (4.44a) and (4.44b) are preferable.

If displacement equations of motion are used instead of action equations, Eq. (4.30) is replaced by

$$\mathbf{FM\ddot{D}} + \mathbf{D} = \mathbf{0} \tag{4.45}$$

We transform this equation to principal coordinates by substituting Eq. (4.34) for $\mathbf{D}$ and Eq. (4.35) for $\mathbf{\ddot{D}}$. Then premultiplication by $\mathbf{\Phi}_\mathrm{M}^{-1}$ yields

$$\mathbf{\Phi}_\mathrm{M}^{-1}\mathbf{FM\Phi}_\mathrm{M}\mathbf{\ddot{D}}_\mathrm{P} + \mathbf{D}_\mathrm{P} = \mathbf{0} \tag{r}$$

When the identity matrix $\mathbf{I} = (\mathbf{\Phi}_\mathrm{M}^{-1})^\mathrm{T}\mathbf{\Phi}_\mathrm{M}^\mathrm{T}$ is inserted before $\mathbf{M}$, the coefficient matrix in Eq. (r) becomes

$$\mathbf{\Phi}_\mathrm{M}^{-1}\mathbf{F}(\mathbf{\Phi}_\mathrm{M}^{-1})^\mathrm{T}\mathbf{\Phi}_\mathrm{M}^\mathrm{T}\mathbf{M\Phi}_\mathrm{M} = \mathbf{F}_\mathrm{P}\mathbf{M}_\mathrm{P} \tag{s}$$

The symbol $\mathbf{F}_\mathrm{P}$ in Eq. (s) represents a *principal flexibility matrix*, corresponding to $\mathbf{S}_\mathrm{P}$, and is defined as

$$\mathbf{F}_\mathrm{P} = \mathbf{\Phi}_\mathrm{M}^{-1}\mathbf{F}(\mathbf{\Phi}_\mathrm{M}^{-1})^\mathrm{T} = \mathbf{S}_\mathrm{P}^{-1} \tag{4.46}$$

Of course, this definition applies only when $\mathbf{S}$ (and hence $\mathbf{S}_\mathrm{P}$) is positive-definite. Thus, the displacement equations of motion in principal coordinates may be written as

$$\mathbf{F}_\mathrm{P}\mathbf{M}_\mathrm{P}\mathbf{\ddot{D}}_\mathrm{P} + \mathbf{D}_\mathrm{P} = \mathbf{0} \tag{4.47}$$

Furthermore, the expanded form of the eigenvalue problem in Eq. (g) is replaced by

$$\mathbf{F}\mathbf{M}\boldsymbol{\Phi}_{\mathrm{M}} = \boldsymbol{\Phi}_{\mathrm{M}}\boldsymbol{\lambda} \tag{t}$$

The eigenvalue matrix $\boldsymbol{\lambda}$ in Eq. (t) consists of a diagonal array containing values of λ_i on the diagonal, as follows:

$$\boldsymbol{\lambda} = \begin{bmatrix} \lambda_1 & 0 & 0 & \cdots & 0 \\ 0 & \lambda_2 & 0 & \cdots & 0 \\ 0 & 0 & \lambda_3 & \cdots & 0 \\ \cdots & \cdots & \cdots & \cdots & \cdots \\ 0 & 0 & 0 & \cdots & \lambda_n \end{bmatrix} = (\boldsymbol{\omega}^2)^{-1} \tag{4.48}$$

Premultiplying Eq. (t) by $\boldsymbol{\Phi}_{\mathrm{M}}^{-1}$ and using relationship (s), we find

$$\mathbf{F}_{\mathrm{P}}\mathbf{M}_{\mathrm{P}} = \boldsymbol{\lambda} \tag{4.49}$$

When the modal matrix is normalized with respect to the mass matrix, the principal flexibility matrix, from Eqs. (4.46) and (4.49), is seen to be

$$\boldsymbol{\Phi}_{\mathrm{N}}^{-1}\mathbf{F}(\boldsymbol{\Phi}_{\mathrm{N}}^{-1})^{\mathrm{T}} = \mathbf{F}_{\mathrm{N}} = \boldsymbol{\lambda} = (\boldsymbol{\omega}^2)^{-1} \tag{4.50}$$

Thus, the flexibility matrix in normal coordinates becomes the eigenvalue matrix $\boldsymbol{\lambda}$, which is also equal to $(\boldsymbol{\omega}^2)^{-1}$. From this we conclude that Eq. (4.43) gives the form of the equations of motion in normal coordinates, regardless of the method of formulation in the original coordinates.

As an example of the use of normal coordinates for displacement equations, we again consider the three masses on a stretched wire in Fig. 4.2a. The flexibility matrix for this system is

$$\mathbf{F} = \frac{\ell}{4T}\begin{bmatrix} 3 & 2 & 1 \\ 2 & 4 & 2 \\ 1 & 2 & 3 \end{bmatrix} \tag{u}$$

Inverting $\boldsymbol{\Phi}_{\mathrm{N}}$ from Eq. (m) in accordance with Eq. (4.44b) and substituting the result, along with Eq. (u), into Eq. (4.50) gives

$$\mathbf{F}_{\mathrm{N}} = \frac{\ell m}{2T}\begin{bmatrix} 2+\sqrt{2} & 0 & 0 \\ 0 & 1 & 0 \\ 0 & 0 & 2-\sqrt{2} \end{bmatrix} \tag{v}$$

which contains the values of $\lambda_1 = 1/\omega_1^2$, $\lambda_2 = 1/\omega_2^2$, and $\lambda_3 = 1/\omega_3^2$ on the diagonal (see Example 2 in Sec. 4.2).

As described at the end of the preceding section, the eigenvalue problem is often solved by first transforming it to standard form with a symmetric coefficient matrix. When this approach is used, the eigenvectors are ordinarily normalized to be of unit length. Letting the symbol $\mathbf{V}_i$ represent such a normalized eigenvector, we specify that

$$\mathbf{V}_i^{\mathrm{T}}\mathbf{V}_i = 1 \tag{w}$$

The eigenvector $\mathbf{\Phi}_{\mathrm{U}i}$ from the standard form [see Eq. (4.12a), in Sec. 4.2] is scaled to obtain $\mathbf{V}_i$ as

$$\mathbf{V}_i = \frac{\mathbf{\Phi}_{\mathrm{U}i}}{C_i} \tag{x}$$

where the scalar C_i is

$$C_i = \pm\sqrt{\mathbf{\Phi}_{\mathrm{U}i}^{\mathrm{T}}\mathbf{\Phi}_{\mathrm{U}i}} = \pm\sqrt{\sum_{j=1}^{n} \Phi_{\mathrm{U}ji}^2} \tag{4.51}$$

With this type of normalization, the modal matrix $\mathbf{V}$ has the properties

$$\mathbf{V}^{\mathrm{T}}\mathbf{V} = \mathbf{I} \qquad \mathbf{V}^{-1} = \mathbf{V}^{\mathrm{T}} \tag{4.52}$$

and is called simply an *orthogonal matrix*. Transforming this modal matrix back to the original coordinates [see Eq. (4.12b), Sec. 4.2], we obtain

$$\mathbf{\Phi}_{\mathrm{N}} = \mathbf{U}^{-1}\mathbf{V} \tag{4.53}$$

To prove that this operation produces $\mathbf{\Phi}_{\mathrm{N}}$, we substitute Eq. (4.53) and the definition $\mathrm{M} = \mathbf{U}^{\mathrm{T}}\mathbf{U}$ into Eq. (4.40), giving

$$\mathbf{V}^{\mathrm{T}}(\mathbf{U}^{-1})^{\mathrm{T}}\mathbf{U}^{\mathrm{T}}\mathbf{U}\mathbf{U}^{-1}\mathbf{V} = \mathbf{I}$$

Thus, the eigenvectors in the original coordinates are already normalized with respect to M.

4.4 NORMAL-MODE RESPONSE TO INITIAL CONDITIONS

From Eq. (4.43) in the preceding section, we see that a typical equation of motion for undamped free vibrations in normal coordinates is

$$\ddot{u}_{\mathrm{N}i} + \omega_i^2 u_{\mathrm{N}i} = 0 \qquad (i = 1, 2, 3, \ldots, n) \tag{4.54}$$

Each quation of this type is uncoupled from all of the others, and we will treat the expression as if it pertained to a one-degree-of-freedom system [see Eq. (1.1) in Sec. 1.1]. If we can provide initial conditions of displacement $u_{\text{N}0i}$ and velocity $\dot{u}_{\text{N}0i}$ in each normal coordinate (at time $t = 0$), we will be able to calculate the free-vibrational response of the ith mode as

$$u_{\text{N}i} = u_{\text{N}0i} \cos \omega_i t + \frac{\dot{u}_{\text{N}0i}}{\omega_i} \sin \omega_i t \qquad (i = 1, 2, 3, \ldots, n) \tag{4.55}$$

This expression is drawn from Eq. (1.5) in Sec. 1.1 for an undamped one-degree system.

Using Eq. (4.32), we obtain the initial displacements in normal coordinates as follows:

$$\mathbf{D}_{\text{N}0} = \mathbf{\Phi}_{\text{N}}^{-1} \mathbf{D}_0 \tag{4.56}$$

The symbols $\mathbf{D}_0$ and $\mathbf{D}_{\text{N}0}$ in Eq. (4.56) represent vectors of initial displacements in the original coordinates and in normal coordinates, respectively. Thus,

$$\mathbf{D}_0 = \begin{bmatrix} u_{01} \\ u_{02} \\ u_{03} \\ \cdots \\ u_{0n} \end{bmatrix} \qquad \mathbf{D}_{\text{N}0} = \begin{bmatrix} u_{\text{N}01} \\ u_{\text{N}02} \\ u_{\text{N}03} \\ \cdots \\ u_{\text{N}0n} \end{bmatrix} \tag{a}$$

Similarly, the initial velocities of the system can be transformed to normal coordinates by the operation

$$\dot{\mathbf{D}}_{\text{N}0} = \mathbf{\Phi}_{\text{N}}^{-1} \dot{\mathbf{D}}_0 \tag{4.57}$$

where $\dot{\mathbf{D}}_0$ and $\dot{\mathbf{D}}_{\text{N}0}$ are vectors of initial velocities in the original coordinates and in normal coodinates, respectively. Equation (4.57) is obtained from Eq. (4.56) by differentiation with respect to time, and the form of each velocity vector is the same as that for each displacement vector in Eqs. (a).

Having the required initial conditions in normal coordinates, we can apply Eq. (4.55) repetitively to calculate the terms in the vector of normal-mode displacements $\mathbf{D}_{\text{N}} = \{u_{\text{N}i}\}$. These results are then transformed back to the original coordinates, using the operation given by Eq. (4.34). Thus,

$$\mathbf{D} = \mathbf{\Phi}_{\text{N}} \mathbf{D}_{\text{N}} \tag{4.58}$$

This sequence of operations is the same regardless of whether the original equations of motions are written as action equations or displacement equations. However, for the action-equation approach, there exists the possibility of one or more *rigid-body modes*. For such a principal mode the eigenvalue ω_i^2 is zero, and Eq. (4.54) becomes

$$\ddot{u}_{\mathrm{N}i} = 0 \tag{4.59}$$

Integration of this equation twice with respect to time yields

$$u_{\mathrm{N}i} = u_{\mathrm{N}0i} + \dot{u}_{\mathrm{N}0i} t \tag{4.60}$$

Equation (4.60) is used in place of Eq. (4.55) to evaluate the response of a rigid-body mode in normal coordinates.

Example 1. In Sec. 3.5 we calculated the free-vibrational response of the two-mass system in Fig. 3.1a to the initial conditions $u_{01} = u_{02} = 1$ and $\dot{u}_{01} = \dot{u}_{02} = 0$ by evaluating arbitrary constants. We shall now determine the same results using the normal-mode method.

Solution. Assuming that $m_1 = m_2 = m$ and $k_1 = k_2 = k$, we previously found the eigenvalues of the system to be $\omega_1^2 = 0.382\,k/m$ and $\omega_2^2 = 2.618\,k/m$. In addition, the amplitude ratios were obtained as $r_1 = 0.618$ and $r_2 = -1.618$. Therefore, the modal matrix is

$$\mathbf{\Phi}_{\mathrm{M}} = \begin{bmatrix} 0.618 & -1.618 \\ 1.000 & 1.000 \end{bmatrix} \tag{b}$$

To normalize this array with respect to $\mathbf{M} = m\mathbf{I}$, we evaluate from Eq. (4.39) the scalars

$$C_1 = \sqrt{m(0.618)^2 + m(1.000)^2} = 1.175\sqrt{m}$$
$$C_2 = \sqrt{m(-1.618)^2 + m(1.000)^2} = 1.902\sqrt{m}$$

When the columns of $\mathbf{\Phi}_{\mathrm{M}}$ are divided by these values, the results are

$$\mathbf{\Phi}_{\mathrm{N}} = \frac{1}{\sqrt{m}} \begin{bmatrix} 0.526 & -0.851 \\ 0.851 & 0.526 \end{bmatrix} \tag{c}$$

The inverse of $\mathbf{\Phi}_{\mathrm{N}}$, required for transforming the initial conditions to normal coordinates, is obtained from Eq. (4.44b) as

$$\mathbf{\Phi}_{\mathrm{N}}^{-1} = \mathbf{\Phi}_{\mathrm{N}}^{\mathrm{T}}\mathrm{M} = \sqrt{m} \begin{bmatrix} 0.526 & 0.851 \\ -0.851 & 0.526 \end{bmatrix} \tag{d}$$

Initial conditions in vector form are

$$\mathbf{D}_0 = \begin{bmatrix} 1 \\ 1 \end{bmatrix} \qquad \dot{\mathbf{D}}_0 = \begin{bmatrix} 0 \\ 0 \end{bmatrix} \tag{e}$$

In accordance with Eq. (4.56), the vector of nonzero initial displacements is transformed to normal coordinates, as follows:

$$\mathbf{D}_{N0} = \mathbf{\Phi}_N^{-1}\mathbf{D}_0 = \sqrt{m}\begin{bmatrix} 1.377 \\ -0.325 \end{bmatrix} \tag{f}$$

Using Eq. (4.55) twice, we form the solution vector in normal coordinates.

$$\mathbf{D}_N = \sqrt{m}\begin{bmatrix} 1.377 \cos \omega_1 t \\ -0.325 \cos \omega_2 t \end{bmatrix} \tag{g}$$

Then the response in the original coordinates is obtained by the back-transformation in Eq. (4.58) to be

$$\mathbf{D} = \mathbf{\Phi}_N \mathbf{D}_N = \begin{bmatrix} 0.724 \cos \omega_1 t + 0.276 \cos \omega_2 t \\ 1.71 \cos \omega_1 t - 0.171 \cos \omega_2 t \end{bmatrix} \tag{h}$$

These results are the same as those in Sec. 3.5.

Example 2. For an example with rigid-body motion, we reconsider the three mass system in Fig. 4.1*a* and let $k_1 = 0$. In addition, we shall assume that $m_1 = m_2 = m_3 = m$ and $k_2 = k_3 = k$. Under these conditions, the eigenvalues were found to be $\omega_1^2 = 0$, $\omega_2^2 = k/m$, and $\omega_3^2 = 3\,k/m$ (see Example 1 in Sec. 4.2). The eigenvectors were also obtained, and the modal matrix is

$$\mathbf{\Phi}_M = \begin{bmatrix} 1 & 1 & 1 \\ 1 & 0 & -2 \\ 1 & -1 & 1 \end{bmatrix} \tag{i}$$

Suppose that the system is at rest when the first mass is struck in such a manner that it suddenly acquires a velocity v. Determine the response of the system due to this impact.

Solution. Normalization of $\mathbf{\Phi}_M$ with respect to the mass matrix yields

$$\mathbf{\Phi}_N = \frac{1}{\sqrt{6m}}\begin{bmatrix} \sqrt{2} & \sqrt{3} & 1 \\ \sqrt{2} & 0 & -2 \\ \sqrt{2} & -\sqrt{3} & 1 \end{bmatrix} \tag{j}$$

and the inverse of $\mathbf{\Phi}_N$ is

$$\mathbf{\Phi}_N^{-1} = \mathbf{\Phi}_N^T \mathbf{M} = \sqrt{\frac{m}{6}} \begin{bmatrix} \sqrt{2} & \sqrt{2} & \sqrt{2} \\ \sqrt{3} & 0 & -\sqrt{3} \\ 1 & -2 & 1 \end{bmatrix} \tag{k}$$

The vectors of initial conditions are

$$\mathbf{D}_0 = \begin{bmatrix} 0 \\ 0 \\ 0 \end{bmatrix} \qquad \dot{\mathbf{D}}_0 = \begin{bmatrix} v \\ 0 \\ 0 \end{bmatrix} \tag{l}$$

Transforming the nonzero initial-velocity vector to normal coordinates, we find

$$\dot{\mathbf{D}}_{N0} = \mathbf{\Phi}_N^{-1} \dot{\mathbf{D}}_0 = v\sqrt{\frac{m}{6}} \begin{bmatrix} \sqrt{2} \\ \sqrt{3} \\ 1 \end{bmatrix} \tag{m}$$

Equation (4.60) must be applied for the rigid-body mode, while Eq. (4.55) provides the response for the vibrational modes. Thus, the vector of normal-mode responses becomes

$$\mathbf{D}_N = v\sqrt{\frac{m}{6}} \begin{bmatrix} \sqrt{2}\,t \\ (\sqrt{3} \sin \omega_2 t)/\omega_2 \\ (\sin \omega_3 t)/\omega_3 \end{bmatrix} \tag{n}$$

Transforming these results to the original coordinates gives

$$\mathbf{D} = \mathbf{\Phi}_N \mathbf{D}_N = \frac{v}{6} \begin{bmatrix} 2t + (3 \sin \omega_2 t)/\omega_2 + (\sin \omega_3 t)/\omega_3 \\ 2t - (2 \sin \omega_3 t)/\omega_3 \\ 2t - (3 \sin \omega_2 t)/\omega_2 + (\sin \omega_3 t)/\omega_3 \end{bmatrix} \tag{o}$$

The rigid-body component of each response in Eq. (o) is equal to $vt/3$.

If all of the masses have the same initial velocity v, the initial-velocity vector becomes $\dot{\mathbf{D}}_0 = \{v, v, v\}$; and Eqs. ($m$), ($n$), and ($o$) are replaced by

$$\dot{\mathbf{D}}_{N0} = v\sqrt{3m} \begin{bmatrix} 1 \\ 0 \\ 0 \end{bmatrix} \qquad \mathbf{D}_N = v\sqrt{3m} \begin{bmatrix} t \\ 0 \\ 0 \end{bmatrix} \qquad \mathbf{D} = vt \begin{bmatrix} 1 \\ 1 \\ 1 \end{bmatrix} \tag{p}$$

In this case, the motion consists of pure rigid-body translation, without vibrations.

Example 3. Determine the free-vibrational response of the system in Fig. 4.2*a* to the sudden release of a static force P, applied in the x direction at the second mass. Assume that $m_1 = m_2 = m_3 = m$ and $\ell_1 = \ell_2 = \ell_3 = \ell_4 = \ell$.
Solution. From Example 2 of Sec. 4.2, we have the modal matrix

$$\mathbf{\Phi}_{\mathrm{M}} = \begin{bmatrix} 1 & 1 & 1 \\ \sqrt{2} & 0 & -\sqrt{2} \\ 1 & -1 & 1 \end{bmatrix} \qquad (q)$$

Normalizing $\mathbf{\Phi}_{\mathrm{M}}$ with respect to M produces

$$\mathbf{\Phi}_{\mathrm{N}} = \frac{1}{2\sqrt{m}} \begin{bmatrix} 1 & \sqrt{2} & 1 \\ \sqrt{2} & 0 & -\sqrt{2} \\ 1 & -\sqrt{2} & 1 \end{bmatrix} \qquad \mathbf{\Phi}_{\mathrm{N}}^{-1} = \frac{\sqrt{m}}{2} \begin{bmatrix} 1 & \sqrt{2} & 1 \\ \sqrt{2} & 0 & -\sqrt{2} \\ 1 & -\sqrt{2} & 1 \end{bmatrix} \qquad (r)$$

The initial-condition vectors are

$$\mathbf{D}_0 = \frac{P\ell}{2T} \begin{bmatrix} 1 \\ 2 \\ 1 \end{bmatrix} \qquad \dot{\mathbf{D}}_0 = \begin{bmatrix} 0 \\ 0 \\ 0 \end{bmatrix} \qquad (s)$$

In normal coordinates the initial displacements become

$$\mathbf{D}_{\mathrm{N}0} = \mathbf{\Phi}_{\mathrm{N}}^{-1}\mathbf{D}_0 = \frac{P\ell\sqrt{m}}{2T} \begin{bmatrix} 1+\sqrt{2} \\ 0 \\ 1-\sqrt{2} \end{bmatrix} \qquad (t)$$

Applying Eq. (4.55) repetitively, we obtain the normal-mode responses to be

$$\mathbf{D}_{\mathrm{N}} = \frac{P\ell\sqrt{m}}{2T} \begin{bmatrix} (1+\sqrt{2}) \cos \omega_1 t \\ 0 \\ (1-\sqrt{2}) \cos \omega_3 t \end{bmatrix} \qquad (u)$$

Then the results in the original coordinates are calculated as

$$\mathbf{D} = \mathbf{\Phi}_{\mathrm{N}}\mathbf{D}_{\mathrm{N}} = \frac{P\ell}{4T} \begin{bmatrix} (1+\sqrt{2}) \cos \omega_1 t + (1-\sqrt{2}) \cos \omega_3 t \\ (\sqrt{2}+2) \cos \omega_1 t - (\sqrt{2}-2) \cos \omega_3 t \\ (1+\sqrt{2}) \cos \omega_1 t + (1-\sqrt{2}) \cos \omega_3 t \end{bmatrix} \qquad (v)$$

The antisymmetric second mode is not excited, and only the symmetric first and third modes contribute terms in Eq. (v). Consequently, masses 1 and 3 have the same response.

4.5 NORMAL-MODE RESPONSE TO APPLIED ACTIONS

We now consider the case of a multi-degree system subjected to applied actions that correspond to the displacement coordinates. The action equations of motion in matrix form are

$$\mathbf{M\ddot{D}} + \mathbf{SD} = \mathbf{Q} \tag{4.61}$$

where the symbol $\mathbf{Q}$ represents a column matrix (or vector) of time-varying applied actions, as follows:

$$\mathbf{Q} = \begin{bmatrix} Q_1 \\ Q_2 \\ Q_3 \\ \cdots \\ Q_n \end{bmatrix} = \begin{bmatrix} F_1(t) \\ F_2(t) \\ F_3(t) \\ \cdots \\ F_n(t) \end{bmatrix} \tag{a}$$

Equation (4.61) will be transformed to principal coordinates by premultiplying both sides with $\mathbf{\Phi}_{\mathrm{M}}^{\mathrm{T}}$ and substituting Eqs. (4.34) and (4.35) for $\mathbf{D}$ and $\mathbf{\ddot{D}}$ to produce

$$\mathbf{\Phi}_{\mathrm{M}}^{\mathrm{T}}\mathbf{M}\mathbf{\Phi}_{\mathrm{M}}\mathbf{\ddot{D}}_{\mathrm{P}} + \mathbf{\Phi}_{\mathrm{M}}^{\mathrm{T}}\mathbf{S}\mathbf{\Phi}_{\mathrm{M}}\mathbf{D}_{\mathrm{P}} = \mathbf{\Phi}_{\mathrm{M}}^{\mathrm{T}}\mathbf{Q}$$

This equation may also be written as

$$\mathbf{M}_{\mathrm{P}}\mathbf{\ddot{D}}_{\mathrm{P}} + \mathbf{S}_{\mathrm{P}}\mathbf{D}_{\mathrm{P}} = \mathbf{Q}_{\mathrm{P}} \tag{4.62}$$

in which the matrices $\mathbf{M}_{\mathrm{P}}$ and $\mathbf{S}_{\mathrm{P}}$ are given by Eqs. (4.28) and (4.29). The symbol $\mathbf{Q}_{\mathrm{P}}$ in Eq. (4.62) denotes a vector of applied actions in principal coordinates, which is computed by the operation

$$\mathbf{Q}_{\mathrm{P}} = \mathbf{\Phi}_{\mathrm{M}}^{\mathrm{T}}\mathbf{Q} \tag{4.63a}$$

In expanded form, the results of this multiplication are

$$\begin{bmatrix} Q_{\mathrm{P}1} \\ Q_{\mathrm{P}2} \\ Q_{\mathrm{P}3} \\ \cdots \\ Q_{\mathrm{P}n} \end{bmatrix} = \begin{bmatrix} \Phi_{\mathrm{M}11}Q_1 + \Phi_{\mathrm{M}21}Q_2 + \Phi_{\mathrm{M}31}Q_3 + \cdots + \Phi_{\mathrm{M}n1}Q_n \\ \Phi_{\mathrm{M}12}Q_1 + \Phi_{\mathrm{M}22}Q_2 + \Phi_{\mathrm{M}32}Q_3 + \cdots + \Phi_{\mathrm{M}n2}Q_n \\ \Phi_{\mathrm{M}13}Q_1 + \Phi_{\mathrm{M}23}Q_2 + \Phi_{\mathrm{M}33}Q_3 + \cdots + \Phi_{\mathrm{M}n3}Q_n \\ \cdots \\ \Phi_{\mathrm{M}1n}Q_1 + \Phi_{\mathrm{M}2n}Q_2 + \Phi_{\mathrm{M}3n}Q_3 + \cdots + \Phi_{\mathrm{M}nn}Q_n \end{bmatrix} \tag{4.63b}$$

If the modal matrix is normalized with respect to the mass matrix, Eq.

(4.63a) becomes

$$\mathbf{Q}_{\mathrm{N}} = \mathbf{\Phi}_{\mathrm{N}}^{\mathrm{T}}\mathbf{Q} \tag{4.64}$$

and the ith equation of motion in normal coordinates takes the form

$$\ddot{u}_{\mathrm{N}i} + \omega_i^2 u_{\mathrm{N}i} = q_{\mathrm{N}i} \qquad (i = 1, 2, 3, \ldots, n) \tag{4.65}$$

where the ith *normal-mode load* is

$$q_{\mathrm{N}i} = \Phi_{\mathrm{N}1i}Q_1 + \Phi_{\mathrm{N}2i}Q_2 + \Phi_{\mathrm{N}3i}Q_3 + \cdots + \Phi_{\mathrm{N}ni}Q_n \tag{4.66}$$

The term $q_{\mathrm{N}i}$, determined by Eq. (4.66), constitutes an applied action in the ith normal coordinate. It appears to have units of acceleration because the generalized mass is unity.

Each of the n equations represented by (4.65) is uncoupled from all of the others, and we see that it has the same form as that for a one-degree system. Therefore, we can calculate the response of the ith normal coordinate to applied actions, using Duhamel's integral, as follows:

$$u_{\mathrm{N}i} = \frac{1}{\omega_i}\int_0^t q_{\mathrm{N}i} \sin \omega_i(t - t')\, dt' \tag{4.67}$$

This expression is drawn from Eq. (1.64) in Sec. 1.12 and was derived for an undamped one-degree system that is initially at rest. It is applied repetitively to calculate the terms in the vector of normal-mode displacements $\mathbf{D}_{\mathrm{N}} = \{u_{\mathrm{N}i}\}$. Then the results are transformed to the original coordinates, using Eq. (4.58) from the preceding section.

For a normal mode corresponding to rigid-body motion, the eigenvalue ω_i^2 is zero; and Eq. (4.65) becomes

$$\ddot{u}_{\mathrm{N}i} = q_{\mathrm{N}i} \tag{4.68}$$

In this case, the normal-mode response (with the system initially at rest) is

$$u_{\mathrm{N}i} = \int_0^t \int_0^{t'} q_{\mathrm{N}i}\, dt''\, dt' \tag{4.69}$$

Equation (4.69) replaces Eq. (4.67) whenever a rigid-body mode is encountered.

In summary, we calculate the dynamic response of a multidegree system to applied actions by first transforming those actions to normal coordinates, using Eq. (4.64). Then the response of each vibrational mode is obtained from the integral in Eq. (4.67), and that for each rigid-body mode is determined by the integral in Eq. (4.69). Finally, the values of the actual displacement coordinates are found with the back-transformation operation of Eq. (4.58). If applied actions do not correspond to the displacement coordinates, the appropriate *equivalent loads* can always be calculated as a preliminary step (see Example 3 at the end of this section).

Before proceeding further, let us examine the effect of a load $Q_j = F_j(t)$, corresponding to the jth displacement coordinate, in producing dynamic response of the kth displacement coordinate. From Eq. (4.66), the ith normal-mode load due to Q_j is

$$q_{Ni} = \Phi_{Nji} Q_j \qquad (b)$$

If the system has only vibrational modes, the response of the ith mode is obtained from Eq. (4.67) to be

$$u_{Ni} = \frac{\Phi_{Nji}}{\omega_i} \int_0^t Q_j \sin \omega_i (t - t')\, dt' \qquad (c)$$

Transformation of the response back to the original coordinates by Eq. (4.58) yields the response of the kth displacement coordinate as

$$(u_k)_{Q_j} = \sum_{i=1}^{n} \left[\frac{\Phi_{Nki} \Phi_{Nji}}{\omega_i} \int_0^t Q_j \sin \omega_i (t - t')\, dt' \right] \qquad (d)$$

Similarly, the response of the jth displacement coordinate caused by a load $Q_k = F_k(t)$, corresponding to the kth displacement coordinate, may be written

$$(u_j)_{Q_k} = \sum_{i=1}^{n} \left[\frac{\Phi_{Nji} \Phi_{Nki}}{\omega_i} \int_0^t Q_k \sin \omega_i (t - t')\, dt' \right] \qquad (e)$$

If $Q_j = Q_k = F(t)$, the right-hand sides of Eqs. (d) and (e) are equal, and we may equate the left-hand sides to obtain

$$(u_k)_{Q_j} = (u_j)_{Q_k} \qquad [Q_j = Q_k = F(t)] \qquad (4.70)$$

Equation (4.70) constitutes a *reciprocal theorem for dynamic loads* [3] that is similar to Maxwell's reciprocal theorem for static loads. It states that the dynamic response of the kth displacement coordinate due to any time-varying action corresponding to the jth coordinate is equal to the response of the jth coordinate due to the the same action applied at the kth coordinate. The theorem holds for systems with rigid-body modes as well as vibrational modes, as can be seen by using Eq. (4.69) in place of Eq. (4.67) in Eq. (c).

If we analyze a vibrational system using displacement equations of motion instead of action equations, Eq. (4.61) is replaced by

$$\mathbf{FM\ddot{D}} + \mathbf{D} = \mathbf{FQ} = \mathbf{\Delta} \qquad (4.71)$$

In this equation, the symbol $\boldsymbol{\Delta}$ represents a vector of time-varying displacements that are found by static analysis. Since this vector contains functions of time, it is more general than the vector $\boldsymbol{\Delta}_{st}$, which was defined in Eq. (3.37) of Sec. 3.6. The latter symbol denotes a vector of constants that are static displacements due to the maximum values of a harmonic forcing function. The array $\boldsymbol{\Delta}$ in Eq. (4.71) may contain terms due to disturbances other than actions corresponding to the displacement coordinates. Cases for which such time-varying displacements arise from support motions are discussed in the next section.

Equation (4.71) is transformed to normal coordinates by substituting Eqs. (4.34) and (4.35) for $\mathbf{D}$ and $\ddot{\mathbf{D}}$, respectively (with $\boldsymbol{\Phi}_{\mathrm{M}}$ normalized to become $\boldsymbol{\Phi}_{\mathrm{N}}$). Then premultiplication by $\boldsymbol{\Phi}_{\mathrm{N}}^{-1}$ yields

$$\boldsymbol{\Phi}_{\mathrm{N}}^{-1}\mathbf{F}\mathbf{M}\boldsymbol{\Phi}_{\mathrm{N}}\ddot{\mathbf{D}}_{\mathrm{N}} + \mathbf{D}_{\mathrm{N}} = \boldsymbol{\Phi}_{\mathrm{N}}^{-1}\mathbf{F}\mathbf{Q} = \boldsymbol{\Phi}_{\mathrm{N}}^{-1}\boldsymbol{\Delta}$$

Or,

$$\mathbf{F}_{\mathrm{N}}\ddot{\mathbf{D}}_{\mathrm{N}} + \mathbf{D}_{\mathrm{N}} = \boldsymbol{\Delta}_{\mathrm{N}} \tag{4.72}$$

in which $\mathbf{F}_{\mathrm{N}}$ is given by Eq. (4.50), and $\boldsymbol{\Delta}_{\mathrm{N}}$ is defined as

$$\boldsymbol{\Delta}_{\mathrm{N}} = \boldsymbol{\Phi}_{\mathrm{N}}^{-1}\boldsymbol{\Delta} \tag{4.73}$$

Each of the n equations represented by Eq. (4.72) has the form

$$\lambda_i \ddot{u}_{\mathrm{N}i} + u_{\mathrm{N}i} = \delta_{\mathrm{N}i} \qquad (i = 1, 2, 3, \ldots, n) \tag{4.74}$$

where $\delta_{\mathrm{N}i}$ is a time-varying displacement in the ith normal coordinate.

If both sides of Eq. (4.72) are premultiplied by $\mathbf{S}_{\mathrm{N}} = \mathbf{F}_{\mathrm{N}}^{-1}$, the expression becomes

$$\ddot{\mathbf{D}}_{\mathrm{N}} + \mathbf{S}_{\mathrm{N}}\mathbf{D}_{\mathrm{N}} = \mathbf{S}_{\mathrm{N}}\boldsymbol{\Delta}_{\mathrm{N}} \tag{4.75}$$

and each of the n equations takes the form

$$\ddot{u}_{\mathrm{N}i} + \omega_i^2 u_{\mathrm{N}i} = q_{\mathrm{N}\delta i} \qquad (i = 1, 2, 3, \ldots, n) \tag{4.76}$$

in which

$$q_{\mathrm{N}\delta i} = \omega_i^2 \delta_{\mathrm{N}i} \tag{4.77}$$

The term $q_{\mathrm{N}\delta i}$, defined by Eq. (4.77), constitutes an *equivalent normal-mode load* due to the time-varying displacement $\delta_{\mathrm{N}i}$. It takes the place of $q_{\mathrm{N}i}$ when displacement equations of motion are utilized.

The response of the ith normal mode to the equivalent load $q_{N\delta i}$ can again be calculated by the Duhamel integral. In this case we have

$$u_{Ni} = \frac{1}{\omega_i}\int_0^t q_{N\delta i} \sin \omega(t-t')\, dt' = \omega_i \int_0^t \delta_{Ni} \sin \omega_i(t-t')\, dt' \qquad (4.78)$$

which corresponds to Eq. (1.70) in Sec. 1.13. After repetitive applications of Eq. (4.78), the results are transformed back to the original coordinates by the operation in Eq. (4.58).

A *reciprocal theorem for dynamic displacements* can be derived in a manner analogous to that for loads in Eq. (4.70). Thus, we write the equality

$$(u_k)_{\Delta_j} = (u_j)_{\Delta_k} \qquad [\Delta_j = \Delta_k = f(t)] \qquad (4.79)$$

This expression means that the dynamic response of the kth displacement coordinate due to any time-varying displacement of the jth coordinate is equal to the response of the jth coordinate due to the same displacement of the kth coordinate.

Example 1. We shall consider again the two-mass system in Fig. 3.1*a*, for which the response to initial conditions was calculated in Example 1 of the preceding section. Suppose that a step function $Q_1 = P$ is applied to the first mass. Determine the response of the system to this forcing function, starting from rest.

Solution. The vector of applied actions for this simple case is $\mathbf{Q} = \{P, 0\}$. In accordance with Eq. (4.64), we transform this vector to normal coordinates by the operation

$$\mathbf{Q}_N = \mathbf{\Phi}_N^T \mathbf{Q} = \frac{1}{\sqrt{m}}\begin{bmatrix} 0.526 & 0.851 \\ -0.851 & 0.526 \end{bmatrix}\begin{bmatrix} P \\ 0 \end{bmatrix} = \frac{P}{\sqrt{m}}\begin{bmatrix} 0.526 \\ -0.851 \end{bmatrix} \qquad (f)$$

From the result of the Duhamel integral for a step function [see Eq. (1.66) in Sec. 1.12], we obtain the form of the response for the ith normal coordinate as

$$u_{Ni} = q_{Ni}(1 - \cos \omega_i t)/\omega_i^2 \qquad (g)$$

Thus, the vector of normal-mode displacements is

$$\mathbf{D}_N = \frac{P}{\sqrt{m}}\begin{bmatrix} 0.526(1 - \cos \omega_1 t)/\omega_1^2 \\ -0.851(1 - \cos \omega_2 t)/\omega_2^2 \end{bmatrix} \qquad (h)$$

Substitution of $\omega_1^2 = 0.382k/m$ and $\omega_2^2 = 2.618k/m$ into Eq. (h) yields

$$\mathbf{D}_{\mathrm{N}} = \frac{P\sqrt{m}}{k}\begin{bmatrix} 1.377(1 - \cos\omega_1 t) \\ -0.325(1 - \cos\omega_2 t) \end{bmatrix} \tag{i}$$

Transforming the solution to the original coordinates by Eq. (4.58), we obtain

$$\mathbf{D} = \mathbf{\Phi}_{\mathrm{N}}\mathbf{D}_{\mathrm{N}} = \frac{P}{k}\begin{bmatrix} 1 - 0.724\cos\omega_1 t - 0.276\cos\omega_2 t \\ 1 - 1.171\cos\omega_1 t + 0.171\cos\omega_2 t \end{bmatrix} \tag{j}$$

Inspection of these results shows that the masses oscillate about the displaced positions $(u_1)_{\mathrm{st}} = (u_2)_{\mathrm{st}} = P/k$, due to the load applied statically.

Proceeding in a similar manner, we can also calculate the response of the system to a step function $Q_2 = P$ applied to the second mass. In this case, we find

$$\mathbf{D} = \frac{P}{k}\begin{bmatrix} 1 - 1.171\cos\omega_1 t + 0.171\cos\omega_2 t \\ 2 - 1.895\cos\omega_1 t - 0.105\cos\omega_2 t \end{bmatrix} \tag{k}$$

Equation (k) shows that the first mass oscillates about the displaced position $(u_1)_{\mathrm{st}} = P/k$, while the second mass oscillates about the position $(u_2)_{\mathrm{st}} = 2P/k$ (due to the load applied statically). Comparing Eqs. (j) and (k), we also observe that the dynamic response of mass 2 due to the step function $Q_1 = P$ applied to mass 1 is equal to the response of mass 1 due to the step function $Q_2 = P$ applied to mass 2. This equality confirms the reciprocal theorem in Eq. (4.70).

Example 2. Assume that the semidefinite system of Example 2 in the preceding section is subjected to the ramp function $Q_2 = Rt$, applied to the second mass. (The symbol R denotes the rate of change of force with respect to time). Calculate the response of the three-mass system to this excitation.

Solution. Premultiplying the vector of applied actions $\mathbf{Q} = \{0, Rt, 0\}$ by $\mathbf{\Phi}_{\mathrm{N}}^{\mathrm{T}}$ for this system, we transform it to normal coordinates, as follows:

$$\mathbf{Q}_{\mathrm{N}} = \mathbf{\Phi}_{\mathrm{N}}^{\mathrm{T}}\mathbf{Q} = \frac{1}{\sqrt{6m}}\begin{bmatrix} \sqrt{2} & \sqrt{2} & \sqrt{2} \\ \sqrt{3} & 0 & -\sqrt{3} \\ 1 & -2 & 1 \end{bmatrix}\begin{bmatrix} 0 \\ Rt \\ 0 \end{bmatrix} = \frac{Rt}{\sqrt{6m}}\begin{bmatrix} \sqrt{2} \\ 0 \\ -2 \end{bmatrix} \tag{l}$$

The response of the first normal mode (the rigid-body mode) is obtained from Eq. (4.69) to be

$$u_{\mathrm{N1}} = \frac{Rt^3\sqrt{2}}{6\sqrt{6m}} \tag{m}$$

Because the second mode is symmetric, it does not respond to this anti-symmetric loading. However, the third mode responds in accordance with the solution of Eq. (4.67). Evaluation of Duhamel's integral for the ramp function yields

$$u_{N3} = -2R\left(t - \frac{1}{\omega_3}\sin\omega_3 t\right) \Big/ \omega_3^2\sqrt{6m} \tag{n}$$

which is drawn from Example 1 in Sec. 1.12. Using the previously calculated value of $\omega_3^2 = 3k/m$, we write the normal-mode responses as

$$\mathbf{D}_N = \frac{R}{6\sqrt{6m}}\begin{bmatrix} t^3\sqrt{2} \\ 0 \\ -\dfrac{4m}{k}\left(t - \dfrac{1}{\omega_3}\sin\omega_3 t\right) \end{bmatrix} \tag{o}$$

Transformation of these displacements to the original coordinates gives

$$\mathbf{D} = \mathbf{\Phi}_N\mathbf{D}_N = \frac{R}{18m}\begin{bmatrix} t^3 - \dfrac{2m}{k}\left(t - \dfrac{1}{\omega_3}\sin\omega_3 t\right) \\ t^3 + \dfrac{6m}{k}\left(t - \dfrac{1}{\omega_3}\sin\omega_3 t\right) \\ t^3 - \dfrac{2m}{k}\left(t - \dfrac{1}{\omega_3}\sin\omega_3 t\right) \end{bmatrix} \tag{p}$$

The rigid-body component of each response in Eq. (p) is equal to $Rt^3/18m$, while only the third principal mode contributes vibratory motions.

Example 3. Referring again to the system in Fig. 4.2a, we assume that the masses and lengths are as specified in Example 3 of the preceding section. Suppose that a harmonic forcing function $P \sin \Omega t$, acting in the x direction, is applied to the wire midway between the first and second masses. The resulting steady-state forced vibrations of this system are to be determined by both the action-equation approach and the displacement-equation method.

Solution. By inspection, we see that the equivalent forces applied to the masses are $\mathbf{Q} = \{P(\sin \Omega t)/2,\ P(\sin \Omega t)/2,\ 0\}$. Premultiplying this vector by $\mathbf{\Phi}_N^T$ produces the following forcing functions in normal coordinates:

$$\mathbf{Q}_N = \mathbf{\Phi}_N^T\mathbf{Q} = \frac{P\sin\Omega t}{4\sqrt{m}}\begin{bmatrix} 1 & \sqrt{2} & 1 \\ \sqrt{2} & 0 & -\sqrt{2} \\ 1 & -\sqrt{2} & 1 \end{bmatrix}\begin{bmatrix} 1 \\ 1 \\ 0 \end{bmatrix} = \frac{P\sin\Omega t}{4\sqrt{m}}\begin{bmatrix} 1+\sqrt{2} \\ \sqrt{2} \\ 1-\sqrt{2} \end{bmatrix} \tag{q}$$

Omitting transients, we write the normal-mode responses as

$$\mathbf{D}_{\mathrm{N}} = \frac{P \sin \Omega t}{4\sqrt{m}} \begin{bmatrix} (1+\sqrt{2})\beta_1/\omega_1^2 \\ \sqrt{2}\beta_2/\omega_2^2 \\ (1-\sqrt{2})\beta_3/\omega_3^2 \end{bmatrix} \tag{r}$$

where the magnification factor for the ith normal mode is given by

$$\beta_i = \frac{1}{1-\Omega^2/\omega_i^2} \qquad (i = 1, 2, 3) \tag{s}$$

Substitution of $1/\omega_1^2 = (2+\sqrt{2})\ell m/2T$, $1/\omega_2^2 = \ell m/2T$, and $1/\omega_3^2 = (2-\sqrt{2})\ell m/2T$ into Eq. (r) changes it to

$$\mathbf{D}_{\mathrm{N}} = \frac{P\ell\sqrt{m} \sin \Omega t}{8T} \begin{bmatrix} (4+3\sqrt{2})\beta_1 \\ \sqrt{2}\beta_2 \\ (4-3\sqrt{2})\beta_3 \end{bmatrix} \tag{t}$$

Then the response in the original coordinates becomes

$$\mathbf{D} = \mathbf{\Phi}_{\mathrm{N}}\mathbf{D}_{\mathrm{N}} = \frac{P\ell \sin \Omega t}{16T} \begin{bmatrix} (4+3\sqrt{2})\beta_1 + 2\beta_2 + (4-3\sqrt{2})\beta_3 \\ 2(3+2\sqrt{2})\beta_1 + 2(3-2\sqrt{2})\beta_3 \\ (4+3\sqrt{2})\beta_1 - 2\beta_2 + (4-3\sqrt{2})\beta_3 \end{bmatrix} \tag{u}$$

To apply the displacement-equation method to this problem, we first develop the displacement vector $\mathbf{\Delta}$ as the product

$$\mathbf{\Delta} = \mathbf{FQ} = \frac{P\ell \sin \Omega t}{8T} \begin{bmatrix} 3 & 2 & 1 \\ 2 & 4 & 2 \\ 1 & 2 & 3 \end{bmatrix} \begin{bmatrix} 1 \\ 1 \\ 0 \end{bmatrix} = \frac{P\ell \sin \Omega t}{8T} \begin{bmatrix} 5 \\ 6 \\ 3 \end{bmatrix} \tag{v}$$

These displacements are transformed to normal coordinates, using Eq. (4.73), as follows:

$$\mathbf{\Delta}_{\mathrm{N}} = \mathbf{\Phi}_{\mathrm{N}}^{-1}\mathbf{\Delta} = \frac{P\ell\sqrt{m} \sin \Omega t}{16T} \begin{bmatrix} 1 & \sqrt{2} & 1 \\ \sqrt{2} & 0 & -\sqrt{2} \\ 1 & -\sqrt{2} & 1 \end{bmatrix} \begin{bmatrix} 5 \\ 6 \\ 3 \end{bmatrix}$$

$$= \frac{P\ell\sqrt{m} \sin \Omega t}{8T} \begin{bmatrix} 4+3\sqrt{2} \\ \sqrt{2} \\ 4-3\sqrt{2} \end{bmatrix} \tag{w}$$

This vector causes the same normal-mode responses as those in Eq. (t); and the final results are again given by Eq. (u).

4.6 NORMAL-MODE RESPONSE TO SUPPORT MOTIONS

In many instances we are interested in the response of a multidegree system caused by support motions instead of applied actions. For example, if the ground in Fig. 4.1*a* translates in the *x* direction in accordance with the function

$$u_g = F_g(t) \tag{a}$$

the action equations of motion may be written as

$$\mathbf{M\ddot{D}} + \mathbf{SD}^* = \mathbf{0} \tag{b}$$

in which the vector $\mathbf{D}^*$ contains the displacements of the masses relative to the ground, as follows:

$$\mathbf{D}^* = \mathbf{D} - \mathbf{1}u_g \tag{c}$$

The symbol **1** in expression (*c*) denotes a vector that is filled with values of unity, and it has the effect of reproducing u_g n times. This manner of handling ground displacements is analogous to the technique used previously for systems with one or two degrees of freedom [e.g., see Eq. (*j*), Sec. 1.6]. However, a more general approach consists of writing the equations of motion in the equivalent form

$$\mathbf{M\ddot{D}} + \mathbf{SD} + \mathbf{S}_g u_g = \mathbf{0} \tag{d}$$

where, in this case,

$$\mathbf{S}_g = -\mathbf{S1} \tag{e}$$

The symbol $\mathbf{S}_g$ represents a vector of stiffness influence coefficients that are defined as holding actions corresponding to the free displacement coordinates† when a unit value of u_g is induced. Such holding actions may be generated directly (by static analysis of the system with $u_g = 1$); but for the case under consideration, they can also be calculated from Eq. (*e*), which shows that they are equal to the negatives of the row sums of matrix **S**.

We shall put Eq. (*d*) into the form of Eq. (4.61) in the preceding section by moving the product $\mathbf{S}_g u_g$ to the right-hand side of the equal sign. Thus, we have

$$\mathbf{M\ddot{D}} + \mathbf{SD} = \mathbf{Q}_g \tag{4.80}$$

† To avoid ambiguities in this section, we use the term "free displacement coordinate" to distinguish a displacement coordinate that can displace freely from one that is imposed. The latter type is referred to as a support motion, or a restraint motion.

where

$$\mathbf{Q}_g = -\mathbf{S}_g u_g = \mathbf{S1} u_g \tag{4.81}$$

The vector $\mathbf{Q}_g$ contains equivalent loads corresponding to the free displacement coordinates due to ground motions. Such equivalent loads are transformed to normal coordinates by the same operation as that for actual loads. Hence, from Eq. (4.64) we write

$$\mathbf{Q}_{Ng} = \mathbf{\Phi}_N^T \mathbf{Q}_g \tag{4.82}$$

The ith equation of motion in normal coordinates becomes

$$\ddot{u}_{Ni} + \omega_i^2 u_{Ni} = q_{Ngi} \qquad (i = 1, 2, 3, \ldots, n) \tag{4.83}$$

where the term q_{Ngi} is the equivalent action in the ith normal coordinate due to ground motions.

To calculate the response of the ith mode, we use Duhamel's integral, as follows:

$$u_{Ni} = \frac{1}{\omega_i} \int_0^t q_{Ngi} \sin \omega_i (t - t')\, dt' \tag{4.84}$$

This equation is of the same mathematical form as Eq. (4.67) in the preceding section, but the term q_{Ni} has been replaced by the term q_{Ngi}. As before, the results obtained from repetitive applications of Eq. (4.84) are transformed to the original coordinates by the operation $\mathbf{D} = \mathbf{\Phi}_N \mathbf{D}_N$.

If the ground acceleration $\ddot{u}_g$ is specified (instead of the ground translation u_g) for the system in Fig. 4.1*a*, we change coordinates to the relative motions given in Eq. (*c*). Accelerations corresponding to $\mathbf{D}^*$ are

$$\ddot{\mathbf{D}}^* = \ddot{\mathbf{D}} - \mathbf{1}\ddot{u}_g \tag{f}$$

Substitution of $\ddot{\mathbf{D}}$ from Eq. (*f*) into Eq. (*b*) gives the equations of motion in relative coordinates as

$$\mathbf{M}\ddot{\mathbf{D}}^* + \mathbf{S}\mathbf{D}^* = \mathbf{Q}_g^* \tag{4.85}$$

where

$$\mathbf{Q}_g^* = -\mathbf{M1}\ddot{u}_g \tag{4.86}$$

Because the mass matrix is diagonal for the system in Fig. 4.1*a*, the expanded form of $\mathbf{Q}_g^*$ in this particular case is

$$\mathbf{Q}_g^* = \begin{bmatrix} -m_1 \ddot{u}_g \\ -m_2 \ddot{u}_g \\ -m_3 \ddot{u}_g \end{bmatrix} \tag{g}$$

Thus, the equivalent loads corresponding to the relative free displacement coordinates are equal to the negatives of the masses times the ground acceleration $\ddot{u}_g$. Having determined these equivalent actions, we may calculate the response of the system relative to ground, using the vector $\mathbf{Q}_g^*$ in place of $\mathbf{Q}_g$. Because the coefficient matrices in Eq. (4.85) are the same as those in Eq. (4.80), the same transformation operator $\mathbf{\Phi}_N$ can be used to relate relative coordinates to normal coordinates.

If we work with displacement equations of motion instead of action equations, the effect of ground translation in Fig. 4.1*a* may be expressed as

$$\mathbf{FM\ddot{D}} + \mathbf{D} - \mathbf{1}u_g = \mathbf{0} \tag{h}$$

which can be obtained from Eq. (*b*) by premultiplication with $\mathbf{F} = \mathbf{S}^{-1}$. Putting this expression into the form of Eq. (4.71) in the preceding section, we obtain

$$\mathbf{FM\ddot{D}} + \mathbf{D} = \mathbf{\Delta}_g \tag{4.87}$$

where

$$\mathbf{\Delta}_g = \mathbf{1}u_g \tag{4.88}$$

The vector $\mathbf{\Delta}_g$ consists of time-varying displacements of the free displacement coordinates due to the ground motions, and its terms are found by static analysis. Of course, for the case under consideration each term in this vector is simply equal to the translation u_g. These displacements are transformed to normal coordinates by the operation given previously in Eq. (4.73). Thus, we have

$$\mathbf{\Delta}_{Ng} = \mathbf{\Phi}_N^{-1}\mathbf{\Delta}_g \tag{4.89}$$

In this case the ith equation of motion in normal coordinates (see Eq. (4.74)) becomes

$$\lambda_i \ddot{u}_{Ni} + u_{Ni} = \delta_{Ngi} \qquad (i = 1, 2, 3, \ldots, n) \tag{4.90}$$

in which the term δ_{Ngi} is a time-varying displacement of the ith normal coordinate due to ground motions.

For the purpose of calculating the response of the ith mode, we apply the second form of the Duhamel integral in Eq. (4.78) of the preceding section. Thus, we have

$$u_{Ni} = \omega_i \int_0^t \delta_{Ngi} \sin \omega_i (t - t')\, dt' \tag{4.91}$$

where $\delta_{\mathrm{N}gi}$ has replaced $\delta_{\mathrm{N}i}$. Following the repetitive use of Eq. (4.91), the results are transformed to the original coordinates in the usual manner.

If the ground acceleration $\ddot{u}_g$ is specified instead of u_g, the displacement equations of motion in relative coordinates take the form

$$\mathbf{FM\ddot{D}}^* + \mathbf{D}^* = \mathbf{FQ}_g^* \tag{4.92}$$

which is found by premultiplication of Eq. (4.85) with $\mathbf{F} = \mathbf{S}^{-1}$. In this case we have from Eq. (4.86)

$$\mathbf{\Delta}_g^* = \mathbf{FQ}_g^* = -\mathbf{FM1}\ddot{u}_g \tag{4.93}$$

Then the response of the system relative to ground may be calculated using $\mathbf{\Delta}_g^*$ in place of $\mathbf{\Delta}_g$.

We have discussed four methods for dealing with support motions of a particular type of multi-degree system (Fig. 4.1*a*). When action equations of motion are used, equivalent loads are determined by Eq. (4.81) for specified displacements and by Eq. (4.86) for specified accelerations. The latter operation is usually simpler than the former, but the calculated response is relative to the moving support. On the other hand, when displacement equations of motion are written, time-varying values of the free displacement coordinates due to ground translations are given by Eq. (4.88); and those due to ground accelerations are found by Eq. (4.93). Comparing these two expressions, we see that the former is more convenient than the latter. Furthermore, Eq. (4.88) is also simpler than either Eq. (4.81) or Eq. (4.86) for the action-equation approaches. Therefore, if support displacements are specified and flexibilities are not difficult to obtain, the displacement-equation approach can be advantageous. This is certaintly true for the statically determinate system in Fig. 4.1*a* when it is subjected to rigid-body translations of the ground. However, if the system is statically indeterminate, the action-equation methods usually offer greater convenience.

Example 1. Suppose that the ground restraining the two-mass system in Fig. 3.1*a* suddenly moves to the right (as a rigid body) in accordance with the displacement step function $u_g = d$. Determine the response of the system to this instantaneous support displacement, assuming that $m_1 = m_2 = m$ and $k_1 = k_2 = k$.

Solution. For the action-equation approach, we calculate the equivalent loads corresponding to the free displacement coordinates from Eq. (4.81), as follows:

$$\mathbf{Q}_g = -\mathbf{S}_g u_g = \mathbf{S1}u_g = \begin{bmatrix} 2k & -k \\ -k & k \end{bmatrix} \begin{bmatrix} 1 \\ 1 \end{bmatrix} d = \begin{bmatrix} kd \\ 0 \end{bmatrix} \tag{i}$$

Transformation of this vector to normal coordinates by Eq. (4.82) yields

$$\mathbf{Q}_{\mathrm{Ng}} = \mathbf{\Phi}_{\mathrm{N}}^{\mathrm{T}} \mathbf{Q}_{\mathrm{g}} = \frac{1}{\sqrt{m}} \begin{bmatrix} 0.526 & 0.851 \\ -0.851 & 0.526 \end{bmatrix} \begin{bmatrix} kd \\ 0 \end{bmatrix} = \frac{kd}{\sqrt{m}} \begin{bmatrix} 0.526 \\ -0.851 \end{bmatrix} \tag{j}$$

Applying Eq. (4.84) twice, we obtain the responses to the step functions in normal coordinates to be

$$\mathbf{D}_{\mathrm{N}} = \frac{kd}{\sqrt{m}} \begin{bmatrix} 0.526(1 - \cos \omega_1 t)/\omega_1^2 \\ -0.851(1 - \cos \omega_2 t)/\omega_2^2 \end{bmatrix} = d\sqrt{m} \begin{bmatrix} 1.377(1 - \cos \omega_1 t) \\ -0.325(1 - \cos \omega_2 t) \end{bmatrix} \tag{k}$$

In the original coordinates the solution becomes

$$\mathbf{D} = \mathbf{\Phi}_{\mathrm{N}} \mathbf{D}_{\mathrm{N}} = d \begin{bmatrix} 1 - 0.724 \cos \omega_1 t - 0.726 \cos \omega_2 t \\ 1 - 1.171 \cos \omega_1 t + 0.171 \cos \omega_2 t \end{bmatrix} \tag{l}$$

which is similar to that for a step force applied at the first mass (see Example 1 of the preceding section), except that P/k is replaced by d.

To solve this problem by the displacement-equation approach, we evaluate from Eq. (4.88) the translations of the free displacement coordinates, due to the step displacement, as

$$\mathbf{\Delta}_{\mathrm{g}} = \mathbf{1} \mathbf{u}_{\mathrm{g}} = \begin{bmatrix} 1 \\ 1 \end{bmatrix} d = \begin{bmatrix} d \\ d \end{bmatrix} \tag{m}$$

In normal coordinates this vector transforms to

$$\mathbf{\Delta}_{\mathrm{Ng}} = \mathbf{\Phi}_{\mathrm{N}}^{-1} \mathbf{\Delta}_{\mathrm{g}} = \sqrt{m} \begin{bmatrix} 0.526 & 0.851 \\ -0.851 & 0.526 \end{bmatrix} \begin{bmatrix} d \\ d \end{bmatrix} = d\sqrt{m} \begin{bmatrix} 1.377 \\ -0.325 \end{bmatrix} \tag{n}$$

Two applications of Eq. (4.91) produce the same normal-mode responses as those in Eq. (k). Therefore, the final results are again given by Eq. (l).

Example 2. For the three-mass system in Fig. 4.1a, let $m_1 = m_2 = m_3 = m$ and $k_1 = k_2 = k_3 = k$. Determine the response of this system to the parabolic function $\ddot{u}_{\mathrm{g}} = a_1 t^2 / t_1^2$ for support acceleration, where a_1 is the rigid-body ground acceleration at time t_1.

Solution. Considering first the action-equation method, we have the equivalent loads $\mathbf{Q}_{\mathrm{g}}^*$ for this system given by Eq. (g). Thus,

$$\mathbf{Q}_{\mathrm{g}}^* = \frac{-a_1 t^2 m}{t_1^2} \begin{bmatrix} 1 \\ 1 \\ 1 \end{bmatrix} \tag{o}$$

In normal coordinates this vector becomes

$$\mathbf{Q}_{\mathrm{Ng}}^{*} = \mathbf{\Phi}_{\mathrm{N}}^{\mathrm{T}}\mathbf{Q}_{\mathrm{g}}^{*} = \frac{-a_1 t^2 \sqrt{m}}{t_1^2}\begin{bmatrix} 0.328 & 0.591 & 0.737 \\ 0.737 & 0.328 & -0.591 \\ 0.591 & -0.737 & 0.328 \end{bmatrix}\begin{bmatrix} 1 \\ 1 \\ 1 \end{bmatrix}$$

$$= \frac{-a_1 t^2 \sqrt{m}}{t_1^2}\begin{bmatrix} 1.656 \\ 0.474 \\ 0.182 \end{bmatrix} \tag{p}$$

From three applications of Duhamel's integral, we obtain

$$\mathbf{D}_{\mathrm{N}}^{*} = \frac{-a_1 \sqrt{m}}{t_1^2}\begin{bmatrix} 1.656[t^2 - 2(1 - \cos\omega_1 t)/\omega_1^2]/\omega_1^2 \\ 0.474[t^2 - 2(1 - \cos\omega_2 t)/\omega_2^2]/\omega_2^2 \\ 0.182[t^2 - 2(1 - \cos\omega_3 t)/\omega_3^2]/\omega_3^2 \end{bmatrix} \tag{q}$$

in which the form of each solution is given by the answer to Prob. 1.13-6 at the end of the book. Substitution of $1/\omega_1^2 = 5.05m/k$, $1/\omega_2^2 = 0.643m/k$, and $1/\omega_3^2 = 0.308m/k$ into Eq. (q) produces

$$\mathbf{D}_{\mathrm{N}}^{*} = \frac{-a\sqrt{m^3}}{t_1^2 k}\begin{bmatrix} 8.363[t^2 - 10.10m(1 - \cos\omega_1 t)/k] \\ 0.305[t^2 - 1.286m(1 - \cos\omega_2 t)/k] \\ 0.056[t^2 - 0.616m(1 - \cos\omega_3 t)/k] \end{bmatrix} \tag{r}$$

Then the response in the original relative coordinates becomes

$$\mathbf{D}^{*} = \mathbf{\Phi}_{\mathrm{N}}\mathbf{D}_{\mathrm{N}}^{*} = \frac{-a_1 m}{t_1^2 k}\begin{bmatrix} 3t^2 - 27.70f_1(t) - 0.289f_2(t) - 0.020f_3(t) \\ 5t^2 - 49.92f_1(t) - 0.129f_2(t) + 0.025f_3(t) \\ 6t^2 - 62.26f_1(t) + 0.232f_2(t) - 0.011f_3(t) \end{bmatrix} \tag{s}$$

where $f_1(t) = m(1 - \cos\omega_1 t)/k$; $f_2(t) = m(1 - \cos\omega_2 t)/k$; and $f_3(t) = m(1 - \cos\omega_3 t)/k$.

Turning to the displacement-equation approach, we determine the vector $\mathbf{\Delta}_{\mathrm{g}}^{*}$ from Eq. (4.93) to be

$$\mathbf{\Delta}_{\mathrm{g}}^{*} = -\mathbf{FM1}\ddot{u}_{\mathrm{g}} = \frac{-a_1 t^2 m}{t_1^2 k}\begin{bmatrix} 1 & 1 & 1 \\ 1 & 2 & 2 \\ 1 & 2 & 3 \end{bmatrix}\begin{bmatrix} 1 \\ 1 \\ 1 \end{bmatrix} = \frac{-a_1 t^2 m}{t_1^2 k}\begin{bmatrix} 3 \\ 5 \\ 6 \end{bmatrix} \tag{t}$$

Transformation of this vector to normal coordinate gives

$$\mathbf{\Delta}_{\mathrm{Ng}}^{*} = \mathbf{\Phi}_{\mathrm{N}}^{-1}\mathbf{\Delta}_{\mathrm{g}}^{*}$$

$$= \frac{-a_1 t^2 \sqrt{m^3}}{t_1^2 k}\begin{bmatrix} 0.328 & 0.591 & 0.737 \\ 0.737 & 0.328 & -0.591 \\ 0.591 & -0.737 & 0.328 \end{bmatrix}\begin{bmatrix} 3 \\ 5 \\ 6 \end{bmatrix} = \frac{-a_1 t^2 \sqrt{m^3}}{t_1^2 k}\begin{bmatrix} 8.363 \\ 0.305 \\ 0.056 \end{bmatrix} \tag{u}$$

Three applications of Duhamel's integral yield the normal-mode responses in Eq. (r), and the final results are the same as those in Eq. (s). From these results it is apparent that the fundamental mode contributes most of the response, the second mode contributes much less than the first, and the third mode contributes much less than the second.

The preceding examples illustrate the various techniques for analyzing systems for which only one type of rigid-body ground translation is specified. In more complicated problems, three components of rigid-body translation as well as three rigid-body rotations of ground are possible [4]. In such cases the term u_g must be expanded into a vector containing the six types of displacements, and the vector $\mathbf{S}_g$ becomes an $n \times 6$ matrix. Furthermore, the ground rotations must be restricted to small rotations in order to maintain the linear characteristics upon which the normal-mode method is based. The only large motions permissible in linear analysis are rigid-body translations. Problems involving such large motions should be analyzed using relative coordinates in order to avoid loss of numerical accuracy in the vibrational response. Then the absolute motion of the system may be obtained by adding the relative displacements to the ground displacements.

For a system with multiple connections to ground, it is also possible to calculate the response to *independent motions of the support points* by generating the appropriate stiffness or flexibility coefficients [5]. In such a case the relative displacements of the supports must be small in order to retain linear behavior. If a system with n degrees of freedom has r support restraints that can move independently, the action equation (d) generalizes to

$$\mathbf{M}\ddot{\mathbf{D}} + \mathbf{S}\mathbf{D} + \mathbf{S}_{\mathrm{R}}\mathbf{D}_{\mathrm{R}} = \mathbf{0}$$

Or,

$$\mathbf{M}\ddot{\mathbf{D}} + \mathbf{S}\mathbf{D} = \mathbf{Q}_{\mathrm{R}} \tag{4.94}$$

where

$$\mathbf{Q}_{\mathrm{R}} = -\mathbf{S}_{\mathrm{R}}\mathbf{D}_{\mathrm{R}} \tag{4.95}$$

In these expressions $\mathbf{D}_{\mathrm{R}}$ denotes a vector of restraint displacements, $\mathbf{S}_{\mathrm{R}}$ is an $n \times r$ stiffness matrix coupling the free displacement coordinates with the support restraints, and $\mathbf{Q}_{\mathrm{R}}$ is a vector of equivalent loads due to motions of the restraints.

Alternatively, we convert Eq. (4.94) to displacement-equation form by premultiplying both sides with $\mathbf{F} = \mathbf{S}^{-1}$ to obtain

$$\mathbf{F}\mathbf{M}\ddot{\mathbf{D}} + \mathbf{D} = \mathbf{\Delta}_{\mathrm{R}} \tag{4.96}$$

in which

$$\mathbf{\Delta}_{\mathrm{R}} = -\mathbf{F}\mathbf{S}_{\mathrm{R}}\mathbf{D}_{\mathrm{R}} = \mathbf{T}_{\mathrm{R}}\mathbf{D}_{\mathrm{R}} \tag{4.97}$$

In this approach, the vector $\mathbf{\Delta}_{\mathrm{R}}$ consists of time-varying displacements of the free displacement coordinates due to independent motions of support restraints. As with other vectors of this nature, its terms are found by static analysis. Equation (4.97) shows that such terms may be generated by premultiplication of $\mathbf{D}_{\mathrm{R}}$ with the transformation operator

$$\mathbf{T}_{\mathrm{R}} = -\mathbf{F}\mathbf{S}_{\mathrm{R}} = -\mathbf{S}^{-1}\mathbf{S}_{\mathrm{R}} \tag{4.98}$$

The symbol $\mathbf{T}_{\mathrm{R}}$ represents an $n \times r$ matrix of *displacement influence coefficients,* defined as displacements of the free displacement coordinates due to unit values of restraint motions. While Eq. (4.98) provides a useful formula for calculating such terms in complicated systems, it is sometimes possible to develop them directly. The following example demonstrates the application of these techniques to a system with independent restraint motions.

Example 3. Considering once more the system in Fig. 4.2*a*, assume that the support points A and B can translate independently in the x direction. Let u_{R1} and u_{R2} represent small displacements of points A and B, respectively. Determine the steady-state response of the system due to the restraint motion $u_{\mathrm{R2}} = d \sin \Omega t$, while $u_{\mathrm{R1}} = 0$. As before, take $m_1 = m_2 = m_3 = m$ and $\ell_1 = \ell_2 = \ell_3 = \ell_4 = \ell$, so that previously calculated system characteristics may be used.
Solution. The vector of restraint displacements for this problem takes the form

$$\mathbf{D}_{\mathrm{R}} = \begin{bmatrix} u_{\mathrm{R1}} \\ u_{\mathrm{R2}} \end{bmatrix} = \begin{bmatrix} 0 \\ d \sin \Omega t \end{bmatrix} \tag{v}$$

To apply the action-equation method of Eq. (4.94), we develop the matrix $\mathbf{S}_{\mathrm{R}}$ to be

$$\mathbf{S}_{\mathrm{R}} = \begin{bmatrix} S_{\mathrm{R11}} & S_{\mathrm{R12}} \\ S_{\mathrm{R21}} & S_{\mathrm{R22}} \\ S_{\mathrm{R31}} & S_{\mathrm{R32}} \end{bmatrix} = \frac{T}{\ell} \begin{bmatrix} -1 & 0 \\ 0 & 0 \\ 0 & -1 \end{bmatrix} \tag{w}$$

The first column in this array is derived as the holding actions at the masses in Fig. 4.2*e* for a displacement $u_{\mathrm{R1}} = 1$. Similarly, the second column of $\mathbf{S}_{\mathrm{R}}$ consists of the holding actions in Fig. 4.2*f* for $u_{\mathrm{R2}} = 1$. In accordance with Eq. (4.95), we obtain the equivalent loads corresponding to the free displacement coordinates as

$$\mathbf{Q}_{\mathrm{R}} = -\mathbf{S}_{\mathrm{R}}\mathbf{D}_{\mathrm{R}} = \frac{Td \sin \Omega t}{\ell}\begin{bmatrix} 0 \\ 0 \\ 1 \end{bmatrix} \tag{x}$$

Transformation of this vector to normal coordinates yields

$$\mathbf{Q}_{\mathrm{NR}} = \mathbf{\Phi}_{\mathrm{N}}^{\mathrm{T}}\mathbf{Q}_{\mathrm{R}} = \frac{Td \sin \Omega t}{2\ell\sqrt{m}}\begin{bmatrix} 1 & \sqrt{2} & 1 \\ \sqrt{2} & 0 & -\sqrt{2} \\ 1 & -\sqrt{2} & 1 \end{bmatrix}\begin{bmatrix} 0 \\ 0 \\ 1 \end{bmatrix}$$

$$= \frac{Td \sin \Omega t}{2\ell\sqrt{m}}\begin{bmatrix} 1 \\ -\sqrt{2} \\ 1 \end{bmatrix} \tag{y}$$

The steady-state responses of principal modes are

$$\mathbf{D}_{\mathrm{N}} = \frac{Td \sin \Omega t}{2\ell\sqrt{m}}\begin{bmatrix} \beta_1/\omega_1^2 \\ -\sqrt{2}\beta_2/\omega_2^2 \\ \beta_3/\omega_3^2 \end{bmatrix} = \frac{d\sqrt{m} \sin \Omega t}{4}\begin{bmatrix} (2+\sqrt{2})\beta_1 \\ -\sqrt{2}\beta_2 \\ (2-\sqrt{2})\beta_3 \end{bmatrix} \tag{z}$$

Thus, in the original coordinates we find the results to be

$$\mathbf{D} = \mathbf{\Phi}_{\mathrm{N}}\mathbf{D}_{\mathrm{N}} = \frac{d \sin \Omega t}{8}\begin{bmatrix} (2+\sqrt{2})\beta_1 - 2\beta_2 + (2-\sqrt{2})\beta_3 \\ 2(1+\sqrt{2})\beta_1 + 2(1-\sqrt{2})\beta_3 \\ (2+\sqrt{2})\beta_1 + 2\beta_2 + (2-\sqrt{2})\beta_3 \end{bmatrix} \tag{a'}$$

For the displacement-equation approach of Eq. (4.96), we can generate the array $\mathbf{T}_{\mathrm{R}}$ of displacement influence coefficients by Eq. (4.98) as follows:

$$\mathbf{T}_{\mathrm{R}} = -\mathbf{F}\mathbf{S}_{\mathrm{R}} = \frac{-\ell}{4T}\begin{bmatrix} 3 & 2 & 1 \\ 2 & 4 & 2 \\ 1 & 2 & 3 \end{bmatrix}\begin{bmatrix} -1 & 0 \\ 0 & 0 \\ 0 & -1 \end{bmatrix}\frac{T}{\ell} = \frac{1}{4}\begin{bmatrix} 3 & 1 \\ 2 & 2 \\ 1 & 3 \end{bmatrix} \tag{b'}$$

Alternatively, we derive terms in the first and second columns of $\mathbf{T}_{\mathrm{R}}$ as the displacements of the masses in Figs. 4.2*g* and 4.2*h* due to displacements $u_{\mathrm{R1}} = 1$ and $u_{\mathrm{R2}} = 1$, respectively. When this operator is used as a premultiplier of $\mathbf{D}_{\mathrm{R}}$ in Eq. (v), we obtain the time-varying displacements of the free displacement coordinates. Thus,

$$\mathbf{\Delta}_{\mathrm{R}} = \mathbf{T}_{\mathrm{R}}\mathbf{D}_{\mathrm{R}} = \frac{d \sin \Omega t}{4}\begin{bmatrix} 3 & 1 \\ 2 & 2 \\ 1 & 3 \end{bmatrix}\begin{bmatrix} 0 \\ 1 \end{bmatrix} = \frac{d \sin \Omega t}{4}\begin{bmatrix} 1 \\ 2 \\ 3 \end{bmatrix} \tag{c'}$$

Transforming this vector to normal coordinates, we find

$$\mathbf{\Delta}_{\mathrm{NR}} = \mathbf{\Phi}_{\mathrm{N}}^{-1}\mathbf{\Delta}_{\mathrm{R}} = \frac{d\sqrt{m}\sin\Omega t}{8}\begin{bmatrix} 1 & \sqrt{2} & 1 \\ \sqrt{2} & 0 & -\sqrt{2} \\ 1 & -\sqrt{2} & 1 \end{bmatrix}\begin{bmatrix} 1 \\ 2 \\ 3 \end{bmatrix}$$

$$= \frac{d\sqrt{m}\sin\Omega t}{4}\begin{bmatrix} 2+\sqrt{2} \\ -\sqrt{2} \\ 2-\sqrt{2} \end{bmatrix} \qquad (d')$$

At this point we see that the solution in normal coordinates corresponds to Eq. (z), and the results in the original coordinates are given by Eq. (a').

4.7 ITERATION METHOD FOR FREQUENCIES AND MODE SHAPES

The method described in Sec. 4.2 for determining the eigenvalues and eigenvectors of a linear system is suitable only when the roots of the characteristic equation are easy to obtain. For most systems having more than two degrees of freedom, the characteristic equation is not factorable, so a trial-and-error procedure is required to extract the roots. Other numerical methods are also available for solving the algebraic eigenvalue problem [6], and they are usually more efficient for systems with many degrees of freedom. The technique to be discussed in this section is known as *forward* or *direct iteration.*† For matrices of small order, this approach may be applied using a desk calculator, but for larger problems it should be programmed on a digital computer.

The iteration method is most useful in vibration problems when only a few of the lowest frequencies and their corresponding mode shapes are to be determined. If all of the eigenvalues and eigenvectors are desired for a system with many degrees of freedom, other numerical approaches are preferable because they require fewer arithmetic operations. In addition, the iteration process converges faster if the mode shapes can be estimated in advance. The configuration of the fundamental mode can usually be approximated fairly well, but the shapes of higher modes are more difficult to estimate. Nevertheless, the accuracy with which mode shapes can be predicted affects only the rate of convergence, not the final results.

To develop the iteration procedure, without particular reference to vibrations, we begin with the eigenvalue problem expressed in standard form as

$$\mathbf{A}\mathbf{X}_{\mathrm{M}i} = \lambda_i \mathbf{X}_{\mathrm{M}i} \qquad (4.99)$$

† *Reverse* or *inverse iteration* is described in Ref. [4].

A first approximation to one of the eigenvalues λ_i may be obtained by substituting a trial eigenvector $(\mathbf{X})_1$ into both sides of Eq. (4.99) and then solving for λ_i. For this purpose, let the product of the matrix **A** and the trial eigenvector $(\mathbf{X})_1$ on the left-hand side of the equation be represented by the vector $(\mathbf{Y})_1$, as follows:

$$(\mathbf{Y})_1 = \mathbf{A}(\mathbf{X})_1 \tag{a}$$

If the vector $(\mathbf{X})_1$ is not the true eigenvector, its substitution into Eq. (4.99) will only satisfy the equation approximately. Thus, we have

$$(\mathbf{Y})_1 \approx \lambda_i(\mathbf{X})_1 \tag{b}$$

A first approximation $(\lambda_i)_1$ for the eigenvalue may be obtained by dividing any of the n elements of the vector $(\mathbf{Y})_1$ by the corresponding element of the vector $(\mathbf{X})_1$. [Note that if $(\mathbf{X})_1$ were the true eigenvector all such ratios would be equal.] We shall express all of these possible choices as the ratio

$$(\lambda_i)_1 = (y_j)_1/(x_j)_1 \tag{c}$$

where $1 \leqq j \leqq n$.

In preparation for a second step of iteration, the vector $(\mathbf{Y})_1$ is usually normalized in some manner (such as dividing all of its elements by the first, largest, or last element). In general, we divide, $(\mathbf{Y})_1$ by an arbitrary constant b_1 to produce a second trial vector $(\mathbf{X})_2$.

$$(\mathbf{X})_2 = (\mathbf{Y})_1/b_1 \tag{d}$$

This vector is then premultiplied by matrix **A** to obtain a new vector $(\mathbf{Y})_2$, as follows:

$$(\mathbf{Y})_2 = \mathbf{A}(\mathbf{X})_2 \tag{e}$$

Next, a second approximation $(\lambda_i)_2$ for the eigenvalue is calculated as

$$(\lambda_i)_2 = (y_j)_2/(x_j)_2 \tag{f}$$

Then the vector $(\mathbf{Y})_2$ is scaled by dividing it with the arbitrary constant b_2 to produce a third trial vector $(\mathbf{X})_3$.

$$(\mathbf{X})_3 = (\mathbf{Y})_2/b_2 \tag{g}$$

This procedure is repeated until the eigenvalue and the associated eigenvector are determined to the desired degree of accuracy.

In the kth iteration, the recurrence equations for the steps described above are

$$(\mathbf{Y})_k = \mathbf{A}(\mathbf{X})_k \tag{4.100}$$

$$(\lambda_i)_k = (y_j)_k/(x_j)_k \tag{4.101}$$

$$(\mathbf{X})_{k+1} = (\mathbf{Y})_k/b_k \tag{4.102}$$

in which b_k is an arbitrarily selected divisor. It will be shown that when these equations are applied repeatedly, the procedure converges to the numerically largest eigenvalue and the associated eigenvector. Therefore, for the purpose of aiding convergence, the first trial vector $(\mathbf{X})_1$ should be an approximation to the eigenvector associated with the numerically largest eigenvalue.

In order to prove that the iteration technique converges to the largest eigenvalue, we expand the first trial vector $(\mathbf{X})_1$ in terms of the true eigenvectors of the system as

$$(\mathbf{X})_1 = \sum_{i=1}^{n} a_i \mathbf{\Phi}_{\mathrm{M}i} \tag{h}$$

where $a_1, a_2, \ldots, a_i, \ldots, a_n$ are scale factors. The validity of this expression depends upon the existence of n linearly independent (albeit unknown) eigenvectors, a criterion that is satisfied for vibratory systems. We shall also assume that the corresponding eigenvalues are arranged in descending order and that the largest eigenvalue λ_1 is not repeated. That is,

$$\lambda_1 > \lambda_2 \geqq \cdots \geqq \lambda_i \geqq \cdots \geqq \lambda_n \tag{i}$$

Substituting Eq. (h) into Eq. (a) and using Eq. (4.99) in the result, we develop the following expansion of $(\mathbf{Y})_1$ in terms of the true eigenvectors:

$$(\mathbf{Y})_1 = \sum_{i=1}^{n} a_i \mathbf{A} \mathbf{\Phi}_{\mathrm{M}i} = \sum_{i=1}^{n} a_i \lambda_i \mathbf{\Phi}_{\mathrm{M}i} \tag{j}$$

In the second step of the iteration we have

$$(\mathbf{Y})_2 = \mathbf{A}(\mathbf{X})_2 = \mathbf{A}(\mathbf{Y}_1)/b_1 \tag{k}$$

Substituting Eq. (j) into Eq. (k) and again using Eq. (4.99), we find the expansion of $(\mathbf{Y})_2$ to be

$$(\mathbf{Y})_2 = \sum_{i=1}^{n} a_i \lambda_i \mathbf{A} \mathbf{\Phi}_{\mathrm{M}i}/b_1 = \sum_{i=1}^{n} a_i \lambda_i^2 \mathbf{\Phi}_{\mathrm{M}i}/b_1 \tag{l}$$

In the kth iteration the expansion of $(\mathbf{Y})_k$ becomes

$$(\mathbf{Y})_k = \sum_{i=1}^{n} a_i \lambda_i^k \mathbf{\Phi}_{\mathrm{M}i} / b_1 b_2 \cdots b_{k-1} \tag{m}$$

where λ_i is raised to the kth power. Factoring λ_1^k and separating the term involving $\mathbf{\Phi}_{\mathrm{M}1}$ from the others, we obtain

$$(\mathbf{Y})_k = \lambda_1^k \left[a_1 \mathbf{\Phi}_{\mathrm{M}1} + \sum_{i=2}^{n} a_i (\lambda_i / \lambda_1)^k \mathbf{\Phi}_{\mathrm{M}i} \right] \Big/ b_1 b_2 \cdots b_{k-1}$$

From the assumed ordering of the eigenvalues [see Eq. (i)], it is apparent that the factors $(\lambda_i/\lambda_1)^k$ approach zero as k increases. Therefore, we conclude that

$$(\mathbf{Y})_k \approx \lambda_1^k a_1 \mathbf{\Phi}_{\mathrm{M}1} / b_1 b_2 \cdots b_{k-1} \tag{n}$$

This expression shows that $(\mathbf{Y}_k)$ approaches the form of $\mathbf{\Phi}_{\mathrm{M}1}$, because all other terms are constants. If we factor λ_1 from the right-hand side of Eq. (n), we find

$$(\mathbf{Y})_k \approx \lambda_1 (\lambda_1^{k-1} a_1 \mathbf{\Phi}_{\mathrm{M}1} / b_1 b_2 \cdots b_{k-1}) = \lambda_1 (\mathbf{X})_k \tag{o}$$

Thus, the kth approximation of λ_i, given by Eq. (4.101), is seen from Eq. (o) to be an approximation to λ_1. Because of this condition of convergence, the numerically largest eigenvalue λ_1 is said to be the *dominant eigenvalue*; and the corresponding vector $\mathbf{\Phi}_{\mathrm{M}1}$ is called the *dominant eigenvector.*

We shall now use the iteration method to determine the eigenvalue and eigenvector for the fundamental mode of vibration of a system with several degrees of freedom. Because the method converges to the largest eigenvalue, we must use the displacement-equation approach, where the largest eigenvalue is equal to the reciprocal of the square of the lowest angular frequency. Thus, from Eq. (4.9) in Sec. 4.2, we have

$$\mathbf{A}_1 \mathbf{\Phi}_{\mathrm{M}1} = \mathbf{FM\Phi}_{\mathrm{M}1} = \lambda_1 \mathbf{\Phi}_{\mathrm{M}1} \tag{4.103}$$

in which

$$\lambda_1 = 1/\omega_1^2 \tag{4.104}$$

Equation (4.103) is in the standard form of the eigenvalue problem in Eq. (4.99). For this application we see that

$$\mathbf{A}_1 = \mathbf{FM} \tag{4.105}$$

which is called the *dynamical matrix*. It is unsymmetric unless **M** is a diagonal matrix with equal terms on the diagonal. However, such a lack of symmetry is of no consequence when determining the fundamental mode of vibration by the method of iteration.

For a numerical example, let us consider the displacement-equation approach to the three-mass system in Fig. 4.1*a*, which was analyzed by the action-equation method in Example 1 of Sec. 4.2. If $k_1 = k_2 = k_3 = k$ and $m_1 = m_2 = m_3 = m$, the flexibility and mass matrices for this system are

$$\mathbf{F} = \delta \begin{bmatrix} 1 & 1 & 1 \\ 1 & 2 & 2 \\ 1 & 2 & 3 \end{bmatrix} \qquad \mathbf{M} = m \begin{bmatrix} 1 & 0 & 0 \\ 0 & 1 & 0 \\ 0 & 0 & 1 \end{bmatrix}$$

where $\delta = 1/k$, as before. Therefore, the matrix $\mathbf{A}_1$ in Eq. (4.105) becomes

$$\mathbf{A}_1 = m\delta \begin{bmatrix} 1 & 1 & 1 \\ 1 & 2 & 2 \\ 1 & 2 & 3 \end{bmatrix} \tag{p}$$

A good approximation for the shape of the fundamental mode could be obtained by taking the row sums of matrix $\mathbf{A}_1$. This approach produces a vector of displacements due to statically applied forces that are proportional to the masses, as in Rayleigh's method (see Sec. 1.5). An indirect technique for accomplishing the same objective consists of taking

$$(\mathbf{X}_1)_1 = \{1, 1, 1\}$$

as the first trial vector. Premultiplication of $(\mathbf{X}_1)_1$ by matrix $\mathbf{A}_1$, in accordance with Eq. (4.100), yields the vector

$$(\mathbf{Y}_1)_1 = m\delta\{3, 5, 6\}$$

The first approximation for λ_1, as indicated by Eq. (4.101), could be calculated in three different ways. For convenience in later calculations, we shall divide the last element of $(\mathbf{Y}_1)_1$ by the last element of $(\mathbf{X}_1)_1$ to obtain

$$(\lambda_1)_1 = (y_n)_1/(x_n)_1 = 6m\delta$$

In preparation for the second cycle of iteration, the vector $(\mathbf{Y}_1)_1$ is normalized by dividing each element by the last element (see Eq. 4.102), producing the second trial vector

$$(\mathbf{X}_1)_2 = \{0.500, 0.833, 1.000\}$$

When this type of normalization is used, the divisor $b_1 = 6m\delta$ coincides with the approximate eigenvalue.

The second cycle of iteration consists of applying Eqs. (4.100), (4.101), and (4.102) to obtain values for $(\mathbf{Y}_1)_2$, $(\lambda_1)_2$, and $(\mathbf{X}_1)_3$. This process of iteration is repeated for additional cycles until two successive iterations yield eigenvalues and eigenvectors that agree within a predetermined level of accuracy. Table 4.1 summarizes the results, and in this case the terms obtained in the fifth cycle agrees with those for the fourth cycle to an accuracy of approximately three significant digits. Thus, we have

$$\lambda_1 \approx 5.049m\delta \qquad \omega_1^2 = 1/\lambda_1 \approx 0.198k/m$$
$$\mathbf{\Phi}_{M1} \approx \{0.445, 0.802, 1.000\} \tag{q}$$

These results are the same as those given in Example 1 of Sec. 4.2, except for the manner of normalizing $\mathbf{\Phi}_{M1}$.

After the fundamental mode has been determined, it is also possible to proceed further and obtain eigenvalues and eigenvectors for higher modes by the method of iteration. If the first mode is suppressed by the introduction of appropriate constraints, the second mode will become dominant. If both the first and second modes are constrained, the third mode will become dominant, and so on. Because the number of natural modes is equal to the number of degrees of freedom, the introduction of *modal constraints* reduces the number of degrees of freedom. Thus, we should expect that the order of the coefficient matrix will be $n-1$ when iterating for the second mode, $n-2$ when iterating for the third mode, and so on. However, for tabular and digital computations it is advantageous to retain a matrix of order n, using a simple artifice that will soon be explained.

Modal constraints can be introduced by specifying that certain principal-mode displacements are equal to zero. From Eqs. (4.32) and (4.44a) in Sec. 4.3, we have

$$\mathbf{D}_P = \mathbf{\Phi}_M^{-1}\mathbf{D} = \mathbf{M}_P^{-1}\mathbf{\Phi}_M^T\mathbf{M}\mathbf{D} \tag{r}$$

TABLE 4.1 Iteration of Fundamental Mode

Trial Vector $(\mathbf{X}_1)_k$	$(\mathbf{X}_1)_1$	$(\mathbf{X}_1)_2$	$(\mathbf{X}_1)_3$	$(\mathbf{X}_1)_4$	$(\mathbf{X}_1)_5$
$\frac{\mathbf{A}_1}{m\delta} = \begin{bmatrix} 1 & 1 & 1 \\ 1 & 2 & 2 \\ 1 & 2 & 3 \end{bmatrix}$	1 1 1	0.500 0.833 1.000	0.452 0.806 1.000	0.446 0.803 1.000	0.445 0.802 1.000
Eigenvalue $(\lambda_1)_k/m\delta$	6	5.167	5.065	5.051	5.049

which relates the principal coordinates in vector $\mathbf{D}_\mathrm{P}$ to the original coordinates in vector $\mathbf{D}$. In order to eliminate the first mode, we set the expression for $u_{\mathrm{P}1}$ from Eq. (r) equal to zero, as follows:

$$u_{\mathrm{P}1} = \frac{1}{M_{\mathrm{P}1}} \mathbf{\Phi}_{\mathrm{M}1}^{\mathrm{T}} \mathbf{M} \mathbf{D} = 0 \tag{s}$$

If the vector $\mathbf{D}$ is taken to be an eigenvector $\mathbf{\Phi}_{\mathrm{M}i}$ (where $i = 2, 3, \ldots, n$), we see that this constraint condition coincides with the orthogonality of the first mode and the higher modes, with respect to $\mathbf{M}$. For simplicity, the mass matrix is assumed to be diagonal; then the expansion of Eq. (s) becomes

$$M_{11}\Phi_{\mathrm{M}11}u_1 + M_{22}\Phi_{\mathrm{M}21}u_2 + M_{33}\Phi_{\mathrm{M}31}u_3 + \cdots + M_{nn}\Phi_{\mathrm{M}n1}u_n = 0$$

As an arbitrary choice, we solve for u_1 in terms of the other displacements, as follows:

$$u_1 = -\frac{M_{22}\Phi_{\mathrm{M}21}}{M_{11}\Phi_{\mathrm{M}11}} u_2 - \frac{M_{33}\Phi_{\mathrm{M}31}}{M_{11}\Phi_{\mathrm{M}11}} u_3 - \cdots - \frac{M_{nn}\Phi_{\mathrm{M}n1}}{M_{11}\Phi_{\mathrm{M}11}} u_n \tag{t}$$

Substitution of this expression for u_1 into the original equations of the eigenvalue problem [see Eq. (4.99)] produces n equations in $n-1$ unknowns. The first of these equations will be a linear combination of the remaining $n-1$ equations and could be discarded, leaving a reduced set of $n-1$ equations in $n-1$ unknowns. However, the full set of n linearly dependent equations (devoid of the first mode) can be iterated for the second mode without reducing the size of the matrix. For this purpose we cast expression (t), together with the trivial relationships $u_2 = u_2$, $u_3 = u_3$, etc., into the matrix format

$$\begin{bmatrix} u_1 \\ u_2 \\ u_3 \\ \cdots \\ u_n \end{bmatrix} = \left[\begin{array}{c|cccc} 0 & c_{12} & c_{13} & \cdots & c_{1n} \\ 0 & 1 & 0 & \cdots & 0 \\ 0 & 0 & 1 & \cdots & 0 \\ \cdots & \cdots & \cdots & \cdots & \cdots \\ 0 & 0 & 0 & \cdots & 1 \end{array}\right] \begin{bmatrix} u_1' \\ \hline u_2 \\ u_3 \\ \cdots \\ u_n \end{bmatrix} \tag{u}$$

where

$$c_{12} = -\frac{M_{22}\Phi_{\mathrm{M}21}}{M_{11}\Phi_{\mathrm{M}11}} \qquad c_{13} = -\frac{M_{33}\Phi_{\mathrm{M}31}}{M_{11}\Phi_{\mathrm{M}11}} \qquad \cdots \qquad c_{1n} = -\frac{M_{nn}\Phi_{\mathrm{M}n1}}{M_{11}\Phi_{\mathrm{M}11}} \tag{v}$$

The first element u_1' of the right-hand vector in Eq. (u) is taken to be a dummy displacement that is always multiplied by zero. This matrix equation may be written concisely as

$$\mathbf{D} = \mathbf{T}_{S1}\mathbf{D}' \tag{w}$$

in which $\mathbf{T}_{S1}$ expresses the dependence of $u_1, u_2, u_3, \ldots, u_n$ upon $u_2, u_3, \ldots, u_n$. Substituting $\mathbf{A}_1$ and $\mathbf{\Phi}_{Mi} = \mathbf{T}_{S1}\mathbf{\Phi}'_{Mi}$ for $\mathbf{X}_{Mi}$ on the left-hand side of Eq. (4.99) gives

$$\mathbf{A}_1\mathbf{T}_{S1}\mathbf{\Phi}'_{Mi} = \lambda_i\mathbf{\Phi}_{Mi} \tag{x}$$

The presence of $\mathbf{T}_{S1}$ in this expression provides the linear dependency that is necessary for iterating the second mode. In each cycle of iteration this operator premultiplies a trial vector and assures its orthogonality with the first mode. For greater convenience, however, it is ordinarily used only once, as a postmultiplier upon matrix $\mathbf{A}_1$. The array $\mathbf{T}_{S1}$ is called a *sweeping matrix*, because its effect is to "sweep out" the first-mode characteristics and allow the second mode to become dominant. Thus, we have

$$\mathbf{A}_2\mathbf{\Phi}'_{Mi} = \lambda_i\mathbf{\Phi}_{Mi} \tag{4.106}$$

where

$$\mathbf{A}_2 = \mathbf{A}_1\mathbf{T}_{S1} \tag{4.107}$$

and $\mathbf{T}_{S1}$ has the form given in Eq. (u).

To show the application of this technique, we shall now determine the eigenvalue and eigenvector for the second mode of the system in Fig. 4.1a. Having already obtained the first eigenvector $\mathbf{\Phi}_{M1}$, we use its components to form the sweeping matrix $\mathbf{T}_{S1}$, as follows:

$$\mathbf{T}_{S1} = \left[\begin{array}{c:cc} 0 & -1.802 & -2.247 \\ 0 & 1 & 0 \\ 0 & 0 & 1 \end{array}\right] \tag{y}$$

Postmultiplication of matrix $\mathbf{A}_1$ [see Eq. (p)] by $\mathbf{T}_{S1}$ produces

$$\mathbf{A}_2 = m\delta\left[\begin{array}{c:cc} 0 & -0.802 & -1.247 \\ 0 & 0.198 & -0.247 \\ 0 & 0.198 & 0.753 \end{array}\right] \tag{z}$$

Assuming that we know nothing about the second mode shape, we take for the first trial vector

$$(\mathbf{X}_2)_1 = \{1, 1, 1\}$$

which represents a poor approximation of the true eigenvector. A more informed estimate would converge to the correct solution in fewer cycles of iteration.

Premultiplication of $(\mathbf{X}_2)_1$ by matrix $\mathbf{A}_2$ [see Eq. (4.100)] yields the vector

$$(\mathbf{Y}_2)_1 = m\delta\{-2.049,\ -0.049,\ 0.951\}$$

Dividing the last element of $(\mathbf{Y}_2)_1$ by the last element of $(\mathbf{X}_2)_1$ [see Eq. (4.101)], we obtain the first approximation for λ_2 to be

$$(\lambda_2)_1 = 0.951 m\delta$$

Then the vector $(\mathbf{Y}_2)_1$ is normalized by dividing each element by the last [see Eq. (4.102)], which gives the second trial vector

$$(\mathbf{X}_2)_2 = \{-2.155,\ -0.052,\ 1.000\}$$

These and subsequent results appear in Table 4.2, and after five cycles of iteration the procedure has not yet converged. Upon continuing the calculations through eleven cycles of iteration, we find the converged values to be

$$\lambda_2 \approx 0.643 m\delta \qquad \omega_2^2 = \frac{1}{\lambda_2} \approx 1.555\frac{k}{m} \tag{a'}$$

$$\mathbf{\Phi}_{M2} \approx \{-1.247,\ -0.555,\ 1.000\}$$

Except for the method of normalizing $\mathbf{\Phi}_{M2}$, these results agree with those given in Example 1 of Sec. 4.2.

After the second mode has been determined, it can be eliminated from the system of equations by a procedure similar to that used for eliminating the first mode. Using the constraint conditions $u_{P1} = 0$ and $u_{P2} = 0$, we solve for u_2 in terms of $u_3, \ldots, u_n$ and cast the displacement relationships into

TABLE 4.2 Iteration of Second Mode

Trial Vector $(\mathbf{X}_2)_k$	$(X_2)_1$	$(X_2)_2$	$(X_2)_3$	$(X_2)_4$	$(X_2)_5$
$\dfrac{\mathbf{A}_2}{m\delta} = \begin{bmatrix} 0 & -0.802 & -1.247 \\ 0 & 0.198 & -0.247 \\ 0 & 0.198 & 0.753 \end{bmatrix}$	1 1 1	−2.155 −0.052 1.000	−1.623 −0.346 1.000	−1.416 −0.461 1.000	−1.326 −0.511 1.000
Eigenvalue $(\lambda_2)_k/m\delta$	0.951	0.743	0.684	0.662	0.652

the matrix format

$$\begin{bmatrix} u_1' \\ \hdashline u_2 \\ u_3 \\ \cdots \\ u_n \end{bmatrix} = \left[\begin{array}{cc:cccc} 1 & 0 & 0 & \cdots & 0 \\ \hdashline 0 & 0 & d_{23} & \cdots & d_{2n} \\ 0 & 0 & 1 & \cdots & 0 \\ \cdots & \cdots & \cdots & \cdots & \cdots \\ 0 & 0 & 0 & \cdots & 1 \end{array}\right] \begin{bmatrix} u_1' \\ u_2' \\ \hdashline u_3 \\ \cdots \\ u_n \end{bmatrix} \qquad (b')$$

where

$$d_{23} = -\frac{M_{33}(\Phi_{M11}\Phi_{M32} - \Phi_{M12}\Phi_{M31})}{M_{22}(\Phi_{M11}\Phi_{M22} - \Phi_{M12}\Phi_{M21})} \quad \cdots$$

$$d_{2n} = -\frac{M_{nn}(\Phi_{M11}\Phi_{Mn2} - \Phi_{M12}\Phi_{Mn1})}{M_{22}(\Phi_{M11}\Phi_{M22} - \Phi_{M12}\Phi_{M21})} \qquad (c')$$

Expressions within the parentheses are minors of the first two columns of $\mathbf{\Phi}_M$ that arise in the process of solving for u_2 in terms of $u_3, \ldots, u_n$. An additional dummy displacement u_2' appears in the right-hand vector of Eq. (b'). This equation can be written more concisely as

$$\mathbf{D}' = \mathbf{T}_{S2}\mathbf{D}'' \qquad (d')$$

The matrix $\mathbf{T}_{S2}$ accounts for the dependence of $u_2, u_3, \ldots, u_n$ upon $u_3, \ldots, u_n$. Substitution of Eq. (d') for $\mathbf{\Phi}'_{Mi}$ on the left-hand side of Eq. (4.106) results in

$$\mathbf{A}_3\mathbf{\Phi}''_{Mi} = \lambda_i\mathbf{\Phi}_{Mi} \qquad (4.108)$$

where

$$\mathbf{A}_3 = \mathbf{A}_2\mathbf{T}_{S2} \qquad (4.109)$$

Then Eq. (4.108), which is devoid of the first two modes, can be iterated for the third mode.

Applying this technique to the system in Fig. 4.1a, we evaluate the first of expressions (c') to be

$$d_{23} = -\frac{(0.445)(1.000) - (-1.247)(1.000)}{(0.445)(-0.555) - (-1.247(0.802)} = -\frac{1.692}{0.753} = -2.247$$

and form the sweeping matrix $\mathbf{T}_{S2}$ as

$$\mathbf{T}_{S2} = \left[\begin{array}{cc:c} 1 & 0 & 0 \\ \hdashline 0 & 0 & -2.247 \\ 0 & 0 & 1 \end{array}\right] \qquad (e')$$

Postmultiplication of matrix $\mathbf{A}_2$ [see Eq. (z)] by $\mathbf{T}_{S2}$ gives

$$\mathbf{A}_3 = m\delta \begin{bmatrix} 0 & 0 & 0.555 \\ 0 & 0 & -0.692 \\ 0 & 0 & 0.308 \end{bmatrix} \tag{f'}$$

The third column of $\mathbf{A}_3$ is proportional to the third eigenvector, and no iteration is required for the last mode. Moreover, the third eigenvalue appears in the last position of this vector, and we have

$$\lambda_3 \approx = 0.308 m\delta \qquad \omega_3^2 = 1/\lambda_3 \approx 3.247 k/m$$

$$\mathbf{\Phi}_{M3} \approx \{1.802, -2.247, 1.000\} \tag{g'}$$

These values are the same as those of obtained in Example 1 of Sec. 4.2, except for the manner of normalizing $\mathbf{\Phi}_{M3}$.

If a system has two or more *repeated eigenvalues*, they will be equally dominant, and the eigenvector to which the iteration method converges depends upon the form of the first trial vector. The sweeping matrix will cause each successive eigenvector to be orthogonal to those obtained previously, even though the eigenvalue may be repeated. Because eigenvectors corresponding to repeated eigenvalues frequently contain zero elements, special care must be exercised in the normalization process to avoid division by zero.

As a simple example of a system with repeated roots, assume that the three springs supporting the mass in Prob. 4.2-9 (see Prob. Set. 4.2) lie along the x, y, and z axes. Taking $k_1 = k_2 = k_3 = k$, we find that the matrix $\mathbf{A}_1$ for this problem becomes

$$\mathbf{A}_1 = \mathbf{FM} = m\delta \begin{bmatrix} 1 & 0 & 0 \\ 0 & 1 & 0 \\ 0 & 0 & 1 \end{bmatrix}$$

It can be seen by inspection that the eigenvalues of this system are $\lambda_1 = \lambda_2 = \lambda_3 = m\delta$. However, the eigenvectors are unknown and will be determined by the same approach as that used for the preceding example. The use of a trial vector $(\mathbf{X}_1)_1 = \{1, 1, 1\}$ satisfies Eq. (4.103); so that vector becomes the first eigenvector of the system. The first sweeping matrix $\mathbf{T}_{S1}$ [see Eqs. (u), (v), and (w)] is then found to be

$$\mathbf{T}_{S1} = \left[\begin{array}{c:cc} 0 & -1 & -1 \\ 0 & 1 & 0 \\ 0 & 0 & 1 \end{array}\right]$$

and matrix A_2 is calculated from Eq. (4.107) as

$$\mathbf{A}_2 = \mathbf{A}_1\mathbf{T}_{S1} = m\delta \begin{bmatrix} 0 & -1 & -1 \\ 0 & 1 & 0 \\ 0 & 0 & 1 \end{bmatrix}$$

To determine the second mode shape, we again use the trial vector $(\mathbf{X}_2)_1 = \{1, 1, 1\}$; and in two cycles of iteration we obtain the second eigenvector to be $\mathbf{\Phi}_{M2} = \{-2, 1, 1\}$. Then the second sweeping matrix $\mathbf{T}_{S2}$ [see Eqs. (b'), (c'), and (d')] is formed as the following array:

$$\mathbf{T}_{S2} = \begin{bmatrix} 1 & 0 & 0 \\ 0 & 0 & -1 \\ 0 & 0 & 1 \end{bmatrix}$$

Using this operator, we find matrix $\mathbf{A}_3$ from Eq. (4.109).

$$\mathbf{A}_3 = \mathbf{A}_2\mathbf{T}_{S2} = m\delta \begin{bmatrix} 0 & 0 & 0 \\ 0 & 0 & -1 \\ 0 & 0 & 1 \end{bmatrix}$$

Thus, the third eigenvector is seen to be $\mathbf{\Phi}_{M3} = \{0, -1, 1\}$, which is orthogonal to the other two eigenvectors (with respect to M). This set of eigenvectors is not unique, but it fulfills the orthogonality conditions required for eigenvectors in the normal-mode method of analysis.

The iteration-reduction process described in the preceding discussion can theoretically be applied repeatedly until all of the frequencies and mode shapes of a multidegree system have been determined. However, each eigenvalue and eigenvector obtained in this manner is only approximate. For this reason the orthogonalization inherent in each reduction will be imperfect. Furthermore, every reduction involves roundoff errors that accumulate from one reduction to another. Related to the matter of accuracy is the fact that an inordinate number of arithmetic operations is required to obtain very many frequencies and mode shapes by this approach. Therefore, as stated at the beginning of the section, the iteration method is best applied when only a few of the lowest modes are to be determined. In addition, the numerous arithmetic operations required for systems with many degrees of freedom should be relegated to a digital computer, especially when mode shapes are difficult to estimate. Appendix B.4 describes a computer program named **EIGIT** for calculating the first several eigenvalues and eigenvectors of a matrix by the method of iteration.

The fact that only a few of the eigenvalues and eigenvectors of a system are determined by the method of iteration does not preclude the use of the normal-mode method for response calculations. If n_1 of the modes have been found, where $n_1 \leqq n$, the modal matrix $\mathbf{\Phi}_\mathrm{M}$ (or $\mathbf{\Phi}_\mathrm{N}$) is composed of only n_1 columns instead of n columns. Such a rectangular array does not have an inverse; so the relationships in Eqs. (4.44a) and (4.44b), involving the transposed arrays $\mathbf{\Phi}_\mathrm{M}^\mathrm{T}$ and $\mathbf{\Phi}_\mathrm{N}^\mathrm{T}$, must always be used in place of the inverse. Furthermore, only n_1 of the normal-mode responses are calculated, whereas contributions of the other modes to the total reponse are omitted. This adaptation of the normal-mode method will be referred to as *modal truncation*, a concept of great value in problems where only a few of the natural modes of a system are significantly excited.

We have not considered applying the iteration method to the eigenvalue problem for the action-equation approach [see Eq. (4.17), Sec. 4.2], because the highest eigenvalue ω_n^2 would be dominant. In a problem where stiffness coefficients are more easily developed than flexibilities, a nonsingular stiffness matrix $\mathbf{S}$ can always be inverted to obtain the flexibility matrix $\mathbf{F}$ required in Eq. (4.103). On the other hand, a *semidefinite system,* for which the stiffness matrix is singular, requires special treatment. In such a case the stiffness and mass matrices may be reduced by changing coordinates to a new set that is devoid of the rigid-body modes. Fortunately, the mode shapes for rigid-body motion can usually be determined by inspection, and a simple procedure for eliminating them can be devised.

Suppose that a vibratory system is known to have only one rigid-body mode, which is defined by the first eigenvector $\mathbf{\Phi}_\mathrm{M1}$. This mode can be eliminated from a set of action equations by specifying the constraint condition $u_\mathrm{P1} = 0$. Then we form a *reduction-transformation matrix* $\mathbf{T}_\mathrm{R1}$, consisting of the $n \times (n-1)$ submatrix of the sweeping matrix $\mathbf{T}_\mathrm{S1}$ to the right of the partitioning line in Eq. (u). In this case we use the relationship

$$\mathbf{D} = \mathbf{T}_\mathrm{R1}\mathbf{D}_\mathrm{R1} \tag{4.110}$$

which expresses the dependence of $\mathbf{D} = \{u_1, u_2, u_3, \ldots, u_n\}$ upon $\mathbf{D}_\mathrm{R1} = \{u_2, u_3, \ldots, u_n\}$. (The latter vector does not contain the dummy displacement u_1' that was used with the sweeping matrix.) Since the array $\mathbf{T}_\mathrm{R1}$ is not a function of time, we can also write

$$\ddot{\mathbf{D}} = \mathbf{T}_\mathrm{R1}\ddot{\mathbf{D}}_\mathrm{R1} \tag{h'}$$

where $\ddot{\mathbf{D}}_\mathrm{R1} = \{\ddot{u}_2, \ddot{u}_3, \ldots, \ddot{u}_n\}$. Substituting Eqs. (4.110) and (h') into the action equations of motion for free vibrations [Eq. (4.30), Sec. 4.3] and premultiplying the result by $\mathbf{T}_\mathrm{R1}^\mathrm{T}$, we obtain

$$\mathbf{T}_\mathrm{R1}^\mathrm{T}\mathbf{M}\mathbf{T}_\mathrm{R1}\ddot{\mathbf{D}}_\mathrm{R1} + \mathbf{T}_\mathrm{R1}^\mathrm{T}\mathbf{S}\mathbf{T}_\mathrm{R1}\mathbf{D}_\mathrm{R1} = \mathbf{0}$$

Or

$$\mathbf{M}_\mathrm{R1}\ddot{\mathbf{D}}_\mathrm{R1} + \mathbf{S}_\mathrm{R1}\mathbf{D}_\mathrm{R1} = \mathbf{0} \tag{4.111}$$

in which

$$\mathbf{M}_{R1} = \mathbf{T}_{R1}^{T}\mathbf{M}\mathbf{T}_{R1} \qquad \mathbf{S}_{R1} = \mathbf{T}_{R1}^{T}\mathbf{S}\mathbf{R}_{R1} \tag{4.112}$$

The arrays $\mathbf{M}_{R1}$ and $\mathbf{S}_{R1}$ are both square, symmetric matrices of order $n-1$, and Eq. (4.111) represents the equations of motion transformed to a reduced set of coordinates devoid of the rigid-body mode. In these coordinates the stiffness matrix $\mathbf{S}_{R1}$ can be inverted to produce the flexibility matrix

$$\mathbf{F}_{R1} = \mathbf{S}_{R1}^{-1} \tag{4.113}$$

Then the product

$$\mathbf{A}_{R1} = \mathbf{F}_{R1}\mathbf{M}_{R1} \tag{4.114}$$

is formed in preparation for applying the iteration method to the eigenvalue problem associated with Eq. (4.111). The eigenvectors found by this procedure are transformed back to the original coordinates using Eq. (4.110).

If a second rigid-body mode exists and is defined by the eigenvector $\mathbf{\Phi}_{M2}$, a second constraint condition $u_{P2} = 0$ is required. The reduction-transformation matrix $\mathbf{T}_{R2}$ associated with this condition consists of the $(n-1) \times (n-2)$ submatrix of the sweeping matrix $\mathbf{T}_{S2}$, below and to the right of the partitioning lines in Eq. (b'). In this case we have the relationship

$$\mathbf{D}_{R1} = \mathbf{T}_{R2}\mathbf{D}_{R2} \tag{4.115}$$

which expresses $\mathbf{D}_{R1} = \{u_2, u_3, \ldots, u_n\}$ in terms of $\mathbf{D}_{R2} = \{u_3, \ldots, u_n\}$. Following the same line of reasoning as before, we can reduce the matrices $\mathbf{M}_{R1}$ and $\mathbf{S}_{R1}$ to $\mathbf{M}_{R2}$ and $\mathbf{S}_{R2}$, as follows:

$$\mathbf{M}_{R2} = \mathbf{T}_{R2}^{T}\mathbf{M}_{R1}\mathbf{T}_{R2} \qquad \mathbf{S}_{R2} = \mathbf{T}_{R2}^{T}\mathbf{S}_{R1}\mathbf{T}_{R2} \tag{4.116}$$

Alternatively, substitution of Eq. (4.115) into Eq. (4.110) produces the relationship

$$\mathbf{D} = \mathbf{T}_{R2}^{*}\mathbf{D}_{R2} \tag{4.117}$$

where

$$\mathbf{T}_{R2}^{*} = \mathbf{T}_{R1}\mathbf{T}_{R2} \tag{4.118}$$

The combined operator $\mathbf{T}^*_{R2}$ [of size $n \times (n-2)$] expresses the dependency of $\mathbf{D}$ directly upon $\mathbf{D}_{R2}$, by virtue of multiplying $\mathbf{T}_{R1}$ [$n \times (n-1)$] and $\mathbf{T}_{R2}$ [$(n-1) \times (n-2)$]. With such a combined operator, the reduction of $\mathbf{M}$ and $\mathbf{S}$ directly to $\mathbf{M}_{R2}$ and $\mathbf{S}_{R2}$ becomes

$$\mathbf{M}_{R2} = (\mathbf{T}^*_{R2})^T \mathbf{M} \mathbf{T}^*_{R2} \qquad \mathbf{S}_{R2} = (\mathbf{T}^*_{R2})^T \mathbf{S} \mathbf{T}^*_{R2} \tag{4.119}$$

This technique can be extended to eliminate whatever number of rigid-body modes exist in a given system.

To illustrate the method, suppose that the stiffness constant k_1 for the first spring in Fig. 4.1*a* is zero. If $k_2 = k_3 = k$ and $m_1 = m_2 = m_3 = m$, the stiffness and mass matrices are

$$\mathbf{S} = k \begin{bmatrix} 1 & -1 & 0 \\ -1 & 2 & -1 \\ 0 & -1 & 1 \end{bmatrix} \qquad \mathbf{M} = m \begin{bmatrix} 1 & 0 & 0 \\ 0 & 1 & 0 \\ 0 & 0 & 1 \end{bmatrix}$$

By inspection we can see that the system is semidefinite and that the shape of the rigid-body mode is defined by the vector

$$\mathbf{\Phi}_{M1} = \{1, 1, 1\}$$

From this information, the 3×2 reduction-transformation matrix $\mathbf{T}_{R1}$ is found to be

$$\mathbf{T}_{R1} = \begin{bmatrix} -1 & -1 \\ 1 & 0 \\ 0 & 1 \end{bmatrix} \tag{i'}$$

Using this operator, we transform the mass and stiffness matrices, in accordance with Eqs. (4.112), as follows:

$$\mathbf{M}_{R1} = m \begin{bmatrix} -1 & 1 & 0 \\ -1 & 0 & 1 \end{bmatrix} \begin{bmatrix} 1 & 0 & 0 \\ 0 & 1 & 0 \\ 0 & 0 & 1 \end{bmatrix} \begin{bmatrix} -1 & -1 \\ 1 & 0 \\ 0 & 1 \end{bmatrix} = m \begin{bmatrix} 2 & 1 \\ 1 & 2 \end{bmatrix} \tag{j'}$$

$$\mathbf{S}_{R1} = k \begin{bmatrix} -1 & 1 & 0 \\ -1 & 0 & 1 \end{bmatrix} \begin{bmatrix} 1 & -1 & 0 \\ -1 & 2 & -1 \\ 0 & -1 & 1 \end{bmatrix} \begin{bmatrix} -1 & -1 \\ 1 & 0 \\ 0 & 1 \end{bmatrix} = k \begin{bmatrix} 5 & 1 \\ 1 & 2 \end{bmatrix} \tag{k'}$$

Inversion of the latter result produces the flexibility matrix

$$\mathbf{F}_{\mathrm{R1}} = \mathbf{S}_{\mathrm{R1}}^{-1} = \frac{\delta}{9}\begin{bmatrix} 2 & -1 \\ -1 & 5 \end{bmatrix} \tag{l'}$$

Next, we form the product given by Eq. (4.114) as

$$\mathbf{A}_{\mathrm{R1}} = \mathbf{F}_{\mathrm{R1}}\mathbf{M}_{\mathrm{R1}} = \frac{m\delta}{3}\begin{bmatrix} 1 & 0 \\ 1 & 3 \end{bmatrix} \tag{m'}$$

It is easy to verify (by iteration or direct solution) that the eigenvalues associated with this matrix are $\lambda_2 = m\delta$ and $\lambda_3 = m\delta/3$; and the corresponding eigenvectors are $(\mathbf{\Phi}_{\mathrm{R1}})_{\mathrm{M2}} = \{0, -1\}$ and $(\mathbf{\Phi}_{\mathrm{R1}})_{\mathrm{M3}} = \{-2, 1\}$. Transforming these vectors back to the original coordinates by Eq. (4.110), we obtain

$$\mathbf{\Phi}_{\mathrm{M2}} = \mathbf{T}_{\mathrm{R1}}\begin{bmatrix} 0 \\ -1 \end{bmatrix} = \begin{bmatrix} 1 \\ 0 \\ -1 \end{bmatrix} \qquad \mathbf{\Phi}_{\mathrm{M3}} = \mathbf{T}_{\mathrm{R1}}\begin{bmatrix} -2 \\ 1 \end{bmatrix} = \begin{bmatrix} 1 \\ -2 \\ 1 \end{bmatrix} \tag{n'}$$

which are the same as those given in the latter part of Example 1, Sec. 4.2.

4.8 DAMPING IN MULTIDEGREE SYSTEMS

The normal-mode method, as presented in Secs. 4.3–4.6, applies only to multidegree systems without damping. Often the effects of damping upon the response of a vibratory system are minor and can be ignored. For example, the influence of a small amount of damping upon the response of a system during an excitation of short duration is not likely to be significant. In addition, damping plays a minor role in the steady-state response of a system to a periodic forcing function when the frequency of the excitation is not near a resonance. However, for a periodic excitation with a frequency at or near a natural frequency, damping is of primary importance and must be taken into account. Because its effects are usually not known in advance, damping should ordinarily be included in a vibrational analysis until its importance is ascertained.

In Chapter 3 we discussed free and forced vibrations of two-degree systems with viscous damping, and we shall now consider damped systems having n degrees of freedom. When dissipative forces are provided by dashpot dampers, as in the three-mass system of Fig. 4.3, the action equations of motion may be written as

$$\mathbf{M}\ddot{\mathbf{D}} + \mathbf{C}\dot{\mathbf{D}} + \mathbf{S}\mathbf{D} = \mathbf{Q} \tag{4.120}$$

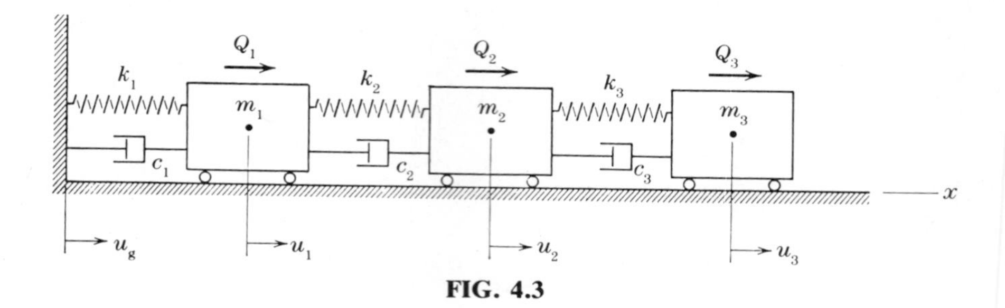

FIG. 4.3

in which the damping matrix **C** has the general form

$$\mathbf{C} = \begin{bmatrix} C_{11} & C_{12} & C_{13} & \cdots & C_{1n} \\ C_{21} & C_{22} & C_{23} & \cdots & C_{2n} \\ C_{31} & C_{32} & C_{33} & \cdots & C_{3n} \\ \cdots & \cdots & \cdots & \cdots & \cdots \\ C_{n1} & C_{n2} & C_{n3} & \cdots & C_{nn} \end{bmatrix} \tag{a}$$

The influence coefficients in this symmetric array were defined previously in Sec. 3.7.

We consider first the special systems for which the damping matrix happens to be linearly related to the mass and stiffness matrices. That is, we take

$$\mathbf{C} = a\mathbf{M} + b\mathbf{S} \tag{4.121}$$

where a and b are constants. This type of damping is called *proportional damping* because **C** is proportional to a linear combination of **M** and **S**. In such a case the equations of motion (4.120) are uncoupled by the same transformation as that for the undamped system [3]. Thus, in principal coordinates we have

$$\mathbf{M}_{\mathrm{P}}\ddot{\mathbf{D}}_{\mathrm{P}} + \mathbf{C}_{\mathrm{P}}\dot{\mathbf{D}}_{\mathrm{P}} + \mathbf{S}_{\mathrm{P}}\mathbf{D}_{\mathrm{P}} = \mathbf{Q}_{\mathrm{P}} \tag{4.122}$$

where

$$\mathbf{C}_{\mathrm{P}} = \mathbf{\Phi}_{\mathrm{M}}^{\mathrm{T}}\mathbf{C}\mathbf{\Phi}_{\mathrm{M}} = a\mathbf{M}_{\mathrm{P}} + b\mathbf{S}_{\mathrm{P}} \tag{b}$$

The symbol $\mathbf{C}_{\mathrm{P}}$ represents a diagonal array that will be referred to as a *principal damping matrix*, and it consists of a linear combination of $\mathbf{M}_{\mathrm{P}}$ and $\mathbf{S}_{\mathrm{P}}$. When the modal matrix is normalized with respect to **M**, the damping matrix in normal coordinates becomes

$$\mathbf{C}_{\mathrm{N}} = \mathbf{\Phi}_{\mathrm{N}}^{\mathrm{T}}\mathbf{C}\mathbf{\Phi}_{\mathrm{N}} = a\mathbf{I} + b\boldsymbol{\omega}^2 \tag{4.123}$$

The diagonal matrix $\boldsymbol{\omega}^2$ in this expression contains the characteristic values ω_i^2 for the undamped case [see Eq. (4.36), Sec. 4.3]. Therefore, the ith equation of motion in normal coordinates is

$$\ddot{u}_{Ni} + (a + b\omega_i^2)\dot{u}_{Ni} + \omega_i^2 u_{Ni} = q_{Ni} \qquad (i = 1, 2, 3, \ldots, n) \tag{c}$$

To make this expression analogous to that for a one-degree system [see Eq. (1.61), Sec. 1.12], we introduce the notations

$$C_{Ni} = 2n_i = a + b\omega_i^2 \qquad \gamma_i = \frac{n_i}{\omega_i} \tag{d}$$

where $C_{Ni} = 2n_i$ is defined as the *modal damping constant* for the ith normal mode, and γ_i represents the corresponding *modal damping ratio.* Using the first of these definitions in Eq. (c), we obtain

$$\ddot{u}_{Ni} + 2n_i\dot{u}_{Ni} + \omega_i^2 u_{Ni} = q_{Ni} \qquad (i = 1, 2, 3, \ldots, n) \tag{4.124}$$

Each of the n equations represented by this expression is uncoupled from all of the others. Therefore, we can determine the response of the ith mode in the same manner as that for a one-degree system with viscous damping.

From the definitions (d) we may express the modal damping ratio γ_i in terms of the constants a and b, as follows:

$$\gamma_i = \frac{a + b\omega_i^2}{2\omega_i} \tag{4.125}$$

This relationship is useful for studying the effects upon the modal damping of varying the constants a and b in Eq. (4.121). For example, setting the constant a equal to zero (while b is nonzero) implies that the damping matrix is proportional to the stiffness matrix. This type of damping is sometimes referred to as *relative damping* because it is associated with relative velocities of displacement coordinates. Thus, under the condition that $a = 0$, eq. (4.125) becomes

$$\gamma_i = \frac{b}{2}\omega_i \tag{e}$$

which means that the damping ratio in each principal mode is proportional to the undamped angular frequency of that mode. Therefore, the responses of the higher modes of a system will be damped out more rapidly than those of the lower modes.

On the other hand, setting b equal to zero (while a is nonzero) implies that the damping matrix is proportional to the mass matrix. This type of damping is sometimes called *absolute damping* because it is associated with absolute velocities of displacement coordinates. In this case Eq. (4.125) simplifies to

$$\gamma_i = \frac{a}{2\omega_i} \tag{f}$$

so that the damping ratio in each mode is inversely proportional to the undamped frequency. Under this condition the lower modes of a system will be suppressed more strongly than the higher modes.

It has been shown [7] that the condition given by Eq. (4.121) is sufficient but not necessary for the existence of principal modes in damped systems. The essential condition for the existence of principal modes is that the transformation that diagonalizes the damping matrix also uncouples the equations of motion. This condition is less restrictive than Eq. (4.121) and encompasses more possibilities.

However, in the most general case, the damping influence coefficients are such that the damping matrix cannot be diagonalized simultaneously with the mass and stiffness matrices. As was shown in Sec. 3.7, the natural modes which do exist have phase relationships that complicate the analysis. The eigenvalues for this type of system are either real and negative or complex with negative real parts. The complex eigenvalues occur as conjugate pairs [see Eqs. (3.42a) and (3.42b), Sec. 3.7], and the corresponding eigenvectors also consist of complex conjugate pairs. In highly damped systems, where the imaginary terms due to dissipative forces are significant, the method of Foss [8] may be utilized. This approach involves transformation of the n second-order equations of motion into $2n$ uncoupled first-order equations.

Lightly damped systems need not be treated in such a complicated manner, especially in view of the fact that the nature of damping in physical systems is not well understood. The simplest approach consists of assuming that the equations of motion are uncoupled by the modal matrix obtained for the system without damping. In other words, the matrix $\mathbf{\Phi}_{\mathrm{M}}$ is assumed to be orthogonal with respect to not only $\mathbf{M}$ and $\mathbf{S}$ (see Eqs. 4.23 and 4.24) but also $\mathbf{C}$, as follows:

$$\mathbf{\Phi}_{\mathrm{M}j}^{\mathrm{T}}\mathbf{C}\mathbf{\Phi}_{\mathrm{M}i} = \mathbf{\Phi}_{\mathrm{M}i}^{\mathrm{T}}\mathbf{C}\mathbf{\Phi}_{\mathrm{M}j} = 0 \qquad (i \neq j) \tag{4.126}$$

This assumption implies that any off-diagonal terms resulting from the operation $\mathbf{C}_{\mathrm{P}} = \mathbf{\Phi}_{\mathrm{M}}^{\mathrm{T}}\mathbf{C}\mathbf{\Phi}_{\mathrm{M}}$ are small and can be neglected. In addition, it is usually more convenient to obtain experimentally (or to assume) the damping ratio γ_i for the natural modes of vibration than it is to determine the damping influence coefficients in matrix $\mathbf{C}$. Therefore, we rewrite Eq. (4.124) in terms of γ_i as

$$\ddot{u}_{\mathrm{N}i} + 2\gamma_i\omega_i\dot{u}_{\mathrm{N}i} + \omega_i^2 u_{\mathrm{N}i} = q_{\mathrm{N}i} \qquad (i = 1, 2, 3, \ldots, n) \tag{4.127}$$

In order that this expression may pertain to a lightly damped system, we shall specify that $0 \leqq \gamma_i \leqq 0.20$ for all modes. The type of damping associated with this set of assumptions is of great practical value, and it will be referred to simply as *modal damping*. It should be remembered that this concept is based on the normal coordinates for the undamped system and that damping ratios are specified in those coordinates.

When modal damping is assumed in the normal coordinates of a system, it may also be of interest to determine the damping matrix in the original coordinates. This array can be found by means of the reverse transformation

$$\mathbf{C} = (\mathbf{\Phi}_{\mathrm{N}}^{-1})^{\mathrm{T}} \mathbf{C}_{\mathrm{N}} \mathbf{\Phi}_{\mathrm{N}}^{-1} \tag{g}$$

Instead of attempting to invert $\mathbf{\Phi}_{\mathrm{N}}$, however, we use the relationship $\mathbf{\Phi}_{\mathrm{N}}^{-1} = \mathbf{\Phi}_{\mathrm{N}}^{\mathrm{T}} \mathbf{M}$ [see Eq. (4.44b), Sec. 4.3] and rewrite Eq. (g) as

$$\mathbf{C} = \mathbf{M} \mathbf{\Phi}_{\mathrm{N}} \mathbf{C}_{\mathrm{N}} \mathbf{\Phi}_{\mathrm{N}}^{\mathrm{T}} \mathbf{M} \tag{4.128}$$

This form of the transformation is necessary when not all of the natural modes are included in the analysis (modal truncation).

4.9 DAMPED RESPONSE TO PERIODIC FORCING FUNCTIONS

As mentioned in the preceding section, damping is of greatest importance when a periodic excitation has a frequency that is close to one of the natural frequencies of a multidegree system. The topic of steady-state forced vibrations of two-degree systems was treated in Sec. 3.8 by the transfer-function method. This approach can easily be extended to systems with n degrees of freedom, in which case the basic relationships [see Eqs. (3.51) and (3.52)] remain symbolically the same. However, the solution by this technique requires the inversion of a matrix of order $n \times n$, containing complex terms. If the eigenvalues and eigenvectors of a system have already been obtained, the normal-mode method provides a useful alternative to the transfer-function approach. Knowing the frequency of the excitation and the natural frequencies of the system, we can calculate in a direct manner the steady-state responses of the modes having frequencies in the vicinity of the imposed frequency. Both simple harmonic and general periodic forcing functions will be discussed; and either proportional damping or modal damping will be assumed, as described in the preceding section.

If a lightly damped system is subjected to a set of actions that are all proportional to the simple harmonic function $\cos \Omega t$, the action vector $\mathbf{Q}$ may be written as

$$\mathbf{Q} = \mathbf{P} \cos \Omega t \tag{a}$$

where

$$\mathbf{P} = \{P_1, P_2, P_3, \ldots, P_n\} \tag{b}$$

In Eq. (a) the values of P act as scale factors upon the function $\cos \Omega t$. Transformation of the action equations of motion to normal coordinates produces the typical modal equation

$$\ddot{u}_{\mathrm{N}i} + 2n_i\dot{u}_{\mathrm{N}i} + \omega_i^2 u_{\mathrm{N}i} = q_{\mathrm{N}i} \cos \Omega t \qquad (i = 1, 2, 3, \ldots, n) \tag{4.129}$$

in which $q_{\mathrm{N}i}$ is a constant. This equation has the same form as Eq. (1.42) in Sec. 1.9. Therefore, we can take the damped steady-state response of the ith mode to be

$$u_{\mathrm{N}i} = \frac{q_{\mathrm{N}i}}{\omega_i^2} \beta_i \cos(\Omega t - \theta_i) \tag{4.130}$$

in which the magnification factor β_i is

$$\beta_i = \frac{1}{\sqrt{(1 - \Omega^2/\omega_i^2)^2 + (2\gamma_i \Omega/\omega_i)^2}} \tag{4.131}$$

and the phase angle θ_i is

$$\theta_i = \tan^{-1}\left(\frac{2\gamma_i \Omega/\omega_i}{1 - \Omega^2/\omega_i^2}\right) \tag{4.132}$$

Equations (4.130), (4.131), and (4.132) are drawn from Eqs. (1.46), (1.47), and (1.48), respectively. The response given by Eq. (4.130) may then be transformed back to the original coordinates in the usual manner.

To determine the response of the mode having its angular frequency ω_i closest to the impressed angular frequency, we need only use the modal column $\mathbf{\Phi}_{\mathrm{N}i}$ in the transformations to and from normal coordinates. That is, Eq. (4.64) in Sec. 4.5 is specialized to

$$q_{\mathrm{N}i} = \mathbf{\Phi}_{\mathrm{N}i}^{\mathrm{T}} \mathbf{P} \tag{4.133}$$

and Eq. (4.58) in Sec. (4.4) becomes

$$\mathbf{D} = \mathbf{\Phi}_{\mathrm{N}i} u_{\mathrm{N}i} \tag{4.134}$$

If desired, this process can be repeated for other modes with frequencies in the vicinity of Ω.

When displacement equations of motion are used instead of action equations, the vector of harmonic displacements $\mathbf{\Delta}$ becomes

$$\mathbf{\Delta} = \mathbf{FQ} = \mathbf{FP} \cos \Omega t = \mathbf{\Delta}_{st} \cos \Omega t \tag{c}$$

where $\mathbf{\Delta}_{st}$ represents a vector of displacements due to static application of the actions in $\mathbf{P}$. Because displacements are transformed to normal coordinates using the inverse operator $\mathbf{\Phi}_N^{-1} = \mathbf{\Phi}_N^T\mathbf{M}$, the counterpart of Eq. (4.133) becomes

$$\delta_{Ni} = \mathbf{\Phi}_{Ni}^T \mathbf{M} \mathbf{\Delta}_{st} \tag{4.135}$$

where δ_{Ni} is a constant. The ith equivalent normal-mode load, as given by Eq. (4.77) in Sec. 4.5, is

$$q_{N\delta i} = \omega_i^2 \delta_{Ni} \tag{d}$$

Therefore, the damped steady-state response of the ith mode [see Eq. (4.130)] becomes

$$u_{Ni} = \delta_{Ni} \beta_i \cos(\Omega t - \theta_i) \tag{4.136}$$

and transformation of this response back to the original coordinates remains as in Eq. (4.134).

Now let us consider a lightly damped system subjected to a set of actions that are all proportional to the general periodic function $f(t)$. In this case the action vector $\mathbf{Q}$ may be written as

$$\mathbf{Q} = \mathbf{F}(t) = \mathbf{P} f(t) \tag{e}$$

where the vector $\mathbf{P}$ is given by Eq. (b). Proceeding as described in Sec. 1.11, we express $f(t)$ in the form of a Fourier series [see Eq. (1.58)], as follows:

$$f(t) = a_0 + \sum_{j=1}^{\infty} (a_j \cos j\Omega t + b_j \sin j\Omega t) \tag{4.137}$$

The coefficients a_j, b_j, and a_0 in this expression may be evaluated in the manner indicated by Eqs. (1.59a, b, and c).

Transformation of the action equations of motion to normal coordinates produces the typical modal equation

$$\ddot{u}_{Ni} + 2n_i \dot{u}_{Ni} + \omega_i^2 u_{Ni} = q_{Ni} f(t) \qquad (i = 1, 2, 3, \ldots, n) \tag{4.138}$$

where q_{Ni} is a constant. From the solution of Prob. 1.11-6 at the end of the book, we take the damped steady-state response of the ith mode to be

$$u_{Ni} = \frac{q_{Ni}}{\omega_i^2}\left\{a_0 + \sum_{j=1}^{\infty} \beta_{ij}[a_j \cos(j\Omega t - \theta_{ij}) + b_j \sin(j\Omega t - \theta_{ij})]\right\} \tag{4.139}$$

in which the magnification factor β_{ij} is

$$\beta_{ij} = \frac{1}{\sqrt{(1 - j^2\Omega^2/\omega_i^2)^2 + (2\gamma_i j\Omega/\omega_i)^2}} \tag{4.140}$$

and the phase angle θ_{ij} is

$$\theta_{ij} = \tan^{-1}\left(\frac{2\gamma_i j\Omega/\omega_i}{1 - j^2\Omega^2/\omega_i^2}\right) \tag{4.141}$$

Because a multiplicity of terms contribute to the response of the ith mode in Eq. (4.139), the possibility of resonance ($j\Omega \approx \omega_i$) is much greater for a general periodic function than for a simple harmonic function. Therefore, it becomes somewhat more difficult to predict in advance which of the natural modes will be strongly affected. However, after the forcing function has been expressed as a Fourier series, each of the $j\Omega$ frequencies can be compared with each of the ω_i frequencies for the purpose of predicting large forced vibrations.

If displacement equations of motion are used, the vector of periodic displacements $\boldsymbol{\Delta}$ is given by

$$\boldsymbol{\Delta} = \boldsymbol{\Delta}_{\text{st}} f(t) \tag{f}$$

where $\boldsymbol{\Delta}_{\text{st}}$ retains its previous definition. Following the same line of reasoning as before, we conclude that the constant q_{Ni}/ω_i^2 in Eq. (4.139) is replaced by δ_{Ni}. In any case, the results are transformed back to the original coordinates in accordance with Eq. (4.134).

Example 1. Suppose that the system in Fig. 4.3 is subjected to the simple harmonic forcing functions $Q_1 = Q_2 = Q_3 = P \cos \Omega t$, where $\Omega = 1.25\sqrt{k/m}$. Calculate the steady-state response of the masses, assuming that $m_1 = m_2 = m_3 = m$, $k_1 = k_2 = k_3 = k$, and the damping ratio in each principal mode is $\gamma_i = 0.01$ ($i = 1, 2, 3$).

Solution. The square of the impressed angular frequency ($\Omega^2 = 1.5625k/m$) is very close to the second eigenvalue of the system ($\omega_2^2 = 1.555k/m$), obtained in Example 1 of Sec. 4.2. Therefore, we should expect the second mode of the system to be the primary contributor to the response, even though the pattern of the loads is similar to the shape of the first mode. Using Eq. (4.133), we determine the second normal-mode load to be

$$q_{N2} = \mathbf{\Phi}_{N2}^{T}\mathbf{P} = \frac{1}{\sqrt{m}}[0.737 \quad 0.328 \quad -0.591]\begin{bmatrix} 1 \\ 1 \\ 1 \end{bmatrix} P = 0.474\frac{P}{\sqrt{m}}$$

The magnification factor for the second mode is obtained from Eq. (4.131) as

$$\beta_2 = \frac{1}{\sqrt{(1 - 1.5625/1.555)^2 + (0.02)^2(1.5625/1.555)}}$$

$$= \frac{1}{\sqrt{(0.004823)^2 + (0.02006)^2}} = 48.50$$

From Eq. (4.130), we find the damped steady-state response of the second mode to be

$$u_{N2} = \frac{(0.474)(48.50)}{(1.555)}\frac{P}{k}\sqrt{m}\cos(\Omega t - \theta_2)$$

$$= 14.77\frac{P}{k}\sqrt{m}\cos(\Omega t - \theta_2)$$

where

$$\theta_2 = \tan^{-1}\left(\frac{0.02006}{0.004823}\right) = \tan^{-1} 4.159 = 76°29'$$

as given by Eq. (4.132).

Transforming the response of the second mode back to the original coordinates with Eq. (4.134), we obtain

$$\mathbf{D} = \mathbf{\Phi}_{N2}u_{N2} = \begin{bmatrix} 10.89 \\ 4.84 \\ -8.73 \end{bmatrix}\frac{P}{k}\cos(\Omega t - \theta_2) \tag{g}$$

It is seen that the small amount of damping in this system has a great effect on the response of the second mode. If the system were undamped, the magnification factor would be $\beta_2 = 1/0.004823 = 207.3$, and the phase angle would be $\theta_2 = 0$.

Proceeding in a similar manner, we can determine the response contributed by the first mode as

$$\mathbf{D} = \begin{bmatrix} 0.398 \\ 0.717 \\ 0.895 \end{bmatrix}\frac{P}{k}\cos(\Omega t - \theta_1) \tag{h}$$

and that due to the third mode is found to be

$$\mathbf{D} = \begin{bmatrix} 0.0637 \\ -0.0794 \\ 0.0353 \end{bmatrix} \frac{P}{k} \cos(\Omega t - \theta_3) \tag{i}$$

The amplitudes in both of these vectors are small compared to those in Eq. (g). Furthermore, the influence of damping upon the results in Eqs. (h) and (i) is negligible.

Example 2. Figure 4.4 shows a periodic forcing function in the shape of a square wave. If this function is applied to the first mass in Fig. 4.3, determine the damped steady-state response for each of the normal modes.
Solution. Expanding the square wave as a Fourier series (see Prob. 1.11-2, Sec. 1.11), we obtain

$$F(t) = Pf(t) = \frac{4P}{\pi}\left(\sin \Omega t + \frac{1}{3} \sin 3\Omega t + \cdots\right) \tag{j}$$

Transformation of the load vector to normal coordinates produces

$$\mathbf{Q}_{\mathrm{N}} = \mathbf{\Phi}_{\mathrm{N}}^{\mathrm{T}}\mathbf{Q} = \mathbf{\Phi}_{\mathrm{N}}^{\mathrm{T}} \begin{bmatrix} F(t) \\ 0 \\ 0 \end{bmatrix} = \begin{bmatrix} \Phi_{\mathrm{N}11} \\ \Phi_{\mathrm{N}12} \\ \Phi_{\mathrm{N}13} \end{bmatrix} F(t) \tag{k}$$

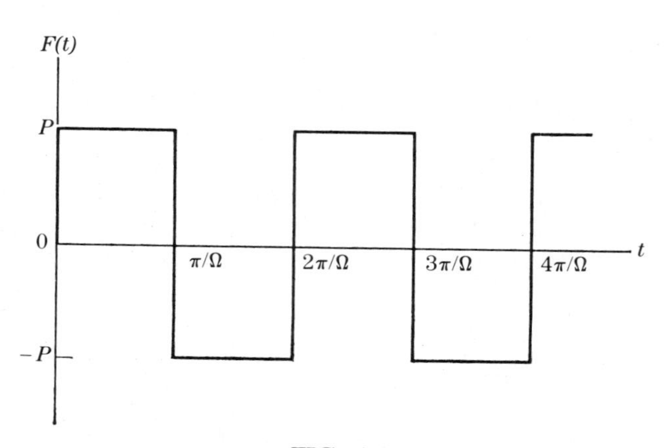

FIG. 4.4

In accordance with Eq. (4.139), the normal-mode responses become

$$\mathbf{D}_{\mathrm{N}} = \frac{4P}{\pi} \begin{bmatrix} \Phi_{\mathrm{N}11}\left[\beta_{11}\sin(\Omega t - \theta_{11}) + \dfrac{\beta_{13}}{3}\sin(3\Omega t - \theta_{13}) + \cdots\right] \Big/ \omega_1^2 \\ \Phi_{\mathrm{N}12}\left[\beta_{21}\sin(\Omega t - \theta_{21}) + \dfrac{\beta_{23}}{3}\sin(3\Omega t - \theta_{23}) + \cdots\right] \Big/ \omega_2^2 \\ \Phi_{\mathrm{N}13}\left[\beta_{31}\sin(\Omega t - \theta_{31}) + \dfrac{\beta_{33}}{3}\sin(3\Omega t - \theta_{33}) + \cdots\right] \Big/ \omega_3^2 \end{bmatrix} \tag{l}$$

where the magnification factors and phase angles are given by Eqs. (4.140) and (4.141), respectively.

4.10 TRANSIENT RESPONSE OF DAMPED SYSTEMS

The effects of damping should be included in calculations for transient response whenever the time of interest is relatively long in comparison with the natural periods of a system. If the time of interest is short but the modal damping ratios are relatively high ($\gamma_i > 0.05$), the presence of damping still might be of some significance. Therefore, we shall modify the formulations in Secs. 4.4, 4.5, and 4.6 to account for the influence of damping upon transient response in normal coordinates. As in the preceding sections, either proportional damping or modal damping will be assumed throughout the discussion.

In Sec. 4.4, we formulated the normal-mode responses of a multi-degree system to intitial conditions of displacement and velocity. In the presence of damping, the free-vibrational response of the ith mode, given by Eq. (4.55), must be changed to

$$u_{\mathrm{N}i} = e^{-n_i t}\left(u_{\mathrm{N}0i}\cos\omega_{\mathrm{d}i}t + \frac{\dot{u}_{\mathrm{N}0i} + n_i u_{\mathrm{N}0i}}{\omega_{\mathrm{d}i}}\sin\omega_{\mathrm{d}i}t\right) \tag{4.142}$$

which is drawn from Eq. (1.35) in Sec. 1.8. The angular frequency of damped vibration in Eq. (4.142) is

$$\omega_{\mathrm{d}i} = \sqrt{\omega_i^2 - n_i^2} = \omega_i\sqrt{1 - \gamma_i^2} \tag{a}$$

wherein ω_i represents the undamped angular frequency. Transformation of the initial-condition vectors $\mathbf{D}_0$ and $\dot{\mathbf{D}}_0$ to normal coordinates remains the same as in Eqs. (4.56) and (4.57), and back-transformation of the response is still given by Eq. (4.58).

Rigid-body motions will not be affected by either modal damping or relative damping (in proportion to the stiffness matrix). However, absolute damping (in proportion to the mass matrix) will influence rigid-body modes, and for this particular type of damping Eq. (4.59) in Sec. 4.4 is replaced by

$$\ddot{u}_{Ni} + a\dot{u}_{Ni} = 0 \tag{4.143}$$

The solution of Eq. (4.143) is

$$u_{Ni} = u_{N0i} + \dot{u}_{N0i}\left(\frac{1 - e^{-at}}{a}\right) \tag{4.144}$$

This expression may be used in place of Eq. (4.60) to evaluate rigid-body motions for systems with absolute damping. Note that if a is set equal to zero (for no damping), the term in parentheses in Eq. (4.144) becomes 0/0. In the limit, this indeterminate form is

$$\lim_{a \to 0}\left(\frac{1 - e^{-at}}{a}\right) = t \tag{b}$$

and the response expression reverts to Eq. (4.60).

Similarly, the calculation of normal-mode responses to applied actions, as described in Sec. 4.5, requires only a few modifications in the presence of either proportional or modal damping. Transformation of applied actions to normal coordinates remains the same as in Eq. (4.64), but Duhamel's integral in Eq. (4.67) must be changed to

$$u_{Ni} = \frac{e^{-n_i t}}{\omega_{di}} \int_0^t e^{n_i t'} q_{Ni} \sin \omega_{di}(t - t')\, dt' \tag{4.145}$$

which is taken from Eq. (1.62) in Sec. 1.12. If absolute damping is assumed, the rigid-body equation of motion (4.68) is replaced by

$$\ddot{u}_{Ni} + a\dot{u}_{Ni} = q_{Ni} \tag{4.146}$$

For the system initially at rest, the solution of Eq. (4.146) has the form

$$u_{Ni} = \int_0^t q_{Ni}\left(\frac{1 - e^{-a(t-t')}}{a}\right) dt' \tag{4.147}$$

which replaces Eq. (4.69).

If displacement equations of motion are used instead of action equations, the transformation of the displacement vector $\mathbf{\Delta}$ to normal coordinates is given by Eq. (4.73) in Sec. 4.5. In addition, the ith equivalent normal-mode load is

$$q_{\mathrm{N}\delta i} = \omega_i^2 \delta_{\mathrm{N}i} \tag{c}$$

which is taken from Eq. (4.77). Therefore, the Duhamel integral in Eq. (4.145) becomes

$$u_{\mathrm{N}i} = \frac{\omega_i^2}{\omega_{\mathrm{d}i}} e^{-n_i t} \int_0^t e^{n_i t'} \, \delta_{\mathrm{N}i} \sin \omega_{\mathrm{d}i}(t - t') \, dt' \tag{4.148}$$

This form of the integral is particularly useful for evaluating the damped response of systems for which time-varying displacements $\mathbf{\Delta}$ are easily found by static analysis, as was demonstrated in Sec. 4.5.

Normal-mode responses to support motions, as described in Sec. 4.6, may also be modified to account for either proportional or modal damping. In many cases it is only necessary to replace $q_{\mathrm{N}i}$ and δ_{Ni} in Eqs. (4.145) and (4.148) by their counterparts $q_{\mathrm{N}gi}$, $q^*_{\mathrm{N}gi}$, $\delta_{\mathrm{N}gi}$, $\delta^*_{\mathrm{N}gi}$, $q_{\mathrm{NR}i}$, or $\delta_{\mathrm{NR}i}$, which were defined in Sec. 4.6. However, in certain cases where support displacements are imposed, the vectors $\mathbf{Q}_{\mathrm{g}}$, $\mathbf{\Delta}_{\mathrm{g}}$, $\mathbf{Q}_{\mathrm{R}}$, or $\mathbf{\Delta}_{\mathrm{R}}$ in the original coordinates must be modified to include velocity coupling between free displacement coordinates and support restraints. For example, the system in Fig. 4.3 has a dashpot connection to ground. If the ground translates in the x direction in accordance with $u_{\mathrm{g}} = F_{\mathrm{g}}(t)$, the action equations of motion become

$$\mathbf{M}\ddot{\mathbf{D}} + \mathbf{C}\dot{\mathbf{D}} + \mathbf{C}_{\mathrm{g}}\dot{u}_{\mathrm{g}} + \mathbf{S}\mathbf{D} + \mathbf{S}_{\mathrm{g}}u_{\mathrm{g}} = \mathbf{0} \tag{d}$$

in which $\mathbf{C}_{\mathrm{g}}$ denotes a vector of damping influence coefficients coupling the masses with ground. Putting Eq. (d) into the form of Eq. (4.120), we obtain

$$\mathbf{M}\ddot{\mathbf{D}} + \mathbf{C}\dot{\mathbf{D}} + \mathbf{S}\mathbf{D} = \mathbf{Q}_{\mathrm{g}} = \mathbf{Q}_{\mathrm{g}1} + \mathbf{Q}_{\mathrm{g}2} \tag{4.149}$$

where

$$\mathbf{Q}_{\mathrm{g}1} = -\mathbf{S}_{\mathrm{g}}u_{\mathrm{g}} \qquad \mathbf{Q}_{\mathrm{g}2} = -\mathbf{C}_{\mathrm{g}}\dot{u}_{\mathrm{g}} \tag{4.150}$$

The vectors $\mathbf{Q}_{\mathrm{g}1}$ and $\mathbf{Q}_{\mathrm{g}2}$ contain equivalent loads (corresponding to the displacement coordinates) due to ground displacement and ground velocity, respectively. These terms can be kept separated throughout the analysis if desired, in which case the transformation to normal coordinates produces

$$\mathbf{Q}_{\mathrm{Ng}} = \mathbf{\Phi}_{\mathrm{N}}^{\mathrm{T}}\mathbf{Q}_{\mathrm{g}} = \mathbf{\Phi}_{\mathrm{N}}^{\mathrm{T}}\mathbf{Q}_{\mathrm{g}1} + \mathbf{\Phi}_{\mathrm{N}}^{\mathrm{T}}\mathbf{Q}_{\mathrm{g}2} = \mathbf{Q}_{\mathrm{Ng}1} + \mathbf{Q}_{\mathrm{Ng}2} \tag{4.151}$$

Then the ith equation of motion in normal coordinates takes the form

$$\ddot{u}_{\mathrm{N}i} + 2n_i\dot{u}_{\mathrm{N}i} + \omega_i^2 u_{\mathrm{N}i} = q_{\mathrm{Ng}i} = q_{\mathrm{Ng}i1} + q_{\mathrm{Ng}i2} \qquad (i = 1, 2, 3, \ldots, n) \quad (4.152)$$

which is analogous to Eq. (1.68) in Sec. 1.13. Finally, the Duhamel integral in Eq. (4.145) becomes

$$u_{\mathrm{N}i} = u_{\mathrm{N}i1} + u_{\mathrm{N}i2} = \frac{e^{-n_i t}}{\omega_{\mathrm{d}i}} \int_0^t e^{n_i t'} (q_{\mathrm{Ng}i1} + q_{\mathrm{Ng}i2}) \sin \omega_{\mathrm{d}i}(t - t')\, dt' \quad (4.153)$$

which may be compared with Eq. (1.69) in Sec. 1.13.

The complication explained in the example above for velocity-coupled ground motions can be avoided by working with ground accelerations and changing coordinates to the relative displacements $\mathbf{D}^* = \mathbf{D} - \mathbf{1}u_{\mathrm{g}}$. As demonstrated for the one-degree system in Sec. 1.13 [see Eqs. (1.71) and (1.72)], there is neither displacement coupling nor velocity coupling between the masses and the ground in the relative coordinates. There exists only inertial coupling with ground, which is the same as that for the system without damping. On the other hand, if r independent motions of support restraints are involved, the concept of rigid-body ground motion must be abandoned. In such a case, any velocity coupling that exists between free displacement coordinates and support restraints must be taken into account directly.

Example 1. Let the the third mass of the system in Fig. 4.3 be subjected to the step function $Q_3 = P$, while $Q_1 = Q_2 = 0$. Assuming that the system is initially at rest, determine the damped responses of its normal modes due to this applied action.

Solution. Transforming the load to normal coordinates, we obtain

$$\mathbf{Q}_{\mathrm{N}} = \mathbf{\Phi}_{\mathrm{N}}^{\mathrm{T}}\mathbf{Q} = \mathbf{\Phi}_{\mathrm{N}}^{\mathrm{T}} \begin{bmatrix} 0 \\ 0 \\ P \end{bmatrix} = \begin{bmatrix} \Phi_{\mathrm{N}31} \\ \Phi_{\mathrm{N}32} \\ \Phi_{\mathrm{N}33} \end{bmatrix} P \qquad (e)$$

Due to the step function, the normal-mode responses (see Example 3, Sec. 1.12) are

$$\mathbf{D}_{\mathrm{N}} = P \begin{bmatrix} \Phi_{\mathrm{N}31}\left[1 - e^{-n_1 t}\left(\cos \omega_{\mathrm{d}1}t + \dfrac{n_1}{\omega_{\mathrm{d}1}} \sin \omega_{\mathrm{d}1}t\right)\right] \Big/ \omega_1^2 \\ \Phi_{\mathrm{N}32}\left[1 - e^{-n_2 t}\left(\cos \omega_{\mathrm{d}2}t + \dfrac{n_2}{\omega_{\mathrm{d}2}} \sin \omega_{\mathrm{d}2}t\right)\right] \Big/ \omega_2^2 \\ \Phi_{\mathrm{N}33}\left[1 - e^{-n_3 t}\left(\cos \omega_{\mathrm{d}3}t + \dfrac{n_3}{\omega_{\mathrm{d}3}} \sin \omega_{\mathrm{d}3}t\right)\right] \Big/ \omega_3^2 \end{bmatrix} \qquad (f)$$

Example 2. Suppose that the ground in Fig. 4.3 translates in accordance with the ramp function $u_g = d_1 t/t_1$, where d_1 is the rigid-body ground translation at time t_1. Formulate the damped responses of the normal modes, assuming that the system is initially at rest.
Solution. The equivalent load vectors, as given by Eqs. (4.150), are

$$\mathbf{Q}_{g1} = -\mathbf{S}_g u_g = \begin{bmatrix} k_1 \\ 0 \\ 0 \end{bmatrix} \frac{d_1 t}{t_1} \qquad \mathbf{Q}_{g2} = -\mathbf{C}_g \dot{u}_g = \begin{bmatrix} c_1 \\ 0 \\ 0 \end{bmatrix} \frac{d_1}{t_1} \tag{g}$$

Transformation of these vectors to normal coordinates (see Eq. 4.151) produces

$$\mathbf{Q}_{Ng} = \mathbf{Q}_{Ng1} + \mathbf{Q}_{Ng2} = \frac{k_1 d_1 t}{t_1} \begin{bmatrix} \Phi_{N11} \\ \Phi_{N12} \\ \Phi_{N13} \end{bmatrix} + \frac{c_1 d_1}{t_1} \begin{bmatrix} \Phi_{N11} \\ \Phi_{N12} \\ \Phi_{N13} \end{bmatrix} \tag{h}$$

The normal-mode responses due to the ramp functions in $\mathbf{Q}_{Ng1}$ are represented by

$$u_{Ni1} = \frac{k_1 d_1 \Phi_{N1i}}{t_1 \omega_i^2} \left[t - \frac{2n_i}{\omega_i^2} + e^{-n_i t} \left(\frac{2n_i}{\omega_i^2} \cos \omega_{di} t - \frac{\omega_{di}^2 - n_i^2}{\omega_i^2 \omega_{di}} \sin \omega_{di} t \right) \right] \tag{i}$$

where $i = 1, 2, 3$ (see Prob. 1.12-9, Sec. 1.12). In addition, the step functions in $\mathbf{Q}_{Ng2}$ cause the normal-mode responses given by (see Example 1)

$$u_{Ni2} = \frac{c_1 d_1 \Phi_{N1i}}{t_1 \omega_i^2} \left[1 - e^{-n_i t} \left(\cos \omega_{di} t + \frac{n_i}{\omega_{di}} \sin \omega_{di} t \right) \right] \tag{j}$$

where $i = 1, 2, 3$.

4.11 STEP-BY-STEP RESPONSE CALCULATIONS

In Sec. 1.15 we discussed step-by-step solutions for systems with one degree of freedom subjected to forcing functions that are not analytical expressions. The approach in that section was to use piecewise-linear interpolation of the loads. This technique will now be incorporated into the normal-mode method for calculating responses of multidegree systems. We shall assume only modal damping, as described previously in Sec. 4.8. Because of the extensive calculations required, it is implied that the method of this section is to be programmed on a digital computer.

Consider again the piecewise-linear type of interpolation illustrated by Fig. 1.55. Without loss of generality, we will handle only one such forcing function at a time. Then the piecewise-linear action vector $\mathbf{Q}_{\ell j}$ may be expressed as

$$\mathbf{Q}_{\ell j} = \mathbf{P} f_{\ell}(\Delta t_j) \qquad (j+1 = 1, 2, \ldots, n_1) \tag{4.154}$$

where Δt_j represents a small but finite time step, and n_1 is the number of steps. In this form the values of $\mathbf{P}$ act as scale factors on the dimensionless piecewise-linear forcing function $f_{\ell}(\Delta t_j)$. If more than one function is applied simultaneously, the responses for each of them handled separately can be superimposed.

Transformation of the action equations of motion to normal coordinates produces the typical modal equation

$$\ddot{u}_{\mathrm{N}i} + 2n_i \dot{u}_{\mathrm{N}i} + \omega_i^2 u_{\mathrm{N}i} = q_{\mathrm{N}i,j} + \Delta q_{\mathrm{N}i,j} \frac{t'}{\Delta t_j} \qquad (i = 1, 2, \ldots, n) \quad \text{and} \quad (j+1 = 1, 2, \ldots, n_1) \tag{4.155}$$

where $t' = t - t_j$. The symbol $q_{\mathrm{N}i,j}$ in Eq. (4.155) represents the ith normal-mode load at time t_j. Thus,

$$q_{\mathrm{N}i,j} = \mathbf{\Phi}_{\mathrm{N}i}^{\mathrm{T}} \mathbf{Q}_{\ell j} = \mathbf{\Phi}_{\mathrm{N}i}^{\mathrm{T}} \mathbf{P} f_{\ell}(\Delta t_j) \tag{4.156}$$

In addition, we have the change in the ith modal load during the time step Δt_j, defined as follows:

$$\Delta q_{\mathrm{N}i,j} = q_{\mathrm{N}i,j+1} - q_{\mathrm{N}i,j} \tag{4.157}$$

where the symbol $q_{\mathrm{N}i,j+1}$ denotes the action at time t_{j+1}.

In a manner similar to that in Sec. 1.15, we can express the response of the ith mode at the end of the jth time step as the sum of three parts, which are

$$u_{\mathrm{N}i,j+1} = (u_{\mathrm{N}1} + u_{\mathrm{N}2} + u_{\mathrm{N}3})_{i,j+1} \tag{4.158}$$

The first part of the response consists of the ith free-vibrational motion caused by the conditions of displacement and velocity at time t_j (the beginning of the interval). Therefore, we have

$$(u_{\mathrm{N}1})_{i,j+1} = e^{-n_i \Delta t_j}\left(u_{\mathrm{N}i,j} \cos \omega_{\mathrm{d}i} \, \Delta t_j + \frac{\dot{u}_{\mathrm{N}i,j} + n_i u_{\mathrm{N}i,j}}{\omega_{\mathrm{d}i}} \sin \omega_{\mathrm{d}i} \, \Delta t_j \right) \tag{4.159a}$$

This formula represents an extension of Eq. (1.79a).

The other two parts of the response in Eq. (4.158) are due to the linear forcing function within the time step. The rectangular portion of this impulse yields

$$(u_{N2})_{i,j+1} = \frac{q_{Ni,j}}{\omega_i^2}\left[1 - e^{-n_i \Delta t_j}\left(\cos \omega_{di}\,\Delta t_j + \frac{n_i}{\omega_{di}}\sin \omega_{di}\,\Delta t_j\right)\right] \quad (4.159b)$$

which is taken from Eq. (1.79b). And that associated with the triangular portion becomes

$$(u_{N3})_{i,j+1} = \frac{\Delta q_{Ni,j}}{\omega_i^4\,\Delta t_j}\left[\omega_i^2\,\Delta t_j - 2n_i + e^{-n_i \Delta t_j}\left(2n_i \cos \omega_{di}\,\Delta t_j - \frac{\omega_{di}^2 - n_i^2}{\omega_{di}}\sin \omega_{di}\,\Delta t_j\right)\right] \quad (4.159c)$$

which is drawn from Eq. (1.79c).

We can also write the velocity of the ith mode at the end of the jth time step in three parts, as follows:

$$\dot{u}_{Ni,j+1} = (\dot{u}_{N1} + \dot{u}_{N2} + \dot{u}_{N3})_{i,j+1} \quad (4.160)$$

These three contributions may be formed by extending the notation in Eqs. (1.80a, b, and c) to obtain

$$(\dot{u}_{N1})_{i,j+1} = e^{-n_i \Delta t_j}\left[-\left(u_{Ni,j}\omega_{di} + n_i \frac{\dot{u}_{Ni,j} + n_i u_{Ni,j}}{\omega_{di}}\right)\sin \omega_{di}\,\Delta t_j + \dot{u}_{Ni,j}\cos \omega_{di}\,\Delta t_j\right] \quad (4.161a)$$

$$(\dot{u}_{N2})_{i,j+1} = \frac{q_{Ni,j}}{\omega_{di}}\, e^{-n_i \Delta t_j}\sin \omega_{di}\,\Delta t_j \quad (4.161b)$$

$$(\dot{u}_{N3})_{i,j+1} = \frac{\Delta q_{Ni,j}}{\omega_i^2\,\Delta t_j}\left[1 - e^{-n_i \Delta t_j}\left(\cos \omega_{di}\,\Delta t_j + \frac{n_i}{\omega_{di}}\sin \omega_{di}\,\Delta t_j\right)\right] \quad (4.161c)$$

Equations (4.158) through (4.161) constitute recurrence formulas for calculating the damped response of each normal mode at the end of the jth time step. They also provide the initial conditions of displacement and velocity at the beginning of step $j+1$. These expressions may be applied repetitively to obtain the time history of response for each of the normal modes. Then the results for each time station are transformed back to the original coordinates in the usual manner.

If the ith mode of a structure is a rigid body motion, appropriate expressions for rigid-body response must be used instead of the recurrence formulas given above. That is, the displacements in Eqs. (4.158) and (4.159) are replaced by

$$u_{\mathrm{N}i,j+1} = u_{\mathrm{N}i,j} + \dot{u}_{\mathrm{N}i,j}\,\Delta t_j + \tfrac{1}{2}(q_{\mathrm{N}i,j} + \tfrac{1}{3}\Delta q_{\mathrm{N}i,j})(\Delta t_j)^2 \tag{4.162}$$

And the velocities in Eqs. (4.160) and (4.161) are supplanted by

$$\dot{u}_{\mathrm{N}i,j+1} = \dot{u}_{\mathrm{N}i,j} + (q_{\mathrm{N}i,j} + \tfrac{1}{2}\Delta q_{\mathrm{N}i,j})\,\Delta t_j \tag{4.163}$$

These equations pertain to rigid-body motions with no absolute damping.

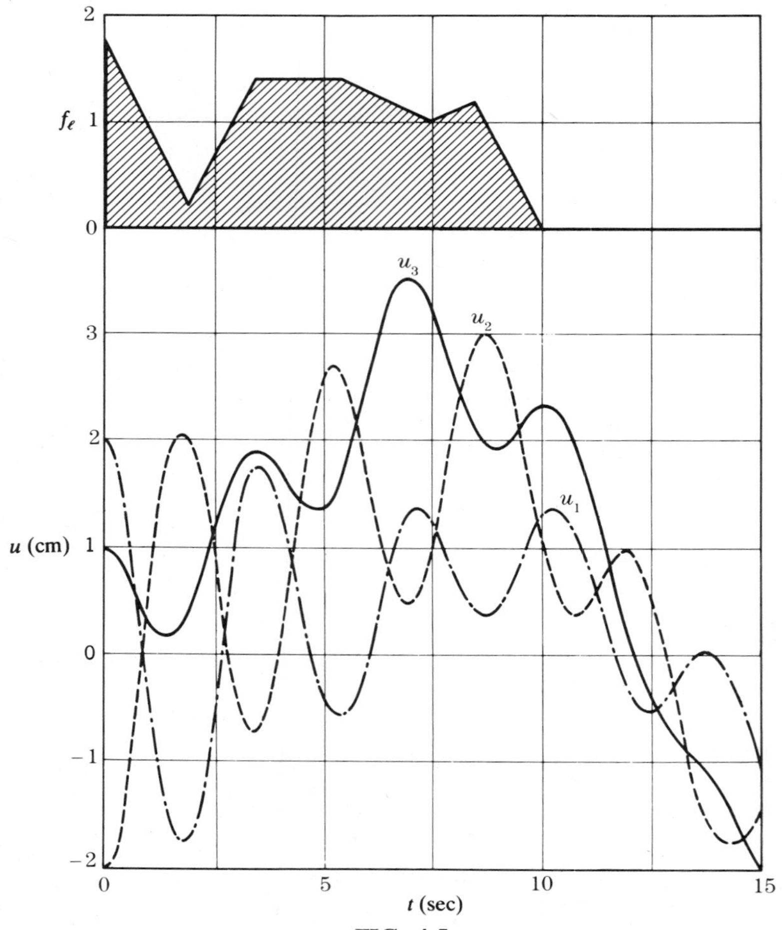

FIG. 4.5

When displacement equations of motion are used instead of action equations, the vector of piecewise-linear displacement functions $\mathbf{\Delta}_{\ell j}$ becomes

$$\mathbf{\Delta}_{\ell j} = \mathbf{F}\mathbf{Q}_{\ell j} = \mathbf{F}\mathbf{P}f_{\ell}(\Delta t_j) = \mathbf{\Delta}_{\mathrm{st}} f_{\ell}(\Delta t_j) \qquad (j+1 = 1, 2, \ldots, n_1) \quad (4.164)$$

In such an analysis, the terms $q_{\mathrm{N}i,j}/\omega_i^2$ and $\Delta q_{\mathrm{N}i,j}/\omega_i^2$ in Eqs. (4.159) and (4.161) would be replaced by $\delta_{\mathrm{N}i,j}$ and $\Delta\delta_{\mathrm{N}i,j}$, respectively.

For the purpose of applying the recurrence formulas given by Eqs. (4.158) through (4.161), we coded a computer program named **NOMOLIN**,

TABLE 4.3 Results from Program NOMOLIN

j	t_j (sec)	Δt_j (sec)	$f_{\ell j}$	u_{1j} (cm)	u_{2j} (cm)	u_{3j} (cm)
0	0	—	1.8	2.000	−2.000	1.000
1	0.5	0.5	1.4	1.260	−1.205	0.781
2	1.0	0.5	1.0	−0.348	0.519	0.344
3	1.5	0.5	0.6	−1.615	1.868	0.135
4	2.0	0.5	0.2	−1.671	1.945	0.425
5	2.5	0.5	0.6	−0.553	0.880	1.080
6	3.0	0.5	1.0	0.885	−0.358	1.718
7	3.5	0.5	1.4	1.668	−0.735	1.977
8	4.0	0.5	1.4	1.393	0.073	1.798
9	4.5	0.5	1.4	0.444	1.502	1.455
10	5.0	0.5	1.4	−0.379	2.569	1.358
11	5.5	0.5	1.4	−0.484	2.623	1.748
12	6.0	0.5	1.3	0.130	1.791	2.501
13	6.5	0.5	1.2	0.950	0.826	3.217
14	7.0	0.5	1.1	1.386	0.513	3.501
15	7.5	0.5	1.0	1.214	1.095	3.239
16	8.0	0.5	1.1	0.698	2.135	2.652
17	8.5	0.5	1.2	0.321	2.878	2.138
18	9.0	0.5	0.8	0.398	2.815	1.975
19	9.5	0.5	0.4	0.859	2.019	2.110
20	10.0	0.5	0.0	1.310	1.038	2.250
21	10.5	0.5	0.0	1.352	0.452	2.093
22	11.0	0.5	0.0	0.875	0.461	1.555
23	11.5	0.5	0.0	0.138	0.782	0.786
24	12.0	0.5	0.0	−0.435	0.906	0.048
25	12.5	0.5	0.0	−0.576	0.503	−0.468
26	13.0	0.5	0.0	−0.347	−0.347	−0.755
27	13.5	0.5	0.0	−0.072	−1.237	−0.953
28	14.0	0.5	0.0	−0.076	−1.745	−1.219
29	14.5	0.5	0.0	−0.444	−1.734	−1.596
30	15.0	0.5	0.0	−0.976	−1.414	−1.981

which is described in Appendix B.5. This program uses the normal-mode method to calculate the dynamic response of a damped multi-degree system subjected to a piecewise-linear forcing function. The upper part of Fig. 4.5 shows such a function f_ℓ, for which the numerical values (at 0.5 sec time intervals) are listed in the fourth column of Table 4.3. This function, multiplied by the vector of load factors $\mathbf{P} = \{-3.0, 0, 6.0\}$ kN, in combination with the initial-condition vectors $\mathbf{D}_0 = \{2, -2, 1\}$ cm and $\dot{\mathbf{D}}_0 = \mathbf{0}$, was applied to the system in Fig. 4.3. For this analysis we assumed that $m_1 = m_2 = m_3 = 1$ kg, $k_1 = k_2 = k_3 = 1$ kN/mm, and $\gamma_1 = \gamma_2 = \gamma_3 = 0.05$. The transient responses obtained from Program **NOMOLIN** appear in the last three columns of Table 4.3.

The lower part of Fig. 4.5 shows plots of the displacements u_1, u_2, and u_3. We see that the initial displacement pattern results in a conspicuous excitation of the third mode, which has a natural period of approximately 3.5 sec. Because the duration of the forcing function is 10 sec, the response thereafter consists only of free vibrations.

REFERENCES

[1] Gere, J. M., and Weaver, W., Jr., *Matrix Algebra for Engineers,* 2nd edn., Wadsworth, Belmont, CA, 1983.

[2] Weaver, W., Jr., and Gere, J. M., *Matrix Analysis of Framed Structures,* 2nd edn., Van Nostrand-Reinhold, New York, 1980.

[3] Rayleigh, J. W. S., *Theory of Sound,* Vol. 1, Dover, New York, 1945.

[4] Weaver, W., Jr., and Johnston, P. R., *Structural Dynamics by Finite Elements,* Prentice-Hall, Englewood Cliffs, NJ, 1987.

[5] Weaver, W., Jr., "Dynamics of Discrete-Parameter Structures," *Developments in Theoretical and Applied Mechanics,* Vol. 2, Pergamon Press, New York, 1965, pp. 629–651.

[6] Wilkinson, J. H., *The Algebraic Eigenvalue Problem,* Oxford University Press, London, 1965.

[7] Caughey, T. K., "Classical Normal Modes in Damped Linear Dynamic Systems," *Jour. Appl. Mech.,* ASME, **27,** 1960, pp. 269–271; also **32,** 1965, pp. 583–588.

[8] Foss, K. A., "Coordinates which Uncouple the Equations of Motion of Damped Linear Dynamic Systems," *Jour. Appl. Mech., ASME,* **25,** 1958, pp. 361–364.

PROBLEMS

4.2-1. Determine the eigenvalues ω_i^2 and the eigenvectors $\mathbf{\Phi}_{Mi}$ $(i = 1, 2, 3)$ for the system in Fig. 4.2*a* by the action-equation approach, using stiffness influence coefficients. Assume that $m_1 = m_2 = m_3 = m$ and $\ell_1 = \ell_2 = \ell_3 = \ell_4 = \ell$.

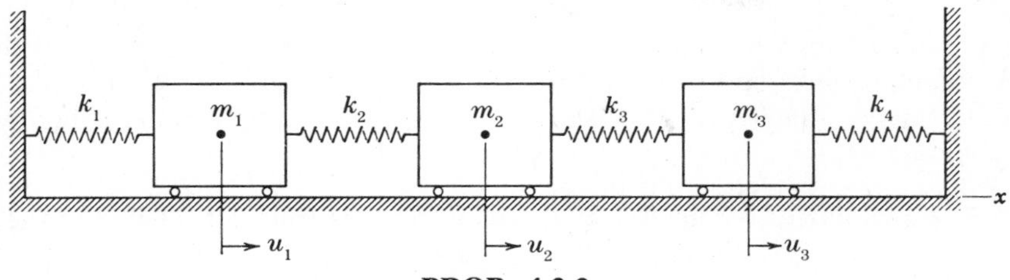

PROB. 4.2-2

4.2-2. The figure shows a system with three masses and four springs. Assuming that $m_1 = m_2 = m_3 = m$ and $k_1 = k_2 = k_3 = k_4 = k$, find the eigenvalues and eigenvectors by the displacement-equation method, using flexibility influence coefficients.

4.2-3. Three simple pendulums are connected by two springs, as shown in the figure. Let $m_1 = m_2 = m_3 = m$ and $k_1 = k_2 = k$; and find the eigenvalues and eigenvectors by the action-equation approach, using the small angles θ_1, θ_2, and θ_3 as displacement coordinates.

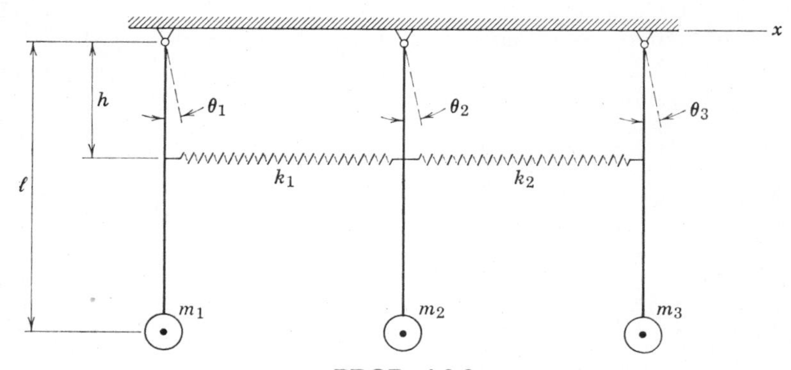

PROB. 4.2-3

4.2-4. The figures shows three disks attached to a shaft that is fixed at

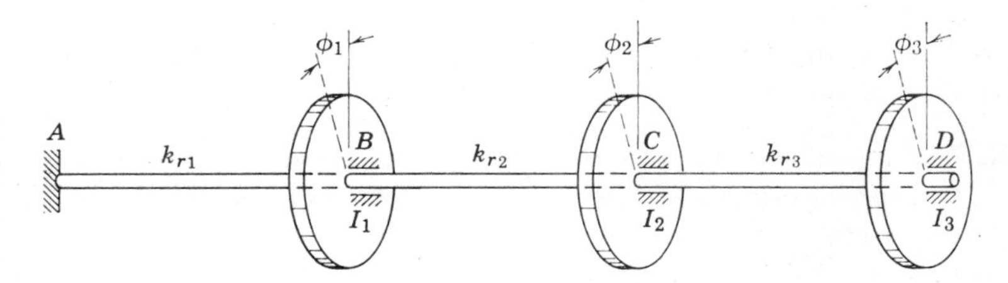

PROB. 4.2-4

point A but free to rotate in bearings at points B, C, and D. Assuming that $I_1 = I_2 = I_3 = I$ and $k_{r1} = k_{r2} = k_{r3} = k_r$, determine the eigenvalues and eigenvectors by the displacement-equation method. Use the angles ϕ_1, ϕ_2, and ϕ_3 as displacement coordinates for this system.

4.2-5. Four masses connected by three springs are free to translate in the x direction, as shown in the figure. Assuming that $m_1 = m_2 = m_3 = m_4 = m$ and $k_1 = k_2 = k_3 = k$, calculate the eigenvalues and eigenvectors by the action-equation method.

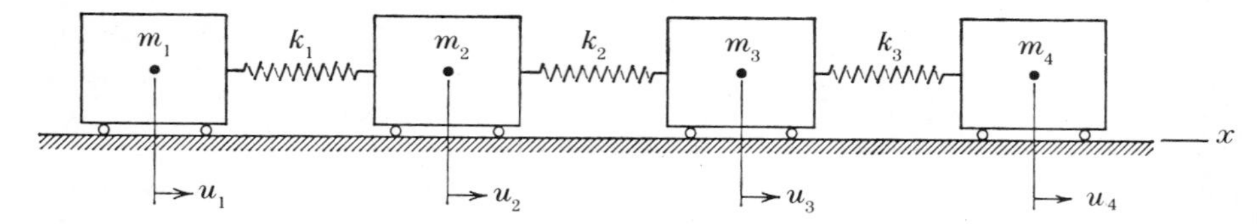

PROB. 4.2-5

4.2-6. The simply supported beam in the figure has three masses attached at its quarter points. Using the small translations v_1, v_2, and v_3 as displacement coordinates, determine the eigenvalues and eigenvectors for this system by the displacement-equation approach. Assume that $m_1 = m_2 = m_3 = m$ and that the massless prismatic beam has flexural rigidity EI.

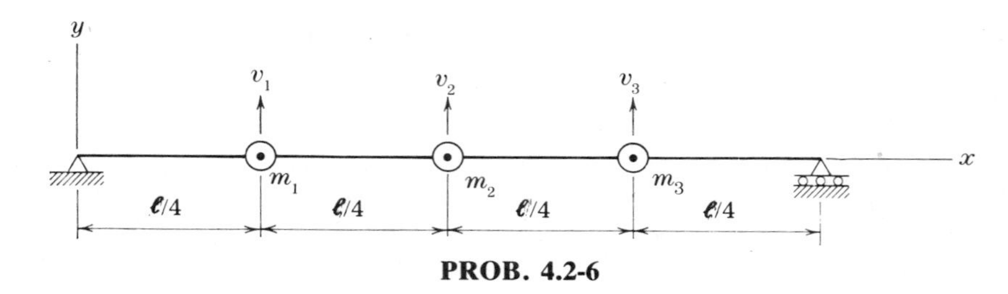

PROB. 4.2-6

4.2-7. For the triple pendulum in the figure, find the eigenvalues and eigenvectors by the action-equation method. Use the small translations u_1, u_2, and u_3 as displacement coordinates, and let $m_1 = m_2 = m_3 = m$ and $\ell_1 = \ell_2 = \ell_3 = \ell$.

4.2-8. The figure shows a three-story building frame with rigid girders and flexible columns. Assume that $m_1 = m_2 = m_3 = m$, $h_1 = h_2 = h_3 = h$, $EI_1 = 3EI$, $EI_2 = 2EI$, and $EI_3 = EI$. Using the small horizontal translations u_1, u_2, and u_3 as displacement coordinates, calculate the

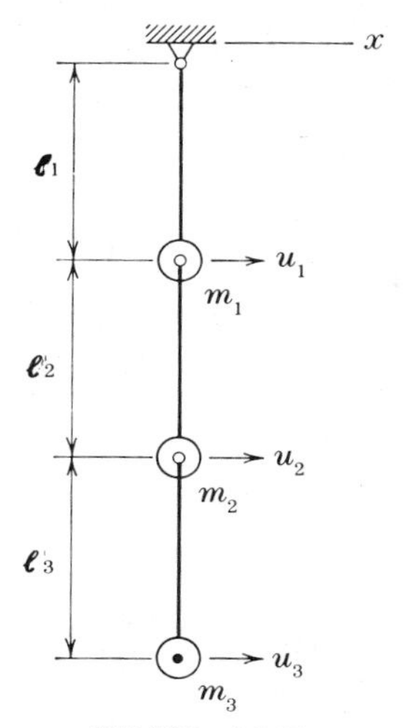

PROB. 4.2-7

eigenvalues and eigenvectors by the displacement-equation approach.

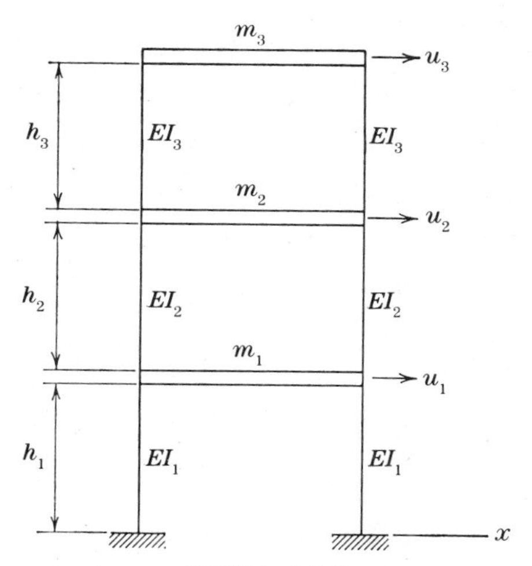

PROB. 4.2-8

4.2-9. Small orthogonal translations u_1, v_1, and w_1 are to be used as displacement coordinates for the spring-suspended mass in the figure. Unit vectors in the directions of the springs are given as $\mathbf{e}_1 = 0.8\mathbf{i} - 0.6\mathbf{j}$, $\mathbf{e}_2 = 0.6\mathbf{j} + 0.8\mathbf{k}$, and $\mathbf{e}_3 = 0.6\mathbf{j} - 0.8\mathbf{k}$ (where $\mathbf{i}$, $\mathbf{j}$, and $\mathbf{k}$ denote unit vectors in the x, y, and z directions, respectively).

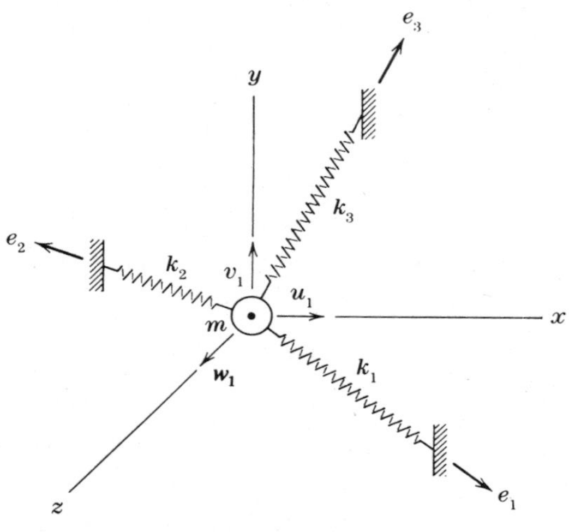

PROB. 4.2-9

Determine the eigenvalues and eigenvectors by the action-equation approach, assuming that the stiffness constants for all of the springs are equal.

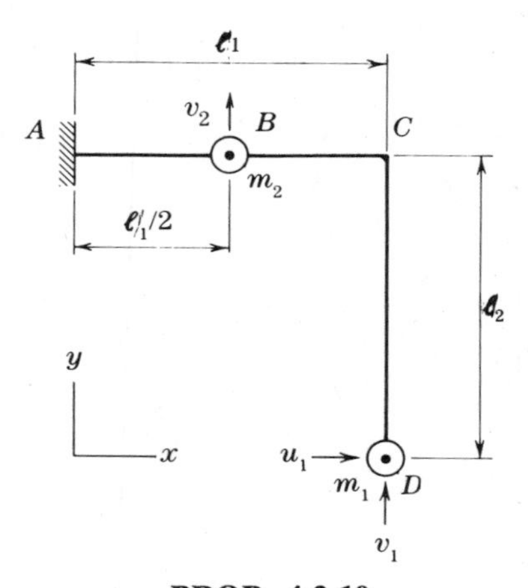

PROB. 4.2-10

4.2-10. The plane frame in the figure has two prismatic members with flexural rigidities EI. It is fixed at point A, continuous at C, and masses are attached at points B and D. Small displacements u_1, v_1, and v_2 are to be taken as displacement coordinates. Assuming that $m_1 = m_2 = m$ and $\ell_1 = \ell_2 = \ell$, find the eigenvalues and eigenvectors by the displacement-equation method.

4.2-11. Assume that a beam with three masses attached (see figure) is free to translate only in the y direction. Under the conditions that $m_1 = m_2 = m_3 = m$ and $\ell_1 = \ell_2 = \ell$, find the eigenvalues and eigenvectors by the action-equation approach. The beam is prismatic and has flexural rigidity EI.

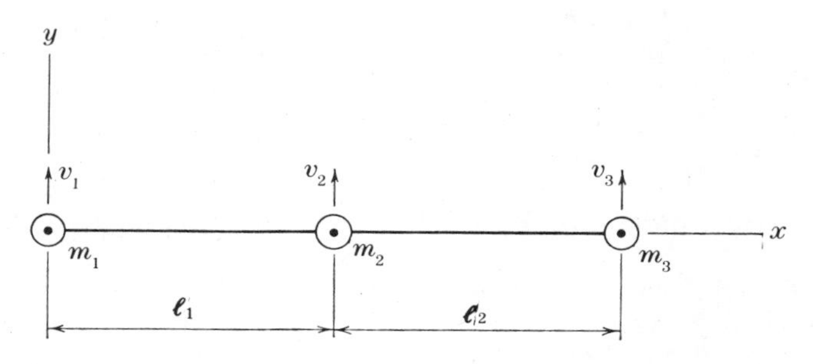

PROB. 4.2-11

4.2-12. The figure shows two spring-supported rigid bars that are hinged together at point B. Assume that $\ell_1 = \ell_2 = \ell$, $k_1 = k_2 = k_3 = k$, $m_1 = m_2 = m$, and that the masses are uniformly distributed along the lengths of the bars. The small translation v_1 of point B and the small rotations θ_1 and θ_2 of the bars about point B are to be used as displacement coordinates. Determine the eigenvalues and eigenvectors of this sytem by the action-equation method.

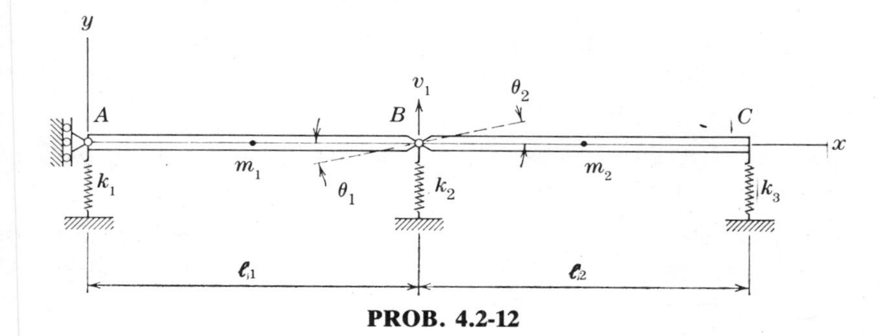

PROB. 4.2-12

4.4-1. Determine the free-vibrational response of the system in Fig. 4.1*a* to the sudden release of a static force P, applied in the x direction at the third mass. If we take $m_1 = m_2 = m_3 = m$ and $k_1 = k_2 = k_3 = k$, the eigenvalues and eigenvectors are the first set calculated in Example 1 of Sec. 4.2.

4.4-2. For the three-mass system in Prob. 4.2-2, determine the free-vibrational response to the initial conditions $\mathbf{D}_0 = \{0, 0, 0\}$ and $\dot{\mathbf{D}}_0 = \{\dot{u}_{01}, 0, -\dot{u}_{01}\}$.

4.4-3. For the spring-connected pendulums in Prob. 4.2-3, calculate the free-vibrational response to the initial conditions $\mathbf{D}_0 = \{0, \phi, 0\}$ and $\dot{\mathbf{D}}_0 = \{0, 0, 0\}$.

4.4-4. Determine the free-vibrational response of the rotational system in Prob. 4.2-4 to the initial conditions $\mathbf{D}_0 = \{0, 0, 0\}$ and $\dot{\mathbf{D}}_0 = \{\dot{\theta}, \dot{\theta}, \dot{\theta}\}$.

4.4-5. Calculate the response of the four-mass system in Prob. 4.2-5 to the initial conditions $\mathbf{D}_0 = \{0, 0, 0, 0\}$ and $\dot{\mathbf{D}}_0 = \{\dot{u}_{01}, 0, 0, \dot{u}_{01}\}$.

4.4-6. Suppose that the massless beam in Prob. 4.2-6 is rotating with uniform angular velocity $\dot{\theta}$ about the left-hand support when it is caught at the right-hand support. Determine the response of the system to this condition of initial velocity.

4.4-7. For the triple pendulum in Prob. 4.2-7, assume the initial conditions $\mathbf{D}_0 = \{\Delta, \Delta, \Delta\}$ and $\dot{\mathbf{D}}_0 = \{0, 0, 0\}$; and calculate the ensuing response.

4.4-8. Determine the response of the three-story building frame in Prob. 4.2-8 due to suddenly releasing a static load $Q_3 = P$, applied at the third level.

4.4-9. Let the mass in Prob. 4.2-9 have an initial velocity $\dot{u}_{01}$ in the x direction, while all other initial velocity and displacement components are zero. Determine the resulting motion of the mass.

4.4-10. Assume that a static force P is applied in the y direction at point C on the frame in Prob. 4.2-10. Find the response of the system due to suddenly releasing this force.

4.4-11. Calculate the response of the system in Prob. 4.2-11 to the initial conditions $\mathbf{D}_0 = \{0, 0, 0\}$ and $\dot{\mathbf{D}}_0 = \{\dot{v}_{01}, 2\dot{v}_{01}, \dot{v}_{01}\}$.

4.4-12. For the system in Prob. 4.2-12, find the response to the initial conditions $v_{01} = \Delta$, $\theta_{01} = \theta_{02} = 0$, $\dot{v}_{01} = 0$, and $\dot{\theta}_{01} = \dot{\theta}_{02} = 0$. Let $\ell = 3$ ft.

4.5-1. For the three-mass system in Fig. 4.1a, assume that $m_1 = m_2 = m_3 = m$ and $k_1 = k_2 = k_3 = k$. Determine the steady-state response of this system to the harmonic forcing function $Q_3 = P \cos \Omega t$, applied to the third mass.

4.5-2. Calculate the response of the three-mass system in Prob. 4.2-2 to the step function $Q_1 = P$, applied to the first mass.

4.5-3. Suppose that a ramp force Rt, acting horizontally and to the right, is applied to the mass of the central pendulum in Prob. 4.2-3. Determine the response of the system for small angles of rotation.

4.5-4. For the system in Prob. 4.2-4, calculate the steady-state response to the torque $T \sin \Omega t$, applied to the shaft midway between the second and third disks.

4.5-5. Determine the response of the four-mass system in Prob. 4.2-5 to the step functions $Q_1 = Q_4 = P$, applied to the first and fourth masses.

4.5-6. Assuming that ramp functions $Q_1 = Q_3 = Rt$ act upon masses m_1 and m_3, attached to the beam in Prob. 4.2-6, find the response of the system.

4.5-7. Let the triple pendulum in Prob. 4.2-7 be subjected to the force $P \cos \Omega t$, acting in the x direction and applied midway between the first and second masses. Calculate the steady-state response of the system.

4.5-8. For the three-story building frame in Prob. 4.2-8, find the response to the step functions $Q_1 = Q_2 = Q_3 = P$, applied simultaneously at each level.

4.5-9. Determine the response of the spring-suspended mass in Prob. 4.2-9 due to the ramp function Rt, applied to the mass in the z direction.

4.5-10. Calculate the steady-state response of the frame in Prob. 4.2-10 to the force $P \sin \Omega t$, applied in the y direction at point B.

4.5-11. Let the system in Prob. 4.2-11 be subjected to the ramp functions $Q_1 = Q_3 = Rt$, applied in the y direction at the first and third masses. Evaluate the response of the system.

4.5-12. For the system in Prob. 4.2-12, find the response due to a step force P, applied in the y direction at the mass center of the right-hand bar. Let $\ell = 3$ ft.

4.6-1. Suppose that the ground in Fig. 4.1a translates as specified by the ramp function $u_g = d_1 t/t_1$, where d_1 is the rigid-body ground translation at time t_1. Using the action-equation approach, determine the response of this system, assuming that $m_1 = m_2 = m_3 = m$ and $k_1 = k_2 = k_3 = k$.

4.6-2. Find the response of the three-mass system in Prob. 4.2-2 due to a rigid-body step displacement $u_g = d$ of the ground. For this problem use the displacement-equation method of analysis.

4.6-3. By the action-equation method, determine the steady-state rotational responses of the spring-connected pendulums in Prob. 4.2-3 due to the rigid-body harmonic ground acceleration $\ddot{u}_g = a \sin \Omega t$. In this case the rotations are absolute, not relative.

4.6-4. Consider the system in Prob. 4.2-4, and assume that the shaft at point A is given the rotational acceleration $\ddot{\phi}_A = \alpha_1 t^2/t_1^2$, where α_1 is the angular acceleration at time t_1. Find the rotational responses of the disks relative to the rotation of the shaft at point A, using the displacement-equation approach.

4.6-5. Let the fourth mass of the system in Prob. 4.2-5 translate in accordance with the ramp function $u_4 = d_1 t/t_1$, where d_1 is its displacement at time t_1. Determine the responses of the other three masses, using the action-equation method.

4.6-6. Suppose that the left-hand support of the beam in Prob. 4.2-6 suddenly translates an amount d in the y direction. By the displacement-equation approach, calculate the responses of the masses attached to the beam.

4.6-7. Using the displacement-equation method, find the response of the triple pendulum in Prob. 4.2-7 due to a step translation $u_g = d$ of the support point.

4.6-8. Assume that the building frame in Prob. 4.2-8 is subjected to the horizontal ground acceleration given by $\ddot{u}_g = a \sin \Omega t$. Determine the steady-state horizontal motions at the three levels, relative to the ground, using the action-equation method.

4.6-9. Consider the spring-supported mass in Prob. 4.2-9, and let the ground connection point of the first (or lower) spring suddenly displace an amount d in the x direction. Calculate the response of the mass by applying the action-equation approach.

4.6-10. Let point A of the frame in Prob. 4.2-10 undergo a small harmonic rotation $\theta_A = \phi \cos \Omega t$ about an axis perpendicular to the x–y plane. Find the steady-state responses of the attached masses, using the displacement-equation method.

4.6-11. Suppose that the central mass in the system of Prob. 4.2-11 is caused to accelerate as specified by the parabolic function $\ddot{v}_2 = a_1 t^2/t_1^2$, where a_1 is the acceleration at time t_1. Determine the responses of masses m_1 and m_3 relative to the motion of m_2, using the action-equation approach.

4.6-12. Assume that the support point at the lower end of the spring beneath point C in Prob. 4.2-12 translates harmonically in the y direction as specified by $v_R = d \sin \Omega t$. Use the displacement-equation method to find the steady-state response to this disturbance, assuming that $\ell = 3$ ft.

4.7-1. Using the method of iteration, determine the eigenvalues and eigenvectors for the system in Fig. 4.2a. Assume that $m_1 = m_3 = m$, that $m_2 = 2m$, and that $\ell_1 = \ell_2 = \ell_3 = \ell_4 = \ell$.

4.7-2. Repeat Prob. 4.2-2 by the iteration method, assuming that $m_1 = m_3 = 2m$ and that $m_2 = m$.

4.7-3. Repeat Prob. 4.2-6 by the iteration method, assuming that $m_1 = m_2 = m$ and that $m_3 = 3m$.

4.7-4. Repeat Prob. 4.2-7 by the iteration method, assuming that $m_1 = m_2 = 3m$ and that $m_3 = m$.

4.7-5. Repeat Prob. 4.2-8 by the iteration method, assuming that $m_1 = m_2 = 2m$ and that $m_3 = m$.

4.7-6. Repeat Prob. 4.2-10 by the iteration method, assuming that $m_1 = m$ and that $m_2 = 3m$.

4.7-7. Repeat Prob. 4.2-11, using the method for constraining rigid-body modes described at the end of Sec. 4.7. Assume that $m_1 = m_3 = m$ and that $m_2 = 2m$.

CHAPTER 5

CONTINUA WITH INFINITE DEGREES OF FREEDOM

5.1 INTRODUCTION

All structures and machines consist of parts that have both mass and flexibility. In many cases these parts can be idealized as point masses, rigid bodies, or deformable members without mass. Such systems have a finite number of degrees of freedom and can be analyzed by the methods described in the preceding chapters. However, it is also possible to treat certain systems more rigorously, without discretization of the analytical model. In this chapter we will analyze elastic bodies in which the mass and deformation properties are continuously distributed. Elements that can be handled in this manner include bars, shafts, cables, beams, simple frames, rings, arches, membranes, plates, shells, and three-dimensional solids. Many of these problems will be discussed in detail, but the topics of shells and three-dimensional solids are considered to be outside the scope of this book.† Furthermore, geometrically complex structures, such as general frames, arches, plates with cut-outs, aircraft fuselages, ship hulls, and so on, are very difficult (if not impossible) to analyze as elastic continua. In such cases it is necessary to use a discretized analytical model with a large but finite number of degrees of freedom. For this purpose, we shall describe the finite-element method in Chapter 6.

When analyzing a body as an elastic continuum, we consider it to be composed of an infinite number of particles. To specify the position of every point in the body, an infinite number of displacement coordinates is

† A more complete discussion of the vibrations of elastic bodies can be found in Refs. [1] and [2].

required; and the system is said to ahve an infinite number of degrees of freedom. These coordinates are manipulated as continuous functions, for which the first and second derivatives with respect to time represent velocity and acceleration of a general point. Because its mass is distributed, an elastic body has an infinite number of natural modes of vibration; and its dynamic response may be calculated as the sum of an infinite number of normal-mode contributions.

For vibrations of elastic bodies, we will assume that the material is homogeneous and isotropic and that it follows Hooke's law. Displacements will be assumed to be sufficiently small that the response to dynamic excitations is always linearly elastic. Although damping is not discussed in this chapter, it may easily be included by assuming modal damping ratios, as described in Sec. 4.8.

5.2 FREE LONGITUDINAL VIBRATIONS OF PRISMATIC BARS

Among the types of natural vibrations exhibited by an elastic bar, its longitudinal vibrations are the simplest to analyze. Torsional and lateral vibrations can also occur, and these types of motions are covered in subsequent sections. In dealing with longitudinal vibrations, we will assume that cross sections of a bar remain plane and that the particles in every cross section move only in the axial direction of the bar. The longitudinal extensions and compressions that take place during such a vibration of the bar will be accompanied by some lateral deformation. However, in the following discussion only those cases will be considered where the length of the longitudinal waves is large in comparison with the cross-sectional dimensions of the bar. In these cases the effects of lateral displacements upon the longitudinal motions can be neglected without substantial errors.†

Figure 5.1*a* shows an unrestrained prismatic bar of length ℓ with a typical segment of infinitesimal length dx located at the distance x from the left-hand end. Let the symbol u denote the longitudinal displacement of a point on the cross section at x. When the bar vibrates longitudinally, the axial forces on each segment (see Fig. 5.1*b*) may be summed in accordance with d'Alembert's principle to obtain

$$S + \frac{\partial S}{\partial x} dx - S - \rho A \, dx \frac{\partial^2 u}{\partial t^2} = 0 \tag{a}$$

in which the symbol S denotes the internal axial stress resultant (or force)

† A complete solution of the problem of longitudinal vibrations of a cylindrical bar with a circular cross section, in which the lateral displacements are also taken into consideration, was given by L. Pochhammer, *Jour. Math.* (Crelle), **81,** 1876, p. 324; see also E. Giebe and E. Blechschmidt, *Ann. Physik,* Ser. 5, **18,** 1933, p. 457.

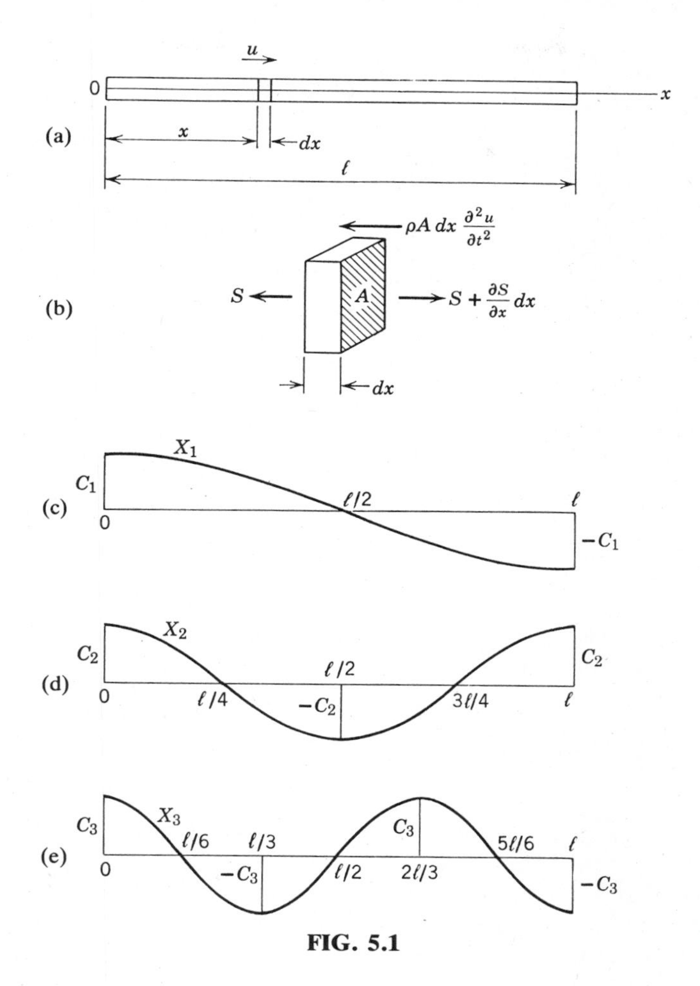

FIG. 5.1

on the cross section at x. The inertial force in this expression consists of the product of the *mass density* ρ of the material, the volume of the segment $A\,dx$ (where A is the cross-sectional area of the bar), and the acceleration $\partial^2 u/\partial t^2$. Using Hooke's law, we can express the axial force S in terms of the axial stress σ, and then in terms of the axial strain $\epsilon = \partial u/\partial x$, as follows:

$$S = A\sigma = EA\epsilon = EA\,\frac{\partial u}{\partial x} \tag{b}$$

where E is the modulus of elasticity. Substitution of expression (b) into Eq. (a) and rearrangement of terms produces

$$\frac{\partial^2 u}{\partial x^2} = \frac{1}{a^2}\frac{\partial^2 u}{\partial t^2} \tag{5.1}$$

in which

$$a = \sqrt{\frac{E}{\rho}} \tag{5.2}$$

Equation (5.1) is often called the *one-dimensional wave equation* to indicate the fact that during longitudinal vibrations displacement patterns are propagated in the axial direction at the velocity a, which is the speed of sound within the material. The *wave solution* for this problem is expressed in the form

$$u = f(x - at) \tag{c}$$

representing any function of x traveling at the velocity a. It can be seen that this expression satisfies Eq. (5.1) by evaluating the necessary derivatives, as follows:

$$\frac{\partial u}{\partial x} = f'(x - at) \qquad \frac{\partial^2 u}{\partial x^2} = f''(x - at)$$

$$\frac{\partial u}{\partial t} = -af'(x - at) \qquad \frac{\partial^2 u}{\partial t^2} = a^2 f''(x - at)$$

Substitution of the second derivatives into Eq. (5.1) yields identical results on both sides, thus satisfying the equation. A more general form of the wave solution is expressed as

$$u = f_1(x - at) + f_2(x + at) \tag{d}$$

in which the first term represents the function $f_1(x)$ traveling in the positive x direction and the second term consists of the function $f_2(x)$ traveling in the negative x direction. Although the wave solution is suitable for certain impact problems with impulses of very short duration, this type of analysis is not as useful as the *vibration solution,* which will now be developed in detail.

When the bar in Fig. 5.1*a* is vibrating in one of its natural modes, the solution of Eq. (5.1) may be taken in the form

$$u = X(A \cos \omega t + B \sin \omega t) \tag{e}$$

where A and B are constants and ω is the angular frequency. The symbol X represents a function of x that defines the shape of the natural mode of vibration and is called a *principal function,* or a *normal function.* Substitution of Eq. (*e*) into Eq. (5.1) results in

$$\frac{d^2 X}{dx^2} + \frac{\omega^2}{a^2} X = 0 \tag{f}$$

for which the solution is

$$X = C \cos \frac{\omega x}{a} + D \sin \frac{\omega x}{a} \tag{g}$$

The constants C and D in this expression for the function X are determined (to within an arbitrary constant) by satisfying the *boundary conditions* at the ends of the bar. Since the bar in Fig. 5.1*a* has free ends, the axial force, which is proportional to dX/dx, must be zero at each extremity. Thus, the boundary conditions for this problem may be written as

$$\left(\frac{dX}{dx}\right)_{x=0} = 0 \qquad \left(\frac{dX}{dx}\right)_{x=\ell} = 0 \tag{h}$$

In order to satisfy the first of these conditions, it is necessary to put $D = 0$ in Eq. (g). The second condition will be satisfied for $C \neq 0$ (the nontrivial solution) only if

$$\sin \frac{\omega \ell}{a} = 0 \tag{5.3}$$

This is the *frequency equation* for the case under consideration, from which the frequencies of the natural modes of the longitudinal vibrations of a bar with free ends can be calculated. This equation will be satisfied by putting

$$\frac{\omega_i \ell}{a} = i\pi \tag{i}$$

where i is an integer. Taking $i = 0, 1, 2, 3, \ldots, \infty$ we can obtain the frequencies of the various modes of longitudinal motion. The value of $i = 0$ yields a zero frequency, which connotes a rigid-body translation of the bar in the x direction. The frequency of the fundamental vibrational mode will be found by putting $i = 1$ in Eq. (i), which gives

$$\omega_1 = \frac{a\pi}{\ell} = \frac{\pi}{\ell}\sqrt{\frac{E}{\rho}} \tag{5.4}$$

The corresponding period of vibration will be

$$\tau_1 = \frac{1}{f_1} = \frac{2\pi}{\omega_1} = 2\ell\sqrt{\frac{\rho}{E}} \tag{5.5}$$

This mode of vibration has the shape illustrated in Fig. 5.1*c*, which [from Eq. (*g*)] is expressed as

$$X_1 = C_1 \cos \frac{\omega_1 x}{a} = C_1 \cos \frac{\pi x}{\ell}$$

In Figs. 5.1*d* and 5.1*e*, the second and third modes of vibration are represented, for which

$$\frac{\omega_2 \ell}{a} = 2\pi \qquad X_2 = C_2 \cos \frac{2\pi x}{\ell}$$

and

$$\frac{\omega_3 \ell}{a} = 3\pi \qquad X_3 = C_3 \cos \frac{3\pi x}{\ell}$$

The general form of the vibrational solution (*e*) of Eq. (5.1) may be written as

$$u_i \cos \frac{i\pi x}{\ell} \left(A_i \cos \frac{i\pi a t}{\ell} + B_i \sin \frac{i\pi a t}{\ell} \right) \tag{j}$$

By superimposing such solutions, we can represent any longitudinal vibration of the bar in the following form:

$$u = \sum_{i=1}^{\infty} \cos \frac{i\pi x}{\ell} \left(A_i \cos \frac{i\pi a t}{\ell} + B_i \sin \frac{i\pi a t}{\ell} \right) \tag{5.6}$$

The constants A_i and B_i in Eq. (5.6) can always be chosen in such a manner as to satisfy any initial conditions. Assume, for instance, that at the initial moment (when $t = 0$), the displacements u are given by the equation $(u)_{t=0} = f_1(x)$ and that the initial velocities are expressed by the equation $(\dot{u})_{t=0} = f_2(x)$. Substituting $t = 0$ into Eq. (5.6), we obtain

$$f_1(x) = \sum_{i=1}^{\infty} A_i \cos \frac{i\pi x}{\ell} \tag{k}$$

By taking the derivative of Eq. (5.6) with respect to t and substituting $t = 0$, we obtain

$$f_2(x) = \sum_{i=1}^{\infty} \frac{i\pi a}{\ell} B_i \cos \frac{i\pi x}{\ell} \tag{l}$$

The coefficients A_i and B_i in Eqs. (*k*) and (*l*) can now be calculated, as explained before [see Eq. (1.59a), Sec. 1.11] by using the formulas

$$A_i = \frac{2}{\ell} \int_0^\ell f_1(x) \cos \frac{i\pi x}{\ell} \, dx \tag{m}$$

$$B_i = \frac{2}{i\pi a} \int_0^\ell f_2(x) \cos \frac{i\pi a}{\ell} \, dx \tag{n}$$

As an example, consider now the case when a prismatic bar compressed by forces applied at the ends is suddenly relieved of this compression at the time $t = 0$. Assuming that the middle of the bar remains stationary, we take

$$(u)_{t=0} = f_1(x) = \frac{\epsilon_0 \ell}{2} - \epsilon_0 x \qquad f_2(x) = 0$$

where ϵ_0 denotes the amount of compressive strain at the moment $t = 0$. We obtain from Eqs. (m) and (n)

$$A_i = \frac{4\epsilon_0 \ell}{\pi^2 i^2} \text{ (for } i \text{ odd)} \qquad A_i = 0 \text{ (for } i \text{ even)} \qquad B_i = 0$$

and the general solution (5.6) becomes

$$u = \frac{4\epsilon_0 \ell}{\pi^2} \sum_{i=1,3,5,\ldots}^{\infty} \frac{1}{i^2} \cos \frac{i\pi x}{\ell} \cos \frac{i\pi a t}{\ell} \tag{o}$$

Only odd integers $i = 1, 3, 5, \ldots$ enter into this solution, and the vibration is symmetric with respect to the middle cross section of the bar.

As a second example, let us consider free longitudinal vibrations of a bar, one end of which is fixed and the other free (see Fig. 5.2a). The end conditions in this case are

$$(u)_{x=0} = 0 \qquad \left(\frac{du}{dx}\right)_{x=\ell} = 0 \tag{p}$$

To satisfy the first of these conditions, we take $C = 0$ in the general expression (g) for normal functions. The second condition yields the frequency equation

$$\cos \frac{\omega \ell}{a} = 0$$

from which the frequencies and the periods of various modes of vibration are

$$\omega_i = \frac{i\pi a}{2\ell} \qquad \tau_i = \frac{2\pi}{\omega_i} = \frac{4\ell}{ia} \qquad (i = 1, 3, 5, \ldots, \infty) \tag{q}$$

Then the general expression (*e*) for various modes of vibration becomes:

$$u_i = \sin\frac{i\pi x}{2\ell}\left(A_i \cos\frac{i\pi at}{2\ell} + B_i \sin\frac{i\pi at}{2\ell}\right) \tag{r}$$

The total solution for longitudinal vibrations is then obtained by superposition, which gives

$$u = \sum_{i=1,3,5,\ldots}^{\infty} \sin\frac{i\pi x}{2\ell}\left(A_i \cos\frac{i\pi at}{2\ell} + B_i \sin\frac{i\pi at}{2\ell}\right) \tag{5.7}$$

The constants A_i, B_i are to be determined in each particular case from the initial conditions (at the time $t = 0$).

For example, assume that the bar is initially stretched by an axial force P_0, applied at the free end (see Fig. 5.2*a*), and that at the time $t = 0$ this force is suddenly removed. Using the notation ϵ_0 for the initial strain P_0/EA, we have the initial conditions

$$(u)_{t=0} = \epsilon_0 x \qquad (\dot{u})_{t=0} = 0$$

The second of these conditions will be satisfied by making the constant B_i vanish in expression (*r*). For determining the constant A_i, we then have the equation

$$\sum_{i=1,3,5,\ldots}^{\infty} A_i \sin\frac{i\pi x}{2\ell} = \epsilon_0 x$$

Using Eq. (1.59b) from Sec. 1.11, we find

$$A_i = \frac{2\epsilon_0}{\ell}\int_0^{\ell} x \sin\frac{i\pi x}{2\ell}\,dx = \frac{8\epsilon_0 \ell}{i^2\pi^2}(-1)^{(i-1)/2}$$

and expression (5.7) becomes

$$u = \frac{8\epsilon_0 \ell}{\pi^2}\sum_{i=1,3,5,\ldots}^{\infty} \frac{(-1)^{(i-1)/2}}{i^2}\sin\frac{i\pi x}{2\ell}\cos\frac{i\pi at}{2\ell} \tag{s}$$

Figures 5.2*b*, *c*, and *d* show the contributions of the first three modes to the total response of the bar. It may be appreciated that the amplitudes of

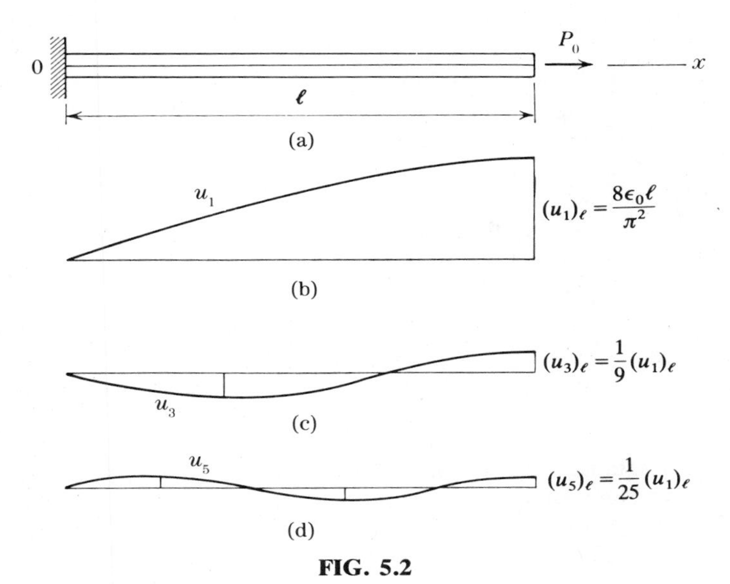

FIG. 5.2

various modes of vibration are rapidly decreasing as i increases. The displacements of the free end of the bar are obtained by substituting $x = \ell$ into expression (s). For time $t = 0$, we find

$$(u)_{\substack{x=\ell \\ t=0}} = \frac{8\epsilon_0\ell}{\pi^2}\left(1 + \frac{1}{9} + \frac{1}{25} + \cdots\right) = \frac{8\epsilon_0\ell}{\pi^2}\left(\frac{\pi^2}{8}\right) = \epsilon_0\ell$$

as it should be.

Example 1. Find the normal functions for longitudinal vibrations of a bar of length ℓ if both ends are fixed.
Solution. The end conditions in this case are

$$(u)_{x=0} = (u)_{x=\ell} = 0$$

To satisfy these conditions, we put $C = 0$ in expression (g) and obtain the frequency equation $\sin \omega_i\ell/a = 0$, from which $\omega_i = i\pi a/\ell$. Hence, the normal functions will be

$$X_i = A_i \sin\frac{i\pi x}{\ell} \qquad (i = 1, 2, 3, \ldots, \infty) \tag{t}$$

Example 2. A bar with built-in ends is acted upon by a concentrated axial force P_0, applied at the middle, as shown in Fig. 5.3a. What vibrations will be produced if the force P_0 is suddenly removed?

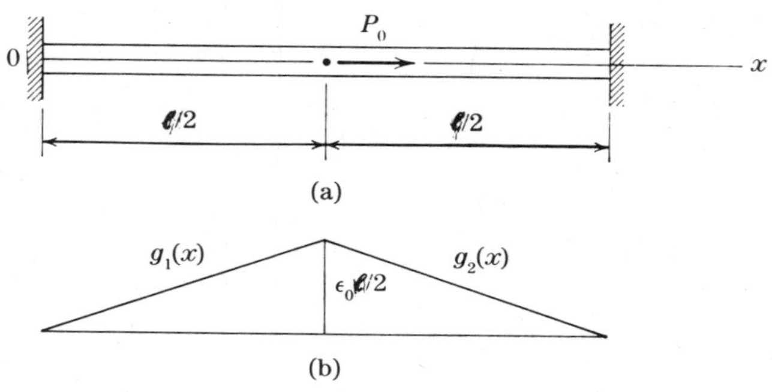

FIG. 5.3

Solution. The tensile strain in the left-hand part of the bar, which is numerically equal to the compressive strain in the right-hand part, is $\epsilon_0 = P_0/2EA$. The initial displacements, $(u)_{t=0}$, are given by $g_1(x) = \varepsilon_0 x$ (for $0 \leqq x \leqq \ell/2$) and $g_2(x) = \epsilon_0(\ell - x)$ for $\ell/2 \leqq x \leqq \ell$, as represented by Fig. 5.3*b*. In the preceding example, we found the normal functions [Eq. (*t*)] for this case; and the general expression for vibration, satisfying the initial condition $(\dot{u})_{t=0} = 0$, is

$$u = \sum_{i=1}^{\infty} A_i \sin \frac{i\pi x}{\ell} \cos \frac{i\pi a t}{\ell} \tag{u}$$

The constants A_i are found from the initial displacement configuration, which leads to

$$A_i = \frac{2}{\ell} \left[\int_0^{\ell/2} \epsilon_0 x \sin \frac{i\pi x}{\ell} dx + \int_{\ell/2}^{\ell} \epsilon_0(\ell - x) \sin \frac{i\pi x}{\ell} dx \right]$$

$$= \left(\frac{4\epsilon_0 \ell}{\pi^2} \right) \frac{(-1)^{(i-1)/2}}{i^2} \qquad \text{(for } i = 1, 3, 5, \ldots, \infty)$$

$$A_i = 0 \qquad \text{(for } i = 2, 4, \ldots, \infty)$$

Thus,

$$u = \frac{4\epsilon_0 \ell}{\pi^2} \sum_{i=1,3,5,\ldots}^{\infty} \frac{(-1)^{(i-1)/2}}{i^2} \sin \frac{i\pi x}{\ell} \cos \omega_i t \tag{v}$$

Example 3. A bar moving along the x axis with constant velocity v is suddenly stopped at the end $x = 0$, so that the initial conditions are $(u)_{t=0} = 0$ and $(\dot{u})_{t=0} = v$. Determine the ensuing vibrations.

Solution. The general expression for displacements in this case is given by Eq. (5.7). Since the initial displacements vanish, we put $A_i = 0$ in that expression. The constants B_i are then obtained from the equation

$$(\dot{u})_{t=0} = \sum_{i=1,3,5,\ldots}^{\infty} B_i \frac{i\pi a}{2\ell} \sin \frac{i\pi x}{2\ell} = v$$

which gives

$$B_i = \frac{8v\ell}{\pi^2 i^2 a}$$

Hence

$$u = \frac{8v\ell}{\pi^2 a} \sum_{i=1,3,5,\ldots}^{\infty} \frac{1}{i^2} \sin \frac{i\pi x}{2\ell} \sin \omega_i t \tag{w}$$

By using this formula, we can calculate the displacement of any cross section of the bar at any given time. Taking, for example, the free end of the bar ($x = \ell$) and assuming that $t = \ell/a$ (which is the time required for sound to travel the distance ℓ), we obtain

$$(u)_{\substack{x=\ell \\ t=l/a}} = \frac{8v\ell}{\pi^2 a}\left(1 + \frac{1}{9} + \frac{1}{25} + \cdots\right) = \frac{v\ell}{a}$$

The strain in the bar during vibration is

$$\frac{du}{dx} = \frac{8v\ell}{\pi^2 a} \sum_{i=1,3,5,\ldots}^{\infty} \frac{1}{i^2} \frac{i\pi}{2\ell} \cos \frac{i\pi x}{2\ell} \sin \omega_i t$$

At the fixed end ($x = 0$) we obtain

$$\left(\frac{du}{dx}\right)_{x=0} = \frac{4v}{\pi a} \sum_{i=1,3,5,\ldots}^{\infty} \frac{1}{i} \sin \frac{i\pi a t}{2\ell} = \frac{v}{a} \qquad \left(\text{for } 0 < \frac{\pi a t}{2\ell} < \frac{\pi}{2}\right)$$

The tension wave, which was originated at the left end of the bar at the instant of stopping ($t = 0$), is moving along the bar with speed a; and at the instant $t = \ell/a$ it reaches the free end of the bar. At this time the velocities of all particles of the bar vanish; and at the same time it is uniformly extended, so that the tensile strain is $\epsilon = v/a$.

5.3 FORCED LONGITUDINAL RESPONSE OF PRISMATIC BARS

Let us now consider a forcing function $P = F(t)$ applied to the right-hand end of the prismatic bar in Fig. 5.2*a*. Free vibrations of this bar were investigated in the preceding section, and we found that normal functions for this case are given by

$$X_i = D_i \sin \frac{i\pi x}{2\ell} \qquad (i = 1, 3, 5, \ldots, \infty) \tag{a}$$

Any displacement $u = f(x)$ can be obtained by superposition of displacements corresponding to the normal modes of vibration (a). Therefore, vibrations produced by the force P will be represented by the series

$$u = \phi_1 \sin\frac{\pi x}{2\ell} + \phi_3 \sin\frac{3\pi x}{2\ell} + \phi_5 \sin\frac{5\pi x}{2l} + \cdots = \sum_{i=1,3,5,\ldots}^{\infty} \phi_i \sin\frac{i\pi x}{2\ell} \tag{5.8}$$

in which $\phi_1, \phi_3, \phi_5, \ldots$ are some unknown functions of time. In the case of free vibration, these functions are represented by the expressions within the parentheses of Eq. (5.7) in the preceding section. To find these functions for the case of forced vibrations, we will apply the principle of virtual work. We must consider here three kinds of forces: The inertial force within each element of the vibrating bar, the elasticity force on each element due to deformation of the bar, and the force P applied at the end. As a virtual displacement, we can take any longitudinal displacement δu satisfying the condition of continuity and the condition at the fixed end $(\delta u_{x=0} = 0)$. It will be found advantageous to take virtual displacements in the form of normal functions, as given by Eq. (a). thus,

$$\delta u_i = X_i = D_i \sin\frac{i\pi x}{2\ell} \tag{b}$$

Observing that the mass of an element between two adjacent cross sections of the bar is $\rho A\, dx$, we find that the work δW_I done by inertial forces on the assumed virtual displacement is

$$\delta W_I = \int_0^\ell (-\rho A\, dx)\ddot{u}\, \delta u_i = -\rho A \int_0^\ell \ddot{u} D_i \sin\frac{i\pi x}{2\ell}\, dx$$

substituting for u the series (5.8) and observing that

$$\int_0^\ell \sin\frac{i\pi x}{2\ell} \sin\frac{j\pi x}{2\ell}\, dx = 0 \qquad \int_0^\ell \sin^2\frac{i\pi x}{2\ell}\, dx = \frac{\ell}{2}$$

we obtain

$$\delta W_I = -\frac{\rho A \ell}{2} D_i \ddot{\phi}_i \tag{c}$$

To calculate the virtual work δW_E produced by elasticity forces, we observe that the force on each element is $EAu''\, dx$. Thus,

$$\delta W_E = \int_0^\ell (EAu''\, dx)\, \delta u_i$$

Substituting into this expression the second derivative of Eq. (5.8) with respect to x and using Eq. (b) for δu_i, we obtain

$$\delta W_E = -\frac{i^2\pi^2 EA}{8\ell} D_i\phi_i \tag{d}$$

In later sections it will be advantageous to obtain the virtual work of elasticity forces by starting with the *strain energy* of the body. The expression for strain energy in the present case of the elastic bar is

$$U = \frac{1}{2}\int_0^\ell EA\left(\frac{\partial u}{\partial x}\right)^2 dx \tag{e}$$

Substituting for u the series (5.8) and observing that

$$\int_0^\ell \cos\frac{i\pi x}{2\ell}\cos\frac{j\pi x}{2\ell}\,dx = 0 \qquad \int_0^\ell \cos^2\frac{i\pi x}{2\ell}\,dx = \frac{\ell}{2}$$

we find the strain energy to be

$$U = \frac{\pi^2 EA}{16\ell}\sum_{i=1,3,5,\ldots}^{\infty} i^2\phi_i^2 \tag{f}$$

We see that the amount of strain energy in the bar at any instant depends upon the quantities ϕ_i defining the displacement of the bar. If we give to one of these quantities an increment $\delta\phi_i$, the corresponding displacement is

$$\delta u_i = \delta\phi_i \sin\frac{i\pi x}{2\ell} \tag{g}$$

and the corresponding increment of the strain energy is

$$\delta U = \frac{\partial U}{\partial \phi_i}\delta\phi_i = \frac{i^2\pi^2 EA}{8\ell}\phi_i\,\delta\phi_i \tag{h}$$

The same quantity taken with negative sign will represent the work of elasticity forces on the displacement (g). To get the work of elasticity forces on the virtual displacement (b), we need only replace $\delta\phi_i$ by D_i, as can be seen by comparing expressions (b) and (g). In this way we obtain for that work the expression

$$\delta W_E = -\frac{\partial U}{\partial \phi_i}\delta\phi_i = -\frac{i^2\pi^2 EA}{8\ell}\phi_i D_i \tag{i}$$

which is the same as Eq. (d).

To determine the virtual work δW_P of the disturbing force P applied at the end, we observe that the virtual displacement of this end is obtained by substituting ℓ for x in expression (b). The corresponding virtual work is

$$\delta W_P = PD_i \sin\frac{i\pi}{2} = PD_i(-1)^{(i-1)/2} \tag{j}$$

Summation of expressions (c), (i), and (j) gives us the total virtual work. Equating it to zero, we obtain

$$\frac{\rho A\ell}{2}\ddot{\phi}_i + \frac{i^2\pi^2 EA}{8\ell}\phi_i = P(-1)^{(i-1)/2}$$

or

$$\ddot{\phi}_i + \omega_i^2\phi_i = \frac{2}{\rho A\ell}P(-1)^{(i-1)/2} \tag{k}$$

where $\omega_i = i\pi a/2\ell$ and $i = 1, 3, 5, \ldots, \infty$. Note that the constant D_i, defining the instantaneous magnitude of the virtual displacement (b), cancels out of Eq. (k), as it should.

We see that each quantity ϕ_i in the series (5.8) can be readily obtained from Eq. (k) if P is known as a function of time. If the initial displacements and velocities are zero, we need only consider vibration produced by the force P. Writing the solution of Eq. (k) in the form of the Duhamel integral, we have

$$\phi_i = \frac{4(-1)^{(i-1)/2}}{i\pi a\rho A}\int_0^t P\sin\left[\frac{i\pi a}{2\ell}(t-t')\right]dt' \tag{l}$$

Substituting Eq. (l) into expression (5.8), we obtain the response produced by the force P in the following form:

$$u = \frac{4}{\pi a\rho A}\sum_{i=1,3,5,\ldots}^{\infty}\frac{(-1)^{(i-1)/2}}{i}\sin\frac{i\pi x}{2\ell}\int_0^t P\sin\left[\frac{i\pi a}{2\ell}(t-t')\right]dt' \tag{5.9}$$

As a particular example, let us take the case of vibrations produced in the bar by a constant force P, suddenly applied at the time $t = 0$. In such a case the integral in Eq. (5.9) can be readily evaluated, and we obtain

$$u = \frac{8\ell P}{\pi^2 a^2\rho A}\sum_{i=1,3,5,\ldots}^{\infty}\frac{(-1)^{(i-1)/2}}{i^2}\sin\frac{i\pi x}{2\ell}\left(1-\cos\frac{i\pi at}{2\ell}\right) \tag{m}$$

Substituting $x = \ell$ into this series, we find the displacement of the end of the bar to be

$$(u)_{x=\ell} = \frac{8\ell P}{\pi^2 a^2 \rho A} \sum_{i=1,3,5,\ldots}^{\infty} \frac{1}{i^2}\left(1 - \cos\frac{i\pi a t}{2\ell}\right) \tag{n}$$

It is seen that by a sudden application of the force P all modes of vibration of the bar are excited. The maximum deflection occurs when $t = 2\ell/a$, because at that instant

$$1 - \cos\frac{i\pi a t}{2\ell} = 2$$

and we obtain

$$(u)_{t=2\ell/a} = \frac{16\ell P}{\pi^2 a^2 \rho A} \sum_{i=1,3,5,\ldots}^{\infty} \frac{1}{i^2}$$

Observing that

$$\sum_{i=1,3,5,\ldots}^{\infty} \frac{1}{i^2} = \frac{\pi^2}{8} \qquad \text{and} \qquad a^2 = \frac{E}{\rho}$$

we find

$$(u)_{t=2\ell/a} = \frac{2\ell P}{EA}$$

Thus, we arrive at the conclusion that the suddenly applied force P produces twice as great a deflection as it would if it were applied statically.

As a second example, let us consider the longitudinal response of a bar with free ends (see Fig. 5.1a), subjected to a suddenly applied force P at the end $x = \ell$.† Proceeding as in the above example and using normal functions for a bar with free ends [see Eq. (5.6), Sec. 5.2], we can represent longitudinal displacements of the vibrating bar by the series

$$\begin{aligned} u &= \phi_0 + \phi_1 \cos\frac{\pi x}{\ell} + \phi_2 \cos\frac{2\pi x}{\ell} + \phi_3 \cos\frac{3\pi x}{\ell} + \cdots \\ &= \phi_0 + \sum_{i=1}^{\infty} \phi_i \cos\frac{i\pi x}{\ell} \end{aligned} \tag{5.10}$$

The first term, ϕ_0, represents the motion of the bar as a rigid body. On this motion various modes of longitudinal vibrations of the bar are superimposed. For determining the function ϕ_0, we have the equation

$$\rho A \ell \ddot{\phi}_0 = P \tag{o}$$

† A similar problem is encountered in investigation of the vibrations produced during the lifting of a long drill stem, as used in deep oil wells. This problem was discussed by B. F. Langer and E. H. Lamberger, *Jour. Appl. Mech.*, **10**, 1943, p. 1.

The functions $\phi_1, \phi_2, \phi_3, \ldots$ will be found as before by using the principle of virtual displacements. Taking a virtual displacement

$$\delta u_i = C_i \cos \frac{i\pi x}{\ell} \tag{p}$$

we find that the work of inertia forces on this displacement is

$$\delta W_1 = -\int_0^\ell \rho A \ddot{u} C_i \cos \frac{i\pi x}{\ell}\, dx = -\frac{1}{2}\rho A \ell C_i \ddot{\phi}_i \tag{q}$$

The strain energy of the vibrating bar at any instant is

$$U = \frac{1}{2}\int_0^\ell EA\left(\frac{\partial u}{\partial x}\right)^2 dx = \frac{\pi^2 EA}{4\ell}\sum_{i=1}^{\infty} i^2 \phi_i^2 \tag{r}$$

and the work of the elasticity forces on the displacement (p) will be

$$\delta W_E = -\frac{\partial U}{\partial \phi_i}\,\delta\phi_i = -\frac{i^2\pi^2 EA}{2\ell}\, C_i \phi_i \tag{s}$$

Finally, the work of the force P on the displacement (p) is

$$\delta W_P = PC_i \cos i\pi = C_i P(-1)^i \tag{t}$$

Equating to zero the sum of expressions (q), (s), and (t), we obtain the equation

$$\ddot{\phi}_i + \omega_i^2 \phi_i = \frac{2}{\rho A \ell} P(-1)^i \tag{u}$$

where $\omega_i = i\pi a/\ell$. From this equation and Eq. (o), we obtain (assuming that the bar is initially at rest)

$$\phi_0 = \frac{Pt^2}{2\rho A \ell} \tag{v}$$

$$\begin{aligned}\phi_i &= (-1)^i \frac{2}{i\pi a \rho A}\int_0^t P \sin\left[\frac{i\pi a}{\ell}(t - t')\right] dt' \\ &= \frac{(-1)^i 2\ell P}{i^2\pi^2 a^2 \rho A}\left(1 - \cos\frac{i\pi a t}{\ell}\right)\end{aligned} \tag{w}$$

Substituting these expressions into Eq. (5.10), we find

$$u = \frac{Pt^2}{2\rho A \ell} + \frac{2\ell P}{\pi^2 a^2 \rho A} \sum_{i=1}^{\infty} \frac{(-1)^i}{i^2} \cos \frac{i\pi x}{\ell} \left(1 - \cos \frac{i\pi at}{\ell}\right) \tag{x}$$

To obtain the displacement of the end of the bar where the force P is applied, we substitute $x = \ell$ into the solution (x), which gives

$$(u)_{x=\ell} = \frac{Pt^2}{2\rho A \ell} + \frac{2\ell P}{\pi^2 a^2 \rho A} \sum_{i=1}^{\infty} \frac{1}{i^2} \left(1 - \cos \frac{i\pi at}{\ell}\right) \tag{y}$$

For $t = \ell/a$, this becomes

$$(u)_{t=\ell/a} = \frac{P\ell}{2EA} + \frac{4P\ell}{\pi^2 EA} \left(1 + \frac{1}{9} + \frac{1}{25} + \cdots\right) = \frac{P\ell}{EA} \tag{z}$$

At this instant the displacement is equal to the extension of the bar under the action of the uniform tensile force P.

Example 1. Determine the steady-state forced vibrations of a bar with one end fixed and the other free (see Fig. 5.2*a*), produced by a harmonic axial force $P = P_1 \sin \Omega t$ acting at the free end of the bar.
Solution. Equation (k) in this case becomes

$$\ddot{\phi}_i + \omega_i^2 \phi_i = \frac{2(-1)^{(i-1)/2}}{\rho A \ell} P_1 \sin \Omega t$$

and the steady-state forced vibration is

$$\phi_i = \frac{2P_1(-1)^{(i-1)/2}}{\rho A \ell(\omega_i^2 - \Omega^2)} \sin \Omega t$$

Substituting this expression into Eq. (5.8), we can obtain the required forced vibrations of the bar. It is seen that if Ω approaches the value of one of the natural frequencies of the bar, the amplitude of the corresponding type of vibration becomes large.

Example 2. A drill stem is a steel tube 4000 ft long. Considering it as a bar with free ends, find the period τ_1 of the fundamental mode of vibration. In addition, determine the displacement δ of the end ($x = \ell$) at time $t = \tau_1/2$, produced by a tensile stress $\sigma = P/A = 3000$ psi suddenly applied to its end. Take $E = 30 \times 10^6$ psi and $\rho = 0.735 \times 10^{-3}$ lb-sec^2/in.^{4}.
Solution. The velocity of sound in this bar is

$$a = \sqrt{\frac{E}{\rho}} = 202 \times 10^3 \text{ in./sec}$$

and the period of the fundamental mode of vibration is $\tau_1 = 2\ell/a = 0.475$ sec. From Eq. (z) the required displacement is $\delta = (3000)(4000)/(30 \times 10^6) = 0.40$ ft.

5.4 NORMAL-MODE METHOD FOR PRISMATIC BARS

The analyses in the preceding sections have shown certain similarities with the normal-mode method for multidegree systems discussed in Chapter 4. We shall now develop the normal-mode technique for prismatic bars with distributed mass and an infinite number of degrees of freedom. Although the formulations will have particular reference to longitudinal vibrations of prismatic bars, the general concepts of the normal-mode method discussed herein may be extended to the analysis of any type of elastic body.

Let us reconsider the free longitudinal vibrations of the prismatic bar in Fig. 5.1a. The differential equation of motion for a typical element of the bar [see Eqs. (a) and (b) in Sec. 5.2] may be written as

$$m\ddot{u}\, dx - ru''\, dx = 0 \tag{a}$$

where the dots and primes signify differentiation of the displacement u with respect to t and x, respectively. The term $m = \rho A$ represents the *mass of the bar per unit length,* and the quantity $r = EA$ is its *axial rigidity*. When the bar vibrates in its ith natural mode, it has the harmonic motion

$$u_i = X_i(A_i \cos \omega_i t + B_i \sin \omega_i t) \tag{b}$$

Substitution of Eq. (b) into Eq. (a) and rearrangement of terms produces

$$rX_i'' + m\omega_i^2 X_i = 0 \tag{c}$$

for which the solution has the form

$$X_i = C_i \cos \frac{\omega_i x}{a} + D_i \sin \frac{\omega_i x}{a} \tag{d}$$

(where $a = \sqrt{r/m}$), as was discussed previously in Sec. 5.2.

We shall now restate Eq. (c) in the alternative manner

$$X_i'' = \lambda_i X_i \tag{5.11}$$

where

$$\lambda_i = -\frac{m\omega_i^2}{r} = -\left(\frac{\omega_i}{a}\right)^2 \tag{e}$$

Equation (5.11) is in the form of an eigenvalue problem, where the *eigenvalues* λ_i and the *eigenfunctions* X_i are determined from the boundary conditions. This type of eigenvalue problem may be characterized as one in which the second derivative (with respect to x) of the function X_i is equal to the same function multiplied by the constant λ_i.

Let us examine the orthogonality properties of the eigenfunctions by considering modes i and j of the eigenvalue problem, as follows:

$$X_i'' = \lambda_i X_i \tag{f}$$

$$X_j'' = \lambda_j X_j \tag{g}$$

Multiplying Eq. (f) by X_j and Eq. (g) by X_i and integrating the products over the length of the bar, we obtain

$$\int_0^{\ell} X_i'' X_j \, dx = \lambda_i \int_0^{\ell} X_i X_j \, dx \tag{h}$$

$$\int_0^{\ell} X_j'' X_i \, dx = \lambda_j \int_0^{\ell} X_i X_j \, dx \tag{i}$$

Integration of the left-hand sides of these equations by parts yields

$$[X_i' X_j]_0^{\ell} - \int_0^{\ell} X_i' X_j' \, dx = \lambda_i \int_0^{\ell} X_i X_j \, dx \tag{j}$$

$$[X_j' X_i]_0^{\ell} - \int_0^{\ell} X_i' X_j' \, dx = \lambda_j \int_0^{\ell} X_i X_j \, dx \tag{k}$$

For either free or fixed conditions at the ends of the bar, the integrated terms are zero. Therefore, subtraction of Eq. (k) from Eq. (j) produces

$$(\lambda_i - \lambda_j) \int_0^{\ell} X_i X_j \, dx = 0 \tag{l}$$

To satisfy Eq. (l) when $i \neq j$ and the eigenvalues are distinct ($\lambda_i \neq \lambda_j$), the following relationship must be true:

$$\int_0^{\ell} X_i X_j \, dx = 0 \qquad (i \neq j) \tag{5.12}$$

Using this relationship in Eq. (j), we find

$$\int_0^{\ell} X_i' X_j' \, dx = 0 \qquad (i \neq j) \tag{5.13}$$

From Eq. (h) it is also seen that

$$\int_0^\ell X_i'' X_j \, dx = 0 \qquad (i \neq j) \tag{5.14}$$

Thus, the eigenfunctions of the prismatic bar are not only orthogonal with each other, but orthogonality relationships also exist among their derivatives.

For the case of $i = j$, the integral in Eq. (l) may be any constant. If the constant is designated as α_i, we have

$$\int_0^\ell X_i^2 \, dx = \alpha_i \qquad (i = j) \tag{5.15}$$

When the eigenfunctions are normalized in the manner implied by this expression, Eqs. (h) and (j) yield

$$\int_0^\ell X_i'' X_i \, dx = -\int_0^\ell (X_i')^2 \, dx = \lambda_i \alpha_i = -\frac{m\omega_i^2}{r} \alpha_i = -\left(\frac{\omega_i}{a}\right)^2 \alpha_i \tag{5.16}$$

In the following discussion it will become apparent what value of α_i should be selected for response calculations.

As in the preceding section, we shall express the longitudinal motion of a bar in terms of time functions ϕ_i and displacement functions X_i, as follows:

$$u = \sum_i \phi_i X_i \qquad (i = 1, 2, 3, \ldots, \infty) \tag{5.17}$$

Substitution of Eq. (5.17) into the equation of motion (a) for free vibration produces

$$\sum_{i=1}^{\infty} (m\ddot{\phi}_i X_i - r\phi_i X_i'') \, dx = 0$$

Multiplying this expression by the normal function X_j and integrating over the length of the bar, we obtain

$$\sum_{i=1}^{\infty} \left(m\ddot{\phi}_i \int_0^\ell X_i X_j \, dx - r\phi_i \int_0^\ell X_i'' X_j \, dx \right) = 0 \tag{m}$$

From the orthogonality relationships given by Eqs. (5.12) and (5.14), we conclude that (for $i = j$) the equation of motion (m) reduces to the form

$$m_{\mathrm{P}i}\ddot{\phi}_i + r_{\mathrm{P}i}\phi_i = 0 \qquad (i = 1, 2, 3, \ldots, \infty) \tag{5.18}$$

in which

$$m_{\mathrm{P}i} = m\int_0^{\ell} X_i^2\,dx = m\alpha_i \tag{5.19}$$

and

$$r_{\mathrm{P}i} = -r\int_0^{\ell} X_i''X_i\,dx = r\int_0^{\ell} (X_i')^2\,dx = m\omega_i^2\alpha_i \tag{5.20}$$

The symbol $m_{\mathrm{P}i}$ represents a *principal mass* (or *generalized mass*) for the ith mode, and $r_{\mathrm{P}i}$ is a *principal rigidity* (also called a *generalized stiffness*). Thus, Eq. (5.18) constitutes a typical equation of motion for free vibration in *principal coordinates.*

If the eigenfunctions X_i are normalized in such a manner that

$$m_{\mathrm{P}i} = m\int_0^{\ell} X_i^2\,dx = 1 \tag{n}$$

they are said to be normalized with respect to the mass per unit length. With this type of normalization, the principal mass $m_{\mathrm{P}i}$ is unity; the constant α_i [see Eq. (5.15)] is equal to $1/m$; and Eq. (5.20) shows that the principal rigidity becomes

$$r_{\mathrm{P}i} = \omega_i^2 \tag{o}$$

Then the equation of motion (5.18) simplifies to

$$\ddot{\phi}_i + \omega_i^2\phi_i = 0 \qquad (i = 1, 2, 3, \ldots, \infty) \tag{5.21}$$

which is now said to be expressed in *normal coordinates.* If the arbitrary constant α_i is selected as unity, we will have $m_{\mathrm{P}i} = m$ and $r_{\mathrm{P}i} = m\omega_i^2$. Then the equation of motion (5.21) will contain the common factor m, which can be divided out. Thus, for convenience we shall take $\alpha_i = 1$ (instead of $1/m$).

In summary, the differential equation of motion (a) has been transformed to normal coordinates by substituting Eq. (5.17) for u, multiplying by X_j, and integrating over the length of the bar. When the eigenfunctions are normalized so that

$$\int_0^{\ell} X_i^2\,dx = 1 \qquad \int_0^{\ell} X_i''X_i\,dx = -\int_0^{\ell} (X_i')^2\,dx = -\left(\frac{\omega_i}{a}\right)^2 \tag{5.22}$$

the generalized mass for each of the principal coordinates is equal to m; and the generalized stiffness is $m\omega_i^2$. However, the common factor m can be divided out to produce Eq. (5.21).

We shall now use the normal-mode approach for determining the longitudinal response of a bar to initial conditions of displacement and velocity. As in Sec. 5.2, we assume that when $t=0$ the initial displacements are expressed as $u_0 = f_1(x)$ and that the initial velocities are given by $\dot{u}_0 = f_2(x)$. Expanding u_0 and $\dot{u}_0$ in the form of Eq. (5.17), we have

$$\sum_{i=1}^{\infty} \phi_{0i} X_i = f_1(x) \tag{p}$$

$$\sum_{i=1}^{\infty} \dot{\phi}_{0i} X_i = f_2(x) \tag{q}$$

Multiplying these expressions by X_j and integrating over the length of the bar, we obtain

$$\sum_{i=1}^{\infty} \phi_{0i} \int_0^{\ell} X_i X_j \, dx = \int_0^{\ell} f_1(x) X_j \, dx \tag{r}$$

$$\sum_{i=1}^{\infty} \dot{\phi}_{0i} \int_0^{\ell} X_i X_j \, dx = \int_0^{\ell} f_2(x) X_j \, dx \tag{s}$$

From the orthogonality and normalization relationships given by Eqs. (5.12) and (5.22), it is seen that (for $i=j$) Eqs. (r) and (s) yield the following initial conditions in normal coordinates:

$$\phi_{0i} = \int_0^{\ell} f_1(x) X_i \, dx \tag{5.23}$$

$$\dot{\phi}_{0i} = \int_0^{\ell} f_2(x) X_i \, dx \tag{5.24}$$

Therefore, the free-vibrational responses of the normal modes are

$$\phi_i = \phi_{0i} \cos \omega_i t + \frac{\dot{\phi}_{0i}}{\omega_i} \sin \omega_i t \qquad (i = 1, 2, 3, \ldots, \infty) \tag{t}$$

Substitution of this expression into Eq. (5.17) produces the combined response of all modes as

$$u = \sum_{i=1}^{\infty} X_i \left(\phi_{0i} \cos \omega_i t + \frac{\dot{\phi}_{0i}}{\omega_i} \sin \omega_i t \right) \tag{5.25}$$

which represents the general form of the specific solutions given by Eqs. (5.6) and (5.7) in Sec. 5.2.

Next, we shall apply the normal-mode method to the calculation of forced longitudinal response of prismatic bars. For this purpose it will be assumed that the bar is subjected to a distributed force per unit length,

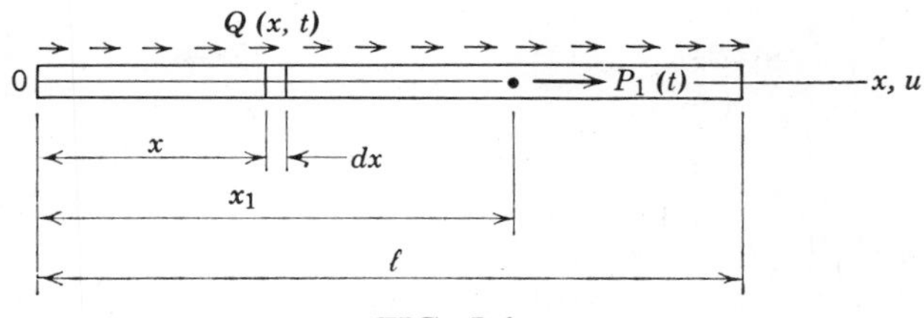

FIG. 5.4

$Q(x, t)$ as indicated in Fig. 5.4. In this case the differential equation of motion for a typical element of the bar becomes

$$m\ddot{u}\,dx - ru''\,dx = Q(x, t)\,dx \tag{u}$$

For convenience, we divide both sides of this equation by $m = \rho A$ (the mass per unit length) to obtain

$$\ddot{u}\,dx - a^2u''\,dx = q(x, t)\,dx \tag{v}$$

where $a^2 = r/m = E/\rho$ and $q(x, t) = Q(x, t)/m$. Equation (*v*) will be transformed to normal coordinates by substituting expression (5.17) for u, multiplying by X_j, and integrating over the length of the bar, as follows:

$$\sum_{i=1}^{\infty}\left(\ddot{\phi}_i\int_0^{\ell} X_iX_j\,dx - a^2\phi_i\int_0^{\ell} X_i''X_j\,dx\right) = \int_0^{\ell} X_jq(x, t)\,dx \tag{w}$$

Using the conditions of orthogonality and normalization given by Eqs. (5.12), (5.14), and (5.22), we obtain (for $i = j$)

$$\ddot{\phi}_i + \omega_i^2\phi_i = \int_0^{\ell} X_iq(x, t)\,dx \qquad (i = 1, 2, 3, \ldots, \infty) \tag{5.26}$$

This expression represents a typical equation of motion in normal coordinates, and the integral on the right is the ith *normal-mode load*.

The response of the ith vibrational mode is found by the Duhamel integral to be

$$\phi_i = \frac{1}{\omega_i}\int_0^{\ell} X_i\int_o^{t} q(x, t')\sin\omega_i(t - t')\,dt'\,dx \tag{5.27}$$

Substitution of this time function into Eq. (5.17) gives the total vibrational response as

$$u = \sum_{i=1}^{\infty} \frac{X_i}{\omega_i} \int_0^{\ell} X_i \int_0^t q(x, t') \sin \omega_i(t - t')\, dt'\, dx \tag{5.28}$$

If a load $P_1(t)$ is concentrated at point x_1, as shown in Fig. 5.4, no integration over the length is required. The response to this type of loading is calculated from the simpler expression

$$u = \sum_{i=1}^{\infty} \frac{X_i X_{i1}}{\omega_i} \int_0^t q_1(t') \sin \omega_i(t - t')\, dt' \tag{5.29}$$

in which the symbol X_{i1} represents the normal function X_i evaluated at point x_1; and $q_1(t) = P_1(t)/m$.

The normal-mode method for calculating the response to applied loads is equivalent to the virtual-work approach described in the preceding section. The following examples demonstrate the use of Eqs. (5.28) and (5.29) for distributed and concentrated forces.

Example 1. Assuming that the bar in Fig. 5.4 is fixed at the left end and free at the right end, determine its response to the sudden application of a uniformly distributed longitudinal force of intensity Q per unit length.
Solution. Because the load intensity $q = Q/m$ does not vary with either x or t, it may be taken outside of the integals in Eq. (5.28). From free-vibration analysis of this bar, we have

$$\omega_i = \frac{i\pi a}{2\ell} \qquad X_i = D_i \sin \frac{\omega_i x}{a} \qquad (i = 1, 3, 5, \ldots, \infty)$$

To normalize X_i in accordance with Eq. (5.22), we must take $D_i = \sqrt{2/\ell}$. Then Eq. (5.28) yields

$$u = \frac{2Q}{\ell m} \sum_{i=1,3,5,\ldots}^{\infty} \frac{1}{\omega_i} \sin \frac{\omega_i x}{a} \int_0^{\ell} \sin \frac{\omega_i x}{a} \int_0^t \sin \omega_i(t - t')\, dt'\, dx$$

$$= \frac{4Q}{\pi m} \sum_{i=1,3,5,\ldots}^{\infty} \frac{1}{i\omega_i^2} \sin \frac{\omega_i x}{a} (1 - \cos \omega_i t)$$

$$= \frac{16\ell^2 Q}{\pi^3 a^2 m} \sum_{i=1,3,5,\ldots}^{\infty} \frac{1}{i^3} \sin \frac{i\pi x}{2\ell} \left(1 - \cos \frac{i\pi a t}{2\ell}\right) \tag{x}$$

Example 2. For the bar with end conditions as in Example 1, calculate the response to a concentrated force P suddenly applied at the right-hand end ($x_1 = \ell$).

Solution. In this case we use Eq. (5.29) to obtain the response, as follows:

$$u = \frac{2P}{\ell m} \sum_{i=1,3,5,\ldots}^{\infty} \frac{1}{\omega_i} \sin \frac{\omega_i x}{a} \sin \frac{\omega_i \ell}{a} \int_0^t \sin \omega_i (t - t')\, dt'$$

$$= \frac{2P}{\ell m} \sum_{i=1,3,5,\ldots}^{\infty} \frac{1}{\omega_i^2} \sin \frac{\omega_i x}{a} \sin \frac{\omega_i \ell}{a} (1 - \cos \omega_i t)$$

$$= \frac{8\ell P}{\pi^2 a^2 m} \sum_{i=1,3,5,\ldots}^{\infty} \frac{(-1)^{(i-1)/2}}{i^2} \sin \frac{i\pi x}{2\ell} \left(1 - \cos \frac{i\pi a t}{2\ell}\right) \qquad (y)$$

This expression is the same as Eq. (m) in Sec. 5.3.

5.5 PRISMATIC BAR WITH A MASS OR SPRING AT THE END

In addition to the fixed or free end conditions discussed in the previous sections, we may encounter a concentrated mass or an elastic restraint, both of which are present at the right-hand end of the bar in Fig. 5.5. these two types of end conditions will be treated in this section, using the normal-mode method.

Let us consider first the case where the stiffness constant k of the spring in Fig. 5.5 is equal to zero, and we have only the lumped mass M at the right-hand end of the bar. In this instance the inertial force exerted by the concentrated mass upon the end of the bar during vibration will be equal to $-M(\ddot{u})_{x=\ell}$. Thus, the boundary conditions for the bar may be written as

$$(u)_{x=0} = 0 \qquad r(u')_{x=\ell} = -M(\ddot{u})_{x=\ell} \qquad (a)$$

Because the second of these conditions involves the motion of the lumped mass, the analysis will be somewhat more complicated than that for the bar alone. However, the motion will still be harmonic, and we may again take for the ith mode

$$u_i = X_i(A_i \cos \omega_i t + B_i \sin \omega_i t) \qquad (b)$$

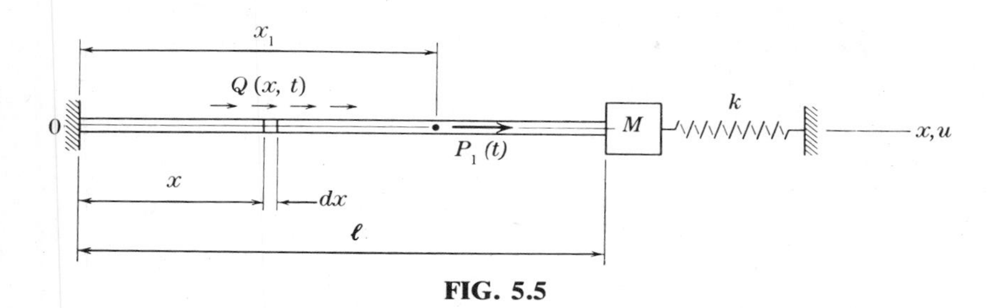

FIG. 5.5

Substitution of Eq. (*b*) into the boundary conditions (*a*) produces

$$X_{i0} = 0 \qquad rX'_{i\ell} = M\omega_i^2 X_{i\ell} \tag{c}$$

where the subscripts 0 and ℓ denote the locations $x = 0$ and $x = \ell$, respectively. As before, the normal functions have the form

$$X_i = C_i \cos\frac{\omega_i x}{a} + D_i \sin\frac{\omega_i x}{a} \tag{d}$$

The first boundary condition in Eq. (*c*) requires that $C_i = 0$, and the second condition yields the relationship

$$\frac{r\omega_i}{a}\cos\frac{\omega_i \ell}{a} = M\omega_i^2 \sin\frac{\omega_i \ell}{a} \tag{e}$$

We may rewrite this expression more succinctly as

$$\xi_i \tan \xi_i = \eta \tag{5.30}$$

in which $\xi_i = \omega_i \ell / a$ and $\eta = m\ell/M$ (the ratio of the mass of the bar to that of the concentrated mass).

Equation (5.30) is the frequency equation for the case under consideration. Because this expression is transcendental, the angular frequencies ω_i must be found by trial. The fundamental mode of vibration is usually of greatest interest; and for various values of the mass ratio η the corresponding values of ξ_1 (for the first mode) are given as follows:

η =	0.01	0.10	0.30	0.50	0.70	0.90	1.00	1.50
ξ_1 =	0.10	0.32	0.52	0.65	0.75	0.82	0.86	0.98

η =	2.00	3.00	4.00	5.00	10.0	20.0	100.0	∞
ξ_1 =	1.08	1.20	1.27	1.32	1.42	1.52	1.57	$\pi/2$

If the mass of the bar is small compared to that of the attached mass, the vaues of η and ξ_1 will both be small; and Eq. (5.30) can be simplified by taking $\tan \xi_1 \approx \xi_1$. Then we have

$$\xi_1^2 \approx \eta = \frac{m\ell}{M} \qquad \xi_1 = \frac{\omega_1 \ell}{a} \approx \sqrt{\frac{m\ell}{M}}$$

Hence,

$$\omega_1 \approx \frac{a}{\ell}\sqrt{\frac{m\ell}{m}} = \sqrt{\frac{EA}{M\ell}}$$

where EA/ℓ is the *axial stiffness* of the bar. This result coincides with that obtained with treating the bar and mass as a one-degree system. On the other hand, if the mass ratio η is large, the frequency equation becomes

$$\tan \frac{\omega_i \ell}{a} = \infty$$

This expression yields the angular frequencies

$$\omega_i = \frac{i\pi a}{2\ell} \qquad (i = 1, 3, 5, \ldots, \infty)$$

which are the same as those obtained in Sec. 5.2.

In order to develop orthogonality relationships for the bar with a mass at the end, we rewrite the eigenvalue problem [see Eq. (5.11), Sec. 5.4] for two distinct modes i and j, as follows:

$$rX_i'' = -m\omega_i^2 X_i \tag{f}$$

$$rX_j'' = -m\omega_j^2 X_j \tag{g}$$

Multiplying the first of these expressions by X_j and the second by X_i and integrating over the length of the bar, we have

$$r\int_0^\ell X_i''X_j\,dx = -m\omega_i^2\int_0^\ell X_iX_j\,dx \tag{h}$$

$$r\int_0^\ell X_j''X_i\,dx = -m\omega_j^2\int_0^\ell X_iX_j\,dx \tag{i}$$

The mass located at $x = \ell$ must also be included in the orthogonality relationships, and the second boundary condition in Eqs. (c) for modes i and j may be written as

$$rX_{i\ell}'X_{j\ell} = M\omega_i^2 X_{i\ell}X_{j\ell} \tag{j}$$

$$rX_{j\ell}'X_{i\ell} = M\omega_j^2 X_{i\ell}X_{j\ell} \tag{k}$$

where the first expression has been multiplied by $X_{j\ell}$ and the second by $X_{i\ell}$. Subtraction of Eqs. (j) and (k) from Eqs. (h) and (i) produces the combined relationships

$$r\int_0^\ell X_i''X_j\,dx - rX_{i\ell}'X_{j\ell} = -\omega_i^2\left(m\int_0^\ell X_iX_j\,dx + MX_{i\ell}X_{j\ell}\right) \tag{l}$$

$$r\int_0^\ell X_j''X_i\,dx - rX_{j\ell}'X_{i\ell} = -\omega_j^2\left(m\int_0^\ell X_iX_j\,dx + MX_{i\ell}X_{j\ell}\right) \tag{m}$$

Integration of the expressions on the left by parts results in

$$-rX'_{i0}X_{j0} - r\int_0^\ell X'_iX'_j\,dx = -\omega_i^2\left(m\int_0^\ell X_iX_j\,dx + MX_{i\ell}X_{j\ell}\right) \qquad (n)$$

$$-rX'_{j0}X_{i0} - r\int_0^\ell X'_iX'_j\,dx = -\omega_j^2\left(m\int_0^\ell X_iX_j\,dx + MX_{i\ell}X_{j\ell}\right) \qquad (o)$$

Because the integrated terms on the left are zero, subtraction of Eq. (n) from Eq. (o) yields

$$(\omega_i^2 - \omega_j^2)\left(m\int_0^\ell X_iX_j\,dx + MX_{i\ell}X_{j\ell}\right) = 0 \qquad (p)$$

When $i \neq j$ (and $\omega_i^2 \neq \omega_j^2$), Eq. ($p$) gives the orthogonality relationship

$$m\int_0^\ell X_iX_j\,dx + MX_{i\ell}X_{j\ell} = 0 \qquad (i \neq j) \qquad (5.31)$$

From Eq. (n), we also see that

$$r\int_0^\ell X'_iX'_j\,dx = 0 \qquad (i \neq j) \qquad (5.32)$$

and Eq. (l) gives

$$r\int_0^\ell X''_iX_j\,dx - rX'_{i\ell}X_{j\ell} = 0 \qquad (i \neq j) \qquad (5.33)$$

Comparing these expressions with their counterparts in Sec. 5.2 (see Eqs. (5.12), (5.13), and (5.14)], we see that additional terms are present in Eqs. (5.31) and (5.33).

For the case of $i = j$, the second factor in Eq. (p) may be any constant. We shall arbitrarily select the constant to be m, which results in

$$m\int_0^\ell X_i^2\,dx + MX_{i\ell}^2 = m \qquad (i = j) \qquad (5.34)$$

When the eigenfunctions are normalized to satisfy this expression, Eqs. (l) and (n) give

$$r\int_0^\ell X''_iX_i\,dx - rX'_{i\ell}X_{i\ell} = -r\int_0^\ell (X'_i)^2\,dx = -m\omega_i^2 \qquad (5.35)$$

To determine the longitudinal response of the system to the initial conditions $u_0 = f_1(x)$ and $\dot{u}_0 = f_2(x)$ at time $t = 0$, we first evaluate the initial displacement $u_{0\ell} = f_1(\ell)$ and velocity $\dot{u}_{0\ell} = f_2(\ell)$ of the attached mass at $x = \ell$. Then the initial conditions for the bar and the mass are expanded in terms of time and displacement functions [see Eq. (5.17), Sec. 5.4], as follows:

$$\sum_{i=1}^{\infty} \phi_{0i} X_i = f_1(x) \qquad \sum_{i=1}^{\infty} \dot{\phi}_{0i} X_i = f_2(x) \tag{q}$$

$$\sum_{i=1}^{\infty} \phi_{0i} X_{i\ell} = f_1(\ell) \qquad \sum_{i=1}^{\infty} \dot{\phi}_{0i} X_{i\ell} = f_2(\ell) \tag{r}$$

Next, we multiply Eqs. (q) by mX_j and integrate over the length of the bar. In addition, Eqs. (r) are multiplied by $MX_{j\ell}$ and added to the results from Eqs. (q) to obtain

$$\sum_{i=1}^{\infty} \phi_{0i}\left(m\int_0^\ell X_i X_j\, dx + M X_{i\ell} X_{j\ell}\right) = m\int_0^\ell f_1(x) X_j\, dx + M f_1(\ell) X_{j\ell} \tag{s}$$

$$\sum_{i=1}^{\infty} \dot{\phi}_{0i}\left(m\int_0^\ell X_i X_j\, dx + M X_{i\ell} X_{j\ell}\right) = m\int_0^\ell f_2(x) X_j\, dx + M f_2(\ell) X_{j\ell} \tag{t}$$

Using the orthogonality and normalization relationships given by Eqs. (5.31) and (5.34), we find that (for $i = j$) Eqs. (s) and (t) produce the following initial conditions in normal coordinates:

$$\phi_{0i} = \int_0^\ell f_1(x) X_i\, dx + \frac{\ell}{\eta} f_1(\ell) X_{i\ell} \tag{5.36}$$

$$\dot{\phi}_{0i} = \int_0^\ell f_2(x) X_i\, dx + \frac{\ell}{\eta} f_2(\ell) X_{i\ell} \tag{5.37}$$

With these expressions for ϕ_{0i} and $\dot{\phi}_{0i}$, the response to initial conditions is the same as in Eq. (5.25) of Sec. 5.4.

To show how the response of the system to applied longitudinal forces may be calculated, we set up the differential equation of motion for a typical element of the bar (see Fig. 5.5) as

$$m\ddot{u}\, dx - ru''\, dx = Q(x, t)\, dx \tag{u}$$

At the right-hand end of the bar, we have from Eq. (a) the relationship

$$M\ddot{u}_\ell + ru'_\ell = 0 \tag{v}$$

Expanding Eq. (u) in accordance with Eq. (5.17), we then multiply it by X_j and integrate over the length to obtain

$$\sum_{i=1}^{\infty}\left(m\ddot{\phi}_i\int_0^{\ell} X_iX_j\,dx - r\phi_i\int_0^{\ell} X_i''X_j\,dx\right) = \int_0^{\ell} X_jQ(x, t)\,dx \tag{w}$$

Similarly, expansion of Eq. (v) and multiplication by $X_{j\ell}$ gives

$$\sum_{i=1}^{\infty}(M\ddot{\phi}_iX_{i\ell}X_{j\ell} + r\phi_iX'_{i\ell}X_{j\ell}) = 0 \tag{x}$$

Addition of Eqs. (w) and (x) results in

$$\sum_{i=1}^{\infty}\left[\ddot{\phi}_i\left(m\int_0^{\ell} X_iX_j\,dx + MX_{i\ell}X_{j\ell}\right) - r\phi_i\left(\left(\int_0^{\ell} X_i''X_j\,dx - X'_{i\ell}X_{j\ell}\right)\right)\right] = \int_0^{\ell} X_jQ(x, t)\,dx \tag{y}$$

From the conditions of orthogonality and normalization in Eqs. (5.31), (5.33), (5.34), and (5.35), we see that (for $i = j$)

$$m\ddot{\phi}_i + m\omega_i^2\phi_i = \int_0^{\ell} X_iQ(x, t)\,dx \tag{z}$$

When this equation is divided by m, it becomes the same as Eq. (5.26) in Sec. 5.4, in which $q(x, t) = Q(x, t)/m$. Therefore, the response of the ith mode is again given by Eq. (5.27), and the total response is found from Eq. (5.28). For a concentrated force $P_1(t)$ located at point x_1 (see Fig. 5.5), the response may be calculated from Eq. (5.29). When $x_1 = \ell$, the concentrated force is applied directly to the mass M; and no special treatment is required for that condition.

All of the preceding discussion in this section pertains to the case in which the stiffness constant k of the spring in Fig. 5.5 is equal to zero and the concentrated mass M is nonzero. We shall now consider the opposite situation, in which $k \neq 0$ and $M = 0$. In this instance the force exerted by the spring upon the end of the bar during vibration will be equal to $-k(u)_{x=\ell}$. Therefore, the boundary conditions for the bar become

$$(u)_{x=0} = 0 \qquad r(u')_{x=\ell} = -k(u)_{x=\ell} \tag{a'}$$

Following a line of reasoning similar to that for the case of the lumped mass, we conclude that the normal functions are again given by

$$X_i = D_i \sin\frac{\omega_i x}{a} \tag{b'}$$

and that the second boundary condition in Eq. (a') leads to the relationship

$$\frac{r\omega_i}{a}\cos\frac{\omega_i\ell}{a} = -k\sin\frac{\omega_i\ell}{a} \tag{c'}$$

If we define the dimensionless parameter $\zeta_i = m\ell\omega_i^2/k$, the compact form of the frequency equation becomes

$$\xi_i \tan \xi_i = -\zeta_i \tag{5.38}$$

in which $\xi_i = \omega_i\ell/a$, as before. Thus, the series of numerical values given previously for the first mode still apply if the parameter η is replaced by $-\zeta_i$. When the stiffness constant k of the spring is small ($k \to 0$), Eq. (c') becomes the frequency equation for a bar with no restraint at the right-hand end (but fixed at the left end). On the other hand, when k is large ($k \to \infty$), Eq. (c') divided by k becomes the frequency equation for a bar with both ends fixed.

To develop orthogonality relationships for the bar with a spring at the end, we proceed in a manner similar to that for the bar with a mass at the end. In this case, however, Eqs. (j) and (k) are replaced by

$$rX'_{i\ell}X_{j\ell} = -kX_{i\ell}X_{j\ell} \tag{d'}$$

and

$$rX'_{j\ell}X_{i\ell} = -kX_{i\ell}X_{j\ell} \tag{e'}$$

Subtraction of these expressions from Eqs. (h) and (i) yields the combined relationships

$$r\int_0^\ell X_i''X_j\,dx - rX'_{i\ell}X_{j\ell} - kX_{i\ell}X_{j\ell} = -m\omega_i^2\int_0^\ell X_iX_j\,dx \tag{f'}$$

$$r\int_0^\ell X_j''X_i\,dx - rX'_{j\ell}X_{i\ell} - kX_{i\ell}X_{j\ell} = -m\omega_j^2\int_0^\ell X_iX_j\,dx \tag{g'}$$

Integration of terms on the left by parts and subtraction of Eq. (f') from Eq. (g') leads to the following orthogonality relationships for this system:

$$m\int_0^\ell X_iX_j\,dx = 0 \qquad (i \neq j) \tag{5.39}$$

$$r\int_0^\ell X_i'X_j'\,dx + kX_{i\ell}X_{j\ell} = 0 \qquad (i \neq j) \tag{5.40}$$

$$r\int_0^\ell X_i''X_j\,dx - rX'_{i\ell}X_{j\ell} - kX_{i\ell}X_{j\ell} = 0 \qquad (i \neq j) \tag{5.41}$$

Comparison of these expressions with Eqs. (5.12), (5.13), and (5.14) in Sec. 5.4 shows that Eq. (5.39) is the same as for the system without the spring. However, additional terms appear in Eqs. (5.40) and (5.41).

For the case of $i = j$, the normalization procedure is arbitrarily selected to give

$$m \int_0^\ell X_i^2 \, dx = m \tag{5.42}$$

and

$$r \int_0^\ell X_i'' X_i \, dx - r X_{i\ell}' X_{i\ell} - k X_{i\ell}^2 = -r \int_0^\ell (X_i')^2 \, dx - k X_{i\ell}^2 = -m\omega_i^2 \tag{5.43}$$

Because Eq. (5.42) coincides with the type of normalization used in Sec. 5.4 [see Eq. (5.22)], the formulas derived in that section for response to initial conditions [Eqs. (5.23), (5.24), and (5.25)] apply also to the present case. Furthermore, the response of the system to applied longitudinal forces can again be calculated using Eqs. (5.28) and (5.29) in Sec. 5.4. Thus, we see that although the presence of the spring affects the frequencies and mode shapes of the bar, the normal-mode method for determining its dynamic response is unchanged.

If both the mass and the spring are present ($M \neq 0$ and $k \neq 0$), as shown in Fig. 5.5, the boundary conditions become

$$(u)_{x=0} = 0 \qquad r(u')_{x=\ell} = -M(\ddot{u})_{x=\ell} - k(u)_{x=\ell} \tag{h'}$$

In this combined case the normal functions are still represented by Eq. (c'), and the second boundary condition in Eq. (h') results in

$$\frac{r\omega_i}{a} \cos \frac{\omega_i \ell}{a} = (M\omega_i^2 - k) \sin \frac{\omega_i \ell}{a} \tag{i'}$$

from which the brief form of the frequency equation is found to be

$$\xi_i \tan \xi_i = \frac{\eta \zeta_i}{\zeta_i - \eta} \tag{5.44}$$

The orthogonality relationships for this case consist of Eqs. (5.31), (5.40), and (5.41), and the normalization requirements are covered by Eqs. (5.34) and (5.43). Initial displacements and velocities in normal coordinates are represented by Eqs. (5.36) and (5.37), and the solutions for response to initial conditions and applied forces are given by Eqs. (5.25), (5.28), and (5.29) in Sec. 5.4.

The techniques in this section can be extended to the more complicated case of a mass and a spring at both ends of the member. In such a case the normal functions will contain two nonzero parts, as represented by Eq. (d); and more terms will appear in the frequency equation. In additon, the orthogonality and normalization relationships will contain mass and stiffness terms for both ends of the member, whereas the initial conditions in normal coordinates can be calculated on the basis of mass effects only. It is left as an exercise for the reader to develop these more complicated (but straightforward) expressions pertaining to longitudinal vibrations of prismatic bars. The mathematically similar case of a shaft with disks at both ends will be discussed in Sec. 5.7, and a stretched wire with lateral restraining springs at both ends is treated in Sec. 5.8.

5.6 BARS SUBJECTED TO LONGITUDINAL SUPPORT MOTIONS

We shall now consider the longitudinal response of prismatic bars caused by support motions instead of applied forces. for example, if the ground in Fig. 5.6 translates in the x direction in accordance with the function

$$u_g = g(t) \tag{a}$$

the differential equation of motion for a typical element of the bar may be written as

$$m\ddot{u}\, dx - r(u - u_g)''\, dx = 0 \tag{b}$$

To solve this equation, we introduce the notation

$$u^* = u - u_g \tag{c}$$

which represents the displacement of any point on the bar relative to the rigid-body translation of the ground. In addition, the absolute acceleration u of any point may be expressed as

$$\ddot{u} = \ddot{u}^* + \ddot{u}_g \tag{d}$$

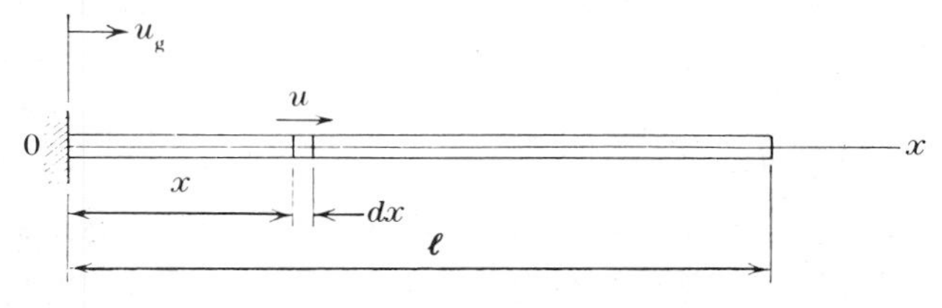

FIG. 5.6

Substitution of Eqs. (c) and (d) into Eq. (b) produces

$$m(\ddot{u}^* + \ddot{u}_g)\,dx - r(u^*)''\,dx = 0$$

or

$$m\ddot{u}^*\,dx - r(u^*)''\,dx = -m\ddot{u}_g\,dx = -m\ddot{g}(t)\,dx \tag{e}$$

Comparison of this equation of motion with Eq. (u) in Sec. 5.4 shows that the equivalent distributed loading in the relative coordinate is $-m\ddot{g}(t)$. This formulation is analogous to those developed in previous chapters for discrete-parameter systems where the acceleration of the ground is specified [e.g., see Eq. (l) in Sec. 1.6].

For convenience, we divide Eq. (e) by the mass m per unit length to obtain

$$\ddot{u}^*\,dx - a^2(u^*)''\,dx = -\ddot{g}(t)\,dx \tag{f}$$

Comparing this equation with Eq. (v) in Sec. 5.4, we see that $q(x, t)$ is replaced by $-\ddot{g}(t)$; and Eq. (5.26) becomes

$$\ddot{\phi}_i + \omega_i^2\phi_i = -\ddot{g}(t)\int_0^\ell X_i\,dx \qquad (i = 1, 2, 3, \ldots, \infty) \tag{5.45}$$

where the right-hand term is the ith *equivalent normal-mode load.* The response of the ith vibrational mode is calculated by the Duhamel integral as

$$\phi_i = -\frac{1}{\omega_i}\int_0^\ell X_i\,dx \int_0^t \ddot{g}(t')\sin\omega_i(t - t')\,dt' \tag{5.46}$$

Superimposing all of the normal-mode responses in accordance with Eq. (5.17), we obtain

$$u^* = -\sum_{i=1}^{\infty}\frac{X_i}{\omega_i}\int_0^\ell X_i\,dx \int_0^t \ddot{g}(t')\sin\omega_i(t - t')\,dt' \tag{5.47}$$

which represents the vibrational motion of any point on the bar relative to the motion of the ground. The total solution is determined by adding the relative (vibrational) motion to the ground motion, as follows:

$$u = u_g + u^* = g(t) + u^* \tag{5.48}$$

Thus, by using the second derivative (with respect to time) of $u_0 = g(t)$ in Eq. (5.47) and the function itself in Eq. (5.48), we can calculate the response of the bar to longitudinal rigid-body translations of the ground.

The function $u_g = g(t)$ in Eq. (5.48) may be characterized as the *rigid-body motion of the massless bar* (as well as the ground), and the relative motion u^* accounts for the inertia forces distributed over the length of the bar.

In a manner that is similar in certain aspects to the above derivation for rigid-body ground motion, we shall formulate the longitudinal response of a bar to independent translations of restraints at both ends. For this purpose let us consider Fig. 5.7 and the functions

$$u_{g1} = g_1(t) \qquad u_{g2} = g_2(t) \tag{g}$$

which represent independent translations of the left and right ends, respectively. Although these translations may be large, their difference at any time t is assumed to be small.

When analyzing the longitudinal response of a bar to independent support motions, it is advantageous to consider the absolute displacement u as the sum

$$u = u_{st} + u^* \tag{h}$$

In this expression the symbol u_{st} denotes the displacement of any point on the massless bar (with both ends restrained) due to the given support motons. Such a function is determined by static analysis, and for a prismatic bar it takes the form

$$u_{st} = \frac{\ell - x}{\ell} g_1(t) + \frac{x}{\ell} g_2(t) = (u_{st})_1 + (u_{st})_2 \tag{5.49}$$

This part of the displacement will be characterized as the *flexible-body motion of the massless bar.* The symbol u^* in Eq. (h) now represents the displacement of any point on the bar relative to u_{st}. Thus, the relative motion u^* will again be associated with the inertia forces distributed over the length of the bar.

Similarly, the acceleration $\ddot{u}$ of any point on the bar may be written as

$$\ddot{u} = \ddot{u}_{st} + \ddot{u}^* \tag{i}$$

which is obtained by differentiating Eq. (h) twice with respect to time. The equation of motion for a typical element of the bar in Fig. 5.7 will now be

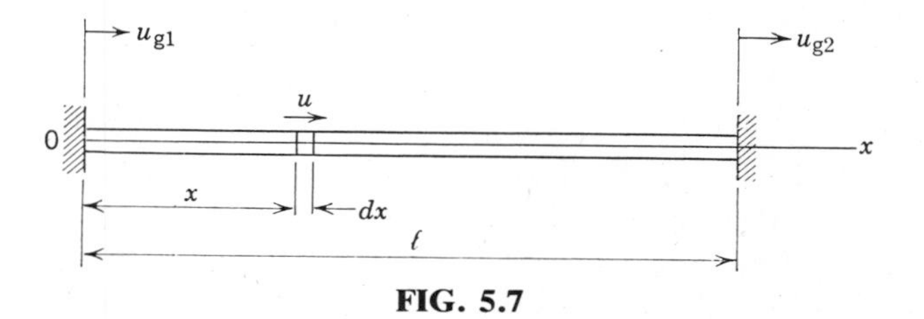

FIG. 5.7

written in terms of the definitions in Eqs. (*h*) and (*i*), as follows:

$$m(\ddot{u}_{st} + \ddot{u}^*)\,dx - r(u_{st} + u^*)''\,dx = 0 \tag{j}$$

For a prismatic bar the term u''_{st} vanishes, and Eq. (*j*) may be rewritten as

$$m\ddot{u}^*\,dx - r(u^*)''\,dx = -m\ddot{u}_{st}(x, t)\,dx \tag{k}$$

which is comparable to Eq. (*e*) for rigid-body ground motion. We see from Eq. (*k*) that in the case under consideration the equivalent distributed loading in relative coordinates is $-m\ddot{u}_{st}(x, t)$. Dividing Eq. (*k*) by the mass m per unit length, we obtain

$$\ddot{u}^*\,dx - a^2(u^*)''\,dx = -\ddot{u}_{st}(x, t)\,dx \tag{l}$$

and in this instance the term $-\ddot{u}_{st}(x, t)$ appears in place of $q(x, t)$. Therefore, the ith equation of motion in normal coordinates becomes

$$\ddot{\phi}_i + \omega_i^2\phi_i = -\int_0^\ell X_i\ddot{u}_{st}(x, t)\,dx \qquad (i = 1, 2, 3, \ldots, \infty) \tag{5.50}$$

in which the ith equivalent normal-mode load is given by the right-hand expression. Using Duhamel's integral, we determine the response of the ith vibrational mode as

$$\phi_i = -\frac{1}{\omega_i}\int_0^\ell X_i \int_0^t \ddot{u}_{st}(x, t')\sin\omega_i(t - t')\,dt'\,dx \tag{5.51}$$

and superposition of all modes yields

$$u^* = -\sum_{i=1}^{\infty}\frac{X_i}{\omega_i}\int_0^\ell X_i \int_0^t \ddot{u}_{st}(x, t')\sin\omega_i(t - t')\,dt'\,dx \tag{5.52}$$

This expression represents the response of any point on the bar relative to the motion u_{st} of the massless bar. To obtain the total response, we add the two types of motion, as follows:

$$u = u_{st} + u^* = \frac{\ell - x}{\ell}g_1(t) + \frac{x}{\ell}g_2(t) + u^* \tag{5.53}$$

In summary, the longitudinal response of a bar to independent translations of end restraints may be calculated by adding the relative motion u^* (which may also be called the vibrational motion) to the flexible-body motion u_{st} of the massless bar. Although the displacement u_{st} is determined by static analysis, it is a function of both x and t. The displacement u^* represents the deviation of the total response u from the displacement u_{st} and accounts for the nonzero mass of the bar. However, the equivalent distributed loading $-m\ddot{u}_{st}$ in Eq. (k) is not the same as either the distributed inertia force $-m\ddot{u}$ in the original coordinates or the distributed inertia force $-m\ddot{u}^*$ in the relative coordinates. It is a term-that can be treated as if it were an applied loading only by virtue of the change of coordinates from absolute to relative motions. The eigenvalues and eigenfunctions for the relative coordinates are the same as for the original coordinates (with both ends of the bar fixed), because the coefficients m and r in Eq. (k) have the same values as before.

If the functions $g_1(t)$ and $g_2(t)$ at the two ends of the bar are equal, so that

$$g_1(t) = g_2(t) = g(t)$$

then Eq. (5.49) yields $u_{st} = u_g = g(t)$. In this case the flexible-body motion simplifies to rigid-body motion (with both ends restrained), and Eqs. (5.50) through (5.53) become the same as Eqs. (5.45) through (5.48).

Example 1. Suppose that the support in Fig. 5.6 translates in accordance with the parabolic function $u_g = g(t) = u_1(t/t_1)^2$, where u_1 is the displacement at time t_1. Assuming that the bar is initially at rest, determine its response to this ground motion.

Solution. From previous analyses of this bar, we have

$$\omega_i = \frac{i\pi a}{2\ell} \qquad X_i = \sqrt{\frac{2}{\ell}}\sin\frac{\omega_i x}{a} \qquad (i = 1, 3, 5, \ldots, \infty)$$

where X_i has been normalized to satisfy Eq. (5.22) in Sec. 5.4. Differentiation of the rigid-body motion u_g twice with respect to time gives

$$\ddot{u}_g = \ddot{g}(t) = \frac{2u_1}{t_1^2}$$

Then Eq. (5.47) yields the relative motion

$$\begin{aligned} u^* &= -\frac{4u_1}{\ell t_1^2}\sum_{i=1,3,5,\ldots}^{\infty}\frac{1}{\omega_i}\sin\frac{\omega_i x}{a}\int_0^\ell \sin\frac{\omega_i x}{a}\,dx\int_0^t \sin\omega_i(t-t')\,dt' \\ &= -\frac{8u_1}{\pi t_1^2}\sum_{i=1,3,5,\ldots}^{\infty}\frac{1}{i\omega_i^2}\sin\frac{\omega_i x}{a}(1-\cos\omega_i t) \\ &= -\frac{32\ell^2 u_1}{\pi^3 a^2 t_1^2}\sum_{i=1,3,5,\ldots}^{\infty}\frac{1}{i^3}\sin\frac{i\pi x}{2\ell}\left(1-\cos\frac{i\pi a t}{2\ell}\right) \end{aligned} \qquad (m)$$

and from Eq. (5.48) the total response is

$$u = \frac{u_1}{t_1^2}\left[t^2 - \frac{32\ell^2}{\pi^3 a^2}\sum_{i=1,3,5,\ldots}^{\infty}\frac{1}{i^3}\sin\frac{i\pi x}{2\ell}\left(1-\cos\frac{i\pi a t}{2\ell}\right)\right] \tag{n}$$

Example 2. Considering the bar in Fig. 5.7, let us assume that the supports are oscillating with the simple harmonic motions

$$u_{g1} = g_1(t) = u_1 \sin \Omega_1 t \qquad u_{g2} = g_2(t) = u_2 \sin \Omega_2 t$$

In these expressions the symbols u_1 and u_2 represent the amplitudes of the oscillations at the left- and right-hand supports, respectively; and Ω_1 and Ω_2 are the impressed angular frequencies. Determine the steady-state forced vibrations of any point on the bar due to these independent support motions.

Solution. Because both ends of the bar in Fig. 5.7 are restrained, the natural frequencies and normalized mode shapes to be used in this case are

$$\omega_i = \frac{i\pi a}{\ell} \qquad X_i = \sqrt{\frac{2}{\ell}}\sin\frac{\omega_i x}{a} \qquad (i = 1, 2, 3, \ldots, \infty)$$

From Eq. (5.49) we see that the flexible-body response of the massless bar is given by

$$u_{\mathrm{st}} = \frac{\ell - x}{\ell} u_1 \sin \Omega_1 t + \frac{x}{\ell} u_2 \sin \Omega_2 t \tag{o}$$

and the second derivative of this expression with respect to time is

$$\ddot{u}_{\mathrm{st}} = -\frac{\ell - x}{\ell}\Omega_1^2 u_1 \sin \Omega_1 t - \frac{x}{\ell}\Omega_2^2 u_2 \sin \Omega_2 t \tag{p}$$

The steady-state response u^* of any point on the bar, relative to the displacement u_{st}, is found [after integration of Eq. (5.52) with respect to time] as follows:

$$\begin{aligned} u^* &= \frac{2}{\ell}\sum_{i=1}^{\infty}\frac{1}{\omega_i^2}\sin\frac{i\pi x}{\ell}\left(\Omega_1^2 u_1 \beta_{i1}\sin\Omega_1 t\int_0^{\ell}\frac{\ell - x}{\ell}\sin\frac{i\pi x}{\ell}\,dx\right. \\ &\quad \left. + \Omega_2^2 u_2 \beta_{i2}\sin\Omega_2 t\int_0^{\ell}\frac{x}{\ell}\sin\frac{i\pi x}{\ell}\,dx\right) \\ &= \frac{2\ell^2}{\pi^3 a^2}\sum_{i=1}^{\infty}\frac{1}{i^3}\sin\frac{i\pi x}{\ell}\left[\Omega_1^2 u_1 \beta_{i1}\sin\Omega_1 t - (-1)^i \Omega_2^2 u_2 \beta_{i2}\sin\Omega_2 t\right] \end{aligned} \tag{q}$$

In this result the magnification factors are defined as

$$\beta_{i1} = \frac{1}{1 - \Omega_1^2/\omega_i^2} \qquad \beta_{i2} = \frac{1}{1 - \Omega_2^2/\omega_i^2}$$

Substituting Eqs. (o) and (q) into Eq. (5.53), we find the total response to be

$$u = \left[\frac{\ell - x}{\ell} + \frac{2\ell^2\Omega_1^2}{\pi^3 a^2} \sum_{i=1}^{\infty} \frac{\beta_{i1}}{i^3} \sin\frac{i\pi x}{\ell}\right] u_1 \sin\Omega_1 t$$
$$+ \left[\frac{x}{\ell} - \frac{2\ell^2\Omega_2^2}{\pi^3 a^2} \sum_{i=1}^{\infty} (-1)^i \frac{\beta_{i2}}{i^3} \sin\frac{i\pi x}{\ell}\right] u_2 \sin\Omega_2 t \qquad (r)$$

5.7 TORSIONAL VIBRATIONS OF CIRCULAR SHAFTS

Figure 5.8a shows a straight shaft with a circular cross section, which we will analyze for torsional vibrations. For this purpose, let the symbol θ denote the rotation (about the axis of the shaft) of any cross section located at the distance x from the left-hand end. When the shaft vibrates in a torsional mode, the elasticity and inertial torques on a typical segment (see Fig. 5.8b) may be summed in accordance with d'Alembert's principle, as follows:

$$T + \frac{\partial T}{\partial x} dx - T - \rho I_p \, dx \frac{\partial^2 \theta}{\partial t^2} = 0 \qquad (a)$$

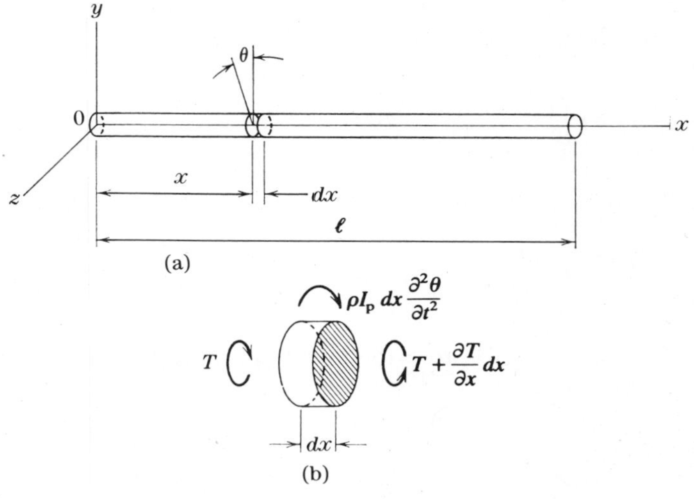

FIG. 5.8

In this differential equation the internal torque on the cross section at x is represented by the symbol T, for which the positive sense is shown in Fig. 5.8*b*. In addition, the polar moment of inertia of the cross section is signified by I_p. With this notation the mass moment of inertia of the segment is $\rho I_p\, dx$, and its rotational acceleration is $\partial^2\theta/\partial t^2$. From elementary torsion theory we have the relationship

$$T = GI_p \frac{\partial \theta}{\partial x} \tag{b}$$

where G is the shear modulus of elasticity. Substituting expression (b) into Eq. (a) and rearranging terms, we obtain

$$\frac{\partial^2 \theta}{\partial x^2} = \frac{1}{b^2}\frac{\partial^2 \theta}{\partial t^2} \tag{5.54}$$

This expression has the form of the one-dimensional wave equation, for which the velocity of torsional wave propagation is

$$b = \sqrt{\frac{G}{\rho}} \tag{5.55}$$

Equations (5.54) and (5.55) have the same mathematical forms as Eqs. (5.1) and (5.2) in Sec. 5.2, except that the symbols u, a, and E have been replaced by θ, b, and G, respectively. Therefore, all previous derivations for longitudinal vibrations of prismatic bars may also be utilized for torsional vibrations of circular shafts by merely changing the notation. For example, in the case of a shaft with free ends, the frequencies and normal functions for the natural modes of torsional vibration are

$$\omega_i = \frac{i\pi b}{\ell} \qquad X_i = C_i \cos\frac{\omega_i x}{b} \qquad (i = 1, 2, 3, \ldots, \infty) \tag{c}$$

and the solution for free vibtations [see Eq. (5.6)] has the form

$$\theta = \sum_{i=1}^{\infty} \cos\frac{i\pi x}{\ell}\left(A_i \cos\frac{i\pi bt}{\ell} + B_i \sin\frac{i\pi bt}{\ell}\right) \tag{5.56}$$

Similarly, the solutions for free torsional vibrations of shafts with one or both ends fixed can also be drawn from previously developed formulas in Sec. 5.2; and the general form of the normal-mode solution is given by Eq. (5.25) in Sec. 5.4.

To analyze the forced rotational response of a shaft subjected to a distributed torque, we can use Eq. (5.28) in Sec. 5.4. In this case the symbol

$q(x, t)$ will represent the distributed torque per unit length divided by the mass moment of inertia ρI_p per unit length. Similarly, Eq. (5.29) can be adapted to the case of a concentrated torque $T_1(t)$ applied at point $x = x_1$, provided that the load term in that expression is taken to be $q_1(t) = T_1(t)/\rho I_p$.

In the preceding section we analyzed the response of prismatic bars to longitudinal support translations. The methods developed therein can be readily applied to the situation in which a shaft is subjected to rotations of restraints about its own axis. If the restraints rotate as a rigid body, Eqs. (5.47) and (5.48) in Sec. 5.6 may be used; but the displacement function becomes $\theta_g = g(t)$. On the other hand, if a shaft that is restrained at both ends is subjected to independent rotations of the two ends, Eqs. (5.52) and (5.53) apply. In this case the support rotations are specified as the functions $\theta_{g1} = g_1(t)$ and $\theta_{g2} = g_2(t)$ for the left and right ends, respectively.

We shall now consider the case of a shaft with disks at the ends, as shown in Fig. 5.9. The shaft is free to rotate, and mass moments of inertia of the disks with respect to the x axis are indicated as I_1 (at $x = 0$) and I_2 (at $x = \ell$). This arrangement was previously studied in Sec. 1.3 as a system with only one vibrational mode, by virtue of the fact that the distributed mass of the shaft was neglected. When this mass is taken into account, the system has an infinite number of natural modes, and more accurate results can be obtained. The approach described previously in Sec. 5.5 for a prismatic bar with a mass or spring at the end will be applied now to the shaft with disks at both ends.

When the system in Fig. 5.9 undergoes torsional vibrations, the inertial torques exerted by the disks upon the ends of the shaft constitute the boundary conditions, as follows:

$$GI_p(\theta')_{x=0} = I_1(\ddot{\theta})_{x=0} \qquad GI_p(\theta')_{x=\ell} = -I_2(\ddot{\theta})_{x=\ell} \tag{d}$$

As before, we assume for the ith natural mode of vibration the harmonic motion

$$\theta_i = X_i(A_i \cos \omega_i t + B_i \sin \omega_i t) \tag{e}$$

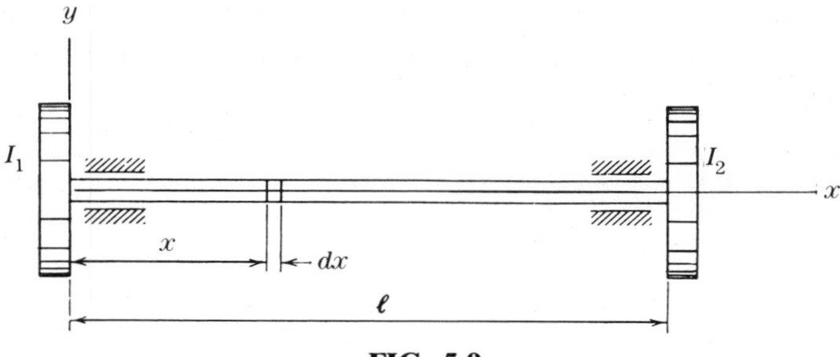

FIG. 5.9

Substituting this expression into Eqs. (d), we obtain

$$GI_p X'_{i0} = -I_1 \omega_i^2 X_{i0} \qquad GI_p X'_{i\ell} = I_2 \omega_i^2 X_{i\ell} \tag{f}$$

where the subscripts 0 and ℓ signify the locations $x = 0$ and $x = \ell$, respectively. In this instance the normal functions may be written as

$$X_i = C_i \cos \frac{\omega_i x}{b} + D_i \sin \frac{\omega_i x}{b} \tag{g}$$

Substitution of Eq. (g) into Eqs. (f) produces

$$GI_p \frac{\omega_i}{b} D_i = -I_1 \omega_i^2 C_i \tag{h}$$

and

$$GI_p \frac{\omega_i}{b} \left(-C_i \sin \frac{\omega_i \ell}{b} + D_i \cos \frac{\omega_i \ell}{b} \right) = I_2 \omega_i^2 \left(C_i \cos \frac{\omega_i \ell}{b} + D_i \sin \frac{\omega_i \ell}{b} \right) \tag{i}$$

Equations (h) and (i) constitute a pair of homogeneous algebraic equations involving the unknown constants C_i and D_i. By eliminating these constants from the equations, we can obtain the frequency equation

$$-GI_p \frac{\omega_i}{b} \left(\sin \frac{\omega_i \ell}{b} + \frac{I_1 b \omega_i}{GI_p} \cos \frac{\omega_i \ell}{b} \right) = I_2 \omega_i^2 \left(\cos \frac{\omega_i \ell}{b} - \frac{I_1 b \omega_i}{GI_p} \sin \frac{\omega_i \ell}{b} \right) \tag{j}$$

Since ω_i appears as a common factor on both sides of this expression, it is apparent that the value $\omega_0 = 0$ represents the frequency for the rigid-body rotation of the system. For convenience in determining the frequencies of the vibrational modes, we let

$$\xi_i = \frac{\omega_i \ell}{b} \qquad \eta_1 = \frac{\rho I_p \ell}{I_1} = \frac{I_o}{I_1} \qquad \eta_2 = \frac{I_o}{I_2} \tag{k}$$

where $I_o = \rho I_p \ell$ is the mass moment of inertia of the shaft about its own axis. Using notations (k), we rewrite the frequency equation (j) in the simpler form

$$-\left(\tan \xi_i + \frac{\xi_i}{\eta_1} \right) = \frac{\xi_i}{\eta_2} \left(1 - \frac{\xi_i}{\eta_1} \tan \xi_i \right)$$

or

$$\left(\frac{\xi_i^2}{\eta_1 \eta_2} - 1 \right) \tan \xi_i = \left(\frac{1}{\eta_1} + \frac{1}{\eta_2} \right) \xi_i \tag{5.57}$$

If we let ξ_1, ξ_2, ξ_3, . . . , etc., represent the nonzero positive roots (in ascending order) of this transcendental equation, then the corresponding normal functions from Eqs. (g) and (h) may be written as

$$X_i = C_i\left(\cos\frac{\xi_i x}{\ell} - \frac{\xi_i}{\eta_1}\sin\frac{\xi_i x}{\ell}\right) \tag{5.58}$$

Suppose that the mass moments of inertia I_1 and I_2 of the disks are small in comparison with the mass moment of inertia I_o of the shaft. In this instance the parameters η_1 and η_2 will be large; the consecutive roots of interest obtained from Eq. (5.57) are π, 2π, 3π, . . . , etc.; and the normal functions in Eq. (5.58) become those for a shaft with free ends [see Eq. (c)]. On the other hand, if the values of I_1 and I_2 are large in comparison with I_o, the ratios η_1 and η_2 will be small; and the value of unity in the factor on the left-hand side of Eq. (5.57) can be neglected in comparison with the term $\xi_i^2/\eta_1\eta_2$. In this case the frequency equation becomes

$$\xi_i \tan \xi_i = \eta_1 + \eta_2 \tag{l}$$

This equation is of the same form as Eq. (5.30) in Sec. 5.5, pertaining to longitudinal vibrations. For the first torsional mode of vibration, all of the terms in Eq. (l) will be small, and the relationship can be simplified by taking $\tan \xi_1 \approx \xi_1$. Then we have

$$\xi_1^2 \approx \eta_1 + \eta_2 = \frac{I_o(I_1 + I_2)}{I_1 I_2}$$

Hence,

$$\omega_1 = \frac{b\xi_1}{\ell} \approx \frac{b}{\ell}\sqrt{\frac{I_o(I_1 + I_2)}{I_1 I_2}} = \sqrt{\frac{GI_p(I_1 + I_2)}{\ell I_1 I_2}}$$

and the fundamental period becomes

$$\tau_1 = 2\pi\sqrt{\frac{\ell I_1 I_2}{GI_p(I_1 + I_2)}}$$

This result coincides with Eq. (1.11) in Sec. 1.3, which was obtained by neglecting the mass of the shaft and analyzing the system for only one vibrational mode.

By the approach used previously in Sec. 5.5, we can develop the following orthogonality relationships for the shaft with disks at both ends:

$$\rho I_{\mathrm{p}} \int_0^{\ell} X_i X_j \, dx + I_1 X_{i0} X_{j0} + I_2 X_{i\ell} X_{j\ell} = 0 \qquad (i \neq j) \tag{5.59}$$

$$GI_{\mathrm{p}} \int_0^{\ell} X_i' X_j' \, dx = 0 \qquad (i \neq j) \tag{5.60}$$

$$GI_{\mathrm{p}} \left(\int_0^{\ell} X_i'' X_j \, dx + X_{i0}' X_{j0} - X_{i\ell}' X_{j\ell} \right) = 0 \qquad (i \neq j) \tag{5.61}$$

In addition, the normalization condition is selected (for $i = j$) to be

$$\rho I_{\mathrm{p}} \int_0^{\ell} X_i^2 \, dx + I_1 X_{i0}^2 + I_2 X_{i\ell}^2 = \rho I_{\mathrm{p}} \tag{5.62}$$

and it follows that

$$GI_{\mathrm{p}} \left(\int_0^{\ell} X_i'' X_i \, dx + X_{i0}' X_{i0} - X_{i\ell}' X_{i\ell} \right) = -GI_{\mathrm{p}} \int_0^{\ell} (X_i')^2 \, dx = -\rho I_{\mathrm{p}} \omega_i^2 \tag{5.63}$$

Equations (5.59) through (5.63) are similar to Eqs. (5.31) through (5.35); but $m = \rho A$ has been replaced by ρI_{p}, and $r = EA$ has been replaced by GI_{p}. Furthermore, terms associated with the disk at location $x = 0$ appear in the present expressions, together with those for the disk at location $x = \ell$.

Let us consider the rotational response of the system in Fig. 5.9 to the initial conditions $\theta_0 = f_1(x)$ and $\dot{\theta}_0 = f_2(x)$ at time $t = 0$. For this purpose we must evaluate the initial displacements $f_1(0)$, $f_1(\ell)$ and velocities $f_2(0)$, $f_2(\ell)$ of the disks at $x = 0$ and $x = \ell$. Then the initial conditions for the shaft and disks are expanded in terms of time and displacement functions, as follows:

$$\sum_{i=1}^{\infty} \phi_{0i} X_i = f_1(x) \qquad \sum_{i=1}^{\infty} \dot{\phi}_{0i} X_i = f_2(x) \tag{m}$$

$$\sum_{i=1}^{\infty} \phi_{0i} X_{i0} = f_1(0) \qquad \sum_{i=1}^{\infty} \dot{\phi}_{0i} X_{i0} = f_2(0) \tag{n}$$

$$\sum_{i=1}^{\infty} \phi_{0i} X_{i\ell} = f_1(\ell) \qquad \sum_{i=1}^{\infty} \dot{\phi}_{0i} X_{i\ell} = f_2(\ell) \tag{o}$$

In combining these expressions, we multiply Eqs. (m) by $\rho I_{\mathrm{p}} X_j$ and integrate over the length of the shaft. Equations (n) and (o) are then multiplied by $I_1 X_{j0}$ and $I_2 X_{j\ell}$, respectively, and added to the results from

Eqs. (m). By this means we obtain

$$\sum_{i=1}^{\infty} \phi_{0i}\left(\rho I_p \int_0^{\ell} X_i X_j \, dx + I_1 X_{i0} X_{j0} + I_2 X_{i\ell} X_{j\ell}\right)$$
$$= \rho I_p \int_0^{\ell} f_1(x) X_j \, dx + I_1 f_1(0) X_{j0} + I_2 f_1(\ell) X_{j\ell} \quad (p)$$

$$\sum_{i=1}^{\infty} \dot{\phi}_{0i}\left(\rho I_p \int_0^{\ell} X_i X_j \, dx + I_1 X_{i0} X_{j0} + I_2 X_{i\ell} X_{j\ell}\right)$$
$$= \rho I_p \int_0^{\ell} f_2(x) X_j \, dx + I_1 f_2(0) X_{j0} + I_2 f_2(\ell) X_{j\ell} \quad (q)$$

From the orthogonality and normalization relationships given by Eqs. (5.59) and (5.62), we see that (for $i = j$) Eqs. (p) and (q) yield the following initial displacements and velocities in normal coordinates:

$$\phi_{0i} = \int_0^{\ell} f_1(x) X_i \, dx + \frac{\ell}{\eta_1} f_1(0) X_{i0} + \frac{\ell}{\eta_2} f_1(\ell) X_{i\ell} \quad (5.64)$$

$$\dot{\phi}_{0i} = \int_0^{\ell} f_2(x) X_i \, dx + \frac{\ell}{\eta_1} f_2(0) X_{i0} + \frac{\ell}{\eta_2} f_2(\ell) X_{i\ell} \quad (5.65)$$

These expressions are similar to Eqs. (5.36) and (5.37) in Sec. 5.5, but terms for both of the disks appear in the present formulas. With these definitions for ϕ_{0i} and $\dot{\phi}_{0i}$, we can determine the response of the system under consideration, using Eq. (5.25) in Sec. 5.4.

If the eigenfunctions are normalized to satisfy Eq. (5.62), the vibrational response of the system in Fig. 5.9 to applied torques can be calculated using Eqs. (5.28) and (5.29) in Sec. 5.4. To this response can be added the rigid-body motion, which is determined from the equation

$$J \ddot{\phi}_0 = R \quad (r)$$

In this expression the symbol $J = I_o + I_1 + I_2$ represents the total mass moment of inertia of the system; $\ddot{\phi}_0$ is the acceleration of the rigid-body mode; and R is the total torque applied to the shaft and the disks.

Example 1. Suppose that the unrestrained shaft in Fig. 5.8a is subjected to a torque $R = R_1 t / t_1$ (where R_1 is the torque at time t_1), applied at the left-hand end. Assuming that the shaft is initially at rest, determine its response to this ramp function.

Solution. Substituting $J = I_o = \rho I_p \ell$ and $R = R_1 t / t_1$ into Eq. (r) and integrating it twice with respect to t, we obtain the rigid-body rotation

$$\phi_0 = \frac{R_1 t^3}{6 \rho I_p \ell t_1} = \frac{R_1 t^3}{6 I_o t_1} \quad (s)$$

The vibrational response to be added to this rigid-body motion is found from Eq. (5.29) to be

$$\theta = \frac{2R_1}{\rho I_p \ell t_1} \sum_{i=1}^{\infty} \frac{1}{\omega_i} \cos \frac{\omega_i x}{b} \int_0^t t' \sin \omega_i (t - t')\, dt'$$

$$= \frac{2R_1}{I_o t_1} \sum_{i=1}^{\infty} \frac{1}{\omega_i^2} \cos \frac{\omega_i x}{b} \left(t - \frac{1}{\omega_i} \sin \omega_i t \right)$$

$$= \frac{2\ell^2 R_1}{\pi^2 b^2 I_o t_1} \sum_{i=1}^{\infty} \frac{1}{i^2} \cos \frac{i\pi x}{\ell} \left(t - \frac{\ell}{i\pi b} \sin \frac{i\pi b t}{\ell} \right) \tag{t}$$

and the total motion of the shaft consists of the sum of Eqs. (s) and (t). For instance, the total displacement of the end of the bar where the torque is applied (at $x = 0$) is

$$(\theta)_{x=0} = \frac{R_1}{I_o t_1} \left[\frac{t^3}{6} + \frac{2\ell^2}{\pi^2 b^2} \sum_{i=1}^{\infty} \frac{1}{i^2} \left(t - \frac{\ell}{i\pi b} \sin \frac{i\pi b t}{\ell} \right) \right] \tag{u}$$

Letting $t = t_1 = \ell / b$, we find

$$(\theta)_{x=0} = \frac{R_1 \ell^2}{I_o b^2} \left[\frac{1}{6} + \frac{2}{\pi^2} \left(1 + \frac{1}{4} + \frac{1}{9} + \cdots \right) \right] = \frac{R_1 \ell}{2 G I_p} \tag{v}$$

Example 2. Assuming that the mass moments of inertia of the three parts of the system in Fig. 5.9 are equal ($I_o = I_1 = I_2$), determine an expression for the response to the initial conditions

$$\theta_0 = f_1(x) = \frac{\alpha_0}{\ell}(2x - \ell) \qquad \dot{\theta}_0 = f_2(x) = 0 \tag{w}$$

The initial displacement function θ_0 is due to equal and opposite torques applied to the disks (causing a relative rotation $2\alpha_0$), which are then released at the instant $t = 0$.

Solution. For this case $\eta_1 = \eta_2 = 1$, and the transcendental frequency equation (5.57) simplifies to

$$(\xi_i^2 - 1) \tan \xi_i = 2\xi_i \tag{x}$$

In this instance the functions X_i are to be normalized [see Eq. (5.62)] such that

$$\int_0^\ell X_i^2\, dx + X_{i0}^2 + X_{i\ell}^2 = 1 \tag{y}$$

For the given initial conditions (w), Eqs. (5.64) and (5.65) become

$$\phi_{0i} = \frac{\alpha_0}{\ell}\left[\int_0^{\ell} (2x - \ell)X_i\,dx - \ell^2(X_{i0} - X_{i\ell})\right]$$

and

$$\dot{\phi}_{0i} = 0$$

Thus, the expression for free-vibrational response [see Eq. (5.25)] may be written as

$$\theta = \frac{\alpha_0}{\ell}\sum_{i=1}^{\infty} X_i\left[\int_0^{\ell} (2x - \ell)X_i\,dx - \ell^2(X_{i0} - X_{i\ell})\right]\cos\left(\frac{\xi_i bt}{\ell}\right) \tag{z}$$

5.8 TRANSVERSE VIBRATIONS OF STRETCHED WIRES

Another elastic system for which the equation of motion has the form of the one-dimensional wave equation is shown in Fig. 5.10*a*. It consists of a tightly stretched wire (having no flexural stiffness) that is free to vibrate transversely. The tensile force S in the wire is assumed to remain constant during small vibrations in the x–y plane. We shall use the symbol v to denote the transverse displacement of any point on the wire, located at the distance x from the left-hand end. Figure 5.10*b* shows a free-body diagram of a typical segment of length dx, for which the forces in the y direction are of primary interest. During free vibrations, the inertia force is counteracted by the difference between the y components of the tensile forces at the ends of the segment. For small slopes this condition of dynamic equilibrium may be written as

$$S\left(\frac{\partial v}{\partial x} + \frac{\partial^2 v}{\partial x^2}dx\right) - S\frac{\partial v}{\partial x} - m\,dx\frac{\partial^2 v}{\partial t^2} = 0 \tag{a}$$

where m is the mass of the wire per unit length. Hence, the differential equation of motion for this system is

$$\frac{\partial^2 v}{\partial x^2} = \frac{1}{c^2}\frac{\partial^2 v}{\partial t^2} \tag{5.66}$$

in which

$$c = \sqrt{\frac{S}{m}} \tag{5.67}$$

denotes the longitudinal velocity of transverse waves.

We see that Eqs. (5.66) and (5.67) have the same mathematical forms as Eqs. (5.1) and (5.2) in Sec. 5.2, except that the symbols u, a, E, and ρ have

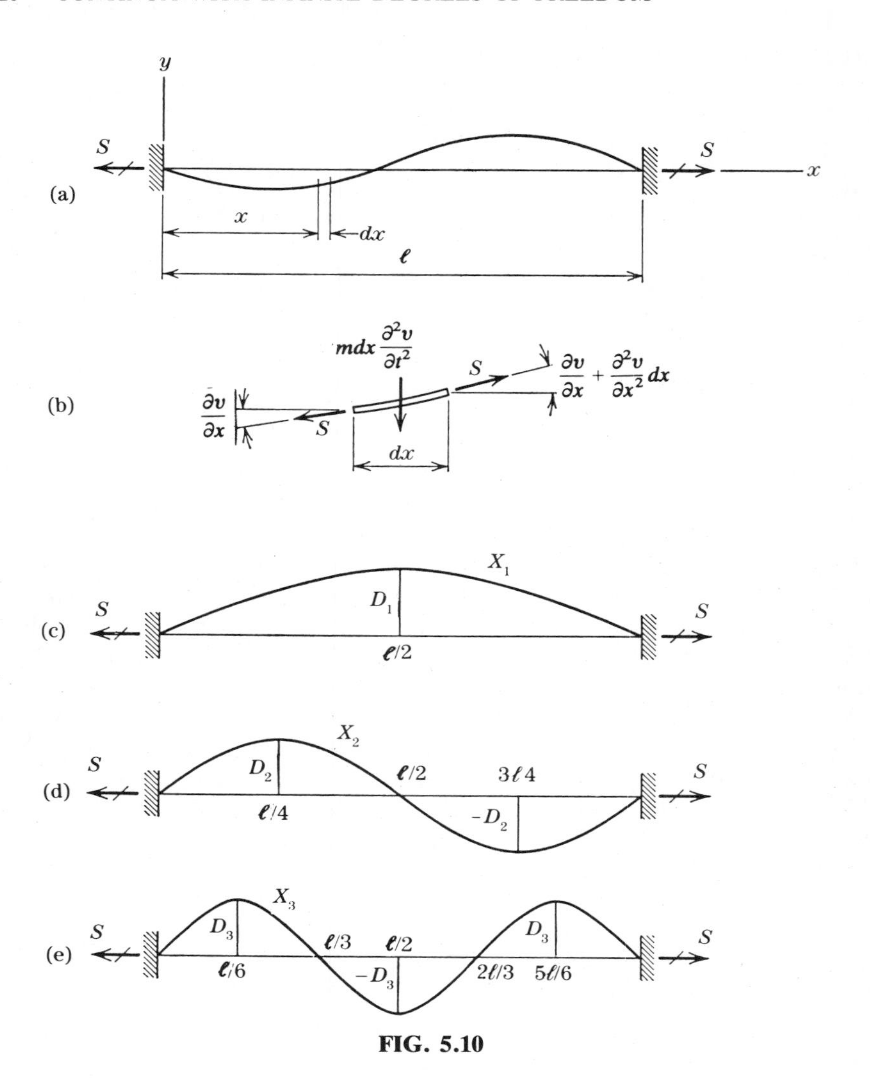

FIG. 5.10

been replaced by v, c, S, and m, respectively. Therefore, many of the previously developed formulations for longitudinal and torsional vibrations of bars and shafts can be applied to lateral vibrations of stretched wires by simply changing the notations. However, in this case the boundary conditions are somewhat restricted by the necessity to maintain the tensile force S in the wire. The simplest boundary conditions appear in Fig. 5.10*a*, where both ends of the wire are fixed. For this case the conditions are

$$(v)_{x=0} = 0 \qquad (v)_{x=\ell} = 0 \tag{b}$$

and the angular frequencies and normal functions become

$$\omega_i = \frac{i\pi c}{\ell} \qquad X_i = D_i \sin\frac{\omega_i x}{c} \qquad (i = 1, 2, 3, \ldots, \infty) \tag{c}$$

Figures 5.10*c*, *d*, and *e* illustrate the normal functions for modes, 1, 2, and 3, respectively. When these functions are normalized in accordance with Eq. (5.22) [see Sec. 5.4], we have $D_i = \sqrt{2/\ell}$.

Let us represent the initial transverse displacement of any point on the wire (at time $t = 0$) as $v_0 = f_1(x)$ and the initial velocity as $\dot{v}_0 = f_2(x)$. From Eqs. (5.23) and (5.24) in Sec. 5.4, we see that the initial conditions in the normal coordinates are

$$\phi_{0i} = \sqrt{\frac{2}{\ell}} \int_0^\ell f_1(x) \sin\frac{i\pi x}{\ell}\, dx \qquad \dot{\phi}_{0i} = \sqrt{\frac{2}{\ell}} \int_0^\ell f_2(x) \sin\frac{i\pi x}{\ell}\, dx \tag{d}$$

and Eq. (5.25) gives the response to these conditions as

$$v = \sqrt{\frac{2}{\ell}} \sum_{i=1}^{\infty} \sin\frac{i\pi x}{\ell} \left(\phi_{0i} \cos\frac{i\pi ct}{\ell} + \frac{\dot{\phi}_{0i}}{\omega_i} \sin\frac{i\pi ct}{\ell}\right) \tag{5.68}$$

Furthermore, the response to a distributed transverse force $Q(x, t)$ is found from Eq. (5.28) to be

$$v = \frac{2}{\ell} \sum_{i=1}^{\infty} \frac{1}{\omega_i} \sin\frac{i\pi x}{\ell} \int_0^\ell \sin\frac{i\pi x}{\ell} \int_0^t q(x, t') \sin\omega_i(t - t')\, dt'\, dx \tag{5.69}$$

in which $q(x, t) = Q(x, t)/m$. If a transverse load $P_1(t)$ is concentrated at point x_1, the response given by Eq. (5.29) becomes

$$v = \frac{2}{\ell} \sum_{i=1}^{\infty} \frac{1}{\omega_i} \sin\frac{i\pi x}{\ell} \sin\frac{i\pi x_1}{\ell} \int_0^t q_1(t') \sin\omega_i(t - t')\, dt' \tag{5.70}$$

where $q_1(t) = P_1(t)/m$.

In Sec. 5.6 we discussed the case of a prismatic bar subjected to independent longitudinal translations of restraints at both ends. The formulas developed for that purpose will be applied in a very direct manner to the situation in which the ends of a stretched wire translate laterally (in the y direction). Figures 5.11*a* and 5.11*b* show independent lateral translations of the supports, specified as

$$v_{g1} = g_1(t) \qquad v_{g2} = g_2(t) \tag{e}$$

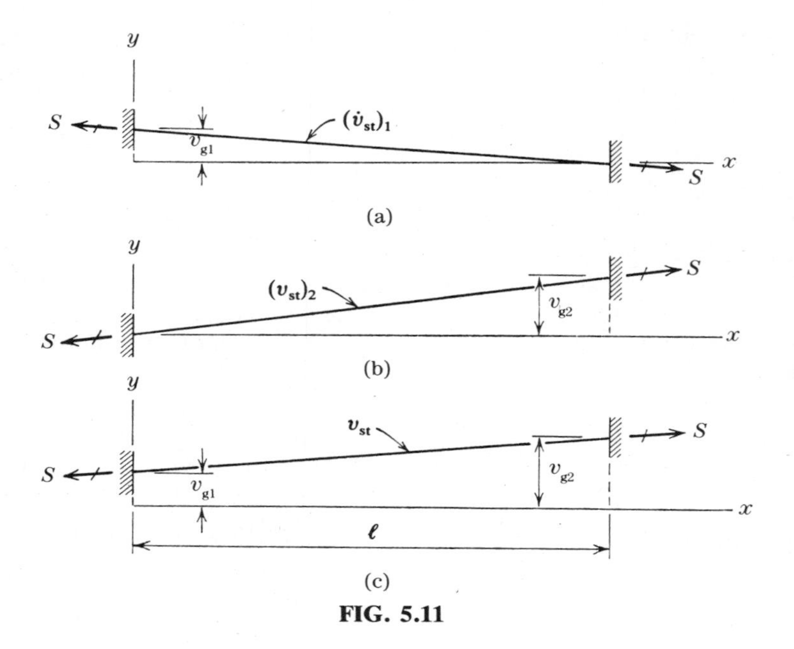

FIG. 5.11

at the left and right ends, respectively. Due to these displacements the transverse motion of any point on the corresponding massless wire is

$$v_{st} = (v_{st})_1 + (v_{st})_2 = \frac{\ell - x}{\ell} g_1(t) + \frac{x}{\ell} g_2(t) \tag{5.71}$$

In this case the displacement v_{st} (see Fig. 5.11c) is rigid-body motion, consisting of a translation in the y direction and a small rotation about an axis perpendicular to the x–y plane. Furthermore, each of the component motions $(v_{st})_1$ and $(v_{st})_2$ in Figs. 5.11*a* and 5.11*b* have this same character. The combined motion is pure translation when $g_1(t) = g_2(t) = g(t)$ and pure rotation about the midpoint when $g_1(t) = -g_2(t)$. Thus, we see that the concept of flexible-body motions for massless systems, which was introduced in Sec. 5.6, coincides with the concept of rigid-body motions in the case of a stretched wire with transverse end displacements.

Substitution of the second derivative of Eq. (5.71) with respect to time into Eq. (5.52) of Sec. 5.6, together with the normalized eigenfunctions from Eqs. (*c*), gives the vibrational response of the wire as

$$v^* = -\frac{2}{\ell} \sum_{i=1}^{\infty} \frac{1}{\omega_i} \sin \frac{i\pi x}{\ell} \int_0^{\ell} \sin \frac{i\pi x}{\ell} \times \int_0^t \left[\frac{\ell - x}{\ell} \ddot{g}_1(t') + \frac{x}{\ell} \ddot{g}_2(t') \right] \sin \omega_i (t - t')\, dt'\, dx \tag{5.72}$$

To obtain the total response, we add the vibrational and rigid-body motions, as follows:

$$v = v_{st} + v^* = \frac{\ell - x}{\ell} g_1(t) + \frac{x}{\ell} g_2(t) + v^* \tag{5.73}$$

The above discussion pertains to a stretched wire that is rigidly restrained at both ends. We shall now investigate the case of a wire with elastic transverse restraints, as shown in Fig. 5.12. It is assumed that the stiffness constants k_1 and k_2 for the springs at $x = 0$ and $x = \ell$ are known and that the ends of the wire are free to translate only in the y direction. The approach used earlier (see Sec. 5.5) for a prismatic bar with a mass or spring at one end will be applied to the wire with springs at both ends.

The equation of motion for a typical element of the wire in Fig. 5.12 [see Eq. (*a*)] may be written as

$$m\ddot{v}\,dx - Sv''\,dx = 0 \tag{f}$$

At each end of the wire, the spring force must be in equilibrium with the y component of the tensile force S in the wire. Therefore, the boundary conditions are

$$S(v')_{x=0} = k_1(v)_{x=0} \qquad S(v')_{x=\ell} = -k_2(v)_{x=\ell} \tag{g}$$

As usual, we assume for the ith natural mode of vibration the harmonic motion

$$v_i = X_i(A_i \cos \omega_i t + B_i \sin \omega_i t) \tag{h}$$

Substitution of this expression into Eqs. (*g*) yields

$$SX'_{i0} = k_1 X_{i0} \qquad SX'_{i\ell} = -k_2 X_{i\ell} \tag{i}$$

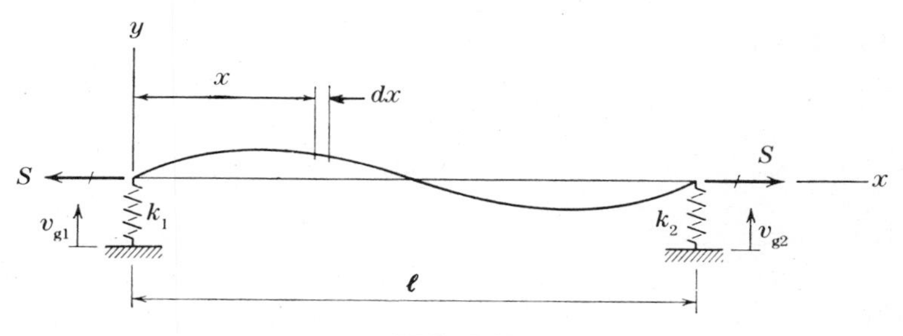

FIG. 5.12

Writing the normal functions as before, we have

$$X_i = C_i \cos \frac{\omega_i x}{c} + D_i \sin \frac{\omega_i x}{c} \tag{j}$$

By substituting Eq. (j) into Eqs. (i), we obtain

$$\frac{S\omega_i}{c} D_i = k_1 C_i \tag{k}$$

and

$$\frac{S\omega_i}{c}\left(-C_i \sin \frac{\omega_i \ell}{c} + D_i \cos \frac{\omega_i \ell}{c}\right) = -k_2\left(C_i \cos \frac{\omega_i \ell}{c} + D_i \sin \frac{\omega_i \ell}{c}\right) \tag{l}$$

Elimination of the constants C_i and D_i from Eqs. (k) and (l) results in the frequency equation

$$\frac{S\omega_i}{c}\left(\sin \frac{\omega_i \ell}{c} - \frac{k_1 c}{S\omega_i} \cos \frac{\omega_i \ell}{c}\right) = k_2\left(\cos \frac{\omega_i \ell}{c} + \frac{k_1 c}{S\omega_i} \sin \frac{\omega_i \ell}{c}\right) \tag{m}$$

To simplify this expression, we let

$$\xi_i = \frac{\omega_i \ell}{c} \qquad \zeta_{i1} = \frac{m\ell\omega_i^2}{k_1} \qquad \zeta_{i2} = \frac{m\ell\omega_i^2}{k_2} \tag{n}$$

and rewrite Eq. (m) as

$$\tan \xi_i - \frac{\xi_i}{\zeta_{i1}} = \frac{\xi_i}{\zeta_{i2}}\left(1 + \frac{\xi_i}{\zeta_{i1}} \tan \xi_i\right)$$

or

$$\left(1 - \frac{\xi_i^2}{\zeta_{i1}\zeta_{i2}}\right) \tan \xi_i = \left(\frac{1}{\zeta_{i1}} + \frac{1}{\zeta_{i2}}\right)\xi_i \tag{5.74}$$

This expression is the transcendental frequency equation for the wire with springs at both ends. In this case the normal functions from Eqs. (j) and (k) may be written as

$$X_i = C_i\left(\cos \frac{\xi_i x}{\ell} + \frac{\xi_i}{\zeta_{i1}} \sin \frac{\xi_i x}{\ell}\right) \tag{5.75}$$

If the values of k_1 and k_2 are both large, Eq. (m) approaches

$$\sin \frac{\omega_i \ell}{c} = 0 \tag{o}$$

which is the frequency equation for the wire with both ends fixed. In this case Eq. (5.75) approaches the form of Eq. (*c*). On the other hand, when k_1 and k_2 are both small, Eqs. (5.74) and (5.75) result in

$$\omega_i = \frac{i\pi c}{\ell} \qquad X_i = C_i \cos\frac{\omega_i x}{c} \qquad (i = 1, 2, 3, \ldots, \infty) \tag{p}$$

which implies no lateral restraints. Finally, if only k_1 is large and k_2 is small, Eq. (*m*) approaches

$$\cos\frac{\omega_i \ell}{c} = 0 \tag{q}$$

for which the frequencies and normal functions (from Eq. 5.75) are

$$\omega_i = \frac{i\pi c}{2\ell} \qquad X_i = D_i \sin\frac{\omega_i x}{c} \qquad (i = 1, 3, 5, \ldots, \infty) \tag{r}$$

These expressions imply that the left end of the wire is fixed, while the right end can translate freely.

By the method used before in Sec. 5.5, we can derive the following orthogonality relationships for the stretched wire with springs at both ends (for $i \neq j$):

$$m\int_0^\ell X_i X_j \, dx = 0 \tag{5.76}$$

$$S\int_0^\ell X_i' X_j' \, dx + k_1 X_{i0} X_{j0} + k_2 X_{i\ell} X_{j\ell} = 0 \tag{5.77}$$

$$S\int_0^\ell X_i'' X_j \, dx + S X_{i0}' X_{j0} - k_1 X_{i0} X_{j0} - S X_{i\ell}' X_{j\ell} - k_2 X_{i\ell} X_{j\ell} = 0 \tag{5.78}$$

Also, the normalization condition is selected as

$$m\int_0^\ell X_i^2 \, dx = m \qquad (i = j) \tag{5.79}$$

and it follows that

$$S\int_0^\ell X_i'' X_i \, dx + S X_{i0}' X_{i0} - k_1 X_{i0}^2 - S X_{i\ell}' X_{i\ell} - k_2 X_{i\ell}^2$$
$$= -S\int_0^\ell (X_i')^2 \, dx - k_1 X_{i0}^2 - k_2 X_{i\ell}^2 = -m\omega_i^2 \tag{5.80}$$

Equations (5.76) through (5.80) are similar to Eqs. (5.39) through (5.43), but the rigidity $r = EA$ has been replaced by the tensile force S. In addition, terms pertaining to the spring at the left end are included, along with those for the spring at the right end.

Because Eq. (5.79) coincides with the type of normalization used in Sec. 5.4, the expressions derived therein for response to initial conditions [Eqs. (5.23), (5.24), and (5.25)] can also be used for the case of an elastically restrained wire. In addition, the response of this system to applied transverse forces can be calculated using Eqs. (5.28) and (5.29). Furthermore, the response to independent support translations v_{g1} and v_{g2} (see Fig. 5.12) may be determined from Eqs. (5.52) and (5.53) in Sec. 5.6. Thus, we see that the elastic restraints affect the frequencies and mode shapes of the wire but not the subsequent steps in the solution for dynamic response.

5.9 TRANSVERSE VIBRATIONS OF PRISMATIC BEAMS

Let us now consider the transverse vibrations of a prismatic beam in the x–y plane (see Fig. 5.13*a*), which is assumed to be a plane of symmetry for any cross section. As in the preceding section for a stretched wire, we shall use the symbol v to represent the transverse displacement of a typical segment of the beam, located at the distance x from the left-hand end. Whereas the flexural rigidity EI of the wire was assumed to be small, that for the beam must be taken into account. Figure 5.13*b* shows a free-body diagram of an element of length dx with internal and inertial actions upon it. In this figure the senses of the shearing force V and the bending moment M conform to the beam sign convention [3]. When the beam is vibrating transversely, the

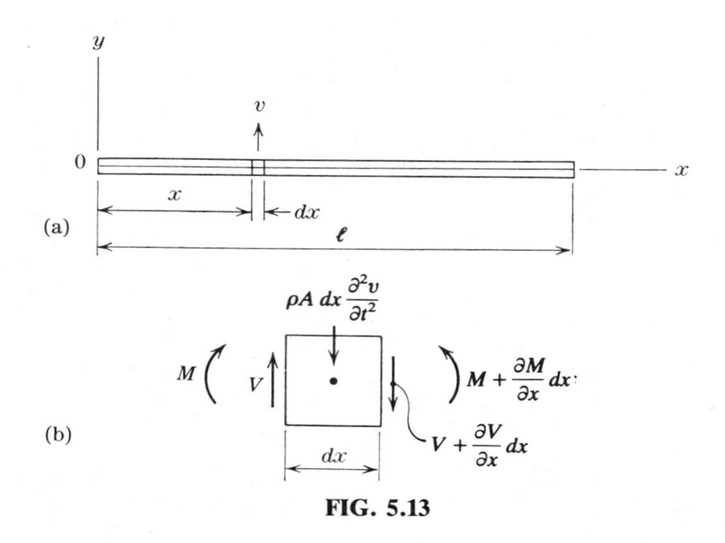

FIG. 5.13

dynamic equilibrium condition for forces in the y direction is

$$V - V - \frac{\partial V}{\partial x} dx - \rho A \, dx \frac{\partial^2 v}{\partial t^2} = 0 \tag{a}$$

and the moment equilibrium condition gives

$$-V \, dx + \frac{\partial M}{\partial x} dx \approx 0 \tag{b}$$

Substitution of V from Eq. (b) into Eq. (a) produces

$$\frac{\partial^2 M}{\partial x^2} dx = -\rho A \, dx \frac{\partial^2 v}{\partial t^2} \tag{c}$$

From elementary flexural theory we have the relationship

$$M = EI \frac{\partial^2 v}{\partial x^2} \tag{d}$$

Using this expression in Eq. (c), we obtain

$$\frac{\partial^2}{\partial x^2}\left(EI \frac{\partial^2 v}{\partial x^2}\right) dx = -\rho A \, dx \frac{\partial^2 v}{\partial t^2} \tag{5.81}$$

which is the general equation for transverse free vibrations of a beam. In the particular case of a prismatic beam the flexural rigidity EI does not vary with x, and we have

$$EI \frac{\partial^4 v}{\partial x^4} dx = -\rho A \, dx \frac{\partial^2 v}{\partial t^2} \tag{5.82}$$

This equation may also be written as

$$\frac{\partial^4 v}{\partial x^4} = -\frac{1}{a^2} \frac{\partial^2 v}{\partial t^2} \tag{5.83}$$

In this application the symbol a has the definition

$$a = \sqrt{\frac{EI}{\rho A}} \tag{5.84}$$

When a beam vibrates transversely in one of its natural modes, the deflection at any location varies harmonically with time, as follows:

$$v = X(A \cos \omega t + B \sin \omega t) \tag{e}$$

where the subscript i for the ith mode has been omitted for notational convenience. Substitution of Eq. (e) into Eq. (5.83) results in

$$\frac{d^4X}{dx^4} - \frac{\omega^2}{a^2} X = 0 \tag{f}$$

As an aid in solving this fourth-order ordinary differential equation, we introduce the notation

$$\frac{\omega^2}{a^2} = k^4 \tag{g}$$

and rewrite Eq. (f) as

$$\frac{d^4X}{dx^4} - k^4 X = 0 \tag{h}$$

To satisfy Eq. (h), we let $X = e^{nx}$ and obtain

$$e^{nx}(n^4 - k^4) = 0 \tag{i}$$

Thus, the values of n are found to be $n_1 = k$, $n_2 = -k$, $n_3 = jk$, and $n_4 = -jk$, where $j = \sqrt{-1}$. The general form of the solution for Eq. (h) becomes

$$X = Ce^{kx} + De^{-kx} + Ee^{jkx} + Fe^{-jkx} \tag{j}$$

which may also be written in the equivalent form

$$X = C_1 \sin kx + C_2 \cos kx + C_3 \sinh kx + C_4 \cosh kx \tag{5.85}$$

This expression represents a typical normal function for transverse vibrations of a prismatic beam.

The constants C_1, C_2, C_3, and C_4 in Eq. (5.85) must be determined in each particular case from the boundary conditions at the ends of the beam. For example, at an end that is simply supported, the deflection and bending moment are equal to zero, and we have

$$X = 0 \qquad X'' = 0 \tag{k}$$

At a fixed end the deflection and slope are equal to zero. In this case the conditions are

$$X = 0 \qquad X' = 0 \tag{l}$$

At a free end the bending moment and the shearing force both vanish, and we obtain

$$X'' = 0 \qquad X''' = 0 \tag{m}$$

For the two ends of a beam, we will always have four such end conditions that can be used to find the values of C_1, C_2, C_3, and C_4, which lead to the frequencies and mode shapes for free vibrations. Then the normal modes can be superimposed to produce the total response as

$$v = \sum_{i=1}^{\infty} X_i(A_i \cos \omega_i t + B_i \sin \omega_i t) \tag{5.86}$$

Particular cases of beams with various end conditions will be considered in subsequent sections.

We shall now re-examine Eq. (h) as an eigenvalue problem of the form

$$X_i'''' = \lambda_i X_i \tag{5.87}$$

where

$$\lambda_i = k_i^4 = \left(\frac{\omega_i}{a}\right)^2 \tag{n}$$

This type of problem may be characterized as one in which the fourth derivative (with respect to x) of the eigenfunction X_i is equal to the same function multiplied by the eigenvalue λ_i. Orthogonality properties of the eigenfunctions will be studied by considering modes i and j, as follows:

$$X_i'''' = \lambda_i X_i \tag{o}$$

$$X_j'''' = \lambda_j X_j \tag{p}$$

Multiplying Eq. (o) by X_j and Eq. (p) by X_i and integrating the products over the length of the beam, we obtain

$$\int_0^\ell X_i'''' X_j \, dx = \lambda_i \int_0^\ell X_i X_j \, dx \tag{q}$$

$$\int_0^\ell X_j'''' X_i \, dx = \lambda_j \int_0^\ell X_i X_j \, dx \tag{r}$$

Integration of the left-hand sides of these equations by parts yields

$$[X_i''' X_j]_0^\ell - [X_i'' X_j']_0^\ell + \int_0^\ell X_i'' X_j'' \, dx = \lambda_i \int_0^\ell X_i X_j \, dx \tag{s}$$

$$[X_j''' X_i]_0^\ell - [X_j'' X_i']_0^\ell + \int_0^\ell X_i'' X_j'' \, dx = \lambda_j \int_0^\ell X_i X_j \, dx \tag{t}$$

The end conditions (k), (l), and (m) require that the integrated terms in Eqs. (s) and (t) vanish. Therefore, subtraction of Eq. (t) from Eq. (s) produces

$$(\lambda_i - \lambda_j)\int_0^\ell X_i X_j \, dx = 0 \tag{u}$$

To satisfy Eq. (u) when $i \neq j$ and the eigenvalues are distinct ($\lambda_i \neq \lambda_j$), we must have

$$\int_0^\ell X_i X_j \, dx = 0 \qquad (i \neq j) \tag{5.88}$$

Substituting this expression into Eq. (s), we find

$$\int_0^\ell X_i'' X_j'' \, dx = 0 \qquad (i \neq j) \tag{5.89}$$

and from Eq. (q) we also see that

$$\int_0^\ell X_i'''' X_j \, dx = 0 \qquad (i \neq j) \tag{5.90}$$

Equations (5.88), (5.89), and (5.90) constitute the orthogonality relationships for transverse vibrations of a prismatic beam.

For the case of $i = j$, the integral in Eq. (u) may be any constant α_i, as follows:

$$\int_0^\ell X_i^2 \, dx = \alpha_i \qquad (i = j) \tag{5.91}$$

When the eigenfunctions are normalized in this manner, Eqs. (q) and (s) yield

$$\int_0^\ell X_i'''' X_i \, dx = \int_0^\ell (X_i'')^2 \, dx = \lambda_i \alpha_i = k_i^4 \alpha_i = \left(\frac{\omega_i}{a}\right)^2 \alpha_i \tag{5.92}$$

For the purpose of transforming the equation of motion (5.82) to principal coordinates, we rewrite it as

$$m\ddot{v} \, dx + rv'''' \, dx = 0 \tag{v}$$

in which $m = \rho A$ is the mass per unit length of the beam, and $r = EI$ is its flexural rigidity. Expansion of the transverse motion in terms of time functions ϕ_i and displacement functions X_i gives

$$v = \sum_i \phi_i X_i \qquad (i = 1, 2, 3, \ldots, \infty) \tag{5.93}$$

Substituting Eq. (5.93) into the equation of motion (v), we obtain

$$\sum_{i=1}^{\infty} (m\ddot{\phi}_i X_i + r\phi_i X_i'''') \, dx = 0 \tag{w}$$

Multiplication of Eq. (w) by the normal function X_j, followed by integration over the length, produces

$$\sum_{i=1}^{\infty} \left(m\ddot{\phi}_i \int_0^{\ell} X_i X_j \, dx + r\phi_i \int_0^{\ell} X_i'''' X_j \, dx \right) = 0 \tag{x}$$

From the relationships given by Eqs. (5.88) and (5.90) through (5.92), we see that (for $i = j$) the equation of motion in principal coordinates becomes

$$m_{Pi}\ddot{\phi}_i + r_{Pi}\phi_i = 0 \qquad (i = 1, 2, 3, \ldots, \infty) \tag{5.94}$$

in which

$$m_{Pi} = m \int_0^{\ell} X_i^2 \, dx = m\alpha_i \tag{5.95}$$

and

$$r_{Pi} = r \int_0^{\ell} X_i'''' X_i \, dx = r \int_0^{\ell} (X_i'')^2 \, dx = m\omega_i^2 \alpha_i \tag{5.96}$$

Thus, the principal mass m_{Pi} for flexural vibrations is calculated in the same manner as that for axial vibrations [see Eq. (5.19), Sec. 5.4]. However, the principal rigidity r_{Pi} in Eq. (5.96) is not obtained in the same way as that for axial vibrations (see Eq. 5.20).

As before, we may set the normalization constant equal to unity so that Eqs. (5.91) and (5.92) give

$$\int_0^{\ell} X_i^2 \, dx = 1 \qquad \int_0^{\ell} X_i'''' X_i \, dx = \int_0^{\ell} (X_i'')^2 \, dx = k_i^4 = \left(\frac{\omega_i}{a} \right)^2 \tag{5.97}$$

Then Eq. (5.94), divided by m, takes the familiar form

$$\ddot{\phi}_i + \omega_i^2 \phi_i = 0 \qquad (i = 1, 2, 3, \ldots, \infty) \tag{5.98}$$

We see that the normal-mode method for flexural vibrations of beams is similar to that for axial vibrations, described previously in Sec. 5.4. Because of this similarity, we need not rederive expressions for response to initial conditions and applied actions. The response formulas for flexural vibrations will be the same as those for axial vibrations [see Eqs. (5.23) through (5.29) in Sec. 5.4], except that the longitudinal displacement u is replaced by the transverse displacement v.

5.10 TRANSVERSE VIBRATIONS OF A SIMPLE BEAM

As the first particular case of transverse vibrations of beams, we shall analyze the simply supported prismatic beam shown in Fig. 5.14. The boundary conditions for this case are

$$(X)_{x=0}=0 \qquad \left(\frac{d^2X}{dx^2}\right)_{x=0}=0 \qquad (X)_{x=\ell}=0 \qquad \left(\frac{d^2X}{dx^2}\right)_{x=\ell}=0 \qquad (a)$$

which connote the fact that the translation and the bending moment are zero at each end.

It is useful to write the general expression for a normal function [Eq. (5.85)] in the following equivalent form:

$$X = C_1(\cos kx + \cosh kx) + C_2(\cos kx - \cosh kx) + C_3(\sin kx + \sinh kx) + C_4(\sin kx - \sinh kx) \qquad (5.99)$$

From the first two conditions in Eqs. (a), we conclude that the constants C_1 and C_2 in Eq. (5.99) must be equal to zero. From the third and fourth conditions, we obtain $C_3 = C_4$ and

$$\sin k\ell = 0 \qquad (5.100)$$

which is the frequency equation for the case under consideration. The nonzero positive consecutive roots of this equation are $k_i\ell = i\pi$ for $i = 1, 2, 3, \ldots, \infty$. Hence, we have

$$k_i = \frac{i\pi}{\ell} \qquad (i = 1, 2, 3, \ldots, \infty) \qquad (5.101)$$

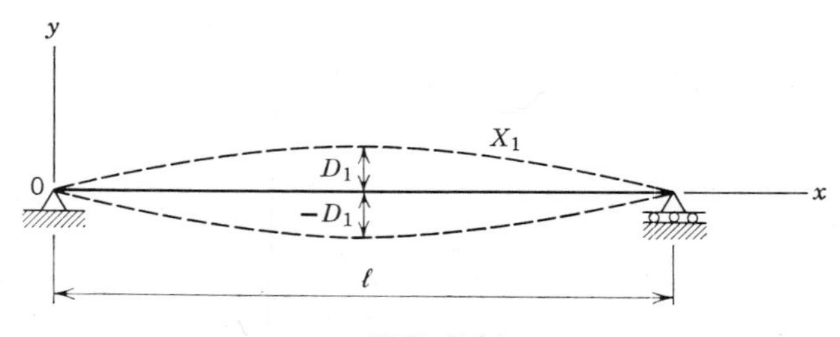

FIG. 5.14

The angular frequencies corresponding to these values of k_i are obtained as

$$\omega_i = k_i^2 a = \frac{i^2 \pi^2 a}{\ell^2} = \frac{i^2 \pi^2}{\ell^2} \sqrt{\frac{EI}{\rho A}} \tag{5.102}$$

and the natural periods are

$$\tau_i = \frac{1}{f_i} = \frac{2\pi}{\omega_i} = \frac{2\ell^2}{i^2 \pi} \sqrt{\frac{\rho A}{EI}} \tag{5.103}$$

It is seen that the period of vibration for any mode is proportional to the square of the length and inversely proportional to the radius of gyration of the cross section. Thus, for geometrically similar beams of the same material, the natural periods vary in direct proportion to the dimensions.

The shapes of the deflection curves for the various modes of vibration are given by the normal function [Eq. (5.99)], with $C_1 = C_2 = 0$ and $C_3 = C_4 = D/2$, as follows:

$$X_i = D_i \sin k_i x = D_i \sin \frac{i\pi x}{\ell} \qquad (i = 1, 2, 3, \ldots, \infty) \tag{5.104}$$

Thus, the mode shapes are sine curves, the first of which is indicated by dashed lines in Fig. 5.14. We see that the normal functions for a simple beam are the same as those for a stretched wire with fixed ends (see Figs. 5.10*c*, *d*, and *e*). To satisfy the normalization requirement of Eq. (5.97) in the preceding section, we would use the constant $D_i = \sqrt{2/\ell}$.

We shall now determine the transverse response of a simple beam to initial conditions of displacement and velocity. As for a stretched wire, we represent the initial transverse displacement of any point on the beam (at time $t = 0$) as $v_0 = f_1(x)$ and the initial velocity as $\dot{v}_0 = f_2(x)$. The general form of the solution is given by Eq. (5.86) in the preceding section, which is equivalent to Eq. (5.25) for the normal-mode method in Sec. 5.4. If we substitute the normalized function from Eq. (5.104) into Eqs. (5.23) and (5.24), the results are

$$\phi_{0i} = \int_0^\ell f_1(x) X_i \, dx = \sqrt{\frac{2}{\ell}} \int_0^\ell f_1(x) \sin \frac{i\pi x}{\ell} \, dx \tag{b}$$

$$\dot{\phi}_{0i} = \int_0^\ell f_2(x) X_i \, dx = \sqrt{\frac{2}{\ell}} \int_0^\ell f_2(x) \sin \frac{i\pi x}{\ell} \, dx \tag{c}$$

Substitution of these expressions into Eq. (5.25), with u replaced by v, produces

$$v = \frac{2}{\ell}\sum_{i=1}^{\infty} \sin\frac{i\pi x}{\ell}\left[\cos\omega_i t\int_0^\ell f_1(x)\sin\frac{i\pi x}{\ell}\,dx + \frac{1}{\omega_i}\sin\omega_i t\int_0^\ell f_2(x)\sin\frac{i\pi x}{\ell}\,dx\right] \tag{5.105}$$

Comparing this formula with Eq. (5.86), we see that the constants A_i and B_i are given by

$$A_i = \frac{2}{\ell}\int_0^\ell f_1(x)\sin\frac{i\pi x}{\ell}\,dx \tag{d}$$

$$B_i = \frac{2}{\ell\omega_i}\int_0^\ell f_2(x)\sin\frac{i\pi x}{\ell}\,dx \tag{e}$$

As an example, let us assume that due to impact an initial velocity $\dot{v}_1$ is given to a short portion δ of the beam at the distance x_1 from the left-hand support. In this case we have $f_1(x) = 0$; and $f_2(x)$ is also equal to zero at all locations except the point $x = x_1$, for which $f_2(x_1) = \dot{v}_1$. Substituting these conditions into Eqs. (*d*) and (*e*), we obtain (without integration)

$$A_i = 0 \qquad B_i = \frac{2\dot{v}_1\delta}{\ell\omega_i}\sin\frac{i\pi x_1}{\ell}$$

and the total response becomes

$$v = \frac{2\dot{v}_1\delta}{\ell}\sum_{i=1}^{\infty}\frac{1}{\omega_i}\sin\frac{i\pi x}{\ell}\sin\frac{i\pi x_1}{\ell}\sin\omega_i t \tag{f}$$

If the impact occurs at the middle of the span ($x_1 = \ell/2$), we have

$$\begin{aligned} v &= \frac{2\dot{v}_1\delta}{\ell}\left(\frac{1}{\omega_1}\sin\frac{\pi x}{\ell}\sin\omega_1 t - \frac{1}{\omega_3}\sin\frac{3\pi x}{\ell}\sin\omega_3 t + \frac{1}{\omega_5}\sin\frac{5\pi x}{\ell}\sin\omega_5 t - \cdots\right) \\ &= \frac{2\dot{v}_1\delta\ell}{a\pi^2}\left(\sin\frac{\pi x}{\ell}\sin\omega_1 t - \frac{1}{9}\sin\frac{3\pi x}{\ell}\sin\omega_3 t + \frac{1}{25}\sin\frac{5\pi x}{\ell}\sin\omega_5 t - \cdots\right) \end{aligned} \tag{g}$$

For this case, only symmetric modes of vibration are excited, and the amplitudes of consecutive modal contributions in Eq. (*g*) decrease in proportion to $1/i^2$.

5.11 VIBRATIONS OF BEAMS WITH OTHER END CONDITIONS

Beam with Free Ends

In this case we have the following boundary conditions:

$$\left(\frac{d^2X}{dx^2}\right)_{x=0}=0 \qquad \left(\frac{d^3X}{dx^3}\right)_{x=0}=0 \qquad \left(\frac{d^2X}{dx^2}\right)_{x=\ell}=0 \qquad \left(\frac{d^3X}{dx^3}\right)_{x=\ell}=0 \quad (a)$$

To satisfy the first two conditions, we must take in the general solution [Eq. (5.99)] $C_2=C_4=0$, so that

$$X = C_1(\cos kx + \cosh kx) + C_3(\sin kx + \sinh kx) \tag{5.106}$$

From the third and fourth conditions, we find

$$C_1(-\cos k\ell + \cosh k\ell) + C_3(-\sin k\ell + \sinh k\ell) = 0 \quad (b)$$

$$C_1(\sin k\ell + \sinh k\ell) + C_3(-\cos k\ell + \cosh k\ell) = 0 \quad (c)$$

A solution for the constants C_1 and C_3 (different from zero) can be obtained only when the determinant of the coefficients in Eqs. (b) and (c) vanishes. In this manner we determine the following frequency equation:

$$(-\cos k\ell + \cosh k\ell)^2 - (-\sin^2 k\ell + \sinh^2 k\ell) = 0 \quad (d)$$

Or, remembering that

$$\cos^2 k\ell + \sin^2 k\ell = 1 \qquad \cosh^2 k\ell - \sinh^2 k\ell = 1$$

we rewrite Eq. (d) as

$$\cos k\ell \cosh k\ell = 1 \tag{5.107}$$

A few of the lowest consecutive roots of this equation are

$k_0\ell$	$k_1\ell$	$k_2\ell$	$k_3\ell$	$k_4\ell$	$k_5\ell$
0	4.730	7.853	10.996	14.137	17.279

the first of which is a repeated root representing rigid-body motions of two types. The nonzero roots may be approximated with the formula

$$k_i\ell \approx (i + \tfrac{1}{2})\pi$$

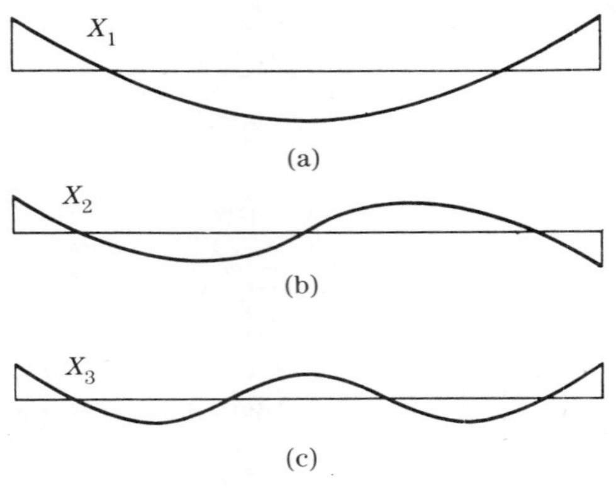

FIG. 5.15

The frequencies for the beam under consideration can be calculated by using the equation $f_i = \omega_i/2\pi = k_i^2 a/2\pi$, which gives

$$f_0 = 0 \qquad f_1 = \frac{\omega_1}{2\pi} = \frac{k_1^2 a}{2\pi} \qquad f_2 = \frac{\omega_2}{2\pi} = \frac{k_2^2 a}{2\pi} \qquad \cdots \tag{e}$$

By substituting the consecutive roots of Eq. (5.107) into Eqs. (*b*)and (*c*), we can find C_1/C_3 for each mode of vibration. Then the shape of the deflection curve during vibration is obtained from Eq. (5.106). The first three modes of vibration, corresponding to the frequencies f_1, f_2, and f_3, are shown in Figs. 5.15*a*, *b*, and *c*, respectively. On these vibrations the displacements of the beam as a rigid body can be superimposed. The combined rigid-body motion may be characterized as

$$X_0 = c_1 + c_2 x \tag{f}$$

This expression represents a translation and a rotation, which can be superimposed on the free vibrations.

Beam with Fixed Ends

The boundary conditions for this case are

$$(X)_{x=0} = 0 \qquad \left(\frac{dX}{dx}\right)_{x=0} = 0 \qquad (X)_{x=\ell} = 0 \qquad \left(\frac{dX}{dx}\right)_{x=\ell} = 0 \tag{g}$$

The first two conditions will be satisfied if in the general solution [Eq. (5.99)] we take $C_1 = C_3 = 0$, so that

$$X = C_2(\cos kx - \cosh kx) + C_4(\sin kx - \sinh kx) \tag{5.108}$$

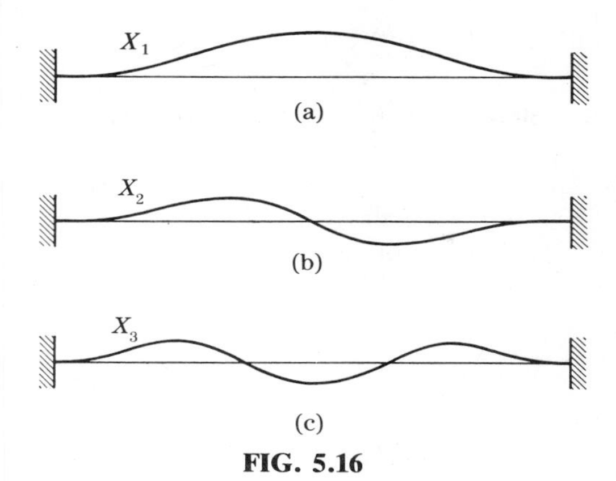

FIG. 5.16

From the other two condisitons the following equations are obtained:

$$C_2(\cos k\ell - \cosh k\ell) + C_4(\sin k\ell - \sinh k\ell) = 0 \tag{h}$$

$$C_2(\sin k\ell + \sinh k\ell) + C_4(-\cos k\ell + \cosh k\ell) = 0 \tag{i}$$

from which the same frequency equation as in the preceding case [Eq. (5.107)] can be deduced. This result means that the consecutive nonzero frequencies of vibration of a beam with fixed ends are the same as for the beam with free ends. Figures 5.16*a, b,* and *c* illustrate the first three mode shapes for this type of beam.

Beam with One End Fixed, Other End Free

Assuming that the left end ($x = 0$) is built in, we have the boundary conditions

$$(X)_{x=0} = 0 \qquad \left(\frac{dX}{dx}\right)_{x=0} = 0 \qquad \left(\frac{d^2X}{dx^2}\right)_{x=\ell} = 0 \qquad \left(\frac{d^3X}{dx^3}\right)_{x=\ell} = 0 \tag{j}$$

From the first two conditions we conclude that $C_1 = C_3 = 0$ in the general solution [Eq. (5.99)], so that the form of the mode shapes is again given by Eq. (5.108). The remaining two conditions give us the following frequency equation:

$$\cos k\ell \cosh k\ell = -1 \tag{5.109}$$

The consecutive roots of this equation are as follows:

$k_1\ell$	$k_2\ell$	$k_3\ell$	$k_4\ell$	$k_5\ell$	$k_6\ell$
1.875	4.694	7.855	10.996	14.137	17.279

Approximate values of these roots may be calculated with the formula

$$k_i\ell \approx (i - \tfrac{1}{2})\pi$$

We see that with increasing frequency the roots of Eq. (5.109) approach the roots of Eq. (5.107), obtained above for a beam with free ends.

The frequency of vibration of any mode will be

$$f_i = \frac{\omega_i}{2\pi} = \frac{ak_i^2}{2\pi} \tag{k}$$

Taking, for instance, the fundamental mode of vibration, we obtain

$$f_1 = \frac{a}{2\pi}\left(\frac{1.875}{\ell}\right)^2 \tag{l}$$

The corresponding period of vibration is

$$\tau_1 = \frac{1}{f_1} = \frac{2\pi}{a}\frac{\ell^2}{(1.875)^2} = \frac{2\pi}{3.515}\sqrt{\frac{\rho A\ell^4}{EI}} \tag{m}$$

The first three mode shapes for this beam are depicted in Figs. 5.17*a*, *b*, and *c*.

Beam with One End Fixed, Other End Simply Supported

In this case the frequency equation becomes

$$\tan k\ell = \tanh k\ell \tag{5.110}$$

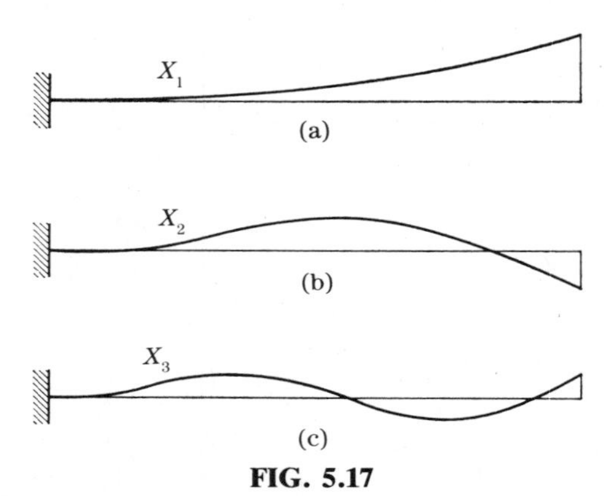

FIG. 5.17

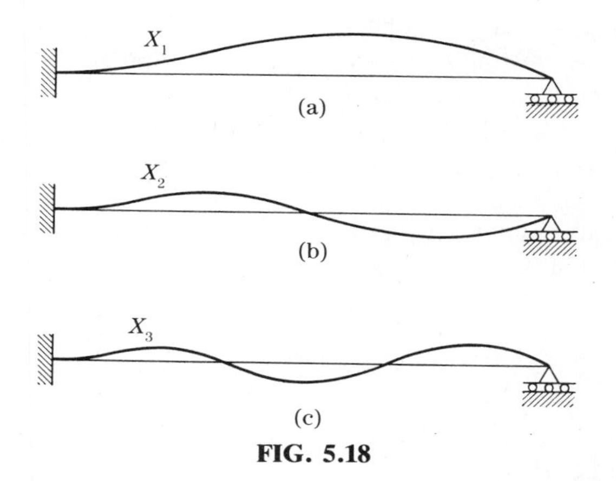

FIG. 5.18

The consecutive roots of this equation are

$k_1\ell$	$k_2\ell$	$k_3\ell$	$k_4\ell$	$k_5\ell$
3.927	7.069	10.210	13.352	16.493

These roots are given with satisfactory accuracy by the formula

$$k_i\ell \approx \left(i + \frac{1}{4}\right)\pi$$

Figures 5.18*a*, *b*, and *c* show the first three mode shapes for such a beam.

For all end conditions that we have considered, the normal functions and their consecutive derivatives have been tabulated.† By the use of these tables, the solution of vibration problems for beams can be greatly simplified. We shall now show how the values in such tables can be used for the purpose of calculating the response of beams to initial conditions. A similar technique for calculating the response to applied actions will be covered in a later section.

The method used previously for determining the response of an elastic body to initial conditions involves the evaluation of integrals of the forms

$$\int_0^\ell f_1(x)X_i\,dx \qquad \int_0^\ell f_2(x)X_i\,dx$$

Direct integrations of such expressions become difficult when the normal functions X_i are complicated. Among the types of beams that we have studied, only the simply supported beam has simple mode shapes. Beams with other end conditions involve hyperbolic functions, which usually

† See "Tables of Characteristic Functions Representing Normal Modes of Vibration of a Beam," by Dana Young and R. P. Felgar, *Univ. Texas Publ.*, No. 4913, 1949.

necessitate numerical integration. However, an alternative approach will be found to be advantageous, especially when the initial conditions are caused by a concentrated force or a concentrated moment. The case of an initial displacement $v_0 = f_1(x)$ due to a concentrated force P_0 (suddenly removed at time $t = 0$) is discussed below, and that for a concentrated moment may be handled in a similar manner.

Any initial displacement curve $v_0 = f_1(x)$ for a beam can be expressed in terms of the normal functions X_i, as follows:

$$v_0 = b_1 X_1 + b_2 X_2 + b_3 X_3 + \cdots = \sum_{i=1}^{\infty} b_i X_i \qquad (n)$$

where the constants b_i are scale factors that are to be determined. The strain energy within a prismatic beam in the deflected position may be written as

$$U_0 = \frac{EI}{2} \int_0^{\ell} (v_0'')^2 \, dx \qquad (o)$$

Substitution of the second derivative (with respect to x) of v_0 from Eq. (n) into Eq. (o) gives

$$U_0 = \frac{EI}{2} \sum_{i=1}^{\infty} b_i^2 \int_0^{\ell} (X_i'')^2 \, dx \qquad (p)$$

From Eqs. (5.91) and (5.92) of Sec. 5.9, we have

$$\int_0^{\ell} (X_i'')^2 \, dx = k_i^4 \int_0^{\ell} X_i^2 \, dx \qquad (q)$$

Using this relationship in Eq. (p), we obtain

$$U_0 = \frac{EI}{2} \sum_{i=1}^{\infty} b_i^2 k_i^4 \int_0^{\ell} X_i^2 \, dx \qquad (r)$$

Suppose that the initial displacement v_0 in Eq. (n) is caused by the concentrated force P_0, acting in the y direction at the point $x = x_1$. We shall determine the constants b_i in Eq. (n) required for this case, using the principle of virtual work. Let us consider the virtual displacement $\delta b_i X_i$ and set the virtual work of the applied force equal to the increment of the strain energy, as follows:

$$P_0 \delta b_i X_{i1} = \frac{\partial U_0}{\partial b_i} \delta b_i = EI b_i k_i^4 \delta b_i \int_0^{\ell} X_i^2 \, dx \qquad (s)$$

in which X_{i1} represents X_i evaluated at $x = x_1$. Solving for b_i in Eq. (s), we obtain

$$b_i = \frac{P_0 X_{i1}}{EIk_i^4 \int_0^\ell X_i^2\, dx} \tag{t}$$

Substitution of this expression for b_i into Eq. (n) results in

$$v_0 = \frac{P_0}{EI} \sum_{i=1}^{\infty} \left(\frac{X_i X_{i1}}{k_i^4 \int_0^\ell X_i^2\, dx} \right) \tag{u}$$

We see that the manner of normalizing X_i will not affect the value of v_0. In the tables mentioned above, the normalization procedure is given by

$$\int_0^\ell X_i^2\, dx = \ell \tag{v}$$

Under this condition, Eq. (t) may be rewritten as

$$b_i = \frac{P_0 \ell^3 X_{i1}}{EI(k_i \ell)^4} \tag{5.111}$$

and Eq. (u) becomes

$$v_0 = \frac{P_0 \ell^3}{EI} \sum_{i=1}^{\infty} \frac{X_i X_{i1}}{(k_i \ell)^4} \tag{5.112}$$

which could be used in static analysis for representing the deflection curve of a beam.

Remembering that the free-vibrational response of an elastic beam displaced initially in the configuration $b_i X_i$ is

$$v_i = b_i X_i \cos \omega_i t \tag{w}$$

we can calculate the total response due to v_0 as

$$v = \sum_{i=1}^{\infty} b_i X_i \cos \omega_i t \tag{x}$$

Substitution of b_i from Eq. (5.111) into Eq. (x) gives

$$v = \frac{P_0 \ell^3}{EI} \sum_{i=1}^{\infty} \frac{X_i X_{i1}}{(k_i \ell)^4} \cos \omega_i t \tag{5.113}$$

To use this expression, we obtain values of X_{i1} from the tables for as many modes as desired.

This method can be used for a distributed load as well as for a concentrated force; however, from a practical viewpoint it is not especially advantageous to do so. Calculation of the virtual work of a distributed loading [see Eq. (s)] requires an integration of the product of the load intensity and each normal function over the length of the beam. This integration with the load function is similar to that for the displacement function, which is the type of operation we wished to avoid in the first place. However, in most cases it is simpler to integrate with the load function than with the displacement function.

Example. Suppose that a beam with fixed ends is subjected to a transverse force P_0 applied at midspan. Assuming that the force is removed instantaneously at the time $t = 0$, determine the free-vibrational response at the center of the beam.

Solution. The normal functions for a beam with fixed ends [see Eq. (5.108)] may be expressed as

$$X_i = \cosh k_i x - \cos k_i x - \alpha_i (\sinh k_i x - \sin k_i x) \tag{y}$$

where

$$\alpha_i = \frac{\cosh k_i \ell - \cos k_i \ell}{\sinh k_i \ell - \sin k_i \ell}$$

Thus, $\alpha_1 = 0.9825$, $\alpha_2 = 1.0008$, $\alpha_3 \approx 1$, $\alpha_4 \approx 1$, etc. From the above-mentioned tables, we find for the odd modes of a fixed-end beam

$$(X_1)_{x=\ell/2} = 1.588 \qquad (X_3)_{x=\ell/2} = -1.406 \qquad (X_5)_{x=\ell/2} = 1.415$$

and so on. Using these values in Eq. (5.113), together with the values of $k_i \ell$ given earlier, we obtain the following response at the center of the beam:

$$\begin{aligned}(v)_{x=\ell/2} &= \frac{P_0 \ell^3}{EI} \left[\frac{(1.588)^2}{(4.730)^4} \cos \omega_1 t + \frac{(1.406)^2}{(10.996)^4} \cos \omega_3 t + \frac{(1.415)^2}{(17.279)^4} \cos \omega_5 t + \cdots \right] \\ &= \frac{P_0 \ell^3}{EI} (5038 \cos \omega_1 t + 135 \cos \omega_3 t + 22 \cos \omega_5 t + \cdots) \times 10^{-6}\end{aligned} \tag{z}$$

Thus, the series converges rapidly for the case under consideration.

5.12 EFFECTS OF ROTARY INERTIA AND SHEARING DEFORMATIONS

In previous discussions of flexural vibrations, we tacitly assumed that the cross-sectional dimensions of the beam were small in comparison with its length. Corrections to the theory will now be given for the purpose of taking into account the effects of the cross-sectional dimensions on the frequencies. These corrections may be of considerable importance for studying the modes of vibration of higher frequencies when a vibrating bar is subdivided by *nodal cross sections* into comparatively short portions.

It is easy to see that during vibration a typical element of a beam (see Fig. 5.13*b*) performs not only a translatory motion but also rotates [1]. The angle of rotation, which is equal to the slope of the deflection curve, is expressed by $\partial v/\partial x$; and the corresponding angular velocity and angular acceleration will be given by

$$\frac{\partial^2 v}{\partial x\, \partial t} \qquad \text{and} \qquad \frac{\partial^3 v}{\partial x\, \partial t^2}$$

Therefore, the inertial moment of the element about an axis through its center of mass and perpendicular to the x–y plane will be

$$-\rho I \frac{\partial^3 v}{\partial x\, \partial t^2}\, dx$$

which is positive in the counterclockwise sense. This moment should be included when writing the equations of dynamic equilibrium for a typical element. Thus, instead of Eq. (*b*) in Sec. 5.9, we will have

$$-V\, dx + \frac{\partial M}{\partial x}\, dx - \rho I \frac{\partial^3 v}{\partial x\, \partial t^2}\, dx = 0 \tag{a}$$

Substituting the shearing force V from this expression into the equation for equilibrium of forces in the y direction [Eq. (*a*), Sec. 5.9], we have

$$\frac{\partial V}{\partial x}\, dx = \frac{\partial}{\partial x}\left(\frac{\partial M}{\partial x} - \rho I \frac{\partial^3 v}{\partial x\, \partial t^2}\right) dx = -\rho A\, dx \frac{\partial^2 v}{\partial t^2} \tag{b}$$

and using Eq. (*d*), Sec. 5.9, we obtain

$$EI \frac{\partial^4 v}{\partial x^4} = -\rho A \frac{\partial^2 v}{\partial t^2} + \rho I \frac{\partial^4 v}{\partial x^2\, \partial t^2} \tag{5.114}$$

This is the differential equation for the transverse vibrations of prismatic beams, in which the second term on the right side represents the effect of rotary inertia.

A still more accurate differential equation is obtained if not only the rotary inertia but also the deflection due to shear is taken into account.† The slope of the deflection curve depends not only on the rotation of cross sections of the beam but also on the shearing deformations. Let ψ denote the slope of the deflection curve when the shearing force is neglected and β the angle of shear at the neutral axis in the same cross section. Then we find for the total slope

$$\frac{dv}{dx} = \psi + \beta \tag{c}$$

From elementary flexure theory we have for bending moment and shearing force the following equations:

$$M = EI\frac{d\psi}{dx} \qquad V = -k'\beta AG = -k'\left(\frac{dv}{dx} - \psi\right)AG \tag{d}$$

in which k' is a numerical factor depending on the shape of the cross section, A is the cross-sectional area, and G is the modulus of elasticity in shear. The differential equation for rotation of an element will now be

$$-V\,dx + \frac{\partial M}{\partial x}\,dx - \rho I\frac{\partial^2 \psi}{\partial t^2}\,dx = 0 \tag{e}$$

Substituting Eqs. (d) into Eq. (e), we obtain

$$EI\frac{\partial^2 \psi}{\partial x^2} + k'\left(\frac{\partial v}{\partial x} - \psi\right)AG - \rho I\frac{\partial^2 \psi}{\partial t^2} = 0 \tag{f}$$

The differential equation for the translatory motion of the same element in the y direction will still be

$$-\frac{\partial V}{\partial x}\,dx - \rho A\,\frac{\partial^2 v}{\partial t^2}\,dx = 0 \tag{g}$$

† See the paper by *S. Timoshenko, Phil. Mag.* Ser. 6, **41,** p. 744; and **43,** p. 125, 1921. An experimental verification of shear effect was made by E. Goens, *Ann. Physik,* Ser. 5, **11,** 1931, p. 649; see also R. M. Davies, *Phil. Mag., Ser.* 7, **23,** 1937, p. 1129. The need to consider shear deformation in the case of impact on a beam was discussed by W. Flügge, *Z. angew. Math. u. Mech.,* **22,** 1942, p. 312.

Substitution of the second relationship of Eqs. (d) into Eq. (g) gives

$$k'\left(\frac{\partial^2 v}{\partial x^2} - \frac{\partial \psi}{\partial x}\right)G - \rho\frac{\partial^2 v}{\partial t^2} = 0 \tag{h}$$

Eliminating ψ from Eqs. (f) and (h), we obtain the following more complete differential equation for the transverse vibrations of prismatic beams:

$$EI\frac{\partial^4 v}{\partial x^4} + \rho A\frac{\partial^2 v}{\partial t^2} - \rho I\left(1 + \frac{E}{k'G}\right)\frac{\partial^4 v}{\partial x^2\,\partial t^2} + \frac{\rho^2 I}{k'G}\frac{\partial^4 v}{\partial t^4} = 0 \tag{5.115}$$

The application of this equation in calculating the frequencies of a beam will be shown in the following discussion.

Let us reconsider the simply supported beam that was analyzed in Sec. 5.10 (see Fig. 5.14). In order to find the values of the frequencies more accurately, we should use Eq. (5.115) instead of Eq. (5.83). Dividing Eq. (5.115) by ρA and using Eq. (5.84) and the notation

$$r_g^2 = \frac{I}{A} \tag{i}$$

we obtain

$$a^2\frac{\partial^4 v}{\partial x^4} + \frac{\partial^2 v}{\partial t^2} - r_g^2\left(1 + \frac{E}{k'G}\right)\frac{\partial^4 v}{\partial x^2\,\partial t^2} + r_g^2\frac{\rho}{k'G}\frac{\partial^4 v}{\partial t^4} = 0 \tag{5.116}$$

This equation and the end conditions will be satisfied by taking

$$v_i = \left(\sin\frac{i\pi x}{\ell}\right)(A_i \cos\omega_i t + B_i \sin\omega_i t) \tag{j}$$

which involves the same normal function as that for the beam without rotary inertia and shearing deformation. Substituting Eq. (j) into Eq. (5.116), we obtain the following equation for calculating the frequencies:

$$a^2\frac{i^4\pi^4}{\ell^4} - \omega_i^2 - \omega_i^2\frac{i^2\pi^2 r_g^2}{\ell^2} - \omega_i^2\frac{i^2\pi^2 r_g^2}{\ell^2}\frac{E}{k'G} + \frac{r_g^2\rho}{k'G}\omega_i^4 = 0 \tag{5.117}$$

Considering only the first two terms in this equation, we have

$$\omega_i = a\frac{i^2\pi^2}{\ell^2} = \frac{a\pi^2}{\lambda_i^2} \tag{k}$$

in which $\lambda_i = (\ell/i)$ is the length of the half waves into which the bar is subdivided during vibration. This coincides with the result in Eq. (5.102) obtained before. By taking the first three terms in Eq. (5.117) and using the binomial expansion, we obtain

$$\omega_i = \frac{a\pi^2}{\lambda_i^2}\left(1 - \frac{\pi^2 r_g^2}{2\lambda_i^2}\right) \tag{l}$$

In this manner the *effect of rotary inertia* is taken into account; and we see that this correction becomes more and more important with a decrease of λ_i, that is, with an increase in the frequency of vibration.

In order to obtain the *effect of shearing deformations,* we should take all of the terms in Eq. (5.117) into consideration. By substituting the first approximation (k) for ω_i into the last term of this equation, it can be shown that this term is a small quantity of the second order compared with the quantity $\pi^2 r_g^2/\lambda_i^2$. Neglecting the last term, we obtain

$$\omega_i = \frac{a\pi^2}{\lambda_i^2}\left[1 - \frac{1}{2}\frac{\pi^2 r_g^2}{\lambda_i^2}\left(1 + \frac{E}{k'G}\right)\right] \tag{5.118}$$

Assuming that $G = 3E/8$ and taking a beam of rectangular cross section, for which $k' = 0.833$,† we have

$$\frac{E}{k'G} = 3.2$$

Thus, the correction due to shear is 3.2 times larger than the correction due to rotary inertia.‡ Assuming that the wave length λ_i for the ith mode is 10 times larger than the depth of the beam, we obtain

$$\frac{1}{2}\left(\frac{\pi^2 r_g^2}{\lambda_i^2}\right) = \frac{1}{2}\left(\frac{\pi^2}{12}\right)\left(\frac{1}{100}\right) \approx 0.004$$

and the total correction for rotary inertia and shearing deformation will be about 1.7%.

5.13 FORCED RESPONSE OF A SIMPLE BEAM

In this section we shall consider the transverse response of a simply supported beam to a distributed force $Q(x, t)$, a concentrated force $P_1(t)$, or a concentrated moment $M_1(t)$ applied at $x = x_1$ (see Fig. 5.19). As

† A somewhat different value for k' was suggested by R. G. Olsson, see *Z. angew. Math. u. Mech.*, **15,** 1935.

‡ Regarding the solution of Eq. (5.116), see R. A. Anderson, "Flexural Vibrations in Uniform Beams According to the Timoshenko Theory," *Trans. ASME,* **75,** 1953, pp. APM 504–514.

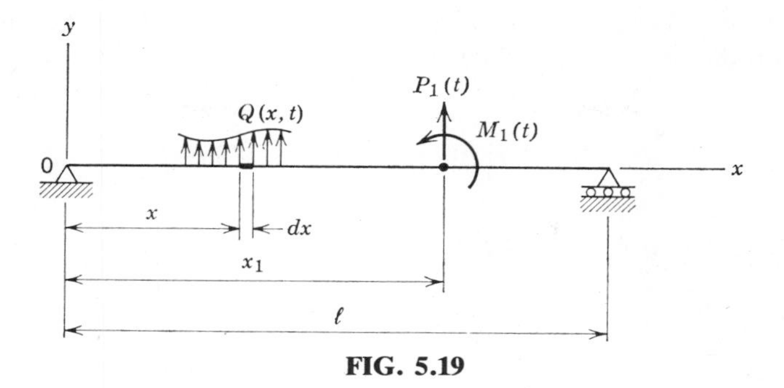

FIG. 5.19

mentioned earlier (Sec. 5.9), there is no need to derive the general expressions for the response of a beam to the first two types of loads. Referring to Eq. (5.28) in Sec. 5.4, we write the transverse response v, due to the distributed force $Q(x, t)$, as

$$v = \sum_{i=1}^{\infty} \frac{X_i}{\omega_i} \int_0^{\ell} X_i \int_0^t q(x, t') \sin \omega_i(t - t')\, dt'\, dx \tag{5.119}$$

where $q(x, t) = Q(x, t)/m$. Similarly, from Eq. (5.29) we obtain the expression for the response caused by the concentrated force $P_1(t)$ as

$$v = \sum_{i=1}^{\infty} \frac{X_i X_{i1}}{\omega_i} \int_0^t q_1(t') \sin \omega_i(t - t')\, dt' \tag{5.120}$$

in which X_{i1} denotes the normal function X_i evaluated at $x = x_1$; and $q_1(t) = P_1(t)/m$.

Because the moment $M_1(t)$ does not correspond to a generic translation, it must be handled indirectly. For this purpose we shall use the method of virtual work in the manner shown before (see Sec. 5.3). In this approach the deflection curve is expanded in the series

$$v = \sum_{j=1}^{\infty} \phi_j X_j \tag{a}$$

and we take the virtual displacement $\delta v_i = \delta\phi_i X_i$ for the ith mode. Then the virtual work of the distributed inertia forces through the virtual displacement of the ith mode is

$$\delta W_{Ii} = \int_0^{\ell} (-\rho A\, dx\, \ddot{v})\, \delta v_i = -m\, \delta\phi_i \int_0^{\ell} \ddot{v} X_i\, dx \tag{b}$$

Substituting Eq. (a) into Eq. (b) and using the orthogonality and normalization relationships in Eqs. (5.88) and (5.97), we obtain (for $i = j$)

$$\delta W_{Ii} = -m\ddot{\phi}_i\, \delta\phi_i \int_0^\ell X_i^2\, dx = -m\ddot{\phi}_i\, \delta\phi_i \tag{c}$$

The strain energy associated with bending in the beam is

$$U = \int_0^\ell \frac{EI}{2} (v'')^2\, dx = \frac{r}{2} \int_0^\ell (v'')^2\, dx \tag{d}$$

Substitution of Eq. (a) into Eq. (d) and use of the relationships (5.89) and (5.97) yields

$$U = \frac{r}{2} \sum_{j=1}^{\infty} \phi_j^2 \int_0^\ell (X_j'')^2\, dx = \frac{r}{2} \sum_{j=1}^{\infty} k_j^4 \phi_j^2 \tag{e}$$

Then the virtual work of elasticity forces is (for $i = j$)

$$\delta W_{Ei} = -\frac{\partial U}{\partial \phi_i}\, \delta\phi_i = -rk_i^4 \phi_i\, \delta\phi_i = -m\omega_i^2 \phi_i\, \delta\phi_i \tag{f}$$

To determine the virtual work of the concentrated moment, we observe that it does work through the rotation $\delta v_i'$ at its point of application. Thus, we write the work of M_1 through a virtual displacement of the ith mode as

$$\delta W_{M_1 i} = M_1\, \delta v_{i1}' = M_1\, \delta\phi_i X_{i1}' \tag{g}$$

where X_{i1}' is the first derivative of X_i (with respect to x), evaluated at $x = x_1$.

Summing the expressions (c), (f), and (g) and equating the result to zero, we obtain

$$m\ddot{\phi}_i + m\omega_i^2 \phi_i = M_1 X_{i1}' \tag{h}$$

Division of Eq. (h) by m produces

$$\ddot{\phi}_i + \omega_i^2 \phi_i = \frac{M_1 X_{i1}'}{m} \qquad (i = 1, 2, 3, \ldots, \infty) \tag{5.121}$$

This expression represents a typical equation of motion in normal coordinates, and the term on the right is the ith normal-mode load for the case under consideration.

The response of the ith vibrational mode is found by the Duhamel integral to be

$$\phi_i = \frac{X'_{i1}}{m\omega_i} \int_0^t M_1(t') \sin \omega_i(t - t')\, dt' \tag{5.122}$$

and the total vibrational response is

$$v = \sum_{i=1}^{\infty} \frac{X_i X'_{i1}}{m\omega_i} \int_0^t M_1(t') \sin \omega_i(t - t')\, dt' \tag{5.123}$$

Thus, the virtual-work technique, combined with the normal-mode concept, produces a response expression for the concentrated moment which is similar to that for a concentrated force [see Eq. (5.120)]. However, the terms P_1 and X_{i1} are replaced by M_1 and X'_{i1}, respectively.

In the particular case of a simply supported beam, the angular frequencies and normalized shape functions are (see Sec. 5.10)

$$\omega_i = k_i^2 a = \frac{i^2\pi^2 a}{\ell^2} \qquad X_i = \sqrt{\frac{2}{\ell}} \sin \frac{i\pi x}{\ell} \qquad (i = 1, 2, 3, \ldots, \infty) \tag{i}$$

Substitution of the normal functions into Eq. (5.119) for the distributed force gives

$$v = \frac{2}{\ell} \sum_{i=1}^{\infty} \frac{1}{\omega_i} \sin \frac{i\pi x}{\ell} \int_0^{\ell} \sin \frac{i\pi x}{\ell} \int_0^t q(x, t') \sin \omega_i(t - t')\, dt'\, dx \tag{5.124}$$

Similarly, Eq. (5.120) for the concentrated force becomes

$$v = \frac{2}{\ell} \sum_{i=1}^{\infty} \frac{1}{\omega_i} \sin \frac{i\pi x}{\ell} \sin \frac{i\pi x_1}{\ell} \int_0^t q_1(t') \sin \omega_i(t - t')\, dt' \tag{5.125}$$

and Eq. (5.123) for the concentrated moment yields

$$v = \frac{2\pi}{m\ell^2} \sum_{i=1}^{\infty} \frac{i}{\omega_i} \sin \frac{i\pi x}{\ell} \cos \frac{i\pi x_1}{\ell} \int_0^t M_1(t') \sin \omega_i(t - t')\, dt' \tag{5.126}$$

Equations (5.124) and (5.125) are the same as Eqs. (5.69) and (5.70) for a stretched wire (see Sec. 5.8), but Eq. (5.126) is feasible only for a member with flexural rigidity.

As an example, let us consider the case of a harmonically varying force $P_1 = P \sin \Omega t$, applied at $x = x_1$. In this instance the response given by Eq.

(5.125) is

$$
\begin{aligned}
v &= \frac{2P}{m\ell} \sum_{i=1}^{\infty} \frac{1}{\omega_i} \sin \frac{i\pi x}{\ell} \sin \frac{i\pi x_1}{\ell} \int_0^t \sin \Omega t' \sin \omega_i (t - t')\, dt' \\
&= \frac{2P}{m\ell} \sum_{i=1}^{\infty} \frac{1}{\omega_i^2} \sin \frac{i\pi x}{\ell} \sin \frac{i\pi x_1}{\ell} \left(\sin \Omega t - \frac{\Omega}{\omega_i} \sin \omega_i t \right) \beta_i \\
&= \frac{2P\ell^3}{m\pi^4 a^2} \sum_{i=1}^{\infty} \frac{1}{i^4} \sin \frac{i\pi x}{\ell} \sin \frac{i\pi x_1}{\ell} \left(\sin \Omega t - \frac{\Omega \ell^2}{i^2 \pi^2 a} \sin \omega_i t \right) \beta_i
\end{aligned}
\tag{5.127}
$$

where the magnification factor β_i is given by

$$
\beta_i = \frac{1}{1 - \Omega^2 / \omega_i^2} \tag{j}
$$

The first part of Eq. (5.127) represents steady-state forced vibrations of the beam, whereas the second part consists of transient free vibrations. The latter vibrations will die out in the presence of damping, and only the steady-state response will be of practical importance.

If the harmonic force $P \sin \Omega t$ is varying very slowly, Ω is a very small quantity; and $\beta_i \approx 1$. Then the steady-state response becomes

$$
v = \frac{2P\ell^3}{m\pi^4 a^2} \sum_{i=1}^{\infty} \frac{1}{i^4} \sin \frac{i\pi x}{\ell} \sin \frac{i\pi x_1}{\ell} \sin \Omega t
$$

Or, using $ma^2 = EI$, we have

$$
v = \frac{2P\ell^3}{EI\pi^4} \sum_{i=1}^{\infty} \frac{1}{i^4} \sin \frac{i\pi x}{\ell} \sin \frac{i\pi x_1}{\ell} \sin \Omega t \tag{k}
$$

This expression represents the static deflection of the beam produced by the load $P \sin \Omega t$. In the particular case when the force P is applied at the middle ($x_1 = \ell/2$), we obtain

$$
v = \frac{2P\ell^3}{EI\pi^4} \left(\sin \frac{\pi x}{\ell} - \frac{1}{3^4} \sin \frac{3\pi x}{\ell} + \frac{1}{5^4} \sin \frac{5\pi x}{\ell} - \cdots \right) \sin \Omega t \tag{l}
$$

The series (l) converges rapidly, and a satisfactory approximation for the deflections will be obtained by taking the first term only. In this manner we find for the amplitude v_m at the middle

$$
(v_m)_{x=\ell/2} = \frac{2P\ell^3}{EI\pi^4} = \frac{P\ell^3}{48.7EI}
$$

The error of this approximation is about 1.5%.

Denoting by α the ratio of the frequency of the disturbing force to the frequency of the fundamental mode of free vibration, we obtain from (i)

$$\alpha = \frac{\Omega}{\omega_1} = \frac{\Omega \ell^2}{a\pi^2}$$

and the steady-state forced vibrations from Eq. (5.127) become

$$v = \frac{2P\ell^3 \sin \Omega t}{EI\pi^4} \sum_{i=1}^{\infty} \frac{\sin(i\pi x/\ell)\sin(i\pi x_1/\ell)}{i^4 - \alpha^2} \tag{m}$$

If the harmonic force is applied at the middle, we obtain

$$v = \frac{2P\ell^3 \sin \Omega t}{EI\pi^4} \left[\frac{\sin(\pi x/\ell)}{1-\alpha^2} - \frac{\sin(3\pi x/\ell)}{3^4 - \alpha^2} + \frac{\sin(5\pi x/\ell)}{5^4 - \alpha^2} - \cdots \right] \tag{n}$$

For small values of α the first term of this series represents the deflection with good accuracy; and comparing Eq. (n) with Eq. (l), we can conclude that the ratio of the dynamic deflection to the static deflection is approximately equal to

$$\beta_1 = \frac{1}{1-\alpha^2} \tag{o}$$

For example, if the frequency of the disturbing force is one-fourth of the frequency of the fundamental mode of vibration, the dynamic deflection will be about 6% greater than the static deflection.

From Eq. (5.126) we can obtain the response produced by a harmonic moment $M_1 = M \sin \Omega t$ applied at the end ($x = 0$). Following a line of reasoning similar to the above, we find

$$v = \frac{2M\ell^2 \sin \Omega t}{EI\pi^3} \sum_{i=1}^{\infty} \frac{\beta_i}{i^3} \sin \frac{i\pi x}{\ell} \tag{5.128}$$

which represents steady-state forced vibrations produced by the moment $M \sin \Omega t$.

Because problems on vibration of beams are represented by linear differential equations, the principle of superposition holds; and if there are several harmonic forces or moments acting on a beam, the resulting vibration may be found by superimposing the vibrations produced by the individual actions.

The case of continuously distributed harmonic forces also can be solved in a similar manner, using Eq. (5.124). For instance, assume that the beam is loaded by a uniformly distributed force of the intensity $Q(t) = w \sin \Omega t$.

Then from Eq. (5.124) we obtain

$$v = \frac{4w\ell^4 \sin \Omega t}{EI\pi^5} \sum_{i=1}^{\infty} \frac{\beta_i}{i^5} \sin \frac{i\pi x}{\ell} \tag{5.129}$$

If the frequency of the load is very small in comparison with the frequency of the fundamental mode of vibration, we find the approximation

$$v = \frac{4w\ell^4}{EI\pi^5}\left(\sin \frac{\pi x}{\ell} + \frac{1}{3^5}\sin \frac{3\pi x}{\ell} + \frac{1}{5^5}\sin \frac{5\pi x}{\ell} + \cdots\right)\sin \Omega t \tag{p}$$

This very rapidly converging series represents the static deflection of the beam produced by a uniformly distributed load $w \sin \Omega t$. By taking $x = \ell/2$, we obtain for the deflection at the middle

$$(v)_{x=\ell/2} = \frac{4w\ell^4}{EI\pi^5}\left(1 - \frac{1}{3^5} + \frac{1}{5^5} - \cdots\right)\sin \Omega t \tag{q}$$

If only the first term in this series is used, the error in the deflection at the middle will be about 0.25%.

5.14 FORCED RESPONSE OF BEAMS WITH OTHER END CONDITIONS

Equation (5.119) in the preceding section implies that in general the distributed loading is a function of both x and t. However, if the loading $Q(x, t)$ can be expressed as the product

$$Q(x, t) = f(x)Q(t) \tag{a}$$

then Eq. (5.119) may be written in the simpler form

$$v = \sum_{i=1}^{\infty} \frac{X_i}{\omega_i} \int_0^{\ell} f(x)X_i \, dx \int_0^{\ell} q(t') \sin \omega_i(t - t') \, dt' \tag{5.130}$$

in which $q(t) = Q(t)/m$. The first integral in Eq. (5.130) involves the product of the load function $f(x)$ and the ith normal function. This type of operation was discussed previously in Sec. 5.11, and it constitutes a difficult step for beams that are not simply supported.

On the other hand, if the loads consist of concentrated forces or moments, Eqs. (5.120) and (5.123) may be used for calculating the vibrational responses of a beam, regardless of how it is supported. As an

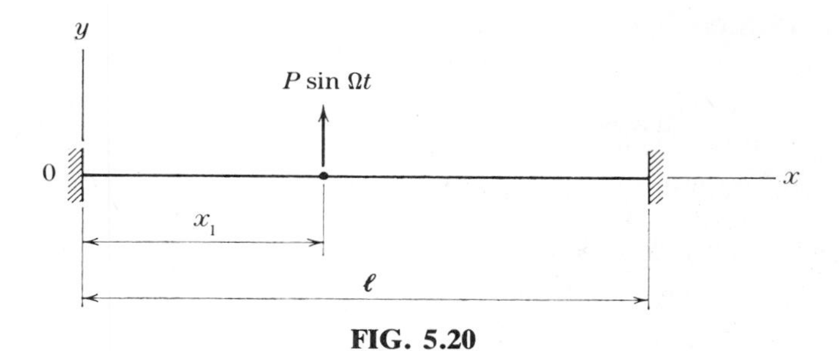

FIG. 5.20

example, let us consider the case of a beam with fixed ends and assume that vibrations are produced by a harmonic force $P_1(t) = P \sin \Omega t$, applied at the distance $x = x_1$ from the left end (see Fig. 5.20). In this case, Eq. (5.120) gives

$$v = \frac{P}{m\ell}\sum_{i=1}^{\infty}\frac{X_i X_{i1}}{\omega_i}\int_0^t \sin \Omega t' \sin \omega_i (t - t')\, dt'$$

$$= \frac{P}{m\ell}\sum_{i=1}^{\infty}\frac{X_i X_{i1}}{\omega_i^2}\left(\sin \Omega t - \frac{\Omega}{\omega_i}\sin \omega_i t\right)\beta_i \tag{5.131}$$

The length ℓ appearing in this expression is due to the type of normalization described in Sec. 5.11. Equation (5.131) holds for any type of beam; and if we wish to implement it in the case under consideration, we must use the frequencies ω_i and normal functions X_i for a member with fixed ends.

Suppose that the harmonic force in Fig. 5.20 is applied at the middle of the beam and that we wish to compute the steady-state response at the point of application. For this purpose the first part of Eq. (5.131) yields

$$(v)_{x=\ell/2} = \frac{P \sin \Omega t}{m\ell}\sum_i \frac{\beta_i}{\omega_i^2}(X_i)^2_{x=\ell/2} \qquad (i = 1, 3, 5, \ldots, \infty) \tag{b}$$

Substituting the relationships $\omega_i^2 = a^2 k_i^4$ and $ma^2 = EI$ into Eq. (b), we obtain

$$(v)_{x=\ell/2} = \frac{P\ell^3 \sin \Omega t}{EI}\sum_i \frac{\beta_i}{(k_i\ell)^4}(X_i)^2_{x=\ell/2} \qquad (i = 1, 3, 5, \ldots, \infty) \tag{c}$$

From the Example in Sec. 5.11, we have the values of terms in the series, as follows:

$$(v)_{x=\ell/2} \frac{P\ell^3 \sin \Omega t}{EI}(5038\beta_1 + 135\beta_3 + 22\beta_5 + \cdots) \times 10^{-6} \tag{d}$$

5.15 BEAMS SUBJECTED TO SUPPORT MOTIONS

In Sec. 5.6 we discussed the longitudinal response of a prismatic bar, due to either rigid-body translation of the ground or independent translations of the support restraints (in the axial direction). For beams there are two types of rigid-body motions to be considered. They are usually taken to be a pure translation in the y direction and a small rotation about the z axis, which is perpendicular to the x and y axes at the origin (see Fig. 5.13). With these two types of ground motions, the y translation of any point on a beam will be given by

$$v_g = g_1(t) + xg_2(t) \tag{a}$$

where $g_1(t)$ and $g_2(t)$ are the rigid-body translation and rotation, respectively. Referring to Eq. (5.47) in Sec. 5.6, we write the relative response of a beam due to such rigid-body ground motions as

$$v^* = -\sum_{i=1}^{\infty} \frac{X_i}{\omega_i} \left[\int_0^\ell X_i \, dx \int_0^t \ddot{g}_1(t') \sin \omega_i(t-t') \, dt' + \int_0^\ell xX_i \, dx \int_0^t \ddot{g}_2(t') \sin \omega_i(t-t') \, dt' \right] \tag{5.132}$$

in which the functions X_i pertain to beams and are normalized as indicated by Eq. (5.97). This expression represents the vibrational motion of any point on the beam, relative to the rigid-body motion. The total solution is obtained by adding the vibrational motion to the ground motion, as follows:

$$v = v_g + v^* = g_1(t) + xg_2(t) + v^* \tag{5.133}$$

It is seen from Eq. (5.132) that the integrations over the length are relatively simple and are separable from the integrations with respect to time.

A more general type of support-motion problem involves independent displacements of individual restraints. Figures (5.21a and 5.21b show the effects of unit translations (in the y direction) of the restraints for a simply supported beam. For this type of beam the displacement functions

$$\delta_1(x) = 1 - \frac{x}{\ell} \qquad \delta_2(x) = \frac{x}{\ell} \tag{b}$$

as well as the normal functions X_i are the same as those for the stretched wire discussed in Sec. 5.8. Therefore, the relative response for the beam

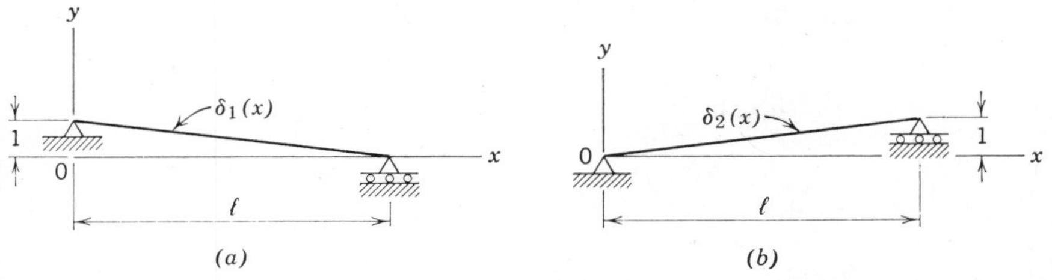

FIG. 5.21

may be written as [see (Eq. 5.72)]

$$v^* = -\frac{2}{\ell}\sum_{i=1}^{\infty}\frac{1}{\omega_i}\sin\frac{i\pi x}{\ell}\left[\int_0^{\ell}\delta_1(x)\sin\frac{i\pi x}{\ell}\,dx\int_0^t \ddot{g}_1(t')\sin\omega_i(t-t')\,dt'\right.$$
$$\left.+\int_0^{\ell}\delta_2(x)\sin\frac{i\pi x}{\ell}\,dx\int_0^t \ddot{g}_2(t')\sin\omega_i(t-t')\,dt'\right] \quad (5.134)$$

and the total response is [see Eq. (5.73)]

$$v = v_{\text{st}} + v^* = \delta_1(x)g_1(t) + \delta_2(x)g_2(t) + v^* \quad (5.135)$$

where $g_1(t)$ and $g_2(t)$ are the specified translations at the left and right ends, respectively.

The displacement functions in Eqs. (b) for the simple beam are examples of expressions that we will refer to as *displacement influence functions*. We define this type of function as the displacement of a generic point due to a unit displacement of a support restraint. Figures 5.22*a, b, c,* and *d* show

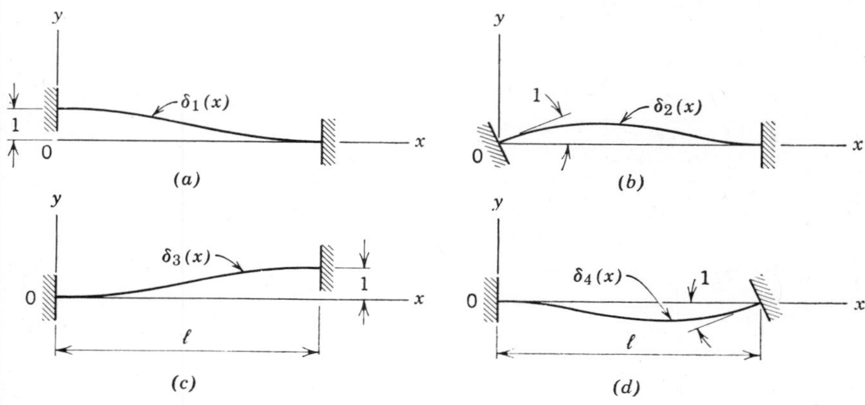

FIG. 5.22

four such influence functions for a beam with fixed ends. They are

$$\delta_1(x) = 1 - \frac{3x^2}{\ell^2} + \frac{2x^3}{\ell^3} \qquad \delta_2(x) = x - \frac{2x^2}{\ell} + \frac{x^3}{\ell^2}$$
$$\delta_3(x) = \frac{3x^2}{\ell^2} - \frac{2x^3}{\ell^3} \qquad \delta_4(x) = -\frac{x^2}{\ell} + \frac{x^3}{\ell^2} \tag{c}$$

When these influence functions are multiplied by the specified motions of the support restraints, they produce the flexible-body motions of the beam, as follows:

$$v_{\mathrm{st}} = \delta_1(x)g_1(t) + \delta_2(x)g_2(t) + \delta_3(x)g_3(t) + \delta_4(x)g_4(t) \tag{d}$$

The symbols $g_1(t)$ and $g_2(t)$ in this expression denote the translation and rotation of the left end, and $g_3(t)$ and $g_4(t)$ are the translation and rotation of the right end.

Figures 5.23*a* and 5.23*b* depict unit values of independent restraint motions for a cantilever beam that is fixed at the left end (the origin of the coordinates). In this case the displacement influence functions are

$$\delta_1(x) = 1 \qquad \delta_2(x) = x \tag{e}$$

which coincide with the rigid-body ground motions discussed earlier [see Eq. (*a*)]. On the other hand, a propped cantilever (see Figs. 5.24*a*, *b*, and *c*) has the displacement influence functions

$$\delta_1(x) = 1 - \frac{3x^2}{2\ell^2} + \frac{x^3}{2\ell^3} \qquad \delta_2(x) = x - \frac{3x^2}{2\ell} + \frac{x^3}{2\ell^2} \qquad \delta_3(x) = \frac{3x^2}{2\ell^2} - \frac{x^3}{2\ell^3} \tag{f}$$

When these functions are multiplied by the corresponding support motions, they produce the flexible-body motions of the propped cantilever, as follows:

$$v_{\mathrm{st}} = \delta_1(x)g_1(t) + \delta_2(x)g_2(t) + \delta_3(x)g_3(t) \tag{g}$$

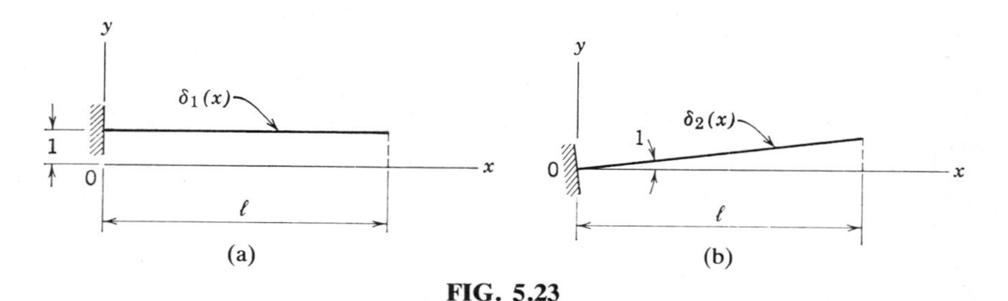

FIG. 5.23

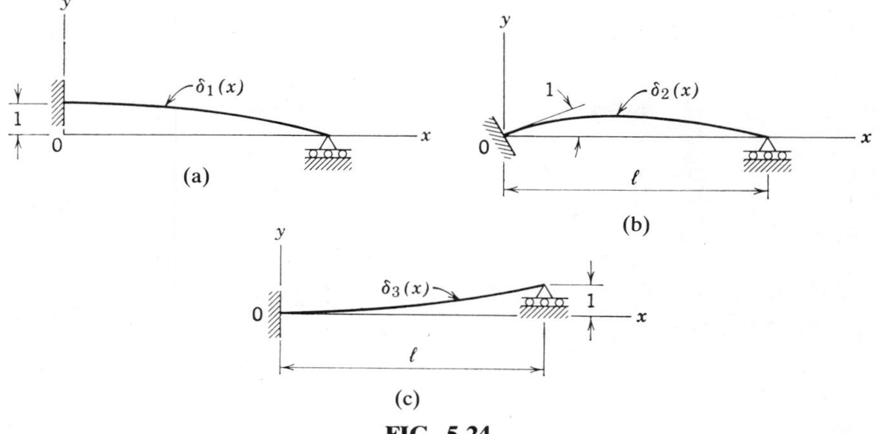

FIG. 5.24

Thus, for any independent restraint motion $g(t)$ for a supported beam, we may determine the corresponding displacement influence function $\delta(x)$ and evaluate the vibrational response as

$$v^* = -\sum_{i=1}^{\infty} \frac{X_i}{\omega_i} \int_0^{\ell} \delta(x) X_i \, dx \int_0^t \ddot{g}(t') \sin \omega_i(t - t') \, dt' \tag{5.136}$$

Then the total response is obtained from

$$v = v_{\text{st}} + v^* = \delta(x) g(t) + v^* \tag{5.137}$$

If more than one type of support motion is involved, the results for each type may be superimposed, as was demonstrated for the simple beam [see Eqs. (5.134) and (5.135)].

Example. Suppose that a beam is simply supported at the left end and fixed at the right end. Formulate an expression for the response of the beam due to a specified translation $g(t)$ of the left end in the y direction.
Solution. The normal functions for this case are found to be

$$X_i = \sinh k_i\ell \sin k_i x - \sin k_i\ell \sinh k_i x \tag{h}$$

where the values of k_i are calculated from the transcendental frequency equation

$$\tan k_i\ell = \tanh k_i\ell \tag{i}$$

as discussed in Sec. 5.11. To normalize the functions in the manner indicated by Eq. (5.97), we use Eq. (h) to obtain

$$\int_0^\ell X_i^2\,dx = \frac{\ell}{2}(\sinh^2 k_i\ell - \sin^2 k_i\ell) = \alpha_i \tag{j}$$

Then the normalized form of X_i becomes

$$X_i = (\sinh k_i\ell \sin k_i x - \sin k_i\ell \sinh k_i x)/\sqrt{\alpha_i} \tag{k}$$

The displacement influence function for translation of the left-hand support in the y direction is

$$\delta(x) = 1 - \frac{3x}{2\ell} + \frac{x^3}{2\ell^3} \tag{l}$$

In this case Eq. (5.136) for the relative response becomes

$$v^* = -\sum_{i=1}^{\infty} \frac{X_i}{\omega_i} \int_0^\ell \left(1 - \frac{3x}{2\ell} + \frac{x^3}{2\ell^3}\right) X_i\,dx \int_0^t \ddot{g}(t') \sin \omega_i(t-t')\,dt' \tag{m}$$

in which the normalized functions X_i are given by Eq. (k). The complete solution for this case [see Eq. (5.137)] is

$$v = \left(1 - \frac{3x}{2\ell} + \frac{x^3}{2\ell^3}\right) g(t) + v^* \tag{n}$$

5.16 BEAMS TRAVERSED BY MOVING LOADS

It is well known that a rolling load on a bridge or a girder produces a greater deflection and greater stresses than does the same load acting statically. This effect of live loads on bridges is of great practical importance, and many engineers have worked on the solution of the problem.† In this section we shall discusse the case of a moving load that exerts either a constant or a harmonic force on the beam. The distributed mass of the beam will be taken into account, but the mass of the load itself will be omitted. Systems in which the mass of the load (either sprung or

† The problem of a massless beam traversed by an unsprung mass was first considered by R. Willis, "Appendix to the Report of the Commissioners Appointed to Inquire into the Application of Iron to Railway Structures," H.M. Stationery Office, London, 1849. A series solution of this problem (for a mass moving with constant velocity) was reported by G. G. Stokes, "Discussions of a Differential Equation Related to the Breaking of Railway Bridges," *Trans. Cambridge Phil. Soc.*, **8,** Part 5, 1867, pp. 707–735. Stokes also showed a solution for a beam of uniform mass traversed by a constant-velocity moving force.

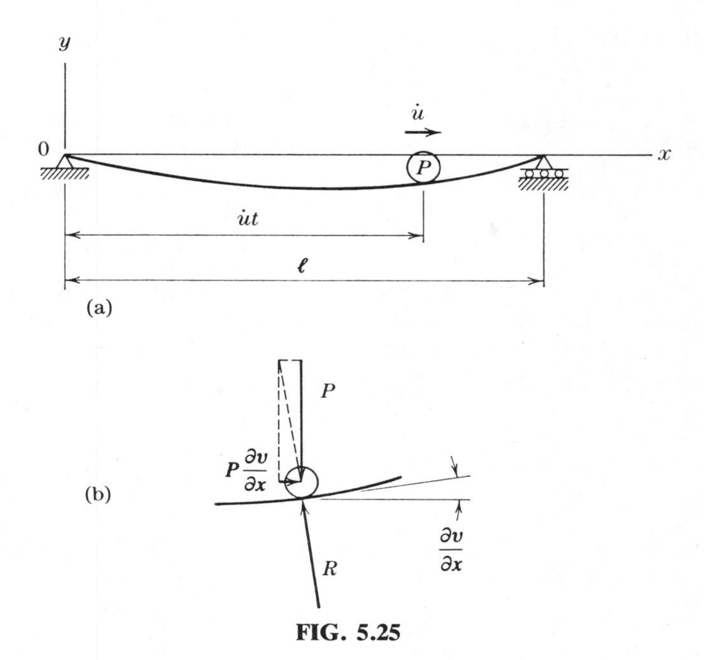

FIG. 5.25

unsprung) is included have differential equations with variable coefficients, because the position of the load is continuously changing. The analysis of such a system becomes rather complicated and is considered to be outside the scope of this book.

Let us consider first a rolling load P, acting downward on a simple beam and moving to the right at a constant velocity $\dot{u}$, as indicated in Fig. 5.25a.† Assuming that at time $t = 0$ the load is at the left-hand support, we observe that at any later time t the distance of the load from the left support will be $\dot{u}t$. The virtual work of the vertical force through the virtual displacement $\delta v_i = \delta\phi_i X_i$ for the ith natural mode of the beam will be

$$\delta W_{\mathrm{P}i} = -P\,\delta\phi_i X_{i1} = -P\,\delta\phi_i \sin\frac{i\pi\dot{u}t}{\ell} \qquad (a)$$

Using this expression for the virtual work of the moving load and

† See the papers by A. N. Krylov, "Uber die Erzwungenen Schwingungen von Gleich-formigen Elastischen Staben," *Math. Ann.*, **61,** 1905, p. 211; S. Timoshenko, "Erzwungene Schwingungen Prismatishe Stabe," *Z. Math. u. Phys.*, **59,** 1911, p. 163; and C. E. Inglis, "A Mathematical Treatise on Vibrations in Railway Bridges," Cambridge University Press, London, 1934.

proceeding as in Sec. 5.13, we obtain the following solution:

$$v = -\frac{2P\ell^3}{m\pi^2} \sum_{i=1}^{\infty} \frac{\sin(i\pi x/\ell)}{i^2(i^2\pi^2 a^2 - \dot{u}^2\ell^2)} \sin\frac{i\pi\dot{u}t}{\ell} + \frac{2P\ell^4\dot{u}}{m\pi^3 a} \sum_{i=1}^{\infty} \frac{\sin(i\pi x/\ell)}{i^3(i^2\pi^2 a^2 - \dot{u}^2\ell^2)} \sin\frac{i^2\pi^2 at}{\ell^2} \tag{5.138}$$

The first series in this solution represents forced vibrations, and the second series pertains to free vibrations of the beam.

If the velocity $\dot{u}$ of the moving force is very small, we can put $\dot{u} = 0$ and $\dot{u}t = x_1$ in the solution above. Then we have, from the first part,

$$v = -\frac{2P\ell^3}{m\pi^4 a^2} \sum_{i=1}^{\infty} \frac{1}{i^4} \sin\frac{i\pi x}{\ell} \sin\frac{i\pi x_1}{\ell} \tag{b}$$

This expression represents the static deflection of the beam produced by the load P applied at the distance x_1 from the left support. By using the notation

$$\alpha^2 = \frac{\dot{u}^2}{a\omega_1} = \frac{\dot{u}^2\ell^2}{a^2\pi^2} \tag{c}$$

and the relationship $ma^2 = EI$, we cast the solution for the forced vibrations in Eq. (5.138) into the following form:

$$v = -\frac{2P\ell^3}{EI\pi^4} \sum_{i=1}^{\infty} \frac{\sin(i\pi x/\ell)\sin(i\pi\dot{u}t/\ell)}{i^2(i^2 - \alpha^2)} \tag{d}$$

It is interesting to note that this deflection coincides with the static deflection of a beam on which (in addition to the lateral load P applied at a distance $x_1 = \dot{u}t$ from the left support) a longitudinal compressive force S is acting, such that

$$\frac{S}{S_{\text{cr}}} = \frac{S\ell^2}{EI\pi^2} = \alpha^2 \tag{e}$$

Here S_{er} denotes the *Euler buckling load* for the beam, From Eqs. (c) and (e) we obtain

$$\frac{S\ell^2}{EI\pi^2} = \frac{\dot{u}^2\ell^2}{a^2\pi^2}$$

or

$$S = m\dot{u}^2 \tag{f}$$

Conversely, the effect of this force on the static deflection of the beam loaded by P is equivalent to the effect of the velocity of a moving force P on the deflection (d) representing forced vibrations.

As the velocity $\dot{u}$ increases, a condition is reached where one of the denominators in Eq. (5.138) becomes equal to zero. For instance, assume that

$$\dot{u}^2\ell^2 = a^2\pi^2 \tag{g}$$

In this case the period of the fundamental vibration of the beam, given by $\tau_1 = 2\pi/\omega_1 = 2\ell^2/a\pi$, becomes equal to $2\ell/\dot{u}$ and is twice as great as the time required for the force P to pass over the beam. The denominators in the first terms of both series in Eq. (5.138) become, under the condition (g), equal to zero, and the sum of these two terms will be

$$v = -\frac{2P\ell^3}{m\pi^2}\left(\sin\frac{\pi x}{\ell}\right)\frac{\sin(\pi\dot{u}t/\ell) - (\ell\dot{u}/\pi a)\sin(\pi^2 at/\ell^2)}{\pi^2a^2 - \dot{u}^2\ell^2} \tag{h}$$

This expression has the form 0/0 and can be evaluated as follows:

$$\lim_{\dot{u}\to a\pi/\ell} v = \frac{Pt}{m\pi\dot{u}}\cos\frac{\pi\dot{u}t}{\ell}\sin\frac{\pi x}{\ell} - \frac{P\ell}{m\pi^2\dot{u}^2}\sin\frac{\pi\dot{u}t}{\ell}\sin\frac{\pi x}{\ell} \tag{i}$$

Equation (i) has its maximum value when $t = \ell/\dot{u}$ and is then equal to

$$v_{\max} = -\frac{P\ell}{m\pi^2\dot{u}^2}\left(\sin\frac{\pi\dot{u}t}{\ell} - \frac{\pi\dot{u}t}{\ell}\cos\frac{\pi\dot{u}t}{\ell}\right)_{t=\ell/\dot{u}}\sin\frac{\pi x}{\ell} = -\frac{P\ell^3}{EI\pi^3}\sin\frac{\pi x}{\ell} \tag{j}$$

Taking into consideration the fact that expression (i) represents a satisfactory approximation for the dynamic deflection given by Eq. (5.138), we conclude that the maximum dynamic deflection at the resonance condition (g) is about 50% greater than the maximum static deflection, which is

$$v_{st} = -\frac{P\ell^3}{48EI} \tag{k}$$

It is interesting to note that the maximum dynamic deflection occurs when the force P is leaving the beam. At this instant the deflection under the force P is equal to zero, hence the work done by this force while passing over the beam is evidently also equal to zero. In order to explain the source of the energy accumulated in the vibrating beam during the passage of the force P, we assume that there is no friction and that the beam produces a reaction R in the direction of the normal to the elastic curve (see Fig.

5.25*b*). From the condition of equilibrium, it follows that there must exist a horizontal force equal to $P(\partial v/\partial x)$. The work done by this force during its passage along the beam will be

$$W = \int_0^{\ell/\dot{u}} P\left(\frac{\partial v}{\partial x}\right)_{x=\dot{u}t} \dot{u}\, dt \tag{l}$$

Substitution of Eq. (*i*) for v in Eq. (*l*) produces

$$W = -\frac{P^2}{m\pi\dot{u}^2}\int_0^{\ell/\dot{u}}\left(\sin\frac{\pi\dot{u}t}{\ell} - \frac{\pi\dot{u}t}{\ell}\cos\frac{\pi\dot{u}t}{\ell}\right)\cos\frac{\pi\dot{u}t}{\ell}\,\dot{u}\,dt = \frac{P^2\ell}{m\pi^2\dot{u}^2}\left(\frac{\pi^2}{4}\right)$$

Using Eq. (*g*) and the relationship $ma^2 = EI$, we obtain

$$W = \frac{P^2\ell^3}{4EI\pi^2} \tag{m}$$

This amount of work is very close to the amount of potential energy of bending in the beam at the moment $t = \ell/\dot{u}$. The potential energy of the beam bent by the force P at the middle is

$$U = \frac{P^2\ell^3}{96EI} \qquad \text{and} \qquad \frac{W}{U} = 2.43$$

This ratio is very close to the square of the ratio of the maximum deflections for the dynamic and static conditions, which is equal to $(48/\pi^3)^2 = 2.38$. The discrepancy should be attributed to the higher modes of vibration.†

In the case of bridges, the time it takes the load to cross the bridge is usually large in comparison with the period of the fundamental mode of vibration; and the quantity α^2, given by Eq. (*c*), is small. Then, by taking only the first term in each series of Eq. (5.138) and assuming that in the most unfavorable case the amplitudes of the forced and free vibrations are added to one another, we obtain for the maximum deflection,

$$\begin{aligned} v_{\max} &= -\frac{2P\ell^3}{m\pi^2}\left(\frac{1}{\pi^2a^2 - \dot{u}^2\ell^2} + \frac{\dot{u}\ell}{a\pi}\frac{1}{\pi^2a^2 - \dot{u}^2\ell^2}\right) \\ &= -\frac{2P\ell^3}{EI\pi^4}\frac{1+\alpha}{1-\alpha^2} = -\frac{2P\ell^3}{EI\pi^4}\frac{1}{1-\alpha} \end{aligned} \tag{5.139}$$

† For a further discussion of the question see the paper by E. H. Lee, "On a Paradox in Beam Vibration Theory," *Quarterly of Appl. Math.* **X,** (3), 1952, p. 290.

This is a somewhat exaggerated value of the maximum dynamic deflection, because damping was completely neglected in the discussion. From the principle of superposition, the solution of the problem in the case of a system of concentrated moving forces and in the case of moving distributed forces can also be made without difficulty.

Consider now that a harmonic force $P_1(t) = -P \cos \Omega t$ is moving along the beam with a constant velocity $\dot{u}$.† Such a condition may occur, for instance, when an unbalanced locomotive wheel passes over a railway bridge. It is assumed that at the initial moment ($t = 0$) the force is maximum and is acting downward. Using the same manner of reasoning as before, we find that the virtual work of the moving harmonic force on the displacement $\delta v_i = \delta\phi_i X_i$ is

$$\delta W_{Pi} = -P \cos \Omega t \left(\delta\phi_i \sin \frac{i\pi\dot{u}t}{\ell} \right) \tag{n}$$

Using this expression for the virtual work of the moving load and proceeding as before, we obtain the solution

$$v = -\frac{P\ell^3}{EI\pi^4} \sum_{i=1}^{\infty} \sin \frac{i\pi x}{\ell} \left[\frac{\sin(i\pi\dot{u}/\ell + \Omega)t}{i^4 - (\psi + i\alpha)^2} + \frac{\sin(i\pi\dot{u}/\ell - \Omega)t}{i^4 - (\psi - i\alpha)^2} \right.$$
$$\left. - \frac{\alpha}{i} \left(\frac{\sin(i^2\pi^2 at/\ell^2)}{-i^2\alpha^2 + (i^2 - \psi)^2} + \frac{\sin(i^2\pi^2 at/\ell^2)}{-i^2\alpha^2 + (i^2 + \psi)^2} \right) \right] \tag{5.140}$$

where $\alpha = \dot{u}\ell/\pi a$ is the ratio of the period $\tau_1 = 2\ell^2/\pi a$ of the fundamental mode of vibration of the beam to twice the time $t = \ell/\dot{u}$ that it takes the force to pass over the beam. Also, $\psi = \tau_1/T$ is the ratio of the period of the fundamental mode of vibration of the beam to the period $T = 2\pi/\Omega$ of the harmonic force.

When the period T of the harmonic force is equal to the period τ_1 of the fundamental mode of vibration of the beam, the term $\psi = 1$, and we obtain the condition of resonance. The amplitude of vibration during motion of the harmonic force will gradually build up and attains its maximum at the moment $t = \ell/\dot{u}$. At this time the first term (for $i = 1$) in the series on the right-hand side of Eq. (5.140), which is the most important part of v, may be reduced to the form

$$v = -\frac{2P\ell^3}{\alpha EI\pi^4} \sin \frac{\pi x}{\ell} \sin \Omega t \tag{o}$$

† See S. Timoshenko, "On the Transverse Vibrations of Bars of Uniform Cross Section," *Phil. Mag.*, **43,** 1922, p. 125.

and the maximum deflection is given by the formula

$$v_{\max} = -\frac{2P\ell^3}{\alpha EI\pi^4} = -\frac{2\ell}{\dot{u}\tau_1}\left(\frac{2P\ell^3}{EI\pi^4}\right) \tag{5.141}$$

Due to the fact that in actual cases the time interval $t = \ell/\dot{u}$ is large in comparison with the period τ_1 of the natural vibration, the maximum dynamic deflection produced by the harmonic force will be many times greater than the deflection $2P\ell^3/EI\pi^4$, which would be produced by the same force if applied statically at the middle of the beam.

5.17 EFFECT OF AXIAL FORCE ON VIBRATIONS OF BEAMS

If a vibrating beam is subjected to a tensile force S, as shown in Fig. 5.26, the differential equation of the deflection curve under static transverse loading is

$$EI\frac{d^2v}{dx^2} = M + Sv \tag{a}$$

where M denotes the bending moment produced by a transverse loading of intensity w (see Fig. 5.26). By double differentiation of Eq. (a) with respect to x, we obtain

$$\frac{d^2}{dx^2}\left(EI\frac{d^2v}{dx^2}\right) = w + S\frac{d^2v}{dx^2} \tag{b}$$

To form the differential equation for transverse vibrations, we substitute the inertial force per unit length for w, which gives

$$\frac{\partial^2}{\partial x^2}\left(EI\frac{\partial^2 v}{\partial x^2}\right) - S\frac{\partial^2 v}{\partial x^2} = -\rho A\frac{\partial^2 v}{\partial t^2} \tag{c}$$

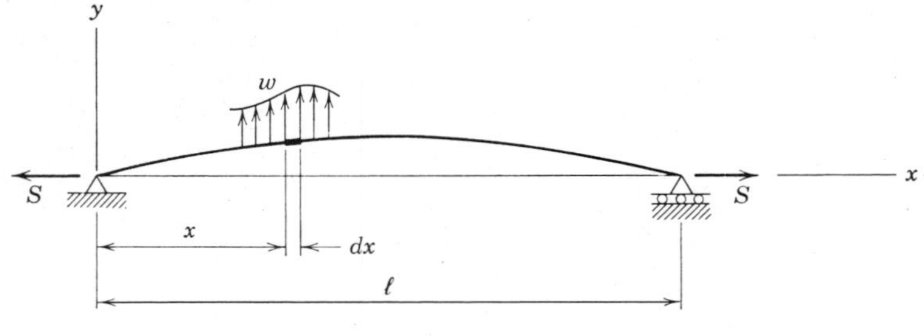

FIG. 5.26

In the case of a prismatic beam, we obtain

$$EI\frac{\partial^4 v}{\partial x^4} - S\frac{\partial^2 v}{\partial x^2} = -\rho A\frac{\partial^2 v}{\partial t^2} \tag{5.142}$$

Assuming that the beam performs one of its natural modes of vibration, we take the solution of Eq. (5.142) in the form

$$v = X(A\cos\omega t + B\sin\omega t) \tag{d}$$

where X is a normal function. Substituting Eq. (d) into Eq. (5.142), we find

$$EI\frac{d^4X}{dx^4} - S\frac{d^2X}{dx^2} = \rho A\omega^2 X \tag{e}$$

The solution of this equation, satisfying the prescribed end conditions, furnishes the corresponding normal functions. We have the simplest case if the ends of the beam are simply supported. These conditions are satisfied by taking

$$X_i = \sin\frac{i\pi x}{\ell} \qquad (i = 1, 2, 3, \ldots, \infty) \tag{f}$$

Substituting this expression into Eq. (e), we obtain the corresponding angular frequency of vibration

$$\omega_i = \frac{i^2\pi^2 a}{\ell^2}\sqrt{1 + \frac{S\ell^2}{i^2 EI\pi^2}} \tag{5.143}$$

where, as before, $a = \sqrt{EI/\rho A}$. This frequency is larger than that obtained before [see Eq. (5.102), Sec. 5.10] when the axial force S was absent.

If we have a very flexible beam (say a wire) under a large tension, the second term under the radical in Eq. (5.143) becomes very large in comparison with unity; and if i^2 is not large, we can put

$$\omega_i \approx \frac{i^2\pi^2 a}{\ell^2}\sqrt{\frac{S\ell^2}{i^2 EI\pi^2}} = \frac{i\pi}{\ell}\sqrt{\frac{S}{\rho A}} \tag{g}$$

which are natural frequencies of a stretched wire (see Sec. 5.8).

Substituting expression (f) into solution (d), we obtain a natural mode of vibration with i sinusoidal half waves. The summation of such modes produces the general solution for free vibrations of a simple beam with axial tension as

$$v = \sum_{i=1}^{\infty} \sin \frac{i\pi x}{\ell} (A_i \cos \omega_i t + B_i \sin \omega_i t) \tag{h}$$

in which ω_i is given by Eq. (5.143). If the initial deflections and the initial velocities are given, the constants A_i an B_i in Eq. (h) can be calculated in the same manner as before (see Sec. 5.10).

If a compressive force instead of a tensile force is acting on the bar, the frequencies of transverse vibrations decrease; and we obtain their values by changing S to $-S$ in Eq. (5.143), which gives

$$\omega_i = \frac{i^2\pi^2 a}{\ell^2} \sqrt{1 - \frac{S\ell^2}{i^2 EI\pi^2}} \tag{5.144}$$

This frequency expression yields smaller values than that for the simple beam without axial compression. Its value depends on the term $S\ell^2/EI\pi^2$, which is the ratio of the axial force to the Euler buckling load. If this ratio approaches unity, the frequency of the lowest mode of vibration approaches zero; and we obtain transverse buckling.

In studying forced response of a simply-supported beam with an axial force S, we proceed as in Sec. 5.13. It is only necessary to use either Eq. (5.143) or Eq. (5.144) in place of the simpler expression (5.102) used previously. All other steps in the analysis remain as before.†

5.18 BEAMS ON ELASTIC SUPPORTS OR ELASTIC FOUNDATIONS

Support restraint conditions at the ends of a beam may be intermediate between the extremes of zero restraint and full restraint. If such restraints against either translations or rotations are linearly elastic, they may be idealized as springs, as shown in Fig. 5.27. Let the symbols k_1 and k_2 represent the stiffness constants for the translational and rotational springs at the left end; k_3 and k_4 are the stiffness constants for the translational and rotational springs at the right end. In this case the boundary conditions may be expressed as

$$\begin{aligned} V_{x=0} &= EI(X''')_{x=0} = -k_1(X)_{x=0} & M_{x=0} &= EI(X'')_{x=0} = k_2(X')_{x=0} \\ V_{x=\ell} &= EI(X''')_{x=\ell} = k_3(X)_{x=\ell} & M_{x=\ell} &= EI(X'')_{x=\ell} = -k_4(X')_{x=\ell} \end{aligned} \tag{a}$$

† If the beam is not simply supported, the mode shapes must satisfy the differential equation (5.142) as well as the known boundary conditions. In such a case, neither the mode shapes nor the natural frequencies will be the same as those in Sec. 5.11.

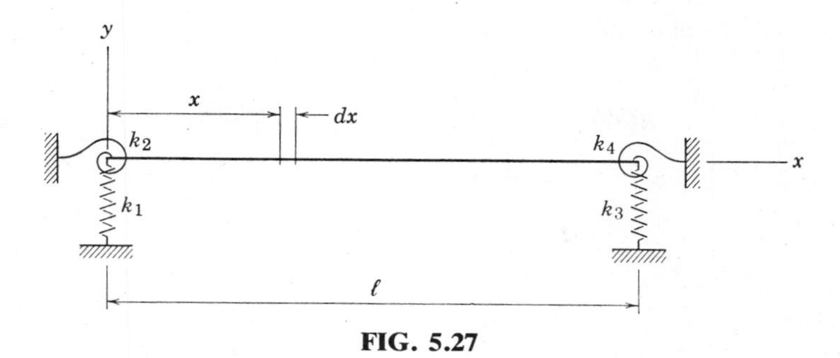

FIG. 5.27

The normal functions and their required derivatives with respect to x are [see Eq. (5.85), Sec. 5.9]

$$\begin{aligned} X &= C_1 \sin kx + C_2 \cos kx + C_3 \sinh kx + C_4 \cosh kx \\ X' &= k(C_1 \cos kx - C_2 \sin kx + C_3 \cosh kx + C_4 \sinh kx) \\ X'' &= k^2(-C_1 \sin kx - C_2 \cos kx + C_3 \sinh kx + C_4 \cosh kx) \\ X''' &= k^3(-C_1 \cos kx + C_2 \sin kx + C_3 \cosh kx + C_4 \sinh kx) \end{aligned} \tag{b}$$

in which $K = \sqrt{\omega/a}$, as before. Substitution of Eqs. (b) into Eqs. (a) produces

$$EIk^3C_1 - k_1C_2 - EIk^3C_3 - k_1C_4 = 0$$
$$-k_2C_1 - EIkC_2 - k_2C_3 + EIkC_4 = 0$$

$$\begin{aligned} &(-EIk^3 \cos k\ell - k_3 \sin k\ell)C_1 + (EIk^3 \sin k\ell - k_3 \cos k\ell)C_2 \\ &\quad + (EIk^3 \cosh k\ell - k_3 \sinh k\ell)C_3 + (EIk^3 \sinh k\ell - k_3 \cosh k\ell)C_4 = 0 \\ &(-EIk \sin k\ell + k_4 \cos k\ell)C_1 + (-EIk \cos k\ell - k_4 \sin k\ell)C_2 \\ &\quad + (EIk \sinh k\ell + k_4 \cosh k\ell)C_3 + (EIk \cosh k\ell + k_4 \sinh k\ell)C_4 = 0 \end{aligned} \tag{5.145}$$

This set of four homogeneous equations will have nontrivial solutions only if the determinant of the coefficients of $C_1, \ldots, C_4$ is zero. Thus, by expansion of this 4×4 determinant, we can obtain the frequency equation for the elastically supported beam in Fig. 5.27. Upon substitution of the roots of this *characteristic equation* back into Eqs. (5.145), it is possible to determine the normal functions.

The determinants corresponding to beams with zero and full restraints can be obtained from Eqs. (5.145) by setting the appropriate spring constants equal to zero or infinity. For example, a cantilever beam that is fixed at the left end and free at the right end will have $k_1 = \infty$, $k_2 = \infty$, $k_3 = 0$, and $k_4 = 0$. In this instance the determinant becomes

$$\begin{vmatrix} 0 & 1 & 0 & 1 \\ 1 & 0 & 1 & 0 \\ -\cos k\ell & \sin k\ell & \cosh k\ell & \sinh k\ell \\ -\sin k\ell & -\cos k\ell & \sinh k\ell & \cosh k\ell \end{vmatrix} = 0 \qquad (c)$$

where the elements of the first row have been divided by $-k_1$ and those of the second row by $-k_2$. Expansion of this determinant produces the frequency equation

$$\cos k\ell \cosh k\ell = -1 \qquad (d)$$

which is the same as Eq. (5.109) in Sec. 5.11. The equation for the normal functions (to within the arbitrary constant C_i) is found to be

$$X_i = C_i\left(\frac{\sin k_i x - \sinh k_i x}{\cos k_i\ell + \cosh k_i\ell} - \frac{\cos k_i x - \cosh k_i x}{\sin k_i\ell - \sinh k_i\ell}\right) \qquad (e)$$

If elastic restraints against transverse motion are continuously distributed along the length of a beam, we refer to the problem as a *beam on an elastic foundation*. Figure 5.28 shows such a beam, under which the elastic foundation is depicted as a large number of closely spaced translational springs. We define the foundation modulus k_f as the load per unit length of the beam necessary to produce a displacement of the foundation equal to unity. When the beam is vibrating transversely, the differential equation of motion for a typical element of length dx is

$$\frac{\partial^2}{\partial x^2}\left(EI\frac{\partial^2 v}{\partial x^2}\right) dx = -k_f v\, dx - \rho A\, dx \frac{\partial^2 v}{\partial t^2} \qquad (f)$$

where the first term on the right represents the restoring force due to the elastic foundation. For a prismatic beam this expression becomes

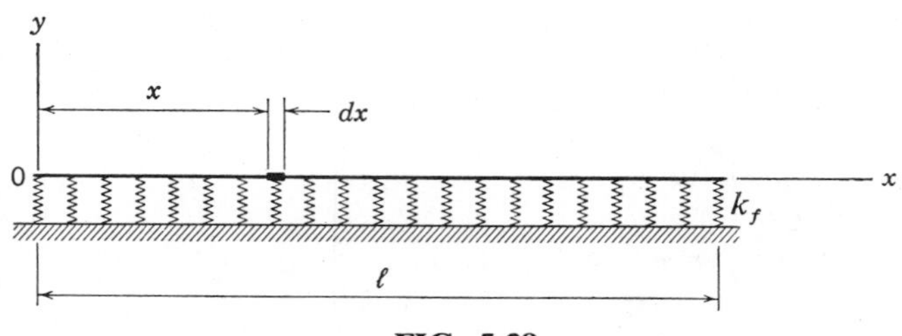

FIG. 5.28

$$EI\frac{\partial^4 v}{\partial x^4} + k_f v = -\rho A \frac{\partial^2 v}{\partial t^2} \tag{5.146}$$

To solve this equation, we take natural modes of vibration in the form

$$v_i = X_i(A_i \cos \omega_i t + B_i \sin \omega_i t) \tag{g}$$

Substituting Eq. (*g*) into Eq. (5.146) gives

$$EI\frac{d^4X_i}{dx^4} - (\rho A \omega_i^2 - k_f)X_i = 0 \tag{h}$$

Dividing this equation by EI, we obtain

$$\frac{d^4X_i}{dx^4} - \left(\frac{\omega_i^2}{a^2} - \frac{k_f}{EI}\right)X_i = 0 \tag{i}$$

For convenience, let

$$\frac{\omega_i^2}{a^2} - \frac{k_f}{EI} = k_i^4 \tag{j}$$

Then Eq. (*i*) becomes

$$\frac{d^4X_i}{dx^4} - k_i^4 X_i = 0 \tag{k}$$

This differential equation has the solution

$$X_i = C_{1i} \sin k_i x + C_{2i} \cos k_i x + C_{3i} \sinh k_i x + C_{4i} \cosh k_i x \tag{l}$$

which is the same as that for a beam without an elastic foundation. Therefore, all formulations previously derived for beams with various end conditions apply also to the case under consideration. The only modification required is that, instead of the relationship $\omega_i = k_i^2 a$, we have from Eq. (*j*)

$$\omega_i = k_i^2 a \sqrt{1 + \frac{k_f}{EIk_i^4}} \tag{5.147}$$

Considering the simplest case of a beam with translational restraints at both ends (a simple beam on an elastic foundation), we have the normal functions

$$X_i = C_i \sin k_i x \qquad (i = 1, 2, 3, \ldots, \infty) \tag{m}$$

and the angular frequencies are

$$\omega_i = \frac{i^2\pi^2 a}{\ell^2}\sqrt{1 + \frac{k_f \ell^4}{EIi^4\pi^4}} = \frac{\pi^2 a}{\ell^2}\sqrt{i^4 + \mu} \tag{5.148}$$

where $\mu = k_f\ell^4/EI\pi^4$. Except for this modification of the angular frequencies, the formulas for responses of a simple beam to various influences (e.g., see Secs. 5.10, 5.13, 5.15, and 5.16) apply also for the beam on an elastic foundation.

In summary, we see that elastic supports at the ends of a beam (see Fig. 5.27) affect both its frequencies and its mode shapes, whereas the presence of an elastic foundation (see Fig. 5.28) influences only the natural frequencies. As was the case for a stretched wire with elastic restraints, subsequent steps in the solutions for dynamic responses of beams on elastic supports or elastic foundations will be the same as those discussed previously for simpler cases.

Example. Consider the case when a harmonic force $P_1(t) = P \sin \Omega t$ is acting at a distance x_1 from the left support of a simple beam on an elastic foundation. Evaluate the steady-state response for this case.

Solution. The vibrations produced by the disturbing force are given by Eq. (5.127) in Sec. 5.13, which for the case under consideration may be written as

$$v = \frac{2P\ell^3}{m}\sum_{i=1}^{\infty} \times \left[\frac{\sin(i\pi x/\ell)\sin(i\pi x_1/\ell)\sin\Omega t}{\pi^4 a^2(i^4+\mu) - \Omega^2\ell^4} - \frac{\Omega \sin(i\pi x/\ell)\sin(i\pi x_1/\ell)\sin\omega_i t}{\ell^4\omega_i(\omega_i^2 - \Omega^2)}\right] \tag{n}$$

The first term in this expression represents the forced vibration, and the second is the free vibration of the beam.

If the harmonic force $P \sin \Omega t$ is varying slowly $(\Omega \to 0)$, the steady-state part of the response from Eq. (n) becomes

$$v = \frac{2P\ell^3}{EI\pi^4}\sum_{i=1}^{\infty}\frac{\sin(i\pi x/\ell)\sin(i\pi x_1/\ell)\sin\Omega t}{i^4+\mu} \tag{o}$$

In the case where $x_1 = \ell/2$, Eq. (o) yields

$$v = \frac{2P\ell^3}{EI\pi^4}\left[\frac{\sin(\pi x/\ell)}{1+\mu} - \frac{\sin(3\pi x/\ell)}{3^4+\mu} + \frac{\sin(5\pi x/\ell)}{5^4+\mu} - \cdots\right]\sin\Omega t \tag{p}$$

Comparing this result with Eq. (l) in Sec. 5.13, we concude that the additional term μ in the denominators represents the effect of the elastic foundation on the deflection of the beam.

5.19 RITZ METHOD FOR CALCULATING FREQUENCIES

In Sec. 1.5 the Rayleigh method was used for approximating the lowest frequency of a beam or a shaft. To use this method it is necessary to make some assumption for the deflected shape of the vibrating elastic body. The corresponding frequency is then found from a consideration of the energy of the system. Choosing a definite shape for the deflection curve in this manner inherently involves introducing some additional constraints, which reduce the system to one having a single degree of freedom. Such additional constraints can only increase the rigidity of the system and make the frequency of vibration (as obtained by Rayleigh's method) somewhat higher than its exact value. Better approximations for calculating the fundamental frequency (and also the frequencies of higher modes of vibration) can be obtained by *Ritz's first method,*† which is a further development of Rayleigh's approach.‡ In using Ritz's first method, we take the assumed deflection curve with several parameters, the magnitudes of which are chosen in such a manner as to reduce to a minimum the frequency of vibration. The manner of choosing the shape of the deflection curve and the procedure for calculating consecutive frequencies will now be shown for the simple case of the vibration of a stretched wire (see Sec. 5.8).

If the transverse deflections of a stretched wire are very small, the change in the tensile force S during vibration can be neglected, and the increase in potential energy of deformation due to the deflection may be obtained by multiplying S with the increase in length of the wire. In a deflected position the length of the wire becomes

$$L = \int_0^\ell \sqrt{1 + \left(\frac{dv}{dx}\right)^2}\, dx$$

For small deflections this expression may be approximated by

$$L \approx \int_0^\ell \left[1 + \frac{1}{2}\left(\frac{dv}{dx}\right)^2\right] dx$$

Then the increase in potential energy will be

$$\Delta U \approx \frac{S}{2}\int_0^\ell \left(\frac{dv}{dx}\right)^2 \mathrm{dx} \tag{a}$$

† W. Ritz, *Gesammelte Werke*, Paris, 1911, p. 265.

‡ Lord Rayleigh used the method only for an approximate calculation of frequency of the fundamental mode of vibration of complicated systems. He was doubtful regarding its application to the investigation of higher modes of vibration (see his papers, *Phil. Mag.*, **47,** Ser. 5, 1899, p. 566; and **22,** Ser. 6, 1911, p. 225).

The maximum potential energy occurs when the vibrating wire occupies its extreme position. In this position $v_{max} = X$, and Eq. (a) becomes

$$\Delta U_{max} \approx \frac{S}{2} \int_0^{\ell} \left(\frac{dX}{dx} \right)^2 dx \tag{b}$$

The kinetic energy of the vibrating wire is

$$T = \frac{m}{2} \int_0^{\ell} (\dot{v})^2 \, dx \tag{c}$$

Its maximum value occurs when the vibrating wire is in its middle position that is, when $\dot{v}_{max} = \omega X$. Thus,

$$T_{max} = \frac{\omega^2 m}{2} \int_0^{\ell} X^2 \, dx \tag{d}$$

Assuming that there are no losses in energy, we may equate Eqs. (b) and (d) to obtain

$$\omega^2 = \frac{S}{m} \frac{\displaystyle\int_0^{\ell} \left(\frac{dX}{dx} \right)^2 dx}{\displaystyle\int_0^{\ell} X^2 \, dx} \tag{5.149}$$

By assuming the shapes of various modes of vibration and substituting into Eq. (5.149) the corresponding expressions for X, we can calculate approximately the frequencies of these modes of vibration.

The first step in Ritz's method is to choose a suitable expression for the deflection curve. Let $\Phi_1(x)$, $\Phi_2(x)$, . . be a series of functions $\Phi_j(x)$ that suitably represent X and satisfy the end conditions. Then we have

$$X = a_1\Phi_1(x) + a_2\Phi_2(x) + a_3\Phi_3(x) + \cdots \tag{e}$$

We know that by taking a finite number of terms in expression (e) we superimpose certain limitations on the possible shapes of the deflection curve of the wire. Therefore, the frequency, as calculated from Eq. (5.149), will usually be higher than the exact value of this frequency. In order that the approximation will be as close as possible to the true value, Ritz proposed to choose the coefficients $a_1, a_2, a_3, \ldots$ in expression (e) so as to make the result from Eq. (5.149) a minimum. In this manner a system of equations, each having the form

$$\frac{\partial}{\partial a_j} \frac{\int_0^\ell \left(\frac{dX}{dx}\right)^2 dx}{\int_0^\ell X^2\, dx} = 0 \tag{5.150}$$

will be obtained.

Performing the differentiation indicated in Eq. (5.150), we have

$$\int_0^\ell X^2\, dx \frac{\partial}{\partial a_j} \int_0^\ell \left(\frac{dX}{dx}\right)^2 dx - \int_0^\ell \left(\frac{dX}{dx}\right)^2 dx \frac{\partial}{\partial a_j} \int_0^\ell X^2\, dx = 0 \tag{f}$$

Noting from Eq. (5.149) that

$$\int_0^\ell \left(\frac{dX}{dx}\right)^2 dx = \frac{\omega^2 m}{S} \int_0^\ell X^2\, dx$$

we find

$$\frac{\partial}{\partial a_j} \int_0^\ell \left[\left(\frac{dX}{dx}\right)^2 - \frac{\omega^2 m}{S} X^2\right] dx = 0 \tag{5.151}$$

Substituting expression (e) for X in Eq. (5.151) and performing the indicated operations, we obtain a system of equations that are homogeneous and linear in $a_1, a_2, a_3, \ldots$. The number of such equations will be equal to the number of coefficients $a_1, a_2, a_3, \ldots$ in the series (e). Such a system of equations can yield solutions different from zero only if the determinant of the coefficients of $a_1, a_2, a_3, \ldots$ is equal to zero. This condition yields the frequency equation from which the angular frequencies of the various modes of vibration can be calculated.

Let us consider the symmetric modes of vibration of a stretched wire of length 2ℓ, as indicated in Figs. 5.29*a*, *b*, and *c*. It is easy to see that a function such as $\ell^2 - x^2$, representing a symmetric parabolic curve and satisfying the end conditions $(v)_{x=\pm\ell} = 0$ is a suitable formula for representing the mode in Fig. 5.29*a*. By multiplying this function with $x^2, x^4, \ldots$ a series of symmetric curves satisfying the end conditions will also be obtained. In this manner we arrive at the following expression for the deflection curve of the vibrating wire:

$$X = a_1(\ell^2 - x^2) + a_2 x^2(\ell^2 - x^2) + a_3 x^4(\ell^2 - x^2) + \cdots \tag{g}$$

In order to show how quickly the accuracy of our calculations improves with an increase in the number of terms in expression (g), we begin with one term only and put

$$X_1 = a_1(\ell^2 - x^2) \tag{h}$$

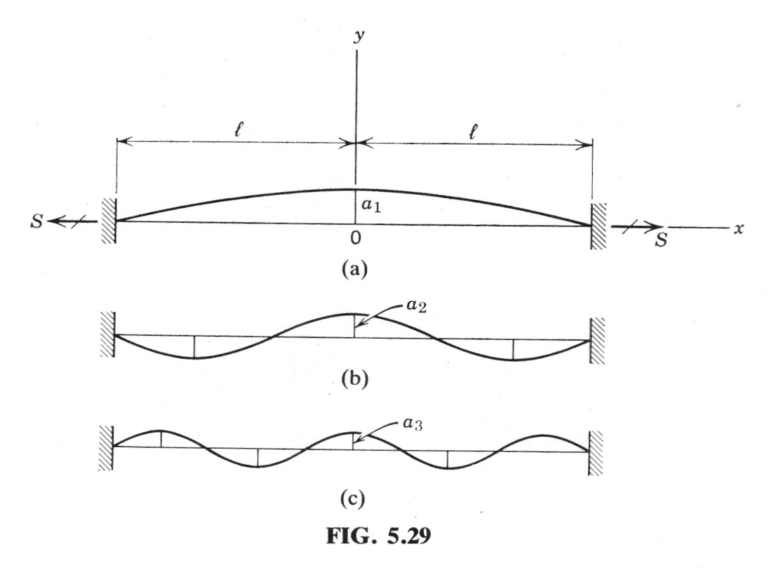

FIG. 5.29

The integrations required for Eq. (5.149) are†

$$\int_0^\ell (X_1)^2\,dx = \frac{8}{15}a_1^2\ell^5 \qquad \int_0^\ell \left(\frac{dX_1}{dx}\right)^2 dx = \frac{4}{3}a_1^2\ell^3$$

Substituting these values into Eq. (5.149), we obtain

$$\omega_1^2 = \frac{5S}{2\ell^2 m} \tag{i}$$

Comparing this result with the exact solution $\omega_1^2 = \pi^2 S/(4\ell^2 m)$, we see that the error in frequency is about 0.66%. Thus, when there is only one term in the expression (*g*), the shape of the curve is completely determined; and the system is reduced to one with a single degree of freedom, as in Rayleigh's method.

To obtain a closer approximation, let us take two terms in the series (*g*). Then we will have two parameters a_1 and a_2, and by changing the ratio of these two quantities we can also change (to a certain extent) the shape of the curve. The best approximation will be obtained when this ratio is such that the result from Eq. (5.149) becomes a minimum, which requires that conditions (5.151) be satisfied. By taking as a second approximation

$$X_2 = a_1(\ell^2 - x^2) + a_2 x^2(\ell^2 - x^2) \tag{j}$$

† For either symmetric or antisymmetric modes, it is sufficient to consider only half of the system.

we obtain

$$\int_0^\ell X_2^2\,dx = \frac{8}{15}a_1^2\ell^5 + \frac{16}{105}a_1a_2\ell^7 + \frac{8}{315}a_2^2\ell^9$$

$$\int_0^\ell \left(\frac{dX_2}{dx}\right)^2 dx = \frac{4}{3}a_1^2\ell^3 + \frac{8}{15}a_1a_2\ell^5 + \frac{44}{105}a_2^2\ell^7$$

Substituting these values into Eq. (5.151) and taking the derivatives with respect to a_1 and a_2, we find

$$\left(1 - \frac{2}{5}k^2\ell^2\right)a_1 + \ell^2\left(\frac{1}{5} - \frac{2}{35}k^2\ell^2\right)a_2 = 0 \qquad (k)$$

$$\left(1 - \frac{2}{7}k^2\ell^2\right)a_1 + \ell^2\left(\frac{11}{7} - \frac{2}{21}k^2\ell^2\right)a_2 = 0 \qquad (l)$$

in which

$$k^2 = \frac{\omega^2 m}{S} \qquad (m)$$

The determinant from Eqs. (k) and (l) will vanish when

$$k^4\ell^4 - 28k^2\ell^2 + 63 = 0$$

The two roots of this equation are

$$k_1^2\ell^2 = 2.46744 \qquad k_2^2\ell^2 = 25.6$$

Remembering that we are considering only modes of vibration that are symmetric about the middle, and using Eq. (m), we obtain for the first and third modes of vibration

$$\omega_1^2 = \frac{2.46744S}{\ell^2 m} \qquad \omega_3^2 = \frac{25.6S}{\ell^2 m}$$

Comparing these results with the following exact solutions

$$\omega_1^2 = \frac{\pi^2 S}{4\ell^2 m} = \frac{2.46740S}{\ell^2 m} \qquad \omega_3^2 = \frac{9\pi^2 S}{4\ell^2 m} = \frac{22.207S}{\ell^2 m}$$

we conclude that the accuracy with which the fundamental frequency is obtained is very high (the error is about 0.00081%). On the other hand, the error in the frequency of the third mode of vibration is about 7.4%. With three terms in the series (g) the frequency of the third mode of vibration will be obtained with an error of less than 0.5%.

It is seen that by using the Ritz method not only the fundamental frequency but also frequencies of higher modes of vibration can be obtained with good accuracy by taking a sufficient number of terms in the expression for the deflection curve. In the next section an application of Ritz's first method to the study of the vibrations of beams of variable cross section will be shown. His second method will also be explained and demonstrated.

5.20 VIBRATIONS OF NONPRISMATIC BEAMS

In our previous discussions, various problems involving the vibrations of prismatic beams were considered. However, several important engineering problems, such as the vibrations of turbine blades, hulls of ships, bridge girders of variable depth, and so on, involve the theory of vibration of a beam of variable cross section. The differential equation of motion for such a vibrating beam has been discussed previously [see Eq. (5.81), Sec. 5.9]. It has the form

$$\frac{\partial^2}{\partial x^2}\left(EI\frac{\partial^2 v}{\partial x^2}\right) + \rho A\frac{\partial^2 v}{\partial t^2} = 0 \tag{a}$$

in which I and A are certain functions of x. Only in some special cases, which will be considered later, can the exact expressions for the normal functions and the frequencies be determined. Thus, approximate methods are often used for calculating the natural frequencies of vibration.

In applying the first method of Ritz to beam vibrations, we write the following expressions for the maximum potential and the maximum kinetic energies:

$$U_{\max} = \frac{1}{2}\int_0^\ell EI\left(\frac{d^2X}{dx^2}\right)^2 dx \tag{b}$$

$$T_{\max} = \frac{\omega^2}{2}\int_0^\ell \rho A X^2\, dx \tag{c}$$

from which

$$\omega^2 = \frac{E}{\rho}\,\frac{\displaystyle\int_0^\ell I\left(\frac{d^2X}{dx^2}\right)^2 dx}{\displaystyle\int_0^\ell AX^2\, dx} \tag{d}$$

In order to obtain an approximate solution, we proceed as in the preceding section and take the shape of the deflection curve in the form of a series

$$X = a_1\Phi_1(x) + a_2\Phi_2(x) + a_3\Phi_3(x) + \cdots \tag{e}$$

in which every one of the functions Φ_j satisfies the conditions at the ends of the beam. The conditions for minimizing the results from Eq. (d) will be

$$\frac{\partial}{\partial a_j} \frac{\int_0^\ell I\left(\frac{d^2X}{dx^2}\right)^2 dx}{\int_0^\ell AX^2\,dx} = 0 \tag{f}$$

or

$$\int_0^\ell AX^2\,dx \frac{\partial}{\partial a_j} \int_0^\ell I\left(\frac{d^2X}{dx^2}\right)^2 dx - \int_0^\ell I\left(\frac{d^2X}{dx^2}\right)^2 dx \frac{\partial}{\partial a_j} \int_0^\ell AX^2\,dx = 0 \tag{g}$$

From Eqs. (d) and (g) we obtain

$$\frac{\partial}{\partial a_j} \int_0^\ell \left[I\left(\frac{d^2X}{dx^2}\right)^2 - \frac{\omega^2 A\rho}{E} X^2 \right] dx = 0 \tag{5.152}$$

Thus, the problem is reduced to finding values for the constants $a_1, a_2, a_3, \ldots$ in Eq. (e) that will make the integral

$$Z = \int_0^\ell \left[I\left(\frac{d^2X}{dx^2}\right)^2 - \frac{\omega^2 A\rho}{E} X^2 \right] dx \tag{h}$$

a minimum. The equations resulting from (5.152) are homogeneous and linear in $a_1, a_2, a_3, \ldots,$ and their number is equal to the number of terms in the series (e). Equating to zero the determinant of the coefficients in these equations, we obtain the frequency equation from which the angular frequencies of the various modes can be calculated.

Vibrations of a Wedge

We shall now apply Ritz's first method to the case of a wedge of unit thickness with one end free and the other fixed, as shown in Fig. 5.30. We

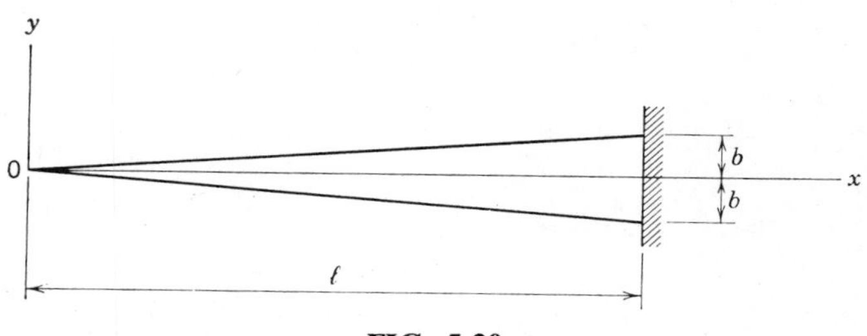

FIG. 5.30

have the geometric relationships

$$A = \frac{2bx}{\ell} \qquad I = \frac{1}{12}\left(\frac{2bx}{\ell}\right)^3 \tag{i}$$

where ℓ is the length of the cantilever and $2b$ is its depth at the fixed end. The boundary conditions for this case are

$$\begin{aligned} \left(EI\frac{d^2X}{dx^2}\right)_{x=0} &= 0 \qquad \frac{d}{dx}\left(EI\frac{d^2X}{dx^2}\right)_{x=0} = 0 \\ (X)_{x=\ell} &= 0 \qquad \left(\frac{dX}{dx}\right)_{x=\ell} = 0 \end{aligned} \tag{j}$$

In order to satisfy these conditions, we take the deflection curve in the form of the series

$$X = a_1\left(1 - \frac{x}{\ell}\right)^2 + x_2\frac{x}{\ell}\left(1 - \frac{x}{\ell}\right)^2 + a_3\frac{x^2}{\ell^2}\left(1 - \frac{x}{\ell}\right)^2 + \cdots \tag{k}$$

It is easy to see that each term, as well as its derivative with respect to x, becomes equal to zero when $x = \ell$. Consequently, the third and fourth end conditions in Eqs. (j) will be satisfied. The first and second conditions are also satisfied, because I and dI/dx are zero for $x = 0$.

Taking as a first approximation

$$X_1 = a_1\left(1 - \frac{x}{\ell}\right)^2 \tag{l}$$

and substituting it into Eq. (d), we find

$$\omega^2 = \frac{10Eb^2}{\rho\ell^4} \qquad f = \frac{\omega}{2\pi} = \frac{5.48b}{2\pi\ell^2}\sqrt{\frac{E}{3\rho}} \tag{m}$$

To get a closer approximation, we take two terms in the series (k), as follows:

$$X_2 = a_1\left(1 - \frac{x}{\ell}\right)^2 + a_2\frac{x}{\ell}\left(1 - \frac{x}{\ell}\right)^2 \tag{n}$$

Substituting this expression into Eq. (h), we obtain

$$Z_2 = \frac{2}{3}\frac{b^3}{\ell^3}\left[(a_1 - 2a_2)^2 + \frac{24}{5}a_2(a_1 - 2a_2) + 6a_2^2\right] - \frac{2b\rho\ell\omega^2}{E}\left[\frac{a_1^2}{30} + \frac{2a_1a_2}{105} + \frac{a_2^2}{280}\right]$$

From the conditions

$$\frac{\partial Z_2}{\partial a_1} = 0 \qquad \frac{\partial Z_2}{\partial a_2} = 0$$

we find the following two linear equations:

$$\left(\frac{E}{\rho}\frac{b^2}{3\ell^4} - \frac{\omega^2}{30}\right)a_1 + \left(\frac{2E}{5\rho}\frac{b^2}{3\ell^4} - \frac{\omega^2}{105}\right)a_2 = 0 \qquad (o)$$

$$\left(\frac{2E}{5\rho}\frac{b^2}{3\ell^4} - \frac{\omega^2}{105}\right)a_1 + \left(\frac{2E}{5\rho}\frac{b^2}{3\ell^4} - \frac{\omega^2}{280}\right)a_2 = 0 \qquad (p)$$

Equating to zero the determinant of the coefficients in these equations, we obtain

$$\left(\frac{E}{\rho}\frac{b^2}{3\ell^4} - \frac{\omega^2}{30}\right)\left(\frac{2E}{5\rho}\frac{b^2}{3\ell^4} - \frac{\omega^2}{280}\right) - \left(\frac{2E}{5\rho}\frac{b^2}{3\ell^4} - \frac{\omega^2}{105}\right)^2 = 0 \qquad (q)$$

From this equation $\omega_{1,2}^2$ can be calculated, and the smaller of the two roots yields

$$f_1 = \frac{\omega_1}{2\pi} = \frac{5.319b}{2\pi\ell^2}\sqrt{\frac{E}{3\rho}} \qquad (r)$$

For the case under consideration an exact solution exists, in which the normal functions are determined as Bessel's functions.† This exact solution gives

$$f_1 = \frac{\omega_1}{2\pi} = \frac{5.315b}{2\pi\ell^2}\sqrt{\frac{E}{3\rho}} \qquad (5.153)$$

Comparing this expression with Eqs. (m) and (r), we conclude that the inaccuracy of the first approximation is about 3.1%, while the error of the second approximation is approximately 0.075%. A further increase in the number of terms from the series (e) is neccessary only if the frequencies of the higher modes of vibration are also to be calculated. For further comparison, we note that in the case of a prismatic cantilever beam having the same section as the wedge at the thick end, the following result was obtained (see Sec. 5.11):

$$f_1 = \frac{\omega_1}{2\pi} = \frac{(1.875)^2 a}{2\pi\ell^2} = \frac{3.515b}{2\pi\ell^2}\sqrt{\frac{E}{3\rho}} \qquad (s)$$

† G. R. Kirchhoff, *Monatsberichte*, Berlin, 1879, p. 815; or *Gesammelte Abhandlungen*, Leipzig, 1882, p. 339.

The first method of Ritz can be applied also in cases where A and I are not represented by continuous functions of x. These functions may have several points of discontinuity or may be defined by different mathematical expressions in different intervals along the length ℓ. In such cases the integrals (h) should be subdivided into intervals such that I and A may be represented by continuous functions in each of these intervals.† If the functions A and I are obtained either graphically or from numerical tables, Ritz's first method can still be used. In such cases it becomes necessary to determine the integrals in (h) by numerical techniques.

The calculations described above can be simplified by applying *Ritz's second method*,‡ in which we use directly the differential equation of vibration instead of energy expressions. As an example, we take the known case of vibration of a cantilever of constant cross section, for which the differential equation defining the normal functions is

$$EI\frac{d^4X}{dx^4} - \rho A\omega^2 X = 0 \tag{t}$$

Assuming that the beam is fixed at the left end and free at the right end, we have the boundary conditions

$$(X)_{x=0} = 0 \qquad \left(\frac{dX}{dx}\right)_{x=0} = 0 \qquad \left(\frac{d^2X}{dx^2}\right)_{x=\ell} = 0 \qquad \left(\frac{d^3X}{dx^3}\right)_{x=\ell} = 0 \tag{u}$$

In using the second Ritz method, we again take X in the form of the series (e). Since it is not the rigorous solution, it will not satisfy Eq. (t); and we will obtain (upon substitution into the left-hand side of the equation) a quantity different from zero, which represents some loading $Q(x)$ distributed along the length of the cantilever. Then the values of the coefficients $a_1, a_2, a_3, \ldots$ in the series (e) may be obtained from the condition that the virtual work of the loading $Q(x)$ on the virtual displacements $\delta v_j = \delta a_j \Phi_j(x)$ vanishes. In this manner we obtain equations of the form

$$\int_0^\ell \left(EI\frac{d^4X}{dx^4} - \rho A\omega^2 X\right)\Phi_j(x)\, dx = 0 \tag{5.154}$$

After substituting the series (e) into this expression and integrating, we have a system of linear equations for $a_1, a_2, a_3, \ldots$; as before, the frequency equation is found by equating to zero the determinant from these equations.

Taking only two terms in the series (e), we shall assume in our case that

$$X = a_1(6\ell^2x^2 - 4\ell x^3 + x^4) + a_2(20\ell^3x^2 - 10\ell^2x^3 + x^5) \tag{v}$$

† Examples of this kind were discussed by K. A. Traenkel, *Ing.-Arch.*, **1,** 1930, p. 499.

‡ As noted in Sec. 2.3, the method is sometimes attributed to Galerkin, but was first intoduced by W. Ritz.

Each of the expressions in the parentheses satisfies the end conditions (u). The first represents (to within a constant factor) deflections of a uniformly loaded cantilever, and the second is the deflection of a cantilever under a triangular loading that is zero at the fixed end and increases linearly toward the free end. Substituting expression (v) into Eq. (5.154) and performing the integrations, we obtain

$$\left(\frac{104}{45}\frac{\omega^2\ell^4}{a^2} - \frac{144}{5}\right)a_1 + \left(\frac{2644}{315}\frac{\omega^2\ell^4}{a^2} - 104\right)a_2 = 0 \tag{w}$$

$$\left(\frac{2644}{315}\frac{\omega^2\ell^4}{a^2} - 104\right)a_1 + \left(\frac{21{,}128}{693}\frac{\omega^2\ell^4}{a^2} - \frac{2640}{7}\right)a_2 = 0 \tag{x}$$

Equating to zero the determinant of the coefficients in these two equations, we find

$$\omega_1 = 3.517\frac{a}{\ell^2} = \frac{3.517}{\ell^2}\sqrt{\frac{EI}{\rho A}} \qquad \omega_2 = \frac{22.78}{\ell^2}\sqrt{\frac{EI}{\rho A}} \tag{y}$$

The value of ω_1 is obtained with a high accuracy, and the error in ω_2 is about 3.4%.

Vibrations of a Conical Beam

The problem of vibrations of a conical beam that has its tip free and the base built in was first treated by Kirchhoff.† For the fundamental mode he obtained in this case

$$f_1 = \frac{\omega_1}{2\pi} = \frac{4.359r}{2\pi\ell^2}\sqrt{\frac{E}{\rho}} \tag{5.155}$$

where r is the radius of the base and ℓ is the length of the beam. For comparison it should be remembered that a cylindrical beam of the same length and base size has the frequency

$$f_1 = \frac{\omega_1}{2\pi} = \frac{(1.875)^2 a}{2\pi\ell^2} = \frac{1.758r}{2\pi\ell^2}\sqrt{\frac{E}{\rho}} \tag{z}$$

Thus, the frequencies of the fundamental modes of a conical and a cylindrical beam are in the ratio $4.359/1.758 \approx 2.5$. More generally, the frequency of any mode of vibration of a conical beam can be calculated

† *Ibid.*

from the equation

$$f_j = \frac{\omega_j}{2\pi} = \frac{\alpha_j r}{2\pi\ell^2}\sqrt{\frac{E}{\rho}} \tag{5.156}$$

in which α_j has the values†

α_1	α_2	α_3	α_4	α_5	α_6
4.359	10.573	19.225	30.339	43.921	59.956

Other Cases of a Cantilever Beam of Variable Cross Section

In the general case the frequencies of transverse vibrations of a cantilever beam can be represented by the equation

$$f_j = \frac{\omega_j}{2\pi} = \frac{\alpha_j r_g}{2\pi\ell^2}\sqrt{\frac{E}{\rho}} \tag{5.157}$$

In this expression r_g is the radius of gyration of the built-in section, ℓ is the length of the cantilever, and α_j is a constant depending on the shape of the beam and on the mode of vibration. Values of the constant α_1 for certain particular cases of practical importance are as follows:

1. If the variations of the cross-sectional area and of the moment of inertia along the x axis can be expressed in the forms

$$A = ax^m \qquad I = bx^m \tag{a'}$$

(x being measured from the free end), then r_g remains constant along the length of the cantilever; and the constant α_1 for the fundamental mode can be represented with sufficient accuracy by the equation‡

$$\alpha_1 = 3.47(1 + 1.05m) \tag{b'}$$

2. If the variation of the cross-sectional area and of the moment of inertia along the x axis can be expressed in the forms

$$A = a\left(1 - c\frac{x}{\ell}\right) \qquad I = b\left(1 - c\frac{x}{\ell}\right) \tag{c'}$$

(x being measured from the built-in end), then r_g remains constant

† See D. Wrinch, *Proc. Roy. Soc. (London)*, **101,** 1922, p. 493.
‡ A. Ono, *Jour. Soc. Mech. Engrs. (Tokyo)*, **27,** 1924, p. 467.

along the length of the beam; and the quantity α_1 will be as given in the following list:†

$c = 0$	0.4	0.6	0.8	1.0
$\alpha_1 = 3.515$	4.098	4.585	5.398	7.16

Beam of Variable Cross Section with Free Ends

Let us consider now the case of a laterally vibrating "free-free" beam consisting of two equal halves joined together at their thick ends (see Fig. 5.31), the left half being generated by revolving the curve

$$y = ax^m \tag{d'}$$

about the x axis. The exact solution in terms of Bessel's functions has been obtained in this case for certain values of m,‡ and the frequency of the fundamental mode can be represented in the form

$$f_1 = \frac{\omega_1}{2\pi} = \frac{\alpha_1 r}{4\pi\ell^2}\sqrt{\frac{E}{\rho}} \tag{5.158}$$

In this formula r is the radius of the thickest cross section, 2ℓ is the length of the beam, and α_1 is a constant depending on the shape of the curve (d'), the values of which are given in the list

$m = 0$	$\frac{1}{4}$	$\frac{1}{2}$	$\frac{3}{4}$	1
$\alpha_1 = 5.593$	6.957	8.203	9.300	10.173

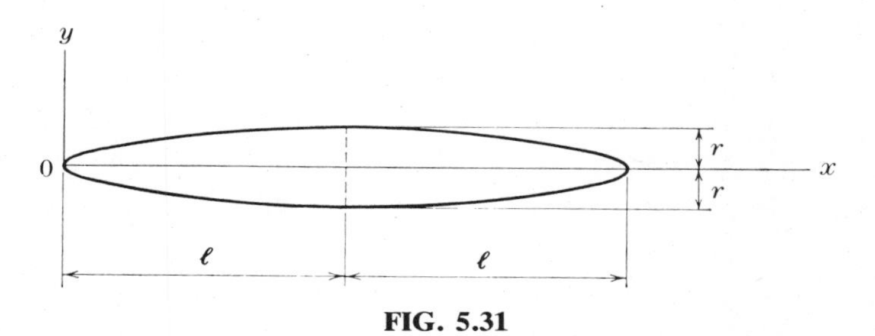

FIG. 5.31

† Ono, *Jour. Soc. Mech. Engrs. (Tokyo)*, **28,** 1925, p. 429.
‡ See J. W. Nicholson, *Proc. Roy. Soc. (London)*, **93,** 1917, p. 506.

5.21 COUPLED FLEXURAL AND TORSIONAL VIBRATIONS OF BEAMS

In previous discussions of transverse vibrations of beams, it was always assumed that the beam vibrates in a plane of symmetry. If this is not the case, the flexural vibrations will usually be coupled with torsional vibrations. As an example, let us consider the vibrations of a channel (see Fig. 5.32*a*) in the x–y plane, which is perpendicular to the plane of symmetry (the x–z plane). If a force is applied in the y direction, bending will occur in the x–y plane and will not be accompanied by torsion only if the load is applied at

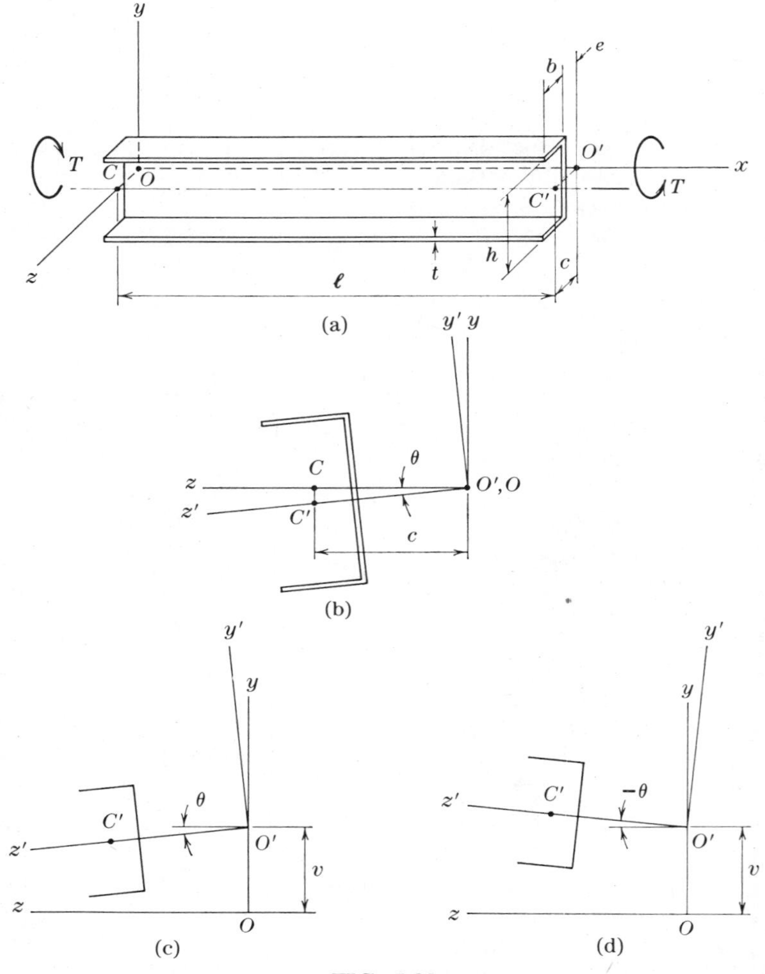

FIG. 5.32

the *shear-center axis* O–O'. This axis is parallel to the centroidal axis C–C' and lies in the plane of symmetry. In Fig. 5.32*a* the shear-center axis is taken to be the x axis. Its distance e from the middle plane of the web and its distance c from the centroid are given by the formulas [3]

$$e = \frac{b^2 h^2 t}{4I_z} \qquad c = e + \frac{b^2}{2b + h} \tag{a}$$

in which b is the width of the flanges, h is the distance between the centers of the flanges, and t is the thickness of the flanges and of the web. For a load in the y direction, the differential equation of the deflection curve is

$$EI_z \frac{d^4 v}{dx^4} = w \tag{b}$$

in which w is the intensity of the distributed load (positive upward) and EI_z is the flexural rigidity of the channel with respect to the z axis.

If the load is distributed along the centroidal axis, we can always replace the given load by the same load and a torque of intensity wc, distributed along the shear-center axis x. In such a case we will have a combination of bending, defined by Eq. (b), and torsion with respect to the shear-center axis x. This torsion is nonuniform; and the relation between the variable torque $T(x)$ and the angle of twist θ is given by the equation [4]

$$T(x) = R\frac{d\theta}{dx} - R_1 \frac{d^3\theta}{dx^3} \tag{c}$$

in which R is the *torsional rigidity* for uniform torsion and R_1 is the *warping rigidity*. We take the positive sense of the angle of twist as shown in Fig. 5.32*b*, which conforms to the right-hand rule. Differentiating Eq. (c) with respect to x and observing that the positive torque has the direction shown in Fig. 5.32*a*, we obtain

$$R\frac{d^2\theta}{dx^2} - R_1 \frac{d^4\theta}{dx^4} = wc \tag{d}$$

Equations (b) and (d) define coupled bending and torsion of the beam when a static load is distributed along the centroidal axis.

When the beam is vibrating (see Fig. 5.32*c*), we must consider transverse inertial forces† of intensity

$$-\rho A \frac{\partial^2}{\partial t^2}(v - c\theta)$$

† Longitudinal inertial forces associated with warping are neglected.

and inertial torques of intensity

$$-\rho I_p \frac{\partial^2 \theta}{\partial t^2}$$

where I_p is the centroidal polar moment of inertia of the cross section. Using the first of these inertial actions in place of the statically applied load in Eqs. (*b*) and (*d*), we obtain the following differential equations for the coupled flexural and torsional vibrations:

$$EI_z \frac{\partial^4 v}{\partial x^4} = -\rho A \frac{\partial^2}{\partial t^2}(v - c\theta) \tag{5.159a}$$

$$R \frac{\partial^2 \theta}{\partial x^2} - R_1 \frac{\partial^4 \theta}{\partial x^4} = -\rho A c \frac{\partial^2}{\partial t^2}(v - c\theta) + \rho I_p \frac{\partial^2 \theta}{\partial t^2} \tag{5.159b}$$

Assuming that the beam performs one of its natural modes of vibration, we put

$$v = X(A \cos \omega t + B \sin \omega t) \qquad \theta = X_1(A_1 \cos \omega t + B_1 \sin \omega t) \tag{e}$$

where ω is the angular frequency of vibration and X and X_1 are the normal functions. Substituting expressions (*e*) into Eqs. (5.159a) and (5.159b), we obtain for finding X and X_1 the following equations:

$$EI_z X'''' = \rho A \omega^2 (X - cX_1) \tag{f}$$

$$R_1 X_1'''' - RX_1'' = -\rho A \omega^2 c(X - cX_1) + \rho I_p \omega^2 X_1 \tag{g}$$

In each particular case we must find for X and X_1 solutions that satisfy the prescribed conditions at the ends of the beam, as well as Eqs. (*f*) and (*g*).

As an example, we take the case of a beam with simply supported ends, for which the boundary conditions are

$$v = \frac{\partial^2 v}{\partial x^2} = \theta = \frac{\partial^2 \theta}{\partial x^2} = 0 \qquad (\text{for } x = 0 \text{ and } x = \ell) \tag{h}$$

These requirements are satisfied by the functions

$$X_i = C_i \sin \frac{i\pi x}{\ell} \qquad X_{1i} = D_i \sin \frac{i\pi x}{\ell} \qquad (i = 1, 2, 3, \ldots, \infty) \tag{i}$$

where C_i and D_i are constants. Substituting these expressions into Eqs. (*f*) and (g) and using the notations

$$\frac{EI_z i^4 \pi^4}{\ell^4 \rho A} = \omega_{bi}^2 \qquad \frac{(Ri^2\pi^2\ell^2 + R_1 i^4 \pi^4)}{\ell^4 \rho (I_p + Ac^2)} = \omega_{ti}^2 \qquad \frac{Ac^2}{I_p + Ac^2} = \lambda \tag{j}$$

we obtain

$$(\omega_{bi}^2 - \omega_i^2)C_i + \omega_i^2 c D_i = 0 \tag{k}$$

$$\frac{\lambda}{c}\,\omega_i^2 C_i + (\omega_{ti}^2 - \omega_i^2)D_i = 0 \tag{l}$$

These equations can give solutions for C_i and D_1 different from zero only if the determinant of their coefficients vanishes, in which case the frequency equation becomes

$$(\omega_{bi}^2 - \omega_i^2)(\omega_{ti}^2 - \omega_i^2) - \lambda\omega_i^4 = 0 \tag{m}$$

from which we find

$$\omega_i^2 = \frac{(\omega_{ti}^2 + \omega_{bi}^2) \mp \sqrt{(\omega_{ti}^2 - \omega_{bi}^2)^2 + 4\lambda\omega_{bi}^2\omega_{ti}^2}}{2(1-\lambda)} \tag{5.160}$$

A similar result will be obtained in all other cases of simply supported beams with one plane of symmetry, which vibrate in a direction perpendicular to that plane.

If the shear center coincides with the centroid, the distance c is zero, $\lambda = 0$, and we obtain

$$\omega_i^2 = \frac{\omega_{ti}^2 + \omega_{bi}^2}{2} \mp \frac{\omega_{ti}^2 - \omega_{bi}^2}{2}$$

which gives us two sets of frequencies

$$\omega_{1i} = \omega_{bi} \qquad \omega_{2i} = \omega_{ti} \tag{n}$$

As can be seen from notations (j), these frequencies correspond to uncoupled flexural and nonuniform torsional vibrations and are independent of each other in this case. If c does not vanish, we obtain from the solution (5.160) two values for ω_i^2, one of which is larger and the other smaller than the values in (n). For the larger value of ω_i^2 we find from Eqs. (k) and (l) that C_i and D_i have the same sign, and for the smaller value of ω_i^2 they are of the opposite sign. The corresponding two configurations are shown in Figs. 5.32c and 5.32d, respectively.

Similar results will also be obtained in cases of beams with other end conditions. The solution of Eqs. (f) and (g) then become more complicated, but we can calculate the approximate values of frequencies of the coupled vibrations by using the Rayleigh–Ritz method.† In the case of a beam having no plane of symmetry the problem becomes even more involved.‡ Torsional vibrations are coupled with flexural vibrations in the two principal planes, and we obtain three simultaneous differential equations instead of two. In practical applications we also encounter the more complicated problem of coupled torsional and bending vibrations of an unsymmetric beam of variable cross section. Such problems occur, for example, in the forms of turbine blades, airplane wings, and propellers. For the solution of these problems, numerical methods are usually applied.

5.22 VIBRATIONS OF CIRCULAR RINGS

The problem of vibrations of a circular ring is encountered in the investigation of the frequencies of various types of circular parts of rotating machinery. In the following discussion, several simple problems involving the vibrations of a circular ring of constant cross section are considered, under the assumption that the cross-sectional dimensions of the ring are small in comparison with the radius r of the center line (see Fig. 5.33*a*). It is also assumed that the plane of the ring (the x–y plane) is a plane of symmetry for each cross section.

Extensional Vibrations

The simplest extensional mode of vibration is that in which the center line of the ring forms a circle of periodically varying radius, and all of the cross sections move radially without rotation (see Fig. 5.33*b*). Let the symbol u denote the radial translational (positive outward) of any point on the ring. Then the unit elongation of the ring in the circumferential direction (extensional strain) is equal to u/r. The potential energy of deformation, consisting in this case of the energy of simple tension, will be given by the equation

$$U = \frac{AEu^2}{2r^2} 2\pi r \tag{a}$$

in which A is the cross-sectional area of the ring. Furthermore, the kinetic energy during vibration will be

$$T = \frac{\rho A}{2} \dot{u}^2 2\pi r \tag{b}$$

† In this manner C. F. Garland investigated coupled bending and torsional vibrations of a cantilever, *Jour. Appl. Mech.*, **7,** 1940, p. 97.

‡ The differential equations for the general case are discussed by K. Federhofer, *Sitzber. Akad. Wiss. Wien, Abt. IIa,* **156,** 1947, p. 343.

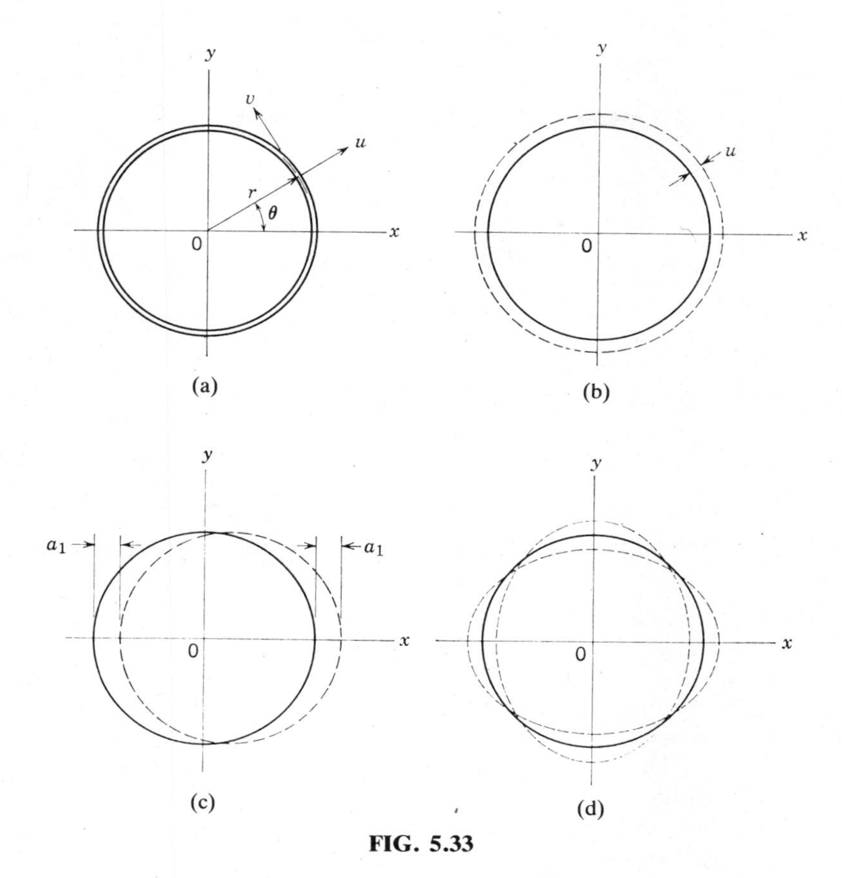

FIG. 5.33

Equating maximum values of the potential and kinetic energies and using $\dot{u}_{max} = \omega u_{max}$, we obtain

$$\omega = \sqrt{\frac{E}{\rho r^2}} \tag{c}$$

Thus, the frequency of the fundamental extensional mode (see Fig. 5.33*b*) is

$$f = \frac{\omega}{2\pi} = \frac{1}{2\pi}\sqrt{\frac{E}{\rho r^2}} \tag{5.161}$$

A circular ring also possesses other extensional modes that resemble the longitudinal vibrations of prismatic bars. If i denotes the number of wavelengths to the circumference, the frequencies of the higher modes of

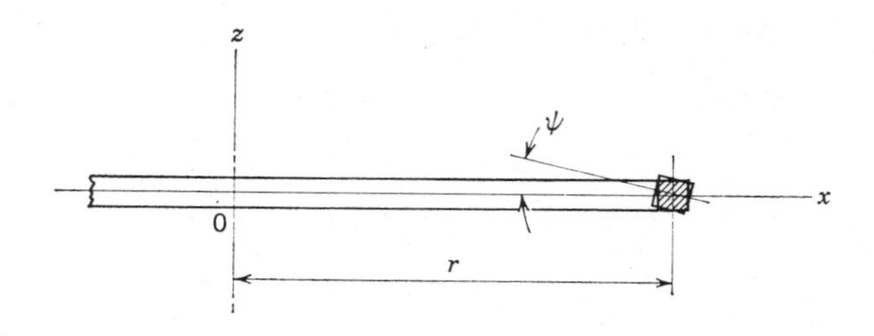

FIG. 5.34

extensional vibration of the ring will be determined from the equation [2]

$$f_i = \frac{1}{2\pi}\sqrt{\frac{E(1+i^2)}{\rho r^2}} \tag{5.162}$$

When $i = 0$, this expression yields Eq. (5.161) for pure radial vibrations.

Torsional Vibrations

Consideration will now be given to the fundamental mode of torsional vibration. In this mode the center line of the ring remains undeformed, and all of the cross sections of the ring rotate through the same small angle ψ (see Fig. 5.34). Due to this rotation, a point at the distance z from the middle plane of the ring will have a radial displacement approximately equal to $z\psi$; and the corresponding circumferential strain can be taken equal to $z\psi/r$. The potential energy of deformation of the ring can now be calculated as follows:

$$U = 2\pi r \int_A \frac{E}{2}\left(\frac{z\psi}{r}\right)^2 dA = \frac{\pi E I_x \psi^2}{r} \tag{d}$$

In this expression I_x is the moment of inertia of the cross section about the x axis.

The kinetic energy of the ring during torsional vibrations will be

$$T = 2\pi r \frac{\rho I_p}{2} \dot{\psi}^2 \tag{e}$$

where I_p is the polar moment of inertia of the cross section.

Equating U_{max} and T_{max} and using $\dot{\psi}_{max} = \omega\psi_{max}$, we find

$$\omega = \sqrt{\frac{EI_x}{\rho r^2 I_p}} \tag{f}$$

The frequency of torsional vibration then becomes

$$f = \frac{1}{2\pi}\sqrt{\frac{EI_x}{\rho r^2 I_p}} \tag{5.163}$$

Comparing this result with formula (5.161), we conclude that the fundamental frequencies of the torsional and pure radial vibrations are in the ratio $\sqrt{I_x/I_p}$.

For rings with circular cross sections, the frequencies of torsional modes of vibration are given by the formula [2]

$$f_i = \frac{1}{2\pi}\sqrt{\frac{E(1+i^2)}{2\rho r^2}} \qquad (i = 1, 2, 3, \ldots) \tag{5.164}$$

Remembering that

$$\sqrt{\frac{E}{\rho r^2}} = \frac{a}{r}$$

where a is the velocity of propagation of sound in the tangential direction of the ring, we see that the extensional and torsional vibrations considered above usually have high frequencies. Much lower frequencies will be encountered when flexural vibrations of the ring are considered.

Flexural Vibrations

Flexural vibrations of a circular ring fall into two classes, as follows: (1), flexural vibrations in the plane of the ring and (2) flexural vibrations involving both translations perpendicular to the plane of the ring and rotations of cross sections. Let us consider the flexural vibrations in the plane of the ring (see Fig. 5.33*a*), using the following notations:

θ = angle determining the position of a point on the ring
u = radial translation (positive outward)
v = tangential translation (positive in the direction of increasing θ)
I_z = moment of inertia of the cross section with respect to a principal axis parallel to the z axis

Due to the translations u and v, the strain of the center line of the ring at any point is

$$\epsilon = \frac{u}{r} + \frac{\partial v}{r\,\partial\theta} \tag{g}$$

and the change in curvature can be represented by the equation†

$$\frac{1}{r+\Delta r}-\frac{1}{r}=-\frac{\partial^2 u}{r^2\,\partial\theta^2}-\frac{u}{r^2} \tag{h}$$

For the most general case of flexural vibrations in the plane of the ring, the radial translation u can be expanded in the form of a trigonometric series‡

$$u=a_1\cos\theta+a_2\cos 2\theta+\cdots+b_1\sin\theta+b_2\sin 2\theta+\cdots \tag{i}$$

in which the coefficients $a_1, a_2, \ldots, b_1, b_2, \ldots$ are functions of time. Considering flexural vibrations without extension,§ we have $\epsilon=0$; and Eq. (g) gives

$$u=-\frac{\partial v}{\partial\theta} \tag{j}$$

Substitution of Eq. (i) into Eq. (j), followed by integration,|| results in

$$v=-a_1\sin\theta-\frac{1}{2}a_2\sin 2\theta-\cdots+b_1\cos\theta+\frac{1}{2}b_2\cos 2\theta+\cdots \tag{k}$$

From Eq. (h) the bending moment at any cross section of the ring is

$$M=-\frac{EI_z}{r^2}\left(\frac{\partial^2 u}{\partial\theta^2}+u\right) \tag{l}$$

Hence, we obtain for the potential energy of bending

$$U=\frac{EI_z}{2r^4}\int_0^{2\pi}\left(\frac{\partial^2 u}{\partial\theta^2}+u\right)^2 r\,d\theta \tag{m}$$

By substituting the series (i) into Eq. (m) and using the formulas

$$\int_0^{2\pi}\cos m\theta\cos n\theta\,d\theta=0 \qquad \int_0^{2\pi}\sin m\theta\sin n\theta\,d\theta=0 \qquad (m\neq n)$$

$$\int_0^{2\pi}\cos m\theta\sin m\theta\,d\theta=0 \qquad \int_0^{2\pi}\cos^2 m\theta\,d\theta=\int_0^{2\pi}\sin^2 m\theta\,d\theta=\pi$$

† This equation was established by J. Boussinesq, *Comptes rend.*, **97**, 1883, p. 843.

‡ The constant term of the series, corresponding to pure radial vibration, is omitted.

§ For discussions of flexural vibrations taking into account extension, see F. W. Waltking, *Ing.-Arch.*, **5**, 1934, p. 429; and K. Federhofer, *Sitzber. Acad. Wiss. Wien, Abt. IIa*, **145**, 1936, p. 29.

|| The constant of integration, representing rigid-body rotation of the ring in its own plane, is omitted.

we obtain

$$U = \frac{EI_z\pi}{2r^3} \sum_{i=1}^{\infty} (1 - i^2)^2(a_i^2 + b_i^2) \tag{n}$$

The kinetic energy of the vibrating ring is

$$T = \frac{\rho A}{2} \int_0^{2\pi} (\dot{u}^2 + \dot{v}^2) r \, d\theta \tag{o}$$

By substituting Eqs. (i) and (k) for u and v, we find that Eq. (o) becomes

$$T = \frac{\pi r \rho A}{2} \sum_{i=1}^{\infty} \left(1 + \frac{1}{i^2}\right)(\dot{a}_i^2 + \dot{b}_i^2) \tag{p}$$

Using *Lagrange's equation* for a conservative system [5] and the generalized coordinates a_i, we obtain the following differential equation of motion:

$$\pi r \rho A\left(1 + \frac{1}{i^2}\right)\ddot{a}_i + \frac{EI_z\pi}{r^3}(1 - i^2)^2 a_i = 0$$

or

$$\ddot{a}_i + \frac{EI_z i^2(1 - i^2)^2}{\rho A r^4(1 + i^2)} a_i = 0 \tag{q}$$

and an equation of the same form is found for the generalized coordinates b_i. Hence, the frequency of the ith mode of vibration is given by

$$f_i = \frac{1}{2\pi} \sqrt{\frac{EI_z i^2(1 - i^2)^2}{\rho A r^4(1 + i^2)}} \tag{5.165}$$

When $i = 1$, we obtain $f_1 = 0$. In this case $u = a_1 \cos\theta$ and $v = -a_1 \sin\theta$; and the ring moves as a rigid body. As shown in Fig. 5.33c, the term a_1 represents rigid-body motion in the x direction. When $i = 2$, the ring performs the fundamental mode of flexural vibration. The extreme positions of the ring during this vibration are shown in Fig. 5.33d by dashed lines.

In the case of flexural vibrations of a ring of circular cross section involving both translations perpendicular to the plane of the ring and rotations of cross sections, the frequencies of the principal modes of vibration can be calculated from the equation [2]

$$f_i = \frac{1}{2\pi} \sqrt{\frac{EI_z i^2(i^2 - 1)^2}{\rho A r^4(i^2 + 1 + \nu)}} \tag{5.166}$$

in which ν denotes Poisson's ratio. Comparing Eqs. (5.165) and (5.166), we conclude that even in the lowest mode ($i = 2$) the frequencies of the two classes of flexural vibrations differ only slightly.

5.23 TRANSVERSE VIBRATIONS OF MEMBRANES

The two-dimensional counterpart of the stretched wire (or string), discussed in Sec. 5.8, consists of a stretched membrane. In the following discussion it is assumed that the membrane is a perfectly flexible thin lamina of constant thickness. It is further assumed that it is uniformly stretched in all directions by a tension so large that the fluctuation in this tension due to small deflections during vibration can be neglected. Taking the plane of the membrane as the x–y plane (see Fig. 5.35), we use the following notations:

w = displacement of any point on the membrane in the direction perpendicular to the x–y plane (the z direction)
S = uniform tension per unit length of the boundary
h = thickness of the membrane

The increase in the potential energy of the deflected membrane can be found by multiplying the uniform tension S by the increase in surface area of the membrane. In a deflected position the surface area of the membrane becomes

$$A = \iint \sqrt{1 + \left(\frac{\partial w}{\partial x}\right)^2 + \left(\frac{\partial w}{\partial y}\right)^2}\, dx\, dy$$

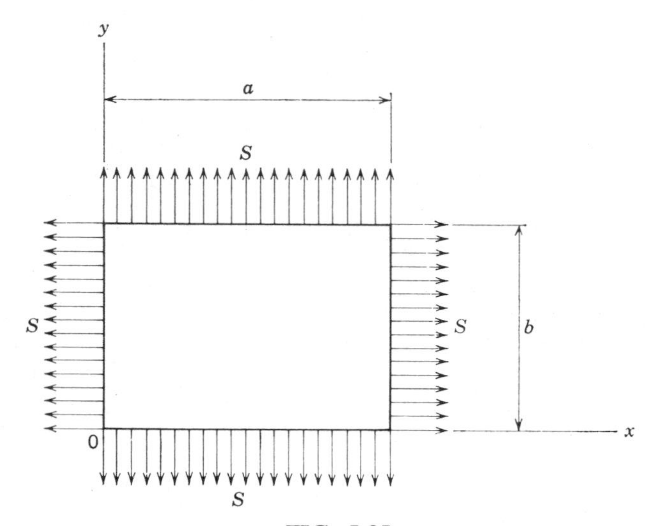

FIG. 5.35

For small deflections this expression may be approximated by

$$A \approx \iint \left\{ 1 + \frac{1}{2}\left(\frac{\partial w}{\partial x}\right)^2 + \frac{1}{2}\left(\frac{\partial w}{\partial y}\right)^2 \right\} dx\, dy$$

Then the increase in potential energy will be

$$\Delta U \approx \frac{S}{2} \iint \left\{ \left(\frac{\partial w}{\partial x}\right)^2 + \left(\frac{\partial w}{\partial y}\right)^2 \right\} dx\, dy \tag{a}$$

Furthermore, the kinetic energy of the membrane during vibration is

$$T = \frac{\rho h}{2} \iint \dot{w}^2\, dx\, dy \tag{b}$$

where ρh is the mass per unit area. The vibrational characteristics of particular types of membranes will now be investigated.

Rectangular Membranes

Let a and b denote the lengths of the sides of a rectangular membrane, as shown in Fig. 5.35. Whatever function of the coordinates w may be, it always can be represented within the limits of the rectangle by the double series

$$w = \sum_{m=1}^{\infty} \sum_{n=1}^{\infty} \phi_{mn} \sin \frac{m\pi x}{a} \sin \frac{n\pi y}{b} \tag{c}$$

in which the coefficients ϕ_{mn} are functions of time. It is easy to see that each term of the series (c) satisfies the boundary conditions. That is, $w = 0$ for $x = 0$ and $x = a$; and $w = 0$ for $y = 0$ and $y = b$. Substituting Eq. (c) into expression (a) for the increase in potential energy, we obtain

$$\Delta U \approx \frac{S\pi^2}{2} \int_0^a \int_0^b \left\{ \left(\sum_{m=1}^{\infty} \sum_{n=1}^{\infty} \phi_{mn} \frac{m}{a} \cos \frac{m\pi x}{a} \sin \frac{n\pi y}{b} \right)^2 + \left(\sum_{m=1}^{\infty} \sum_{n=1}^{\infty} \phi_{mn} \frac{n}{b} \sin \frac{m\pi x}{a} \cos \frac{n\pi y}{b} \right)^2 \right\} dx\, dy$$

Integrating this expression over the area of the membrane, we use the formulas of Sec. 1.11 to find

$$\Delta U \approx \frac{S}{2} \frac{ab\pi^2}{4} \sum_{m=1}^{\infty} \sum_{n=1}^{\infty} \left(\frac{m^2}{a^2} + \frac{n^2}{b^2} \right) \phi_{mn}^2 \tag{d}$$

In a similar manner, the kinetic energy in Eq. (*b*) can be expressed as

$$T = \frac{\rho hab}{8} \sum_{m=1}^{\infty} \sum_{n=1}^{\infty} \dot{\phi}_{mn}^2 \tag{e}$$

The inertial force for a typical element of the membrane is $-(\rho h)\ddot{w}\, dx\, dy$. Proceeding as before and considering a virtual displacement

$$\delta w_{mn} = \delta\phi_{mn} \sin\frac{m\pi x}{a} \sin\frac{n\pi y}{b}$$

we obtain the differential equation of motion for free vibrations in principal coordinates to be

$$\frac{\rho hab}{4}\ddot{\phi}_{mn} + S\frac{ab\pi^2}{4}\left(\frac{m^2}{a^2} + \frac{n^2}{b^2}\right)\phi_{mn} = 0 \tag{f}$$

from which

$$f_{mn} = \frac{1}{2}\sqrt{\frac{S}{\rho h}\left(\frac{m^2}{a^2} + \frac{n^2}{b^2}\right)} \tag{5.167}$$

The fundamental mode of vibration is found by putting $m = n = 1$, which produces

$$f_{11} = \frac{1}{2}\sqrt{\frac{S}{\rho h}\left(\frac{1}{a^2} + \frac{1}{b^2}\right)} \tag{5.168}$$

The deflected surface of the membrane in this case is defined by the first term in the series (*c*) as

$$w = C \sin\frac{\pi x}{a} \sin\frac{\pi y}{b} \tag{g}$$

It is apparent from Eq. (*f*) that such terms in the series (*c*) constitute the normal functions for the case under consideration. If the membrane is square ($a = b$), the lowest frequency given by Eq. (5.168) is

$$f_{11} = \frac{1}{a}\sqrt{\frac{S}{2\rho h}} \tag{5.169}$$

This frequency is directly proportional to the square root of the tension S and inversely proportional to the length a and to the square root of the mass ρh per unit area.

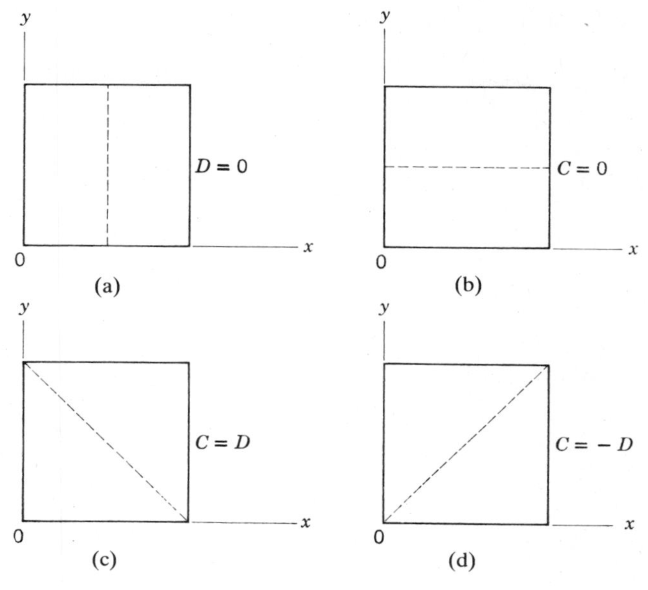

FIG. 5.36

The next two higher modes of vibtation will be obtained by taking one of the numbers m or n equal to 2 and the other equal to 1. These two modes have the same frequency (for $a = b$) but different shapes. The dashed lines in Figs. 5.36*a* and 5.36*b* indicate the *nodal lines* for these two modes of vibration. (Along such lines the deflections during vibration are zero.) Because of the fact that the frequencies are the same, it is of interest to superimpose these two surfaces on each other in various ratios of their maximum deflections. Such a combination is expressed by

$$w = C \sin \frac{2\pi x}{a} \sin \frac{\pi y}{a} + D \sin \frac{\pi x}{a} \sin \frac{2\pi y}{a}$$

where C and D are small but arbitrary quantities. Four particular cases of such a combined vibration are shown in Figs. 5.36*a,b,c,* and *d*. Taking $D = 0$, we obtain the vibration mentioned above and shown in Fig. 5.36*a*. The vibrating membrane is subdivided into two equal parts by a nodal line parallel to the y axis. When $C = 0$, the membrane is subdivided by a nodal line parallel to the x axis, as shown in Fig. 5.36*b*. When $C = D$, we obtain

$$\begin{aligned} w &= C\left(\sin \frac{2\pi x}{a} \sin \frac{\pi y}{a} + \sin \frac{\pi x}{a} \sin \frac{2\pi y}{a}\right) \\ &= 2C \sin \frac{\pi x}{a} \sin \frac{\pi y}{a}\left(\cos \frac{\pi x}{a} + \cos \frac{\pi y}{a}\right) \end{aligned}$$

This expression vanishes when

$$\sin\frac{\pi x}{a} = 0 \qquad \text{or} \qquad \sin\frac{\pi y}{a} = 0$$

and again when

$$\cos\frac{\pi x}{a} + \cos\frac{\pi y}{a} = 0$$

The first two equations give us the sides of the boundary, and from the third equation we obtain

$$\frac{\pi x}{a} = \pi - \frac{\pi y}{a}$$

or

$$x + y = a$$

This expression represents one diagonal of the square, as shown in Fig. 5.36*c*. On the other hand, Fig. 5.36*d* represents the case when $C = -D$. Each half of the membrane in the last two cases can be considered to act separately as a triangular membrane. The frequency for any of the modes in Figs. 5.36*a*, *b*, *c*, and *d* is given by Eq. (5.167) as

$$f = \frac{1}{2}\sqrt{\frac{S}{\rho h}\left(\frac{4}{a^2} + \frac{1}{a^2}\right)} = \frac{1}{2a}\sqrt{\frac{5S}{\rho h}} \tag{5.170}$$

In this manner, higher modes of vibration of a square or rectangular membrane can also be considered [1].

Let us now consider forced response of the membrane, for which the differential equation of motion (*f*) becomes

$$\frac{\rho hab}{4}\ddot{\phi}_{mn} + S\frac{ab\pi^2}{4}\left(\frac{m^2}{a^2} + \frac{n^2}{b^2}\right)\phi_{mn} = Q_{mn} \tag{h}$$

in which Q_{mn} is selected so that $Q_{mn}\,\delta\phi_{mn}$ represents the virtual work of the applied forces in principal coordinates. As an example, we take the case of a harmonic force $P_1(t) = P\cos\Omega t$, acting at the center of the membrane. By inducing a virtual displacement δw_{mn} of the membrane [see Eq. (*c*)], we find for the virtual work done by the force

$$\delta W_P = P\cos\Omega t\,\delta\phi_{mn}\sin\frac{m\pi}{2}\sin\frac{n\pi}{2} = Q_{mn}\,\delta\phi_{mn}$$

From this expression we see that when m and n are both odd, $Q_{mn} = \pm P \cos \Omega t$; otherwise, $Q_{mn} = 0$. Substituting this result into Eq. (h) and using Duhamel's integral, we obtain

$$\phi_{mn} = \pm \frac{4P}{ab\rho h \omega_{mn}} \int_0^t \sin \omega_{mn}(t - t') \cos \Omega t' \, dt'$$

$$= \pm \frac{4P}{ab\rho h(\omega_{mn}^2 - \Omega^2)} (\cos \Omega t - \cos \omega_{mn} t) \tag{i}$$

where m and n are both odd and

$$\omega_{mn}^2 = \frac{S\pi^2}{\rho h} \left(\frac{m^2}{a^2} + \frac{n^2}{b^2} \right)$$

Substitution of Eq. (i) into expression (c) produces the total solution for the case under consideration.

When a distributed dynamic force of intensity $Q(x, y, t)$ is acting on the membrane, we have

$$Q_{mn} = \int_0^b \int_0^a Q \sin \frac{m\pi x}{a} \sin \frac{n\pi y}{b} \, dx \, dy \tag{j}$$

Assume, for instance, that a uniformly distributed pressure Q_0 is suddenly applied to the membrane at the initial moment ($t = 0$). Then from Eq. (j) we have

$$Q_{mn} = \frac{abQ_0}{mn\pi^2} (1 - \cos m\pi)(1 - \cos n\pi)$$

When m and n are both odd, this expression yields

$$Q_{mn} = \frac{4abQ_0}{mn\pi^2} \tag{k}$$

Otherwise, Q_{mn} vanishes. Substituting Eq. (k) into Eq. (h) and assuming that the membrane is initially at rest, we find

$$\phi_{mn} = \frac{16Q_0(1 - \cos \omega_{mn} t)}{\rho h mn\pi^2 \omega_{mn}^2} \tag{l}$$

Hence, the vibrations produced by the suddenly applied pressure Q_0 are given by the expression

$$w = \frac{16Q_0}{\pi^2 \rho h} \sum_m \sum_n \frac{1 - \cos \omega_{mn} t}{mn\omega_{mn}^2} \sin \frac{m\pi x}{a} \sin \frac{n\pi y}{b} \tag{m}$$

where m and n are both odd.

Ritz Method

For calculating the frequencies of the natural modes of vibration of a membrane, Ritz's first method can be very useful. To apply this approach, we assume that the deflections of the vibrating membrane are given by

$$w = Z \cos(\omega t - \alpha) \tag{n}$$

where Z is a suitable function of the coordinates x and y that approximates the shape of the deflected membrane, that is, the mode of vibration. substituting Eq. (n) into expression (a) for the increase in potential energy, we find the maximum value to be

$$\Delta U_{\max} \approx \frac{S}{2} \iint \left\{ \left(\frac{\partial Z}{\partial x} \right)^2 + \left(\frac{\partial Z}{\partial y} \right)^2 \right\} dx\, dy \tag{o}$$

and for the maximum kinetic energy we obtain from Eq. (b)

$$T_{\max} = \frac{\rho h}{2} \omega^2 \iint Z^2\, dx\, dy \tag{p}$$

Equating expressions (o) and (p), we find

$$\omega^2 = \frac{S}{\rho h} \frac{\displaystyle\iint \left\{ \left(\frac{\partial Z}{\partial x} \right)^2 + \left(\frac{\partial Z}{\partial y} \right)^2 \right\} dx\, dy}{\displaystyle\iint Z^2\, dx\, dy} \tag{q}$$

With the Ritz method we take the expression Z for the deflected surface of the membrane in the form of a series

$$Z = a_1 \Phi_1(x, y) + a_2 \Phi_2(x, y) + a_3 \Phi_3(x, y) + \cdots \tag{r}$$

each term of which satisfies the conditions at the boundary. (The deflections at the boundary of the membrane must be equal to zero.) The coefficients $a_1, a_2, a_3, \ldots$ in this series should be chosen in such a manner that expression (q) yields a minimum value for ω^2. Thus,

$$\frac{\partial}{\partial a_j} \frac{\iint \left\{ \left(\frac{\partial Z}{\partial x} \right)^2 + \left(\frac{\partial Z}{\partial y} \right)^2 \right\} dx\, dy}{\iint Z^2\, dx\, dy} = 0$$

or

$$\iint Z^2\, dx\, dy \frac{\partial}{\partial a_j} \iint \left\{ \left(\frac{\partial Z}{\partial x} \right)^2 + \left(\frac{\partial Z}{\partial y} \right)^2 \right\} dx\, dy$$
$$- \iint \left\{ \left(\frac{\partial Z}{\partial x} \right)^2 + \left(\frac{\partial Z}{\partial y} \right)^2 \right\} dx\, dy \frac{\partial}{\partial a_j} \iint Z^2\, dx\, dy = 0$$

By using the relationship (q) in this latter equation, we find

$$\frac{\partial}{\partial a_j} \iint \left\{ \left(\frac{\partial Z}{\partial x} \right)^2 + \left(\frac{\partial Z}{\partial y} \right)^2 - \frac{\omega^2 \rho h}{S} Z^2 \right\} dx\, dy = 0 \tag{s}$$

In this manner we obtain as many equations of the type (s) as there are coefficients in the series (r). All of these equations will be linear in a_1, a_2, a_3, . . . , and the frequency equation for the membrane will be obtained by setting the determinant of the coefficients in such equations equal to zero.

Considering, for example, the modes of vibration of a square membrane that is symmetric with respect to the x and y axes (see Fig. 5.37), we can take the series (r) in the following form:

$$Z = (a^2 - x^2)(a^2 - y^2)(a_1 + a_2 x^2 + a_3 y^2 + a_4 x^2 y^2 + \cdots)$$

Each term of this series becomes equal to zero when $x = y = \pm a$. Hence, the conditions at the boundary are satisfied.

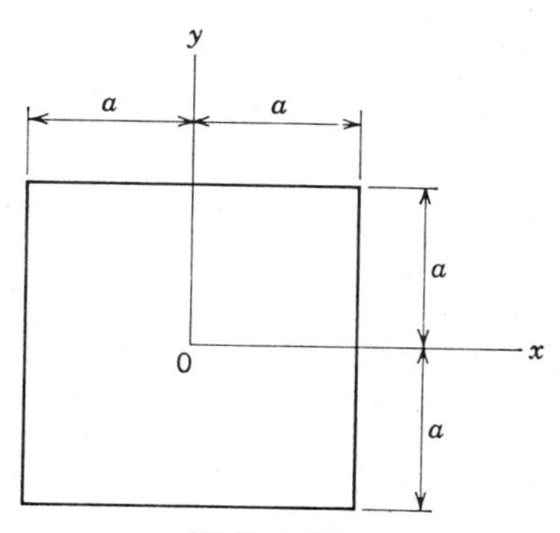

FIG. 5.37

For a membrane in the shape of a convex polygon, the boundary conditions will be satisfied by taking

$$Z = [(a_1x + b_1y + c_1)(a_2x + b_2y + c_2)\cdots]\sum_m \sum_n a_{mn}x^m y^n$$

where $a_1x + b_1y + c_1 = 0, \ldots,$ and so on, are the equations of the sides of the polygon. By taking only the first term ($m = 0, n = 0$) of this series, we will usually find a satisfactory approximation for the fundamental mode of vibration. It is necessary to take more terms if the frequencies of higher modes of vibration are required.

Circular Membranes

We will consider the simplest case of vibration of a circular membrane, where the deflected surface of the membrane is symmetric with respect to the center of the circle. In this case the deflections depend only on the radial distance r, and the boundary condition will be satisfied by the series

$$Z = a_1 \cos\frac{\pi r}{2R} + a_2 \cos\frac{3\pi r}{2R} + \cdots \tag{t}$$

in which R denotes the radius of the boundary.

For convenience, we use polar coordinates; and in this case Eq. (o) must be replaced by the following equation:

$$\Delta U_{\max} \approx \frac{S}{2}\int_0^R \left(\frac{\partial Z}{\partial r}\right)^2 2\pi r\,dr \tag{u}$$

Furthermore, instead of Eq. (p) we have

$$T_{\max} = \frac{\rho h}{2}\omega^2 \int_0^R Z^2 2\pi r\,dr \tag{v}$$

and Eq. (s) is replaced by

$$\frac{\partial}{\partial a_j}\int_0^R \left\{\left(\frac{\partial Z}{\partial r}\right)^2 - \frac{\omega^2 \rho h}{S}Z^2\right\} 2\pi r\,dr = 0 \tag{w}$$

By taking only the first term in the series (t) and substituting $Z = a_1 \times \cos \pi r/2R$ into Eq. (w), we obtain

$$\frac{\pi^2}{4R^2}\int_0^R \sin^2\frac{\pi r}{2R}\,r\,dr = \frac{\omega^2 \rho h}{S}\int_0^R \cos^2\frac{\pi r}{2R}\,r\,dr$$

from which

$$\frac{\pi^2}{4R^2}\left(\frac{1}{2}+\frac{2}{\pi^2}\right)=\frac{\omega^2\rho h}{S}\left(\frac{1}{2}+\frac{2}{\pi^2}\right)$$

or

$$\omega=\frac{2.415}{R}\sqrt{\frac{S}{\rho h}}$$

The exact solution [1] for this case is

$$\omega=\frac{2.404}{R}\sqrt{\frac{S}{\rho h}} \tag{5.171}$$

The error of the first approximation is less than 0.5%.

In order to obtain a better approximation for the fundamental mode and also for the frequencies of the higher modes of vibration, we must take a larger number of terms in the series (t). These higher modes of vibration will have one, two, three, . . . *nodal circles,* at which the displacements w are zero during vibration.

In addition to the modes of vibration that are symmetric with respect to the center, a circular membrane may also have modes in which one, two, three, . . . diameters of the circle are *nodal diameters,* along which the deflections during vibration are zero. Several modes of vibration of a circular membrane are shown in Fig. 5.38, where the nodal circles and nodal diameters are indicated by dashed lines.

In all cases the quantity ω_{ns}, determining the frequencies, can be

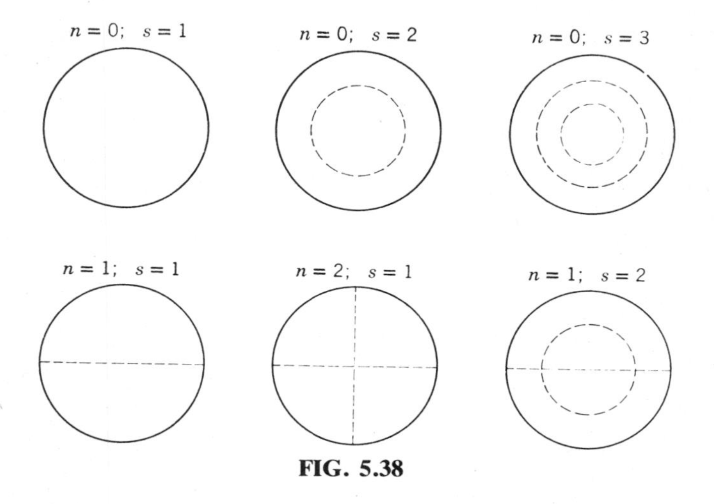

FIG. 5.38

TABLE 5.1

s	$n=0$	$n=1$	$n=2$	$n=3$	$n=4$	$n=5$
1	2.404	3.832	5.135	6.379	7.586	8.780
2	5.520	7.016	8.417	9.760	11.064	12.339
3	8.654	10.173	11.620	13.017	14.373	15.700
4	11.792	13.323	14.796	16.224	17.616	18.982
5	14.931	16.470	17.960	19.410	20.827	22.220
6	18.071	19.616	21.117	22.583	24.018	25.431
7	21.212	22.760	24.270	25.749	27.200	28.628
8	24.353	25.903	27.421	28.909	30.371	31.813

expressed as

$$\omega_{ns} = \frac{\alpha_{ns}}{R}\sqrt{\frac{S}{\rho h}} \tag{5.172}$$

The constants α_{ns} for this formula are given in Table 5.1,† in which n denotes the number of nodal diameters and s the number of nodal circles. (The boundary circle is included in the latter number.)

Membranes with Other Shapes

In the preceding discussion it was assumed that the membrane has a complete circular area and that it is fixed only at the circular boundary; but the results already obtained also include the solution of other problems, such as membranes bounded by two concentric circles and two radii, or membranes in the form of sectors. For example, consider a semicircular membrane. All possible modes of vibration of this membrane will be included among the modes that the circular membrane may perform. It is only necessary to consider one of the nodal diameters of the circular membrane as a fixed boundary.

When the boundary of a membrane is approximately circular, its lowest mode is nearly the same as that of a circular membrane having the same area and the same value of $S/\rho h$. In general, the equation determining the frequency of the fundamental mode of vibration of a membrane may be taken in the form

$$\omega = \alpha\sqrt{\frac{S}{\rho h A}} \tag{5.173}$$

where A is the area of the membrane. The constant α in this equation will

† Values in Table 5.1 were calculated by Bourget, *Ann. l'école normale,* **3,** 1866.

be given by the following list, which shows the effects of departures from the crcular form. [1]

Circle	$\alpha = 2.404\sqrt{\pi}$	$= 4.261$
Square	$\alpha = \pi\sqrt{2}$	$= 4.443$
Quadrant of a circle	$\alpha = (5.135/2)\sqrt{\pi}$	$= 4.551$
60° Sector of a circle	$\alpha = 6.379\sqrt{\pi/6}$	$= 4.616$
Rectangle ($a/b = 3/2$)	$\alpha = \pi\sqrt{13/6}$	$= 4.624$
Equilateral triangle	$\alpha = 2\pi\sqrt{\tan 30°}$	$= 4.774$
Semicircle	$\alpha = 3.832\sqrt{\pi/2}$	$= 4.803$
Rectangle ($a/b = 2/1$)	$\alpha = \pi\sqrt{5/2}$	$= 4.967$
Rectangle ($a/b = 3/1$)	$\alpha = \pi\sqrt{10/3}$	$= 5.736$

In cases where the boundary is different from those discussed above, the investigation of vibrations presents greater mathematical difficulties. However, the case of an elliptical boundary has been completely solved.†

5.24 TRANSVERSE VIBRATIONS OF PLATES

Figure 5.39*a* shows a plate with a uniform thickness h that is assumed to be small in comparison with its other dimensions. We take the x–y plane as the *middle plane* of the plate and assume that deflections in the z direction are small in comparison with the thickness h. In addition, the normals to the middle plane of the plate are assumed to remain normal to the deflected *middle surface* during vibrations.

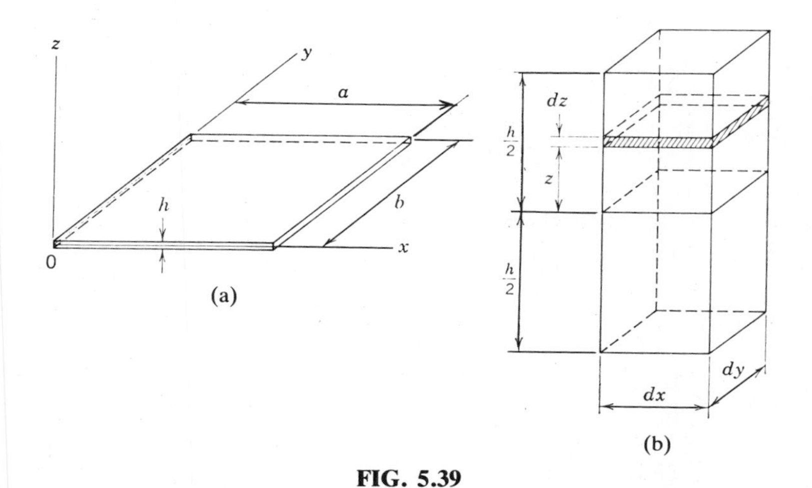

FIG. 5.39

† E. Mathieu, *J. Math.* (*Liouville*), **13,** 1868.

Let us consider the strains in a thin layer of a typical element, as indicated by the shaded area at the distance z from the middle surface in Fig. 5.39*b*. These strains will be represented by the following equations:†

$$\epsilon_x = -z \frac{\partial^2 w}{\partial x^2} \qquad \epsilon_y = -z \frac{\partial^2 w}{\partial y^2} \qquad \gamma_{xy} = -2z \frac{\partial^2 w}{\partial x \, \partial y} \tag{a}$$

In these expressions w denotes the deflection of the plate in the z direction; and ϵ_x, ϵ_y, and γ_{xy} are the normal strains and the shear strain in the thin layer. The stresses corresponding to these strains are determined by the known relationships [6]

$$\begin{aligned} \sigma_x &= \frac{E}{1-\nu^2}(\epsilon_x + \nu\epsilon_y) = -\frac{Ez}{1-\nu^2}\left(\frac{\partial^2 w}{\partial x^2} + \nu \frac{\partial^2 w}{\partial y^2}\right) \\ \sigma_y &= \frac{E}{1-\nu^2}(\epsilon_y + \nu\epsilon_x) = -\frac{Ez}{1-\nu^2}\left(\frac{\partial^2 w}{\partial y^2} + \nu \frac{\partial^2 w}{\partial x^2}\right) \\ \tau_{xy} &= G\gamma_{xy} = -\frac{Ez}{(1+\nu)} \frac{\partial^2 w}{\partial x \, \partial y} \end{aligned} \tag{b}$$

in which ν denotes Poisson's ratio.

The potential energy accumulated in the shaded layer of the element during deformation will be

$$dU = \left(\frac{\epsilon_x \sigma_x}{2} + \frac{\epsilon_y \sigma_y}{2} + \frac{\gamma_{xy}\tau_{xy}}{2}\right) dx \, dy \, dz$$

Substitution of Eqs. (a) and (b) into this expression yields

$$dU = \frac{Ez^2}{2(1-\nu^2)} \left\{ \left(\frac{\partial^2 w}{\partial x^2}\right)^2 + \left(\frac{\partial^2 w}{\partial y^2}\right)^2 + 2\nu \frac{\partial^2 w}{\partial x^2} \frac{\partial^2 w}{\partial y^2} + 2(1-\nu)\left(\frac{\partial^2 w}{\partial x \, \partial y}\right)^2 \right\} dx \, dy \, dz \tag{c}$$

Integrating Eq. (c) over the volume of the plate, we obtain the potential energy of bending as

$$U = \iiint dU = \frac{D}{2} \iint \left\{ \left(\frac{\partial^2 w}{\partial x^2}\right)^2 + \left(\frac{\partial^2 w}{\partial y^2}\right)^2 + 2\nu \frac{\partial^2 w}{\partial x^2} \frac{\partial^2 w}{\partial y^2} + 2(1-\nu)\left(\frac{\partial^2 w}{\partial x \, \partial y}\right)^2 \right\} dx \, dy \tag{5.174}$$

where $D = Eh^3/[12(1-\nu^2)]$ is the *flexural rigidity* of the plate.

† It is assumed that there is no stretching of the middle plane.

The kinetic energy of a transversely vibrating plate will be

$$T = \frac{\rho h}{2} \iint \dot{w}^2 \, dx \, dy \tag{5.175}$$

where ρh is the mass per unit area. These expressions for U and T will now be applied to particular types of plates.

Rectangular Plates

In the case of a rectangular plate (see Fig. 5.39a) with simply supported edges, we can proceed as in the case of a rectangular membrane and take the deflection of the plate during vibtation as the double series

$$w = \sum_{m=1}^{\infty} \sum_{n=1}^{\infty} \phi_{mn} \sin \frac{m\pi x}{a} \sin \frac{n\pi y}{b} \tag{d}$$

which are the normal functions for the case under consideration. It is easy to see that each term of this series satisfies the conditions at the edges, which require that $w = \partial^2 w/\partial x^2 = 0$ at $x = 0$ and $x = a$; and $w = \partial^2 w/\partial y^2 = 0$ at $y = 0$ and $y = b$. If Eq. (d) is substituted into Eq. (5.174), the following expression for the potential energy will be obtained:

$$U = \frac{\pi^4 ab}{8} D \sum_{m=1}^{\infty} \sum_{n=1}^{\infty} \phi_{mn}^2 \left(\frac{m^2}{a^2} + \frac{n^2}{b^2} \right)^2 \tag{5.176}$$

In addition, the kinetic energy (Eq. 5.175) becomes

$$T = \frac{\rho hab}{8} \sum_{m=1}^{\infty} \sum_{n=1}^{\infty} \dot{\phi}_{mn}^2 \tag{5.177}$$

The inertial force for a typical element of the plate is $-\rho h \ddot{w} \, dx \, dy$. Proceeding as before and considering a virtual displacement

$$\delta w_{mn} = \delta \phi_{mn} \sin \frac{m\pi x}{a} \sin \frac{n\pi y}{b}$$

we obtain the differential equation of motion for free vibrations in principal coordinates to be

$$\rho h \ddot{\phi}_{mn} + \pi^4 D \phi_{mn} \left(\frac{m^2}{a^2} + \frac{n^2}{b^2} \right)^2 = 0$$

The solution of this equation is

$$\phi_{mn} = C_1 \cos \omega t + C_2 \sin \omega t$$

where

$$\omega = \pi^3 \sqrt{\frac{D}{\rho h}} \left(\frac{m^2}{a^2} + \frac{n^2}{b^2} \right) \tag{5.178}$$

From this formula the frequencies of vibration can be easily calculated. For example, in the case of a square plate, we obtain for the lowest mode of vibration

$$f_1 = \frac{\omega_1}{2\pi} = \frac{\pi}{a^2} \sqrt{\frac{D}{\rho h}} \tag{5.179}$$

In considering higher modes of vibration and their nodal lines, we see that the discussion previously given for the vibrations of a square membrane (see Fig. 5.36) apply equally well to a square plate. In addition, the case of forced vibrations of a rectangular plate with simply supported edges can be solved without difficulty. It should be noted that the vibrations of a rectangular plate, for which two opposite edges are supported while the other two edges are free or clamped, can also be analyzed without great mathematical difficulty.†

However, vibrational analyses of plates with all edges free or clamped are much more complicated. For the solution of these problems, Ritz's first method has been found to be very useful.‡ To apply this technique, we assume

$$w = Z \cos(\omega t - \alpha) \tag{e}$$

where Z is a function of x and y that approximates the mode of vibration. Substituting Eq. (e) into Eqs. (5.174) and (5.175), we obtain the following expressions for the maximum potential and kinetic energies of vibration:

$$U_{\max} = \frac{D}{2} \iint \left\{ \left(\frac{\partial^2 Z}{\partial x^2} \right)^2 + \left(\frac{\partial^2 Z}{\partial y^2} \right)^2 + 2\nu \frac{\partial^2 Z}{\partial x^2} \frac{\partial^2 Z}{\partial y^2} + 2(1-\nu) \left(\frac{\partial^2 Z}{\partial x \, \partial y} \right)^2 \right\} dx\, dy$$

$$T_{\max} = \frac{\rho h}{2} \omega^2 \iint Z^2 \, dx\, dy$$

† See W. Voigt, *Nachr. Ges. Wiss. Göttinger,* 1893, p. 225.

‡ See W. Ritz, *Ann. Physik,* **28,** 1909, p. 737. The accuracy of the Ritz method has been discussed by S. Tomatika, *Phil. Mag.*, ser. 7, **21,** 1936, p. 745. See also A. Weinstein and W. Z. Chien, *Quart. Appl. Math.*, **1,** 1943, p. 61.

Equating these expressions and solving for ω^2, we find

$$\omega^2 = \frac{2}{\rho h} \frac{U_{\max}}{\iint Z^2\,dx\,dy} \tag{5.180}$$

Now we take the function Z in the form of a series

$$Z = a_1\Phi_1(x, y) + a_2\Phi_2(x, y) + a_3\Phi_3(x, y) + \cdots \tag{f}$$

each term of which satisfies the conditions at the boundary of the plate. It is then necessary to determine the coefficients $a_1, a_2, a_3, \ldots$ in such a manner as to make the result of Eq. (5.180) a minimum. In this way we arrive at a system of equations of the type

$$\frac{\partial}{\partial a_j} \iint \left\{ \left(\frac{\partial^2 Z}{\partial x^2}\right)^2 + \left(\frac{\partial^2 Z}{\partial y^2}\right)^2 + 2\nu \frac{\partial^2 Z}{\partial x^2}\frac{\partial^2 Z}{\partial y^2} + 2(1-\nu)\left(\frac{\partial^2 Z}{\partial x\,\partial y}\right)^2 - \frac{\omega^2 \rho h}{D} Z^2 \right\} dx\,dy = 0 \tag{5.181}$$

which will be linear with respect to the constants $a_1, a_2, a_3, \ldots$. By equating to zero the determinant of the coefficients in these equations, we find the frequency equation for the plate.

W. Ritz applied this method for studying the vibrations of a square plate with free edges.† The series (f) was taken in this case to be of the form

$$Z = \sum_m \sum_n a_{mn} X_m(x) Y_n(y) \tag{f'}$$

where $X_m(x)$ and $Y_n(y)$ are the normal functions of vibration for a prismatic beam with free ends (see Sec. 5.11). The frequencies of the various modes of vibration can be determined by the formula

$$\omega = \frac{\alpha}{a^2}\sqrt{\frac{D}{\rho h}} \tag{5.182}$$

in which α is a constant depending on the mode. For the three lowest modes the values of this constant are‡

$$\alpha_1 = 14.10 \qquad \alpha_2 = 20.56 \qquad \alpha_3 = 23.91$$

† *Ibid.* Applications of the Ritz method for several other edge conditions were made by Dana Young, *Jour. Appl. Mech.*, **17,** 1950, p. 448.

‡ Poisson's ratio is taken equal to 0.225.

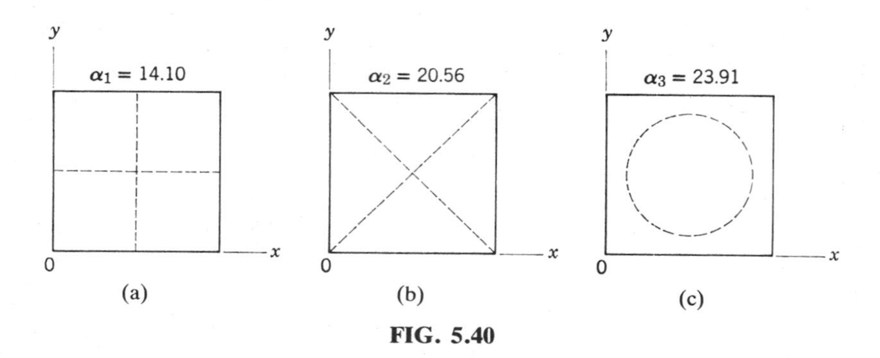

FIG. 5.40

Nodal lines for the corresponding modes of vibration are indicated in Figs. 5.40*a*, *b*, and *c*.

Circular Plates

The problem of vibration of a circular plate was solved by G. R. Kirchhoff,† who calculated the frequencies of several modes of vibration for a plate with a free boundary. The exact solution of this problem involves the use of Bessel functions. In the following, an approximate solution is developed by means of the Ritz method, which usually gives for the lowest mode an accuracy that is sufficient for practical applications. To apply this method it will be useful to transform the expressions (5.174) and (5.175) for the potential and kinetic energies to polar coordinates.

In Fig. 5.41, we see from the elemental triangle ABC that a small increase dx in the coordinate x corresponds to

$$dr = dx \cos \theta \qquad d\theta = -\frac{dx \sin \theta}{r}$$

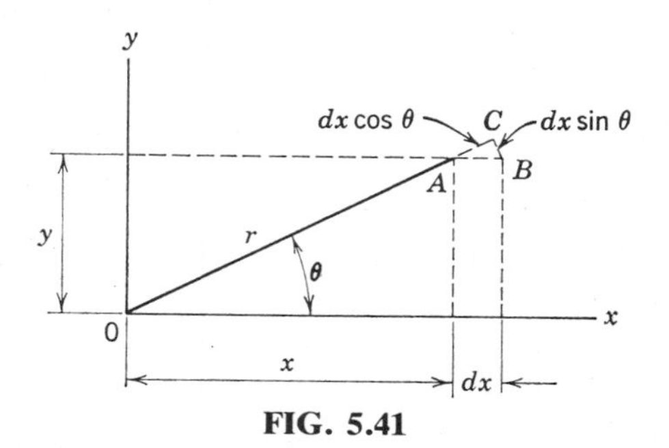

FIG. 5.41

† See *Jour. Math.* (*Crelle*), **40,** 1850; *Gesammelte Abhandlungen*, Leipzig, 1882, p. 237; or *Vorlesungen über mathematische Physik Mechanik*, Leipzig, *Vorlesung* 30, 1876.

Then, considering the deflection w as a function of r and θ, we obtain

$$\frac{\partial w}{\partial x} = \frac{\partial w}{\partial r}\frac{\partial r}{\partial x} + \frac{\partial w}{\partial \theta}\frac{\partial \theta}{\partial x} = \frac{\partial w}{\partial r}\cos\theta - \frac{\partial w}{\partial \theta}\frac{\sin\theta}{r}$$

In a similar manner we also find

$$\frac{\partial w}{\partial y} = \frac{\partial w}{\partial r}\sin\theta + \frac{\partial w}{\partial \theta}\frac{\cos\theta}{r}$$

Repeating the differentiation gives us

$$\frac{\partial^2 w}{\partial x^2} = \frac{\partial^2 w}{\partial r^2}\cos^2\theta - 2\frac{\partial^2 w}{\partial \theta\,\partial r}\frac{\sin\theta\cos\theta}{r} + \frac{\partial w}{\partial r}\frac{\sin^2\theta}{r} + 2\frac{\partial w}{\partial \theta}\frac{\sin\theta\cos\theta}{r^2} + \frac{\partial^2 w}{\partial \theta^2}\frac{\sin^2\theta}{r^2} \tag{g}$$

$$\frac{\partial^2 w}{\partial y^2} = \frac{\partial^2 w}{\partial r^2}\sin^2\theta + 2\frac{\partial^2 w}{\partial \theta\,\partial r}\frac{\sin\theta\cos\theta}{r} + \frac{\partial w}{\partial r}\frac{\cos^2\theta}{r} - 2\frac{\partial w}{\partial \theta}\frac{\sin\theta\cos\theta}{r^2} + \frac{\partial^2 w}{\partial \theta^2}\frac{\cos^2\theta}{r^2} \tag{h}$$

$$\frac{\partial^2 w}{\partial x\,\partial y} = \frac{\partial^2 w}{\partial r^2}\sin\theta\cos\theta + \frac{\partial^2 w}{\partial r\,\partial \theta}\frac{\cos 2\theta}{r} - \frac{\partial w}{\partial \theta}\frac{\cos 2\theta}{r^2} - \frac{\partial w}{\partial r}\frac{\sin\theta\cos\theta}{r} - \frac{\partial^2 w}{\partial \theta^2}\frac{\sin\theta\cos\theta}{r^2} \tag{i}$$

from which we find

$$\frac{\partial^2 w}{\partial x^2} + \frac{\partial^2 w}{\partial y^2} = \frac{\partial^2 w}{\partial r^2} + \frac{1}{r}\frac{\partial w}{\partial r} + \frac{1}{r^2}\frac{\partial^2 w}{\partial \theta^2} \tag{j}$$

and

$$\frac{\partial^2 w}{\partial x^2}\frac{\partial^2 w}{\partial y^2} - \left(\frac{\partial^2 w}{\partial x\,\partial y}\right)^2 = \frac{\partial^2 w}{\partial r^2}\left(\frac{1}{r}\frac{\partial w}{\partial r} + \frac{1}{r^2}\frac{\partial^2 w}{\partial \theta^2}\right) - \left\{\frac{\partial}{\partial r}\left(\frac{1}{r}\frac{\partial w}{\partial \theta}\right)\right\}^2 \tag{k}$$

Substituting Eqs. (j) and (k) into Eq. (5.174) and taking the origin at the center of the plate, we obtain

$$U = \frac{D}{2}\int_0^{2\pi}\int_0^R \left[\left(\frac{\partial^2 w}{\partial r^2} + \frac{1}{r}\frac{\partial w}{\partial r} + \frac{1}{r^2}\frac{\partial^2 w}{\partial \theta^2}\right)^2 - 2(1-\nu)\frac{\partial^2 w}{\partial r^2}\left(\frac{1}{r}\frac{\partial w}{\partial r} + \frac{1}{r^2}\frac{\partial^2 w}{\partial \theta^2}\right) + 2(1-\nu)\left\{\frac{\partial}{\partial r}\left(\frac{1}{r}\frac{\partial w}{\partial \theta}\right)\right\}^2\right] r\,d\theta\,dr \tag{5.183}$$

where R denotes the radius of the plate. When the deflection of the plate is symmetric about the center, w will be a function of r only, and Eq. (5.183) becomes

$$U = \pi D \int_0^R \left\{ \left(\frac{d^2w}{dr^2} + \frac{1}{r}\frac{dw}{dr} \right)^2 - 2(1-\nu)\frac{d^2w}{dr^2}\frac{1}{r}\frac{dw}{dr} \right\} r\, dr \tag{5.184}$$

In the case of a plate that is fixed at the edge, the integral

$$\iint \left[\frac{\partial^2 w}{\partial r^2}\left(\frac{1}{r}\frac{\partial w}{\partial r} + \frac{1}{r^2}\frac{\partial^2 w}{\partial \theta^2} \right) - \left\{ \frac{\partial}{\partial r}\left(\frac{1}{r}\frac{\partial w}{\partial \theta} \right) \right\}^2 \right] r\, d\theta\, dr$$

vanishes, and we obtain from Eq. (5.183)

$$U = \frac{D}{2}\int_0^{2\pi}\int_0^R \left(\frac{\partial^2 w}{\partial r^2} + \frac{1}{r}\frac{\partial w}{\partial r} + \frac{1}{r^2}\frac{\partial^2 w}{\partial \theta^2} \right)^2 r\, d\theta\, dr \tag{5.185}$$

If the deflection of such a plate is symmetric about the center, we have

$$U = \pi D \int_0^R \left(\frac{\partial^2 w}{\partial r^2} + \frac{1}{r}\frac{\partial w}{\partial r} \right)^2 r\, dr \tag{5.186}$$

The expression for the kinetic energy in polar coordinates will be

$$T = \frac{\rho h}{2}\int_0^{2\pi}\int_0^R \dot{w}^2\, r\, d\theta\, dr \tag{5.187}$$

and in symmetric cases we have

$$T = \pi \rho h \int_0^R \dot{w}^2 r\, dr \tag{5.188}$$

By using these expressions for the potential and kinetic energies, we can calculate the frequencies of the natural modes of vibration of a circular plate for various boundary conditions.†

Circular Plate Fixed at the Boundary

The problem of a circular plate fixed at the edge is of particular interest in connection with the application to telephone receivers and similar devices. In using the Ritz method we assume that the solution takes the form of Eq. (*e*), but with Z expressed as a function of r and θ. For the lowest mode of

† Forced vibrations of circular plates were studied by W. Flügge, *Z. tech. Phys.*, **13**, 1932, p. 139.

vibration, the shape of the vibrating plate is symmetric about the center, and Z will be a function of r only. If Z is taken as the series

$$Z = a_1\left(1 - \frac{r^2}{R^2}\right)^2 + a_2\left(1 - \frac{r^2}{R^2}\right)^2 + \cdots \tag{l}$$

the condition of symmetry will be satisfied. Conditions at the boundary will also be satisfied because each term of the series (l), together with its first derivative with respect to r, vanishes when $r = R$.

The equation for minimization, corresponding to Eq. (5.181), in the case under consideration becomes

$$\frac{\partial}{\partial a_j}\int_0^R \left\{\left(\frac{d^2Z}{dr^2} + \frac{1}{r}\frac{dZ}{dr}\right)^2 - \frac{\omega^2 \rho h}{D} Z^2\right\} r\, dr = 0 \tag{5.189}$$

By taking only one term of the series (l) and substituting it into Eq. (5.189), we obtain

$$\frac{96}{9R^2} - \frac{\omega^2 \rho h}{D}\frac{R^2}{10} = 0$$

from which

$$\omega = \frac{10.33}{R^2}\sqrt{\frac{D}{\rho h}} \tag{5.190}$$

In order to get a closer approximation, we take the first two terms of the series (l). Then we have

$$\int_0^R \left(\frac{d^2Z}{dr^2} + \frac{1}{r}\frac{dZ}{dr}\right)^2 r\, dr = \frac{96}{9R^2}\left(a_1^2 + \frac{3}{2}a_1 a_2 + \frac{9}{10}a_2^2\right)$$

and

$$\int_0^R Z^2 r\, dr = \frac{R^2}{10}\left(a_1^2 + \frac{5}{3}a_1 a_2 + \frac{5}{7}a_2^2\right)$$

Equation (5.189) yields

$$a_1\left(\frac{192}{2} - \frac{\lambda}{5}\right) + a_2\left(\frac{144}{9} - \frac{\lambda}{6}\right) = 0$$

and

$$a_1\left(\frac{144}{9} - \frac{\lambda}{6}\right) + a_2\left(\frac{96}{5} - \frac{\lambda}{7}\right) = 0 \tag{m}$$

where

$$\lambda = R^4 \omega^2 \frac{\rho h}{D} \tag{n}$$

Equating to zero the determinant of the coefficients in Eqs. (m), we obtain

$$\lambda^2 - \frac{(204)(48)}{5}\lambda + (768)(36)(7) = 0$$

from which

$$\lambda_1 = 104.3 \qquad \lambda_2 = 1854$$

Substituting these values into Eq. (n), we find

$$\omega_1 = \frac{10.21}{R^2}\sqrt{\frac{D}{\rho h}} \qquad \omega_2 = \frac{43.06}{R^2}\sqrt{\frac{D}{\rho h}} \tag{5.191}$$

Thus, ω_1 gives an improved approximation to the frequency of the lowest mode of vibration of the plate; and ω_2 provides a rough approximation to the frequency of the second mode of vibration, in which the vibrating plate has one nodal circle. By this approach the modes of vibration having nodal diameters can also be investigated.

In all cases the frequency of vibration is determined by the equation

$$\omega = \frac{\alpha}{R^2}\sqrt{\frac{D}{\rho h}} \tag{5.192}$$

Some values of the constant α (for a given number n of nodal diameters and a given number s of nodal circles) are listed in Table 5.2

If a plate is immersed in a fluid, its natural frequencies may be considerably altered. In order to take the mass of the fluid into account for the fundamental mode of vibration, we should replace expression (5.192) by the formula†

$$\omega_1 = \frac{10.21}{R^2\sqrt{1+\eta}}\sqrt{\frac{D}{\rho h}} \tag{5.193}$$

in which

$$\eta = 0.6689\frac{\rho_1}{\rho}\frac{R}{h}$$

TABLE 5.2

s	$n = 0$	$n = 1$	$n = 2$
0	10.21	21.22	34.84
1	39.78	—	—
2	88.90	—	—

† H. Lamb, *Proc. Roy. Soc. (London)*, **98,** 1921, p. 205.

TABLE 5.3

s	$n=0$	$n=1$	$n=2$	$n=3$
0	—	—	5.251	12.23
1	9.076	20.52	35.24	52.91
2	38.52	59.86	—	—

TABLE 5.4

s	0	1	2	3
α	3.75	20.91	60.68	119.7

and ρ_1/ρ is the ratio of the mass density of the fluid to the mass density of the material of the plate. For example, consider a circular steel plate that is clamped at the edge and immersed in water. If $R = 3.5$ in. and $h = 0.125$ in., the value of η becomes

$$\eta = 0.6689\left(\frac{1}{7.8}\right)\left(\frac{3.5}{0.125}\right) = 2.40$$

Then $1/\sqrt{1+\eta} = 0.542$, which indicates that the frequency of the lowest mode of vibration will be reduced to 0.542 times its original value.

Circular Plate with Other Boundary Conditions

In all cases, the frequencies of a vibrating circular plate can be calculated from Eq. (5.192) by changing the value of the constant α. For a free circular plate with n nodal diameters and s nodal circles, α has the values listed in Table 5.3†

For a circular plate with its center fixed and having s nodal circles, values of α are given in Table 5.4.‡ The frequencies of modes having nodal diameters will be the same as in the case of a free plate.

For more comprehensive tables on characteristics of plates, see Appendix A in the book by Szilard [7]. His work covers plates with various shapes, loads, and boundary conditions.

REFERENCES

[1] Rayleigh, J. W. S., *Theory of Sound,* 2nd edn., Vol. 1, Dover, New York, 1945.

[2] Love, A. E. H., *Mathematical Theory of Elasticity,* 4th edn., Cambridge University Press, London, 1934.

† Poisson's ratio is taken equal to 1/3.

‡ See R. V. Southwell, *Proc. Roy. Soc.* (*London*), **101,** 1922, p. 133. Poisson's ratio is taken equal to 0.3.

[3] Gere, J. M., and Timoshenko, S. P., *Mechanics of Materials,* 3rd edn., PWS-Kent, Boston, MA, 1990.

[4] Timoshenko, S. P., *Strength of Materials,* 3rd edn., Vol. 2, Van Nostrand, Princeton, NJ, 1955.

[5] Greenwood, D. T., *Classical Dynamics,* Prentice-Hall, Englewood Cliffs, NJ, 1977.

[6] Timoshenko, S. P., and Goodier, J. N., *Theory of Elasticity,* 3rd edn., McGraw-Hill, New York, 1970.

[7] Szilard, R., *Theory and Analysis of Plates,* Prentice-Hall, Englewood Cliffs, NJ, 1974.

PROBLEMS

5.2-1. Determine the general expression for longitudinal vibrations of a bar that is free at the end where $x = 0$ and fixed at the end where $x = \ell$.

5.2-2. A bar moving along the x axis with constant velocity v is suddenly stopped at the midpoint, where $x = \ell/2$. Find the expression for the resulting free vibrations.

5.2-3. Suppose that the initial force P_0 in Example 2 acts at the quarter point ($x = \ell/4$) of the bar instead of at the midpoint. In addition, assume that an equal and opposite force $-P_0$ acts at the three-quarter point ($x = 3\ell/4$). What vibrations result when these forces are suddenly removed?

5.2-4. Let the initial force P_0 in Fig. 5.2*a* be uniformly distributed along the length of the bar (intensity of load $= P_0/\ell$). Determine the response of the bar due to suddenly removing this distributed load.

5.3-1. Suppose that a constant axial force P is suddenly applied at the midpoint of the fixed-end bar in Fig. 5.3*a*. Determine the longitudinal response of the bar, starting from rest.

5.3-2. A bar with free ends (see Fig. 5.1*a*) is subjected to an axial ramp force $P = P_1 t/t_1$ applied at the end $x = 0$. Assuming that the bar is initialy at rest, determine its longitudinal response.

5.3-3. Find the steady-state forced vibrations of a bar that is fixed at the end $x = 0$ and free at the end $x = \ell$ (see Fig. 5.2*a*) due to a uniformly distributed axial force $(P_1/\ell) \sin \Omega t$ applied along its length.

5.3-4. Consider a bar that is free at the end $x = 0$ and fixed at the end $x = \ell$. Determine its response to a constant axial force P suddenly applied at the midpoint ($x = \ell/2$).

5.4-1. Assume that the initial force P_0 in Fig. 5.3*a* acts at the third point ($x = \ell/3$) instead of at the midpoint of the bar. Using the normal-mode approach, find the free-vibrations due to the sudden removal of this force.

5.4-2. Suppose that the right-hand half of the bar in Fig. 5.2*a* has an initial velocity v in the axial direction at time $t = 0$. Determine the free-vibrations of the bar due to this condition.

5.4-3. A bar with fixed ends is suddenly subjected to a distributed axial load with an intensity that varies linearly from zero at $x = 0$ to Q at $x = \ell$. Find the longitudinal response of this bar by the normal-mode method.

5.4-4. Determine the longitudinal response of a bar with free ends due to an axial force $P = P_1(t/t_1)^2$ applied at its midpoint ($x = \ell/2$).

5.6-1. A bar that is free at the end $x = 0$ and fixed at the end $x = \ell$ is subjected to the harmonic support motion $u_g = g(t) = d \sin \Omega t$, where d is the amplitude of oscillation. Determine the steady-state forced vibrations of the bar due to this motion.

5.6-2. Find the longitudinal response of a fixed-end bar caused by the rigid-body support motion $u_g = g(t) = u_1(t/t_1)^2$.

5.6-3. A fixed-end bar is subjected to the support motion $u_{g1} = g_1(t) = u_1(t/t_1)^2$ at the end $x = 0$, while the end $x = \ell$ remains stationary. Determine the longitudinal response of the bar due to these conditions.

5.6-4. Suppose that the end $x = \ell$ of the fixed-end bar in Prob. 5.6-3 also moves and that this motion is $u_{g2} = g_2(t) = u_2(t/t_2)^3$. Find the additional response of the bar to be added to that from Prob. 5.6-3.

5.10-1. A simply supported beam is deflected by a force P applied at the middle. What vibrations of the beam will occur if the load P is suddenly removed?

5.10-2. Solve the preceding problem, assuming that the force P is applied at the location $x = x_1$.

5.10-3. A simply supported beam carries a uniform load of intensity w. Find the vibrations that result when the load is suddenly removed.

5.10-4. Determine the vibrations of a simply supported beam to which a transverse velocity $\dot{v}_0$ is imparted (at time $t = 0$) at all points except the ends.

5.11-1. Determine numerically the normal functions for a beam with one end fixed and the other simply supported, and plot the deflection curves for the first and second modes of vibration

5.11-2. Solve the preceding problem, assuming that the end $x = 0$ of the beam is fixed and that the end $x = \ell$ is free.

5.11-3. Find the expression for free vibrations of a beam with fixed ends, assuming that it is bent by a concentrated force P_0 applied at $x = \ell/4$ and that at the instant $t = 0$ this force is suddenly removed.

5.11-4. Solve the preceding problem for a beam, one end of which is fixed and the other free. Initial deflection is produced by a load P_0 applied at the free end.

5.13-1. A simply supported beam under the action of a force P, applied at the middle, deflects 1 mm at the point of application. What will be the amplitude of forced vibration produced by a harmonic force $P \sin \Omega t$, applied at the middle, if the frequency Ω is equal to half of the fundamental frequency of the beam?

5.13-2. The beam of the preceding problem is under the action of two harmonic forces $P \sin \Omega t$ applied at the third points. Find the amplitude of forced vibration at the middle if Ω has the same magnitude as in the preceding problem.

5.13-3. Find the amplitude of forced vibration produced at the middle of a simply supported beam by a distributed harmonic load of intensity $w \sin \Omega t$ covering the left-hand half of the span.

5.13-4. Determine the response of a simply supported beam due to a force P suddenly applied at the middle of the span.

5.13-5. Determine the steady-state response produced in a simply supported beam by a sinusoidally distributed harmonic load $Q(x, t) = w \sin(\pi x/\ell) \sin \Omega t$.

5.13-6. By the method of virtual work, derive the general expression for the response of a beam to a distributed moment $M(x, t)$ per unit length. Then give the solution for a simply supported beam.

5.14-1. Determine the steady-state forced vibrations of a cantilever beam with a harmonic force $P \sin \Omega t$ acting at the free end.

5.14-2. Evaluate the response of the free end of the cantilever beam in the preceding problem. From the tables, find

$$(X_i)_{x=\ell} = 2(-1)^{i+1}$$

Then take the values of $k_i\ell$ from Sec. 5.11.

5.14-3. Determine the steady-state forced vibrations of a beam with one end fixed and the other simply supported if a harmonic force $P \sin \Omega t$ is applied at the middle of the span.

5.14-4. Calculate the response at the middle for the beam of the preceding problem by using the following numerical values:

$$(X_1)_{x=\ell/2} = 1.4449 \qquad (X_2)_{x=\ell/2} = 0.5704$$

$$(X_3)_{x=\ell/2} = -1.3005 \qquad (X_4)_{x=\ell/2} = -0.5399 \qquad (X_5)_{x=\ell/2} = 1.3068$$

The values of $k_i\ell$ for this beam are given in Sec. 5.11.

5.15-1. Determine the steady-state response of a simply supported beam subjected to the harmonic motion $g_1(t) = v_1 \sin \Omega t$ of its left-hand support (see Fig. 5.21*a*).

5.15-2. Suppose that the left-hand end of the cantilever beam in Fig. 5.23*b* rotates in accordance with the function $g_2(t) = \theta_2(t/t_2)^2$, where θ_2 is a small angle. Write the general expression for the response of the beam in terms of X_i and ω_i.

5.15-3. Write the general expression for the steady-state response of a fixed-end beam subjected to the harmonic rotation $g_4 = \theta_4 \sin \Omega t$ of its right-hand support (see Fig. 5.22*d*), assuming that θ_4 is a small angle.

5.15-4. Write the general expression for the response of a propped cantilever subjected to the translation $g_3(t) = v_3(t/t_3)^3$ of its right-hand support (see Fig. 5.24*c*).

CHAPTER 6

FINITE-ELEMENT METHOD FOR DISCRETIZED CONTINUA

6.1 INTRODUCTION

The classical solutions discussed in the preceding chapter are feasible only because they are geometrically simple. However, practical problems of framed structures, two- and three-dimensional solids, plates, and shells may have arbitrary shapes and boundary conditions. For such cases we must have a good numerical technique to handle any type of geometry that may be encountered.

In this chapter we introduce the method of *finite elements* [1–3] to discretize elastic continua for dynamic analysis. The basic concept is to divide a continuum into subregions having simpler geometries than the original problem. Each subregion (or finite element) is of finite size (not infinitesimal) and has a number of key points, called *nodes,* that control the behavior of the element. By making the displacements or stresses at any point in an element dependent upon those at the nodes, we need only write a finite number of differential equations of motion for such nodes. This approach enables us to convert a problem with an infinite number of degrees of freedom to one with a finite number, thereby simplifying the solution process. For good accuracy in the solution, the number of nodal degrees of freedom must usually be fairly large, and the details of element formulations are rather complicated. Therefore, it becomes necessary to program this method on a digital computer.

Figure 6.1 shows various examples of solids and structures that are discretized by finite elements, with dots indicating the nodes. In Fig. 6.1*a* we see a continuous beam that is divided into several flexural elements of the type to be described in Sec. 6.4. The space frame with curved members

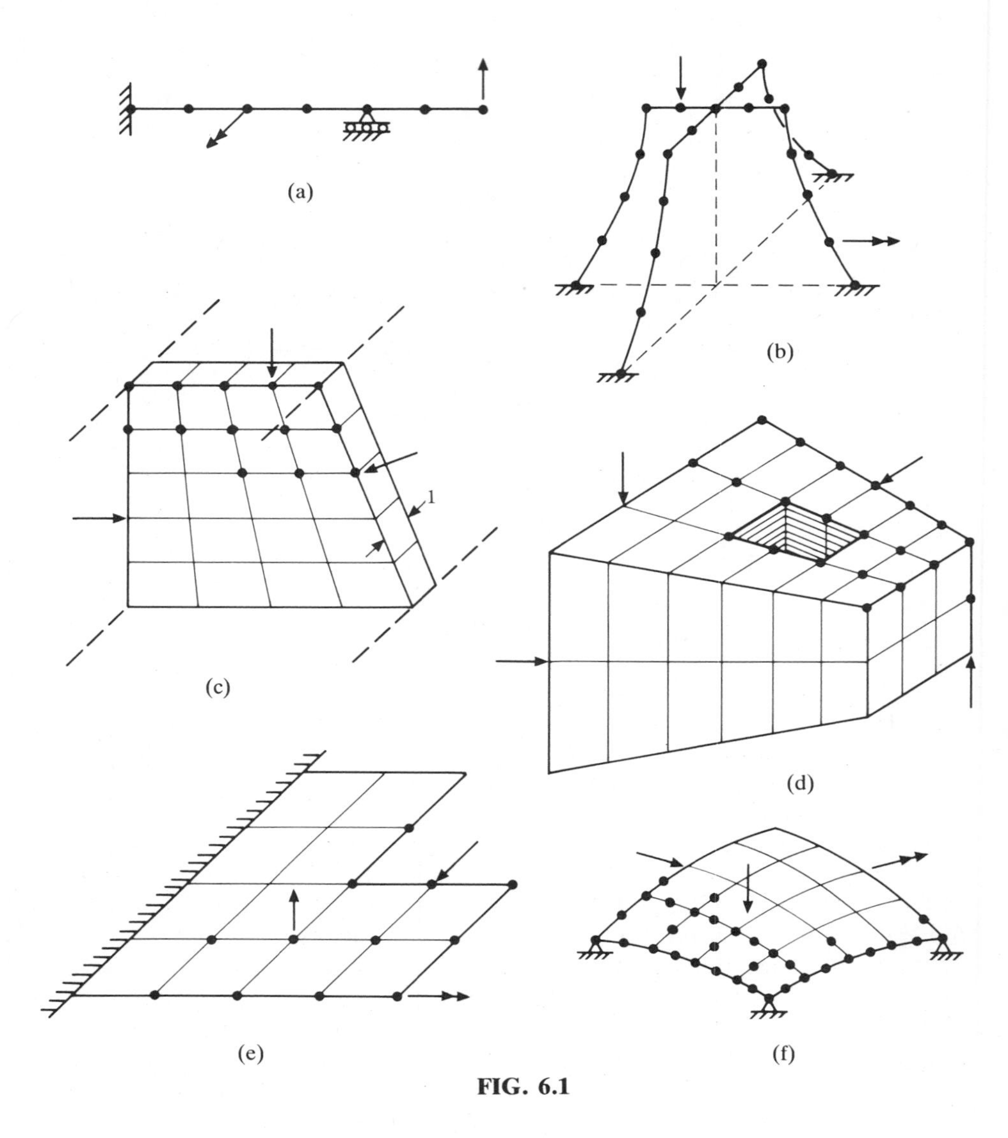

FIG. 6.1

in Fig. 6.1*b* has axial, flexural, and torsional deformations in each of its subdivided members. Figure 6.1*c* depicts a two-dimensional slice of unit thickness, representing the constant state known as plane strain on the cross section of a long, prismatic solid. On the other hand, the discretized general solid in Fig. 6.1*d* has no such restriction. If the thin plate in Fig. 6.1*e* has forces applied in its own plane, it experiences a condition known as plane stress. But if the forces are normal to the plane, it is in a state of flexure, or bending. Finally, a general shell of the type shown in Fig. 6.1*f* can resist any kind of loading. All of the discretized continua in Fig. 6.1 have multiple degrees of freedom and will be referred to as MDOF systems.

The finite-element method to be used in this chapter involves the assumption of *displacement shape functions*† within each element. These functions give approximate results when the element is of finite size and exact results at infinitesimal size. The shape functions make *generic displacements* at any point completely dependent upon *nodal displacements*. Similarly, the local velocities and accelerations are also dependent upon the nodal values. With these dependencies in mind, we can devise a procedure for writing differential equations of motion, as follows.

1. Divide the continuum into a finite number of subregions (or elements) of simple geometry, such as lines, quadrilaterals, or hexahedra.
2. Select key points on the elements to serve as nodes, where conditions of dynamic equilibrium and compatibility with other elements are to be enforced.
3. Assume displacement shape functions within each element so that displacements, velocities, and accelerations at any point are dependent upon nodal values.
4. Satisfy strain–displacement and stress–strain relationships within a typical element for a specific type of problem.
5. Determine equivalent stiffnesses, masses, and nodal loads for each finite element, using a work or energy principle.
6. Develop differential equations of motion for the nodes of the discretized continuum by assembling the finite-element contributions.

From the homogeneous form of the equations of motion, we can perform a vibrational analysis for any linearly elastic structure. This type of analysis consists of finding undamped frequencies and corresponding mode shapes for the discretized analytical model. Such information is often useful by itself, and it is essential for the normal-mode method of dynamic analysis described previously in Chapter 4.

In the present chapter we develop only one-dimensional elements, which are useful for discretizing and analyzing framed structures. Other types of finite elements are discussed in Ref. [4], where the applications include two- and three-dimensional solids, plates in bending, and shells.

6.2 STRESSES AND STRAINS IN CONTINUA

We assume that the continuum to be analyzed consists of a linearly elastic material with small strains and small displacements. Such strains and their

† The *displacement shape functions* in this chapter are similar to the *displacement influence functions* used in Sec. 5.15, but the latter depend upon boundary conditions at the ends of members.

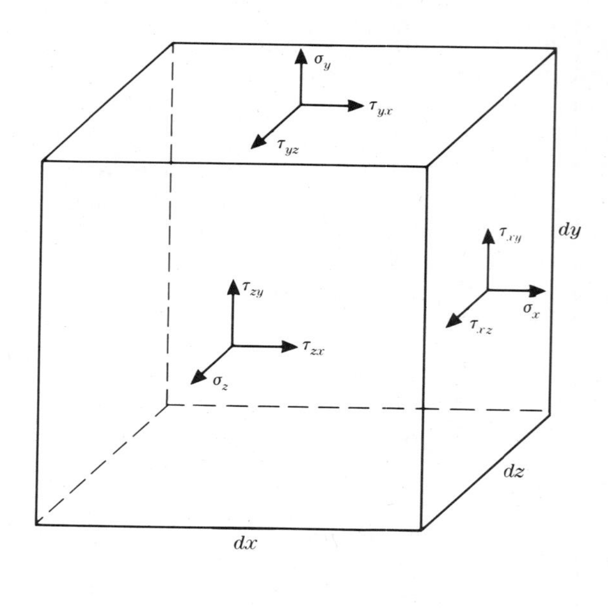

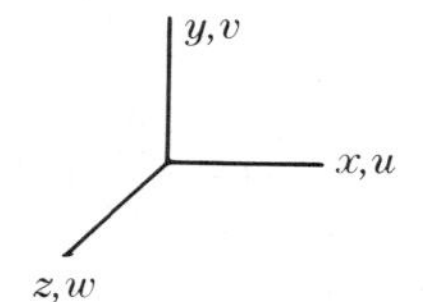

FIG. 6.2

corresponding stresses may be expressed with respect to some right-hand orthogonal coordinate system. For example, in a (rectangular) Cartesian set, the coordinates would be x, y, and z. On the other hand, in a cylindrical coordinate system, the symbols r, θ, and z would serve as the coordinates.

Figure 6.2 shows an infinitesimal element in Cartesian coordinates, where the edges are of lengths dx, dy, and dz. *Normal* and *shearing stresses* are cated by arrows on the faces of the element. The normal stresses are labeled σ_x, σ_y, and σ_z, whereas the shearing stresses are named τ_{xy}, τ_{yz}, and so on. From equilibrium of the element, the following relationships are known:

$$\tau_{xy} = \tau_{yx} \qquad \tau_{yz} = \tau_{zy} \qquad \tau_{zx} = \tau_{xz} \tag{6.1}$$

Thus, only three independent components of shearing stresses need be considered.

Corresponding to the stresses shown in Fig. 6.2 are *normal* and *shearing strains*. Normal strains ϵ_x, ϵ_y, and ϵ_z are defined as

$$\epsilon_x = \frac{\partial u}{\partial x} \qquad \epsilon_y = \frac{\partial v}{\partial y} \qquad \epsilon_z = \frac{\partial w}{\partial z} \tag{6.2}$$

where u, v, and w are translations in the x, y, and z directions. Shearing strains, γ_{xy}, γ_{yz}, and so on, are given by

$$\begin{aligned}
\gamma_{xy} &= \frac{\partial u}{\partial y} + \frac{\partial v}{\partial x} = \gamma_{yx} \\
\gamma_{yz} &= \frac{\partial v}{\partial z} + \frac{\partial w}{\partial y} = \gamma_{zy} \\
\gamma_{zx} &= \frac{\partial w}{\partial x} + \frac{\partial u}{\partial z} = \gamma_{xz}
\end{aligned} \tag{6.3}$$

Hence, only three of the shearing strains are independent.

For convenience, the six independent stresses and the corresponding strains usually will be represented as column matrices (or vectors). Thus,

$$\boldsymbol{\sigma} = \begin{bmatrix} \sigma_1 \\ \sigma_2 \\ \sigma_3 \\ \sigma_4 \\ \sigma_5 \\ \sigma_6 \end{bmatrix} = \begin{bmatrix} \sigma_x \\ \sigma_y \\ \sigma_z \\ \tau_{xy} \\ \tau_{yz} \\ \tau_{zx} \end{bmatrix} \qquad \boldsymbol{\epsilon} = \begin{bmatrix} \epsilon_1 \\ \epsilon_2 \\ \epsilon_3 \\ \epsilon_4 \\ \epsilon_5 \\ \epsilon_6 \end{bmatrix} = \begin{bmatrix} \epsilon_x \\ \epsilon_y \\ \epsilon_z \\ \gamma_{xy} \\ \gamma_{yz} \\ \gamma_{zx} \end{bmatrix} \tag{6.4}$$

where the bold-faced Greek letters $\boldsymbol{\sigma}$ and $\boldsymbol{\epsilon}$ denote the vectors shown.

Strain–stress relationships for an *isotropic material* are drawn from the theory of elasticity [5], as follows:

$$\begin{aligned}
\epsilon_x &= \frac{1}{E}(\sigma_x - \nu\sigma_y - \nu\sigma_z) & \gamma_{xy} &= \frac{\tau_{xy}}{G} \\
\epsilon_y &= \frac{1}{E}(-\nu\sigma_x + \sigma_y - \nu\sigma_z) & \gamma_{yz} &= \frac{\tau_{yz}}{G} \\
\epsilon_z &= \frac{1}{E}(-\nu\sigma_x - \nu\sigma_y + \sigma_z) & \gamma_{zx} &= \frac{\tau_{zx}}{G}
\end{aligned} \tag{6.5}$$

where

$$G = \frac{E}{2(1+\nu)} \tag{6.6}$$

In these expressions E = Young's modulus of elasticity, G = shearing modulus of elasticity, and ν = Poisson's ratio. With matrix format, the relationships in Eqs. (6.5) may be written as

$$\boldsymbol{\epsilon} = \mathbf{C}\,\boldsymbol{\sigma} \tag{6.7}$$

in which

$$\mathbf{C} = \frac{1}{E}\begin{bmatrix} 1 & -\nu & -\nu & 0 & 0 & 0 \\ -\nu & 1 & -\nu & 0 & 0 & 0 \\ -\nu & -\nu & 1 & 0 & 0 & 0 \\ 0 & 0 & 0 & 2(1+\nu) & 0 & 0 \\ 0 & 0 & 0 & 0 & 2(1+\nu) & 0 \\ 0 & 0 & 0 & 0 & 0 & 2(1+\nu) \end{bmatrix} \tag{6.8}$$

Matrix $\mathbf{C}$ is an array that relates the strain vector $\boldsymbol{\epsilon}$ to the stress vector $\boldsymbol{\sigma}$. By the process of inversion (or simultaneous solution), we can also obtain *stress–strain relationships* from Eq. (6.7), as follows:

$$\boldsymbol{\sigma} = \mathbf{E}\,\boldsymbol{\epsilon} \tag{6.9}$$

where

$$\mathbf{E} = \mathbf{C}^{-1} = \frac{E}{(1-\nu)(1-2\nu)}\begin{bmatrix} 1-\nu & \nu & \nu & 0 & 0 & 0 \\ \nu & 1-\nu & \nu & 0 & 0 & 0 \\ \nu & \nu & 1-\nu & 0 & 0 & 0 \\ 0 & 0 & 0 & \dfrac{1-2\nu}{2} & 0 & 0 \\ 0 & 0 & 0 & 0 & \dfrac{1-2\nu}{2} & 0 \\ 0 & 0 & 0 & 0 & 0 & \dfrac{1-2\nu}{2} \end{bmatrix} \tag{6.10}$$

Matrix $\mathbf{E}$ relates the stress vector $\boldsymbol{\sigma}$ to the strain vector $\boldsymbol{\epsilon}$.

For the elements discussed in Sec. 6.4, we will not need the 6×6 stress–strain matrix given by Eq. (6.10). With one-dimensional elements, only one term, such as E or G, is required. If we were to deal with two- and three-dimensional elements, larger matrices [up to the size of that in Eq. (6.10)] would be needed.

6.3 EQUATIONS OF MOTION FOR FINITE ELEMENTS

We shall now introduce definitions and notations that pertain to all types of finite elements for elastic continua. By using the principle of virtual work, we can develop equations of motion that include formulas for stiffnesses, masses, and nodal loads for a typical element. These terms are treated in detail for one-dimensional elements in the next section.

Assume that a three-dimensional finite element with zero damping exists in Cartesian coordinates x, y, and z. Let the time-varying *generic displacements* $\mathbf{u}(t)$ at any point within the element be expressed as the column vector

$$\mathbf{u}(t) = \{u, v, w\} \tag{6.11}$$

where u, v, and w are translations in the x, y, and z directions, respectively.

If the element is subjected to time-varying *body forces*, such forces may be placed into a vector $\mathbf{b}(t)$, as follows:

$$\mathbf{b}(t) = \{b_x, b_y, b_z\} \tag{6.12}$$

Here the symbols b_x, b_y, and b_z represent components of force (per unit of volume, area, or length) acting in the reference directions at a generic point. The time variation for each component of body force is assumed to be the same throughout the element. That is, we may have one time function for b_x, another for b_y, and a third for b_z.

Time-varying *nodal displacements* $\mathbf{q}(t)$ will at first be considered as only translations in the x, y, and z directions. Thus, if n_{en} = number of element nodes,

$$\mathbf{q}(t) = \{\mathbf{q}_i(t)\} \qquad (i = 1, 2, \ldots, n_{\text{en}}) \tag{6.13}$$

where

$$\mathbf{q}_i(t) = \{q_{xi}, q_{yi}, q_{zi}\} = \{u_i, v_i, w_i\} \tag{6.14}$$

However, other types of displacements, such as small rotations ($\partial v/\partial x$, and so on) and curvatures ($\partial^2 v/\partial x^2$, and so on) could also be used.

Similarly, time-varying *nodal actions* $\mathbf{p}(t)$ will temporarily be taken as only forces in the x, y, and z directions at the nodes. That is,

$$\mathbf{p}(t) = \{\mathbf{p}_i(t)\} \qquad (i = 1, 2, \ldots, n_{\text{en}}) \tag{6.15}$$

in which

$$\mathbf{p}_i(t) = \{p_{xi}, p_{yi}, p_{zi}\} \tag{6.16}$$

Time functions for p_{xi}, p_{yi}, and p_{zi} at each node may be independent and arbitrary. Other types of nodal actions, such as moments, rates of change of moments, and so on, could be used as well.

For the type of finite-element method in this chapter, certain assumed *displacement shape functions* relate generic displacements to nodal displacements, as follows:

$$\mathbf{u}(t) = \mathbf{f}\,\mathbf{q}(t) \tag{6.17}$$

In this expression, the symbol $\mathbf{f}$ denotes a rectangular matrix containing geometic functions that make $\mathbf{u}(t)$ completely dependent upon $\mathbf{q}(t)$.

Strain–displacement relationships are obtained by differentiation of the generic displacements. This process may be expressed by forming a matrix $\mathbf{d}$, called a *linear differential operator*, and applying it with the rules of matrix multiplication. Thus,

$$\boldsymbol{\epsilon}(t) = \mathbf{d}\,\mathbf{u}(t) \tag{6.18}$$

In this equation the operator $\mathbf{d}$ expresses the time-varying strain vector $\boldsymbol{\epsilon}(t)$ in terms of generic displacements in the vector $\mathbf{u}(t)$ [see Eqs. (6.2) and (6.3)]. Substitution of Eq. (6.17) into Eq. (6.18) yields

$$\boldsymbol{\epsilon}(t) = \mathbf{B}\,\mathbf{q}(t) \tag{6.19}$$

where

$$\mathbf{B} = \mathbf{d}\,\mathbf{f} \tag{6.20}$$

Matrix $\mathbf{B}$ gives strains at any point within the element due to unit values of nodal displacements.

From Eq. (6.9) we have the matrix form of *stress–strain relationships.* That is,

$$\boldsymbol{\sigma}(t) = \mathbf{E}\,\boldsymbol{\epsilon}(t) \tag{6.21}$$

where $\mathbf{E}$ is a matrix relating time-varying stresses in the vector $\boldsymbol{\sigma}(t)$ to strains in $\boldsymbol{\epsilon}(t)$. Substitution of Eq. (6.19) into (6.21) produces

$$\boldsymbol{\sigma}(t) = \mathbf{E}\,\mathbf{B}\,\mathbf{q}(t) \tag{6.22}$$

in which the matrix product $\mathbf{E}\,\mathbf{B}$ gives stresses at a generic point due to unit values of nodal displacements.

Application of the principle of virtual work to a finite element yields

$$\delta U_{\mathrm{e}} - \delta W_{\mathrm{e}} = 0 \tag{6.23}$$

where δU_{e} is the virtual strain energy of internal stresses and δW_{e} is the virtual work of external actions on the element. To develop both of these quantities in detail, we assume a vector $\delta\mathbf{q}$ of small virtual nodal displacements. Thus,

$$\delta\mathbf{q} = \{\delta\mathbf{q}_i\} \qquad (i = 1, 2, \ldots, n_{\mathrm{en}}\} \tag{6.24}$$

Then the resulting virtual generic displacements become [see Eq. (6.17)]

$$\delta\mathbf{u} = \mathbf{f}\,\delta\mathbf{q} \tag{6.25}$$

Using the strain–displacement relationships in Eq. (6.19), we obtain

$$\delta \boldsymbol{\epsilon} = \mathbf{B}\, \delta \mathbf{q} \tag{6.26}$$

Now the *internal virtual strain energy* can be written as

$$\delta U_e = \int_V \delta \boldsymbol{\epsilon}^{\mathrm{T}} \boldsymbol{\sigma}(t)\, dV \tag{6.27}$$

where integration is over the volume of the element.

For the *external virtual work* we turn to Fig. 6.3, which shows an infinitesimal element with components of *applied body forces* $b_x(t)\,dV$, $b_y(t)\,dV$, and $b_z(t)\,dV$. The figure also indicates *inertial body forces* $\rho\ddot{u}\,dV$, $\rho\ddot{v}\,dV$, and $\rho\ddot{w}\,dV$ due to the accelerations $\ddot{u}$, $\ddot{v}$, and $\ddot{w}$. Note that the inertial forces act in directions that are opposite to the positive senses of the accelerations. Thus, we add the external virtual work of nodal and distributed body forces as follows:

$$\delta W_e = \delta \mathbf{q}^{\mathrm{T}} \mathbf{p}(t) + \int_V \delta \mathbf{u}^{\mathrm{T}} \mathbf{b}(t)\, dV - \int_V \delta \mathbf{u}^{\mathrm{T}} \rho \ddot{\mathbf{u}}\, dV \tag{6.28}$$

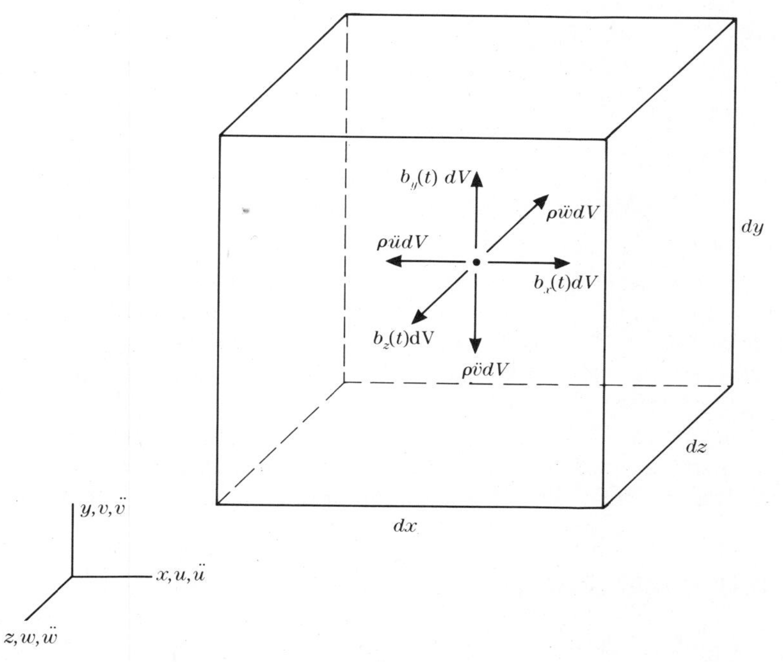

FIG. 6.3

Substitution of Eqs. (6.27) and (6.28) into Eq. (6.23) produces

$$\int_V \delta\boldsymbol{\epsilon}^{\mathrm{T}}\boldsymbol{\sigma}(t)\, dV = \delta\mathbf{q}^{\mathrm{T}}\mathbf{p}(t) + \int_V \delta\mathbf{u}^{\mathrm{T}}\mathbf{b}(t)\, dV - \int_V \delta\mathbf{u}^{\mathrm{T}}\rho\ddot{\mathbf{u}}\, dV \tag{6.29}$$

Now assume that

$$\ddot{\mathbf{u}} = \mathbf{f}\,\ddot{\mathbf{q}} \tag{6.30}$$

Then we can substitute Eqs. (6.22) and (6.30) into Eq. (6.29) and use the transposes of Eqs. (6.25) and (6.26) to obtain

$$\delta\mathbf{q}^{\mathrm{T}}\int_V \mathbf{B}^{\mathrm{T}}\mathbf{E}\,\mathbf{B}\, dV\,\mathbf{q} = \delta\mathbf{q}^{\mathrm{T}}\mathbf{p}(t) + \delta\mathbf{q}^{\mathrm{T}}\int_V \mathbf{f}^{\mathrm{T}}\mathbf{b}(t)\, dV - \delta\mathbf{q}^{\mathrm{T}}\int_V \rho\mathbf{f}^{\mathrm{T}}\mathbf{f}\, dV\,\ddot{\mathbf{q}} \tag{6.31}$$

Cancellation of $\delta\mathbf{q}^{\mathrm{T}}$ and rearrangement of the resulting equations of motion gives

$$\mathbf{M}\,\ddot{\mathbf{q}} + \mathbf{K}\,\mathbf{q} = \mathbf{p}(t) + \mathbf{p}_b(t) \tag{6.32}$$

where

$$\mathbf{K} = \int_V \mathbf{B}^{\mathrm{T}}\mathbf{E}\,\mathbf{B}\, dV \tag{6.33}$$

and

$$\mathbf{M} = \int_V \rho\mathbf{f}^{\mathrm{T}}\mathbf{f}\, dV \tag{6.34}$$

Also,

$$\mathbf{p}_b(t) = \int_V \mathbf{f}^{\mathrm{T}}\mathbf{b}(t)\, dV \tag{6.35}$$

Matrix $\mathbf{K}$ in Eq. (6.33) is the *element stiffness matrix*, which contains *stiffness coefficients* that are fictitious actions at nodes due to unit values of nodal displacements. Equation (6.34) gives the form of the *consistent-mass matrix*, in which the terms are energy-equivalent actions at nodes due to unit values of nodal accelerations. Finally, the vector $\mathbf{p}_b(t)$ in Eq. (6.35) consists of *equivalent nodal loads* due to body forces in the vector $\mathbf{b}(t)$. Other equivalent nodal loads due to initial strains (or stresses) could be derived [1], but analyses for such influences are considered to be statics problems.

6.4 ONE-DIMENSIONAL ELEMENTS

In this section we develop properties of one-dimensional elements subjected to axial, torsional, and flexural deformations, starting with the *axial element*

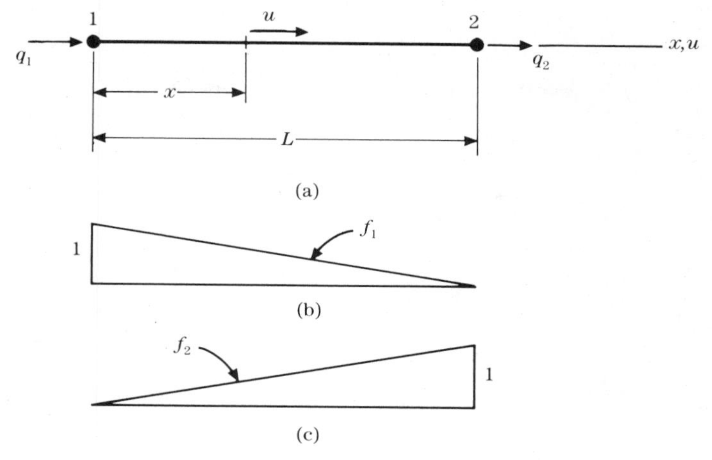

FIG. 6.4

in Fig. 6.4*a*. The figure indicates a single generic translation u in the x direction. Thus, from Eq. (6.11) we have

$$\mathbf{u}(t) = u$$

The corresponding body force is a single component b_x (force per unit length), acting in the x direction. Therefore, Eq. (6.12) gives

$$\mathbf{b}(t) = b_x$$

Nodal displacements q_1 and q_2 consist of translations in the x direction at nodes 1 and 2 (see Fig. 6.4*a*). Hence, Eq. (6.13) becomes

$$\mathbf{q}(t) = \{q_1, q_2\} = \{u_1, u_2\}$$

Corresponding nodal forces at points 1 and 2 are given by Eq. (6.15) as

$$\mathbf{p}(t) = \{p_1, p_2\} = \{p_{x1}, p_{x2}\}$$

Figures 6.4*b* and *c* show linear displacement shape functions f_1 and f_2 assumed for this element. That is, Eq. (6.17) produces

$$u = \mathbf{f}\,\mathbf{q}(t)$$

where

$$\mathbf{f} = [f_1 \quad f_2] = \left[1 - \frac{x}{L} \quad \frac{x}{L}\right] \tag{6.36}$$

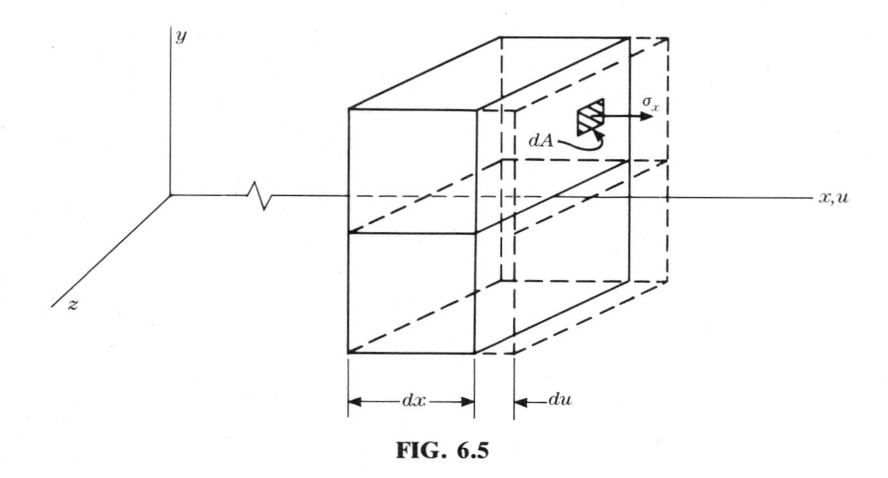

FIG. 6.5

These functions give the variations of u along the length due to unit values of the nodal translations q_1 and q_2. They are exactly correct for a prismatic element.

From Fig. 6.5 we see that the single strain–displacement relationship du/dx for the axial element is constant on the cross section. Thus, Eqs. (6.18), (6.19), and (6.20) yield

$$\boldsymbol{\epsilon}(t) = \epsilon_x = \frac{du}{dx} = \frac{d\mathbf{f}}{dx}\mathbf{q}(t) = \mathbf{B}\,\mathbf{q}(t) \tag{6.37a}$$

where

$$\mathbf{B} = \mathbf{d}\,\mathbf{f} = \frac{d\mathbf{f}}{dx} = \frac{1}{L}[-1 \quad 1] \tag{6.37b}$$

Here the linear differential operator $\mathbf{d}$ is simply d/dx. As shown in Eq. (6.37a), matrix $\mathbf{B}$ expresses the strain ϵ_x in terms of the nodal displacements $\mathbf{q}$. Similarly, the single stress–strain relationship [see Eqs. (6.21) and (6.22)] becomes merely

$$\boldsymbol{\sigma}(t) = \sigma_x = \mathbf{E}\,\boldsymbol{\epsilon}(t) = E\epsilon_x = E\mathbf{B}\,\mathbf{q}(t) \tag{6.38a}$$

Hence,

$$\mathbf{E} = E \qquad \text{and} \qquad E\mathbf{B} = \frac{E}{L}[-1 \quad 1] \tag{6.38b}$$

Equation (6.38a) shows that the product $E\mathbf{B}$ gives the stress σ_x in terms of the nodal displacements.

From Eq. (6.33) the element stiffness matrix **K** may now be evaluated, as follows:

$$\mathbf{K} = \int_V \mathbf{B}^{\mathrm{T}} \mathbf{E} \, \mathbf{B} \, dV = \frac{E}{L^2} \begin{bmatrix} -1 \\ 1 \end{bmatrix} [-1 \quad 1] \int_0^L \int_A dA \, dx$$

Multiplication and integration over the cross section and the length gives

$$\mathbf{K} = \frac{EA}{L} \begin{bmatrix} 1 & -1 \\ -1 & 1 \end{bmatrix} \tag{6.39}$$

assuming that the cross-sectional area A is constant. Similarly, the consistent mass matrix **M** is found from Eq. (6.34) to be

$$\mathbf{M} = \int_V \rho \mathbf{f}^{\mathrm{T}} \mathbf{f} \, dV = \frac{\rho}{L^2} \int_0^L \int_A \begin{bmatrix} L - x \\ x \end{bmatrix} [L - x \quad x] \, dA \, dx$$

$$= \frac{\rho A L}{6} \begin{bmatrix} 2 & 1 \\ 1 & 2 \end{bmatrix} \tag{6.40}$$

assuming also that the mass density ρ is constant.

We see that the stiffness matrix **K** and the consistent mass matrix **M** are unique for a prismatic axial element of uniform mass density. However, an infinite number of equivalent nodal load vectors $\mathbf{p}_b(t)$ may be derived, depending upon the distribution of body forces. For the simplest case, we assume that a uniformly distributed axial load b_x (force per unit length) is suddenly applied to the axial element. Then Eq. (6.35) produces

$$\mathbf{p}_b(t) = \int_0^L \mathbf{f}^{\mathrm{T}} b_x \, dx = \frac{b_x}{L} \begin{bmatrix} L - x \\ x \end{bmatrix} dx = b_x \frac{L}{2} \begin{bmatrix} 1 \\ 1 \end{bmatrix} \tag{6.41}$$

which shows that the equivalent nodal loads are equal forces at the two ends.

Turning now to the *torsional element* in Fig. 6.6a, we use a single generic displacement θ_x, which is a small rotation about the x axis (indicated by a double-headed arrow). Thus,

$$\mathbf{u}(t) = \theta_x$$

Corresponding to this displacement is a single body action

$$\mathbf{b}(t) = m_x$$

which is a moment per unit length acting in the positive x sense. Nodal

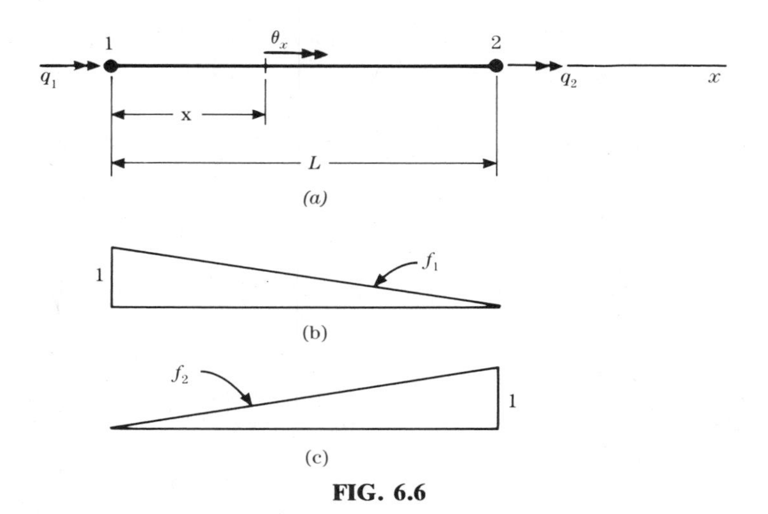

FIG. 6.6

displacements in the figure consist of small axial rotations at nodes 1 and 2. Hence,

$$\mathbf{q}(t) = \{q_1, q_2\} = \{\theta_{x1}, \theta_{x2}\}$$

In addition, the corresponding nodal actions at points 1 and 2 are

$$\mathbf{p}(t) = \{p_1, p_2\} = \{M_{x1}, M_{x2}\}$$

which are moments (or torques) acting in the x direction. As for the axial element, we assume the linear displacement shape functions f_1 and f_2 shown in Figs. 6.6*b* and *c*. Therefore,

$$\theta_x = \mathbf{f}\,\mathbf{q}(t)$$

in which the matrix **f** is again given by Eq. (6.36). However, in this case the functions mean variations of θ_x along the length caused by unit values of the nodal rotations q_1 and q_2.

Strain–displacement relationships can be inferred for a torsional element with a circular cross section by examining Fig. 6.7. Assuming that radii remain straight during torsional deformation, we conclude that the shearing strain γ varies linearly with the radial distance r, as follows:

$$\gamma = r\frac{d\theta_x}{dx} = r\psi \tag{6.42}$$

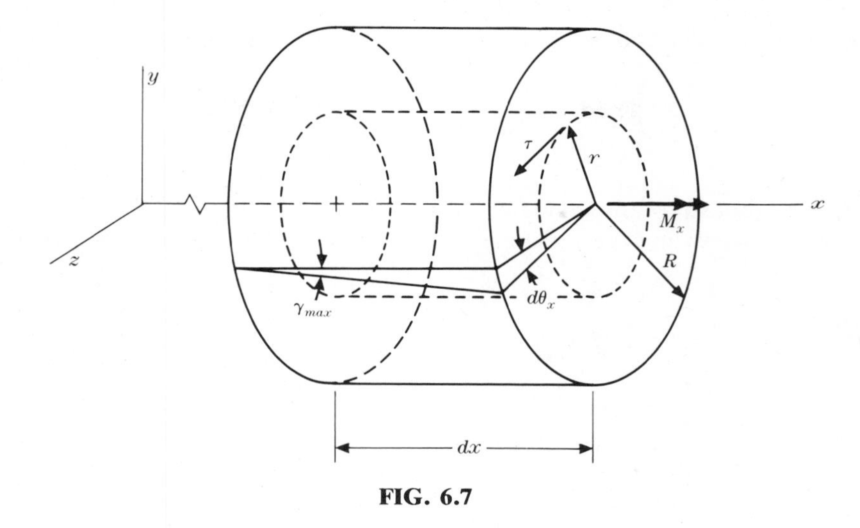

FIG. 6.7

where ψ is the *twist*, or rate of change of angular displacement. Thus,

$$\psi = \frac{d\theta_x}{dx} \tag{6.43}$$

Equation (6.42) shows that the maximum value of the shearing strain occurs at the surface. That is,

$$\gamma_{\max} = R\psi$$

where R is the radius of the cross section (see Fig. 6.7). We also see from Eq. (6.42) that the linear differential operator $\mathbf{d}$ relating γ to θ_x is

$$\mathbf{d} = r\frac{d}{dx} \tag{6.44}$$

Thus, the strain–displacement matrix $\mathbf{B}$ becomes

$$\mathbf{B} = \mathbf{d}\,\mathbf{f} = \frac{r}{L}[-1 \quad 1] \tag{6.45}$$

which is the same as for the axial element, except for the presence of r.

Shearing stress τ (see Fig. 6.7) is related to shearing strain in a torsional element by

$$\tau = G\gamma \tag{6.46a}$$

where the symbol G denotes the shearing modulus of the material. Hence,

$$\mathbf{E} = G \qquad \text{and} \qquad G\mathbf{B} = \frac{Gr}{L}[-1 \quad 1] \tag{6.46b}$$

These relationships are analogous to Eqs. (6.38) for the axial element.

We may now find the torsional stiffness matrix $\mathbf{K}$ by applying Eq. (6.33), as follows:

$$\begin{aligned}\mathbf{K} &= \int_V \mathbf{B}^{\mathrm{T}}\mathbf{E}\,\mathbf{B}\,dV \\ &= \frac{G}{L^2}\begin{bmatrix}-1\\1\end{bmatrix}[-1 \quad 1]\int_0^L \int_0^{2\pi} \int_0^R (r^2) r\,dr\,d\theta\,dx \\ &= \frac{GJ}{L}\begin{bmatrix}1 & -1\\-1 & 1\end{bmatrix}\end{aligned} \tag{6.47}$$

where GJ is constant. The *polar moment of inertia* J for a circular cross section is defined as

$$J = \int_0^{2\pi} \int_0^R r^3\,dr\,d\theta = \frac{\pi R^4}{2} \tag{6.48}$$

If the cross section of a torsional element is not circular, it will warp. Such warping is most severe for elements of open cross sections, such as channel or wide-flanged sections. For most practical cases, the theory of *uniform torsion* described here may be used by substituting the appropriate *torsion constant* [6] for J. If a more precise analysis is desired, the theory of *nonuniform torsion* may be applied (see Sec. 5.21 and [7]).

To obtain the consistent mass matrix $\mathbf{M}$ for a torsional element, we will first integrate over the cross section and then over the length. Due to the small rotation θ_x, the translation of a point on the cross section at distance r from the center is $r\theta_x$. Also, the acceleration of the same point is $r\ddot{\theta}_x$. By integration over the cross section, we find the *inertial moment per unit length* to be $-\rho J\ddot{\theta}_x$, where J is again the polar moment of inertia given in Eq. (6.48). Use of this inertial moment in conjunction with the corresponding virtual rotation $\delta\theta_x$ leads to

$$\mathbf{M} = \int_0^L \rho J \mathbf{f}^{\mathrm{T}}\mathbf{f}\,dx \tag{6.49}$$

which is a specialized version of Eq. (6.34). Integration of Eq. (6.49) over the length yields

$$\mathbf{M} = \frac{\rho J}{L^2} \int_0^L \begin{bmatrix} L - x \\ x \end{bmatrix} [L - x \quad x] \, dx$$

$$= \frac{\rho J L}{6} \begin{bmatrix} 2 & 1 \\ 1 & 2 \end{bmatrix} \tag{6.50}$$

This array is the consistent mass matrix for a torsional element of constant cross section and uniform mass density.

The simplest case of a body force applied to a torsional element consists of a uniformly distributed axial torque (or moment) m_x per unit length. For this loading Eq. (6.35) gives

$$\mathbf{p}_b(t) = \int_0^L \mathbf{f}^{\mathrm{T}} m_x \, dx = \frac{m_x}{L} \int_0^L \begin{bmatrix} L - x \\ x \end{bmatrix} dx = m_x \frac{L}{2} \begin{bmatrix} 1 \\ 1 \end{bmatrix} \tag{6.51}$$

These equivalent nodal loads are equal moments at the two ends of the element.

Figure 6.8*a* shows a *flexural element*, for which the x–y plane is a principal plane of bending. Indicated in the figure is the single generic displacement v, which is a translation in the y direction. Thus,

$$\mathbf{u}(t) = v$$

The corresponding body force is a single component b_y (force per unit length), acting in the y direction. Hence,

$$\mathbf{b}(t) = b_y$$

At node 1 (see Fig. 6.8*a*) the two nodal displacements q_1 and q_2 are a small translation in the y direction and a small rotation in the z sense. The former is indicated by a single-headed arrow, while the latter is shown as a double-headed arrow. Similarly, at node 2 the displacements numbered 3 and 4 are a translation and a rotation, respectively. Therefore, the vector of nodal displacements becomes

$$\mathbf{q}(t) = \{q_1, q_2, q_3, q_4\} = \{v_1, \theta_{z1}, v_2, \theta_{z2}\}$$

in which

$$\theta_{z1} = \left(\frac{dv}{dx}\right)_1 \qquad \theta_{z2} = \left(\frac{dv}{dx}\right)_2$$

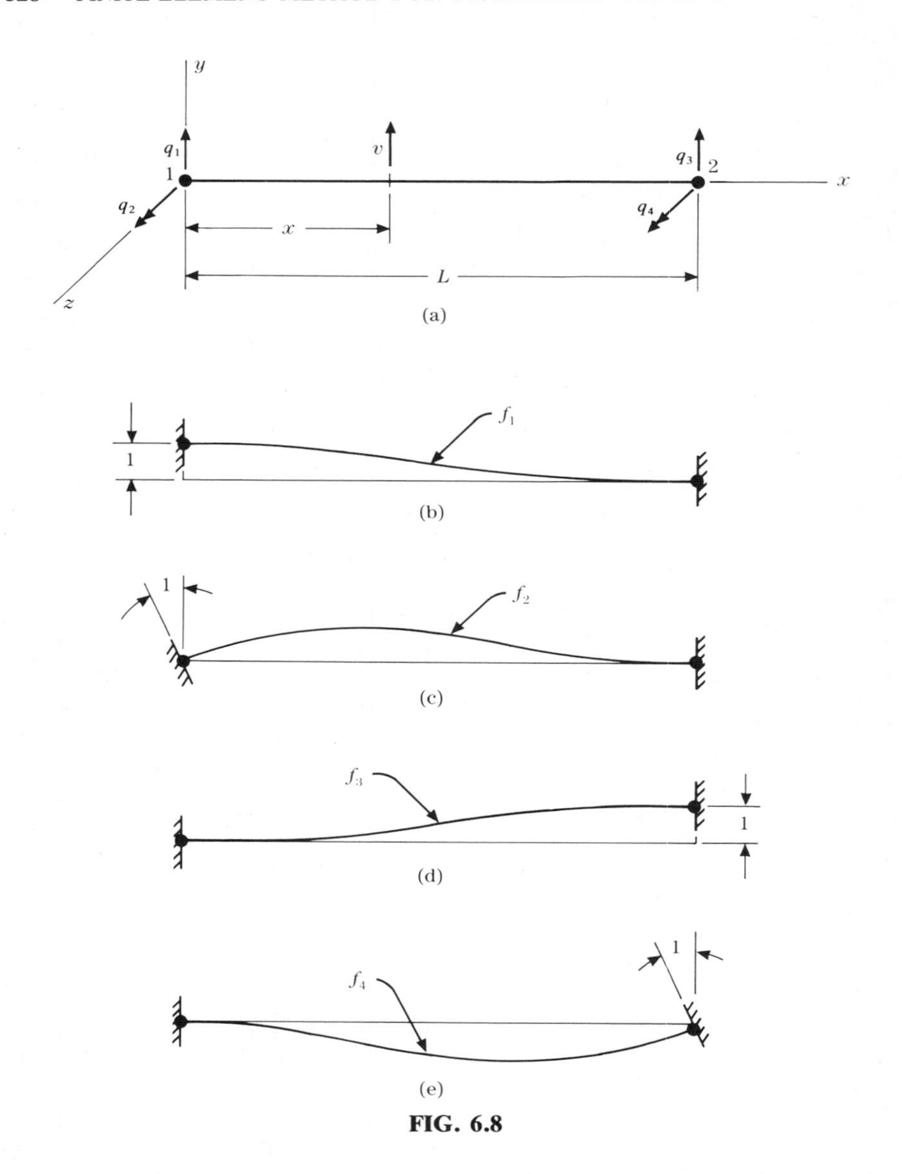

FIG. 6.8

These derivatives (or slopes) may be considered to be small rotations even though they are actually rates of changes of translations at the nodes. Corresponding nodal actions at points 1 and 2 are

$$\mathbf{p}(t) = \{p_1, p_2, p_3, p_4\} = \{p_{y1}, M_{z1}, p_{y2}, M_{z2}\}$$

The terms p_{y1} and p_{y2} denote forces in the y direction at nodes 1 and 2, and the symbols M_{z1} and M_{z2} represent moments in the z sense at those points.

For the flexural element we assume cubic displacement shape functions, and matrix **f** becomes

$$\mathbf{f} = [f_1 \quad f_2 \quad f_3 \quad f_4] \tag{6.52a}$$

where

$$f_1 = \frac{1}{L^3}(2x^3 - 3Lx^2 + L^3) \qquad f_2 = \frac{1}{L^2}(x^3 - 2Lx^2 + L^2x)$$
$$f_3 = \frac{1}{L^3}(-2x^3 + 3Lx^2) \qquad f_4 = \frac{1}{L^2}(x^3 - Lx^2) \tag{6.52b}$$

These four shape functions appear in Figs. 3.8*b*, *c*, *d*, and *e*. They represent the variations of v along the length due to unit values of the four nodal displacements q_1 through q_4. The functions are exact for a prismatic element in which shearing deformations are omitted.

Strain–displacement relationships can be developed for a flexural element by assuming that plane sections remain plane during deformation, as illustrated in Fig. 6.9. The translation u in the x direction at any point on the cross section is

$$u = -y\frac{dv}{dx} \tag{6.53}$$

Using this relationship, we obtain the following expression for flexural

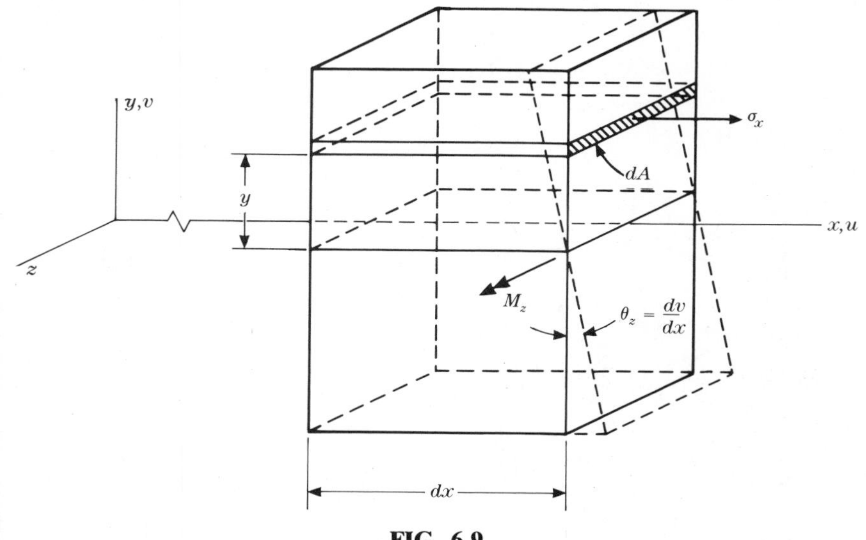

FIG. 6.9

strain:

$$\epsilon_x = \frac{du}{dx} = -y\frac{d^2v}{dx^2} = -y\phi \tag{6.54}$$

in which ϕ represents the *curvature*

$$\phi = \frac{d^2v}{dx^2} \tag{6.55}$$

From Eq. (6.54) we see that the linear differential operator **d** relating ϵ_x to v is

$$\mathbf{d} = -y\frac{d^2}{dx^2} \tag{6.56}$$

Then Eq. (6.20) gives the strain–displacement matrix **B** as

$$\mathbf{B} = \mathbf{df} = -\frac{y}{L^3}[12x - 6L \quad 6Lx - 4L^2 \quad -12x + 6L \quad 6Lx - 2L^2] \tag{6.57}$$

In addition, flexural stress σ_x in Fig. 6.9 is related to flexural strain ϵ_x simply by

$$\sigma_x = E\epsilon_x \tag{6.58a}$$

Hence,

$$\mathbf{E} = E \qquad \text{and} \qquad \mathbf{E\,B} = E\mathbf{B} \tag{6.58b}$$

Element stiffnesses now may be obtained from Eq. (6.33), as follows:

$$\mathbf{K} = \int_V \mathbf{B}^{\mathrm{T}}\mathbf{E\,B}\,dV$$

$$= \int_0^L \int_A \frac{Ey^2}{L^6}\begin{bmatrix} 12x - 6L \\ 6Lx - 4L^2 \\ -12x + 6L \\ 6Lx - 2L^2 \end{bmatrix}[12x - 6L \quad \cdots \quad \cdots \quad 6Lx - 2L^2]\,dA\,dx$$

Multiplication and integration (with EI constant) yields

$$\mathbf{K} = \frac{2EI}{L^3}\begin{bmatrix} 6 & 3L & -6 & 3L \\ 3L & 2L^2 & -3L & L^2 \\ -6 & -3L & 6 & -3L \\ 3L & L^2 & -3L & 2L^2 \end{bmatrix} \tag{6.59}$$

where

$$I = \int_A y^2 \, dA \tag{6.60}$$

represents the *moment of inertia* (second moment of area) of the cross section with respect to the neutral axis. Additional contributions to matrix **K** due to shearing deformations are given in Ref. [6].

The consistent mass matrix **M** for a flexural element will be developed in two parts. A typical cross section for this type of member translates in the y direction, as indicated in Fig. 6.8*a*. However, the section also rotates about its neutral axis, as shown in Fig. 6.9. The *translational inertia* terms are much more important than the rotational terms, so they will be considered first. Using matrix **f** from Eq. (6.52) in Eq. (6.34), we find

$$\mathbf{M}_t = \int_V \rho \mathbf{f}^{\mathrm{T}} \mathbf{f} \, dV = \int_0^L \rho A \mathbf{f}^{\mathrm{T}} \mathbf{f} \, dx$$

Multiplying terms in **f** and integrating over the length gives

$$\mathbf{M}_t = \frac{\rho A L}{420} \begin{bmatrix} 156 & 22L & 54 & -13L \\ 22L & 4L^2 & 13L & -3L^2 \\ 54 & 13L & 156 & -22L \\ -13L & -3L^2 & -22L & 4L^2 \end{bmatrix} \tag{6.61}$$

which is the *consistent-mass matrix for translational inertia* in a prismatic beam.

Rotational inertia (or *rotary inertia*) terms for a beam (see also Sec. 5.12) can be deduced from Fig. 6.9, where the translation u in the x direction of a point on the cross section is

$$u = -y\theta_z \tag{6.62}$$

In this expression,

$$\theta_z = v_{,x} = \mathbf{f}_{,x}\mathbf{q}(t) \tag{6.63}$$

where the symbols $v_{,x}$ and $\mathbf{f}_{,x}$ represent differentiation with respect to x. Similarly, the acceleration of the same point in the x direction is

$$\ddot{u} = -y\ddot{\theta}_z \tag{6.64}$$

where

$$\ddot{\theta}_z = \ddot{v}_{,x} = \mathbf{f}_{,x}\ddot{\mathbf{q}}(t) \tag{6.65}$$

By integrating the moment of inertial force over the cross section, we find the *inertial moment per unit length* to be $-\rho I\ddot{\theta}_z$, where I is again the moment of inertia given in Eq. (6.60). Use of this inertial moment in conjunction with the corresponding virtual rotation $\delta\theta_z$ leads to the formula

$$\mathbf{M}_r = \int_0^L \rho I \mathbf{f}_{,x}^{\mathrm{T}} \mathbf{f}_{,x}\, dx \tag{6.66}$$

which is a modified version of Eq. (6.34). Differentiating matrix $\mathbf{f}$ [see Eq. (6.52)] with respect to x, we find

$$\mathbf{f}_{,x} = \frac{1}{L^3}[6(x^2 - Lx) \quad 3Lx^2 - 4L^2x + L^3 \quad -6(x^2 - Lx) \quad 3Lx^2 - 2L^2x] \tag{6.67}$$

Substitution of this matrix into Eq. (6.66), followed by integration over the length, produces

$$\mathbf{M}_r = \frac{\rho I}{30L}\begin{bmatrix} 36 & 3L & -36 & 3L \\ 3L & 4L^2 & -3L & -L^2 \\ -36 & -3L & 36 & -3L \\ 3L & -L^2 & -3L & 4L^2 \end{bmatrix} \tag{6.68}$$

which is the *consistent matrix for rotational inertia* in a prismatic beam. Additional contributions to matrix $\mathbf{M}$ due to shearing deformations have also been developed and are given in Ref. [8].

We now consider the simple loading case of a uniformly distributed body force b_y (per unit length) applied to a flexural element. Equivalent nodal loads at points 1 and 2 (see Fig. 6.8*a*) may be calculated from Eq. (6.35) as

$$\mathbf{p}_b(t) = \int_0^L \mathbf{f}^{\mathrm{T}} b_y\, dx = b_y \frac{L}{12}\{6, L, 6, -L\} \tag{6.69}$$

For this integration, the displacement shape functions f_1 through f_4 were drawn from Eqs. (6.52b).

By using *generalized stresses and strains*, we can avoid repetitious integrations over the cross sections of one-dimensional elements. While this concept is rather trivial for an axial element, it can be more useful for torsional and flexural elements. Let us reconsider the torsional element in Fig. 6.7 and integrate the moment of the shearing stress τ about the x axis. Thus, we generate the torque M_x, as follows:

$$M_x = \int_0^{2\pi} \int_0^R \tau r^2\, dr\, d\theta \tag{6.70}$$

Substitution of the stress–strain and strain–displacement relationships from Eqs. (6.46a) and (6.42) produces

$$M_x = G\psi \int_0^{2\pi} \int_0^R r^3 \, dr \, d\theta = GJ\psi \tag{6.71}$$

If we take M_x as generalized (or integrated) stress and ψ as generalized strain, the generalized stress–strain (or torque–twist) operator $\bar{G}$ becomes

$$\bar{G} = GJ \tag{6.72}$$

which is the *torsional rigidity* of the cross section. Hence, from Eq. (6.71) we have

$$M_x = \bar{G}\psi \tag{6.73}$$

By this method the operator **d** in Eq. (6.44) does not include the multiplier r. Furthermore, the generalized matrix $\bar{\mathbf{B}}$ [Eq. (6.45) devoid of r] is used instead of matrix **B**. That is,

$$\mathbf{B} = r\bar{\mathbf{B}} \tag{6.74}$$

From this point we can conclude that evaluations of the terms in the stiffness matrix **K** do not require integrations over the cross section. Therefore,

$$\mathbf{K} = \int_0^L \bar{\mathbf{B}}^{\mathrm{T}} \bar{G} \bar{\mathbf{B}} \, dx \tag{6.75}$$

This expression for **K** is equivalent to Eq. (6.33) used previously.

Turning now to the flexural element in Fig. 6.9, we integrate the moment of the normal stress σ_x about the neutral axis to obtain M_z, as follows:

$$M_z = \int_A -\sigma_x y \, dA \tag{6.76}$$

Then substitute the stress–strain and strain–displacement relationships from Eqs. (6.58a) and (6.54) to find

$$M_z = E\phi \int_A y^2 \, dA = EI\phi \tag{6.77}$$

For this element we can take M_z as generalized (or integrated) stress and ϕ as generalized strain. Then the generalized stress–strain (or moment–curvature) operator $\bar{E}$ is

$$\bar{E} = EI \tag{6.78}$$

which is the *flexural rigidity* of the cross section. Thus, from Eq. (6.77) we have

$$M_z = \bar{E}\phi \tag{6.79}$$

With this approach, the operator **d** in Eq. (6.56) is devoid of the multiplier $-y$. In addition, the generalized matrix $\bar{\mathbf{B}}$ [Eq. (6.57) without the factor $-y$] may be used in place of matrix **B**. That is,

$$\mathbf{B} = -y\bar{\mathbf{B}} \tag{6.80}$$

Then integration over the cross section for terms in matrix **K** becomes unnecessary. Hence,

$$\mathbf{K} = \int_0^K \bar{\mathbf{B}}^{\mathrm{T}} \bar{E} \bar{\mathbf{B}} \, dx \tag{6.81}$$

which is analogous to Eq. (6.75).

6.5 VIBRATIONS OF BEAMS BY FINITE ELEMENTS

We could analyze all of the types of framed structures described in Refs. [4] and [6] by using one-dimensional finite elements for the members (or parts of subdivided members). However, two- and three-dimensional structures require rotation-of-axes transformations for action and displacement vectors as well as stiffness and mass matrices of elements. Only the beam type of structure has members that are automatically aligned with reference axes. Therefore, in this chapter we shall examine only beam vibrations, using properties of the flexural element developed in the preceding section.

After stiffnesses, masses, and actual or equivalent nodal loads for individual flexural elements are generated, we can assemble them to form system matrices for a beam using the *direct stiffness method* [6]. With this approach we simply add the contributions from all the elements to obtain stiffnesses, masses, and nodal loads for the whole system. Thus, by summation we have†

$$\mathbf{S}_{\mathrm{s}} = \sum_{i=1}^{n_{\mathrm{e}}} \mathbf{K}_i \qquad \mathbf{M}_{\mathrm{s}} = \sum_{i=1}^{n_{\mathrm{e}}} \mathbf{M}_i \tag{6.82}$$

and

$$\mathbf{A}_{\mathrm{s}}(t) = \sum_{i=1}^{n_{\mathrm{e}}} \mathbf{p}_i(t) \qquad \mathbf{A}_{\mathrm{sb}}(t) = \sum_{i=1}^{n_{\mathrm{e}}} \mathbf{p}_{bi}(t) \tag{6.83}$$

† For the operations in Eqs. (6.82) and (6.83), the matrix or vector on the right must be expanded with zeros to make it the same size as the matrix or vector on the left.

where n_e is the number of elements. In Eqs. (6.82) the symbols $\mathbf{S}_s$ and $\mathbf{M}_s$ represent the *system stiffness matrix* and the *system mass matrix* for all the nodes. Similarly, the action vectors $\mathbf{A}_s(t)$ and $\mathbf{A}_{sb}(t)$ in Eqs. (6.83) are *actual* and *equivalent nodal loads* for the whole system. Then the undamped equations of motion for the assembled system become

$$\mathbf{M}_s\ddot{\mathbf{D}}_s + \mathbf{S}_s\mathbf{D}_s = \mathbf{A}_s(t) + \mathbf{A}_{sb}(t) \tag{6.84}$$

in which $\mathbf{D}_s$ and $\ddot{\mathbf{D}}_s$ are vectors of system displacements and accelerations. Equation (6.84) gives the system equations of motion for all nodal displacements, regardless of whether they are free or restrained.

In preparation for solving Eq. (6.84), we rearrange and partition it, as follows:

$$\begin{bmatrix} \mathbf{M}_{FF} & \mathbf{M}_{FR} \\ \mathbf{M}_{RF} & \mathbf{M}_{RR} \end{bmatrix} \begin{bmatrix} \ddot{\mathbf{D}}_F \\ \ddot{\mathbf{D}}_R \end{bmatrix} + \begin{bmatrix} \mathbf{S}_{FF} & \mathbf{S}_{FR} \\ \mathbf{S}_{RF} & \mathbf{S}_{RR} \end{bmatrix} \begin{bmatrix} \mathbf{D}_F \\ \mathbf{D}_R \end{bmatrix} = \begin{bmatrix} \mathbf{A}_F(t) \\ \mathbf{A}_R(t) \end{bmatrix} \tag{6.85}$$

In this equation actual and equivalent nodal loads have been combined into a single action vector. The subscript F refers to *free nodal displacements*, while the subscript R denotes *restrained nodal displacements*. Writing Eq. (6.85) in two parts gives

$$\mathbf{M}_{FF}\ddot{\mathbf{D}}_F + \mathbf{M}_{FR}\ddot{\mathbf{D}}_R + \mathbf{S}_{FF}\mathbf{D}_F + \mathbf{S}_{FR}\mathbf{D}_R = \mathbf{A}_F(t) \tag{6.86a}$$

and

$$\mathbf{M}_{RF}\ddot{\mathbf{D}}_F + \mathbf{M}_{RR}\ddot{\mathbf{D}}_R + \mathbf{S}_{RF}\mathbf{D}_F + \mathbf{S}_{RR}\mathbf{D}_R = \mathbf{A}_R(t) \tag{6.86b}$$

If support motions (at restraints) are zero, Eqs. (6.86) simplify to

$$\mathbf{M}_{FF}\ddot{\mathbf{D}}_F + \mathbf{S}_{FF}\mathbf{D}_F = \mathbf{A}_F(t) \tag{6.87a}$$

and

$$\mathbf{M}_{RF}\ddot{\mathbf{D}}_F + \mathbf{S}_{RF}\mathbf{D}_F = \mathbf{A}_R(t) \tag{6.87b}$$

These equations can be used for calculating *free displacements* $\mathbf{D}_F$ and *support reactions* $\mathbf{A}_R(t)$.

In many problems it is sufficiently accurate to merely lump tributary masses at the nodes of a discretized continuum [9]. When doing so, we form the *lumped mass matrix* $\mathbf{M}_\ell$ for the whole structure as:

$$\mathbf{M}_\ell = \begin{bmatrix} \mathbf{M}_1 & \mathbf{0} & \cdots & \mathbf{0} & \cdots & \mathbf{0} \\ \mathbf{0} & \mathbf{M}_2 & \cdots & \mathbf{0} & \cdots & \mathbf{0} \\ \cdots & \cdots & \cdots & \cdots & \cdots & \cdots \\ \mathbf{0} & \mathbf{0} & \cdots & \mathbf{M}_j & \cdots & \mathbf{0} \\ \cdots & \cdots & \cdots & \cdots & \cdots & \cdots \\ \mathbf{0} & \mathbf{0} & \cdots & \mathbf{0} & \cdots & \mathbf{M}_{n_n} \end{bmatrix} \tag{6.88}$$

where n_n is the number of nodes. The typical submatrix $\mathbf{M}_j$ in Eq. (6.88) signifies a small diagonal array defined to be

$$\mathbf{M}_j = M_j \mathbf{I}_0 \tag{6.89}$$

In this expression M_j is the tributary mass lumped at node j, and $\mathbf{I}_0$ is an identity matrix with 1 replaced by 0 wherever a nontranslational displacement occurs. Thus, the lumped-mass approach has the advantage that the mass matrix $\mathbf{M}_\ell$ is always diagonal, although not always positive-definite [10].

In many instances the number of degrees of freedom in an assembled system may be reduced. The concept of *matrix condensation* [6] is a well-known procedure for decreasing the number of unknown displacements in a statics problem. With such applications no loss of accuracy results from the reduction, because the method is simply Gaussian elimination of displacements in matrix form. For dynamic analysis, a similar type of condensation was introduced by Guyan [11], which brings in an additional approximation.

Starting with *static reduction*, we write action equations of equilibrium for free displacements in the partitioned form

$$\begin{bmatrix} \mathbf{S}_{\mathrm{AA}} & \mathbf{S}_{\mathrm{AB}} \\ \mathbf{S}_{\mathrm{BA}} & \mathbf{S}_{\mathrm{BB}} \end{bmatrix} \begin{bmatrix} \mathbf{D}_{\mathrm{A}} \\ \mathbf{D}_{\mathrm{B}} \end{bmatrix} = \begin{bmatrix} \mathbf{A}_{\mathrm{A}} \\ \mathbf{A}_{\mathrm{B}} \end{bmatrix} \tag{6.90}$$

Here the subscript A denotes the displacements that are to be eliminated, and the subscript B refers to those that will be retained. Now rewrite Eq. (6.90) as two sets of equations, as follows:

$$\mathbf{S}_{\mathrm{AA}}\mathbf{D}_{\mathrm{A}} + \mathbf{S}_{\mathrm{AB}}\mathbf{D}_{\mathrm{B}} = \mathbf{A}_{\mathrm{A}} \tag{6.91a}$$

$$\mathbf{S}_{\mathrm{BA}}\mathbf{D}_{\mathrm{A}} + \mathbf{S}_{\mathrm{BB}}\mathbf{D}_{\mathrm{B}} = \mathbf{A}_{\mathrm{B}} \tag{6.91b}$$

Solving for the vector of *dependent displacements* $\mathbf{D}_{\mathrm{A}}$ in Eq. (6.91a) yields

$$\mathbf{D}_{\mathrm{A}} = \mathbf{S}_{\mathrm{AA}}^{-1}(\mathbf{A}_{\mathrm{A}} - \mathbf{S}_{\mathrm{AB}}\mathbf{D}_{\mathrm{B}}) \tag{6.92}$$

Substitute Eq. (6.92) into Eq. (6.91b) and collect terms to obtain

$$\mathbf{S}_{\mathrm{BB}}^{*}\mathbf{D}_{\mathrm{B}} = \mathbf{A}_{\mathrm{B}}^{*} \tag{6.93}$$

in which

$$\mathbf{S}_{\mathrm{BB}}^{*} = \mathbf{S}_{\mathrm{BB}} = \mathbf{S}_{\mathrm{BA}}\mathbf{S}_{\mathrm{AA}}^{-1}\mathbf{S}_{\mathrm{AB}} \tag{6.94}$$

and

$$\mathbf{A}_{\mathrm{B}}^{*} = \mathbf{A}_{\mathrm{B}} - \mathbf{S}_{\mathrm{BA}}\mathbf{S}_{\mathrm{AA}}^{-1}\mathbf{A}_{\mathrm{A}} \tag{6.95}$$

From Eq. (6.93) we see that Eqs. (6.91) have been reduced to a smaller set involving only the *independent displacements* in vector $\mathbf{D}_B$. The *reduced stiffness matrix* $\mathbf{S}_{BB}^*$ in Eq. (6.94) is a modified version of the original submatrix $\mathbf{S}_{BB}$. Also, the *reduced action vector* $\mathbf{A}_B^*$ in Eq. (6.95) contains terms modifying the subvector $\mathbf{A}_B$ that are considered to be equivalent loads of type *B* due to actions of type *A*. Furthermore, Eq. (6.92) may now be viewed as the back-substitution formula required to find vector $\mathbf{D}_A$ *exactly* from vector $\mathbf{D}_B$.

Turning next to *dynamic reduction*, we recall from Eq. (6.87a) that the undamped equations of motion for free displacements are

$$\mathbf{M}\ddot{\mathbf{D}} + \mathbf{S}\,\mathbf{D} = \mathbf{A} \tag{6.96}$$

in which the subscript F has been omitted. Then assume as a new approximation that the displacements of type A are dependent upon those of type B, as follows:

$$\mathbf{D}_A = \mathbf{T}_{AB}\mathbf{D}_B \tag{6.97a}$$

where

$$\mathbf{T}_{AB} = -\mathbf{S}_{AA}^{-1}\mathbf{S}_{AB} \tag{6.97b}$$

Even for static analysis, this relationship is correct only when actions of type A do not exist [see Eq. (6.92)]. However, Eq. (6.97a) follows the finite-element theme of "slave" and "master" displacements. Differentiating Eq. (6.97a) twice with respect to time produces

$$\ddot{\mathbf{D}}_A = \mathbf{T}_{AB}\ddot{\mathbf{D}}_B \tag{6.98}$$

For the purpose of reducing the equations of motion to a smaller set, we can form the transformation operator

$$\mathbf{T}_B = \begin{bmatrix} \mathbf{T}_{AB} \\ \mathbf{I}_B \end{bmatrix} \tag{6.99}$$

in which $\mathbf{I}_B$ is an identity matrix of the same order as $\mathbf{S}_{BB}$. Substituting Eqs. (6.97a) and (6.98) into (6.96) and premultiplying the latter by $\mathbf{T}_B^T$ gives

$$\mathbf{M}_{BB}^*\ddot{\mathbf{D}}_B + \mathbf{S}_{BB}^*\mathbf{D}_B = \mathbf{A}_B^* \tag{6.100}$$

In this equation the matrices $\mathbf{S}_{BB}^*$ and $\mathbf{A}_B^*$ still have the definitions given in Eqs. (6.94) and (6.95). However, the *reduced mass matrix* $\mathbf{M}_{BB}^*$ is found to be

$$\mathbf{M}_{BB}^* = \mathbf{T}_B^T\mathbf{M}\mathbf{T}_B = \mathbf{M}_{BB} + \mathbf{T}_{AB}^T\mathbf{M}_{AB} + \mathbf{M}_{BA}\mathbf{T}_{AB} + \mathbf{T}_{AB}^T\mathbf{M}_{AA}\mathbf{T}_{AB} \tag{6.101}$$

As mentioned before, all of the condensed matrices in Eq. (6.100) are approximate. If damping is to be included in the equations of motion, a reduced damping matrix $\mathbf{C}^*_{BB}$ also can be derived, which has a form analogous to $\mathbf{M}^*_{BB}$ in Eq. (6.101).

When applying Guyan reduction to framed structures, we usually choose rotations at the joints of beams, plane frames, grids, and space frames as the dependent set of displacements. However, the method can be used in a much more general manner for various discretized continua. That is, any arbitrarily selected set of displacements may be referred to as type A, while the remaining displacements become type B. The trouble with this generality is that a good choice of "slave" and "master" displacements is not always obvious. Even with framed structures there are cases when joint rotations are more important than translations and should not be eliminated.

Example. The fixed beam in Fig. 6.10*a* is divided into three flexural elements, each of which has the same properties E, I, ρ, and A. By Guyan

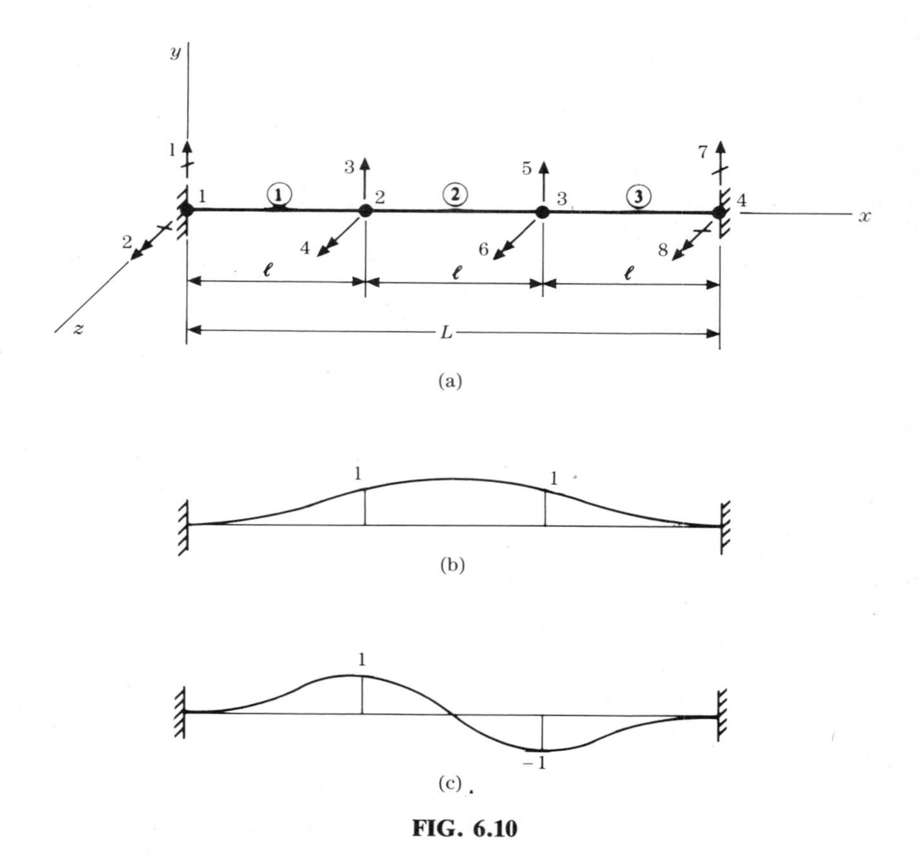

FIG. 6.10

reduction, we shall eliminate the rotations at nodes 2 and 3 and retain the translations. The reduction will be followed by a vibrational analysis that is compared against exact results.

For this example, the assembled system stiffness matrix (without consideration of restraints) is

$$
\mathbf{S}_s = \frac{2EI}{\ell^3}\begin{bmatrix}
6 & & & & & & & \\
3\ell & 2\ell^2 & & & & & \text{Sym.} & \\
-6 & -3\ell & 12 & & & & & \\
3\ell & \ell^2 & 0 & 4\ell^2 & & & & \\
0 & 0 & -6 & -3\ell & 12 & & & \\
0 & 0 & 3\ell & \ell^2 & 0 & 4\ell^2 & & \\
0 & 0 & 0 & 0 & -6 & -3\ell & 6 & \\
0 & 0 & 0 & 0 & 3\ell & \ell^2 & -3\ell & 2\ell^2
\end{bmatrix}
\begin{matrix} 1 \\ 2 \\ 3 \\ 4 \\ 5 \\ 6 \\ 7 \\ 8 \end{matrix}
\qquad (a)
$$

(Column indexes 1 2 3 4 5 6 7 8 below the matrix; dashed boxes ①, ②, ③ enclose element contributions.)

Similarly, the assembled system mass matrix for translational inertias becomes

$$
\mathbf{M}_s = \frac{\rho A\ell}{420}\begin{bmatrix}
156 & & & & & & & \\
22\ell & 4\ell^2 & & & & & \text{Sym.} & \\
54 & 13\ell & 312 & & & & & \\
-13\ell & -3\ell^2 & 0 & 8\ell^2 & & & & \\
0 & 0 & 54 & 13\ell & 312 & & & \\
0 & 0 & -13l & -3\ell^2 & 0 & 8\ell^2 & & \\
0 & 0 & 0 & 0 & 54 & 13\ell & 156 & \\
0 & 0 & 0 & 0 & -13\ell & -3\ell^2 & -22\ell & 4\ell^2
\end{bmatrix}
\begin{matrix} 1 \\ 2 \\ 3 \\ 4 \\ 5 \\ 6 \\ 7 \\ 8 \end{matrix}
\qquad (b)
$$

(Column indexes 1 2 3 4 5 6 7 8 below the matrix; dashed boxes ①, ②, ③ enclose element contributions.)

Dashed boxes in Eqs. (a) and (b) enclose the contributions of elements 1, 2, and 3, which are drawn from Eqs. (6.59) and (6.61). Moreover, the joint displacement indexes for the problem (see Fig. 6.10a) are listed at the right side and below the matrices.

As the first step, we remove the first, second, seventh, and eighth rows and columns from matrices $\mathbf{S}_s$ and $\mathbf{M}_s$, because displacements 1, 2, 7, and 8 are restrained by supports. Then the remaining 4×4 arrays are rearranged to put the rotational terms before the translational terms, as follows:

$$\mathbf{S} = \begin{bmatrix} \mathbf{S}_{AA} & \mathbf{S}_{AB} \\ \mathbf{S}_{BA} & \mathbf{S}_{BB} \end{bmatrix} = \frac{2EI}{\ell^3} \left[\begin{array}{cc:cc} 4\ell^2 & & \text{Sym.} & \\ \ell^2 & 4\ell^2 & & \\ \hdashline 0 & 3\ell & 12 & \\ -3\ell & 0 & -6 & 12 \end{array} \right] \begin{matrix} 4 \\ 6 \\ 3 \\ 5 \end{matrix} \qquad (c)$$
$$\begin{matrix} 4 & 6 & 3 & 5 \end{matrix}$$

$$\mathbf{M} = \begin{bmatrix} \mathbf{M}_{AA} & \mathbf{M}_{AB} \\ \mathbf{M}_{BA} & \mathbf{M}_{BB} \end{bmatrix} = \frac{\rho A \ell}{420} \left[\begin{array}{cc:cc} 8\ell^2 & & \text{Sym.} & \\ -3\ell^2 & 8\ell^2 & & \\ \hdashline 0 & -13\ell & 312 & \\ 13\ell & 0 & 54 & 312 \end{array} \right] \begin{matrix} 4 \\ 6 \\ 3 \\ 5 \end{matrix} \qquad (d)$$
$$\begin{matrix} 4 & 6 & 3 & 5 \end{matrix}$$

From Eq. (c) the inverse of $\mathbf{S}_{AA}$ is

$$\mathbf{S}_{AA}^{-1} = \frac{\ell}{30EI} \begin{bmatrix} 4 & -1 \\ -1 & 4 \end{bmatrix} \qquad (e)$$

Substituting this array and the other submatrices of $\mathbf{S}$ from Eq. (c) into Eq. (6.94) produces

$$\mathbf{S}_{BB}^{*} = \frac{12EI}{\ell^3} \begin{bmatrix} 2 & -1 \\ -1 & 2 \end{bmatrix} - \frac{6EI}{5\ell^3} \begin{bmatrix} 4 & 1 \\ 1 & 4 \end{bmatrix} = \frac{6EI}{5\ell^3} \begin{bmatrix} 16 & -11 \\ -11 & 16 \end{bmatrix} \begin{matrix} 3 \\ 5 \end{matrix} \qquad (f)$$
$$\begin{matrix} 3 & 5 \end{matrix}$$

which is the reduced stiffness matrix.

To reduce the mass matrix, we form the transformation matrix $\mathbf{T}_B$ by evaluating $\mathbf{T}_{AB}$ in Eq. (6.97b) and substituing it into Eq. (6.99). Thus,

$$\mathbf{T}_{AB} = -\mathbf{S}_{AA}^{-1}\mathbf{S}_{AB} = \frac{1}{5\ell} \begin{bmatrix} 1 & 4 \\ -4 & -1 \end{bmatrix} \qquad (g)$$

and

$$\mathbf{T}_B = \begin{bmatrix} \mathbf{T}_{AB} \\ \mathbf{I}_B \end{bmatrix} = \frac{1}{5\ell} \left[\begin{array}{cc} 1 & 4 \\ -4 & -1 \\ \hdashline 5\ell & 0 \\ 0 & 5\ell \end{array} \right] \qquad (h)$$

In this case the submatrix $\mathbf{I}_B$ in the lower partition of matrix $\mathbf{T}_B$ is of order 2 because there are two remaining translational displacements (numbers 3 and

TABLE 6.1 Frequency Coefficients μ_i for Prismatic Beams Modeled by Four Elements

Support Conditions	Mode	Exact[a]	CM-TR	% Error	CM-TO	% Error	LM-TO	% Error
Simple	1	9.870	9.872	+0.020	9.873	+0.030	9.867	−0.030
	2	39.48	39.63	+0.38	39.76	+0.71	39.19	−0.73
	3	88.23	90.45	+2.5	94.03	+6.6	83.21	−5.7
Free	1	22.37	22.41	+0.18	22.46	+0.40	18.91	−15
	2	61.67	62.06	+0.63	63.12	+2.4	48.00	−22
	3	120.9	121.9	+0.83	122.4	+1.2	86.84	−28
Fixed	1	22.37	22.40	+0.13	22.41	+0.18	22.30	−0.31
	2	61.67	62.24	+0.92	62.77	+1.8	59.25	−3.9
	3	120.9	123.5	+2.2	124.8	+3.2	97.40	−19
Cantilever	1	3.516	3.516	+0.00	3.516	+0.00	3.418	−2.8
	2	22.03	22.06	+0.14	22.09	+0.27	20.09	−8.8
	3	61.70	62.18	+0.83	62.97	+2.1	53.20	−14
Propped	1	15.42	15.43	+0.065	15.43	+0.065	15.40	−0.13
	2	49.97	50.28	+0.62	50.56	+1.2	49.05	−1.8
	3	104.2	106.6	+2.3	110.5	+6.0	91.53	−12

$\omega_i = \frac{\mu_i}{L^2}\sqrt{\frac{EI}{\rho A}}$

CM: Consistent Masses
LM: Lumped Masses

TR: Translations and Rotations
TO: Translations Only

[a] See Secs. 5.10 and 5.11

5). Substitution of matrix $\mathbf{M}$ from Eq. (d) and $\mathbf{T}_B$ from Eq. (h) into Eq. (6.101) yields

$$\mathbf{M}^*_{\mathrm{BB}} = \frac{\rho A \ell}{2100} \begin{bmatrix} 1696 & 319 \\ 319 & 1696 \end{bmatrix} \begin{matrix} 3 \\ 5 \end{matrix} \qquad \begin{matrix} 3 & \quad 5 \end{matrix} \tag{i}$$

which is the reduced mass matrix.

Using the terms in matrices $\mathbf{S}^*_{\mathrm{BB}}$ and $\mathbf{M}^*_{\mathrm{BB}}$, we can set up the eigenvalue problem in the form of Eq. (4.17). The angular frequencies found by this method are

$$\omega_{1,2} = 22.51, 63.26 \frac{1}{L^2} \sqrt{\frac{EI}{\rho A}} \tag{j}$$

where $L = 3\ell$. These values are in error by +0.63% and +2.6%, respectively; and they constitute upper bounds of the exact angular frequencies (see Sec. 5.11). The corresponding mode shapes are

$$\mathbf{\Phi} = [\mathbf{\Phi}_1 \quad \mathbf{\Phi}_2] = \begin{bmatrix} 1 & 1 \\ 1 & -1 \end{bmatrix} \tag{k}$$

which appear in Figs. 6.10b and c.

Frequency coefficients μ_i for prismatic beams with various end conditions are summarized in Table 6.1. In each case the beam is modeled by four flexural elements, and the results for the consistent-mass approach (with and without elimination of rotations) are compared with those for the lumped-mass method (with elimination of rotations). The table shows that the consistent-mass model produces much better accuracy than the lumped-mass model in beam analysis.

REFERENCES

[1] Weaver, W., Jr., and Johnston, P. R., *Finite Elements for Structural Analysis,* Prentice-Hall, Englewood Cliffs, NJ, 1984.

[2] Zienkiewicz, O. C., and Taylor, R. L., *The Finite Element Method,* 4th ed., McGraw-Hill, London, 1989.

[3] Cook, R. D., *Concepts and Applications of Finite Element Analysis,* 2nd ed., Wiley, New York, 1981.

[4] Weaver, W., Jr., and Johnston, P. R., *Structural Dynamics by Finite Elements,* Prentice-Hall, Englewood Cliffs, NJ, 1987.

[5] Timoshenko, S. P., and Goodier, J. N., *Theory of Elasticity,* 3rd ed., McGraw-Hill, New York, 1970.

[6] Weaver, W., Jr., and Gere, J. M., *Matrix Analysis of Framed Structures,* 3rd ed., Van Nostrand-Reinhold, New York, 1991.

[7] Oden, J. T., *Mechanics of Elastic Structures,* McGraw-Hill, New York, 1967.

[8] Archer, J. S., "Consistent Matrix Formulations for Structural Analysis Using Finite-Element Techniques," *AIAA J.*, **3,** (10), 1965, pp. 1910–1918.

[9] Clough, R. W., "Analysis of Structural Vibrations and Dynamic Response," *Rec. Adv. Mat. Meth. Struc. Anal. Des.*, ed. by R. H. Gallagher, Y. Yamada, and J. T. Oden, University of Alabama Press, Huntsville, AL, 1971, pp. 25–45.

[10] Gere, J. M., and Weaver, W., Jr., *Matrix Algebra for Engineers,* 2nd ed., Wadsworth, Belmont, CA, 1983.

[11] Guyan, R. J., "Reduction of Stiffness and Mass Matrices," *AIAA J.*, **3** (2), 1965, p. 380.

PROBLEMS

6.4-1. The figure shows an axial element with a linearly distributed load (force per unit length) given by the formula $b_x = b_1 + (b_2 - b_1)x/L$. Find the equivalent nodal loads $\mathbf{p}_b(t) = \{p_{b1}, p_{b2}\}$ due to this influence.

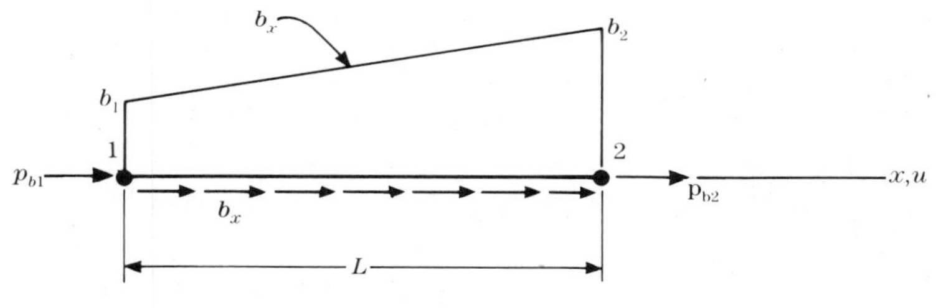

PROB. 6.4-1

6.4-2. A parabolically distributed load (force per unit length) has the formula $b_x = b_2(x/L)^2$, as illustrated in the figure. Determine the equivalent nodal loads $\mathbf{p}_b(t) = \{p_{b1}, p_{b2}\}$ resulting from this body force on an axial element.

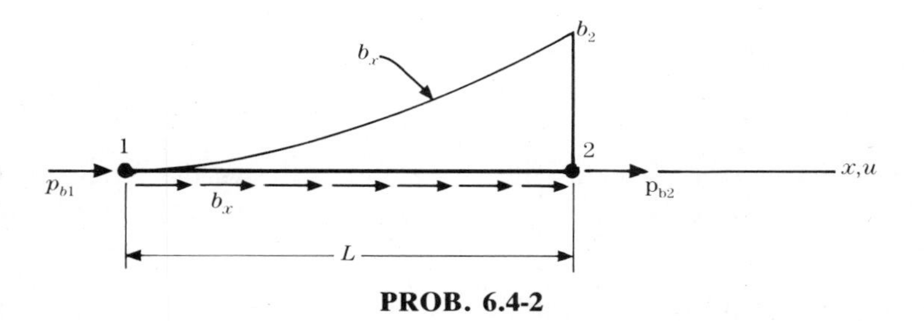

PROB. 6.4-2

6.4-3. Assume that an axial element has three nodes, as shown in part (*a*) of the figure. In terms of the coordinate x measured from node 2, the quadratic displacement shape functions in parts (*b*), (*c*), and (*d*) of the figure are: $f_1 = (2x - L)x/L^2$, $f_2 = (L^2 - 4x^2)/L^2$, and $f_3 = (2x + L)x/L^2$. Derive the 3×3 stiffness matrix $\mathbf{K}$ for this element if the axial rigidity EA is constant along the length.

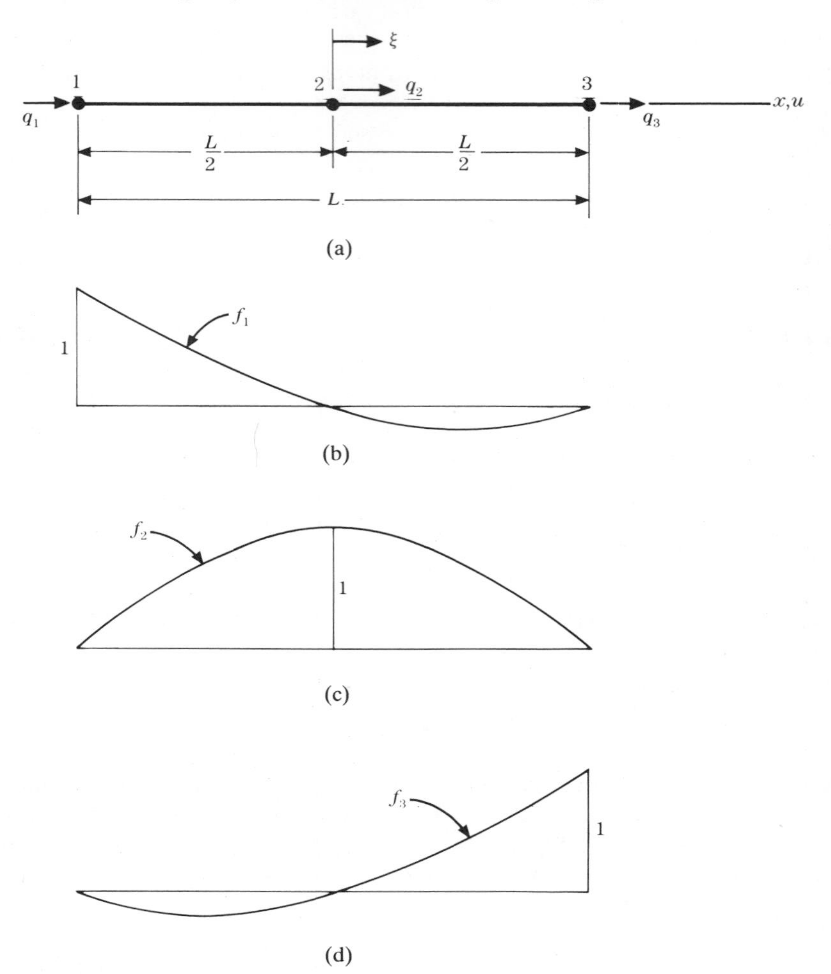

PROB. 6.4-3, 4, and 5

6.4-4. For the axial element with three nodes, derive the 3×3 consistent mass matrix $\mathbf{M}$, assuming that ρ and A are constant along the length.

6.4-5. Let a uniformly distributed load b_x (force per unit length) be applied to the axial element with three nodes. Find the equivalent nodal loads $\mathbf{p}_b(t) = \{p_{b1}, p_{b2}, p_{b3}\}$ due to this body force.

6.4-6. For the torsional element in the figure, obtain the equivalent nodal loads $\mathbf{p}_b(t) = \{p_{b1}, p_{b2}\}$ caused by a parabolically distributed moment (per unit length) given as $m_x = m_{x1}[1 - (x/L)^2]$.

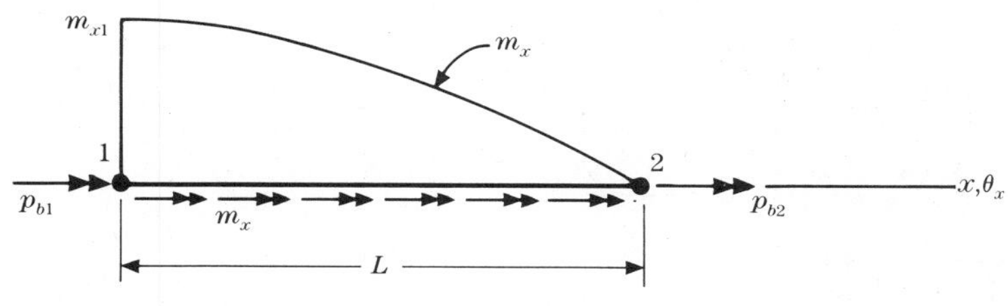

PROB. 6.4-6

6.4-7. Suppose that a concentrated moment M_x is applied to the torsional element at the distance x from node 1, as shown in the figure. Determine the equivalent nodal loads $\mathbf{p}_b(t) = \{p_{b1}, p_{b2}\}$ caused by this moment.

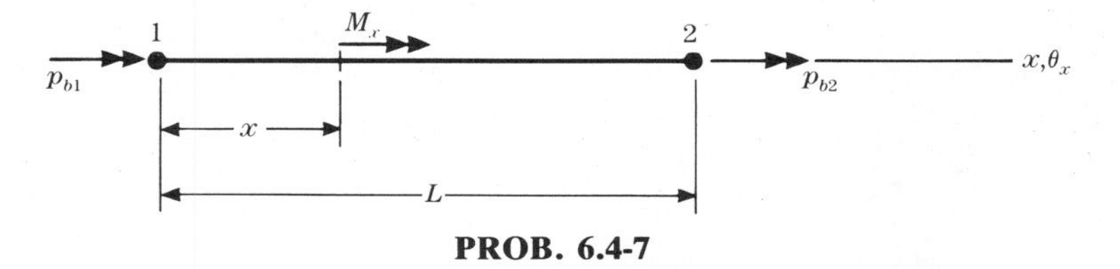

PROB. 6.4-7

6.4-8. The flexural element in the figure is subjected to a triangular load $b_y = b_2 x/L$ (force per unit length). Derive the equivalent nodal loads $\mathbf{p}_b(t) = \{p_{b1}, p_{b2}, p_{b3}, p_{b4}\}$ indicated at points 1 and 2.

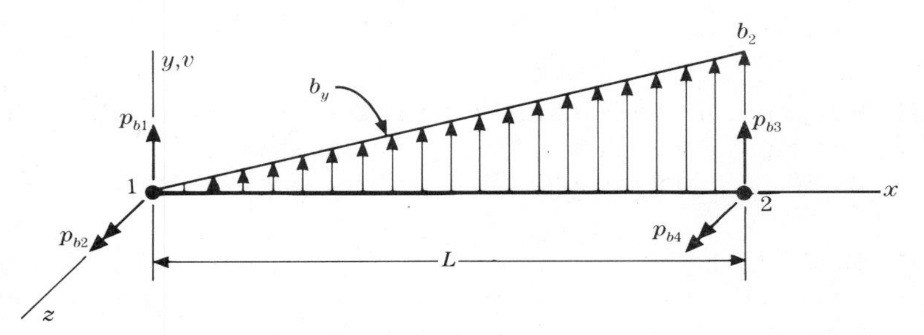

PROB. 6.4-8

6.4-9. The figure depicts a concentrated force P_y and a concentrated moment M_z applied to a flexural element at the distance x from node 1. Obtain the equivalent nodal loads $\mathbf{p}_b(t) = \{p_{b1}, p_{b2}, p_{b3}, p_{b4}\}$ for each of these actions.

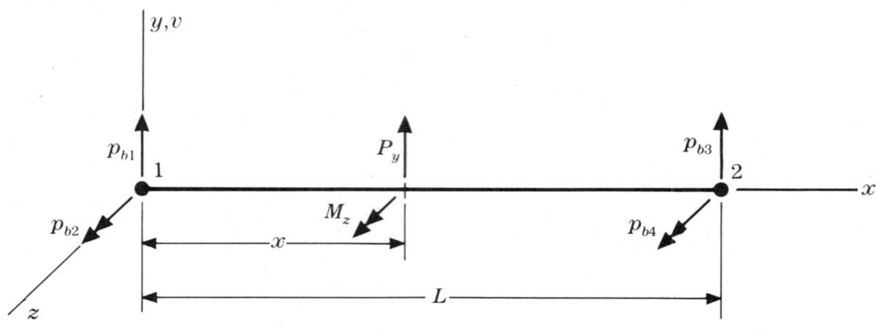

PROB. 6.4-9

6.4-10. In the figure a linearly distributed load $b_y = b_1 + (b_2 - b_1)x/L$ (force per unit length) acts upon a flexural element. Find the equivalent nodal loads $\mathbf{p}_b(t) = \{p_{b1}, p_{b2}, p_{b3}, p_{b4}\}$ due to this influence.

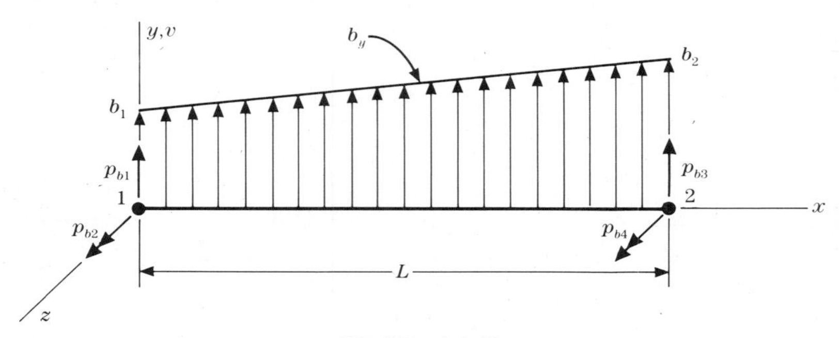

PROB. 6.4-10

6.4-11. Re-derive the 2×2 stiffness matrix $\mathbf{K}$ for a torsional element, using moment M_x and twist ψ as generalized stress and strain. Assume that the torsional rigidity GJ is constant along the length.

6.4-12 Derive again the 4×4 stiffness matrix $\mathbf{K}$ for a flexural element, with moment M_z and curvature ϕ as generalized stress and strain. Let the flexural rigidity EI be constant along the length.

6.5-1. The figure shows a two-element beam for which the parameters E, I, A, and ρ are constant along the length. Assemble the stiffness matrix $\mathbf{S}_s$ and the consistent mass matrix $\mathbf{M}_s$ (for translational inertias) in rearranged and partitioned forms.

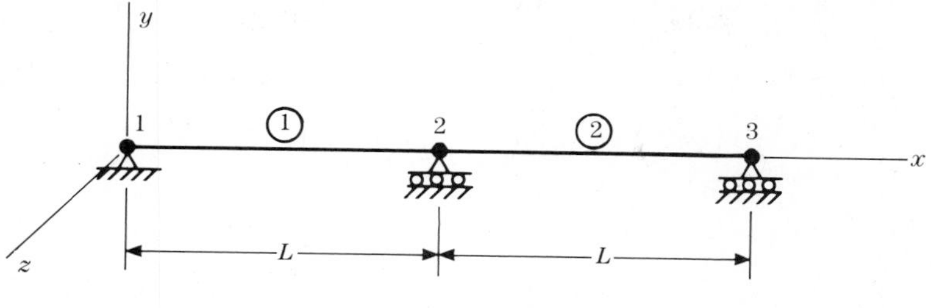

PROB. 6.5-1

6.5-2. Repeat Prob. 6.5-1 for the two-element beam shown in the figure.

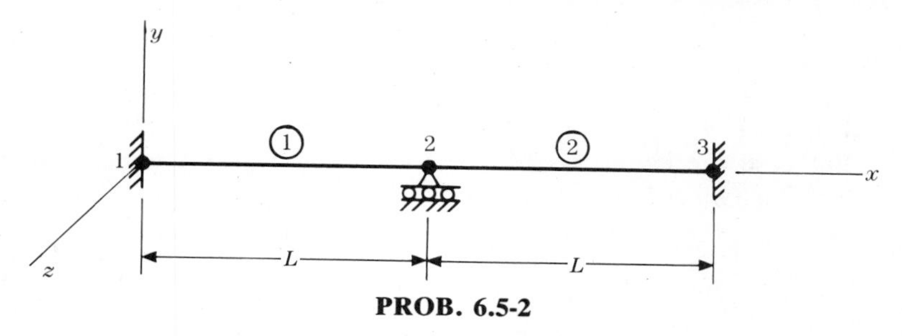

PROB. 6.5-2

6.5-3. Repeat Prob. 6.5-1 for the two-element beam shown in the figure.

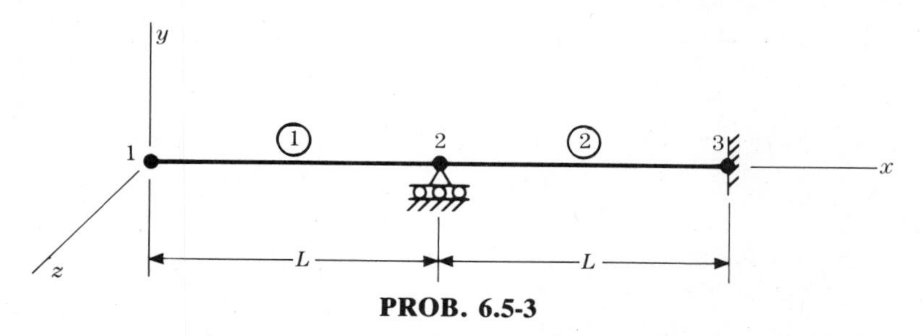

PROB. 6.5-3

6.5-4. Repeat Prob. 6.5-1 for the two-element beam shown in the figure.

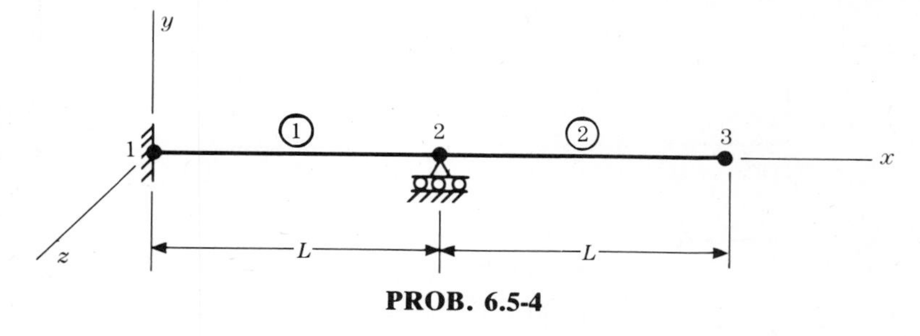

PROB. 6.5-4

6.5-5. The beam in the figure consists of two prismatic flexural elements with two degrees of freedom at node 2. Member 1 has moment of inertia and cross-sectional area equal to I and A, but member 2 has $2I$ and $2A$ for its properties. Determine the angular frequencies and mode shapes for this beam using translational consistent mass terms.

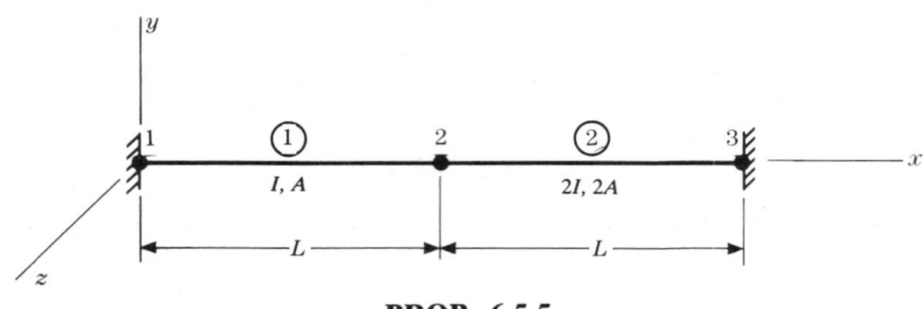

PROB. 6.5-5

6.5-6. Repeat Prob. 6.5-5 for the continuous beam shown in the figure. In this case the beam has constant values of I and A along its length and has two rotational degrees of freedom (at points 2 and 3).

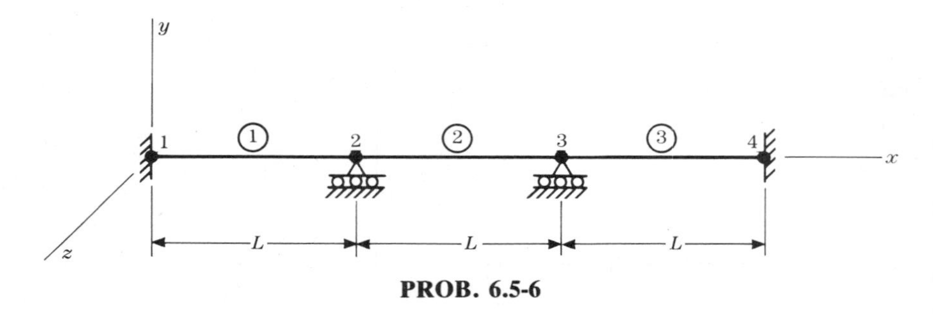

PROB. 6.5-6

6.5-7. Repeat Prob. 6.5-5 for the two-element continuous beam shown in the figure. Cross-sectional properties I and A are constant, and the free displacements at joints 1 and 2 are both rotational.

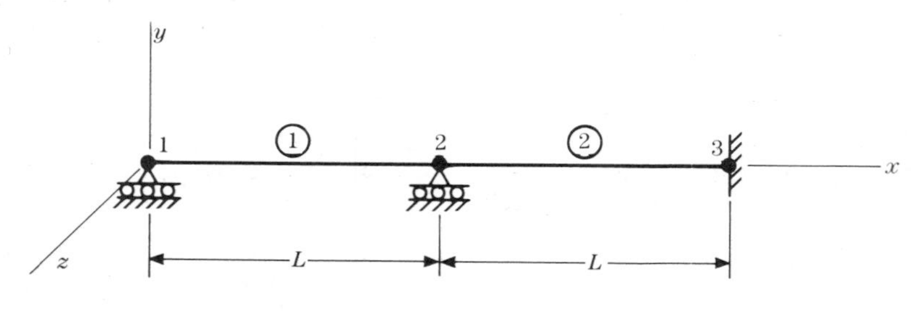

PROB. 6.5-7

6.5-8. The two-element prismatic beam in the figure is fixed at point 3 but free to rotate at point 1. Construct the stiffness and consistent mass matrices **S** and **M** for the three unrestrained displacements. Reduce these matrices by eliminating the rotations and retaining the translation. Then find the angular frequency of vibration for the remaining system, which has only one degree of freedom.

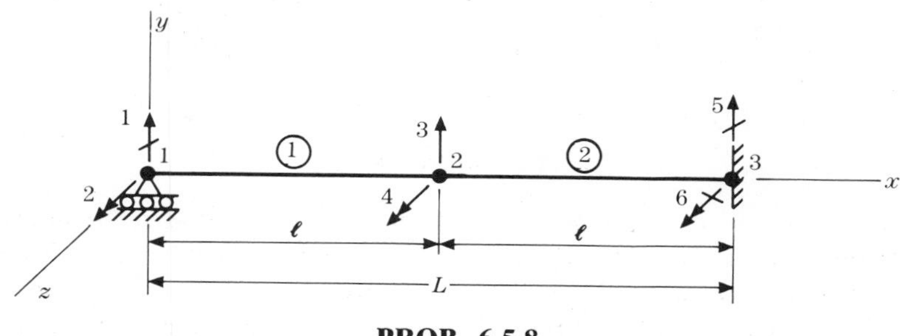

PROB. 6.5-8

6.5-9. The figure shows a two-element prismatic beam that is free to translate (but not rotate) at point 1 and is fixed at point 3. Assemble the stiffness and consistent mass matrices **S** and **M** for the three unrestrained displacements, and reduce them by eliminating the rotation and keeping the translations. Solve the eigenvalue problem to obtain angular frequencies and vibrational mode shapes for the reduced system.

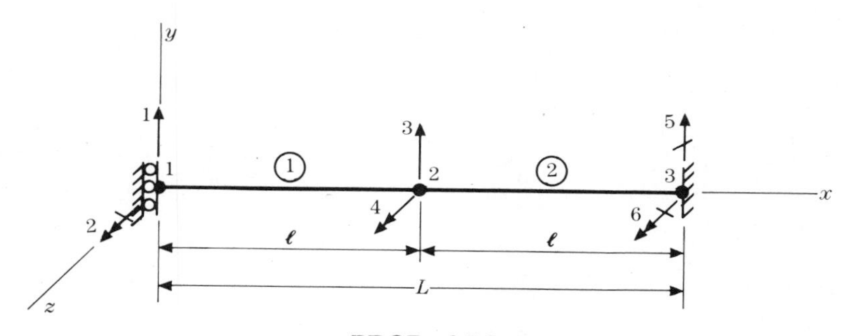

PROB. 6.5-9

6.5-10. The simply-supported beam in the figure is divided into two flexural elements with equal properties. Set up the stiffness and consistent mass matrices **S** and **M** for the four displacements that are unrestrained. Reduce these matrices by eliminating the rotations and retaining the translation. Then find the angular frequency of the remaining system, which has only one degree of freedom.

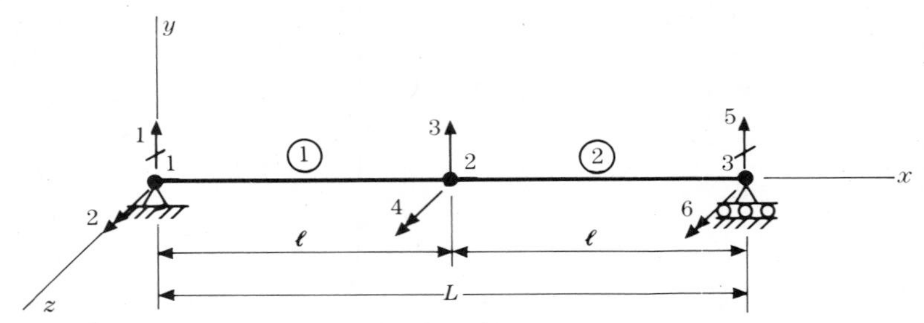

PROB. 6.5-10

6.5-11. A two-element prismatic beam has no restraints whatsoever, as implied in the figure. Construct the stiffness and consistent mass matrices **S** and **M** for the six unrestrained displacements, and eliminate the three rotations. For the three remaining translations, calculate the angular frequencies and mode shapes.

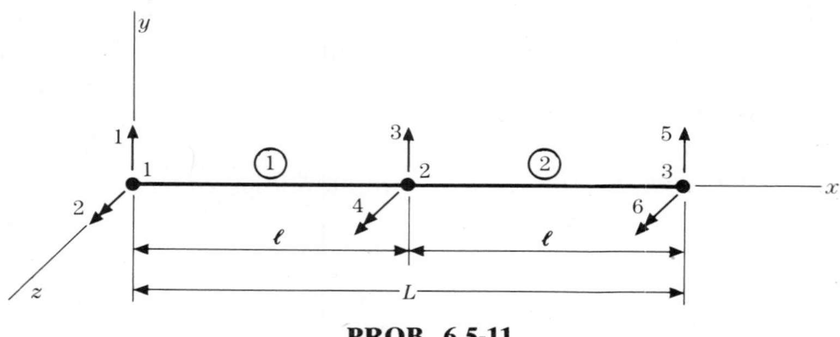

PROB. 6.5-11

BIBLIOGRAPHY

TEXTBOOKS ON VIBRATIONS (CHRONOLOGICAL ORDER)

Rayleigh, J. W. S., 1945, *The Theory of Sound,* Dover, New York.

Den Hartog, J. P., 1956, *Mechanical Vibrations,* 4th ed., McGraw-Hill, New York.

Myklestad, N. O., 1956, *Fundamentals of Vibration Analysis,* McGraw-Hill, New York.

Jacobsen, L. S., and Ayre, R. S., 1958, *Engineering Vibrations,* McGraw-Hill, New York.

Bishop, R. E. D., Gladwell, G. M. L., and Michaelson, S., 1965, *The Matrix Analysis of Vibration,* Cambridge University Press, London.

Anderson, R. A., 1967. *Fundamentals of Vibrations,* Macmillan, New York.

Frýba, L., 1972, *Vibrations of Solids and Structures Under Moving Loads,* Noordhoff, Groningen, The Netherlands.

Newland, D. E., 1975, *An Introduction to Random Vibrations and Spectral Analysis,* Longman, London.

Blevins, R. D., 1977. *Flow-Induced Vibration,* Van Nostrand-Reinhold, New York.

Tse, F. S., Morse, I. E., and Hinkle, R. T., 1978, *Mechanical Vibrations—Theory and Applications,* 2nd ed., Allyn and Bacon, Boston.

Thomson, W. T., 1981, *Theory of Vibration with Applications,* 2nd ed., Prentice-Hall, Englewood Cliffs, NJ.

Harker, R. J., 1983, *Generalized Methods of Vibration Analysis,* Wiley, New York.

Lalanne, M., Berthier, P., and Der Hagopian, J., 1983, *Mechanical Vibrations for Engineers,* Wiley, New York.

Meirovitch, L., 1986, *Elements of Vibration Analysis,* 2nd ed., McGraw-Hill, New York.

Rao, S. S., 1986, *Mechanical Vibrations,* Addison-Wesley, Reading, MA.

Olsson, M., 1986, *Analysis of Structures Subjected to Moving Loads,* LIT Report TVSM-1003, Lund, Sweden.

APPENDIX A

SYSTEMS OF UNITS AND MATERIAL PROPERTIES

A.1 SYSTEMS OF UNITS

The two most commonly-used systems of units are the International System (*SI units*) and the United States Customary (*US units*). The first of these is called an *absolute system* because the fundamental quantity of mass is independent of where it is measured. On the other hand, the US system has force as a fundamental quantity. It is referred to as a *gravitational system* because the unit of force is defined as the weight of a certain mass, which varies with location on Earth.

In the SI system, the three fundamental units required for structural dynamics are mass (*kilogram*), length (*meter*), and time (*second*). Corresponding to mass is a derived force called a *newton*, which is defined as the force needed to accelerate one kilogram by one meter per second squared. Thus, we have:

$$1\ \text{N} = 1\ \text{kg} \cdot \text{m/s}^2 \qquad (a)$$

which is based on Newton's second law that force = mass × acceleration.

In the US system, we use force (*pound*), length (*foot*), and time (*second*). (Note that the unit of time is the same for both systems.) Corresponding to force is a derived mass, which carries the name *slug*. This quantity is defined as the mass that will be accelerated by one foot per second squared when subjected to a force of one pound. Hence,

$$1\ \text{slug} = 1\ \text{lb-s}^2/\text{ft} \qquad (b)$$

which comes from the formula that mass = force/acceleration.

TABLE A.1 Conversion of US Units to SI Units

Quantity	US Units	×Factor	=SI Units
Length	inch (in.)	2.540×10^{-2}	meter (m)
Force	pound (lb)	4.448	newton (N)
Moment	pound-inch (lb-in.)	1.130×10^{-1}	newton · meter (N · m)
Stress	pound/inch^2 (lb/in.^{2} or psi)	6.895×10^3	pascal (Pa)
Mass	pound-sec^2/inch (lb-s^2/in.)	1.751×10^2	kilogram (kg)

Table A.1 presents *conversion factors* for calculating quantities in SI units from those in US units. The factors are given to four significant figures, which usually exceeds the accuracy of the numbers to be converted. Note that stress is defined in SI units as the *pascal*. That is,

$$1 \text{ Pa} = 1 \text{ N/m}^2 \tag{c}$$

For any numerical problem in structural mechanics, we must use a *consistent system of units*. By this we mean that all structural and load parameters must be expressed in the same units within each system. Some examples of consistent units for force, length, and time appear in Table A.2. For instance, in SI(1) we must express an applied force P in newtons (N), a length L in millimeters (mm), the modulus of elasticity E in newtons per square millimeter (N/mm^2), an acceleration $\ddot{u}$ in millimeters per second squared (mm/s^2), and so on.

When programming vibration problems for a digital computer, it is especially important that the system of units for input data be consistent. Otherwise, units would have to be converted within the logic of the program, thereby restricting its usage. For example, if in US units the length L were given in feet and the modulus E were expressed in pounds per square inch, the program would need to convert either L to inches or E to pounds per square foot.

For all of the numerical examples and problems in this book, we use either SI(2) or US(1) in Table A.2. Thus, in SI units we take force P in

TABLE A.2 Consistent Systems of Units

System		Force	Length	Time
SI	(1)	newton	millimeter	second
	(2)	**kilonewton**	**meter**	**second**
	(3)	meganewton	kilometer	second
US	(1)	**pound**	**inch**	**second**
	(2)	kilopound	foot	second
	(3)	megapound	yard	second

kilonewtons (kN), length L in meters (m), modulus E in giganewtons per square meter (GN/m^2 or GPa), acceleration $\ddot{u}$ in meters per second squared (m/s^2), and so on. [Note that the force kilonewton corresponds to the mass megagram (Mg) in Eq. (*a*).] Also, in US units we give force P in pounds (lb), length L in inches (in.), modulus E in pounds per square inch ($lb/in.^2$ or psi), acceleration $\ddot{u}$ in inches per second squared ($in./s^2$), and so on.

In this book, the *gravitational constant* g for conversion of weight to mass is taken to be

$$g = 386 \text{ in.}/\text{s}^2 \qquad \text{(in US units)}$$

or

$$g = 9.80 \text{ m/s}^2 \qquad \text{(in SI units)}$$

Although this constant varies with location on Earth, the three significant digits given here are accurate enough for most engineering applications.

A.2 MATERIAL PROPERTIES

To analyze solids and structures composed of various materials, we need to know certain physical properties. For vibration problems the essential material properties are *modulus of elasticity E, Poisson's ratio ν,* and *mass density ρ*. Table A.3 gives these properties in both US and SI units for some commonly-used materials. Note that the shearing modulus G is not listed in the table because it can be derived from E and ν as $G = E/2(1+\nu)$.

TABLE A.3 Properties of Materials[a]

	Modulus of Elasticity, E			Mass Density, ρ	
Material	$lb/in.^2$	GPa	Poisson's Ratio, ν	$lb\text{-}s^2/in.^4$	Mg/m^3
Aluminum	1.0×10^7	69	0.33	2.45×10^{-4}	2.62
Brass	1.5×10^7	103	0.34	8.10×10^{-4}	8.66
Concrete	3.6×10^6	25	0.15	2.25×10^{-4}	2.40
Steel	3.0×10^7	207	0.30	7.35×10^{-4}	7.85
Titanium	1.7×10^7	117	0.33	4.20×10^{-4}	4.49

[a] Numbers in this table are taken from *Mechanics of Materials,* 3rd edn., by J. M. Gere and S. P. Timoshenko, PWS-Kent, Boston, MA, 1990.

APPENDIX B

COMPUTER PROGRAMS

B.1 INTRODUCTION

For certain topics in this book, the use of a digital computer is highly desirable for expediting the calculations. These topics are numerical in nature and require many arithmetic operations. Therefore, we have devised several useful programs, for which flowcharts are given in this appendix. Descriptions of the required input data are also included in the discussion.

To understand the logic in a flow chart, the reader must be familiar with some algorithmic programming language, such as **FORTRAN**, **BASIC**, or **ALGOL**. A knowledgeable person will see that the flowcharts are oriented toward **FORTRAN**, but they may be coded in any other language if desired. Copies of the **FORTRAN** programs can be purchased from Paul R. Johnston, 838 Mesa Court, Palo Alto, CA 94306.

Various types of *operators* used in computer programs appear in Table B.1. Note that the symbols for multiplication and exponentiation are single and double asterisks, respectively. Furthermore, the replacement operator is an equal sign, as in **FORTRAN**.

Computer programs consist of *declarations* and *statements*. Declarations indicate types of variables, array sizes, formats for printing, and so on. They do not appear in a flowchart because they have no influence on the logic. Only executable statements are shown in a chart, and their arrangement gives the sequence of steps required to solve a problem. Table B.2 contains symbols for the types of statements appearing in our flowcharts. In **FORTRAN** the unconditional control statement is GO TO, the conditional control is IF, and the iterative control is DO.

TABLE B.1 Operators Used in Computer Programs

Operator	Symbol
Arithmetic operators	
Addition	+
Subtraction	−
Multiplication	*
Division	/
Exponentiation	* *
Replacement operator	=
Relational operators	
Less than	$<$
Less or equal	$\leq$
Equal	=
Greater or equal	$\geq$
Greater than	$>$
Not equal	$\neq$

Program notation is defined in conjunction with descriptions of data and flowcharts. In addition, the notation is summarized in Sec. B.6 at the end of this appendix.

B.2 STEP-BY-STEP SOLUTIONS FOR LINEAR ONE-DEGREE SYSTEMS

We described in Sec. 1.15 the topic of step-by-step solutions for linear systems with one degree of freedom. Formulas for the damped response of such a system to a piecewise-linear forcing function were derived as Eqs. (1.79) and (1.80). Because uniform time steps are to be used, the coefficients of u_j, $\dot{u}_j$, Q_j, and ΔQ_j in these equations become constants that need to be calculated only once. Thus, the displacement and velocity at time t_{j+1} may be written in eight parts, as follows:

$$D_{j+1} = C_1 D_j + C_2 V_j + (C_3 A_j + C_4 \, \Delta A_j)/m \tag{B.1}$$

$$V_{j+1} = C_5 D_j + C_6 V_j + (C_7 A_j + C_8 \, \Delta A_j)/m \tag{B.2}$$

In these expressions the symbol D represents displacement, V is velocity, A is actual or equivalent applied action, and ΔA is incremental action. The constant coefficients C_1 through C_8 appearing in Eqs. (B.1) and (B.2) have

TABLE B.2 Statements Used in Computer Programs

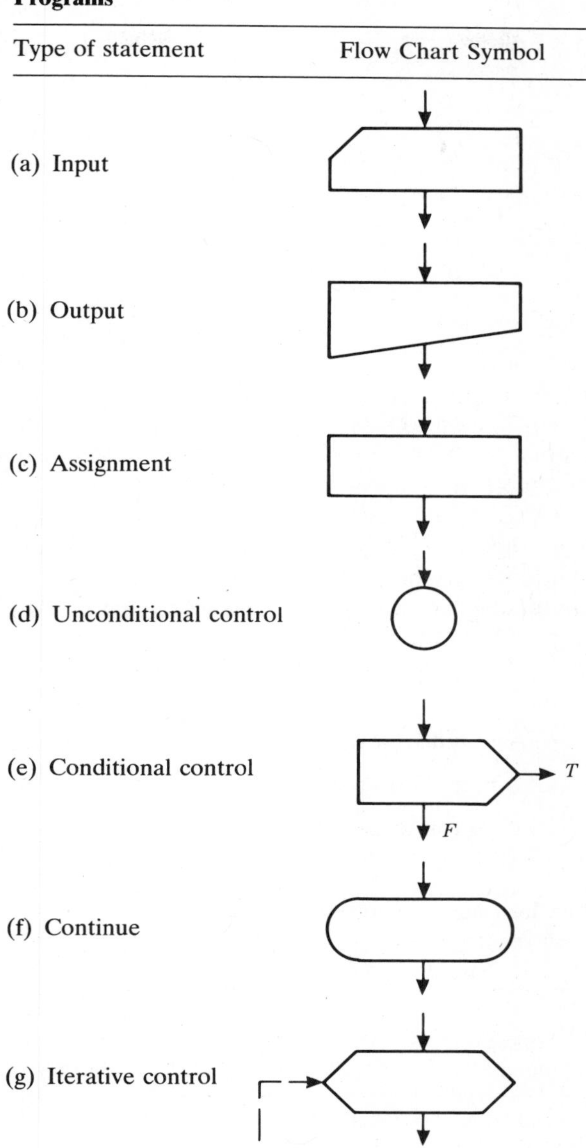

the definitions

$$
\begin{aligned}
C_1 &= e^{-n\Delta t}\left(\cos \omega_d \, \Delta t + \frac{n}{\omega_d} \sin \omega_d \, \Delta t\right) \\
C_2 &= \frac{1}{\omega_d} e^{-n\Delta t} \sin \omega_d \, \Delta t \qquad C_3 = \frac{1}{\omega^2}(1 - C_1) \\
C_4 &= \frac{1}{\omega^2 \, \Delta t}(\Delta t - C_2 - 2nC_3) \qquad C_5 = -\omega^2 C_2 \\
C_6 &= C_1 - 2nC_2 \qquad C_7 = C_2 \qquad C_8 = \frac{1}{\Delta t} C_3
\end{aligned}
\tag{B.3}
$$

Program **LINFORCE** uses Eqs. (B.1), (B.2), and (B.3) to calculate the response of a linear system to a piecewise-linear forcing function.

Table B.3 shows preparation of data for Program **LINFORCE**. In the second line of the table are the system stiffness SK, the system mass SM, the damping ratio DAMPR, and the number of loading systems NLS. Under *dynamic load data* the time parameters consist of the number of time steps NTS and the duration of the uniform time step DT. Next, we have the initial conditions of displacement D0 and velocity V0 at time $t = 0$. Load parameters in the table are defined as follows:

IAA = indicator for applied action (1 or 0)

IGA = indicator for ground acceleration (1 or 0)

AAF = applied-action factor

GAF = ground-acceleration factor

TABLE B.3 Data for Linear One-Degree System

Type of Data	No. of Lines	Items on Data Lines
SYSTEM DATA		
(a) Problem identification	1	Problem description
(b) System parameters	1	SK, SM, DAMPR, NLS
DYNAMIC LOAD DATA		
(a) Time parameters	1	NTS, DT
(b) Initial conditions	1	DØ, VØ
(c) Load parameters	1	IAA, IGA, AAF,GAF
(d) Forcing function		
(1) Function parameter	1	NFO
(2) Function ordinates	NFO	K, T(K), FO(K)

We may put either IAA or IGA equal to unity (indicating existence), but not both. The applied-action and ground-acceleration factors AAF and GAF carry appropriate dimensions and are to be multiplied by a dimensionless forcing function.

At the end of the table we have data for the piecewise-linear forcing function. Line (d)-(1) contains the number of function ordinates NFO. This is followed by NFO lines, each of which gives a subscript K, the time T(K) when the function ordinate occurs, and the value of the function ordinate FO(K) at that time. For simplicity, we restrict the time T(K) to be equal to an even number of time steps DT. If the forcing function changes instantaneously at time $T(K) \neq 0$, two lines of data are required to define FO(K) on both sides of the discontinuity (for the same time). Note again that the function ordinates will receive dimensions only when they are multiplied within the program by either AAF or $-SM*GAF$, thus creating an actual or equivalent load.

Flowchart B.1 shows the outline and executable statements in Program **LINFORCE**. Notes on the right-hand side of the chart indicate the operations or equations being applied. When the data for Examples 1 and 2 in Sec. 1.15 were used with this program, we obtained the responses that are listed in Tables 1.1 and 1.2.

B.3 NUMERICAL SOLUTIONS FOR NONLINEAR ONE-DEGREE SYSTEMS

Section 2.6 contains numerical methods for solving nonlinear equations of motion for one-degree systems. The two iterative approaches discussed in detail consist of the average-acceleration method and the linear-acceleration method. We consider the first to be the better choice because it is unconditionally stable, regardless of the duration of the time step.

To apply the average-acceleration method in a computer program named **AVAC1**, we use Eqs. (2.61) through (2.63) within an iterative loop, with Eqs. (2.64) and (2.65) evaluated beforehand. Also, Eqs. (2.66) through (2.68) serve to start each step, and Eq. (2.69) gives a check on accuracy for each iteration.

For a linear system, the only change required in the data (see Table B.3) is to add the following under SYSTEM DATA:

(c) Solution Parameters NIT, EPS

where NIT = number of iterations allowed

EPS = error limit for displacement accuracy

The partial flowchart B.2 shows response calculations for Program **AVAC1**, which pertains to a linear one-degree system with a piecewise-linear forcing function. This flowchart segment replaces Part 4 in Program **LINFORCE** (see Flowchart B.1). When the data for Example 1 in Sec. 2.6 are processed by Program **AVAC1**, the results are as shown in the first part of Table 2.1.

To convert Program **AVAC1** to handle a nonlinear problem, we need only change the system parameters and the equation for evaluating acceleration. For the pendulum in Example 2 of Sec. 2.6, the system parameters become GOL and NLS, where GOL $= g/\ell$. Also, in a program named **AVAC2A**, the acceleration at each time station is calculated as follows:

```
DDD(J + 1) = -GOL * SIN(D(J + 1))
```

The first part of Table 2.2 gives the results obtained by running Program **AVAC2A** with the data for Example 2. Variants of this program named **AVAC2B**, **2C**, and so on, may be created for the purpose of analyzing other systems with geometric nonlinearities, such as those in Problem Set 2.1.

For the cubically hardening spring in Example 3 of Sec. 2.6, we need only add the scalar multiplier S to the system parameters given in Table B.3. Then in a program named **AVAC3A**, the calculation of acceleration at each time station must include the term

```
-SK * S * D(J + 1) * * 3
```

When the data for Example 3 are run with Program **AVAC3A**, we find the output listed in the first part of Table 2.3. Similar programs, named **AVACB**, **3C**, and so on, may be devised for other systems having nonlinear load–displacement curves, such as those in Problem Set 2.2.

Another series of special-purpose programs, named **AVAC4A**, **4B**, and so on, can also be written for analyzing systems with piecewise-linear characteristics, such as those in Problem Set 2.5. Furthermore, any program for the average-acceleration method may be easily converted to one for the linear-acceleration method by using Eqs. (2.76) and (2.77) in place of Eqs. (2.62) and (2.65). Thus, the conversions of **AVAC1**, **2A**, **3A**, and so on, to **LINAC1**, **2A**, **3A**, and so on, require changing only a few lines in each program. Results from such programs are given in the second parts of Tables 2.1, 2.2, and 2.3. In addition, we could also program the direct linear extrapolation method described in Sec. 2.6, along with the iterative techniques. However, programs of that type are not included in this appendix.

B.4 ITERATION OF EIGENVALUES AND EIGENVECTORS

We described in Sec. 4.7 the iteration method for calculating frequencies and mode shapes of multi-degree linear systems. Recurrence formulas for obtaining the dominant eigenvalue and the corresponding eigenvector are stated as Eqs. (4.100), (4.101), and (4.102). Also, appropriate expressions for the eigenvalue problem in a vibratory system are given by Eqs. (4.103), (4.104), and (4.105). In addition, the introduction of modal constraints and the use of sweeping matrices to eliminate the first and second modes are concepts leading to Eqs. (4.106) through (4.109). To eliminate higher modes, we must formalize the procedure for generating sweeping matrices, because principal minors of order three or more are too complicated to handle with explicit formulas. Furthermore, the mass matrix need not be diagonal, as was assumed in Sec. 4.7.

Let us reconsider the expansion of Eq. (s) in Sec. 4.7, which may be written

$$b_{11}u_1 + b_{12}u_2 + b_{13}u_3 + \cdots + b_{1n}u_n = 0 \tag{a}$$

in which

$$b_{1j} = \mathbf{\Phi}_{\mathrm{M}1}^{\mathrm{T}}\mathbf{M}_j \qquad (j = 1, 2, 3, \ldots, n) \tag{b}$$

and the symbol $\mathbf{M}_j$ denotes the jth column of matrix $\mathbf{M}$. Expressing in matrix form the dependence of the displacement vector $\mathbf{D}$ upon $\mathbf{D}'$, we obtain

$$\begin{bmatrix} b_{11} & b_{12} & b_{13} & \cdots & b_{1n} \\ 0 & 1 & 0 & \cdots & 0 \\ 0 & 0 & 1 & \cdots & 0 \\ \cdots & \cdots & \cdots & \cdots & \cdots \\ 0 & 0 & 0 & \cdots & 1 \end{bmatrix} \begin{bmatrix} u_1 \\ u_2 \\ u_3 \\ \cdots \\ u_n \end{bmatrix} = \begin{bmatrix} 0 & 0 & 0 & \cdots & 0 \\ 0 & 1 & 0 & \cdots & 0 \\ 0 & 0 & 1 & \cdots & 0 \\ \cdots & \cdots & \cdots & \cdots & \cdots \\ 0 & 0 & 0 & \cdots & 1 \end{bmatrix} \begin{bmatrix} u_1' \\ u_2 \\ u_3 \\ \cdots \\ u_n \end{bmatrix} \tag{c}$$

More briefly,

$$\mathbf{B}_1\mathbf{D} = \mathbf{I}_{01}\mathbf{D}' \tag{d}$$

where $\mathbf{B}_1$ is a modified identity matrix with the constant b_{ij} in its first row. Also $\mathbf{I}_{01}$ is another modified identity matrix having a zero as its first diagonal term. Premultiplying both sides of Eq. (d) by $\mathbf{B}_1^{-1}$ yields

$$\mathbf{D} = \mathbf{T}_{\mathrm{S}1}\mathbf{D}' \tag{e}$$

where

$$\mathbf{T}_{\mathrm{S}1} = \mathbf{B}_1^{-1}\mathbf{I}_{01} \tag{f}$$

From here the use of the sweeping matrix $\mathbf{T}_{\mathrm{S}1}$ proceeds as described previously in Sec. 4.7.

If a second deflation is desired, the form of matrix $\mathbf{B}_2$ becomes

$$\mathbf{B}_2 = \begin{bmatrix} b_{11} & b_{12} & b_{13} & \cdots & b_{1n} \\ b_{21} & b_{22} & b_{23} & \cdots & b_{2n} \\ 0 & 0 & 1 & \cdots & 0 \\ \cdots & \cdots & \cdots & \cdots & \cdots \\ 0 & 0 & 0 & \cdots & 1 \end{bmatrix} \qquad (g)$$

in which

$$b_{2j} = \mathbf{\Phi}_{\mathrm{M}2}^{\mathrm{T}} \mathbf{M}_j \qquad (j = 1, 2, 3, \ldots, n) \qquad (h)$$

In this case the matrix $\mathbf{I}_{02}$ has zeros in its first two diagonal positions. Assuming that n_{m} modes are to be iterated in succession, we write the general formula for each sweeping matrix as

$$\mathbf{T}_{\mathrm{S}i} = \mathbf{B}_i^{-1} \mathbf{I}_{0i} \qquad (i = 1, 2, 3, \ldots, n_{\mathrm{m}}) \qquad (\mathrm{B}.4)$$

This notation implies that i rows of matrix $\mathbf{B}_i$ contain constants analogous to those in Eqs. (b) and (h). In addition, the matrix $\mathbf{I}_{0i}$ has zeros for its first i diagonal terms. To avoid undue loss of accuracy by deflations, the value of n_{m} in Eq. (B.4) should not exceed about 6.

Program **EIGIT** incorporates Eqs. (4.100) through (4.105) as well as Eq. (B.4) for determining the first n_{m} eigenvalues and eigenvectors of a linear multi-degree system by the method of iteration. Flowchart B.3 shows the logical steps in this program, where the first part involves reading the system data in Table B.4. System parameters in the table are defined as follows:

ITM = identifier for type of coefficient matrix
(0 for stiffness; 1 for flexibility)
N = number of degrees of freedom
NM = number of modes to be found by iteration
NIT = number of iterations allowed for each mode
EPS = allowable error for terms in eigenvectors

For this program the coefficient matrix **C** may be either a stiffness or a flexibility matrix, as indicated by ITM.

Program **EIGIT** calls the subprogram named **INVERT**, which performs the inversion of a positive-definite matrix by Gauss–Jordan elimination. (For a description of this technique, see Ref. [1] of Chap. 4.) We chose this method because it suits the inversion of matrix **B** in part 4 of this flow chart, where the meanings of arguments in the subprogram-call become obvious. Flowchart B.4 shows the logical steps in subprogram **INVERT**.

TABLE B.4 Data for Linear Multidegree System

Type of Data	No. of Lines	Items on Data Lines
SYSTEM DATA		
(a) Problem identification	1	Problem description
(b) System parameters	1	ITM, N, NM, NIT, EPS
(c) System matrices		
(1) Coefficient matrix[a]	N	C(I, J) J = 1, N
(2) Mass matirix	N	SM(I, J) J = 1, N
DYNAMIC LOAD DATA[b]		
(a) Time parameters	1	NTS, DT, DAMPR
(b) Initial conditions		
(1) Displacements	1	DØ(J) J = 1, N
(2) Velocities	1	VØ(J) J = 1, N
(c) Load information		
(1) Load parameters	1	IAA, IGA, GAF
(2) Applied-action factors	1	AAF(J) J = 1, N
(d) Forcing function		
(1) Function parameter	1	NFO
(2) Function ordinates	NFO	K, T(K), FO(K)

[a] May be either stiffness or flexibility matrix
[b] Not required for Program **EIGIT**.

Program **EIGIT** proves useful for vibration problems having positive-definite coefficient matrices, as do most of those in Problem Set 4.7. This program also serves as a building block for Program **NOMOLIN**, to be described in the next section.

B.5 STEP-BY-STEP SOLUTIONS FOR LINEAR MULTI-DEGREE SYSTEMS

Section 4.11 contains equations for using the normal-mode method to calculate the transient response of damped, linear multi-degree systems to piecewise-linear forcing functions. Here we present Program **NOMOLIN** to apply Eqs. (4.158) through (4.161) for the first n_m vibrational modes of such a system. This program combines the logic in Programs **LINFORCE** and **EIGIT** with the concepts of coordinate transformations required in the normal-mode method of dynamic analysis.

The main program for **NOMOLIN** calls six subprograms (or subroutines), as shown by the double boxes in Flowchart B.5. The first subprogram (**EIGITN**) is a modified version of **EIGIT**, which is the routine for iterating eigenvalues and eigenvectors described in the preceding section. Now it is treated as a subprogram that performs the operations given in Flowchart B.3. However, as the last step in this subprogram, the eigenvectors are normalized with respect to the mass matrix. Then the main program reads the number of loading systems NLS, initializes the loading number LN to zero, and increases LN by one.

Next, Subprogram **DYLOAD** reads and writes the dynamic load data given in the second part of Table B.4. Output from this subprogram includes the loading number LN as well as the number of loading systems NLS. Comparison of the latter part of Table B.4 with Table B.3 shows that a damping ratio DAMPR (for all modes) has been added to the first line. Second, the initial conditions D0 and V0 are now vectors with N terms instead of single values. Third, the applied-action factors AAF are also listed in a vector of N terms, but the ground-acceleration factor GAF remains a single value. The latter factor implies that all free displacements in the system are translational and in the same direction as the ground acceleration. Within Subprogram **DYLOAD** we form a vector of equivalent loads due to ground accelerations. This vector is the negative product of GAF, the row sums of the mass matrix, and the dimensionless forcing function [see Eq. (4.86)]. As in Program **LINFORCE**, we may put either IAA or IGA equal to unity (meaning existence), but not both. Because Subprogram **DYLOAD** is so similar to parts 2 and 3 in Program **LINFORCE**, there is no need to give a flowchart.

The third double box in the main program for **NOMOLIN** carries the name **TRANOR**. This subprogram transforms the vectors of initial conditions and actual or equivalent loads to normal coordinates, using Eqs. (4.56), (4.57), and (4.64). Logical steps for this subprogram appear in Flowchart B.6.

Subprogram **TIHIST** calculates time histories of normal-mode displacements and velocities with the step-by-step method described in Sec. 4.11. As in Program **LINFORCE**, we use uniform time steps; so the coefficients of $u_{\mathrm{N}i,j}$, $\dot{u}_{\mathrm{N}i,j}$, $q_{\mathrm{N}i,j}$, and $\Delta q_{\mathrm{N}i,j}$ in Eqs. (4.158) through (4.161) become constants that need be calculated only once. For vibrational motions, the displacement and velocity at time t_{j+1} may be written as

$$D_{\mathrm{N}i,j+1} = C_1 D_{\mathrm{N}i,j} + C_2 V_{\mathrm{N}i,j} + C_3 A_{\mathrm{N}i,j} + C_4\,\Delta A_{\mathrm{N}i,j} \tag{B.5}$$

$$V_{\mathrm{N}i,j+1} = C_5 D_{\mathrm{N}i,j} + C_6 V_{\mathrm{N}i,j} + C_7 A_{\mathrm{N}i,j} + C_8\,\Delta A_{\mathrm{N}i,j} \tag{B.6}$$

Comparing these expressions with Eqs. (B.1) and (B.2), we see that the action terms are not divided by m because the mass in each normal coordinate is unity. The constants C_1 and C_8 have the definitions given by Eqs. (B.3), except that n_i, ω_i, and $\omega_{\mathrm{d}i}$ replace n, ω, and ω_{d} for each mode. Again, it is not necessary to give a flowchart for Subprogram **TIHIST** because the statements are very similar to those in part 4 of Program **LINFORCE**.

After the response calculations have been completed, the normal-mode displacements are transformed back to physical coordinates by Subprogram **TRABAC**, using Eq. (4.58). Flowchart B.7 shows the logical steps in this procedure.

The last subprogram, named **RESULT**, writes the time histories of displacements in physical coordinates. Then the maximum and minimum values of these quantities and their times of occurrence are written, as in part 5 of Program **LINFORCE**. At the end of the main program for **NOMOLIN**, the test of LN against NLS determines whether to return for another loading system or another vibration problem.

As mentioned in Sec. 4.11, we used Program **NOMOLIN** to produce the results contained in Table 4.3 and illustrated in Fig. 4.5. It is also feasible to include rigid-body modes [see Eqs. (4.162) and (4.163) in Sec. 4.11] and to make the ground-motion capability more versatile (see Sec. 4.6). In addition, some other numerical method for solving the eigenvalue problem might be substituted for the iteration technique.

B.6 PROGRAM NOTATION

Symbol	Definition
A(,)	Dynamical matrix ($\mathbf{A} = \mathbf{F}\,\mathbf{M}$)
AA( )	Applied actions
AAF( )	Applied-action factors
AJ	Action in step j
AN( )	Actions in normal coordinates
B(,)	Modified identity matrix
C(,)	Coefficient matrix
C1, C2, . . .	Constants
D( )	Displacements
DØ( )	Initial displacements
DAMPR	Damping ratio
DD( )	Velocities
DDD( )	Accelerations
DMAX	Maximum displacement
DMIN	Minimum displacement
DN	Displacements in normal coordinates
DT	Duration of time step Δt
EPS	Error limit for accuracy check
EV( )	Eigenvalues
FO( )	Function ordinates

GAF	Ground acceleration factor
I,J,K,L	Indexes
IAA	Indicator for applied actions
IGA	Indicator for ground acceleration
ITM	Indicator for type of matrix (0: stiffness; 1: flexibility)
LN	Loading number
N	Number of degrees of freedom
NFO	Number of function ordinates
NI( )	Number of iterations executed
NIT	Number of iterations allowed
NLS	Number of loading systems
NM	Number of modes
NTS	Number of time steps
OMEGA( )	Angular frequencies ω
PHI(,)	Eigenvectors $\mathbf{\Phi}$ (mode shapes)
QJ, RJ	constants in step j
RHO	Mass density ρ
SC	System damping coefficient
SK	System stiffness coefficient
SM	System mass coefficient (or matrix)
T( )	Times
TIME	Time
TMAX	Time of maximum displacement
TMIN	Time of minimum displacement
TS(,)	Sweeping matrix
VØ( )	Initial velocities
VN( )	Velocities in normal coordinates
W(,)	Working-storage matrix

FLOWCHART B.1 PROGRAM LINFORCE

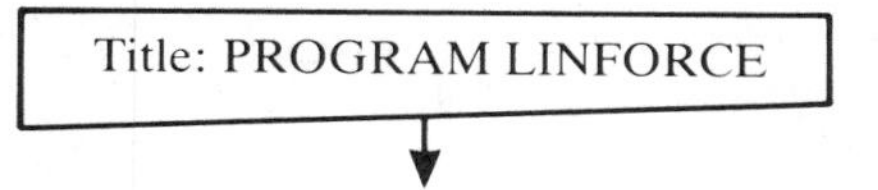

1. System Data

(a) Problem Identification

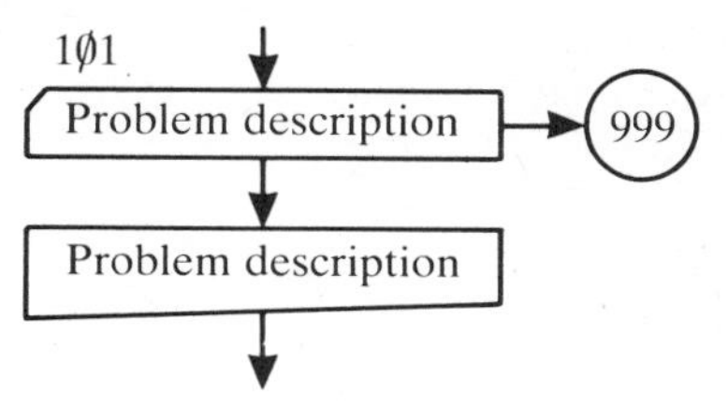

Read and write problem identification.

(b) System Parameters

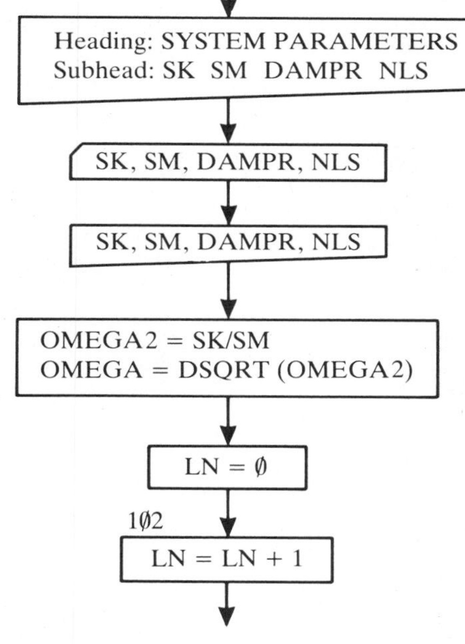

Read and write system parameters.

Compute ω^2 and ω.

Set loading number LN equal to zero.

Increase loading number by 1.

2. Dynamic Load Data

(a) Time Parameters

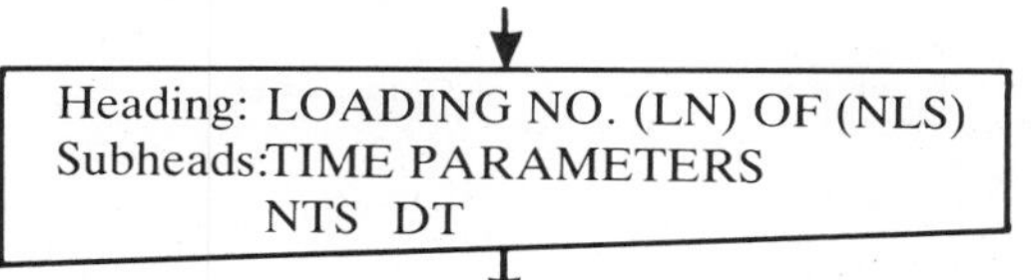

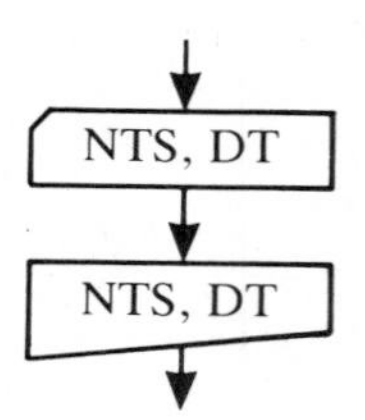

Read and write time parameters NTS and DT.

(b) Initial Condtions

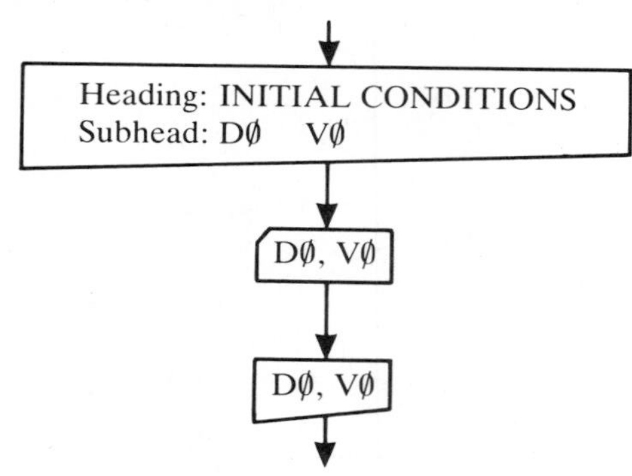

Read and write initial conditions DØ and VØ.

(c) Load Parameters

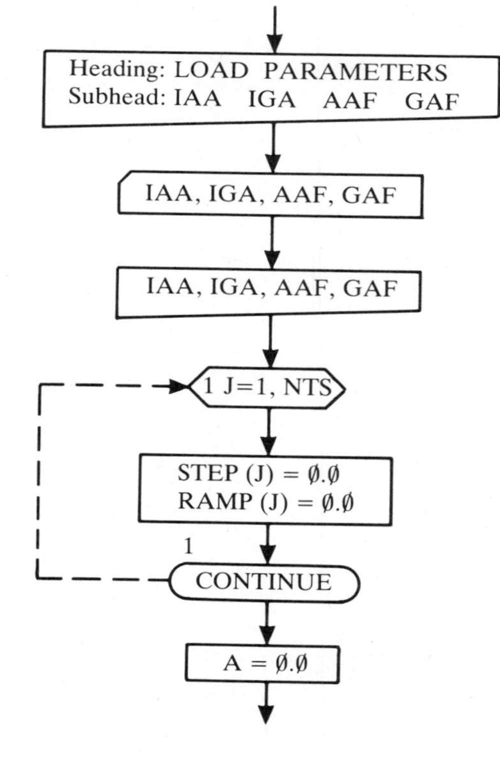

Read and write load parameters.

Clear step and ramp vectors.

Initialize A.

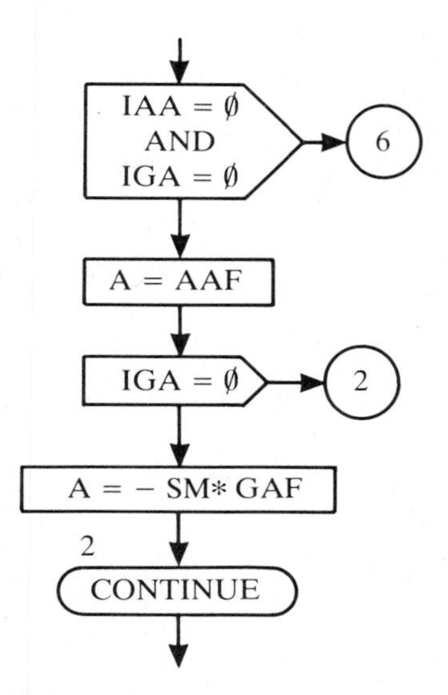

If IAA and IGA are both zero, go to 6 and skip the loads.

By default, let A = AAF.

If IGA = ∅, go to 2.

Otherwise, let A = − SM* GAF.

(d) Forcing Function

(1) Function Parameter

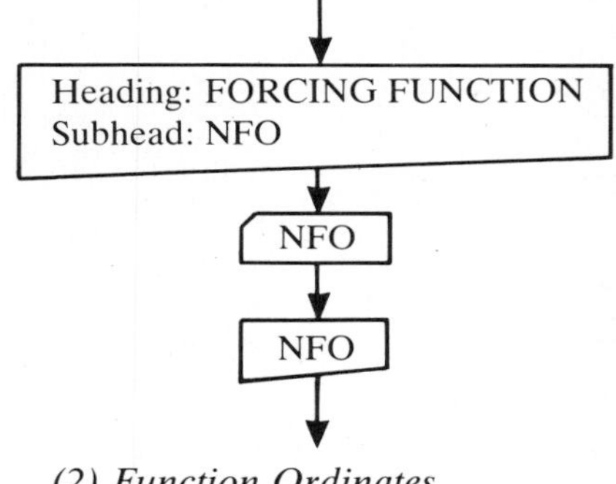

Read and write number of function ordinates NFO.

(2) Function Ordinates

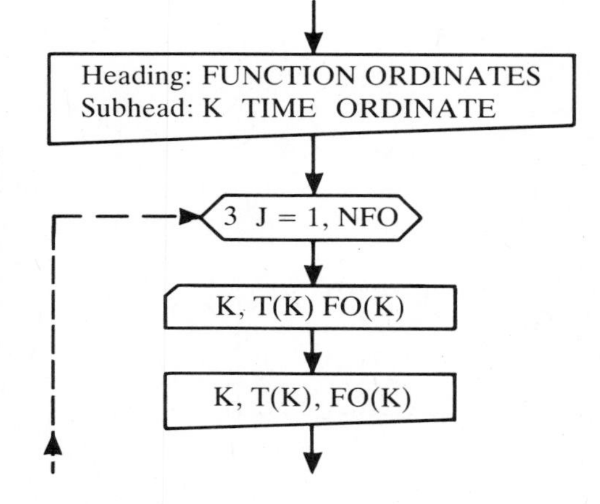

Read and write subscript, time, and function ordinate.

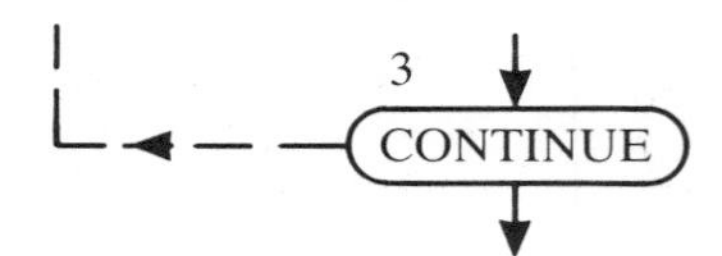

3. Calculate Step and Ramp for Each Time Increment

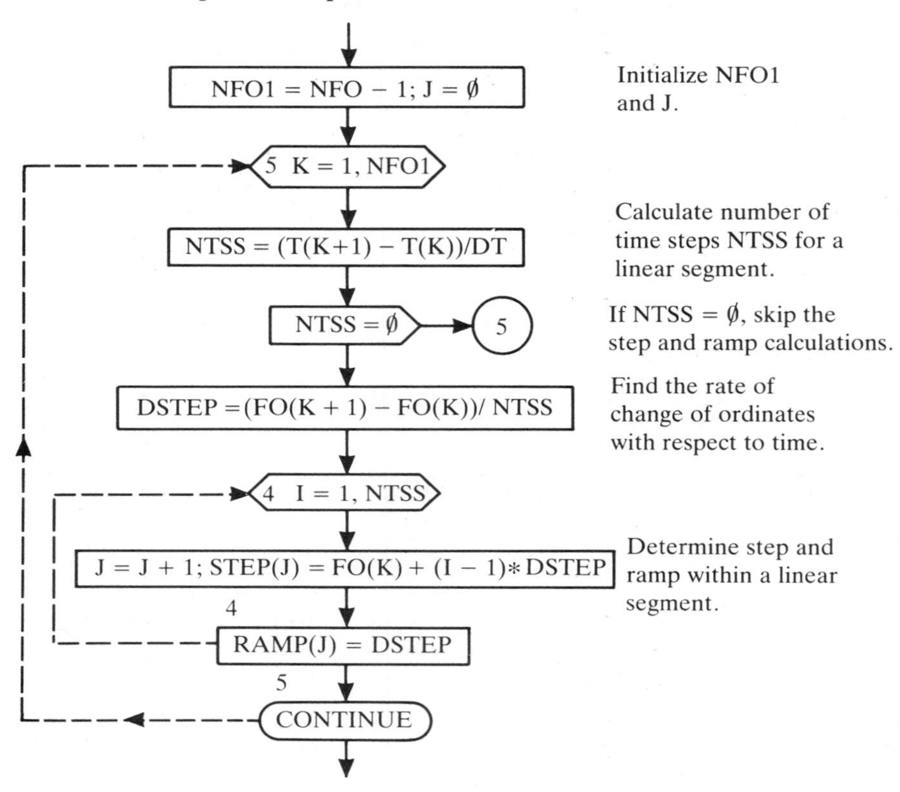

4. Calculate Response

(a) Response Constants

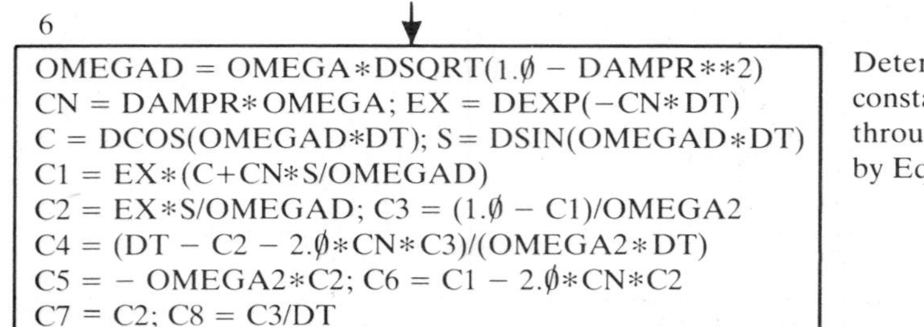

Determine the constants C1 through C8 given by Eq. (B.3).

(b) Step-by-Step Response

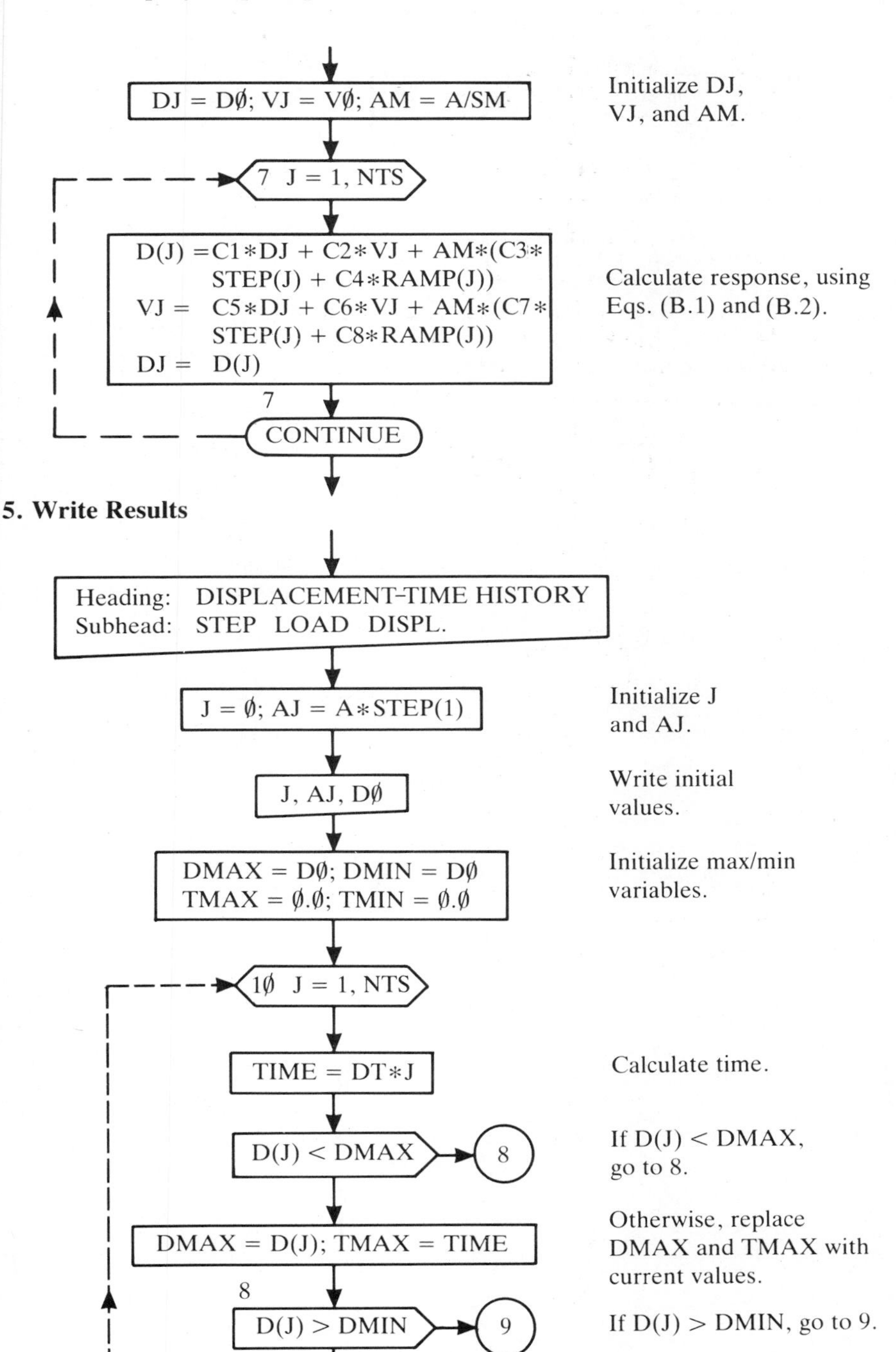

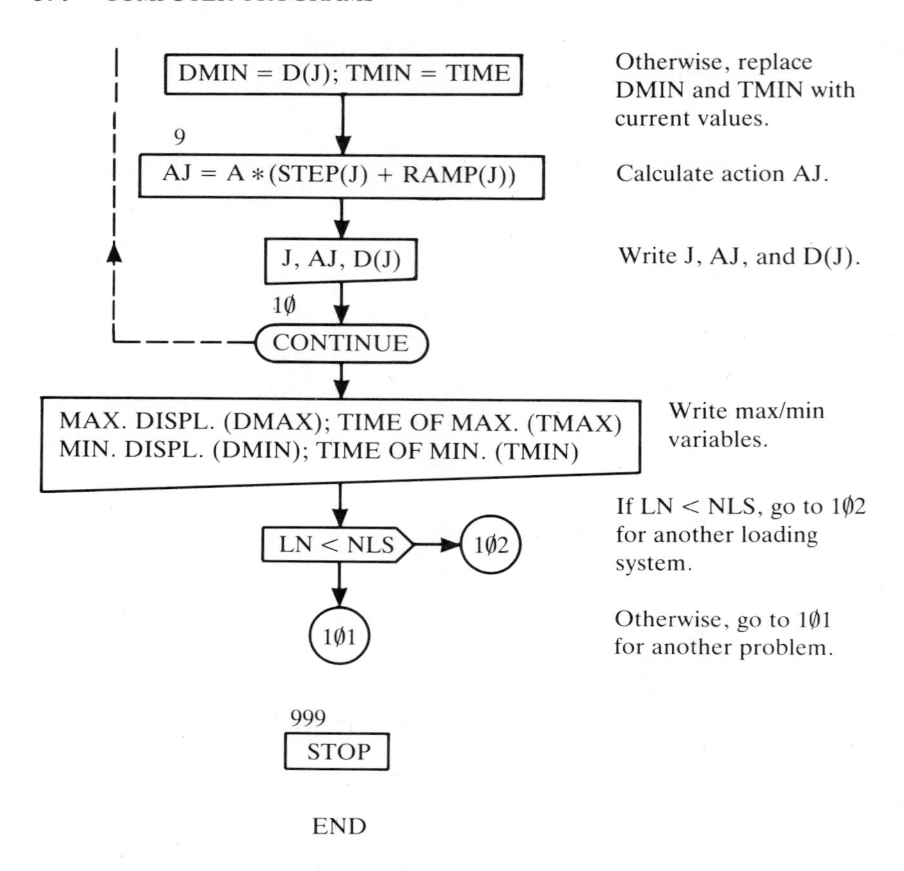
DMIN = D(J); TMIN = TIME
Otherwise, replace DMIN and TMIN with current values.
9
AJ = A * (STEP(J) + RAMP(J))
Calculate action AJ.
J, AJ, D(J)
Write J, AJ, and D(J).
1Ø
CONTINUE
MAX. DISPL. (DMAX); TIME OF MAX. (TMAX)
MIN. DISPL. (DMIN); TIME OF MIN. (TMIN)
Write max/min variables.
LN < NLS
1Ø2
If LN < NLS, go to 1Ø2 for another loading system.
1Ø1
Otherwise, go to 1Ø1 for another problem.
999
STOP
END

FLOWCHART B.2 PROGRAM AVAC1

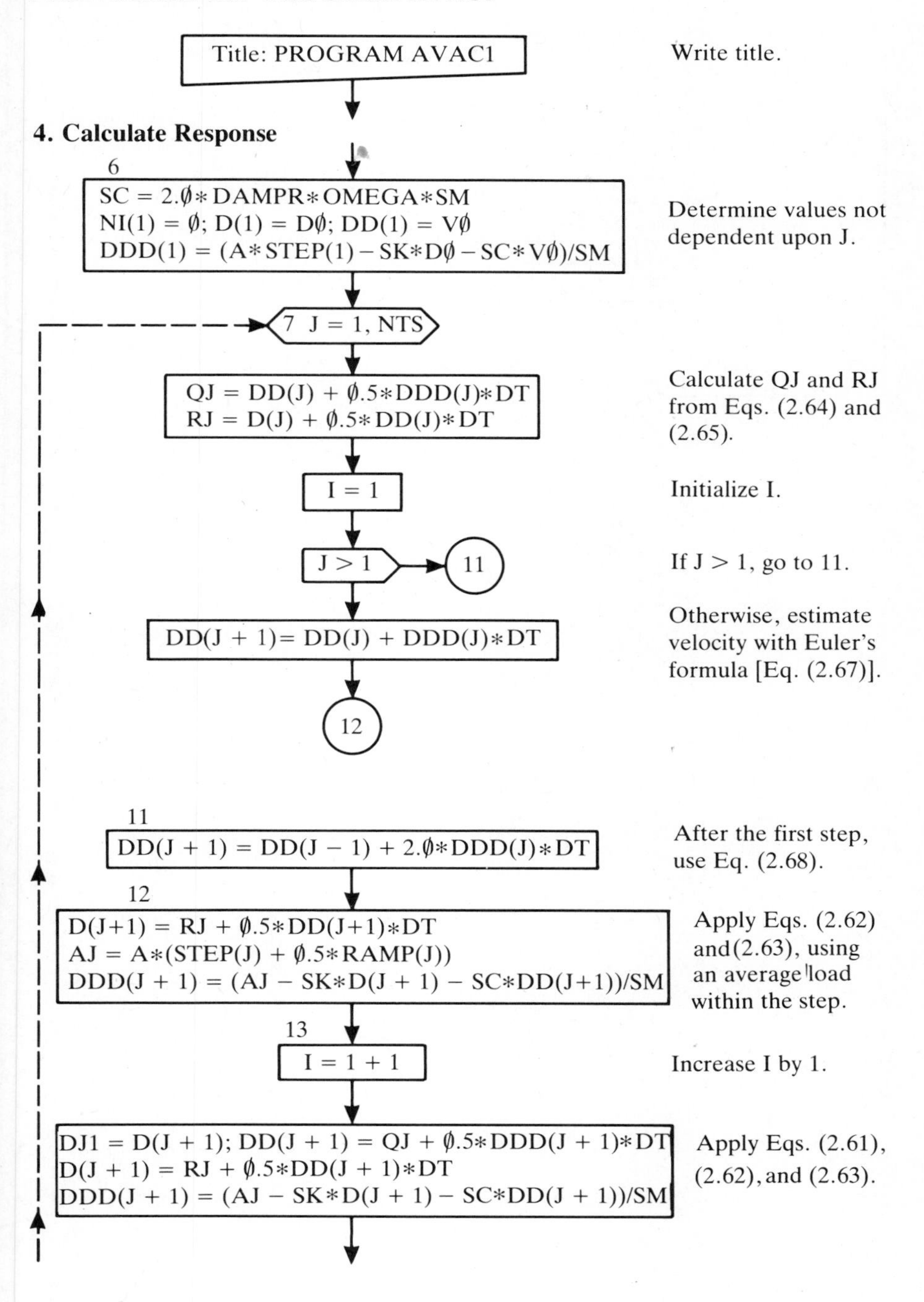

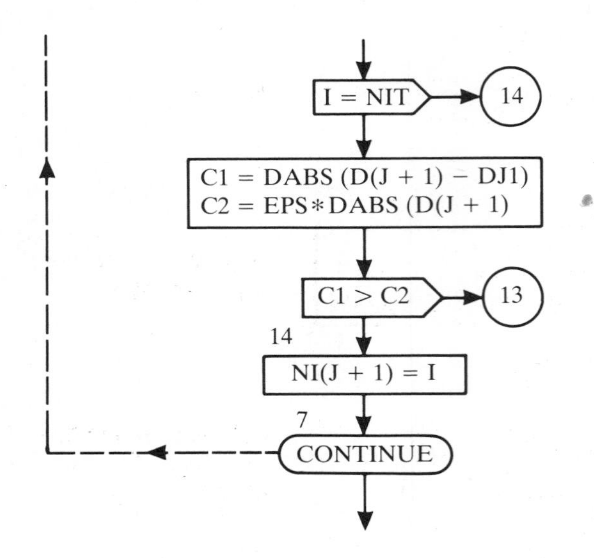

If I = NIT, go to 14.

Compute constants C1 and C2 for accuracy check.

If C1 > C2, go to 13.

Otherwise, put I into NI(J + 1).

Note: When searching for max/min displacements and writing results in Part 5, use the subscript J + 1 in place of J.

FLOWCHART B.3 PROGRAM EIGIT

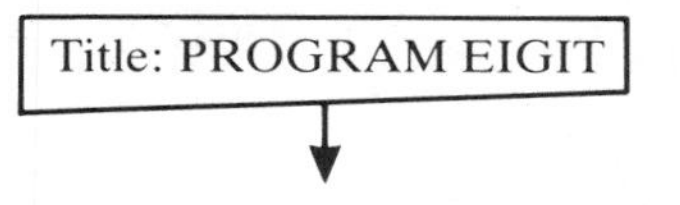

Write title.

1. System Data

(a) Problem Identification

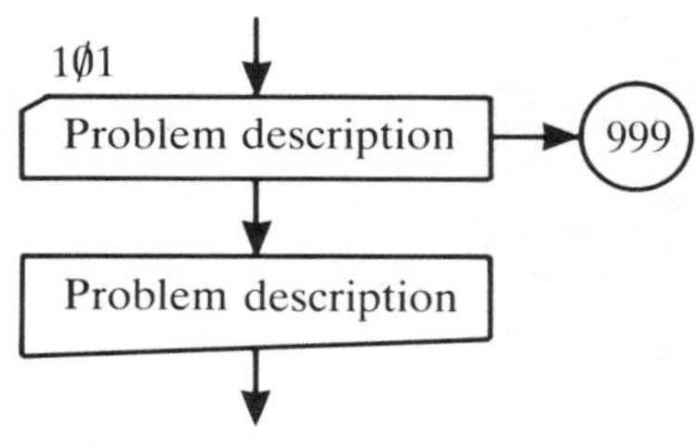

Read and write problem idenfitication.

(b) System Parameters

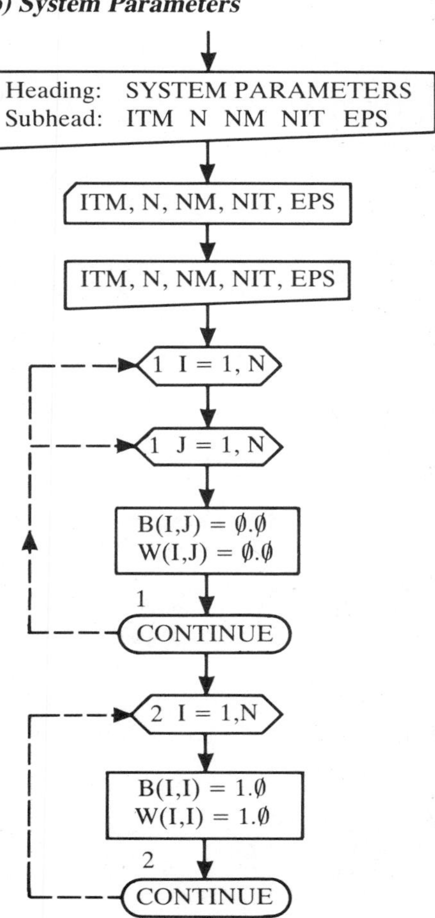

Read and write system and iteration parameters.

Initialize matrices **B** and **W** to zero.

Convert matrices **B** and **W** to identity matrices.

(c) System matrices

(1) Coefficient Matrix

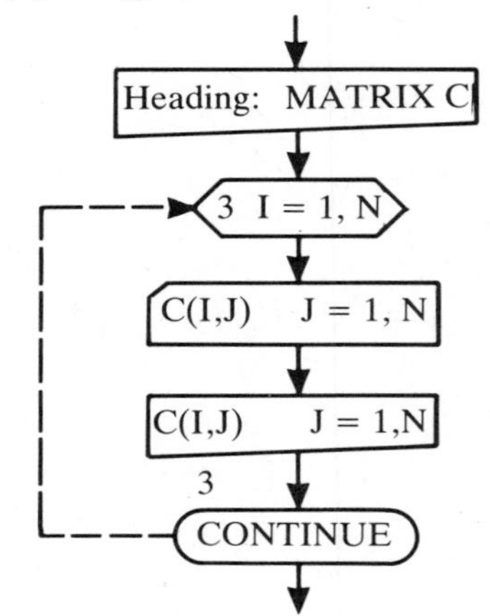

Read and write coefficient matrix **C** by rows.

(2) Mass Matrix

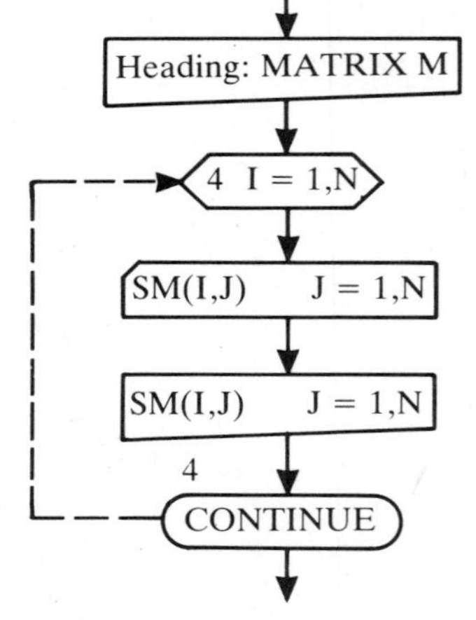

Read and write mass matrix **M** by rows.

2. Calculate Dynamical Matrix

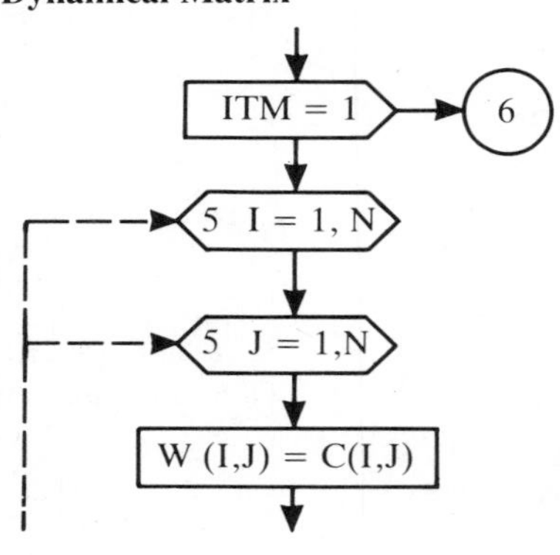

If ITM = 1, the type of matrix is flexibiltiy; so go to 6.

Copy matrix **C** into matrix **W**.

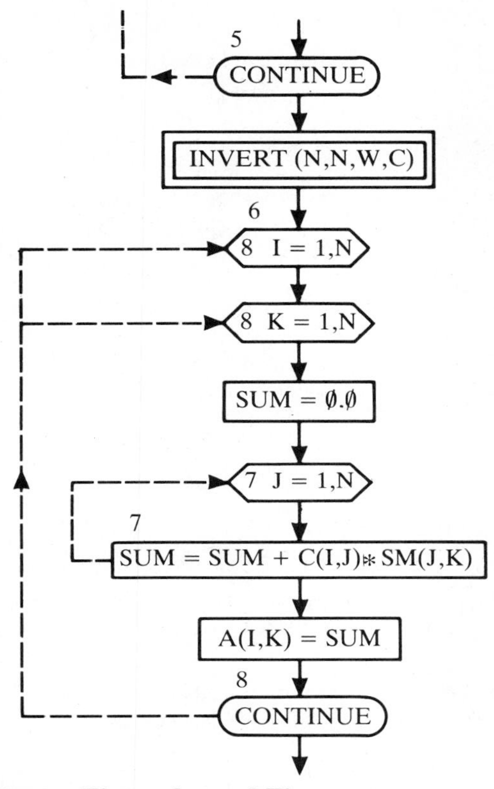

When the type of matrix is stiffness, invert it to obtain flexibilities.

Initialize SUM to zero.

Multiply the flexibility matrix and the mass matrix to find the dynamical matrix **A = FM.**

3. Iterate Eigenvalue and Eigenvector

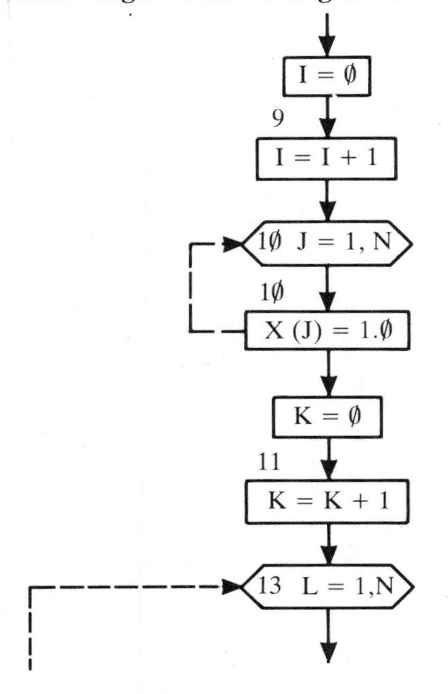

Initialize mode number.

Increment mode number.

Give first trial eigenvector unit values.

Initialize iteration number.

Increment iteration number.

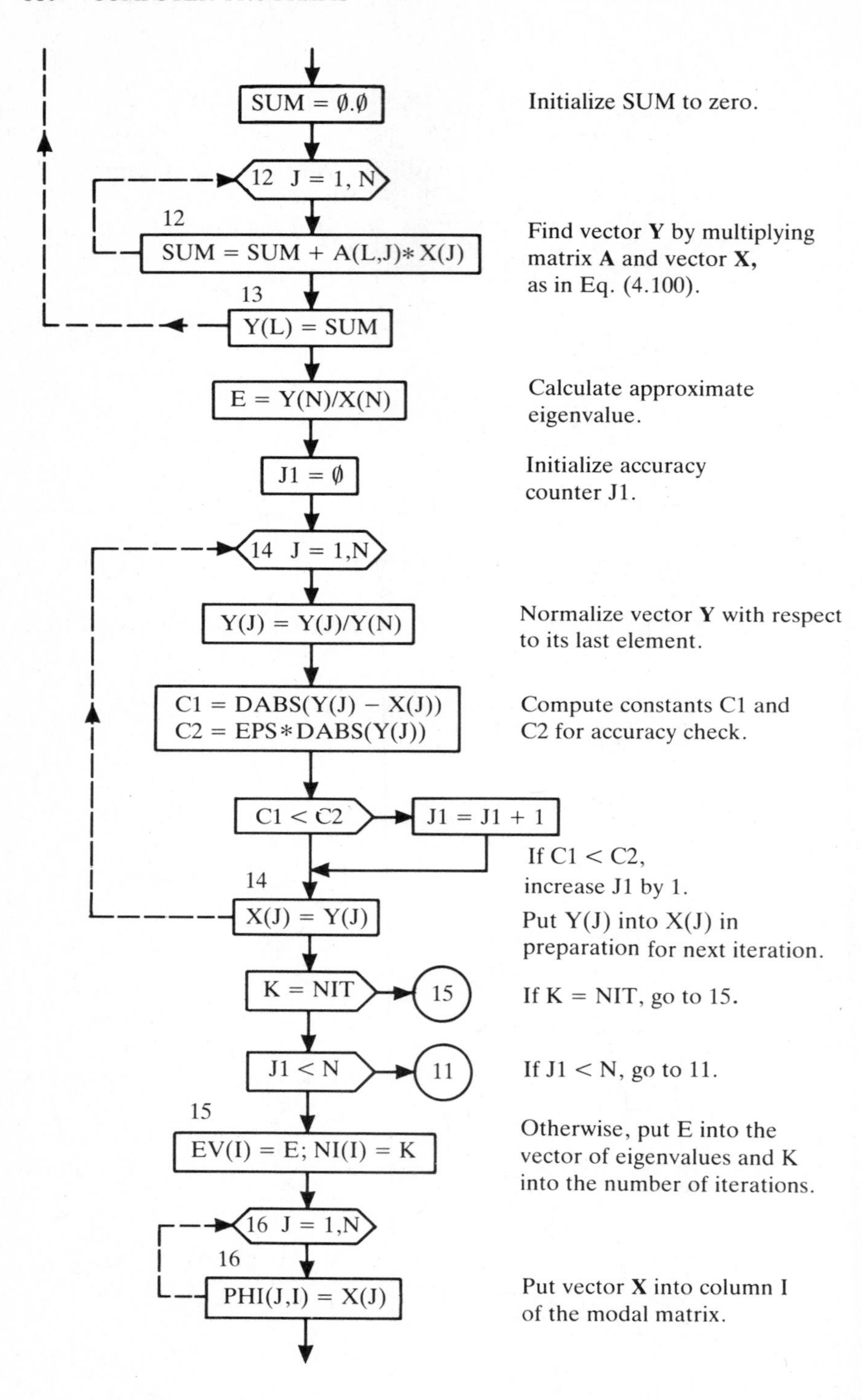

SUM = ∅.∅
Initialize SUM to zero.
12 J = 1, N
12
SUM = SUM + A(L,J)*X(J)
Find vector **Y** by multiplying matrix **A** and vector **X**, as in Eq. (4.100).
13
Y(L) = SUM
E = Y(N)/X(N)
Calculate approximate eigenvalue.
J1 = ∅
Initialize accuracy counter J1.
14 J = 1,N
Y(J) = Y(J)/Y(N)
Normalize vector **Y** with respect to its last element.
C1 = DABS(Y(J) − X(J))
C2 = EPS*DABS(Y(J))
Compute constants C1 and C2 for accuracy check.
C1 < C2
J1 = J1 + 1
If C1 < C2, increase J1 by 1.
14
X(J) = Y(J)
Put Y(J) into X(J) in preparation for next iteration.
K = NIT
15
If K = NIT, go to 15.
J1 < N
11
If J1 < N, go to 11.
15
EV(I) = E; NI(I) = K
Otherwise, put E into the vector of eigenvalues and K into the number of iterations.
16 J = 1,N
16
PHI(J,I) = X(J)
Put vector **X** into column I of the modal matrix.

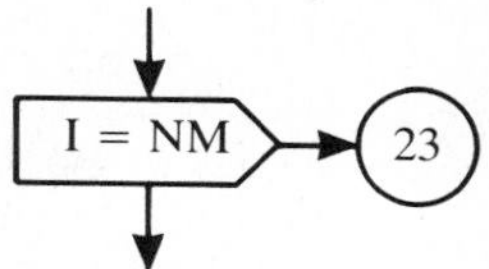

If I = NM, go to 23.

4. Construct and Apply Sweeping Matrix

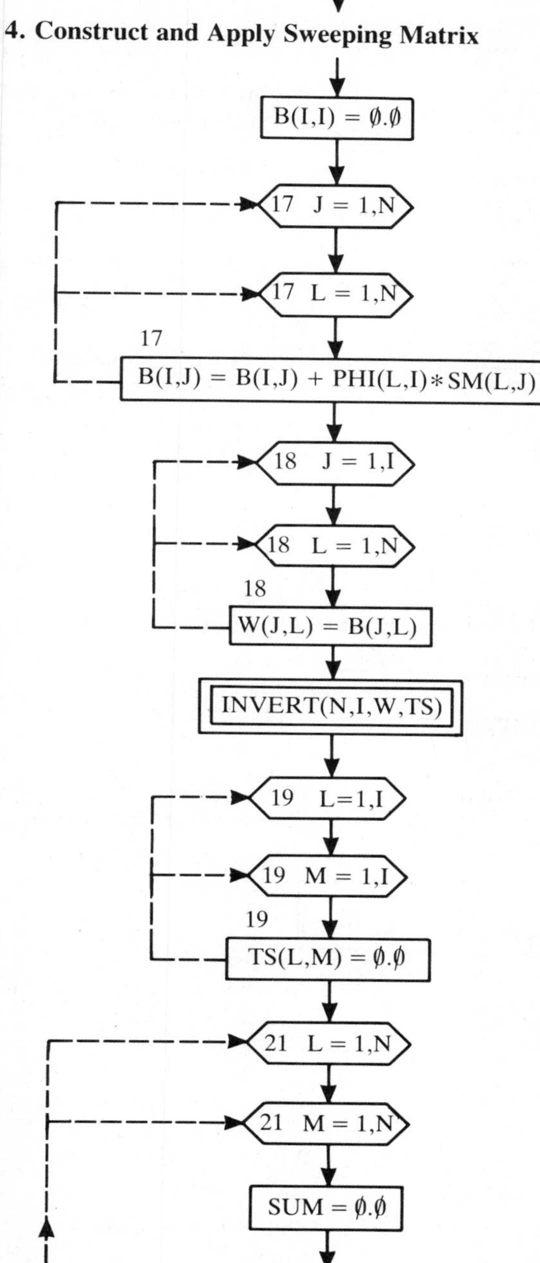

Initialize B(I,I) to zero.

Fill row I of matrix **B**, as in Eqs. (*b*) and (*h*) of Sec. B.4.

Copy I rows of matrix **B** into matrix **W.**

Invert I rows of matrix **W,** and place them into matrix $\mathbf{T}_S$. [Note the meanings of (N,I,W,TS).]

Clear upper left-hand part of matrix $\mathbf{T}_S$, as shown by Eq. (B.4).

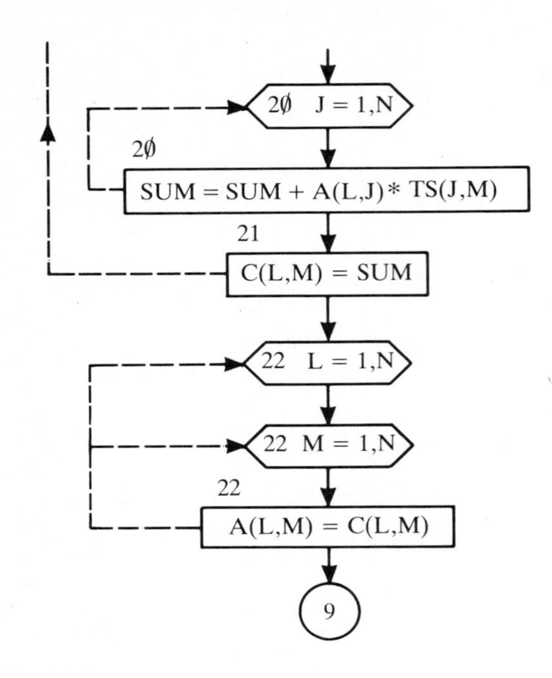

Postmultiply matrix **A** by $\mathbf{T}_S$, and store in matrix **C**.

Copy matrix **C** into matrix **A.**

Go to 9 and iterate for another mode.

5. Write Eigenvalues and Eigenvectors

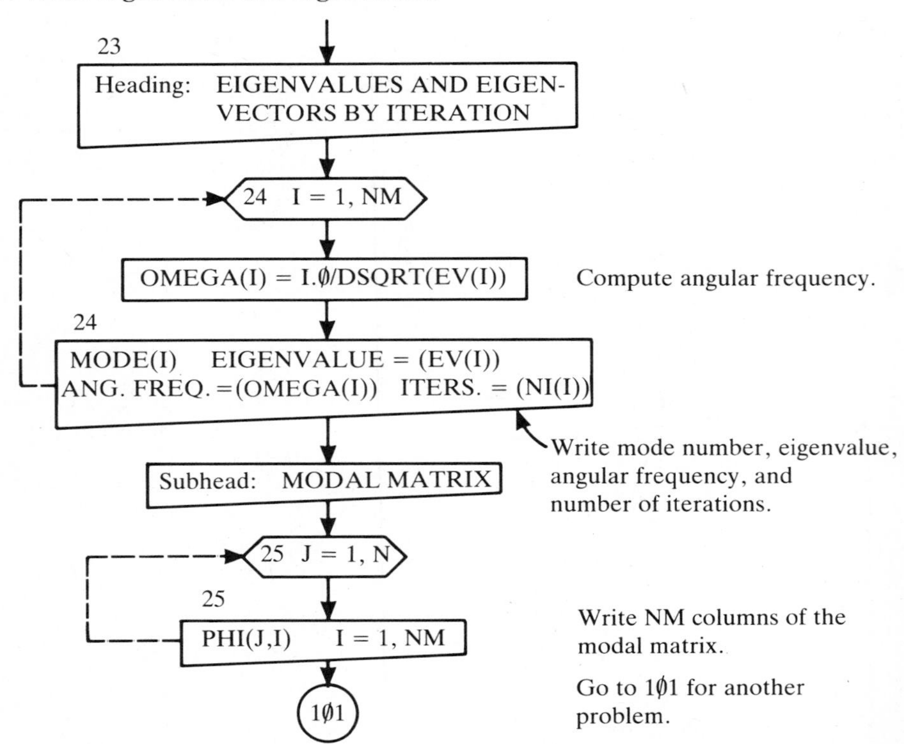

999

FLOWCHART B.4 SUBPROGRAM INVERT

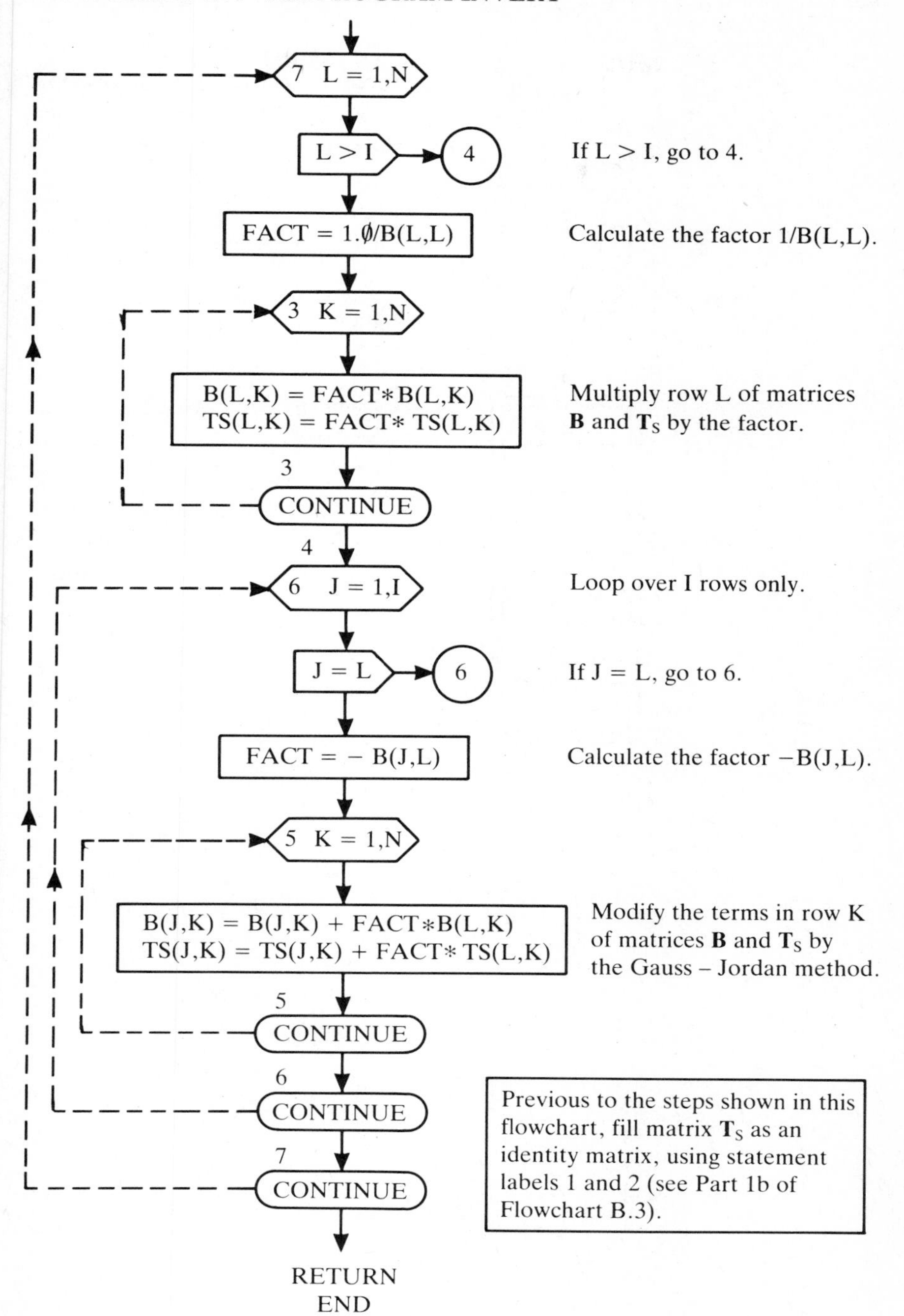

FLOWCHART B.5 MAIN PROGRAM FOR NOMOLIN

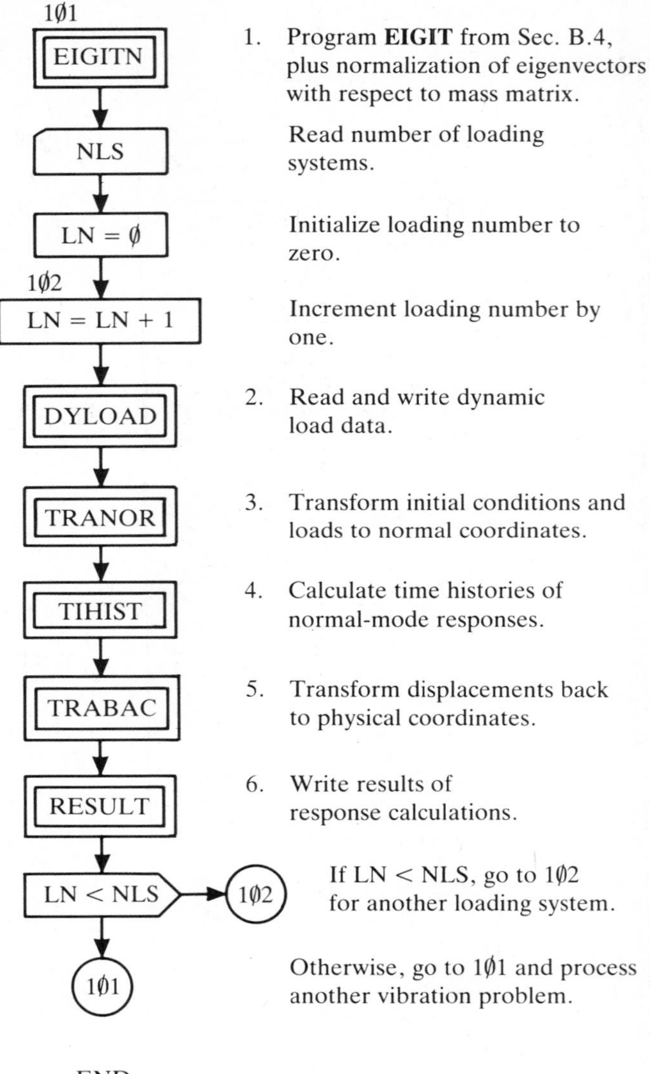

FLOWCHART B.6 SUBPROGRAM TRANOR

1. Calculate Transformation Operator Φ_N^{-1}

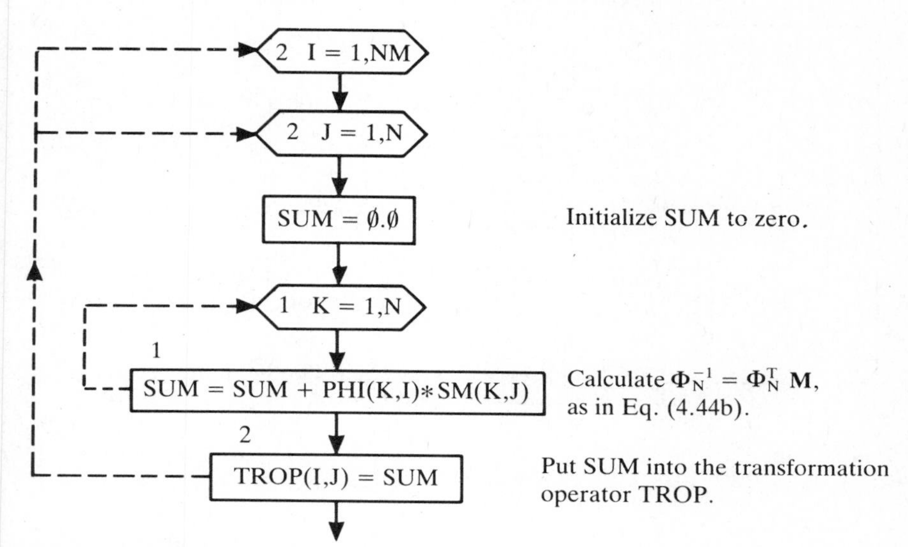

2. Transform Initial Conditions

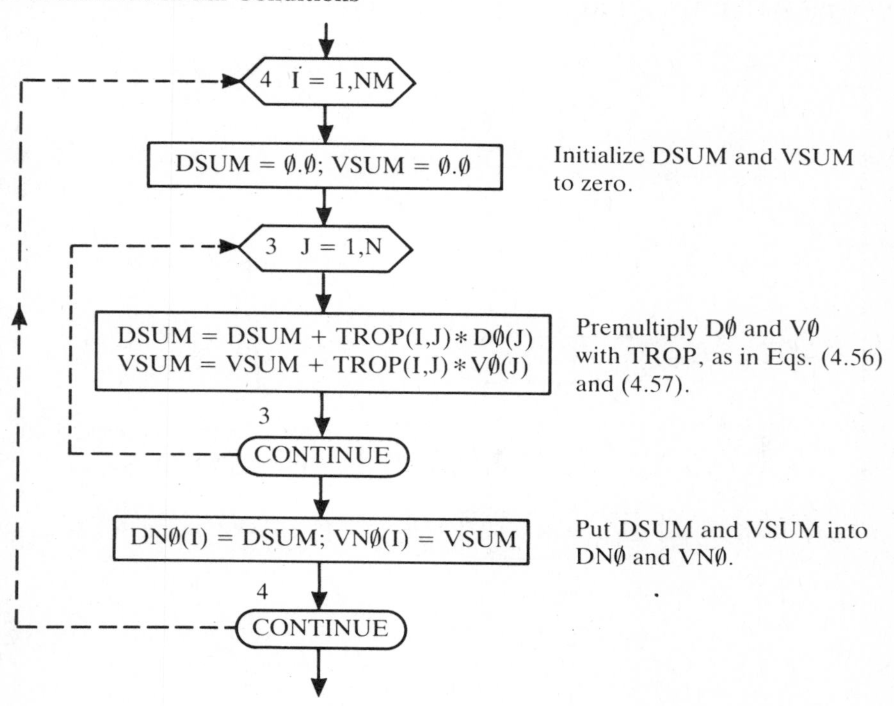

3. Transform Applied Actions

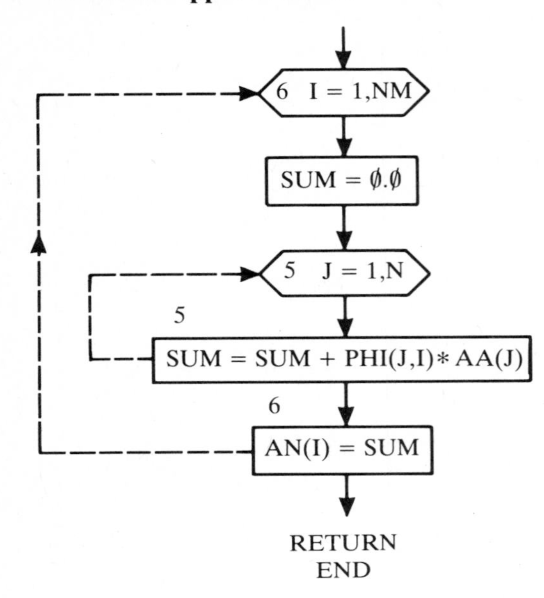

Initialize SUM to zero.

Premultiply vector $\mathbf{A}_A$ with $\mathbf{\Phi}_N^T$, as in Eq. (4.64).

Put SUM into vector $\mathbf{A}_N$.

FLOWCHART B.7 SUBPROGRAM TRABAC

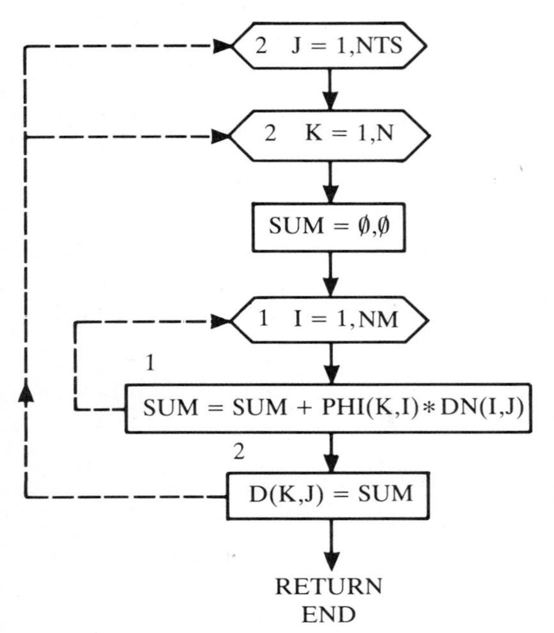

Initialize SUM to zero.

Back-transform displacements with Eq. (4.58).

Place SUM into matrix **D**.

ANSWERS TO PROBLEMS

CHAPTER 1

1.2-1. $\tau = 0.0640$ s

1.2-2. $f = 5.67$ cps

1.2-3. $\tau = 0.533$ s

1.2-4. $\tau = 1.99$ s

1.2-5. $\sigma_{max} = 7230$ psi

1.2-6. $\tau_a = 0.813$ s; $\tau_b = 1.62$ s

1.2-7. $f = 5.88$ cps

1.2-8. $\tau = 2\pi\sqrt{m\ell/(2S)}$

1.2-9. $k = k_1k_2k_3/(k_1k_2 + k_1k_3 + k_2k_3)$

1.3-1. $f = 0.935$ cps

1.3-2. $I_2 = I_1(\tau_2^2 - \tau_0^2)/(\tau_1^2 - \tau_0^2)$

1.3-3. $\tau = 2\pi\sqrt{W/(3kg)}$

1.3-4. $\tau = 2\pi(\ell/a)\sqrt{W/(3kg)}$

1.3-5. $f = \dfrac{1}{2\pi}\sqrt{\dfrac{\pi d^4 G(\ell_1 + \ell_2)}{32I\ell_1\ell_2}}$

1.3-6. $L_1 = 2a + \dfrac{C_1}{C_2}b + 2\dfrac{C_1}{B}r$

1.3-7. $\tau = 2\pi\sqrt{\dfrac{Wr(r - r_0)}{ngS_0 r_0}}$

1.4-1. $\omega = \sqrt{\dfrac{g}{\ell}\sin\beta}$

1.4-2. $f = 2.72$ cps

1.4-3. $W_3 = 27$ N

1.4-4. $\omega = \sqrt{\dfrac{g}{\ell}\left[\dfrac{3ka^2/\ell}{3W + w\ell} - \dfrac{3}{4}\left(\dfrac{4W + 2w\ell}{3W + w\ell}\right)\right]}$

1.4-5. $\omega = \sqrt{\dfrac{g[k_1 a^2 + k_2(a\tan\alpha)^2]}{W\ell^2}}$

1.4-6. $\omega = \sqrt{\dfrac{3ga^2}{\ell b^2}}$

1.4-7. $\omega = \sqrt{\dfrac{cg}{i^2 + (r - c)^2}}$

1.5-1. $\tau = \dfrac{2\pi}{3.58}\sqrt{\dfrac{w\ell^4}{EIg}}$

1.5-2. 13/35

1.5-3. $\tau = \dfrac{2\pi}{22.6}\sqrt{\dfrac{w\ell^4}{EIg}}$

1.5-4. $\tau = 1.73$ s

1.5-5. $239/3360 \approx 1/14$

1.5-6. $\omega = 1.58\sqrt{\dfrac{kg}{w\ell}}$

1.5-7. $\dfrac{w_1\ell_1 k_2^2 + 3w_2\ell_2 k_2(k_1 + k_2) + w_2\ell_2 k_1^2}{3(k_1 + k_2)^2}$

1.5-8. $\dfrac{i_1\ell_1 k_{r_2}^2 + 3i\ell_1 k_{r_1}(k_{r_1} + k_{r_2}) + i_2\ell_2 k_{r_1}^2}{3(k_{r1} + k_{r2})^2}$

1.5-9. $f = \dfrac{1}{2\pi}\sqrt{\dfrac{ngS_0 r_0}{(W + W_s n/3)r(r - r_0)}}$

1.5-10. $f = \dfrac{1}{2\pi}\sqrt{\dfrac{16gB}{[W + (116/105)wr]r^3}}$

1.5-11. $\tau = 0.637\sqrt{\dfrac{w\ell^4}{EIg}}$

1.5-12. $\tau = 1.84\sqrt{\frac{w\ell^4}{EIg}}$

1.5-13. $\omega = 3.07\sqrt{\frac{GJ}{i\ell^2}}$

1.5-14. $\omega = 1.57\sqrt{\frac{EAg}{w\ell^2}}$

1.6-1. 0.0040 in.

1.6-2. 3.3 mm

1.6-3. 0.0078 in.

1.6-4. 5.1 mm

1.6-5. 0.0454 in.

1.6-6. 0.0056 in.

1.7-1. $u_1 = 0.147$ in.

$\dot{u}_1 = -1.22$ in./s

1.7-2. $u_1 = -0.035$ in.

$\dot{u}_1 = -1.81$ in./s

1.7-3. $u = \frac{qt}{2\omega}\sin\omega t$

1.8-1. 0.539/1

1.8-2. $\dot{u}_{max} = -0.24$ m/s

1.8-3. $t = 0.50$ s

$-u_{max} = 0.000496$ in.

1.8-4. $c = 0.00143$ lb-s/in.

1.8-5. $f_d = \sqrt{3/2}\cdot f$

1.9-1. $A = 0.803$ in.

1.9-2. $A = 2.85$ mm

1.9-3. $\beta_{max} = \frac{1}{\sqrt{1 - \Omega^4/\omega^4}}$

1.9-4. $u = \frac{Q}{k}\beta\sin(\Omega t - \theta)$

$\theta = \tan^{-1}\left(\frac{2n\Omega}{\omega^2 - \Omega^2}\right)$

1.9-5. $u = \frac{Bd}{k}\beta\sin(\Omega t - \phi - \theta)$

$\phi = \tan^{-1}\left(\frac{c\Omega}{-k}\right)$

1.9-6. $u^* = \dfrac{-ma}{k}\beta\cos(\Omega t - \theta)$

$$\theta = \tan^{-1}\left(\frac{2n\Omega}{\omega^2 - \Omega^2}\right)$$

1.9-7. $u^* = \dfrac{-ma}{k}\beta\sin(\Omega t - \theta)$

$$\theta = \tan^{-1}\left(\frac{2n\Omega}{\omega^2 - \Omega^2}\right)$$

1.9-8. $u_{tr} = -e^{-nt}\left(-N\cos\omega_{\mathrm{d}}t + \dfrac{M\Omega - Nn}{\omega_{\mathrm{d}}}\sin\omega_{\mathrm{d}}t\right)$

1.11-2. $F(t) = \dfrac{4P}{\pi}(\sin\Omega t + \tfrac{1}{3}\sin 3\Omega t + \cdots)$

1.11-3. $F(t) = \dfrac{4P}{\pi}(\cos\Omega t - \tfrac{1}{3}\cos 3\Omega t + \cdots)$

1.11-4. $F(t) = \dfrac{2P}{\pi}(\sin\Omega t - \tfrac{1}{2}\sin 2\Omega t + \tfrac{1}{3}\sin 3\Omega t - \cdots)$

1.11-5. $F(t) = \dfrac{P}{2} - \dfrac{P}{\pi}(\sin\Omega t + \tfrac{1}{2}\sin 2\Omega t + \tfrac{1}{3}\sin 3\Omega t + \cdots)$

1.11-6. $u = \dfrac{a_0}{k} + \displaystyle\sum_{i=1}^{\infty}\frac{a_i\cos(i\Omega t - \theta_i) + b_i\sin(i\Omega t - \theta_i)}{k\sqrt{(1 - i^2\Omega^2/\omega^2)^2 + (2\gamma i\Omega/\omega)^2}}$

1.12-1. $u = \dfrac{Q_1}{k}(1 - \cos\omega t) \qquad (0 \leqq t \leqq t_1)$

$$u = \frac{Q_1}{k}[\cos\omega(t - t_1) - \cos\omega t] - \frac{Q_2}{k}[1 - \cos\omega(t - t_1)] \qquad (t_1 \leqq t \leqq t_2)$$

$$u = \frac{Q_1}{k}[\cos\omega(t - t_1) - \cos\omega t] - \frac{Q_2}{k}[\cos\omega(t - t_2) - \cos\omega(t - t_1)] \qquad (t_2 \leqq t)$$

1.12-2. $u = \dfrac{Q_1}{k}\left(1 - \cos\omega t - \dfrac{t}{t_1} + \dfrac{\sin\omega t}{\omega t_1}\right) \qquad (0 \leqq t \leqq t_1)$

$$u = \frac{Q_1}{k}\left[-\cos\omega t + \frac{\sin\omega t - \sin\omega(t - t_1)}{\omega t_1}\right] \qquad (t_1 \leqq t)$$

1.12-3. $u = \frac{Q_1}{k}(1 - \cos \omega t) \qquad (0 \leqq t \leqq t_1)$

$$u = \frac{Q_1}{k}\left[1 - \cos \omega t - \frac{t - t_1}{t_2 - t_1} + \frac{\sin \omega(t - t_1)}{\omega(t_2 - t_1)}\right] \qquad (t_1 \leqq t \leqq t_2)$$

$$u = \frac{Q_1}{k}\left[-\cos \omega t + \frac{\sin \omega(t - t_1) - \sin \omega(t - t_2)}{\omega(t_2 - t_1)}\right] \qquad (t_2 \leqq t)$$

1.12-4. $u = \frac{Q_1}{k}\left(\frac{t}{t_1} - \frac{\sin \omega t}{\omega t_1}\right) \qquad (0 \leqq t \leqq t_1)$

$$u = \frac{Q_1}{k}\left[\frac{t}{t_1} - \frac{\sin \omega t}{\omega t_1} - \frac{t_2(t - t_1)}{t_1(t_2 - t_1)} + \frac{t_2 \sin \omega(t - t_1)}{\omega t_1(t_2 - t_1)}\right] \qquad (t_1 \leqq t \leqq t_2)$$

$$u = \frac{Q_1}{k}\left[-\frac{\sin \omega t}{\omega t_1} + \frac{t_2 \sin \omega(t - t_1)}{\omega t_1(t_2 - t_1)} - \frac{\sin \omega(t - t_2)}{\omega(t_2 - t_1)}\right] \qquad (t_2 \leqq t)$$

1.12-5. $u = \frac{Q_1}{k}\left[\left(1 + \frac{2}{\omega^2 t_1^2}\right)(1 - \cos \omega t) - \frac{t^2}{t_1^2}\right] \qquad (0 \leqq t \leqq t_1)$

$$u = \frac{Q_1}{k}\left\{\frac{2}{\omega^2 t_1^2}[\cos \omega(t - t_1) - \cos \omega t] - \frac{2}{\omega t_1}\sin \omega(t - t_1) - \cos \omega t\right\} \qquad (t_1 \leqq t)$$

1.12-6. $u = \frac{Q_1}{k}\left[\left(1 - \frac{2}{\omega^2 t_1^2}\right)(1 - \cos \omega t) - \frac{2t}{t_1} + \frac{t^2}{t_1^2} + \frac{2 \sin \omega t}{\omega t_1}\right] \qquad (0 \leqq t \leqq t_1)$

$$u = \frac{Q_1}{k}\left\{\frac{2}{\omega^2 t_1^2}[\cos \omega t - \cos \omega(t - t_1)] - \cos \omega t + \frac{2 \sin \omega t}{\omega t_1}\right\} \qquad (t_1 \leqq t)$$

1.12-7. $u = \frac{\delta Q}{k}\left[t - \frac{2n}{\omega^2} + e^{-nt}\left(\frac{2n}{\omega^2}\cos \omega_{\mathrm{d}} t - \frac{\omega_{\mathrm{d}}^2 - n^2}{\omega^2 \omega_{\mathrm{d}}}\sin \omega_{\mathrm{d}} t\right)\right]$

1.13-1. $u = d\left(\frac{t}{t_1} - \frac{\sin \omega t}{\omega t_1}\right) \qquad (0 \leqq t \leqq t_1)$

$$u = d\left[\frac{\sin \omega(t - t_1) - \sin \omega t}{\omega t_1} + \cos \omega(t - t_1)\right] \qquad (t_1 \leqq t)$$

1.13-2. $u^* = -\frac{a}{\omega^2}\left(\frac{t}{t_1} - \frac{\sin \omega t}{\omega t_1}\right) \qquad (0 \leqq t \leqq t_1)$

$$u^* = -\frac{a}{\omega^2}\left[1 + \frac{\sin \omega(t - t_1) - \sin \omega t}{\omega t_1}\right] \qquad (t_1 \leqq t)$$

1.13-3. $u = d_1(1 - \cos \omega t) - (d_1 + d_2)\left(\frac{t}{t_1} - \frac{\sin \omega t}{\omega t_1}\right) \qquad (0 \leqq t \leqq t_1)$

$$u = -d_1 \cos \omega t - d_2 \cos \omega(t - t_1) - \frac{d_1 + d_2}{\omega t_1}[\sin \omega(t - t_1) - \sin \omega t] \qquad (t_1 \leqq t)$$

1.13-4. $u^* = -\frac{a_1}{\omega^2}(1 - \cos \omega t) - \frac{a_2 - a_1}{\omega^2}\left(\frac{t}{t_1} - \frac{\sin \omega t}{\omega t_1}\right) \qquad (0 \leqq t \leqq t_1)$

$$u^* = -\frac{a_1}{\omega^2}[\cos \omega(t - t_1) - \cos \omega t] - \frac{a_2 - a_1}{\omega^2}\left[\frac{\sin \omega(t - t_1) - \sin \omega t}{\omega t_1} + \cos \omega(t - t_1)\right] \qquad (t_1 \leqq t)$$

1.13-5. $u = d\left[\frac{2}{\omega^2 t_1^2}(1 - \cos \omega t - \omega t_1 \sin \omega t) + \frac{2t}{t_1} - \frac{t^2}{t_1^2}\right] \qquad (0 \leqq t \leqq t_1)$

$$u = d\left\{\frac{2}{\omega^2 t_1^2}[\cos \omega(t - t_1) - \cos \omega t - \omega t_1 \sin \omega t] + \cos \omega(t - t_1)\right\} \qquad (t_1 \leqq t)$$

1.13-6. $u^* = -\frac{a}{\omega^2}\left[\frac{t^2}{t_1^2} - \frac{2}{\omega^2 t_1^2}(1 - \cos \omega t)\right] \qquad (0 \leqq t \leqq t_1)$

$$u^* = -\frac{a}{\omega^2}\left\{\frac{2}{\omega^2 t_1^2}[\cos \omega t - \cos \omega(t - t_1) + \omega t_1 \sin \omega(t - t_1)] + \cos \omega(t - t_1)\right\} \qquad (t_1 \leqq t)$$

1.13-7. $A^* = \frac{\sqrt{2}a}{\omega^2}\sqrt{1 - \cos\left(\frac{\omega \dot{u}_0}{a}\right)}$

1.15-1. $u_{10} = 0$

1.15-2. $u_{10} = \frac{Q_1}{k}$

1.15-3. Given

1.15-4. Given

1.15-5. $u_{20} = -2.546\frac{Q_1}{k}$

1.15-6. $u_{20} = -4\frac{Q_1}{k}$

1.15-7. $u_{10} = 0.9135\frac{Q_1}{k}$

1.15-8. $u_{10} = 0.3186 \dfrac{Q_1}{k}$

1.15-9. $u_{10} = 1.379 \dfrac{Q_1}{k}$

1.15-10. $u_{10} = 1.102 \dfrac{Q_1}{k}$

CHAPTER 2

2.1-1. **(a)** $m\ddot{u} + 2ku(2 - \ell/\sqrt{\ell^2 + u^2}) = 0$

(b) $m\ddot{u} + 2ku(1 + u^2/2\ell^2) = 0$

(c) $m\ddot{u} + 2ku = 0$

2.1-2. **(a)** $$m\ddot{u} + 2k\left\{2u - \ell\left[\frac{1}{\sqrt{1 + 1/(1 + \sqrt{2}u/\ell)^2}} - \frac{1}{\sqrt{1 + 1/(1 - \sqrt{2}u/\ell)^2}}\right]\right\} = 0$$

(b) $m\ddot{u} + 2ku = 0$

2.1-3. **(a)** $$\ddot{\phi} + \frac{g}{\ell_1}\left[\sin\phi + \frac{F}{W}\cos(\phi - \theta)\right] = 0$$

where

$$F = k\left(\frac{\ell_2 + \ell_1 \sin\phi}{\cos\theta} - \ell_2\right); \qquad \theta = \tan^{-1}\left[\frac{\ell_1(1 - \cos\phi)}{\ell_2 + \ell_1 \sin\phi}\right]$$

(b) $$\ddot{\phi} + \left(\frac{g}{\ell_1} + \frac{kg}{W}\right)\phi = 0$$

2.1-4. **(a)** $$\ddot{\phi} + \frac{g}{\ell_1}\sin\phi + \frac{kg}{W\ell_1}(\ell_1 + \ell_2) \times \left[1 - \frac{\ell_2}{\sqrt{\ell_2^2 + 2\ell_1(\ell_1 + \ell_2)(1 - \cos\phi)}}\right]\sin\phi = 0$$

(b) $$\ddot{\phi} + \frac{g}{\ell_1}\phi + \frac{g}{2}\left[\frac{k}{W}\left(\frac{\ell_1}{\ell_2} + 1\right)^2 - \frac{1}{3\ell_1}\right]\phi^3 = 0$$

(c) $$\ddot{\phi} + \frac{g}{\ell_1}\phi = 0$$

2.1-5. $$m\ddot{u} + k\left[u - \ell\cos\theta + \frac{\ell(\ell\cos\theta - u)}{\sqrt{\ell^2 + u^2 - 2\ell u\cos\theta}}\right] = 0$$

2.1-6. $$\ddot{\phi} + \frac{k_r g}{W\ell^2}\phi - \frac{g}{\ell}\sin\phi = 0$$

2.2-1. $u_m = 2.13$ in.; $P_m = 8563$ lb; $t_m = \tau/4 = 0.308$ s

2.2-2. $u_m = 4.75$ in.; $P_m = 2350$ lb

2.2-3. $u_m = 3.16$ in.; $P_m = 3170$ lb

2.2-4. $u_m = 3.18$ in.; $P_m = 3130$ lb

2.5-1. $t_n = t_1 + \dfrac{1}{\omega}\tan^{-1}[(Q_n/k)/(\dot{u}_1/\omega)]$

$$A = \sqrt{\left(u_1 + \frac{Q_n}{k}\right)^2 + \left(\frac{Q_n}{k}\right)^2 + \left(\frac{\dot{u}_1}{\omega}\right)^2}$$

2.5-2. $t_m = \dfrac{\pi}{2}\left(\sqrt{\dfrac{m}{k_1}} + \sqrt{\dfrac{m}{k_2}}\right); \qquad u_m = u_1\left(1 + \sqrt{\dfrac{k_1}{k_2}}\right)$

2.5-3. $t_n = \dfrac{\pi}{2}\sqrt{\dfrac{m}{k_1}}; \qquad A = u_1\left[1 + \dfrac{k_1}{k_2}\left(1 - \sqrt{1 + \dfrac{k_2}{k_1}}\right)\right]$

2.5-4. $t_m = 3.83\sqrt{m/k}$; $u_m = 3u_1/2$

2.5-5. $u_m = 3u_1/2$; $u_r = u_1/2$

2.5-6. 2.26 s

2.5-7. 0.25

2.5-8. $\tau = \pi(\sqrt{m/k_1} + \sqrt{m/k_2})$; $u_m = \dot{u}_0\sqrt{m/k_2}$

2.5-9. $\tau = \dfrac{2}{\omega}\tan^{-1}\left(\dfrac{k\dot{u}_0}{P_1\omega}\right); \qquad u_m = \sqrt{\left(\dfrac{\dot{u}_0}{\omega}\right)^2 + \left(\dfrac{P_1}{k}\right)^2} - \dfrac{P_1}{k}$

2.5-10. $\tau = 4t_1 + \dfrac{4}{\omega_2}\tan^{-1}\left(\dfrac{k_2\dot{u}_1}{k_1\omega_2 u_1}\right)$

$$u_m = \left(1 - \frac{k_1}{k_2}\right)u_1 + \sqrt{\left(\frac{k_1}{k_2}u_1\right)^2 + \left(\frac{\dot{u}_1}{\omega_2}\right)^2}$$

2.5-11. $u_m = u_1(1 + \sqrt{5})$; $u_r = 4u_1/\sqrt{5}$

2.5-12. $u_m = u_1(6 + 2\sqrt{10})$; $u_r = 4u_1(1 + 2\sqrt{\tfrac{2}{5}})$

2.6-1. Given

2.6-2. Given

2.6-3. $\phi_{10} \approx \pi/2$ rad

2.6-4. $\dot{u}_{min} \approx -1.59$ in./s

2.6-5. $\phi_{max} \approx \pi/2$ rad

2.6-6. $u_m \approx 4.75$ in.; $t_m \approx 0.725$ s

2.6-7. $u_m \approx 3.18$ in.; $t_m \approx 0.50$ s

2.6-8. $u_m \approx 1.50$ in.; $t_m \approx 0.75$ s

2.6-9. $u_m \approx 1.51$ in.; $t_m \approx 1.20$ s

2.6-10. $u_m \approx 3.24$ in.; $t_m \approx 1.60$ s

CHAPTER 3

3.5-1. $r_1 = 1;\ r_2 = -1$

3.5-2. $r_1 = -\sqrt{3}/3;\ r_2 = \sqrt{3}$

3.5-3. $r_1 = 1/(1+\sqrt{2});\ r_2 = 1/(1-\sqrt{2})$

3.5-4. $r_1 = 1/\sqrt{2};\ r_2 = -1/\sqrt{2}$

3.5-5. $r_1 = 0.321;\ r_2 = -3.12$

3.5-6. $r_1 = 2.41;\ r_2 = -0.414$

3.5-7. $r_1 = 1;\ r_2 = -1$

3.5-8. $r_1 = 1/(1+\sqrt{2});\ r_2 = 1/(1-\sqrt{2})$

3.5-11. $r_1 = -13.4$ ft/rad; $r_2 = -2.41$ ft/rad

3.5-12. $r_1 = 0.578\ell;\ r_2 = -0.578\ell$

CHAPTER 4

4.2-1. $\omega^2_{1,2,3} = (2-\sqrt{2})T/m\ell,\ 2T/m\ell,\ (2+\sqrt{2})T/m\ell$

4.2-2. $\lambda_{1,2,3} = (2+\sqrt{2})m/2k,\ m/2k,\ (2-\sqrt{2})m/2k$

4.2-3. $\omega^2_{1,2,3} = \dfrac{g}{\ell},\ \dfrac{g}{\ell} + \dfrac{kh^2}{m\ell^2},\ \dfrac{g}{\ell} + \dfrac{3kh^2}{m\ell^2}$

4.2-4. $\lambda_{1,2,3} = 5.05I/k_r,\ 0.643I/k_r,\ 0.308I/k_r$

4.2-5. $\omega^2_{1,2,3,4} = 0,\ (2-\sqrt{2})k/m,\ 2k/m,\ (2+\sqrt{2})k/m$

4.2-6. $\lambda_{1,2,3} = 31.6\alpha,\ 2\alpha,\ 0.444\alpha\ (\alpha = m\ell^3/768EI)$

4.2-7. $\omega^2_{1,2,3} = 0.416g/\ell,\ 2.29g/\ell,\ 6.29g/\ell$

4.2-8. $\lambda_{1,2,3} = 14.4\alpha,\ 2.62\alpha,\ 0.954\alpha \quad (\alpha = mh^3/144EI)$

4.2-9. $\omega^2_{1,2,3} = 0.332k/m,\ 1.28k/m,\ 1.39k/m$

4.2-10. $\lambda_{1,2,3} = 74.7\alpha,\ 6.99\alpha,\ 0.307\alpha \quad (\alpha = m\ell^3/48EI)$

4.2-11. $\omega^2_{1,2,3} = 0,\ 0,\ 9EI/m\ell^3$

4.2-12. $\omega^2_{1,2,3} = (3-\sqrt{3})k/m,\ 3k/m,\ (3+\sqrt{3})k/m$

4.4-1. $u_1 = P(1.220 \cos \omega_1 t - 0.280 \cos \omega_2 t + 0.060 \cos \omega_3 t)/k$

4.4-2. $u_1 = \dot{u}_{01}(\sin \omega_2 t)\omega_2$

4.4-3. $\theta_1 = \phi(\cos \omega_1 t - \cos \omega_3 t)/3$

4.4-4. $\phi_1 = \dot{\theta}[0.543(\sin \omega_1 t)/\omega_1 + 0.349(\sin \omega_2 t)/\omega_2 + 0.108(\sin \omega_3 t)/\omega_3]$

4.4-5. $u_1 = \dot{u}_{01}[t + (\sin \omega_3 t)/\omega_3]/2$

4.4-6. $v_1 = \dot{\theta}\ell[1.707(\sin \omega_1 t)/\omega_1 - (\sin \omega_2 t)/\omega_2 + 0.293(\sin \omega_3 t)/\omega_3]/4$

4.4-7. $u_1 = \Delta(0.334 \cos \omega_1 t + 0.314 \cos \omega_2 t + 0.352 \cos \omega_3 t)$

4.4-8. $u_1 = Ph^3(2.616 \cos \omega_1 t - 0.702 \cos \omega_2 t + 0.083 \cos {}_3 t)/144EI$

4.4-9. $u_1 = \dot{u}_{01}[0.708(\sin \omega_1 t)/\omega_1 + 0.292(\sin \omega_3 t)/\omega_3]$

4.4-10. $u_1 = P\ell^3(26.33 \cos \omega_1 t - 2.315 \cos \omega_2 t - 0.007 \cos \omega_3 t)/48EI$

4.4-11. $v_1 = \dot{v}_{01}[4t - (\sin \omega_3 t)/\omega_3]/3$

4.4-12. $v_1 = \Delta(1.367 \cos \omega_1 t - 0.367 \cos \omega_3 t)$

4.5-1. $u_1 = (P \cos \Omega t)(0.242\beta_1/\omega_1^2 - 0.436\beta_2/\omega_2^2 + 0.194\beta_3/\omega_3^2)/m$

4.5-2. $u_1 = P[6 - (2 + \sqrt{2}) \cos \omega_1 t - 2 \cos \omega_2 t - (2 - \sqrt{2}) \cos \omega_3 t]/8k$

4.5-3. $\theta_1 = \dfrac{R}{3m\ell}\left[\left(t - \dfrac{1}{\omega_1} \sin \omega_1 t\right)\Big/\omega_1^2 - \left(t - \dfrac{1}{\omega_3} \sin \omega_3 t\right)\Big/\omega_3^2\right]$

4.5-4. $\phi_1 = (T \sin \Omega t)(0.218\beta_1/\omega_1^2 - 0.097\beta_2/\omega_2^2 - 0.121\beta_3/\omega_3^2)/I$

4.5-5. $u_1 = P[t^2 + (1 - \cos \omega_3 t)m/k]/4m$

4.5-6. $v_1 = \dfrac{R}{2m}\left[\left(t - \dfrac{1}{\omega_1} \sin \omega_1 t\right)\Big/\omega_1^2 + \left(t - \dfrac{1}{\omega_3} \sin \omega_3 t\right)\Big/\omega_3^2\right]$

4.5-7. $u_1 = (P \cos \Omega t)(0.077\beta_1/\omega_1^2 + 0.290\beta_2/\omega_2^2 + 0.132\beta_3/\omega_3^2)/m$

4.5-8. $u_1 = Ph^3(5.995 - 4.805 \cos \omega_1 t - 0.871 \cos \omega_2 t$
$- 0.319 \cos \omega_3 t)/144EI$

4.5-9. $w_1 = \dfrac{R}{\omega_2^2 m}\left(t - \dfrac{1}{\omega_2} \sin \omega_2 t\right)$

4.5-10. $u_1 = (P \sin \Omega t)(0.094\beta_1/\omega_1^2 - 0.145\beta_2/\omega_2^2 + 0.052\beta_3/\omega_3^2)/m$

4.5-11. $v_1 = \dfrac{R}{9m}\left[t^3 + 3\left(t - \dfrac{1}{\omega_3} \sin \omega_3 t\right)\Big/\omega_3^2\right]$

4.5-12. $v_1 = P(0.0379 - 0.0647 \cos \omega_1 t + 0.0251 \cos \omega_3 t)/k$

4.6-1. $u_1 = kd_1[0.108f_1(t) + 0.543f_2(t) + 0.349f_3(t)]/t_1 m;$

$f_1(t) = \left(t - \dfrac{1}{\omega_1} \sin \omega_1 t\right)\Big/\omega_1^2$, etc.

4.6-2. $u_1 = d[4 - (2 + \sqrt{2}) \cos \omega_1 t - (2 - \sqrt{2}) \cos \omega_3 t]/4$

4.6-3. $\theta_1 = -(a/g)\beta_1 \sin \Omega t$

4.6-4. $\phi_1^* = -\alpha_1 I[3t^2 - I(28.01 - 27.70 \cos \omega_1 t$
$- 0.289 \cos \omega_2 t - 0.020 \cos \omega_3 t)/k_r]/t_1^2 k_r$

4.6-5. $u_1 = kd_1[0.242f_1(t) - 0.436f_2(t) + 0.194f_3(t)]/t_1 m;$

$$f_1(t) = \left(t - \frac{1}{\omega_1} \sin \omega_1 t\right) \Big/ \omega_1^2, \text{ etc.}$$

4.6-6. $v_1 = d(3 - 1.707 \cos \omega_1 t - \cos \omega_2 t - 0.293 \cos \omega_3 t)/4$

4.6-7. $u_1 = d(1 - 0.334 \cos \omega_1 t - 0.314 \cos \omega_2 t - 0.352 \cos \omega_3 t)$

4.6-8. $u_1^* = -a \sin \Omega t(0.333\beta_1/\omega_1^2 + 0.333\beta_2/\omega_2^2 + 0.334\beta_3/\omega_3^2)$

4.6-9. $u_1 = d(1 - 0.708 \cos \omega_1 t - 0.292 \cos \omega_3 t)$

4.6-10. $u_1 = \phi\ell \cos \Omega t(1.242\beta_1 - 0.249\beta_2 + 0.008\beta_3$

4.6-11. $v_1^* = v_3^* = -a_1[t^2 - 2(1 - \cos \omega t)/\omega^2]\omega^2 t_1^2$, where $\omega^2 = 3EI/\ell^3 m$

4.6-12. $v_1 = d \sin \Omega t(0.096\beta_1 - 0.096\beta_3$

4.7-1. $\lambda_{1,2,3} \approx 2.618m\ell/T,\ 0.500m\ell/T,\ 0.382m\ell/T$

4.7-2. $\lambda_{1,2,3} \approx 2.618m\delta,\ 1.000m\delta,\ 0.382m\delta$

4.7-3. $\lambda_{1,2,3} \approx 39.68\alpha,\ 2.815\alpha,\ 0.501\alpha \qquad (\alpha = m\ell^3/768EI)$

4.7-4. $\omega_{1,2,3}^2 \approx 0.246g/\ell,\ 1.252g/\ell,\ 2.169g/\ell$

4.7-5. $\lambda_{1,2,3} \approx 19.12\alpha,\ 4.000\alpha,\ 1.884\alpha \qquad (\alpha = mh^3/144EI)$

4.7-6. $\lambda_{1,2,3} \approx 76.32\alpha,\ 8.978\alpha,\ 0.700\alpha \qquad (\alpha = m\ell^3/48EI)$

4.7-7. $\omega_{1,2,3}^2 = 0,\ 0,\ 6EI/m\ell^3$

CHAPTER 5

5.2-1. $$u = \sum_{i=1,3,5,\ldots}^{\infty} \cos \frac{i\pi x}{2\ell}\left(A_i \cos \frac{i\pi at}{2\ell} + B_i \sin \frac{i\pi at}{2\ell}\right)$$

5.2-2. $$u = \frac{4v\ell}{\pi^2 a} \sum_{i=1,3,5,\ldots}^{\infty} \frac{1}{i^2} \cos \frac{i\pi x}{\ell} \sin \frac{i\pi at}{\ell}$$

5.2-3. $$u = \frac{P_0\ell}{\pi^2 EA} \sum_{i=2,6,10,\ldots}^{\infty} \frac{(-1)^{(i-2)/4}}{i^2} \sin \frac{i\pi x}{\ell} \cos \frac{i\pi at}{\ell}$$

5.2-4. $$u = \frac{16P_0\ell}{\pi^3 EA} \sum_{i=1,3,5,\ldots}^{\infty} \frac{1}{i^3} \sin \frac{i\pi x}{2\ell} \cos \frac{i\pi at}{2\ell}$$

5.3-1. $u = \dfrac{2\ell P}{\pi^2 a^2 \rho A} \sum_{i=1,3,5,\ldots}^{\infty} \dfrac{(-1)^{(i-1)/2}}{i^2} \sin \dfrac{i\pi x}{\ell}\left(1 - \cos \dfrac{i\pi a t}{\ell}\right)$

5.3-2. $u = \dfrac{P_1 t^3}{6\rho A \ell t_1} + \dfrac{2\ell P_1}{\pi^2 a^2 \rho A t_1} \sum_{i=1}^{\infty} \dfrac{1}{i^2} \cos \dfrac{i\pi x}{\ell}\left(t - \dfrac{\ell}{i\pi a} \sin \dfrac{i\pi a t}{\ell}\right)$

5.3-3. $u = \dfrac{4P_1 \sin \Omega t}{\pi \rho A \ell} \sum_{i=1,3,5,\ldots}^{\infty} \dfrac{\sin(\omega_i x/a)}{i(\omega_i^2 - \Omega^2)}; \qquad \omega_i = \dfrac{i\pi a}{2\ell}$

5.3-4. $u = \dfrac{8\ell P}{\pi^2 a^2 \rho A} \sum_{i=1,3,5,\ldots}^{\infty} \dfrac{\cos(i\pi/4)}{i^2} \cos \dfrac{i\pi x}{2\ell}\left(1 - \cos \dfrac{i\pi a t}{2\ell}\right)$

5.4-1. $u = \dfrac{2P_0}{3\pi E A} \sum_{i=1}^{\infty} \dfrac{1}{i} \sin \dfrac{i\pi x}{\ell}\left[\dfrac{\ell}{i\pi}\left(3 \sin \dfrac{i\pi}{3} - \sin \dfrac{2i\pi}{3}\right) - \dfrac{\ell}{3} \cos \dfrac{2i\pi}{3}\right] \cos \dfrac{i\pi a t}{\ell}$

5.4-2. $u = \dfrac{8v\ell}{\pi^2 a} \sum_{i=1,3,5,\ldots}^{\infty} \dfrac{1}{i^2} \cos \dfrac{i\pi}{4} \sin \dfrac{i\pi x}{2\ell} \sin \dfrac{i\pi a t}{2\ell}$

5.4-3. $u = \dfrac{2\ell^2 Q}{\pi^3 a^2 m} \sum_{i=1}^{\infty} \dfrac{(-1)^{i-1}}{i^3} \sin \dfrac{i\pi x}{\ell}\left(1 - \cos \dfrac{i\pi a t}{\ell}\right)$

5.4-4. $u = \dfrac{P_1 t^4}{12\ell m t_1^2} + \dfrac{2\ell P_1}{\pi^2 a^2 m t_1^2} \sum_{i=2,4,6,\ldots}^{\infty} \dfrac{(-1)^{i/2}}{i^2} \cos \dfrac{i\pi x}{\ell}$
$\times \left[t^2 - \dfrac{2\ell^2}{(i\pi a)^2}\left(1 - \cos \dfrac{i\pi a t}{\ell}\right)\right]$

5.6-1. $u = \left[1 + \dfrac{16\ell^2\Omega^2}{\pi^3 a^2} \sum_{i=1,3,5,\ldots}^{\infty} \dfrac{\beta_i}{i^3}(-1)^{(i-1)/2} \cos \dfrac{i\pi x}{2\ell}\right] d \sin \Omega t$

5.6-2. $u = \dfrac{u_1}{t_1^2}\left[t^2 - \dfrac{8\ell^2}{\pi^3 a^2} \sum_{i=1,3,5,\ldots}^{\infty} \dfrac{1}{i^3} \sin \dfrac{i\pi x}{\ell}\left(1 - \cos \dfrac{i\pi a t}{\ell}\right)\right]$

5.6-3. $u = \dfrac{u_1}{t_1^2}\left[\dfrac{(\ell - x)}{\ell} t^2 - \dfrac{4\ell^2}{\pi^3 a^2} \sum_{i=1}^{\infty} \dfrac{1}{i^3} \sin \dfrac{i\pi x}{\ell}\left(1 - \cos \dfrac{i\pi a t}{\ell}\right)\right]$

5.6-4. $u = \dfrac{u}{t_2^3}\left[\dfrac{x}{\ell} t^3 + \dfrac{12\ell^2}{\pi^3 a^2} \sum_{i=1}^{\infty} \dfrac{(-1)^i}{i^3} \sin \dfrac{i\pi x}{\ell}\left(t - \dfrac{\ell}{i\pi a} \sin \dfrac{i\pi a t}{\ell}\right)\right]$

5.10-1. $v = \dfrac{2P\ell^3}{\pi^4 EI} \sum_i \dfrac{(-1)^{(i-1)/2}}{i^4} \sin \dfrac{i\pi x}{\ell} \cos \omega_i t \qquad (i = 1, 3, 5, \ldots, \infty)$

5.10-2. $v = \dfrac{2P\ell^3}{\pi^4 EI} \sum_i \dfrac{1}{i^4} \sin \dfrac{i\pi x}{\ell} \sin \dfrac{i\pi x_1}{\ell} \cos \omega_i t \qquad (i = 1, 2, 3, \ldots, \infty)$

5.10-3. $v = \dfrac{4w\ell^4}{\pi^5 EI} \sum_i \dfrac{1}{i^5} \sin \dfrac{i\pi x}{\ell} \cos \omega_i t \qquad (i = 1, 3, 5, \ldots, \infty)$

5.10-4. $v = \dfrac{4\dot{v}_0}{\pi} \sum_i \dfrac{1}{i\omega_i} \sin \dfrac{i\pi x}{\ell} \sin \omega_i t \qquad (i = 1, 3, 5, \ldots, \infty)$

5.11-1. $X_i = \cosh k_i x - \cos k_i x - \alpha_i(\sinh k_i x - \sin k_i x)$;

$$\alpha_i = \frac{\cosh k_i \ell - \cos k_i \ell}{\sinh k_i \ell - \sin k_i \ell}$$

$k_1 \ell = 3.927 \qquad k_2 \ell = 7.069 \qquad \alpha_1 = 1.0008 \qquad \alpha_2 = 1.0000$

5.11-2. $X_i = \cosh k_i x - \cos k_i x - \alpha_i(\sinh k_i x - \sin k_i x)$;

$$\alpha_i = \frac{\cosh k_i \ell + \cos k_i \ell}{\sinh k_i \ell + \sin k_i \ell}$$

$k_1 \ell = 1.875 \qquad k_2 \ell = 4.694 \qquad \alpha_1 = 0.7341 \qquad \alpha_2 = 1.0185$

5.11-3. $$v = \frac{P_0 \ell^3}{EI}(1724X_1 \cos \omega_1 t + 380X_2 \cos \omega_2 t + 94X_3 \cos \omega_3 t + \cdots) \times 10^{-6}$$

5.11-4. $$v = \frac{P_0 \ell^3}{EI}(16182X_1 \cos \omega_1 t - 412X_2 \cos \omega_2 t + 53X_3 \cos \omega_3 t - \cdots) \times 10^{-5}$$

5.13-1. $$(v_m)_{x=\ell/2} = \frac{2P\ell^3}{\pi^4 EI}\left(\frac{1}{1-\frac{1}{4}} + \frac{1}{3^4-\frac{1}{4}} + \frac{1}{5^4-\frac{1}{4}} + \cdots\right) \approx 1.328 \text{ mm}$$

5.13-2. $$(v_m)_{x=\ell/2} = \frac{4P\ell^3}{\pi^4 EI}\left[\frac{\sin(\pi/3)}{1-\frac{1}{4}} + \frac{\sin(5\pi/3)}{5^4-\frac{1}{4}} + \frac{\sin(7\pi/3)}{7^4-\frac{1}{4}} + \cdots\right] \approx 2.273 \text{ mm}$$

5.13-3. $$(v_m)_{x=\ell/2} = \frac{2w\ell^4}{\pi^5 EI} \sum_{i=1,3,5,\ldots}^{\infty} \frac{(-1)^{(i-1)/2}}{i(i^4 - \alpha^2)}$$

5.13-4. $$v = \frac{2P\ell^3}{\pi^4 EI} \sum_{i=1,3,5,\ldots}^{\infty} \frac{(-1)^{(i-1)/2}}{i^4} \sin \frac{i\pi x}{\ell}(1 - \cos \omega_i t)$$

5.13-5. $$v = \frac{w\ell^4 \beta_1 \sin \Omega t}{\pi^4 EI} \sin \frac{\pi x}{\ell}; \qquad \beta_1 = \frac{1}{1 - \Omega^2/\omega_1^2}$$

5.13-6. $$v = \frac{2\pi}{m\ell^2} \sum_{i=1}^{\infty} \frac{1}{p_i} \sin \frac{i\pi x}{\ell} \int_0^\ell \cos \frac{i\pi x}{\ell} \int_0^t M(x, t') \sin \omega_i(t - t')\, dt'\, dx$$

5.14-1. $$v = \frac{P\ell^3 \sin \Omega t}{EI} \sum_{i=1}^{\infty} \frac{\beta_i X_i (X_i)_{x=\ell}}{(k_i \ell)^4}; \qquad \beta_i = \frac{1}{1 - \Omega^2/\omega_i^2}$$

5.14-2. $$(v)_{x=\ell} = \frac{4P\ell^3}{EI}(0.08091\beta_1 + 0.00206\beta_2 + 0.00026\beta_3 + \cdots) \sin \Omega t$$

5.14-3. $$v = \frac{P\ell^3 \sin \Omega t}{EI} \sum_{i=1}^{\infty} \frac{\beta_i X_i (X_i)_{x=\ell/2}}{(k_i \ell)^4}$$

5.14-4. $$(v)_{x=\ell/2} = \frac{P\ell^3 \sin \Omega t}{EI}(8829\beta_1 + 130\beta_2 + 156\beta_3 + 9\beta_4 + 23\beta_5 + \cdots)$$

5.15-1. $v = \left[1 - \frac{x}{\ell} + \frac{2\ell^4\Omega^2}{\pi^5 a^2} \sum_{i=1}^{\infty} \frac{\beta_i}{i^5} \sin \frac{i\pi x}{\ell}\right] v_1 \sin \Omega t$

5.15-2. $v = \frac{\theta_2}{t_2^2}\left[xt^2 - \sum_{i=1}^{\infty} \frac{2X_i}{\omega_i^2} \int_0^{\ell} xX_i\, dx(1 - \cos \omega_i t)\right]$

5.15-3. $u = \left[-\frac{x^2}{\ell} + \frac{x^3}{\ell^2} + \sum_{i=1}^{\infty} \frac{\Omega^2 X_i}{(\omega_i^2 - \Omega^2)} \int_0^{\ell} \left(-\frac{x^2}{\ell} + \frac{x^3}{\ell^2}\right) X_i\, dx\right] \theta_4 \sin \Omega t$

5.15-4. $v = \frac{v_3}{t_3^3}\left[\left(\frac{3x^2}{2\ell^2} - \frac{x^3}{2\ell^3}\right)t^3 - \sum_{i=1}^{\infty} \frac{6X_i}{\omega_i^2} \int_0^{\ell} \left(\frac{3x^2}{2\ell^2} - \frac{x^3}{2\ell^3}\right) X_i\, dx\left(t - \frac{1}{\omega_i} \sin \omega_i t\right)\right]$

CHAPTER 6

6.4-1. $\mathbf{p}_b(t) = \{2b_1 + b_2,\ b_1 + 2b_2\} \frac{L}{6}$

6.4-2. $\mathbf{p}_b(t) = \{1,\ 3\} \frac{b_2 L}{12}$

6.4-3. $\mathbf{K} = \frac{EA}{3L}\begin{bmatrix} 7 & -8 & 1 \\ -8 & 16 & -8 \\ 1 & -8 & 7 \end{bmatrix}$

6.4-4. $\mathbf{M} = \frac{\rho AL}{15}\begin{bmatrix} 2 & 1 & -\frac{1}{2} \\ 1 & 8 & 1 \\ -\frac{1}{2} & 1 & 2 \end{bmatrix}$

6.4-5. $\mathbf{p}_b(t) = \{1,\ 4,\ 1\} \frac{b_x L}{6}$

6.4-6. $\mathbf{p}_b(t) = \{5,\ 3\} \frac{m_{x1} L}{12}$

6.4-7. $\mathbf{p}_b(t) = \{L - x,\ x\} \frac{M_x}{L}$

6.4-8. $\mathbf{p}_b(t) = \{9,\ 2L,\ 21,\ -3L\} \frac{b_2 L}{60}$

6.4-9. $\mathbf{p}_b(t)_1 = \{2x^3 - 3x^2L + L^3,\ x^3L - 2x^2L^2 + xL^3,$

$-2x^3 + 3x^2L,\ x^3L - x^2L^2\} \frac{P_y}{L^3}$

$\mathbf{p}_b(t)_2 = \{6x^2 - 6xL,\ 3x^2L - 4xL^2 + L^3,\ -6x^2 + 6xL,\ 3x^2L - 2xL^2\} \frac{M_z}{L^3}$

6.4-10. $\mathbf{p}_b(t) = \{21b_1 + 9b_2,\ (3b_1 + 2b_2)L,\ 9b_1 + 21b_2,\ -(2b_1 + 3b_2)L\} \frac{L}{60}$

6.4-11. Given

6.4-12. Given

6.5-1. $S_{s11} = \dfrac{4EI}{L}; \quad S_{12} = \dfrac{2EI}{L}; \quad S_{s13} = 0; \quad \text{etc.}$

$M_{s11} = \dfrac{\rho AL^3}{105}; \quad M_{s12} = -\dfrac{\rho AL^3}{140}; \quad M_{s13} = 0; \quad \text{etc.}$

6.5-2. $S_{s11} = \dfrac{8EI}{L}; \quad S_{s12} = \dfrac{6EI}{L^2}; \quad S_{s13} = \dfrac{2EI}{L}; \quad \text{etc.}$

$M_{s11} = \dfrac{2\rho AL^3}{105}; \quad M_{s12} = -\dfrac{13\rho AL^2}{420}; \quad M_{s13} = -\dfrac{\rho AL^3}{140}; \quad \text{etc.}$

6.5-3. $S_{s11} = \dfrac{12EI}{L^3}; \quad S_{s12} = \dfrac{6EI}{L^2}; \quad S_{s13} = \dfrac{6EI}{L^2}; \quad \text{etc.}$

$M_{s11} = \dfrac{13\rho AL}{35}; \quad M_{s12} = \dfrac{11\rho AL^2}{210}; \quad M_{s13} = -\dfrac{13\rho AL^2}{420}; \quad \text{etc.}$

6.5-4. $S_{s11} = \dfrac{8EI}{L}; \quad S_{s12} = -\dfrac{6EI}{L^2}; \quad S_{s13} = \dfrac{2EI}{L}; \quad \text{etc.}$

$M_{s11} = \dfrac{2\rho AL^3}{105}; \quad M_{s12} = \dfrac{13\rho AL^2}{420}; \quad M_{s13} = -\dfrac{\rho AL^3}{140}; \quad \text{etc.}$

6.5-5. $\omega_{1,2} = \dfrac{5.546,\ 21.04}{L^2}\sqrt{\dfrac{EI}{\rho A}}; \quad \mathbf{\Phi}_1 = \begin{bmatrix} L \\ -0.3947 \end{bmatrix}; \quad \mathbf{\Phi}_2 = \begin{bmatrix} L \\ -26.61 \end{bmatrix}$

6.5-6. $\omega_{1,2} = \dfrac{15.14,\ 28.98}{L^2}\sqrt{\dfrac{EI}{\rho A}}; \quad \mathbf{\Phi}_1 = \begin{bmatrix} 1 \\ -1 \end{bmatrix}; \quad \mathbf{\Phi}_2 = \begin{bmatrix} 1 \\ 1 \end{bmatrix}$

6.5-7. $\omega_{1,2} = \dfrac{13.32,\ 34.79}{L^2}\sqrt{\dfrac{EI}{\rho A}}; \quad \mathbf{\Phi}_1 = \begin{bmatrix} 1 \\ -0.7071 \end{bmatrix}; \quad \mathbf{\Phi}_2 = \begin{bmatrix} 1 \\ 0.7071 \end{bmatrix}$

6.5-8. $\omega_1 = \dfrac{15.69}{L^2}\sqrt{\dfrac{EI}{\rho A}}$

6.5-9. $\omega_{1,2} = 5.603,\ 31.19\dfrac{1}{L^2}\sqrt{\dfrac{EI}{\rho A}}; \quad \mathbf{\Phi} = \begin{bmatrix} 1.000 & 1.000 \\ 0.5435 & -0.9364 \end{bmatrix}$

6.5-10. $\omega_1 = \dfrac{9.941}{L^2}\sqrt{\dfrac{EI}{\rho A}}$

6.5-11. $\omega_{1,2,3} = 0,\ 0,\ 22.47\dfrac{1}{L^2}\sqrt{\dfrac{EI}{\rho A}}; \quad \mathbf{\Phi} = \begin{bmatrix} 1 & 1 & 1 \\ 1 & 0 & -0.6 \\ 1 & -1 & 1 \end{bmatrix}$

INDEX